SCATTERING OF ELECTROMAGNETIC WAVES

WILEY SERIES IN REMOTE SENSING

Jin Au Kong, Editor

Asrar • THEORY AND APPLICATIONS OF OPTICAL REMOTE SENSING

Crane • ELECTROMAGNETIC WAVE PROPAGATION THROUGH RAIN

Curlander and McDonough • SYNTHETIC APERTURE RADAR: SYSTEMS AND SIGNAL PROCESSING

Elachi • INTRODUCTION TO THE PHYSICS AND TECHNIQUES OF REMOTE SENSING

Haykin, Lewis, Raney, and Rossiter • REMOTE SENSING OF SEA ICE AND ICEBERGS

Haykin and Steinhardt • ADAPTIVE RADAR DETECTION AND ESTIMATION

Hord • REMOTE SENSING: METHODS AND APPLICATIONS

Janssen • ATMOSPHERIC REMOTE SENSING BY MICROWAVE RADIOMETRY

Maffett • TOPICS FOR A STATISTICAL DESCRIPTION OF RADAR CROSS SECTIONS

Steinberg and Subbaram • MICROWAVE IMAGING TECHNIQUES

Szekielda • SATELLITE MONITORING OF THE EARTH

Tsang, Kong, and Shin • THEORY OF MICROWAVE REMOTE SENSING

Tsang, Kong, and Ding • SCATTERING OF ELECTROMAGNETIC WAVES: THEORIES AND APPLICATIONS

Tsang, Kong, Ding, and Ao • SCATTERING OF ELECTROMAGNETIC WAVES: NUMERICAL SIMULATIONS

Tsang and Kong • SCATTERING OF ELECTROMAGNETIC WAVES: ADVANCED TOPICS

SCATTERING OF ELECTROMAGNETIC WAVES

Numerical Simulations

Leung Tsang

Jin Au Kong

Kung-Hau Ding

Chi On Ao

A Wiley-Interscience Publication

JOHN WILEY & SONS, INC.

New York • Chichester • Weinheim • Brisbane • Singapore • Toronto

This text is printed on acid-free paper. ♾

Library of Congress Cataloging in Publication Data

Tsang, Leung.
Scattering of electromagnetic waves: Numerical simulations / L. Tsang . . . [et al.]
p. cm. — (Wiley series in remote sensing)
Includes index.
ISBN 0-471-38800-9 (cloth: alk. paper)
1. Electromagnetic waves—Scattering—Mathematical models. I. Tsang, Leung. II. Series.

QC665.S3 S23 2000
621.36'78'015118—dc21 00-040864

10 9 8 7 6 5 4 3 2

To my family, Hannah, Clarisse, and Kaleb for their love.

— L. Tsang

To our families.

— J. A. Kong, K. H. Ding, C. O. Ao

CONTENTS

CHAPTER 4
RANDOM ROUGH SURFACE SIMULATIONS 111

CHAPTER 5 FAST COMPUTATIONAL METHODS FOR SOLVING ROUGH SURFACE SCATTERING PROBLEMS 177

CHAPTER 10 DENSE MEDIA MODELS AND THREE-DIMENSIONAL SIMULATIONS

CHAPTER 11 ANGULAR CORRELATION FUNCTION AND DETECTION OF BURIED OBJECT

Scattering of Electromagnetic Waves

Volume I: Theories and Applications (Tsang, Kong, and Ding)

Volume II: Numerical Simulations (Tsang, Kong, Ding, and Ao)

Volume III: Advanced Topics (Tsang and Kong)

PREFACE

Electromagnetic wave scattering is an active, interdisciplinary area of research with myriad practical applications in fields ranging from atomic physics to medical imaging to geoscience and remote sensing. In particular, the subject of wave scattering by random discrete scatterers and rough surfaces presents great theoretical challenges due to the large degrees of freedom in these systems and the need to include multiple scattering effects accurately. In the past three decades, considerable theoretical progress has been made in elucidating and understanding the scattering processes involved in such problems. Diagrammatic techniques and effective medium theories remain essential for analytical studies; however, rapid advances in computer technology have opened new doors for researchers with the full power of Monte Carlo simulations in the numerical analysis of random media scattering. Numerical simulations allow us to solve the Maxwell equations exactly without the limitations of analytical approximations, whose regimes of validity are often difficult to assess. Thus it is our aim to present in these three volumes a balanced picture of both theoretical and numerical methods that are commonly used for tackling electromagnetic wave scattering problems. While our book places an emphasis on remote sensing applications, the materials covered here should be useful for students and researchers from a variety of backgrounds as in, for example, composite materials, photonic devices, optical thin films, lasers, optical tomography, and X-ray lithography. Introductory chapters and sections are also added so that the materials can be readily understood by graduate students. We hope that our book would help stimulate new ideas and innovative approaches to electromagnetic wave scattering in the years to come.

The increasingly important role of numerical simulations in solving electromagnetic wave scattering problems has motivated us to host a companion web site that contains computer codes on topics relevant to the book. These computer codes are written in the MATLAB programming language and are available for download from our web site at `www.emwave.com`. They are provided to serve two main purposes. The first is to supply our readers a hands-on laboratory for performing numerical experiments, through which the concepts in the book can be more dynamically relayed. The second is to give new researchers a set of basic tools with which they could quickly build on projects of their own. The fluid nature of the web site would also allow us to regularly update the contents and keep pace with new research developments.

The present volume covers numerical simulation techniques and results for electromagnetic wave scattering in random media and rough surfaces. Due to the large degree of freedom associated with these systems, especially for 3-D scattering problems, fast computational methods are essential for maximizing returns from limited computational resources. Indeed, the subject of numerical electromagnetics has seen explosive growth in recent years. For lack of space, we choose to focus here on methods and techniques which are more directly related to our own research.

We begin in Chapter 1 with Monte Carlo simulations of a simple one-dimensional random medium — a layered medium characterized by permittivity fluctuations. Simulation results are used to explain passive remote sensing measurements of the Antarctic firn. For two- and three-dimensional scattering, it is advantageous to formulate the problem in terms of surface integral equations where the unknowns are confined to a lower dimensionality. Numerical solutions of surface integral equations are often obtained through the method of moments (MoM). We also discuss a useful technique known as the discrete dipole approximation (DDA) for solving volume integral equation. The DDA can be used to model inhomogeneous, irregularily shaped object by discretizing it as a collection of point dipoles. In MoM and DDA, numerical solutions are obtained by approximating the integral equations with a set of linear equations. Thus matrix computation is an essential aspect of numerical electromagnetics. When the size of the system becomes very large, direct matrix inversion becomes inefficient, and iterative methods such as the conjugate gradient methods are often used instead. Iterative methods usually require repeated computations of matrix-vector multiplication, and for problems with translational invariance, it is possible to utilize fast Fourier transform (FFT) to speed up this operation. The use of FFT in conjunction with iterative solvers is the cornerstone of fast computational methods introduced later in this book. Therefore we discuss these topics at some length in Chapter 2.

The remainder of the book is divided into two main parts. Chapters 3–6 deal with simulations of rough surface scattering, while volume scattering simulations involving random discrete scatterers are studied in Chapters 7–13 (except Chapter 11 — which contains aspects of both rough surface and volume scattering). The topic of electromagnetic wave interactions with rough surfaces has important applications in microwave remote sensing of ocean surface, geophysical terrain, and agricultural fields as well as in the design and manufacturing of optical systems and X-ray lithography. In Chapter 3, we discuss scattering and emission by periodic rough surfaces. Two solution methods are used to solve this problem. The first is the T-matrix method, which makes use of Floquet mode expansions and the extended boundary condition. The T-matrix formulation is exact, but the resulting equations become ill-conditioned when the surface is very rough. The second method uses a surface integral equation approach with MoM. Although computationally more intensive than the T-matrix method, the surface integral equation approach is applicable to surfaces with deep corrugation. We also describe Ewald's method for speeding up calculations of the Green's function in periodic medium. This has applications in active research areas such as frequency selective surfaces and photonic bandgap materials.

In Chapter 4, we discuss one-dimensional random rough surface scattering. The core ideas behind rough surface scattering simulations are introduced here. We describe in details the discretization procedure for the surface integral equations in the Dirichlet, Neumann, and two-media problems. Numerical methods for generating Gaussian and fractal rough surface profiles are described. The issue of truncating the rough surface and limiting the computational domain is also an important one. We discuss two popular approaches. The first approach uses a tapered incident wave that illuminates only a part of the entire rough surface, while the second approach uses a periodic boundary condition. As described in Volume I, random rough surfaces are often characterized by their power spectra. This is convenient for theoretical work, but how well does it model reality? We include discussion of wave scattering from real-life rough surface profiles. In addition to simulating bistatic scattering from rough surfaces, we also take an in-depth look at emissivity calculations based on rough surface simulations, which impose much more stringent energy conservation requirement.

Chapters 5 and 6 are devoted respectively to fast computational methods in 1-D and 2-D rough surface scattering simulations. The development of fast computational methods is particularly important in scattering by 2-D rough surfaces (3-D scattering problem) where the number of unknowns can quickly escalate as we increase the surface size. Since real-life surfaces are 2-D, we emphasize in this book fast computational methods that can be applied to scattering by both 1-D and 2-D rough surfaces. We introduce the sparse matrix iterative approach with canonical grid (SMCG). In this method, the impedance matrix is split into a strong part that consists of near-neighbor interactions and a weak part that consists of all the rest. An iterative scheme such as the conjugate gradient method is adopted to solve the matrix equation. The strong matrix is sparse and can be easily handled. However, the weak interactions require the multiplication of the dense weak matrix with successive iterates and could therefore present a major computational bottleneck. To speed up such calculations, the concept of canonical grid (CG) is introduced. The essential nature of CG is that it is translationally invariant. In rough surface scattering problems, the CG is usually taken to be the mean flat surface. By translating the unknowns to the CG, the weak interactions can be performed simultaneously for all unknowns using FFT. This reduces memory requirements from $\mathcal{O}(N^2)$ to $\mathcal{O}(N)$ and operation counts from $\mathcal{O}(N^2)$ to $\mathcal{O}(N \log N)$. We also introduce the physics-based two-grid (PBTG) method for dealing with lossy dielectric surfaces. In this method, a dense grid suitable for the lower half-space and a coarse grid suitable for the upper half-space are chosen. By taking advantage of the attenuative nature of the Green's function in the lower half-space and the slowly varying nature of the Green's function in the upper half-space with respect to the dense grid, one can achieve the accuracy of a single dense grid with the computational efficiency of a single coarse grid. Other fast methods discussed and illustrated in Chapter 5 include the steepest descent fast multipoles method (SDFMM) and the method of ordered multiple interactions (MOMI).

In contrast to rough surface scattering, volume scattering involving dense distributions of discrete scatterers is often a full-fledged 3-D scattering problem. The additional degree of freedom makes direct simulations of scattering coefficients rather difficult. Radiative transfer theory is commonly used for such problems, but the conventional approach fails to take into account of coherent multiple interactions between the scatterers. A better approach is to perform the scattering simulations on a test volume that contains a large number of scatterers but forms only a small part of the whole system. Coherent interactions are captured through the simulated extinction coefficients and phase functions, which can then be used in the dense medium radiative transfer equation (rigorously derived in Volume III) to solve the large-scale problem. These concepts are discussed in Chapter 7, where idealized randomly distributed point scatterers are used to illustrate the methods. The multiple scattering problem is formulated using the Foldy-Lax self-consistent equations.

In a dense medium, the correlation of scatterer positions could significantly affect the scattering results. The pair-distribution function quantifies the two-particle correlation property of the scatterers. In Chapter 8, we introduce the Percus–Yevick equation for the pair-distribution function and give closed-form solutions for hard and sticky spheres. For Monte Carlo simulations, statistical realizations of scatterer configurations are needed. Two methods are commonly employed to generate the particle positions: sequential addition and Metropolis shuffling, the latter method being more efficient when the particles are very closely packed. We show simulation results of the pair distribution functions for hard spheres and spheroids as well as sticky spheres. The simulated pair distribution functions are found to compare well with the Percus–Yevick pair distribution functions. Before dealing with 3-D dense media scattering, it is instructional to first study, in Chapter 9, the simpler problem of 2-D dense media scattering, where the volume scatterers are chosen to be infinitely long cylinders. We describe analytical pair distribution function and Monte Carlo simulations of particle positions in the 2-D case. The Foldy-Lax multiple scattering equations are then used to simulate extinction coefficients for densely packed hard and sticky cylinders. Finally, the SMCG method used in rough surface scattering is generalized to the volume scattering simulations. In Chapter 10, we perform 3-D dense media scattering calculations with dielectric spheres and spheroids. The volume integral equation approach as well as the T-matrix approach based on the Foldy-Lax equations are described in details. Simulation results for the extinction coefficients and phase matrices are shown and compared with analytical approximations.

In Chapter 11, we describe the novel correlation phenomenon in random media scattering known as the memory effect, which manifests itself in wave scattering through the angular correlation function (ACF). ACF has been discussed in Chapter 6 of Volume I in the context of single scattering by point scatterers. Here, we provide a general derivation of the memory effect based on the statistical translational invariance of the random medium. The special property of ACF for random medium makes it a good candidate for the detection of a target embedded in random clutter. We explore such ideas by studying targets buried under rough surface and volume scatterers.

The subject of multiple scattering by finite cylinders has important applications in the remote sensing of vegetation as well as signal coupling among multiple vias in high frequency circuits. In Chapter 12, we consider scattering by vertical cylinders in the presence of reflective boundaries, which introduce additional complications. We discuss Monte Carlo simulations of these systems as well as simple analytical results that take into account of first and second order scattering. In Chapter 13, more realistic modeling of vegetation structures through stochastic Lindenmayer systems are presented. We compare scattering results from such systems obtained using the methods of DDA, the coherent addition approximation, and independent scattering.

This book should provide a good mix of basic principles and current research topics. An introductory course in Monte Carlo simulations can cover most of Chapters 1, 2, 4, 5, 7, and 9.

Acknowledgments

We would like to acknowledge the collaboration with our colleagues and graduate students. In particular, we wish to thank Professor Chi Chan of City University of Hong Kong, Professor Joel T. Johnson of Ohio State University, Dr. Robert T. Shin of MIT Lincoln Laboratory, and Dr. Dale Winebrenner of University of Washington. The graduate students who completed their Ph.D. theses from the University of Washington on random media scattering include Boheng Wen (1989), Kung-Hau Ding (1989), Shu-Hsiang Lou (1991), Charles E. Mandt (1992), Richard D. West (1994), Zhengxiao Chen (1994), Lisa M. Zurk (1995), Kyung Pak (1996), Guifu Zhang (1998), and Qin Li (2000). Much of their dissertation works are included in this book. Financial supports from the Air Force Office of Scientific Research, Army Research Office, National Aeronautics and Space Administration, National Science Foundation, Office of Naval Research, and Schlumberger-Doll Research Center for research materials included in this book are gratefully acknowledged. We also want to acknowledge the current UW graduate students who have helped to develop the numerical codes used throughout this book. These include Chi-Te Chen, Houfei Chen, Jianjun Guo, Chung-Chi Huang, and Lin Zhou. Special thanks are also due to Tomasz Grzegorczyk for proofreading on parts of the manuscript and Bae-Ian Wu for production assistance.

Leung Tsang
Seattle, Washington
Jin Au Kong
Cambridge, Massachusetts
Kung-Hau Ding
Hanscom AFB, Massachusetts
Chi On Ao
Cambridge, Massachusetts

February 2001

SCATTERING OF ELECTROMAGNETIC WAVES

Chapter 1

MONTE CARLO SIMULATIONS OF LAYERED MEDIA

1 One-Dimensional Layered Media with Permittivity Fluctuations

We study Monte Carlo simulations of solutions of the Maxwell equations in Volume II. The simplest case of random medium is one where the permittivity is a random function of positions in a one-dimensional problem. In Fig. 1.1.1, we show a stratified medium of many layers. The permittivity fluctuates from layer to layer. The basic theory of waves in layered medium was covered in Chapter 5 of Volume I. Nevertheless, even in this simple case, there can be two distinct kinds of layering. The first kind is a continuous random medium in which the random medium permittivity $\epsilon(z)$ is a random process that is a continuous function of z. The second kind is discrete layering in which there are abrupt changes of permittivity from layer to layer. To further illustrate the difference, we apply both models to thermal emission of a layered medium and make a comparison with observed brightness temperatures of Antarctica. We found that in order to match the observed brightness temperatures, the two models have to use drastically different physical parameters. The results illustrate the difference between a continuous random medium and a discrete random medium.

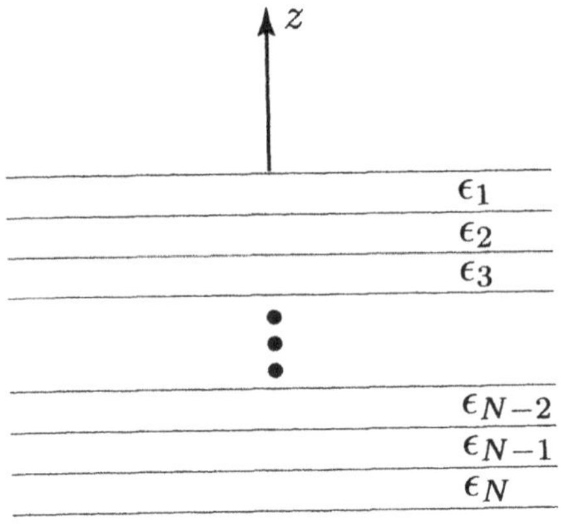

Figure 1.1.1 Stratified medium with permittivity fluctuations from layer to layer.

1.1 Continuous Random Medium

A common approach is to assume a Gaussian random process of the permittivity fluctuations. Figure 1.1.2 illustrates a realization of Gaussian random process as a function of position. The density of snow is used for illustration.

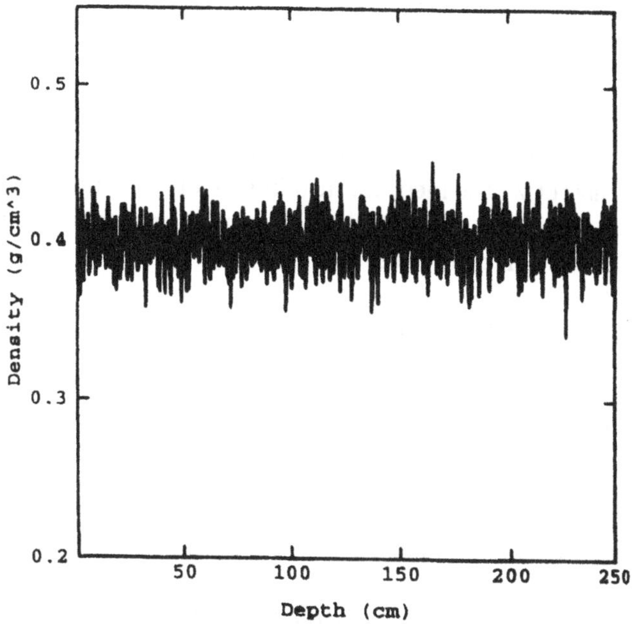

Figure 1.1.2 A single realization of a continuous Gaussian random profile with a mean density of 0.4 g/cm^3, a correlation length of 2 mm, and a standard deviation in density of 0.0156 g/cm^3.

For layered random media, one can assume $\epsilon(z)$ as a one-dimensional Gaussian random process with mean ϵ_m and variance $\sigma^2 = \delta\epsilon_m^2$. The probability density function is

$$p(\epsilon) = \frac{1}{\sqrt{2\pi}\sigma} \exp\left(-\frac{(\epsilon - \epsilon_m)^2}{2\sigma^2}\right) \tag{1.1.1}$$

Let z_1 and z_2 be two positions and let $\epsilon_1 = \epsilon(z_1)$ and $\epsilon_2 = \epsilon(z_2)$. Then the joint probability density function is

$$p(\epsilon_1, \epsilon_2) = \frac{1}{2\pi\sigma^2\sqrt{1-r^2}} \exp\Bigg[-\frac{1}{2\sigma^2(1-r^2)}\Big((\epsilon_1 - \epsilon_m)^2 - 2r(\epsilon_1 - \epsilon_m)(\epsilon_2 - \epsilon_m) + (\epsilon_2 - \epsilon_m)^2\Big)\Bigg] \tag{1.1.2}$$

where r is the correlation coefficient that depends on $|z_1 - z_2|$. If the correlation coefficient is of exponential form, then

$$r(z_1 - z_2) = \exp\left(-\frac{|z_1 - z_2|}{l_z}\right) \tag{1.1.3}$$

where l_z is the correlation length. Let $\epsilon_f(z) = \epsilon(z) - \epsilon_m$ be the fluctuating part of the permittivity. Then the covariance function of the permittivity is

$$\langle \epsilon_f(z)\epsilon_f(z')\rangle = \delta\epsilon_m'^2 \exp\left(-\frac{|z-z'|}{l_z}\right) \tag{1.1.4}$$

where angular bracket denotes average. Given a realization of permittivity profile, we can discretize the medium into fine layers (say up to 30,000 layers). We note that in Fig. 1.1.2 of the Gaussian random process, the permittivity is a continuous function of depth.

1.2 Generation of One-Dimensional Continuous Gaussian Random Medium

If we assume that $f(z)$ is a real Gaussian process with normalized correlation function $C(z)$. Then

$$\langle f(z)f(z')\rangle = \sigma^2 C(z-z') \tag{1.1.5}$$

Let $W(K)$ be the spectral density

$$\sigma^2 C(z) = \int_{-\infty}^{\infty} dK e^{iKz} W(K) \tag{1.1.6}$$

$$W(K) = \frac{1}{2\pi}\int_{-\infty}^{\infty} dz \sigma^2 C(z) e^{-iKz} \tag{1.1.7}$$

For the case that the correlation is exponential

$$C(z) = \exp\left(-\frac{|z|}{l_z}\right) \tag{1.1.8}$$

the spectral density is

$$W(K) = \frac{\sigma^2}{\pi}\frac{l_z}{1+K^2 l_z^2} \tag{1.1.9}$$

Let the sample be generated for $0 \le z \le L$. Outside L, we can assume that $\epsilon_f(z)$ is periodic. Then using a Fourier series,

$$f(z) = \frac{1}{L}\sum_{n=-\infty}^{\infty} b_n e^{\frac{i2\pi n z}{L}} \tag{1.1.10}$$

The discretization is

$$\Delta K = \frac{2\pi}{L} \tag{1.1.11}$$

$$K_n = \frac{2\pi n}{L} = n\Delta K \tag{1.1.12}$$

The b_n's are complex and

$$\langle f(z_1)f(z_2)\rangle = \int_{-\infty}^{\infty} dK e^{iK(z_1-z_2)} W(K)$$

$$= \frac{1}{L^2}\sum_n \sum_m \langle b_n b_m^* \rangle e^{i\frac{2\pi n}{L}z_1 - i\frac{2\pi n}{L}z_2} \tag{1.1.13}$$

$$= \sum_n \frac{2\pi}{L} e^{iK_n(z_1-z_2)} W(K_n) \tag{1.1.14}$$

Thus

$$\langle b_n b_m^* \rangle = \delta_{nm} 2\pi L W(K_n) \tag{1.1.15}$$

Since f is real,

$$b_n = b_{-n}^* \tag{1.1.16}$$

$$\langle |b_n'|^2 \rangle = \langle |b_n''|^2 \rangle = \pi L W(K_n) \tag{1.1.17}$$

where $b_n = b_n' + ib_n''$. The real and imaginary parts are independent Gaussian random variables. Thus $\langle b_n b_{-n}^* \rangle = \langle b_n b_n \rangle = 0$. Let L be divided with N intervals,

$$\Delta z = \frac{L}{N} \tag{1.1.18}$$

and N be a power of 2.

$$f(z_l) = \frac{1}{L}\sum_{n=-N/2+1}^{N/2} b_n \exp\left(i\frac{2\pi n l}{N}\right) \tag{1.1.19}$$

Then b_0 and $b_{N/2}$ are real. We first obtain N independent Gaussian random numbers with zero mean and unit variance. We next multiply the numbers by a normalization factor to b_0, $b_{N/2}$, b_n' and b_n'' with $n = 1, 2, \ldots, N/2 - 1$, such that (1.1.17) holds. We then use

$$b_{-n} = b_n^* \tag{1.1.20}$$

to get b_n with $n = -1, -2, ..., -N/2 + 1$. The permittivity is then calculated by (1.1.19).

1.3 Numerical Results and Applications to Antarctica

In this section we illustrate the numerical results and application to the Antarctic firn. The Antarctic firn has a layering structure. The permittivity of snow is around $1.5\epsilon_o$. Thus if a half-space medium is assumed, the reflectivity at $\theta_o = 0$ is 0.01 so that with $T = 240$ K, the brightness temperature is 237.6 K. However, the measured brightness temperature is significantly less than that. The difference can be attributed to the reflections by the layering structure. Snow is a mixture of ice and air, so that the density of snow ρ indicates the fractional volume of ice in snow. Ice has a density of 0.91 g/cm^3. The density of snow is

$$\rho = 0.91\ f \tag{1.1.21}$$

where f is the fractional volume of ice in snow. We model $\rho(z)$ as a random process. First we assume that $\rho(z)$ is a Gaussian random process with

$$\langle \rho \rangle = \rho_m \tag{1.1.22}$$

$$\rho(z) = \rho_m + \rho_f(z) \tag{1.1.23}$$

$$\langle \rho_f(z_1)\rho_f(z_2) \rangle = \sigma_\rho^2 \exp\left(-\frac{|z_1 - z_2|}{l_z} \right) \tag{1.1.24}$$

In Fig. 1.1.2 we show a simulated density profile of a single realization using $l_z = 2$ mm, $\sigma_\rho = 0.0156$ g/cm^3, and $\rho_m = 0.4$ g/cm^3. The continuous profile of each realization is generated down to a depth of 21 m and is discretized with $2^{14} = 32,768$ layers. This gives a Δz discretization thickness of 0.6 mm. The permittivity of each layer is calculated using the following empirical mixing formula for dry snow

$$\frac{\epsilon'}{\epsilon_o} = 1 + \frac{1.60\,\rho}{1 - 0.35\,\rho} \tag{1.1.25}$$

$$\frac{\epsilon''}{\epsilon_o} = \frac{\epsilon''_{ice}}{\epsilon_o}(0.52\,\rho + 0.62\,\rho^2) \tag{1.1.26}$$

where ϵ''_{ice} is the imaginary part of the permittivity of pure ice and is temperature-dependent. If $\sigma_\rho \ll \rho_m$, we can linearize (1.1.25) and (1.1.26), so that ϵ'_m and ρ_m are related by the same relation as in (1.1.25) and ϵ''_m and ρ_m are related by the same relation as in (1.1.26).

$$\frac{\epsilon'_m}{\epsilon_o} = 1 + \frac{1.60\rho_m}{1 - 0.35\rho_m} \tag{1.1.27a}$$

$$\frac{\epsilon''_m}{\epsilon_o} = \frac{\epsilon''_{ice}}{\epsilon_o}(0.52\rho_m + 0.62\rho_m^2) \tag{1.1.27b}$$

Let $\epsilon = \epsilon_m + \epsilon_f$. Then ϵ_f, the fluctuating part is

$$\frac{\epsilon'_f}{\epsilon_o} = \frac{1.60}{(1 - 0.35\,\rho_m)^2}\rho_f \tag{1.1.28}$$

$$\frac{\epsilon''_f}{\epsilon_o} = \frac{\epsilon''_{ice}}{\epsilon_o}(0.52 + 1.24\,\rho_m)\rho_f \tag{1.1.29}$$

The permittivity fluctuation is also a Gaussian random process.

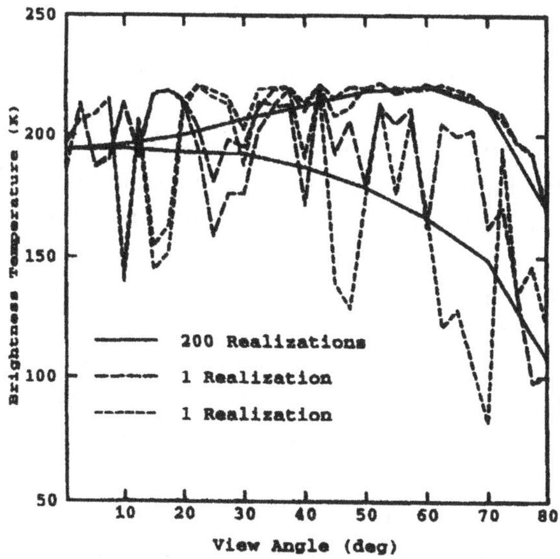

Figure 1.1.3 Computed brightness temperatures using fluctuation dissipation theory and a discretized continuous Gaussian random density profile with a mean density of 0.4 g/cm^3, a correlation length of 2 mm, and a standard deviation in density of 0.0156 g/cm^3. The temperature profile is: $T(z) = 222+34\exp(0.81\,z)$ (z is 0 at the top and negative in the medium). Both profiles are carried down to a depth of 21 m, and are discretized into 32,768 layers, giving a layer thickness of 0.6 mm. Both vertically and horizontally polarized brightness temperatures are shown, with $T_v > T_h$.

The brightness temperature of such a profile is then calculated using the layered medium model of Chapter 5 of Volume I. The layered medium model also accounts for a temperature profile. The temperature profile is assumed to have the form of $T_o + T_h \exp(\gamma z)$. This form of temperature profile is characteristic of Antarctic firn in the summer. We use Eqs. (5.2.36a) and (5.2.36b) of Volume I to compute the brightness temperature. Figure 1.1.3 shows the brightness temperatures of two different realizations which have the same correlation length and variance as the realization shown in Fig. 1.1.2. A single realization of permittivity profile means that a single sample of random permittivity fluctuation is produced. The brightness temperatures are then computed. Since the layered medium model is based on a coherent approach, the results have to be averaged over many realizations. In a real-life situation, there can be built-in incoherence. For example, the interfaces of the layered structure can be rough with rms height larger than 1/16 of a wavelength so that the reflection is rendered incoherent. In Fig. 1.1.3, we also show the brightness temperature averaged over 200 realizations. The averaging shows that all the coherent oscillations have been smoothed out in the average. However, in order to produce a significant amount of decrease in brightness temperature due to reflection, a small correlation length of 2 mm is required as shown in Fig. 1.1.2. The reasons are that the permittivity in a Gaussian random process is a continuous function of z. As we see in some of the examples in Chapter 5 of Volume I, a continuous profile usually does not produce much reflection. In order to produce reflection, the continuous profile must have a substantial change of permittivity over a short distance, say 2 mm. Over a long distance of 2.5 m, this means more than 1200 reflections. Thus to match brightness temperature experimental data, a small standard deviation of 0.0156 g/cm^3 is used for the density fluctuations. This will reduce the amount of power reflections for each reflection. Hence, the choice of parameters of $l_z = 2$ mm and $\sqrt{\langle \rho_f{}^2 \rangle} = 0.0156$ g/cm^3 is dictated by the underlying assumption of Gaussian random process.

By choosing the parameters of the Gaussian random process, we can get reasonable brightness temperatures. However, the question is whether these physical parameters are reasonable. In the next section we discuss the discrete layering model.

2 Random Discrete Layering and Applications

The ground truth measurements of snow density profile have been taken by Rott et al. [1993]. Figure 1.2.1 illustrates the results for the station at Vestestraumen. The profiles indicate discrete layering with layer thickness of the order of 5 cm. Abrupt changes of densities are associated with discrete layering. The ground truth layering geometry does not correspond to the continuous Gaussian random process that is discussed in Section 1. Thus a satisfactory theory needs to (1) give an accurate and correct solution of Maxwell's equations; (2) match experimental measurements, and (3) use a correct medium characterization with parameters that are verified by ground truth measurements.

To incorporate ground truth data on the density profile, we apply a centimeter discrete layering model instead of a millimeter correlation length continuous random medium model. The measured profile in Fig. 1.2.1 shows a background trend of the form $1 - \exp(\delta z)$ $(z < 0)$ which saturates below about 1 m of depth. A theoretical profile similar to the measured profile can be generated using two random variables and a background exponential trend. The two random variables are the layer thickness and the density. The layer thickness is assumed to be exponentially distributed. The layer-to-

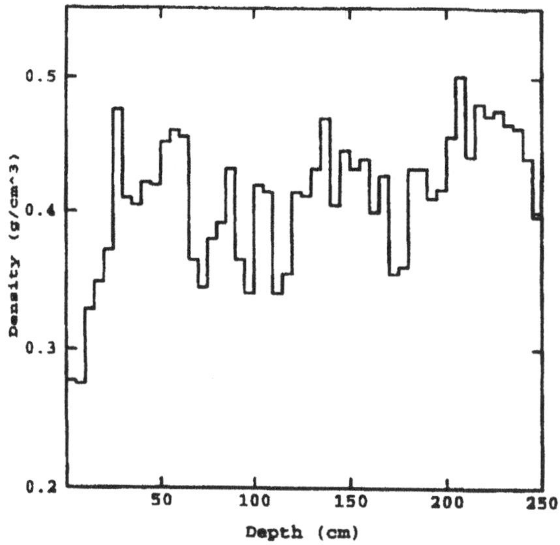

Figure 1.2.1 The snow density profile measured by H. Rott at Vestestraumen. The profile was obtained by measuring the average density in consecutive 5-cm layers.

layer density variations are assumed to be normally distributed and are then superimposed on top of the background exponential profile. The profile is carried down to 10-m depth. Below this depth, layered scattering is expected to drop off. The temperature profile is again assumed to have the form of $T_o + T_h \exp(\gamma z)$. Since the measurements were made in the Antarctic summer, the temperature distribution decays from a high temperature at the surface down to the mean annual temperature at depths greater than 10 m. Figure 1.2.2 shows two sample theoretical profiles generated. The layer thickness is exponentially distributed with a mean layer thickness of 4.1 cm. The density is

$$\rho(z) = 0.25 + 0.16\,[1 - \exp(0.05z)] + \rho_f(z) \tag{1.2.1}$$

with z in centimeters. The fluctuating part $\rho_f(z)$ is normally distributed with standard deviation of of 0.04 g/cm^3. Both parameters of layer thickness and density fluctuation are very different from that of the continuous random medium model. The continuous random medium model used $l_z = 2$ m and density fluctuation of 0.0156 g/cm^3. Figure 1.2.3 shows the corresponding histogram of the densities in a theoretical profile along with the histogram of the actual densities. The background exponential profile was set to provide the correct density level in the theoretical profile. Using the layered medium model, the theoretical brightness temperatures for a range of ob-

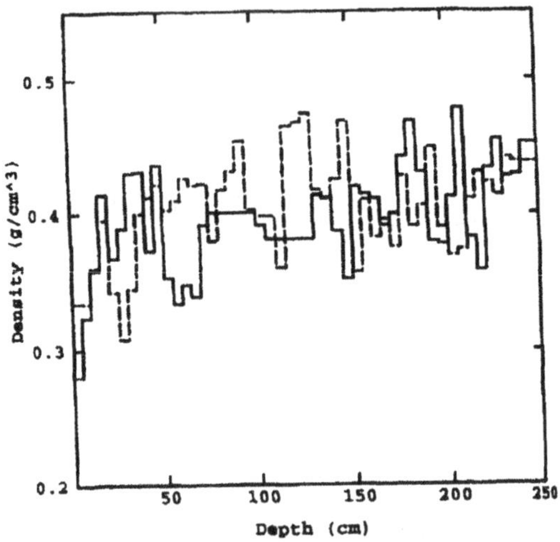

Figure 1.2.2 Two sample realizations of a discrete random profile. The profiles are generated with an exponential distribution of layer thickness and a mean thickness of 4.1 cm. The density in each layer is the sum of a deterministic part given by 0.25 (g/cm^3) + 0.16(1 − exp(z/(20 cm)))(g/cm^3) and a normally distributed random variable with zero mean and a standard deviation of 0.04 g/cm^3 (z is 0 at the top and negative in the medium). For display purposes, the resulting profile is then averaged over 5-cm intervals just as Rott's measured profile was (see Fig. 1.1.2). Calculation of brightness temperatures uses the original profile without the 5-cm averaging.

servation angles are computed and averaged over many realizations. Figure 1.2.4 shows the results when 600 such realizations are averaged. The measured brightness temperatures are also shown in Fig. 1.2.4, along with the results which would be expected from a uniform density profile and the same temperature profile. The layered scattering model gives good agreement of brightness temperature with the measured results out to an observation angle of 50°. Note that the discrete layering used by the theory (Fig. 1.2.2) is statistically similar to the observed profile (Fig. 1.2.1). No ad-hoc correction factors are required. The half-space model of uniform density does not give adequate results. The half-space model results show much less polarization contrast. The half-space model of homogeneous density gives much higher brightness temperatures that does not agree with experimental observations. These simulations show that layer scattering with centimeter layer thickness is an important phenomenon at C-band passive remote sensing of Antarctica.

Extensions have been made to a Poisson process with a modified spectral density. Good agreement with data have also been found for Vestestraumen, Amundsen Ice and Base Camp stations [West, 1994].

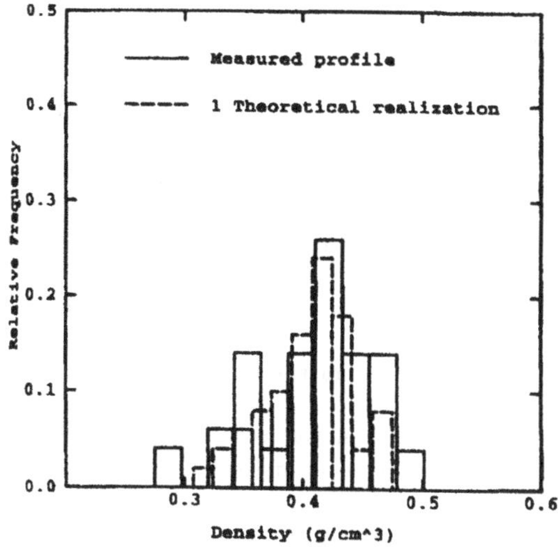

Figure 1.2.3 A histogram of the measured density profile superimposed on a histogram of one realization of a density profile using the discrete random profile described in Fig. 1.2.2.

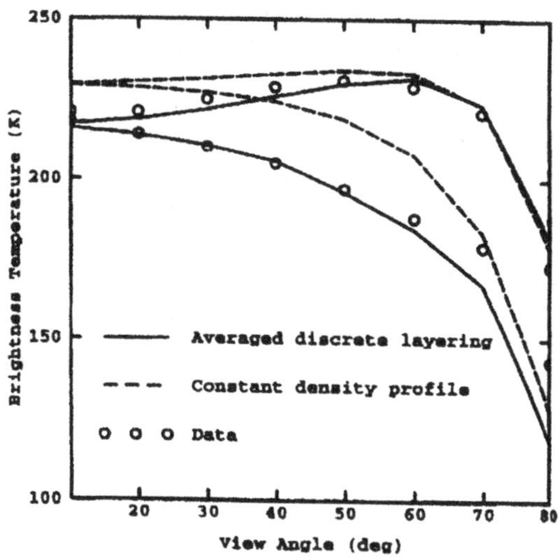

Figure 1.2.4 A comparison of the measured brightness temperatures with theoretical results using the discrete random profile described in Fig. 1.2.2. Average layer thickness is 4.1 cm, standard deviation of density variation is 0.04 g/cm^3, and a background trend is given by 0.25 (g/cm^3)+0.16(1−exp(z/(20 cm)))(g/cm^3). The temperature profile is: $T(z) = 234\ x + 24\exp(0.81\ z)$ (z is 0 at the top and negative in the medium). Both profiles are carried down to a depth of 10 m. Vertically and horizontally polarized brightness temperatures are shown, with $T_v > T_h$ at all viewing angles.

REFERENCES AND ADDITIONAL READINGS

Fung, A. K. (1994), *Microwave Scattering and Emission Models and Their Applications*, Artech House, Norwood, Massachusetts.

Gurvich, A. S., V. L. Kalinin, and D. T. Matveyer (1973), Influence of the internal structure of glaciers on their thermal radio emission, *Atm. Oceanic Phys. USSR*, **9**, 713–717.

Hall, D. K. and J. Martinec (1985), *Remote Sensing of Snow and Ice*, Chapman and Hall, London.

Kong, J. A. (1990), *Electromagnetic Wave Theory*, 2nd edition, John Wiley & Sons, New York.

Mätzler, C. (1987), Applications of the interaction of microwaves with the natural snow cover, *Remote Sens. Rev.*, **2(2)**, 259–387.

Rott, H., K. Sturm, and H. Miller (1993), Active and passive microwave signatures of Antarctic firn by means of field measurements and satellite data, *Annals of Glaciology*, **17**, 337–343.

Sheng, P., B. White, Z.-Q. Zhang, and G. Papanicolaou (1990), Wave localization and multiple scattering in randomly-layered media, in *Scattering and Localization of Classical Waves in Random Media*, edited by P. Sheng, 563–619, World Scientific, Singapore.

Stogryn, A. (1970), The brightness temperature of a vertically structured layer, *J. Geophys. Res.*, **80**, 4484–4496.

Tsang, L. and J. A. Kong (1975), The brightness temperature of a half-space random medium with nonuniform temperature profile, *Radio Sci.*, **10(12)**, 1025–1033.

Tsang, L., J. A. Kong, and R. T. Shin (1985), *Theory of Microwave Remote Sensing*, Wiley-Interscience, New York.

Tsang L., E. Njoku, and J. A. Kong (1975), Microwave thermal emission from a stratified medium with nonuniform temperature distribution, *J. Appl. Phys.*, **46(12)**, 5127–5133.

Ulaby, F. T., R. K. Moore, and A. K. Fung (1981), *Microwave Remote Sensing: Active and Passive*, **1** and **2**, Addison-Wesley, Reading, MA.

West, R. (1994), Microwave emission of polar firn, Ph.D. thesis, University of Washington, Seattle.

West, R., D. P. Winebrenner, L. Tsang, and H. Rott (1996), Microwave emission from density stratified Antarctic firn at 6 cm wavelength, *J. of Glaciology*, **42(140)**, 63–76.

Chapter 2

INTEGRAL EQUATION FORMULATIONS AND BASIC NUMERICAL METHODS

In this chapter, we discuss integral equation formulation of boundary value problems. We will give a brief overview with more detailed analysis to be given in subsequent chapters where the numerical methods are applied to specific problems. We will also describe the basic numerical methods that include the method of moments, discrete dipole approximation, product of Toeplitz matrix and column vector using FFT, and conjugate gradient method. These numerical methods will be used extensively in subsequent chapters.

1 Integral Equation Formulation for Scattering Problems

1.1 Surface Integral Equations

Integral equations can be derived by using Green's functions. The advantage of surface integral equation is that they reduce the dimension of the problem by one. In the following we first consider a two-dimensional scattering problem so that the surface integral equation becomes one-dimensional.

Consider an incident wave impinging upon an object. Let the problem be uniform in the $\hat{y}$-direction so that there is no variation in the $\hat{y}$-direction i.e., $\frac{\partial}{\partial y} = 0$ (Fig. 2.1.1).

We first consider the TE case with electric field in the $\hat{y}$-direction

$$\overline{E}_{inc} = \hat{y}\psi_{inc}(x,z) \tag{2.1.1}$$

$$\overline{E} = \hat{y}\psi(x,z) \tag{2.1.2}$$

The $\overline{H}$ field is

$$i\omega\mu\overline{H} = \hat{x}\left(-\frac{\partial\psi}{\partial z}\right) + \hat{z}\frac{\partial\psi}{\partial x} \tag{2.1.3}$$

For the TM case, the $\overline{H}$ field will be in the $\hat{y}$-direction so that

$$\overline{H}_{inc} = \hat{y}\psi_{inc}(x,z) \tag{2.1.4}$$

$$\overline{H} = \hat{y}\psi(x,z) \tag{2.1.5}$$

Then the electric field is

$$-i\omega\epsilon\overline{E} = \hat{x}\left(-\frac{\partial\psi}{\partial z}\right) + \hat{z}\frac{\partial\psi}{\partial x} \tag{2.1.6}$$

If the scattering object is a perfect electric conductor, then the boundary condition is

$$\hat{n}\times\overline{E} = 0 \tag{2.1.7}$$

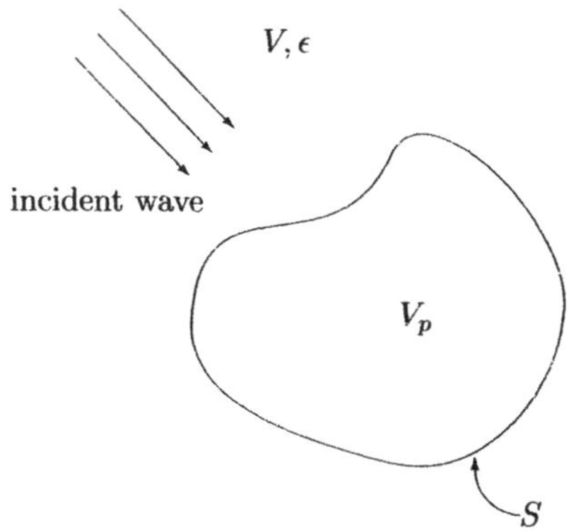

Figure 2.1.1 2-D scattering problem with 1-D boundary.

Let $\hat{n} = n_x\hat{x} + n_z\hat{z}$. We note from (2.1.6) that

$$\hat{n} \times \left(-\hat{x}\frac{\partial \psi}{\partial z} + \hat{z}\frac{\partial \psi}{\partial x}\right) = -\hat{y}\left(n_x\frac{\partial \psi}{\partial x} + n_z\frac{\partial \psi}{\partial z}\right) = -\hat{y}\frac{\partial \psi}{\partial n} \tag{2.1.8}$$

where $\dfrac{\partial \psi}{\partial n} = \hat{n} \cdot \nabla \psi$.

Thus for two-dimensional problem, we can simply use ψ. Then

$$\nabla^2 \psi + k^2 \psi = 0 \tag{2.1.9}$$

(i) For TE problem of perfect electric conductor, the boundary condition is

$$\psi = 0 \qquad \text{at } S \tag{2.1.10}$$

(ii) For TM problem of perfect electric conductor, the boundary condition is

$$\frac{\partial \psi}{\partial n} = 0 \qquad \text{at } S \tag{2.1.11}$$

For two-dimensional problems, the Green's function is

$$g(\overline{\rho}, \overline{\rho}') = \frac{i}{4} H_0^{(1)}\left(k|\overline{\rho} - \overline{\rho}'|\right) \tag{2.1.12}$$

where $\overline{\rho} = x\hat{x} + z\hat{z}$ and $H_0^{(1)}$ is the zeroth order Hankel function of the first kind.

The Green's function obeys the equation

$$\left(\nabla^2 + k^2\right) g(\overline{\rho}, \overline{\rho}') = -\delta\left(\overline{\rho} - \overline{\rho}'\right) \tag{2.1.13}$$

Applying the scalar Green's theorem

$$\int d\overline{\rho}\left(\psi_1 \nabla^2 \psi_2 - \psi_2 \nabla^2 \psi_1\right) = \oint d\overline{s} \cdot \left(\psi_1 \nabla \psi_2 - \psi_2 \nabla \psi_1\right) \tag{2.1.14}$$

Let $\psi_2 = \psi$ and $\psi_1 = g$ in (2.1.14). Then

$$\int d\overline{\rho}\left(g\nabla^2\psi - \psi\nabla^2 g\right)$$

$$= \int d\overline{\rho}\left[g\left(-k^2\psi\right) - \psi\left(-k^2 g - \delta\left(\overline{\rho} - \overline{\rho}'\right)\right)\right]$$

$$= \begin{cases} \psi(\overline{\rho}') & \text{if } \overline{\rho}' \text{ in } V \\ 0 & \text{if } \overline{\rho}' \text{ in } V_p \end{cases} = -\int dS\left(g\frac{\partial\psi}{\partial n} - \psi\frac{\partial g}{\partial n}\right) + \psi_{inc}\left(\overline{\rho}'\right) \quad (2.1.15)$$

where $\hat{n}$ is the outward normal to surface S. In (2.1.14), the surface integral at infinity gives the incident wave $\psi_{inc}(\overline{\rho})$. For TE case of PEC, we apply (2.1.10) to get

$$\begin{cases} \psi(\overline{\rho}') & \text{if } \overline{\rho}' \text{ in } V \\ 0 & \text{if } \overline{\rho}' \text{ in } V_p \end{cases} = -\int dS g(\overline{\rho}, \overline{\rho}')\frac{\partial\psi}{\partial n}(\overline{\rho}) + \psi_{inc}\left(\overline{\rho}'\right) \quad (2.1.16)$$

To obtain a surface integral equation, we let $\overline{\rho}'$ approach surface S, then we have

$$0 = \psi_{inc}(\overline{\rho}') - \int dS g\left(\overline{\rho}, \overline{\rho}'\right)\frac{\partial\psi}{\partial n}(\overline{\rho}) \quad (2.1.17)$$

for $\overline{\rho}' \in S$.

Equation (2.1.17) is a surface integral equation because $\overline{\rho}$ and $\overline{\rho}'$ both belong to the same domain S. Equation (2.1.17) is Fredholm equation of the first kind. The source domain $\overline{\rho}$ and the field domain $\overline{\rho}'$ belong to the same domain. As $\overline{\rho} \to \overline{\rho}'$, $g(\overline{\rho}, \overline{\rho}')$ is large which means the matrix equation will have strong diagonal elements which makes (2.1.17) well conditioned usually. This is unlike Fredholm equation of the first kind in inverse scattering problems. In such inverse scattering problems, the source domain and the field domain are different. The matrix equation does not have large diagonal elements and the equation is often ill-conditioned.

For TM problem of PEC, we apply boundary conditions (2.1.11). Then

$$\begin{cases} \psi(\overline{\rho}') & \text{if } \overline{\rho}' \text{ in } V \\ 0 & \text{if } \overline{\rho}' \text{ in } V_p \end{cases} = \psi_{inc}\left(\overline{\rho}'\right) + \int dS\psi(\overline{\rho})\frac{\partial g}{\partial n}\left(\overline{\rho}, \overline{\rho}'\right) \quad (2.1.18)$$

To get a surface integral equation, we let $\overline{\rho}'$ approach surface S. However, since $\psi(\overline{\rho}')$ experiences a discontinuity of S from V to V_p which is also manifested in the fact that $\frac{\partial g}{\partial n}\left(\overline{\rho}, \overline{\rho}'\right)$ has a non-integrable singularity at $\overline{\rho} = \overline{\rho}'$, care has to be exercised. We will address these problems later when we implement these equations numerically in later chapters. At this point

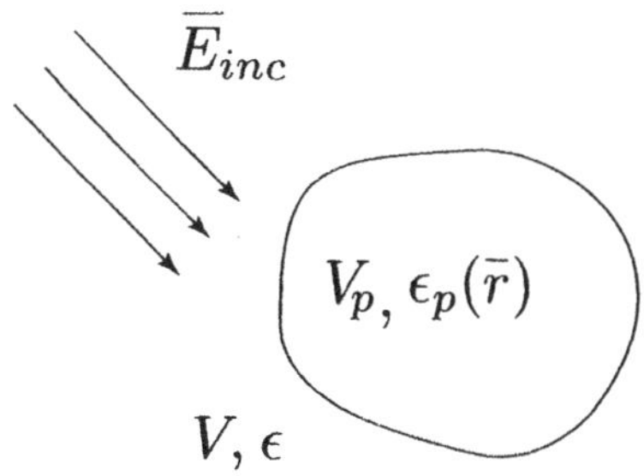

Figure 2.1.2 An incident wave impinges on a scatterer with permittivity $\epsilon_p(\overline{r})$ and volume V_p.

we let $\overline{\rho}' \to \overline{\rho}'_+$ when $+$ stands for the fact that we approach S from V. Then

$$\psi\left(\overline{\rho}'_+\right) = \psi_{inc}\left(\overline{\rho}'\right) + \lim_{\overline{\rho}' \to \overline{\rho}'_+} \int dS \psi\left(\overline{\rho}\right) \frac{\partial g}{\partial n}\left(\overline{\rho}, \overline{\rho}'\right) \tag{2.1.19}$$

for $\overline{\rho}'$ on S.

1.2 Volume Integral Equations

Consider an incident wave $\overline{E}^{inc}(\overline{r})$ impinging upon the scatterer with permittivity $\epsilon_p(\overline{r})$ (Fig. 2.1.2). Maxwell's equations are

$$\nabla \times \overline{E} = +i\omega\mu\overline{H} \tag{2.1.20}$$

$$\nabla \times \overline{H} = -i\omega\epsilon_p(\overline{r})\overline{E} \tag{2.1.21}$$

where

$$\epsilon_p(\overline{r}) = \begin{cases} \epsilon_p(\overline{r}) & \text{if } \overline{r} \in V_p \\ \epsilon & \text{if } \overline{r} \in V \end{cases} \tag{2.1.22}$$

It follows that

$$\nabla \times \nabla \times \overline{E} - k^2\overline{E} = k^2\left(\epsilon_r(\overline{r}) - 1\right)\overline{E} \tag{2.1.23}$$

where $\epsilon_r(\overline{r}) = \epsilon_p(\overline{r})/\epsilon$ is the relative permittivity. Thus the scattered field is

$$\overline{E}_s(\overline{r}) = \int d\overline{r}'\overline{\overline{G}}(\overline{r}, \overline{r}') \cdot k^2\left(\epsilon_r(\overline{r}') - 1\right)\overline{E}\left(\overline{r}'\right) \tag{2.1.24}$$

Adding $\overline{E}_s(\overline{r})$ to $\overline{E}^{inc}(\overline{r})$ gives the total field $\overline{E}(\overline{r})$. Then

$$\overline{E}(\overline{r}) = \overline{E}^{inc}(\overline{r}) + k^2 \int d\overline{r}'\overline{\overline{G}}(\overline{r}, \overline{r}') \cdot \left(\epsilon_r(\overline{r}') - 1\right)\overline{E}(\overline{r}') \tag{2.1.25}$$

We can also regard

$$\overline{P}(\overline{r}) = \epsilon\left(\epsilon_r(\overline{r}) - 1\right)\overline{E}(\overline{r}) \tag{2.1.26}$$

as the induced polarization density. When solving (2.1.25) numerically, we have to take into account the singularity of the dyadic Green's function.

A volume integral equation can also be obtained by using vector and scalar potentials.

$$\nabla \times \overline{E} = +i\omega\mu\overline{H} \tag{2.1.27}$$

$$\begin{aligned}\nabla \times \overline{H} &= -i\omega\epsilon_p(\overline{r})\overline{E} \\ &= -i\omega\epsilon\overline{E} - i\omega\epsilon(\epsilon_r(\overline{r}) - 1)\overline{E}\end{aligned} \tag{2.1.28}$$

$$\nabla \cdot \overline{H} = 0 \tag{2.1.29}$$

$$\nabla \cdot \left(\epsilon_p(\overline{r})\overline{E}\right) = 0 = \nabla \cdot \left[(\epsilon + (\epsilon_r(\overline{r}) - 1)\epsilon)\,\overline{E}\right] \tag{2.1.30}$$

Using equivalent current of

$$\overline{J}(\overline{r}) = -i\omega\epsilon\left(\epsilon_r(\overline{r}) - 1\right)\overline{E}(\overline{r}) \tag{2.1.31}$$

From (2.1.30), $\nabla \cdot (\epsilon\overline{E}) = \nabla \cdot [\epsilon\overline{E} - \epsilon_r(\overline{r})\epsilon\overline{E}]$. Thus

$$\rho_f(\overline{r}) = \frac{1}{i\omega}\nabla \cdot \overline{J}(\overline{r}) = -\epsilon\nabla \cdot \left[\left(\epsilon_r(\overline{r}) - 1\right)\overline{E}(\overline{r})\right] \tag{2.1.32}$$

We can make use of the scalar Green's function

$$g(\overline{r}, \overline{r}') = \frac{e^{ik|\overline{r} - \overline{r}'|}}{4\pi|\overline{r} - \overline{r}'|} \tag{2.1.33}$$

The solutions of scalar and vector potentials are

$$\overline{A}(\overline{r}) = \int d\overline{r}'\, g(\overline{r}, \overline{r}')\mu\overline{J}(\overline{r}') \tag{2.1.34}$$

$$\Phi(\overline{r}) = \int d\overline{r}'\, g(\overline{r}, \overline{r}')\frac{\rho_f(\overline{r}')}{\epsilon} \tag{2.1.35}$$

Since $\overline{E}_s = i\omega\overline{A} - \nabla\Phi$, we have

$$\overline{E}_s(\overline{r}) = i\omega \int d\overline{r}'\, g(\overline{r}, \overline{r}')\mu\overline{J}(\overline{r}') - \nabla \int d\overline{r}'\, g(\overline{r}, \overline{r}')\frac{\rho_f(\overline{r}')}{\epsilon} \tag{2.1.36}$$

Using (2.1.31) and (2.1.32) in (2.1.36), we get

$$\begin{aligned}\overline{E}(\overline{r}) = \overline{E}^{inc}(\overline{r}) &+ k^2 \int d\overline{r}'\, g(\overline{r}, \overline{r}')\left(\epsilon_r(\overline{r}') - 1\right)\overline{E}(\overline{r}') \\ &+ \nabla \int d\overline{r}'\, g(\overline{r}, \overline{r}')\nabla' \cdot \left[\left(\epsilon_r(\overline{r}') - 1\right)\overline{E}(\overline{r}')\right]\end{aligned} \tag{2.1.37}$$

The volume integral over $d\overline{r}'$ is carried over infinite space. Thus if $\epsilon(\overline{r}')$ is discontinuous, the divergence term inside the integrand of (2.1.37) can give rise to surface charge density. This happens when there is a boundary separating two media of different permittivities.

1.3 Dyadic Green's Function Singularity and Electrostatics

From (2.1.25), the volume integral equation is in terms of the dyadic Green's function. Using

$$\overline{\overline{G}}(\overline{r},\overline{r}') = \left(\overline{\overline{I}} + \frac{\nabla\nabla}{k^2}\right) g(\overline{r},\overline{r}') \tag{2.1.38}$$

we have

$$\overline{E}(\overline{r}) = \overline{E}^{inc}(\overline{r}) + k^2 \int d\overline{r}'(\epsilon_r(\overline{r}') - 1)\overline{E}(\overline{r}') + \int d\overline{r}'(\nabla\nabla g(\overline{r},\overline{r}')) \cdot (\epsilon_r(\overline{r}') - 1)\overline{E}(\overline{r}') \tag{2.1.39}$$

To take into account of the singularity of the dyadic Green's function, let

$$\overline{\overline{G}}(\overline{r},\overline{r}') = PV\overline{\overline{G}}(\overline{r},\overline{r}') - \frac{\overline{\overline{L}}\delta(\overline{r}-\overline{r}')}{k^2} \tag{2.1.40}$$

where PV stands for principal value integral and $\delta(\overline{r}-\overline{r}')$ is the three-dimensional Dirac delta function. The volume integration in PV is over the volume with a volume V_δ excluded from the observation point $\overline{r}$ (Fig. 2.1.3). The volume V_δ is infinitesimal. Nevertheless the shape of the exclusion volume has to be specified. The dyad $\overline{\overline{L}}$ depends on the shape of the exclusion volume.

The volume integral equation becomes

$$\overline{E}(\overline{r}) = \overline{E}^{inc}(\overline{r}) + \int_{V-V_\delta} d\overline{r}'\overline{\overline{G}}(\overline{r},\overline{r}')(\epsilon_r(\overline{r}') - 1)k^2\overline{E}(\overline{r}') + \overline{E}_{\text{self}}(\overline{r}) \tag{2.1.41}$$

where

$$\overline{E}_{\text{self}}(\overline{r}) = -(\epsilon_r(\overline{r}) - 1)\,\overline{\overline{L}}\cdot\overline{E}(\overline{r}) \tag{2.1.42}$$

is the self field. The source is an infinitesimal volume V_δ of relative permittivity $\epsilon_r = \epsilon_p/\epsilon$ (i.e. relative to the background ϵ), and with internal field $\overline{E}(\overline{r})$ inside V_δ. On the other hand, the first two terms on the right-hand side of (2.1.41) is the contribution to $\overline{E}(\overline{r})$ from the incident field and its surrounding medium. Since $V_\delta \to 0$, the self field $\overline{E}_{\text{self}}(\overline{r})$ can be interpreted in terms of electrostatic field created by free charge and polarization charge densities and surface charge densities.

On the other hand, if the vector potential and scalar potential are used, we have the volume integral equation of (2.1.37). The subject of singularity of the dyadic Green's function are discussed in [van Bladel, 1961; Livesay and Chen, 1974; Yaghjian, 1980; Tsang et. al. 1985; Chew, 1990]. We note that by comparing (2.1.37) and (2.1.39) that the term due to electric current

$\overline{J}$ is benign. The singularity comes from the electrostatic part of electric field due to charges that are in the third term of (2.1.37) and (2.1.39). Thus we shall use electrostatics and examine how electric charges and polarization charges produce electric field in the source region.

In the electrostatic limit

$$g(\overline{r}, \overline{r}') = \frac{1}{4\pi|\overline{r} - \overline{r}'|} \tag{2.1.43}$$

The electrostatic equations are

$$\nabla \cdot \overline{D} = \rho_f \tag{2.1.44}$$

$$\nabla \times \overline{E} = 0 \tag{2.1.45}$$

The electrostatic potential is Φ such that

$$\overline{E} = -\nabla\Phi \tag{2.1.46}$$

The boundary condition is

$$\hat{n} \cdot (\overline{D}_{\text{out}} - \overline{D}_{\text{in}}) = \rho_s \tag{2.1.47}$$

where "out" and "in" represent the outside and inside of the particle respectively. If we use polarization density, with ϵ being the background permittivity,

$$\overline{D} = \epsilon\overline{E} + \overline{P} \tag{2.1.48}$$

$$\overline{P} = (\epsilon_p(\overline{r}) - \epsilon)\overline{E} \tag{2.1.49}$$

Then

$$\nabla \cdot \epsilon\overline{E} = \rho_f + \rho_p \tag{2.1.50}$$

where

$$\rho_p = -\nabla \cdot P = -\nabla \cdot \left[(\epsilon_p(\overline{r}) - \epsilon)\overline{E}\right] \tag{2.1.51}$$

is the polarization charge density. The surface polarization charge is

$$-\hat{n} \cdot (\overline{P}_{\text{out}} - \overline{P}_{\text{in}}) = \sigma_p \tag{2.1.52}$$

The electrostatic potential can be calculated from the superposition integral of these charges

$$\begin{aligned} \Phi(\overline{r}) &= \frac{1}{\epsilon}\int d\overline{r}'\, g(\overline{r}, \overline{r}')\left[\rho_f(\overline{r}') + \rho_p(\overline{r}')\right] \\ &+ \frac{1}{\epsilon}\int dS'\, g(\overline{r}, \overline{r}')\left[\sigma_s(\overline{r}') + \sigma_p(\overline{r}')\right] \end{aligned} \tag{2.1.53}$$

For the case of an infinitesimal volume V_δ, we can assume that it is of homogeneous permittivity ϵ_p inside V_δ (Fig. 2.1.4). The field inside is $\overline{E}$.

$$\overline{P}_{\text{in}} = (\epsilon_p - \epsilon)\overline{E} \tag{2.1.54}$$

$$\overline{P}_{\text{out}} = 0 \tag{2.1.55}$$

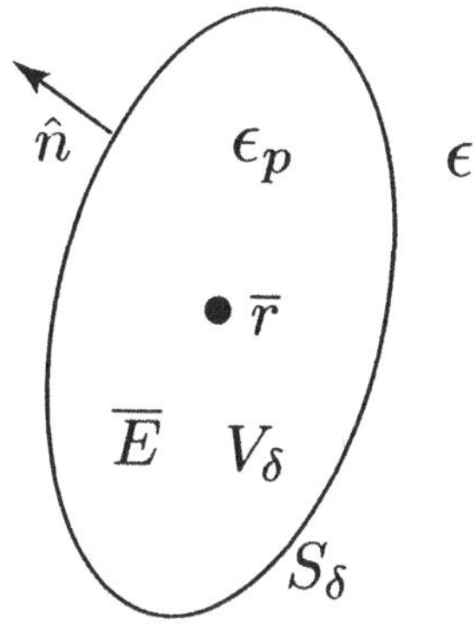

Figure 2.1.4 Field produced by $\overline{E}$ inside V_δ on itself.

Note that $\overline{P}_{\text{in}}$ = constant inside V_δ so that $\rho_p = -\nabla \cdot \overline{P}_{\text{in}} = 0$. For dielectric problem, $\rho_f = \sigma_s = 0$. Thus we only have

$$\sigma_p = -\hat{n} \cdot (\overline{P}_{\text{out}} - \overline{P}_{\text{in}}) = \hat{n} \cdot (\epsilon_r - 1)\epsilon\overline{E} \tag{2.1.56}$$

Thus for $\overline{r}$ inside V_δ, from (2.1.53) and (2.1.56),

$$\begin{aligned}\Phi_{\text{self}}(\overline{r}) &= \frac{1}{\epsilon} \int dS' g(\overline{r}, \overline{r}')\sigma_p(\overline{r}') \\ &= \int dS' \frac{1}{4\pi|\overline{r} - \overline{r}'|} \hat{n}' \cdot (\epsilon_r - 1)\overline{E}(\overline{r}')\end{aligned} \tag{2.1.57}$$

The electric field is $\overline{E}_{\text{self}} = -\nabla\Phi_{\text{self}}(\overline{r})$. Let $\hat{R} = (\overline{r} - \overline{r}')/R$ with $R = |\overline{r} - \overline{r}'|$ and $\hat{R}' = -\hat{R}$. Then $-\nabla(1/R) = \hat{R}/R^2$. Also V_δ is small so that $\overline{E}$ is constant inside V_δ.

$$\overline{E}_{\text{self}}(\overline{r}) = \int_{S_\delta} dS' \frac{\hat{R}\hat{n}'}{4\pi|\overline{r} - \overline{r}'|^2} \cdot (\epsilon_r - 1)\overline{E}(\overline{r}') = -(\epsilon_r - 1)\overline{\overline{L}} \cdot \overline{E} \tag{2.1.58}$$

where

$$\overline{\overline{L}} = \int_{S_\delta} dS' \frac{\hat{R}'\hat{n}'}{4\pi|\overline{r} - \overline{r}'|^2} \tag{2.1.59}$$

Note that $\overline{r}$ is at the center of V_δ and $\hat{R}'$ is a unit vector pointing from $\overline{r}$ to $\overline{r}'$.

Example 1: Without loss of generality, let $\overline{r} = 0$. If V_δ is a sphere of radius δ and $\delta \to 0$, then $\hat{R}' = \hat{n}' = \hat{r} = \sin\theta\cos\phi\hat{x} + \sin\theta\sin\phi\hat{y} + \cos\theta\hat{z}$. We have

$$\overline{\overline{L}} = \frac{1}{4\pi} \int_0^\pi d\theta \sin\theta \int_0^{2\pi} d\phi \ \hat{r}\hat{r} \tag{2.1.60a}$$

and

$$\int d\Omega\ \hat{r}\hat{r} = \int d\Omega(\sin\theta\cos\phi\hat{x} + \sin\theta\sin\phi\hat{y} + \cos\theta\hat{z}) \cdot (\sin\theta\cos\phi\hat{x} + \sin\theta\sin\phi\hat{y} + \cos\theta\hat{z}) = \frac{4\pi}{3}\overline{\overline{I}} \tag{2.1.60b}$$

We have used the result that on integration, the cross terms in (2.1.60) vanish. Hence

$$\overline{\overline{L}} = \frac{\hat{x}\hat{x} + \hat{y}\hat{y} + \hat{z}\hat{z}}{3} = \frac{\overline{\overline{I}}}{3} \tag{2.1.61}$$

Example 2: Let V_δ be a rectangular parallelepiped with sides equal to δ_x, δ_y and δ_z, respectively. Also let $\delta_x = a\delta$, $\delta_y = b\delta$ and $\delta_z = c\delta$ and $\delta \to 0$. Thus a, b and c are finite numbers and the ratios among them are important to determine $\overline{\overline{L}}$. In this case there are six faces for the surface integral of (2.1.60). The sum of the contributions from the top and bottom faces to $\overline{\overline{L}}$ is

$$\begin{aligned}
&\frac{1}{4\pi}\lim_{\delta\to 0}\int_{-\delta_x/2}^{\delta_x/2} dx \int_{-\delta_y/2}^{\delta_y/2} dy \frac{\hat{z}(x\hat{x} + y\hat{y} + (\delta_z/2)\hat{z}) - \hat{z}(x\hat{x} + y\hat{y} - (\delta_z/2)\hat{z})}{(x^2 + y^2 + (\delta_z/2)^2)^{3/2}} \\
&= \hat{z}\hat{z}\frac{\delta_y\delta_z}{2\pi}\int_0^{\delta_x/2} dx \frac{1}{(x^2 + (\delta_z/2)^2)(x^2 + (\delta_y/2)^2 + (\delta_z/2)^2)^{3/2}} \\
&= \hat{z}\hat{z}\frac{2}{\pi}\tan^{-1}\frac{\delta_x\delta_y}{\delta_z(\delta_x^2 + \delta_y^2 + \delta_z^2)^{1/2}}
\end{aligned} \tag{2.1.62}$$

The integration over dy and dx can be found in Gradshteyn and Ryzhik [1965]. Note that in (2.1.62), the result only depends on the ratios of the lengths of the three sides of the rectangular parallelepiped. Similar expressions can be derived for the other four faces.

$$\overline{\overline{L}} = \frac{2}{\pi}\left\{\hat{x}\hat{x}\tan^{-1}\frac{bc}{a(a^2 + b^2 + c^2)^{1/2}} + \hat{y}\hat{y}\tan^{-1}\frac{ca}{b(a^2 + b^2 + c^2)^{1/2}} + \hat{z}\hat{z}\tan^{-1}\frac{ab}{c(a^2 + b^2 + c^2)^{1/2}}\right\} \tag{2.1.63}$$

For the special case of a cube, we set $a = b = c$ in (2.1.63). That gives

$$\overline{\overline{L}} = \frac{\hat{x}\hat{x} + \hat{y}\hat{y} + \hat{z}\hat{z}}{3} = \frac{\overline{\overline{I}}}{3} \tag{2.1.64}$$

Examples of other shapes can be found in Yaghjian [1980].

2 Method of Moments

The method of moments (MoM) is a numerical technique that has been used extensively in the solution of electromagnetic boundary value problems. Many excellent texts have been written on this subject [Harrington, 1968]. The technique is used extensively in this book in Monte Carlo simulations. A characteristic of this technique is that it leads to a full matrix equation which can be solved by matrix inversion. In later chapters, we will describe techniques that can speed up the numerical solution of these matrix equations.

With the use of Green's function, integral equations can be derived. Consider a one dimensional integral equation of the form

$$\int_a^b dx' G(x, x') f(x') = c(x) \tag{2.2.1}$$

where $G(x, x')$ is the Green's function, $f(x)$ is the unknown for the domain $a \le x \le b$, and $c(x)$ is known for $a \le x \le b$. To solve (2.2.1), two sets of functions are used in the MoM: basis functions and weighting functions.

(1) Basis functions. A set of N basis functions in the domain of $a \le x \le b$ is chosen. Let the basis functions be $f_1, f_2, \ldots, f_N$. The unknown function $f(x)$ is expanded in terms of a linear combination of these basis functions.

$$f(x) = \sum_{n=1}^{N} b_n f_n(x) \tag{2.2.2}$$

The linear combination of $f_n(x)$ should well represent the unknown $f(x)$ in the domain. Substitute (2.2.2) into (2.2.1), we have

$$\sum_{n=1}^{N} b_n \int_a^b dx' G(x, x') f_n(x') = c(x) \tag{2.2.3}$$

The unknown coefficients $b_1, b_2, \ldots, b_N$ are to be determined.

(2) Next a set of N weighting functions (testing functions) $w_1(x), w_2(x), \ldots, w_N(x)$ is chosen. Multiply (2.2.3) by $w_m(x)$ and integrate over the domain

$$\sum_{n=1}^{N} b_n \int_a^b dx w_m(x) \int_a^b dx' G(x, x') f_n(x') = \int_a^b dx w_m(x) c(x) \tag{2.2.4}$$

This gives the matrix equation

$$\sum_{n=1}^{N} G_{mn} b_n = c_m \tag{2.2.5}$$

$m = 1, 2, \ldots, N$, where

$$c_m = \int_a^b dx w_m(x) c(x) = \langle w_m, c \rangle \tag{2.2.6}$$

$$G_{mn} = \int_a^b dx w_m(x) \int_a^b dx' G(x, x') f_n(x') = \langle w_m, G f_n \rangle \tag{2.2.7}$$

where the inner product notation is used.

$$\langle f, g \rangle = \int_a^b dx\, f(x)\, g(x) \tag{2.2.8}$$

Computational Considerations

Generally (2.2.5) is a full matrix equation. We note the following

(1) Matrix solution: To solve a full matrix equation of order N by full matrix inversion (e.g., Gaussian elimination) requires $\mathcal{O}(N^3)$ number of operations. This increases rapidly with N.
(2) Matrix filling: To calculate G_{mn}, $m, n = 1, 2, \ldots, N$ can be computationally intensive because there are N^2 values of G_{mn}. Also G_{mn} can require the evaluation of a double integral as given in (2.2.7). The matrix filling can be more computationally intensive than matrix solution because $G(x, x')$ can be of a complicated form. Also since there are N^2 elements of G_{mn}, this can impose a large memory requirement.
(3) The study of f_n, $n = 1, 2, \ldots, N$ is also an important subject as the choice of f_n must well represent the correct solution. Often they have to satisfy differentiation and continuity properties.

Basis Functions

Basis functions can use full domain functions such as sines, cosines, special functions, polynomials, modal solutions, etc. A set that is useful for practical problem is the subsectional basis function. This means that each f_n is only nonzero over a subsection of the domain of f.

A common choice is the pulse function (Fig. 2.2.1a)

$$f_n(x) = \begin{cases} 1 & \text{if } a_n \leq x \leq b_n \\ 0 & \text{otherwise} \end{cases} \tag{2.2.9}$$

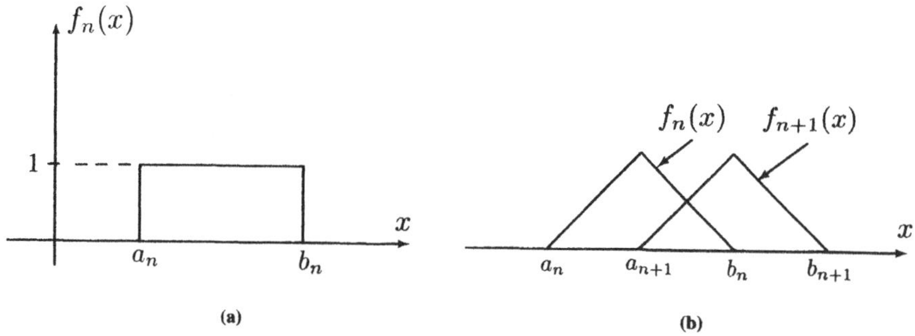

Figure 2.2.1 Common choices of basis functions: (a) pulse functions; (b) triangle functions.

where the interval $a \le x \le b$ have been divided into N intervals with end-points a_n and b_n, $n = 1, 2, \ldots, N$.

Another choice is the triangle basis functions (Fig. 2.2.1b). In Fig. 2.2.1b we show $f_n(x)$ and $f_{n+1}(x)$. Note that $f_n(x)$ and $f_{n+1}(x)$ overlap.

Weighting Functions

Two common choices are

(1) Galerkin's method. In this case the weighting functions, $n = 1, 2, \ldots, N$, are the same as the basis functions, i.e., $\omega_n(x) = f_n(x)$.

(2) Point matching. One can pick a set of points $x = x_1, x_2, \ldots, x_N$ to enforce (2.2.3). Then

$$\sum_{n=1}^{N} b_n \int_a^b dx' G(x_m, x') f_n(x') = c(x_m) \tag{2.2.10}$$

where

$$c_m = c(x_m) \tag{2.2.11}$$

$$G_{mn} = \int_a^b dx' G(x_m, x') f_n(x') \tag{2.2.12}$$

This particular choice of testing procedure is called point matching. In terms of weighting functions, this means that the weighting functions are

$$\omega_m(x) = \delta(x - x_m) \tag{2.2.13}$$

where $m = 1, 2, \ldots, N$ and δ is the Dirac delta function.

In Chapter 1 of Volume I, we have used an infinite cylinder approximation to calculate the scattering by a cylinder of finite length. The surface

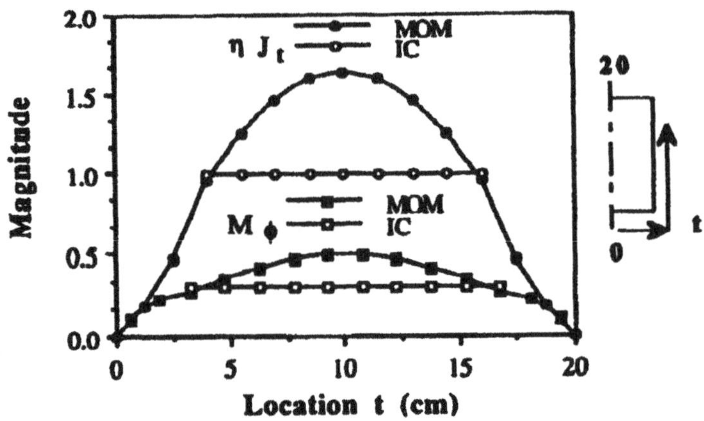

Figure 2.2.2 Comparison of total surface fields for $m = 0$ harmonic computed based on infinite cylinder approximation (IC) and MoM solution using the following parameters: frequency = 1.225 GHz, permittivity of cylinder $\epsilon_p = (6.5+i2)\epsilon_o$, radius of cylinder = 2.5 cm, length of cylinder = 15 cm.

fields are then used to calculate the scattering by integrating the surface fields only over the finite length of the cylinder. Numerically, one can use the method of moment body of revolution code [Glisson and Wilton, 1979] to calculate scattering from a finite length cylinder by solving the surface integral equations. The variations of the unknown electric and magnetic surface fields are approximated by staggered pulse functions in the t-direction and are expanded in Fourier series in the ϕ-direction.

In Fig. 2.2.2 we make a comparison of total surface fields for $m = 0$ harmonic, computed based on infinite cylinder (IC) approximation and MoM solution. The incident wave is of unit amplitude in a direction perpendicular to the axis of the cylinder with polarization parallel to the axis of the cylinder. Equivalent electric surface currents ηJ_t and equivalent magnetic surface current M_φ are shown.

The t coordinate and the t direction is as indicated in Fig. 2.2.2. The t coordinate starts from center of bottom face, radially outward to the edge, along the curved side and then ends at the center of the top face with range 2.5 cm +15 cm +2.5 cm = 20 cm. Thus the infinite cylinder approximation has uniform current densities on the curved surface of the cylinder while the MoM code predicts a maximum at the midpoint.

3 Discrete Dipole Approximation

In this section we discuss how the volume integral equation of Sections 1.2 and 1.3 can be discretized and solved numerically based on the discrete dipole approximation (DDA) [Purcell and Pennypacker, 1973; Goodman et al. 1991].

The volume integral equation is

$$\overline{E}(\overline{r}) = \overline{E}^{inc}(\overline{r}) + k^2 \int_{V-V_\delta} d\overline{r}' \overline{\overline{G}}(\overline{r},\overline{r}') \cdot (\epsilon_r(\overline{r}') - 1)\overline{E}(\overline{r}') - (\epsilon_r(\overline{r}) - 1)\,\overline{\overline{L}} \cdot \overline{E}(\overline{r}) \tag{2.3.1}$$

If we define polarization $\overline{P}$ by

$$(\epsilon_p(\overline{r}) - \epsilon)\,\overline{E}(\overline{r}) = \overline{P}(\overline{r}) \tag{2.3.2}$$

Then

$$\overline{E}(\overline{r}) = \overline{E}^{inc}(\overline{r}) + \frac{k^2}{\epsilon} \int_{V-V_\delta} d\overline{r}' \overline{\overline{G}}(\overline{r},\overline{r}') \cdot \overline{P}(\overline{r}') - (\epsilon_r(\overline{r}) - 1)\,\overline{\overline{L}} \cdot \overline{E}(\overline{r}) \tag{2.3.3}$$

We discretize (2.3.3) into volumes ΔV_j with centers at $\overline{r}_j$, $j = 1, 2, \ldots, N$. Inside ΔV_j, there is uniform field $\overline{E}_j$ and polarization $\overline{P}_j$. Then

$$\overline{E}(\overline{r}) = \overline{E}^{inc}(\overline{r}) + \frac{k^2}{\epsilon} \sum_{\substack{j=1 \\ j\neq i}}^{N} \overline{\overline{G}}(\overline{r}_i,\overline{r}_j) \cdot \overline{P}_j \Delta V_j - (\epsilon_r(\overline{r}_i) - 1)\,\overline{\overline{L}} \cdot \overline{E}_i \tag{2.3.4}$$

Let $\hat{R}$ = unit vector from $\overline{r}'$ to $\overline{r}$, $R = |\overline{r} - \overline{r}'|$ and $\overline{R} = \overline{r} - \overline{r}'$. For $\overline{r} \neq \overline{r}'$, by straightforward differentiation,

$$\overline{\overline{G}}(\overline{r},\overline{r}') = G_1(R)\overline{\overline{I}} + G_2(R)\hat{R}\hat{R} \tag{2.3.5}$$

where

$$G_1(R) = (-1 + ikR + k^2R^2)\frac{e^{ikR}}{4\pi k^2 R^3} \tag{2.3.6}$$

$$G_2(R) = (3 - 3ikR - k^2R^2)\frac{e^{ikR}}{4\pi k^2 R^3} \tag{2.3.7}$$

We let

$$\begin{aligned}\overline{\overline{A}}(\overline{r},\overline{r}') &= -\frac{k^2}{\epsilon}\overline{\overline{G}}(\overline{r},\overline{r}') \\ &= \frac{e^{ikR}}{4\pi\epsilon R^3}\left\{k^2(-R^2\overline{\overline{I}} + \overline{R}\,\overline{R}) + \frac{(1 - ikR)}{R^2}(R^2\overline{\overline{I}} - 3\overline{R}\,\overline{R})\right\}\end{aligned} \tag{2.3.8}$$

3.1 Small Cubes

If the volumes ΔV_j are cubes, then

$$\overline{p}_j = (\Delta V) P_j \tag{2.3.9}$$

where $\Delta V = V/N = d^3$ and d is the length of each volume cube. Then the discretized version of the volume integral equation is, with $\overline{\overline{L}} = \overline{\overline{I}}/3$,

$$\overline{E}_i = \overline{E}_i^{inc} - \sum_{\substack{j=1 \\ j \neq i}}^{N} \overline{\overline{A}}(\overline{r}_i, \overline{r}_j) \cdot \overline{p}_j - \left(\frac{\epsilon_{pi}}{\epsilon} - 1\right) \frac{\overline{E}_i}{3} \tag{2.3.10}$$

where $\overline{E}_i$, $\overline{E}_i^{inc}$, $\overline{p}_i$, ϵ_{pi} are the values in the ith elemental cube. Note that the second term excludes $j = i$ in accordance with principal value and the last term is a result of the $\overline{\overline{L}}$ factor associated with cubic volume. The dipole moment for each cube is

$$\overline{p}_i = d^3(\epsilon_{pi} - \epsilon)\overline{E}_i \tag{2.3.11}$$

Putting the last term in (2.3.10) on the left-hand-side and using (2.3.11) gives the matrix equation for the dipole moment of each cube

$$\overline{p}_i = \alpha_i \overline{E}_i^{inc} - \alpha_i \sum_{\substack{j=1 \\ j \neq i}}^{N} \overline{\overline{A}}_{ij} \cdot \overline{p}_j \tag{2.3.12}$$

where $\overline{\overline{A}}_{ij} = \overline{\overline{A}}(\overline{r}_i, \overline{r}_j)$,

$$\alpha_i = \alpha_i^C = 3\epsilon d^3 \left(\frac{n_i^2 - 1}{n_i^2 + 2}\right) = 3\epsilon \Delta V \frac{\left(\frac{\epsilon_{pi}}{\epsilon} - 1\right)}{\left(\frac{\epsilon_{pi}}{\epsilon} + 2\right)} \tag{2.3.13}$$

and $n = \sqrt{\epsilon_p/\epsilon}$ is the complex relative refractive index. In (2.3.13) α_i^C is the familiar Clausius-Mossoti polarizability. However α_i in (2.3.13) does not obey optical theorem and will subsequently be changed to include more correction terms.

We note that the matrix elements $\overline{\overline{A}}$ is that of the dyadic Green's function which is translational invariant. Thus the product of $\overline{\overline{A}}$ and a column vector can be computed by FFT. That is

$$\sum_{\substack{j=1 \\ j \neq i}}^{N} \overline{\overline{A}}_{ij} \cdot \overline{p}_j$$

where the summation over all cubes except the ith one can be computed by FFT. Thus solution of (2.3.12) using conjugate gradient method combined with FFT makes the solution much faster than that of the Gaussian elimination [Goodman et al. 1991].

However, if ϵ_{pi} is the same as ϵ in most of the the region, it may be more efficient to use the electric field equation from (2.3.11) and (2.3.12)

$$\frac{d^3(\epsilon_{pi}-\epsilon)}{\alpha_i}\overline{E}_i = \overline{E}_i^{inc} - \sum_{\substack{j=1\\ j\neq i}}^{N} \overline{\overline{A}}_{ij}\cdot d^3(\epsilon_{pj}-\epsilon)\overline{E}_j \tag{2.3.14}$$

where N now only need to include those cubes that have $\epsilon_{pj} \neq \epsilon$. Note also that $(\epsilon_{pi}-\epsilon)/\alpha_i$ is finite as can be seen from (2.3.13) even when $\epsilon_{pi} = \epsilon$.

For the case of a single small cube, the scattering solution is calculated by dropping the $\sum \overline{\overline{A}}_{ij}\cdot\overline{p}_j$ term from (2.3.10). The equation can be solved readily to give $\overline{E}_i = \overline{E}_i^{inc}\left[1+\frac{1}{3}\left(\frac{\epsilon_{pi}}{\epsilon}-1\right)\right]^{-1}$. We note the similarity between a small cube and a small sphere. The polarizability of a small sphere is $\alpha_s = 3v_o\epsilon(n^2-1)/(n^2+2)$ where $n = \epsilon_p/\epsilon$ and $v_o = (4\pi/3)a^3$ is the same as (2.3.13) with d^3 replaced by $(4\pi/3)a^3$. The internal field of a small sphere is $(3\epsilon/(\epsilon_p+2\epsilon))\overline{E}^{inc}$ which becomes the same as that of a small cube.

As we examine the optical theorem for a small sphere in Chapter 2, Section 8.2 of Volume I, we have noted that the Clausius-Mossoti internal field is not accurate enough because when one applies the optical theorem using the result forward scattering amplitudes, it does not give the contribution due to the scattering part. For the case of small cube, the same reasoning applies. Thus to ensure that the final scattering obeys energy conservation, one has to take into account radiative correction. In doing so, the expression of α_i in (2.3.14) is modified with the new expression given below.

3.2 Radiative Corrections

In numerical implementations, the cubes are not infinitesimal. One major correction is to improve the self-term impedance matrix element. This will give correction to the Clausius-Mossoti polarizability.

From (2.3.1), the self field $\overline{E}_{\text{self}}$ should be, for volume cube of finite size,

$$\overline{E}_{\text{self}} = k^2\int_{V_c-V_\delta} d\overline{r}'\overline{\overline{G}}(\overline{r},\overline{r}')\cdot(\epsilon_r-1)\overline{E} - (\epsilon_r-1)\overline{\overline{L}}\cdot\overline{E} \tag{2.3.15}$$

where V_c is the self cube, V_δ is the exclusion volume, and $\overline{r}$ is at the center of the cube. Thus

$$\overline{E}_{\text{self}} = k^2\overline{\overline{S}}\cdot(\epsilon_r-1)\overline{E} \tag{2.3.16}$$

where

$$\overline{\overline{S}} = \int_{V_c - V_\delta} d\overline{r}' \overline{\overline{G}}(\overline{r}, \overline{r}') - \frac{\overline{\overline{L}}}{k^2} = -\frac{\epsilon}{k^2} \int_{V_c - V_\delta} d\overline{r}' \overline{\overline{A}}(\overline{r}, \overline{r}') - \frac{\overline{\overline{L}}}{k^2} \tag{2.3.17}$$

and

$$\overline{\overline{A}} = -\frac{k^2}{\epsilon} \overline{\overline{G}} \tag{2.3.18}$$

Then the DDA equation becomes

$$\overline{E}_i = \overline{E}_i^{inc} + \frac{k^2}{\epsilon} \sum_{\substack{j=1 \\ j \neq i}}^{N} \overline{\overline{G}}(\overline{r}_i, \overline{r}_j) \cdot \overline{P}_j \Delta V_j + k^2 \overline{\overline{S}} \cdot (\epsilon_r(\overline{r}_i) - 1) \overline{E}_i \tag{2.3.19}$$

Because of the similarity between a small cube and a small sphere, a small equivalent spherical volume of radius a where $(4\pi/3)a^3 = d^3 = v_o$ will be assumed. Then, using G_1 and G_2 as given by (2.3.6) and (2.3.7)

$$\begin{aligned} \overline{\overline{S}} &= -\frac{\overline{\overline{I}}}{3k^2} + \lim_{\delta \to 0} \int_\delta^a dr' r'^2 \int_{4\pi} d\Omega' [G_1(r') \overline{\overline{I}} + G_2(r') \hat{r}' \hat{r}'] \\ &= -\frac{\overline{\overline{I}}}{3k^2} + \lim_{\delta \to 0} 4\pi \int_\delta^a dr' r'^2 \left[G_1(r') + \frac{G_2(r')}{3} \right] \overline{\overline{I}} \end{aligned} \tag{2.3.20}$$

The second equality in (2.3.20) is a result of (2.1.60b). Since

$$G_1(r') + \frac{G_2(r')}{3} = \frac{e^{ikr'}}{6\pi r'} \tag{2.3.21}$$

we have

$$\overline{\overline{S}} = \overline{\overline{I}} s \tag{2.3.22}$$

where

$$\begin{aligned} s &= -\frac{1}{3k^2} + \frac{2}{3} \lim_{\delta \to 0} \int_\delta^a dr' r' e^{ikr'} = -\frac{1}{3k^2} + \frac{2}{3k^2} \left[-1 + e^{ika}(1 - ika) \right] \\ &\simeq -\frac{1}{3k^2} + \frac{2}{3k^2} \left[\frac{k^2 a^2}{2} + \frac{ik^3 a^3}{3} \right] \end{aligned} \tag{2.3.23}$$

in the limit of small a. Putting (2.3.22) in (2.3.19)

$$\overline{E}_i = \overline{E}_i^{inc} + \left(\frac{\epsilon_{pi}}{\epsilon} - 1 \right) s k^2 \overline{E}_i - \sum_{\substack{j=1 \\ j \neq i}}^{N} \overline{\overline{A}}_{ij} \cdot \overline{p}_j \tag{2.3.24}$$

Thus

$$\overline{E}_i = \frac{\overline{E}_i^{inc}}{1 - \left(\frac{\epsilon_{pi}}{\epsilon} - 1 \right) s k^2} - \frac{1}{1 - \left(\frac{\epsilon_{pi}}{\epsilon} - 1 \right) s k^2} \sum_{\substack{j=1 \\ j \neq i}}^{N} \overline{\overline{A}}_{ij} \cdot \overline{p}_j \tag{2.3.25}$$

Multiply by $\Delta V(\epsilon_{pi} - \epsilon_i)$, and noting that

$$\overline{p}_i = \Delta V(\epsilon_{pi} - \epsilon_i)\overline{E}_i \tag{2.3.26}$$

we have

$$\overline{p}_i = \alpha_i \overline{E}_i^{inc} - \alpha_i \sum_{\substack{j=1 \\ j\neq i}}^{N} \overline{\overline{A}}_{ij} \cdot \overline{p}_j \tag{2.3.27}$$

with

$$\alpha_i = \frac{\Delta V(\epsilon_{pi} - \epsilon)}{1 - \left(\frac{\epsilon_{pi}}{\epsilon} - 1\right) sk^2} \tag{2.3.28}$$

Putting (2.3.23) in (2.3.28) gives

$$\begin{aligned} \alpha_i &= \frac{3\epsilon\Delta V\left(\frac{\epsilon_{pi}}{\epsilon} - 1\right)}{\frac{\epsilon_{pi}}{\epsilon} + 2 - 2\left(\frac{\epsilon_{pi}}{\epsilon} - 1\right)\left(\frac{k^2a^2}{2} + i\frac{k^3a^3}{3}\right)} \\ &= \frac{\alpha_i^C}{1 - \frac{2}{3\epsilon}\frac{\alpha_i^C}{\Delta V}\left[\frac{k^2a^2}{2} + \frac{ik^3a^3}{3}\right]} \end{aligned} \tag{2.3.29}$$

Using $a = (3/4\pi)^{1/3}d$ and $\Delta V = d^3$ in (2.3.29) gives

$$\alpha_i = \frac{\alpha_i^C}{1 - \frac{\alpha_i^C}{4\pi\epsilon d^3}\left[\left(\frac{4\pi}{3}\right)^{1/3} k^2d^2 + \frac{i2k^3d^3}{3}\right]} \tag{2.3.30}$$

The term with imaginary part in the denominator of (2.3.30) is known as radiative correction, which arises for the same reasoning as when scattering by Rayleigh spheres was discussed in Chapter 2, Section 8.2 of Volume I.

3.3 Other Shapes

Let the medium be discretized into rectangular parallelepipeds V_s of sizes $d_x \times d_y \times d_z$. We let $d_x = ad$, $d_y = bd$, $d_z = cd$ where a, b and c are dimensionless quantities and their ratios denote the relative sizes of the three sides of the rectangular parallelepiped. The exclusion volume V_δ for the dyadic Green's function in this case will be an infinitesimal parallelepiped with dimensions $\delta_x = a\delta$, $\delta_y = b\delta$ and $\delta_z = c\delta$ and the $\overline{\overline{L}}$ is as given in (2.1.63). Note that the ratios of the sides of the finite small rectangular parallelepiped V_s is the same as that of the infinitesimal rectangular parallelepiped V_δ. Then it is useful to write a low frequency approximation of $\overline{\overline{G}}_o(\overline{R})$. We note that for $\overline{R} \neq 0$

$$\overline{\overline{G}}(\overline{R}) = -\frac{\epsilon}{k^2}\overline{\overline{A}}(\overline{R}) = -\frac{e^{ikR}}{4\pi k^2 R^3}\left\{k^2(-R^2\overline{\overline{I}}+\overline{R}\,\overline{R})+\frac{(1-ikR)}{R^2}(R^2\overline{\overline{I}}-3\overline{R}\,\overline{R})\right\} \tag{2.3.31}$$

When expanding $\overline{\overline{G}}(R)$ of (2.3.31), it is important to see that there is a singular part of $\mathcal{O}(1/(k^2R^3))$ that is non-integrable over the origin. We also have to expand to the leading term in the imaginary part because that accounts for radiative correction. The Green's function $\overline{\overline{G}}(R)$, on expansion will give $\mathcal{O}(1/(k^2R^3)) + \mathcal{O}(1/R) + i\mathcal{O}(k)$. Thus in (2.3.31), we write $\exp(ikR) \simeq 1 + ikR - k^2R^2/2 - ik^3R^3/6$. We have to include $-ik^3R^3/6$ because this gives a term of order $i\mathcal{O}(k)$ when multiplied with the second term inside the curly bracket of (2.3.31). Thus for $\overline{R} \neq 0$ and $kR \ll 1$

$$\begin{aligned}\overline{\overline{G}}(\overline{R}) \simeq &-\frac{1}{4\pi k^2R^3}\Bigg\{k^2(-R^2\overline{\overline{I}}+\overline{R}\,\overline{R})(1+ikR)\\ &+\frac{(1-ikR)}{R^2}(R^2\overline{\overline{I}}-3\overline{R}\,\overline{R})\left(1+ikR-\frac{k^2R^2}{2}-i\frac{k^3R^3}{6}\right)\Bigg\}\\ &= -\frac{1}{4\pi k^2R^5}(R^2\overline{\overline{I}}-3\overline{R}\,\overline{R}) + \frac{1}{8\pi}\left\{\frac{1}{R^3}(R^2\overline{\overline{I}}+\overline{R}\,\overline{R})+\frac{4ik}{3}\overline{\overline{I}}\right\}\end{aligned} \tag{2.3.32}$$

Note that the imaginary part term of (2.3.32) is just the product of a constant and a unit dyad. Following (2.3.17), let

$$\overline{\overline{S}} = -\frac{\epsilon}{k^2}\int_{V_s-V_\delta} d\overline{r}'\,\overline{\overline{A}}(\overline{r},\overline{r}') - \frac{\overline{\overline{L}}}{k^2} \tag{2.3.33}$$

We use (2.3.32) to write $\overline{\overline{A}}$ as a sum of a regular part $\overline{\overline{A}}_o$ that is integrable over the origin and a singular part $\overline{\overline{A}}_s$ that is non-integrable over the origin. Thus

$$\overline{\overline{A}}(\overline{R}) \simeq \overline{\overline{A}}_o(\overline{R}) + \overline{\overline{A}}_s(\overline{R}) \tag{2.3.34}$$

where

$$\overline{\overline{A}}_o(\overline{R}) = -\frac{k^2}{8\pi\epsilon R^3}(R^2\overline{\overline{I}}+\overline{R}\,\overline{R}) - \frac{ik^3}{6\pi\epsilon}\overline{\overline{I}} \tag{2.3.35a}$$

$$\overline{\overline{A}}_s(\overline{r},\overline{r}') = \frac{1}{4\pi\epsilon R^5}(R^2\overline{\overline{I}}-3\overline{R}\,\overline{R}) \tag{2.3.35b}$$

Thus

$$\overline{\overline{S}} = -\frac{\epsilon}{k^2}\int_{V_s} d\overline{r}'\,\overline{\overline{A}}_o(\overline{r},\overline{r}') - \frac{\epsilon}{k^2}\int_{V_s-V_\delta} d\overline{r}'\,\overline{\overline{A}}_s(\overline{r},\overline{r}') - \frac{\overline{\overline{L}}}{k^2} \tag{2.3.36}$$

Note that $\overline{\overline{A}}_s$ is non-integrable over the origin. However, the origin is excluded in the integration over $V_s - V_\delta$.

The second integral in (2.3.36) can be shown generally to be zero. Here we perform it for the case of rectangular parallelepipeds of V_s and V_δ. The volume $V_s - V_\delta$ can be formed from 8 octants. In integration, the cross terms vanish. The volume integration can also be combined into that of one octant. Thus

$$\begin{aligned}
&-\frac{\epsilon}{k^2}\int_{V_s-V_\delta} d\overline{r}'\overline{\overline{A}}_s(\overline{r},\overline{r}') \\
&= -\frac{2}{\pi k^2}\left\{\int_0^{\frac{\delta_x}{2}} dx \int_0^{\frac{\delta_y}{2}} dy \int_{\frac{\delta_z}{2}}^{\frac{d_z}{2}} dz + \int_0^{\frac{\delta_x}{2}} dx \int_{\frac{\delta_y}{2}}^{\frac{d_y}{2}} dy \int_0^{\frac{d_z}{2}} dz\right. \\
&\left. + \int_{\frac{\delta_x}{2}}^{\frac{d_x}{2}} dx \int_0^{\frac{d_y}{2}} dy \int_0^{\frac{d_z}{2}} dz\right\} \times \frac{1}{(x^2+y^2+z^2)^{5/2}}\left[\hat{x}\hat{x}(-2x^2+y^2+z^2)\right. \\
&\left. + \hat{y}\hat{y}(-2y^2+x^2+z^2) + \hat{z}\hat{z}(x^2+y^2-2z^2)\right] = 0
\end{aligned} \tag{2.3.37}$$

The integral over $\overline{\overline{A}}_o$ can be performed as follows. The dyad $\overline{\overline{S}}$ is diagonal so that

$$\overline{\overline{S}} = S_x\hat{x}\hat{x} + S_y\hat{y}\hat{y} + S_z\hat{z}\hat{z} \tag{2.3.38}$$

$$\overline{\overline{L}} = L_x\hat{x}\hat{x} + L_y\hat{y}\hat{y} + L_z\hat{z}\hat{z} \tag{2.3.39}$$

and

$$\overline{\overline{S}} = \overline{\overline{D}} - \frac{\overline{\overline{L}}}{k^2} \tag{2.3.40}$$

$$\overline{\overline{D}} = D_x\hat{x} + D_y\hat{y} + D_z\hat{z} \tag{2.3.41}$$

where

$$L_x = \frac{2}{\pi}\tan^{-1}\frac{bc}{a(a^2+b^2+c^2)^{1/2}} \tag{2.3.42}$$

$$L_y = \frac{2}{\pi}\tan^{-1}\frac{ca}{b(a^2+b^2+c^2)^{1/2}} \tag{2.3.43}$$

$$L_z = \frac{2}{\pi}\tan^{-1}\frac{ab}{c(a^2+b^2+c^2)^{1/2}} \tag{2.3.44}$$

$$D_x = \frac{1}{\pi}\left\{\left[\frac{d_x}{2}\int_0^{\frac{d_y}{2}} dy \ln \frac{\frac{d_z}{2} + \sqrt{y^2 + \left(\frac{d_z}{2}\right)^2 + \left(\frac{d_x}{2}\right)^2}}{\sqrt{y^2 + \left(\frac{d_x}{2}\right)^2}}\right.\right.$$

$$\left.\left. + \int_0^{\frac{d_x}{2}} dx\, 2x \tan^{-1} \frac{\left(\frac{d_z}{2}\right)\left(\frac{d_y}{2x}\right)}{\sqrt{\left(\frac{d_z}{2}\right)^2 + x^2 + \left(\frac{d_y}{2}\right)^2}}\right] + \frac{ik}{6}\Delta V\right\} \tag{2.3.45}$$

$$D_y = \frac{1}{\pi}\left\{\left[\frac{d_y}{2}\int_0^{\frac{d_z}{2}} dz \ln \frac{\frac{d_x}{2} + \sqrt{z^2 + \left(\frac{d_x}{2}\right)^2 + \left(\frac{d_y}{2}\right)^2}}{\sqrt{z^2 + \left(\frac{d_y}{2}\right)^2}}\right.\right.$$

$$\left.\left. + \int_0^{\frac{d_y}{2}} dy\, 2y \tan^{-1} \frac{\left(\frac{d_x}{2}\right)\left(\frac{d_z}{2y}\right)}{\sqrt{\left(\frac{d_x}{2}\right)^2 + y^2 + \left(\frac{d_z}{2}\right)^2}}\right] + \frac{ik}{6}\Delta V\right\} \tag{2.3.46}$$

$$D_z = \frac{1}{\pi}\left\{\left[\frac{d_z}{2}\int_0^{\frac{d_x}{2}} dx \ln \frac{\frac{d_y}{2} + \sqrt{x^2 + \left(\frac{d_y}{2}\right)^2 + \left(\frac{d_z}{2}\right)^2}}{\sqrt{x^2 + \left(\frac{d_z}{2}\right)^2}}\right.\right.$$

$$\left.\left. + \int_0^{\frac{d_z}{2}} dz\, 2z \tan^{-1} \frac{\left(\frac{d_y}{2}\right)\left(\frac{d_x}{2z}\right)}{\sqrt{\left(\frac{d_y}{2}\right)^2 + z^2 + \left(\frac{d_x}{2}\right)^2}}\right] + \frac{ik}{6}\Delta V\right\} \tag{2.3.47}$$

Substituting (2.3.40) into (2.3.19), we get

$$\frac{\overline{p}_i}{\epsilon\Delta V} = \frac{\overline{\overline{\alpha}}_i}{\epsilon\Delta V}\cdot\overline{E}_i^{inc} - \frac{\overline{\overline{\alpha}}_i}{\epsilon\Delta V}\cdot\sum_{\substack{j=1\\ j\neq i}}^{N}(\epsilon\Delta V)\overline{\overline{A}}_{ij}\cdot\frac{\overline{p}_j}{\epsilon\Delta V} \tag{2.3.48}$$

where

$$\overline{\overline{\alpha}}_i = \alpha_{ix}\hat{x}\hat{x} + \alpha_{iy}\hat{y}\hat{y} + \alpha_{iz}\hat{z}\hat{z} = \left(\frac{\epsilon_{pi}}{\epsilon} - 1\right)\overline{\overline{\beta}}_i \tag{2.3.49}$$

$$\frac{\beta_{ix}}{\epsilon\Delta V} = \frac{1}{1 + \left(\frac{\epsilon_{pi}}{\epsilon} - 1\right)(L_x - D_x k^2)} \tag{2.3.50}$$

$$\frac{\beta_{iy}}{\epsilon\Delta V} = \frac{1}{1+\left(\frac{\epsilon_{pi}}{\epsilon}-1\right)(L_y - D_y k^2)} \tag{2.3.51}$$

$$\frac{\beta_{iz}}{\epsilon\Delta V} = \frac{1}{1+\left(\frac{\epsilon_{pi}}{\epsilon}-1\right)(L_z - D_z k^2)} \tag{2.3.52}$$

and $\Delta V = d_x d_y d_z$.

In the case of cells of circular cylindrical shape of radius a and length l, the corresponding results of L_x, L_y, L_z and D_x, D_y and D_z are

$$L_x = L_y = \frac{l}{2(4a^2 + l^2)^{1/2}} \tag{2.3.53a}$$

$$L_z = 1 - \frac{l}{(4a^2 + l^2)^{1/2}} \tag{2.3.53b}$$

$$D_x = D_y = \frac{ika^2 l}{6} + \frac{l}{8}\left\{\sqrt{l^2+4a^2} - l\right\} + \frac{a^2}{4}\ln\left(\frac{l+\sqrt{l^2+4a^2}}{2a}\right) \tag{2.3.53c}$$

$$D_z = \frac{ika^2 l}{6} + \frac{a^2}{2}\ln\left(\frac{l+\sqrt{l^2+4a^2}}{2a}\right) \tag{2.3.53d}$$

Equation (2.3.48) is the DDA matrix equation and is to be solved numerically. After the solution is obtained, we have the solution of the induced dipole moment $\overline{p}_i$ for every cell. The electric field at cell i is given by

$$\overline{E}_i = \frac{1}{\left(\frac{\epsilon_{pi}}{\epsilon}-1\right)} \frac{\overline{p}_i}{\epsilon\Delta V} \tag{2.3.54a}$$

for $\epsilon_{pi} \neq \epsilon$ for cell i. For the case of $\epsilon_{pi} = \epsilon$,

$$\overline{E}_i = \frac{\overline{\overline{\beta}}_i}{\epsilon\Delta V}\cdot \overline{E}_i^{inc} - \frac{\overline{\overline{\beta}}_i}{\epsilon\Delta V}\cdot \sum_{\substack{j=1\\ j\neq i}}^{N} (\epsilon\Delta V)\overline{\overline{A}}_{ij}\cdot \frac{\overline{p}_j}{\epsilon\Delta V} \tag{2.3.54b}$$

For the case of rectangular parallelepipeds the matrix equation of (2.3.48) is of dimension $3N \times 3N$ where N is the number of rectangular parallelepipes and the factor 3 arises from the x, y and z components of the polarization vector. In the standard form, the matrix equation is

$$\overline{\overline{Z}}\,\overline{x} = \overline{b} \tag{2.3.55}$$

where $\overline{\overline{Z}}$ is the impedance matrix and $\overline{x}$ is the unknown column vector and $\overline{b}$ is the right-hand side. Let the rectangular parallelepipeds be equally spaced in $\hat{x}$, $\hat{y}$ and $\hat{z}$ directions. Then (2.3.48) can be solved by using conjugate gradient method (CGM) combined with fast Fourier transform (FFT) that will be discussed in Section 4. In applying CGM, we need to take the product $\overline{\overline{Z}}\,\overline{d}$ and the product $\overline{\overline{Z}}^{\dagger}\overline{r}$ where $\dagger$ denotes adjoint, $\overline{d}$ is the direction vector and $\overline{r}$ is the residual vector. For example, if there are two rectangular parallelepipeds, $N = 2$, then the impedance matrix $\overline{\overline{Z}}$ is of dimension 6×6 and is

$$\overline{\overline{Z}} = \begin{bmatrix} \overline{\overline{I}} & \dfrac{\overline{\overline{\alpha}}_1}{\epsilon\Delta V}(\epsilon\Delta V)\overline{\overline{A}}_{12} \\ \dfrac{\overline{\overline{\alpha}}_2}{\epsilon\Delta V}(\epsilon\Delta V)\overline{\overline{A}}_{12} & \overline{\overline{I}} \end{bmatrix} \tag{2.3.56}$$

Note that

$$\overline{\overline{A}}_{ij} = \overline{\overline{A}}_{ji} \tag{2.3.57}$$

and

$$\overline{\overline{A}}_{ij}^{t} = \overline{\overline{A}}_{ij} \tag{2.3.58}$$

where t denotes transpose of the dyad. Then the adjoint of the impedance matrix is

$$\overline{\overline{Z}}^{\dagger} = \begin{bmatrix} \overline{\overline{I}} & (\epsilon\Delta V)\overline{\overline{A}}_{12}^{*} \cdot \dfrac{\overline{\overline{\alpha}}_2^{*}}{\epsilon\Delta V} \\ (\epsilon\Delta V)\overline{\overline{A}}_{12}^{*} \cdot \dfrac{\overline{\overline{\alpha}}_1^{*}}{\epsilon\Delta V} & \overline{\overline{I}} \end{bmatrix} \tag{2.3.59}$$

where $*$ denotes complex conjugate.

After the matrix equation (2.3.48) is solved, the far field scattered field in direction $\hat{k}_s$ is

$$\overline{E}^{s}(\overline{r}) = \frac{e^{ikr}}{4\pi r} k^2 \Delta V (\hat{v}_s\hat{v}_s + \hat{h}_s\hat{h}_s) \cdot \sum_i e^{-i\overline{k}_s \cdot \overline{r}_i} \frac{\overline{p}_i}{\epsilon\Delta V} \tag{2.3.60}$$

The time-averaged power absorbed is equal to

$$\langle P_a \rangle = \frac{1}{2}\omega \sum_i \epsilon''_{pi} |\overline{E}_i|^2 \Delta V \tag{2.3.61}$$

where ϵ''_{pi} is the imaginary part of ϵ_{pi} for the ith cell. In terms of dipole moment of each cell, we have

$$\langle P_a \rangle = \frac{1}{2}\omega \sum_i \begin{cases} 0 & \text{if } \epsilon''_{pi} = 0 \\ \epsilon''_{pi}\Delta V \dfrac{1}{\left|\dfrac{\epsilon_{pi}}{\epsilon} - 1\right|^2} \left|\dfrac{\overline{p}_i}{\epsilon\Delta V}\right|^2 & \text{if } \epsilon''_{pi} \neq 0 \end{cases} \tag{2.3.62}$$

Calculation of Matrix Elements by Numerical Integration

In the matrix equation of (2.3.48), the $\overline{\overline{A}}_{ij}$ elements are calculated by taking the value of $\overline{\overline{A}}(\overline{r}, \overline{r}')$ at the point $\overline{r}_i$ and $\overline{r}_j$, for $\overline{r}_i \neq \overline{r}_j$

$$\overline{\overline{A}}_{ij} = \overline{\overline{A}}(\overline{r}_i, \overline{r}_j) \tag{2.3.63}$$

These may not be accurate enough particularly for $\overline{r}_i$ and $\overline{r}_j$ in the neighborhood of each other. Accuracy can be improved by numerical integration over the cell V_j centered at $\overline{r}_j$. Thus we can define a neighborhood distance r_d so that

$$\overline{\overline{A}}_{ij} = \begin{cases} \dfrac{1}{\Delta V} \displaystyle\int_{V_j} d\overline{r}' \overline{\overline{A}}(\overline{r}_i, \overline{r}') & \text{for } |\overline{r}_i - \overline{r}_j| < \overline{r}_d \\ \overline{\overline{A}}(\overline{r}_i, \overline{r}_j) & \text{for } |\overline{r}_i - \overline{r}_j| \geq \overline{r}_d \end{cases} \tag{2.3.64}$$

The expression of $\overline{\overline{A}}_{ij}$ of (2.3.64) will still preserve the translational invariant property so that the FFT can still be taken when the matrix equation is solved by iterative method.

4 Product of Toeplitz Matrix and Column Vector

In matrix equation, the product of a matrix and a column vector is

$$\overline{y} = \overline{\overline{g}}\,\overline{x} \tag{2.4.1}$$

where $\overline{x}$ and $\overline{y}$ are column vectors of dimension N and $\overline{\overline{g}}$ is a matrix of dimension $N \times N$. In matrix notation

$$y(n) = \sum_{m=1}^{N} g(n, m) x(m) \tag{2.4.2}$$

$n = 1, 2, \ldots, N$. The domain has N points. If the kernel $g(n, m)$ is translational invariant, then

$$y(n) = \sum_{m=1}^{N} g(n - m) x(m) \tag{2.4.3}$$

If $g_{nm} = g_{n-m}$, the matrix $\overline{\overline{g}}$ is known as a Toeplitz matrix.

Equation (2.4.3) looks like a discrete convolution. A fast way to do such a computation is to use FFT. However, since $1 \leq n \leq N$ and $1 \leq m \leq N$, the range of $n - m$ is such that $-(N-1) \leq n - m \leq N - 1$ so that the argument of g has $2N - 1$ points. Thus Fourier transform of (2.4.3) should be taken over at least $2N - 1$ points. In the following, we briefly review discrete Fourier transform.

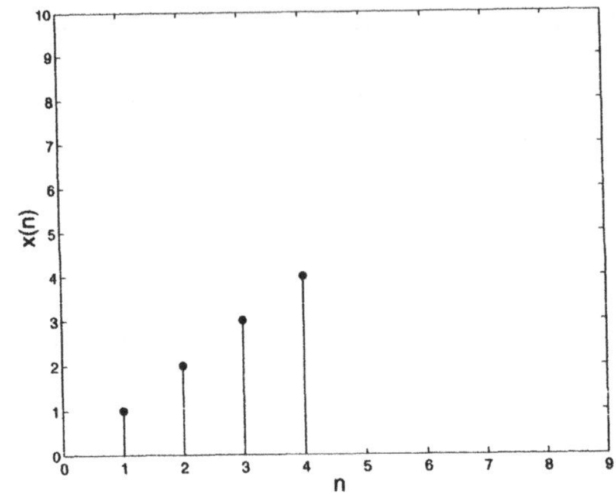

Figure 2.4.1 The $x(n)$ sequence.

4.1 Discrete Fourier Transform and Convolutions

Consider a sequence $x(n)$ that has length N such that $x(n) = 0$ except in the range $1 \le n \le N$. A periodic sequence $\tilde{x}(n)$ can be formed from $x(n)$ by repeating $x(n)$ periodically outside the range $1 \le n \le N$. Suppose $N = 4$ and we have a sequence $x(n)$ as shown in Fig. 2.4.1. We can form a periodic sequence with period N as shown in Fig. 2.4.2.

$$\tilde{x}(n) = \sum_{r=-\infty}^{\infty} x(n + rN) \tag{2.4.4}$$

Let $U_N(n)$ be a rectangular sequence of length N.

$$U_N(n) = \begin{cases} 1 & \text{for } 1 \le n \le N \\ 0 & \text{otherwise} \end{cases}$$

Then

$$x(n) = \tilde{x}(n) U_N(n) \tag{2.4.5}$$

To take discrete Fourier transform, we define the complex number

$$W_N = \exp\left(\frac{2\pi i}{N}\right) \tag{2.4.6a}$$

so that

$$W_N^N = 1 \tag{2.4.6b}$$

$$\sum_{k=1}^{N} W_N^{k(n-n')} = \sum_{k=0}^{N-1} W_N^{k(n-n')} = \begin{cases} N & \text{if } n - n' = \text{integer multiples of } N \\ 0 & \text{otherwise} \end{cases} \tag{2.4.6c}$$

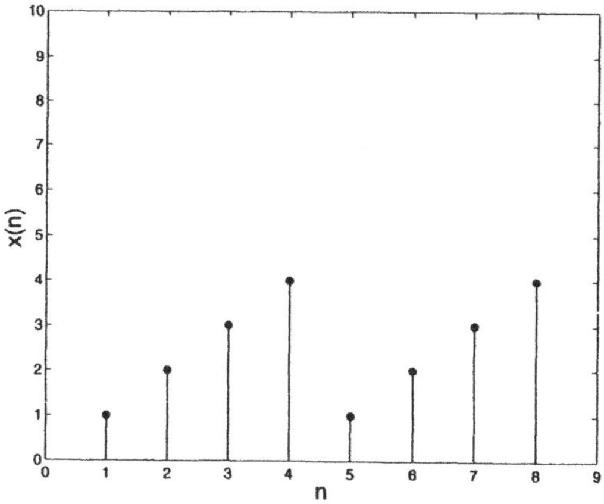

Figure 2.4.2 $x(n)$ with period of $N = 4$.

Then the discrete Fourier transform is, for all k

$$\widetilde{X}(k) = \sum_{n=1}^{N} \tilde{x}(n) W_N^{(k-1)(n-1)} = \sum_{n=1}^{N} x(n) W_N^{(k-1)(n-1)} \tag{2.4.7}$$

From (2.4.7), $\widetilde{X}(k)$ is periodic with period N. It follows that for all n

$$\tilde{x}(n) = \frac{1}{N} \sum_{k=1}^{N} \widetilde{X}(k) W_N^{-(k-1)(n-1)} \tag{2.4.8}$$

The discrete Fourier transform pairs are defined as in (2.4.7) and (2.4.8). Next define $X(k)$ as one period of $\widetilde{X}(k)$.

$$X(k) = \widetilde{X}(k) U_N(k) = \begin{cases} \widetilde{X}(k) & k = 1, \ldots, N \\ 0 & \text{otherwise} \end{cases} \tag{2.4.9}$$

Thus from (2.4.7) and (2.4.9), for $k = 1, \ldots, N$

$$X(k) = \sum_{n=1}^{N} x(n) W_N^{(k-1)(n-1)} \tag{2.4.10}$$

From (2.4.8) and (2.4.10), for $n = 0, 1, \ldots, N-1$

$$x(n) = \frac{1}{N} \sum_{k=1}^{N} X(k) W_N^{-(k-1)(n-1)} \tag{2.4.11}$$

Note that the right-hand sides of (2.4.10) and (2.4.11) are periodic while left-hand sides are only nonzero over one period. The advantage of using fast

Fourier transform (FFT) is that both (2.4.7) and (2.4.8) can be computed in $N \log_2 N$ steps rather than N^2 steps.

The periodic convolution is as follows. Let $\tilde{x}_1(n)$ and $\tilde{x}_2(n)$ be periodic sequences of period N and $\widetilde{X}_1(k)$ and $\widetilde{X}_2(k)$ be their respective DFT. Let

$$\widetilde{X}_3(k) = \widetilde{X}_1(k)\widetilde{X}_2(k) \tag{2.4.12}$$

and $\tilde{x}_3(n)$ be the inverse discrete Fourier transform

$$\begin{aligned}
\tilde{x}_3(n) &= \frac{1}{N}\sum_{k=1}^{N} \widetilde{X}_3(k) W_N^{-(k-1)(n-1)} \\
&= \frac{1}{N}\sum_{k=1}^{N}\sum_{m=1}^{N}\sum_{r=1}^{N} \tilde{x}_1(m)\tilde{x}_2(r) W_N^{(k-1)(m+r-n-1)} \\
&= \begin{cases} \sum_{m=1}^{N}\sum_{r=1}^{N} \tilde{x}_1(m)\tilde{x}_2(r) & m+r-n-1 = \text{integer multiples of } N \\ 0 & \text{otherwise} \end{cases} \\
&= \sum_{m=1}^{N} \tilde{x}_1(m)\tilde{x}_2(n+1-m)
\end{aligned} \tag{2.4.13}$$

Equation (2.4.13) is known as periodic convolution. Note that $\tilde{x}_3(n)$ is also a periodic sequence. However, (2.4.13) is not the usual (linear) convolution. This is because the periodic sequence $\tilde{x}_2(n+1-m)$, when it is "shifted" outside the period N, re-enters on the other side because of the periodic property. Thus

$$x_3^{(c)}(n) = \left[\sum_{m=1}^{N} \tilde{x}_1(m)\tilde{x}_2(n-m+1)\right] U_N(n) \tag{2.4.14}$$

is known as circular convolution.

Circular convolution is not equal to linear convolution of

$$\sum_{n=1}^{N} x_1(n)x_2(m-n+1) \tag{2.4.15}$$

Steps of linear convolution is as shown in Fig. 2.4.3. To use circular convolution to get the result of linear convolution, we need zero padding (Fig. 2.4.4). We also note from Fig. 2.4.3 that to obtain linear convolution from circular convolution, zero padding such that the period becomes $2N$ is sufficient.

Let $x_1(n)$ and $x_2(n)$ be of length N and zero outside $1 \leq n \leq N$. The

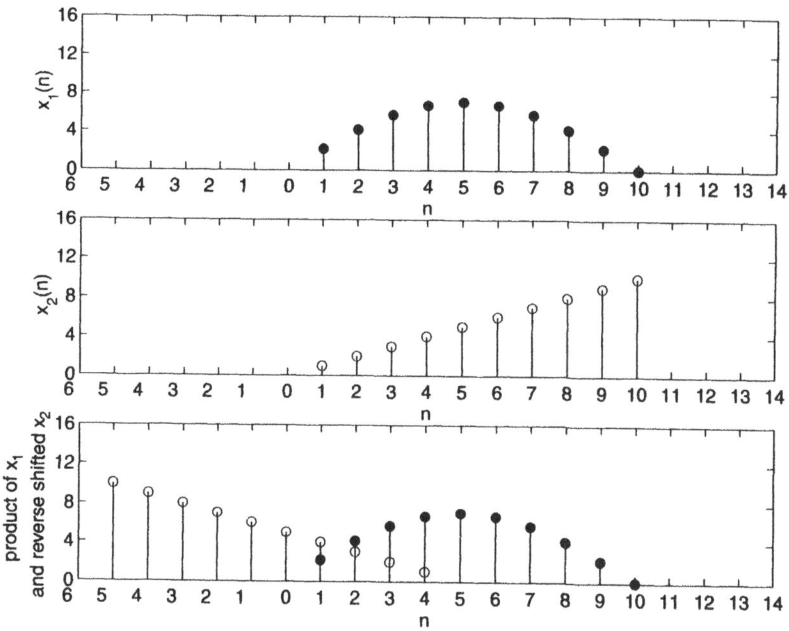

Figure 2.4.3 Steps of linear convolution ($N = 10$).

linear convolution is

$$x_3(n) = \sum_{m=1}^{N} x_1(m)x_2(n-m+1) \tag{2.4.16}$$

and $x_3(n)$ can be nonzero for $1 \leq n \leq 2N-1$. Thus there are $2N-1$ points in $x_3(n)$.

To use DFT to reform the convolution, we need to "pad" $x_1(n)$ and $x_2(n)$ with zeroes to have sequences of length $M \geq 2N-1$

$$x_{1p}(n) = \begin{cases} x_1(n) & 1 \leq n \leq N \\ 0 & N+1 \leq n \leq M \end{cases} \tag{2.4.17}$$

$$x_{2p}(n) = \begin{cases} x_2(n) & 1 \leq n \leq N \\ 0 & N+1 \leq n \leq M \end{cases} \tag{2.4.18}$$

Usually M is chosen as $2N$. Then define $\tilde{x}_{1p}(n)$ and $\tilde{x}_{2p}(n)$ as periodic sequences with period M. That is, they are periodic replicas of $x_{1p}(n)$ and $x_{2p}(n)$ respectively. Then from (2.4.16)–(2.4.18)

$$x_3(n) = \sum_{m=1}^{M} \tilde{x}_{1p}(n)\tilde{x}_{2p}(n-m) \tag{2.4.19}$$

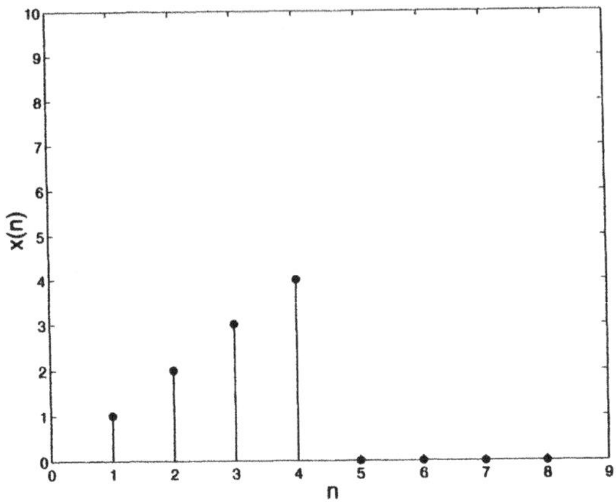

Figure 2.4.4 Zero padding of $x(n)$. ($N = 4$)

for $n = 1, 2, \ldots, 2N$. Thus the DFT pairs that give (2.4.16) are

$$\widetilde{X}_{1p}(k) = \sum_{n=1}^{M} \tilde{x}_{1p}(n) W_M^{(n-1)(k-1)} \tag{2.4.20}$$

$$\widetilde{X}_{2p}(k) = \sum_{n=1}^{M} \tilde{x}_{2p}(n) W_M^{(n-1)(k-1)} \tag{2.4.21}$$

$$\widetilde{X}_{3p}(k) = \widetilde{X}_{1p}(k)\widetilde{X}_{2p}(k) \tag{2.4.22}$$

$$\tilde{x}_{3p}(n) = \frac{1}{M} \sum_{k=1}^{M} \widetilde{X}_{3p}(k) W_M^{-(n-1)(k-1)} \tag{2.4.23}$$

The equation (2.4.23) gives the relation in which $x_3(n)$ as defined by (2.4.16) is computed. Note that the relations (2.4.16)–(2.4.23) are exact. $\tilde{x}_{3p}(n)$ is a periodic sequence of period M and $x_3(n) = \tilde{x}_{3p}(n)$ for $n = 1, 2, \ldots, 2N$.

4.2 FFT for Product of Toeplitz Matrix and Column Vector

Next we address the equation (2.4.3) that describes the product of a matrix and a column vector.

$$y(n) = \sum_{m=1}^{N} g(n-m)x(m) \tag{2.4.24}$$

The equation of (2.4.24) has the following features:

(i) only N values of $y(n)$ are needed. That is, we need to compute $y(n)$, $n = 1, 2, \ldots, N$.

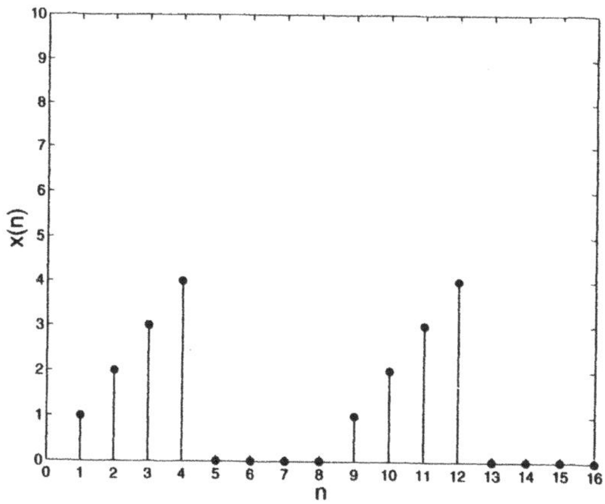

Figure 2.4.5 Zero padding with period of $2N$.

(ii) $g(n)$ is needed for $n = -N+1, -N+2, \ldots, 0, \ldots, N-2, N-1$, a total of $2N-1$ distinct values.
(iii) $x(n)$ is defined for $n = 1, \ldots, N$.
For simplicity, we take $M = 2N$.

To illustrate for the case $N = 4$, we need

(i) $x(n)$, $n = 1, 2, 3, 4$
(ii) $y(n)$, $m = 1, 2, 3, 4$
(iii) $g(n-m)$, $n - m = -3, -2, -1, 0, 1, 2, 3$

First we do zero padding of $x(n)$ (Fig. 2.4.5) to get $x_1(n)$. Let $\tilde{x}_1(n)$ be the periodic version of $x_1(n)$ with period $2N$.

$$x_1(n) = \begin{cases} x(n) & \text{for } n = 1, 2, 3, \ldots, N \\ 0 & \text{for } n = N+1, N+2, \ldots, 2N \end{cases} \tag{2.4.25}$$

Then

$$y(n) = \sum_{m=1}^{2N} g(n-m)\tilde{x}_1(m) \tag{2.4.26}$$

Note that the summation has been changed to $2N$.

Define

$$g'(n) = g(n-1) \tag{2.4.27}$$

See Fig. 2.4.6 for $g'(n)$. Then

$$y(n) = \sum_{m=1}^{2N} g'(n-m+1)\tilde{x}_1(m) \tag{2.4.28}$$

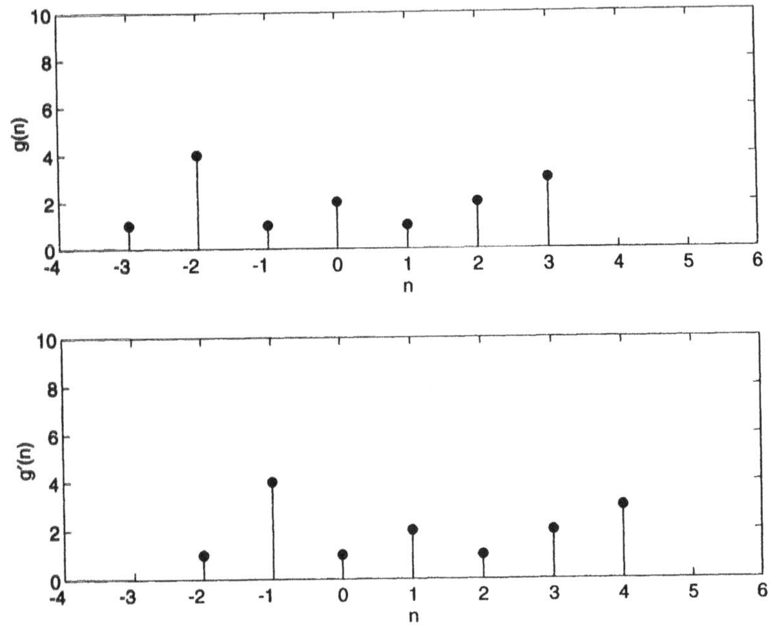

Figure 2.4.6 $g(n)$ and $g(n-1)$.

Define $x_2(n)$ by

$$x_2(n) = \begin{cases} g'(n) & \text{for } 1 \le n \le N \\ 0 & \text{for } n = N+1 \\ g'(n-2N) & \text{for } N+2 \le n \le 2N \end{cases} \tag{2.4.29}$$

for $n = 1, 2, \ldots, 2N$. Let $\tilde{x}_2(n)$ be the periodic version of $x_2(n)$ with period $2N$. Then

$$\tilde{x}_2(n) = g'(n) \quad \text{for } -N+2 \le n \le N \tag{2.4.30}$$

Note that

$$\tilde{x}_2(n) = \begin{cases} g(n-1) & 1 \le n \le N \\ 0 & n = N+1 \\ g(n-2N-1) & N+2 \le n \le 2N \end{cases} \tag{2.4.31}$$

Hence

$$y(n) = \sum_{m=1}^{2N} \tilde{x}_2(n-m+1)\tilde{x}(m) \tag{2.4.32}$$

Define

$$\tilde{y}(n) = \sum_{m=1}^{2N} \tilde{x}_2(n-m+1)\tilde{x}(n) \tag{2.4.33}$$

for all n. The result of $\tilde{y}(n)$ is periodic with period $2N$. Also

$$y(n) = \tilde{y}(n) \qquad \text{for} \quad 1 \leq n \leq N \tag{2.4.34}$$

Now (2.4.33) satisfies the properties of periodic convolution.

We apply periodic convolution.

$$\begin{aligned} \widetilde{X}(k) &= \sum_{n=1}^{2N} \tilde{x}(n) W_{2N}^{(k-1)(n-1)} \\ \widetilde{X}_2(k) &= \sum_{n=1}^{2N} \tilde{x}_2(n) W_{2N}^{(k-1)(n-1)} \end{aligned} \tag{2.4.35}$$

Then

$$\widetilde{Y}(k) = \widetilde{X}_2(k)\widetilde{X}(k) \tag{2.4.36}$$

$$\tilde{y}(n) = \frac{1}{2N} \sum_{k=1}^{2N} \widetilde{Y}(k) W_{2N}^{-(k-1)(n-1)} \tag{2.4.37}$$

We can extend to the case of three-dimensional convolution with three indices as needed for the discrete dipole approximation. Let N_x, N_y and N_z points, respectively, in $\hat{x}$, $\hat{y}$ and $\hat{z}$ directions with N_x, N_y and N_z all equal to powers of 2. Let

$$y(n,m,l) = \sum_{n'=1}^{N_x} \sum_{m'=1}^{N_y} \sum_{l'=1}^{N_z} g(n-n', m-m', l-l')x(n', m', l') \tag{2.4.38}$$

is to be computed. For the sake of simplicity, we illustrate the scalar case. The vector case follows by a simple extension. Then we have 3-D periodic sequences $\tilde{x}$ and $\tilde{x}_2$ with period $M_x = 2N_x$, $M_y = 2N_y$ and $M_z = 2N_z$, respectively in $\hat{x}$, $\hat{y}$ and $\hat{z}$ directions. For one period of $\tilde{x}(n,m,l)$ it is

$$\tilde{x}(n,m,l) = \begin{cases} x(n,m,l) & 1 \leq n \leq N_x \text{ and } 1 \leq m \leq N_y \text{ and } 1 \leq l \leq N_z \\ 0 & \text{when } N_x + 1 \leq n \leq 2N_x \\ & \text{or } N_y + 1 \leq m \leq 2N_y \text{ or } N_z + 1 \leq l \leq 2N_z \end{cases} \tag{2.4.39}$$

The $\tilde{x}_2$ for nonzero values can be computed as in Table 2.4.1. Then the 3-D DFT and inverse DFT can be performed accordingly.

n	m	l	$\tilde{x}_2(n,m,l)$
$[1, N_x]$	$[1, N_y]$	$[1, N_z]$	$g(n-1, m-1, l-1)$
$[N_x+2, 2N_x]$	$[1, N_y]$	$[1, N_z]$	$g(n-2N_x-1, m-1, l-1)$
$[1, N_x]$	$[N_y+2, 2N_y]$	$[1, N_z]$	$g(n-1, m-2N_y-1, l-1)$
$[N_x+2, 2N_x]$	$[N_y+2, 2N_y]$	$[1, N_z]$	$g(n-2N_x-1, m-2N_y-1, l-1)$
$[1, N_x]$	$[1, N_y]$	$[N_z+2, 2N_z]$	$g(n-1, m-1, l-2N_z-1)$
$[N_x+2, 2N_x]$	$[1, N_y]$	$[N_z+2, 2N_z]$	$g(n-2N_x-1, m-1, l-2N_z-1)$
$[1, N_x]$	$[N_y+2, 2N_y]$	$[N_z+2, 2N_z]$	$g(n-1, m-2N_y-1, l-2N_z-1)$
$[N_x+2, 2N_x]$	$[N_y+2, 2N_y]$	$[N_z+2, 2N_z]$	$g(n-2N_x-1, m-2N_y-1, l-2N_z-1)$

Table 2.4.1 Computations for the nonzero values of $\tilde{x}_2(n,m,l)$.

5 Conjugate Gradient Method

Consider a matrix equation of the form

$$\overline{\overline{A}}\,\overline{x} = \overline{b} \tag{2.5.1}$$

where $\overline{\overline{A}}$ is a $N \times N$ nonsingular matrix, x is the unknown column vector and b is the right hand side. Both $\overline{x}$ and $\overline{b}$ are $N \times 1$ column vectors.

In the following we briefly describe the conjugate gradient method. Details can be found in textbooks on matrix computation [Hestenes and Stiefel, 1952; Golub and Van Loan, 1996].

5.1 Steepest Descent Method

Let $\overline{\overline{A}}$ be a real symmetric matrix and positive definite and $\phi(\overline{x})$ be the functional

$$\phi(\overline{x}) = \frac{1}{2}\overline{x}^t\,\overline{\overline{A}}\,\overline{x} - \overline{x}^t\,\overline{b} \tag{2.5.2}$$

where t denotes transpose so that $\overline{x}^t$ is a row vector of dimension $1 \times N$. In index notation

$$\phi(\overline{x}) = \frac{1}{2}\sum_{ij} x_i A_{ij} x_j - \sum_i x_i b_i \tag{2.5.3}$$

Taking the derivatives

$$-\frac{\partial \phi}{\partial x_i} = -\frac{1}{2}\sum_j A_{ij}x_j - \frac{1}{2}\sum_j x_j A_{ji} + b_i$$

$$= b_i - \sum_j A_{ij} x_j \tag{2.5.4}$$

The second equality is due to the fact that

$$A_{ij} = A_{ji} \tag{2.5.5}$$

Thus the gradient is

$$-\nabla\phi = \overline{b} - \overline{\overline{A}}\,\overline{x} \tag{2.5.6}$$

Optimizing ϕ with respect to $\overline{x}$ gives

$$0 = -\nabla\phi = \overline{b} - \overline{\overline{A}}\,\overline{x} \tag{2.5.7}$$

This means that optimizing ϕ is equivalent to solving the matrix equation $\overline{\overline{A}}\,\overline{x} = \overline{b}$.

The residual is the "left over" or the "remainder". Let $\overline{x}_{i-1}$ be the $(i-1)$th iterative solution. The residual is

$$\overline{r}_{i-1} = \overline{b} - \overline{\overline{A}}\,\overline{x}_{i-1} \tag{2.5.8}$$

The direction vector $\overline{d}_i$ gives the next solution

$$\overline{x}_i = \overline{x}_{i-1} + \alpha_i \overline{d}_i \tag{2.5.9}$$

In the method of steepest descent, the direction vector is chosen to be the same as the residual vector.

$$\overline{d}_i = \overline{r}_{i-1} \tag{2.5.10}$$

Then the ith solution is

$$\overline{x}_i = \overline{x}_{i-1} + \alpha_i \overline{r}_{i-1} \tag{2.5.11}$$

Substituting in (2.5.2), we have

$$\begin{aligned}
\phi(\overline{x}_i) &= \phi(\overline{x}_{i-1} + \alpha_i \overline{d}_i) \\
&= \frac{1}{2}\left(\overline{x}_{i-1} + \alpha_i \overline{d}_i\right)^t \overline{\overline{A}}\left(\overline{x}_{i-1} + \alpha_i \overline{d}_i\right) - \left(\overline{x}_{i-1} + \alpha_i \overline{d}_i\right)^t \overline{b} \\
&= \phi(\overline{x}_{i-1}) + \alpha_i \overline{d}_i^t \overline{\overline{A}}\,\overline{x}_{i-1} + \frac{\alpha_i^2}{2}\overline{d}_i^t \overline{\overline{A}}\,\overline{d}_i - \alpha_i \overline{d}_i^t \overline{b} \\
&= \phi(\overline{x}_{i-1}) + \alpha_i \overline{r}_{i-1}^t \overline{\overline{A}}\,\overline{x}_{i-1} + \frac{\alpha_i^2}{2}\overline{r}_{i-1}^t \overline{\overline{A}}\,\overline{r}_{i-1} - \alpha_i \overline{r}_{i-1}^t \overline{b} \\
&= \phi(\overline{x}_{i-1}) - \alpha_i \overline{r}_{i-1}^t \overline{r}_{i-1} + \frac{\alpha_i^2}{2}\overline{r}_{i-1}^t \overline{\overline{A}}\,\overline{r}_{i-1}
\end{aligned} \tag{2.5.12}$$

Optimizing $\phi(\overline{x}_i)$ by taking its derivative with respect to α_i and setting it to zero gives

$$\alpha_i = \frac{\overline{r}_{i-1}^t \overline{r}_{i-1}}{\overline{r}_{i-1}^t \overline{\overline{A}}\,\overline{r}_{i-1}} \tag{2.5.13}$$

In the steepest descent method, the direction vector is in the residual direction. On the other hand, in the conjugate gradient , the direction vector is more general.

5.2 Real Symmetric Positive Definite Matrix

We first summarize the results for the conjugate gradient method for real symmetric positive definite matrix. At the zeroth step, let the solution be

$$\overline{x}_0 = 0 \tag{2.5.14}$$

At the $(i-1)$ th step, let $\overline{x}_{i-1}$ be the approximate solution. Then

$$\overline{r}_{i-1} = \overline{b} - \overline{\overline{A}}\,\overline{x}_{i-1} \tag{2.5.15}$$

is the residual at the $(i-1)$ th step. In particular,

$$\overline{r}_0 = \overline{b} \tag{2.5.16}$$

To get a new solution $\overline{x}_i$, $i = 1, 2, \ldots$, let

$$\overline{x}_i = \overline{x}_{i-1} + \alpha_i \overline{d}_i \tag{2.5.17}$$

where $\overline{d}_i$ is the unknown direction vector and α_i is the unknown scalar to denote the movement in the $\overline{d}_i$ direction. To determine α_i and $\overline{d}_i$, we note that

$$\phi(\overline{x}_{i-1} + \alpha_i \overline{d}_i) = \phi(\overline{x}_{i-1}) + \alpha_i \overline{d}_i^t \overline{\overline{A}}\,\overline{x}_{i-1} + \frac{\alpha_i^2}{2} \overline{d}_i^t \overline{\overline{A}}\,\overline{d}_i - \alpha_i \overline{d}_i^t \overline{b} \tag{2.5.18}$$

Next determine $\overline{d}_i$ by setting the second term to zero

$$\overline{x}_{i-1}^t \overline{\overline{A}}\,\overline{d}_i = 0 \tag{2.5.19}$$

The previous solution $\overline{x}_{i-1}$ for $i = 2, 3, \ldots$ is be a linear combination of previous direction vectors $\overline{d}_1, \overline{d}_2, \ldots, \overline{d}_{i-1}$. That is $\overline{x}_{i-1} \in \text{span}\{\overline{d}_1, \ldots, \overline{d}_{i-1}\}$, for $i = 2, 3, \ldots$.

Thus we have $i-1$ equations,

$$\overline{d}_1^t \overline{\overline{A}}\,\overline{d}_i = 0 \tag{2.5.20}$$

$$\overline{d}_2^t \overline{\overline{A}}\,\overline{d}_i = 0 \tag{2.5.21}$$

$$\cdots\cdots$$

$$\overline{d}_{i-1}^t \overline{\overline{A}}\,\overline{d}_i = 0 \tag{2.5.22}$$

This is called $\overline{\overline{A}}$ conjugation of $\overline{d}_i$ to $\overline{d}_1$, $\overline{d}_2$, $\overline{d}_3$, $\ldots$, $\overline{d}_{i-1}$.

Let $\overline{\overline{D}}_{i-1}$ be the $N \times (i-1)$ matrix containing the column vectors $\overline{d}_1$, $\overline{d}_2$, $\overline{d}_3$, $\ldots$, $\overline{d}_{i-1}$.

$$\overline{\overline{D}}_{i-1}^t \overline{\overline{A}}\,\overline{d}_i = 0 \tag{2.5.23}$$

Then

$$\phi(\overline{x}_{i-1} + \alpha_i \overline{d}_i) = \phi(\overline{x}_{i-1}) + \frac{\alpha_i^2}{2} \overline{d}_i^t \overline{\overline{A}}\, \overline{d}_i - \alpha_i \overline{d}_i^t \overline{b} \tag{2.5.24}$$

Minimizing (2.5.24) with respect to α_i gives

$$\alpha_i = \frac{\overline{d}_i^t \overline{b}}{\overline{d}_i^t \overline{\overline{A}}\, \overline{d}_i} \tag{2.5.25}$$

Next we relate the residual vectors of the ith step and the $(i-1)$th step. Let

$$\overline{d}_1 = \overline{b} \tag{2.5.26}$$

$$\overline{x}_1 = \overline{d}_1 \tag{2.5.27}$$

then

$$\begin{aligned} \overline{r}_1 &= \text{linear combination of } \overline{b} \text{ and } \overline{\overline{A}}\, \overline{b} \\ &= \text{span}\left\{\overline{b}, \overline{\overline{A}}\, \overline{b}\right\} \end{aligned} \tag{2.5.28}$$

The residual vector at the ith step is

$$\overline{r}_i = \overline{b} - \overline{\overline{A}}\, \overline{x}_i = \overline{b} - \overline{\overline{A}}(\overline{x}_{i-1} + \alpha_i \overline{d}_i) \tag{2.5.29}$$

$$\overline{r}_i = \overline{r}_{i-1} - \alpha_i \overline{\overline{A}}\, \overline{d}_i \tag{2.5.30}$$

Since $\overline{d}_i$ is $\overline{\overline{A}}$-orthogonal to previous direction vectors,

$$\overline{d}_i^t \overline{b} = \overline{d}_i^t \left(\overline{b} - \overline{\overline{A}}\, \overline{x}_{i-1}\right) = \overline{d}_i^t \overline{r}_{i-1} \tag{2.5.31}$$

Using (2.5.31) in (2.5.25),

$$\alpha_i = \frac{\overline{d}_i^t \overline{b}}{\overline{d}_i^t \overline{\overline{A}}\, \overline{d}_i} = \frac{\overline{d}_i^t \overline{r}_{i-1}}{\overline{d}_i^t \overline{\overline{A}}\, \overline{d}_i} \tag{2.5.32}$$

The following are three properties concerning the properties of residuals [Golub and van Loan, 1996]:

Property (A):

$$\overline{\overline{D}}_j^t \overline{r}_j = 0 \tag{2.5.33}$$

Property (B):

$$\begin{aligned} \text{span}\left\{\overline{d}_1, \overline{d}_2, \ldots, \overline{d}_i\right\} &= \text{span}\left\{\overline{r}_0, \overline{r}_1, \ldots, \overline{r}_{i-1}\right\} \\ &= \text{span}\left\{\overline{b}, \overline{\overline{A}}\, \overline{b}, \overline{\overline{A}}^2 \overline{b}, \ldots, \overline{\overline{A}}^{i-1} \overline{b}\right\} \\ &= \text{span}\left\{\overline{r}_0, \overline{\overline{A}}\, \overline{r}_0, \ldots, \overline{\overline{A}}^{i-1} \overline{r}_0\right\} = K\left(\overline{r}_0, \overline{\overline{A}}, i\right) \end{aligned} \tag{2.5.34}$$

where $K(\overline{r}_0, \overline{\overline{A}}, i)$ is the Krylov subspace.

Property (C): all the residuals are mutually orthogonal

$$\overline{r}_i^t \overline{r}_{j-1} = 0 \tag{2.5.35}$$

for $i = 0, 1, 2, \ldots, j-2$.

It then follows that [Golub and van Loan, 1996]

$$\overline{d}_k \in \text{span}\{\overline{r}_{k-1}, \overline{d}_{k-1}\} \tag{2.5.36}$$

Let

$$\overline{d}_i = \overline{r}_{i-1} + \beta_i \overline{d}_{i-1} \tag{2.5.37}$$

Then from (2.5.23) and (2.5.37)

$$\overline{d}_{i-1}^t \overline{\overline{A}}\, \overline{d}_i = \overline{d}_{i-1}^t \overline{\overline{A}}\, \overline{r}_{i-1} + \beta_i \overline{d}_{i-1}^t \overline{\overline{A}}\, \overline{d}_{i-1} = 0 \tag{2.5.38}$$

giving

$$\beta_i = -\frac{\overline{d}_{i-1}^t \overline{\overline{A}}\, \overline{r}_{i-1}}{\overline{d}_{i-1}^t \overline{\overline{A}}\, \overline{d}_{i-1}} \tag{2.5.39}$$

Also from (2.5.32) and (2.5.37),

$$\alpha_i = \frac{\overline{d}_i^t \overline{r}_{i-1}}{\overline{d}_i^t \overline{\overline{A}}\, \overline{d}_i} = \frac{\left[\overline{r}_{i-1}^t + \beta_i \overline{d}_{i-1}^t\right] \overline{r}_{i-1}}{\overline{d}_i^t \overline{\overline{A}}\, \overline{d}_i} \tag{2.5.40}$$

On the other hand

$$\overline{d}_{i-1} \in \text{span}\{\overline{r}_0, \overline{r}_1, \ldots, \overline{r}_{i-2}\} \tag{2.5.41}$$

so that

$$\overline{d}_{i-1}^t \overline{r}_{i-1} = 0 \tag{2.5.42}$$

Let

$$\|\overline{x}\|^2 = \overline{x}^t \overline{x} \tag{2.5.43}$$

be the L^2 norm. From (2.5.40) and (2.5.42),

$$\alpha_i = \frac{\|\overline{r}_{i-1}\|^2}{\overline{d}_i^t \overline{\overline{A}}\, \overline{d}_i} \tag{2.5.44}$$

From (2.5.30) and using (2.5.35),

$$\|\overline{r}_i\|^2 = \overline{r}_i^t \overline{r}_i = \overline{r}_i^t \overline{r}_{i-1} - \alpha_i \overline{r}_i^t \overline{\overline{A}}\, \overline{d}_i = -\alpha_i \overline{r}_i^t \overline{\overline{A}}\, \overline{d}_i \tag{2.5.45}$$

Then

$$\|\overline{r}_{i-1}\|^2 = -\alpha_{i-1} \overline{r}_{i-1}^t \overline{\overline{A}}\, \overline{d}_{i-1} = -\alpha_{i-1} \overline{d}_{i-1}^t \overline{\overline{A}}\, \overline{r}_{i-1} \tag{2.5.46}$$

because $\overline{\overline{A}}$ is symmetric. Since $\overline{d}_{i-1}$ is orthogonal to $\overline{r}_{i-1}$, and using (2.5.37),

$$0 = \overline{d}_{i-1}^t \overline{r}_{i-1} = \overline{d}_{i-1}^t \overline{r}_{i-2} - \alpha_{i-1}\overline{d}_{i-1}^t \overline{\overline{A}}\,\overline{d}_{i-1} = \overline{r}_{i-2}^t \overline{r}_{i-2} - \alpha_{i-1}\overline{d}_{i-1}^t \overline{\overline{A}}\,\overline{d}_{i-1} \tag{2.5.47a}$$

Thus

$$\|\overline{r}_{i-2}\|^2 = \alpha_{i-1}\overline{d}_{i-1}^t \overline{\overline{A}}\,\overline{d}_{i-1} \tag{2.5.47b}$$

Using (2.5.46) in (2.5.39), we have

$$\beta_i = \frac{1}{\alpha_{i-1}}\left[\frac{\|\overline{r}_{i-1}\|^2}{\overline{d}_{i-1}^t \overline{\overline{A}}\,\overline{d}_{i-1}}\right] \tag{2.5.48a}$$

Further, use (2.5.47b) in (2.5.48a) to give

$$\beta_i = \frac{\|\overline{r}_{i-1}\|^2}{\|\overline{r}_{i-2}\|^2} \tag{2.5.48b}$$

To summarize, the conjugate gradient algorithm for symmetric positive definite matrix is as follows:

$$\overline{x}_0 = 0 \tag{2.5.49}$$

$$\overline{r}_0 = \overline{b} \tag{2.5.50}$$

$$\overline{d}_1 = \overline{b} \tag{2.5.51}$$

$$\alpha_1 = \frac{\|\overline{r}_0\|^2}{\overline{d}_1^t \overline{\overline{A}}\,\overline{d}_1} \tag{2.5.52}$$

$$\overline{x}_1 = \overline{x}_0 + \alpha_1 \overline{d}_1 \tag{2.5.53}$$

$$\overline{r}_1 = \overline{r}_0 - \alpha_1 \overline{\overline{A}}\,\overline{d}_1 \tag{2.5.54}$$

For $i \geq 2$,

$$\beta_i = \frac{\|\overline{r}_{i-1}\|^2}{\|\overline{r}_{i-2}\|^2} \tag{2.5.55}$$

$$\overline{d}_i = \overline{r}_{i-1} + \beta_i \overline{d}_{i-1} \tag{2.5.56}$$

$$\alpha_i = \frac{\|\overline{r}_{i-1}\|^2}{\overline{d}_i^t \overline{\overline{A}}\,\overline{d}_i} \tag{2.5.57}$$

$$\overline{x}_i = \overline{x}_{i-1} + \alpha_i \overline{d}_i \tag{2.5.58}$$

$$\overline{r}_i = \overline{r}_{i-1} - \alpha_i \overline{\overline{A}}\,\overline{d}_i \tag{2.5.59}$$

The iteration is stopped when the residual becomes small.

5.3 General Real Matrix and Complex Matrix

The procedure can be extended to general nonsymmetric matrix $\overline{\overline{A}}$ that is real and nonsingular (conjugate gradient normal equation residual method). One can solve

$$\overline{\overline{A}}\,\overline{x} = \overline{b} \tag{2.5.60}$$

in the following manner.

Consider the matrix

$$\overline{\overline{B}} = \overline{\overline{A}}^t\,\overline{\overline{A}} \tag{2.5.61}$$

Then $\overline{\overline{B}}^t = \overline{\overline{A}}^t\,\overline{\overline{A}} = \overline{\overline{B}}$, so that $\overline{\overline{B}}$ is symmetric. Also,

$$\overline{y}^t\,\overline{\overline{B}}\,\overline{y} = \overline{y}^t\,\overline{\overline{A}}^t\,\overline{\overline{A}}\,\overline{y} = (\overline{\overline{A}}\,\overline{y})^t(\overline{\overline{A}}\,\overline{y}) = \left\|\overline{\overline{A}}\,\overline{y}\right\|^2 \geq 0 \tag{2.5.62}$$

Hence $\overline{\overline{B}}$ is also positive definite. Since $\overline{\overline{B}}$ is symmetric and positive definite, we can apply the results of (2.5.49)–(2.5.59) to the matrix $\overline{\overline{B}}$. Multiply (2.5.60) by $\overline{\overline{A}}^t$, we have

$$\overline{\overline{B}}\,\overline{x} = \overline{b}' \tag{2.5.63}$$

where

$$\overline{b}' = \overline{\overline{A}}^t\,\overline{b} \tag{2.5.64}$$

In the algorithm, do not calculate $\overline{\overline{B}} = \overline{\overline{A}}^t\overline{\overline{A}}$ because matrix-matrix multiplication is an $\mathcal{O}(N^3)$ procedure for a full matrix. Let

$$\overline{r}' = \overline{b}' - \overline{\overline{B}}\,\overline{x} \tag{2.5.65}$$

$$\overline{r} = \overline{b} - \overline{\overline{A}}\,\overline{x} \tag{2.5.66}$$

be the residuals respectively of the $\overline{\overline{B}}$ and $\overline{\overline{A}}$ matrix equations. Then

$$\overline{r}' = \overline{\overline{A}}^t\,\overline{r} \tag{2.5.67}$$

The steps in the conjugate gradient method will be (2.5.49)–(2.5.59), with $\overline{r}_i$ and $\overline{\overline{A}}$ replaced by $\overline{r}_i' = \overline{\overline{A}}^t\,\overline{r}_i$ and $\overline{\overline{A}}^t\,\overline{\overline{A}}$, respectively.

Suppose that we are solving equation (2.5.63), then $\overline{x}_0 = 0$, $\overline{r}_0' = \overline{b}'$, $\overline{d}_1 = \overline{r}_0'$

$$\beta_i = \frac{\|\overline{r}_{i-1}'\|^2}{\|\overline{r}_{i-2}'\|^2} \tag{2.5.68}$$

$$\overline{d}_i = \overline{r}_{i-1}' + \beta_i\overline{d}_{i-1} \tag{2.5.69}$$

$$\alpha_i = \frac{\|\overline{r}'_{i-1}\|^2}{\overline{d}_i^t\,\overline{\overline{B}}\,\overline{d}_i} = \frac{\|\overline{r}'_{i-1}\|^2}{\left\|\overline{d}_i^t\,\overline{\overline{A}}^t\,\overline{\overline{A}}\,\overline{d}_i\right\|} = \frac{\|\overline{r}'_{i-1}\|^2}{\left\|\overline{\overline{A}}\,\overline{d}_i\right\|^2} \tag{2.5.70}$$

$$\overline{x}_i = \overline{x}_{i-1} + \alpha_i\overline{d}_i \tag{2.5.71}$$

$$\overline{r}'_i = \overline{r}'_{i-1} - \alpha_i\overline{\overline{B}}\overline{d}_i \tag{2.5.72}$$

$$\overline{\overline{A}}^t\,\overline{r}_i = \overline{\overline{A}}^t\,\overline{r}_{i-1} - \alpha_i\,\overline{\overline{A}}^t\,\overline{\overline{A}}\,\overline{d}_i \tag{2.5.73}$$

$$\overline{r}_i = \overline{r}_{i-1} - \alpha_i\,\overline{\overline{A}}\,\overline{d}_i \tag{2.5.74}$$

Then the algorithm is as follows for real non-symmetric matrix:

$$\overline{x}_0 = 0 \tag{2.5.75}$$

$$\overline{r}_0 = \overline{b} \tag{2.5.76}$$

$$\overline{d}_1 = \overline{r}'_0 = \overline{\overline{A}}^t\overline{b} \tag{2.5.77}$$

$$\alpha_1 = \frac{\|\overline{r}'_0\|^2}{d_1^t\overline{\overline{B}}\,\overline{d}_1} = \frac{\left\|\overline{\overline{A}}^t\overline{b}\right\|^2}{\left\|\overline{\overline{A}}\,\overline{d}_1\right\|^2} \tag{2.5.78}$$

$$\overline{x}_1 = \overline{x}_0 + \alpha_1\overline{d}_1 \tag{2.5.79}$$

$$\overline{r}_1 = \overline{r}_0 - \alpha_1\overline{\overline{A}}\,\overline{d}_1 \tag{2.5.80}$$

For $i \geq 2$,

$$\beta_i = \frac{\left\|\overline{r}'_{i-1}\right\|^2}{\left\|\overline{r}'_{i-2}\right\|^2} = \frac{\left\|\overline{\overline{A}}^t\,\overline{r}_{i-1}\right\|^2}{\left\|\overline{\overline{A}}^t\,\overline{r}_{i-2}\right\|^2} \tag{2.5.81}$$

$$\overline{d}_i = \overline{r}'_{i-1} + \beta_i\overline{d}_{i-1} = \overline{\overline{A}}^t\,\overline{r}_{i-1} + \beta_i\overline{d}_{i-1} \tag{2.5.82}$$

$$\alpha_i = \frac{\left\|\overline{r}'_{i-1}\right\|^2}{\overline{d}_i^t\overline{\overline{B}}\,\overline{d}_i} = \frac{\left\|\overline{\overline{A}}^t\,\overline{r}_{i-1}\right\|^2}{\left\|\overline{\overline{A}}\,\overline{d}_i\right\|^2} \tag{2.5.83}$$

$$\overline{x}_i = \overline{x}_{i-1} + \alpha_i\overline{d}_i \tag{2.5.84}$$

$$\overline{r}_i = \overline{r}_{i-1} - \alpha_i\overline{\overline{A}}\,\overline{d}_i \tag{2.5.85}$$

In the algorithm, there is no matrix and matrix products. There are only products of matrix and column vector.

For a general nonsingular complex matrix $\overline{\overline{A}}$, we solve the equation

$$\overline{\overline{A}}\,\overline{x} = \overline{b} \tag{2.5.86}$$

Let superscript † denote adjoint. The L^2 norm is defined by $\|\overline{x}\|^2 = \overline{x}^\dagger \overline{x}$. The algorithm of conjugate gradient method is similar to that of the general real matrix before with transpose replaced by adjoint. One can consider the matrix

$$\overline{\overline{B}} = \overline{\overline{A}}^\dagger \overline{\overline{A}} \tag{2.5.87}$$

where $\overline{\overline{A}}^\dagger = \left(\overline{\overline{A}}^t\right)^*$ is the Hermitian adjoint matrix of $\overline{\overline{A}}$. Then the computational steps become, on replacing t by †

$$\overline{x}_0 = 0 \tag{2.5.88a}$$

$$\overline{r}_0 = \overline{b} \tag{2.5.88b}$$

$$\overline{d}_1 = \overline{\overline{A}}^\dagger \overline{r}_0 \tag{2.5.89a}$$

$$\alpha_1 = \frac{\left\|\overline{\overline{A}}^\dagger \overline{r}_0\right\|^2}{\left\|\overline{\overline{A}}\,\overline{d}_1\right\|^2} \tag{2.5.89b}$$

$$\overline{x}_1 = x_0 + \alpha_1 \overline{d}_1 \tag{2.5.89c}$$

$$\overline{r}_1 = \overline{r}_0 - \alpha_1 \overline{\overline{A}}\,\overline{d}_1 \tag{2.5.89d}$$

and for $i = 2, 3, \ldots$

$$\beta_i = \frac{\left\|\overline{\overline{A}}^\dagger \overline{r}_{i-1}\right\|^2}{\left\|\overline{\overline{A}}^\dagger \overline{r}_{i-2}\right\|^2} \tag{2.5.90a}$$

$$\overline{d}_i = \overline{\overline{A}}^\dagger \overline{r}_{i-1} + \beta_i \overline{d}_{i-1} \tag{2.5.90b}$$

$$\alpha_i = \frac{\left\|\overline{\overline{A}}^\dagger \overline{r}_{i-1}\right\|^2}{\left\|\overline{\overline{A}}\,\overline{d}_i\right\|^2} \tag{2.5.90c}$$

$$\overline{x}_i = \overline{x}_{i-1} + \alpha_i \overline{d}_i \tag{2.5.90d}$$

$$\overline{r}_i = \overline{r}_{i-1} - \alpha_i \overline{\overline{A}}\,\overline{d}_i \tag{2.5.90e}$$

Let $\overline{\overline{A}}$ be a real matrix, then there exist orthogonal matrices of $\overline{\overline{U}}$ and $\overline{\overline{V}}$ and diagonal matrix $\overline{\overline{\Sigma}}$ such that

$$\overline{\overline{A}} = \overline{\overline{U}}\,\overline{\overline{\Sigma}}\,\overline{\overline{V}}^t \tag{2.5.91}$$

where

$$\overline{\overline{\Sigma}} = \mathrm{diag}(\sigma_1, \sigma_2, \ldots, \sigma_n) \qquad \sigma_1 \geq \sigma_2 \geq \ldots \geq \sigma_n \geq 0 \tag{2.5.92}$$

This is known as singular value decomposition. Using the L^2 norm for vectors, let the matrix norm be

$$\left\|\overline{\overline{A}}\right\| = \max_{\|\overline{x}\|=1} \left\|\overline{\overline{A}}\,\overline{x}\right\| \tag{2.5.93}$$

The relations between matrix norm and singular values are

$$\left\|\overline{\overline{A}}\right\| = \sigma_1 \tag{2.5.94}$$

Since

$$\overline{\overline{A}}^{-1} = \overline{\overline{V}}\,\overline{\overline{\Sigma}}^{-1}\overline{\overline{U}}^{t} \tag{2.5.95}$$

the largest singular value of $\overline{\overline{A}}^{-1}$ is $1/\sigma_n$.

$$\left\|\overline{\overline{A}}^{-1}\right\| = \frac{1}{\sigma_n} \tag{2.5.96}$$

The condition number is

$$\kappa\left(\overline{\overline{A}}\right) = \left\|\overline{\overline{A}}\right\|\left\|\overline{\overline{A}}^{-1}\right\| = \frac{\sigma_1}{\sigma_n} \tag{2.5.97}$$

If the condition number is large, the matrix solution can be unstable. To change the condition number, pre-conditioning can be done.

Let

$$\overline{\overline{A}}\,\overline{x} = \overline{b} \tag{2.5.98}$$

The goal is to find a pre-conditioning matrix $\overline{\overline{C}}$

$$\overline{x}' = \overline{\overline{C}}\,\overline{x} \tag{2.5.99}$$

$$\overline{x} = \overline{\overline{C}}^{-1}\overline{x}' \tag{2.5.100}$$

Then

$$\overline{\overline{A}}\,\overline{\overline{C}}^{-1}\overline{x}' = \overline{b} \tag{2.5.101}$$

Multiply by $\overline{\overline{C}}^{-1}$

$$\overline{\overline{C}}^{-1}\overline{\overline{A}}\,\overline{\overline{C}}^{-1}\overline{x}' = \overline{\overline{C}}^{-1}\overline{b} \tag{2.5.102}$$

Let

$$\overline{\overline{A}}' = \overline{\overline{C}}^{-1}\overline{\overline{A}}\,\overline{\overline{C}}^{-1} \tag{2.5.103}$$

$$\overline{b}' = \overline{\overline{C}}^{-1}\overline{b} \tag{2.5.104}$$

Then

$$\overline{\overline{A}}'\,\overline{x}' = \overline{b}' \tag{2.5.105}$$

A judicious choice of the pre-conditioning matrix can drastically change the condition number. To make the computation of $\overline{\overline{A}}'$ efficient, $\overline{\overline{C}}$ and $\overline{\overline{C}}^{-1}$ must be simple matrices.

REFERENCES AND ADDITIONAL READINGS

Anderson, M. G. (1965), Scattering from bodies of revolution, *IEEE Trans. Antennas Propagat.*, **13**, 303–310.

Barrett, R., M. Berry, T. F. Chan, J. Dammel, J. Donato, J. Dongarra, V. Eijkhout, R. Pozo, C. Romine, H. van der Vorst (1993), *Templates for the Solution of Linear Systems: Building Blocks for Iterative Methods*, SIAM Publications, Philadelphia, PA.

Borup, D. T. and O. P. Gandhi (1985), Calculation of high resolution SAR distributions in biological bodies using the FFT algorithm and conjugate gradient method, *IEEE Trans. Microwave Theory Tech.*, **33**, 417–419.

Brigham, E. O. (1988), *The Fast Fourier Transform and its Applications*, Prientice-Hall, Englewood Cliffs, NJ.

Catedra, M. F., E. Gago, and L. Nuno (1989), A numerical scheme to obtain the RCS of three-dimensional bodies of resonant size using the conjugate gradient method and the fast Fourier transform, *IEEE Trans. Antennas Propagat.*, **37**, 528–537.

Chan, C. H. and R. Mittra (1987), Some recent developments in iterative techniques for solving electromagnetic boundary value problems, *Radio Sci.*, **22(6)**, 929–934.

Chew, W. C. (1990), *Waves and Fields in Inhomogeneous Media*, Van Nostrand Reinhold, New York.

Gan, H. and W. C. Chew (1995), A discrete BCG-FFT algorithm for solving 3D inhomogeneous scatterer problems, *J. Electromag. Waves and Appl.*, **9**, 1339–1357.

Glisson, A. W. and D. R. Wilton (1979), Simple and efficient numerical techniques for treating bodies of revolution, *Technical Report 105*, Engineering Experiment Station, The University of Mississippi, University, Mississippi.

Glisson, A. W. and D. R. Wilton (1980), Simple and efficient numerical methods for problems of electromagnetic radiation and scattering from surfaces, *IEEE Trans. Antennas Propagat.*, **28**, 593–603.

Goedecke, G. H. and S. G. O'Brien (1988), Scattering by irregular inhomogeneous particles via the digitized Green's function algorithm, *Appl. Optics*, **27**, 2431–2438.

Golub, G. H. and C. F. Van Loan (1996), *Matrix Computations*, 3rd edition, Johns Hopkins University Press, Baltimore, MD.

Goodman, J. J., B. T. Drain, and P. J. Flatau (1991), Application of fast-Fourier-transform techniques to the discrete-dipole approximation, *Optics Lett.*, **16(15)**, 1198–1200.

Gradshteyn, I. S. and I. M. Ryzhik (1965), *Table of Integrals, Series and Products*, Academic Press, New York.

Harrington, R. F. (1968), *Field Computation by Moment Method*, Macmillan, New York.

Hestenes, M. R. and E. Stiefel (1952), Methods of conjugate gradients for solving linear systems, *J. Res. Nat. Bur. Standards*, **49**, 409–436.

Jackson, J. D. (1975), *Classical Electrodynamics*, John Wiley & Sons, New York.

Jin, J. M. and J. L. Volakis (1992), A biconjugate gradient FFT solution for scattering by planar plates, *Electromagnetics*, **12**, 105–109.

Joseph, J. (1990), Application of integral equation and finite difference method to electromagnetic scattering by two dimensional and boy of revolution geometries, Ph.D. thesis, Department of Electrical Engineering and Computer Science, University of Urbana-Champaign, Urbana, IL.

Kas, A. and E. L. Yip (1987), Preconditioned conjugate gradient methods for solving electromagnetic problems, *IEEE Trans. Antennas Propagat.*, **35**, 147–152.

Lakhtakia, A. (1992), General theory of the Purcell-Pennypacker scattering approach and its extension to bianisotropic scatterers, *Astrophys. J.*, **394(2)**, 494–499.

Lee, S. W., J. Boersma, C. L. Law, and G. A. De Champs (1980), Singularity in Green's function and its numerical evolution, *IEEE Trans. Antennas Propagat.*, **28**, 311–317.

Livesay, D. E. and K. M. Chen (1974), Electromagnetic fields induced inside arbitrarily shaped biological bodies, *IEEE Trans. Microwave Theory Tech.*, **22**, 1273–1280.

Mantz, J. R. and R. F. Harrington (1969), Radiation and scattering from bodies of revolution, *Appl. Sci. Res.*, **20**, 405–435.

Miller, E. K., L. Medgyesi-Mitschang, and E. H. Newman, Eds. (1992), *Computational Electromagnetics: Frequency-Domain Method of Moments*, IEEE Press, New York.

Oppenheim, A. V. and R. W. Schafer (1975), *Digital Signal Processing*, Prentice-Hall, Englewood Cliffs, NJ.

Peterson, A. F. and R. Mittra (1984), Method of conjugate gradient for the numerical solution of large body electromagnetic scattering problems, *J. Opt. Soc. Am.*, **2**, 971–977.

Peterson, A. F. and R. Mittra (1985), The convergence of the conjugate gradient method when applied to matrix equations representing electromagnetic scattering problems, *IEEE Trans. Antennas Propagat.*, **34**, 1447–1454. *J. Opt. Soc. Am.*, **2**, 971–977.

Peterson, A. F., S. L. Ray, C. H. Chan, and R, Mittra (1991), Numerical implementations of the conjugate gradient method and the CG-FFT for electromagnetic scattering, *PIER 5*, T. K. Sarkar, ed., Elsevier, New York.

Peterson, A. F., S. L. Ray, and R. Mittra (1997), *Computational Methods for Electromagnetics*, IEEE Press, New York.

Poggio, A. J. and E. K. Miller (1973), Integral equation solution of three-dimensional scattering problems, *Computer Techniques for Electromagnetics*, R. Mittra, ed., Pergamon, New York.

Purcell, E. M. and C. R. Pennypacker (1973), Scattering and absorption of light by nonspherical dielectric grains, *Astrophys. J.*, **186**, 705–714.

Rao, S. M., D. R. Wilton, and A. W. Glisson (1982), Electromagnetic scattering by surfaces of arbitrary shape, *IEEE Trans. Antennas Propagat.*, **30(3)**, 409–418.

Sarkar, T. K. (1991), editor, *Application of Conjugate Gradient Method to Electromagnetics and Signal Analysis, PIER 5*, Elsevier, New York.

Sarkar, T. K., E. Arvas, and S. M. Rao (1986), Application of the fast Fourier transform and conjugate gradient method for efficient solution of electromagnetic scattering both electrically large and small conducting bodies, *IEEE Trans. Antennas Propagat.*, **34**, 635–640.

Sarkar, T. K., X. Yang, and E. Arvas (1988), A limited survey of various conjugate gradient methods for complex matrix equations answering in electromagnetic wave interactions, *Wave Motion*, **10**, 527–546.

Singham, S. B. and G. C. Salzam (1986), Evaluation of the scattering matrix of an arbitrary particle using the coupled dipole approximation, *J. Chem. Phys.*, **84**, 2658–2667.

Stoer, J. and R. Bulirsch (1992), *Introduction to Numerical Analysis*, 2nd edition, Springer-Verlag, New York.

Tsang, L., J. A. Kong, and R. T. Shin (1985), *Theory of Microwave Remote Sensing*, Wiley-Interscience, New York.

Van Bladel, J. (1961), Some remarks on Green's dyadic for infinite space, *IRE Trans. Ant. and Prop.*, **9**, 563–566.

Van Bladel, J. (1991), *Singular Electromagnetic Fields and Sources*, Oxford, Oxford University Press.

Wang, J. J. H. (1991), *Generalized Moment Methods in Electromagnetics*, John Wiley and Sons, New York.

Yaghjian, A. D. (1980), Electric dyadic Green's functions in the source region, *Proc. IEEE*, **68**, 248–263.

Zwamborn, P. and P. M. van den Berg (1992), The three-dimensional weak form of the conjugate gradient FFT method for solving scattering problems, *IEEE Trans. Microwave Theory Tech.*, **40(9)**, 1757–1766.

Chapter 3

SCATTERING AND EMISSION BY A PERIODIC ROUGH SURFACE

The scattering of waves from a periodic surface has been studied extensively. Two methods of solution will be used in this chapter: (1) the method of moments based on periodic Green's function and (2) the T-matrix method. In Section 1, the problem of a one-dimensional periodic rough surface with Dirichlet boundary condition is illustrated. In Section 2, we consider the case of vector electromagnetic wave obliquely incident on a one-dimensional periodic rough surface. The incident direction is also at a nonzero azimuthal angle with respect to the periodic direction. This problem has applications in polarimetric passive remote sensing of rough surfaces because the third and fourth Stokes parameters are nonzero. The components of the electric and magnetic fields along the row direction are used as unknown scalar functions to reduce the vector nature of the problem to a scalar one. Then, the extended boundary condition (EBC) approach with Fourier series expansion for the surface fields is used to obtain the matrix equations governing the scattered field amplitudes. In Section 3, we study the vector electromagnetic case using integral equation method. In Sections 1 and 3, the periodic surface is one-dimensional, and the Green's function is two-dimensional. A method of speeding up the computation of 2-D Green's function in 1-D lattice is presented. In Section 4, we describe Ewald's method of computing 3-D Green's function in 3-D lattice and 2-D lattice. Besides rough surfaces, periodic structure problems are studied extensively in frequency selective surfaces [Chan, 1995; Munk, 2000] and photonic bandgap materials [Yablonovich, 1987; Joannopoulos et al. 1995].

1 Dirichlet Boundary Conditions

1.1 Surface Integral Equation

Consider a plane wave incident upon a periodic surface with height function $z = f(x)$, such that $f(x+P) = f(x)$. The period of the rough surface is P in the $\hat{x}$-direction. The incident direction is in the x-z plane. The electric field of the incident wave is given by

$$\overline{E}_i = \hat{y}\, e^{i\overline{k}_i \cdot \overline{r}} \tag{3.1.1}$$

where $\overline{k}_i$ denotes the incident wave vector and is given by $\hat{x}k_{ix} - \hat{z}k_{iz}$ with $k_{ix} = k\sin\theta_i$ and $k_{iz} = k\cos\theta_i$. We have $\overline{r} = \hat{x}x + \hat{z}z$. The electric field E_y satisfies the two-dimensional wave equation

$$\left(\frac{\partial^2}{\partial x^2} + \frac{\partial^2}{\partial z^2}\right) E_y + k^2 E_y = 0 \tag{3.1.2}$$

The Green's function is

$$G(\bar{r},\bar{r}') = \frac{i}{4} H_0^{(1)}(k\left|\bar{r}-\bar{r}'\right|) = \frac{i}{4\pi}\int_{-\infty}^{\infty} dk_x \frac{1}{k_z} \exp\left[ik_x(x-x') + ik_z|z-z'|\right] \tag{3.1.3}$$

where $k_z = (k^2 - k_x^2)^{1/2}$. Making use of Green's function of region 0, we have

$$E_{yi}(\bar{r}) - \int_{-\infty}^{\infty} d\sigma' \Big\{G(\bar{r},\bar{r}')\,\hat{n}\cdot\nabla' E_y(\bar{r}') - E_y(\bar{r}')\,\hat{n}\cdot\nabla' G(\bar{r},\bar{r}')\Big\}$$
$$= \begin{cases} E_y(\bar{r}) & z > f(x) \\ 0 & z < f(x) \end{cases} \tag{3.1.4}$$

and

$$d\sigma'\hat{n} = \left[\hat{z} - \hat{x}\,\frac{df(x')}{dx'}\right] dx' \tag{3.1.5}$$

For Dirichlet boundary conditions $E_y = 0$ on the surface, the integral equation in (3.1.4) becomes

$$E_{yi}(x, z = f(x)) = \int_{-\infty}^{\infty} d\sigma' G(x, f(x); x', f(x'))(\hat{n}\cdot\nabla' E_y(\bar{r}'))_{z'=f(x')} \tag{3.1.6}$$

Let the unknown surface variable be denoted by $u(x)$ so that

$$u(x)dx = \Big(d\sigma\hat{n}\cdot\nabla E_y(\bar{r})\Big)_{z=f(x)} \tag{3.1.7}$$

Equation (3.1.6) becomes

$$e^{ik_{ix}x - ik_{iz}f(x)} = \int_{-\infty}^{\infty} dx' G(x, f(x); x', f(x'))u(x') \tag{3.1.8}$$

1.2 Floquet's Theorem and Bloch Condition

The left-hand side of (3.1.8) has the following translational property

$$e^{ik_{ix}(x+nP) - ik_{iz}f(x+nP)} = e^{ik_{ix}nP} e^{ik_{ix}x - ik_{iz}f(x)} \tag{3.1.9}$$

Letting x be replaced by $x + nP$ in (3.1.8) and using (3.1.9), we have also $x' \to x' + nP$.

$$\begin{aligned} e^{ik_{ix}nP} e^{ik_{ix}x - ik_{iz}f(x)} &= \int_{-\infty}^{\infty} dx' G(x + nP, f(x); x' + nP, f(x'))u(x' + nP) \\ &= \int_{-\infty}^{\infty} dx' G(x, f(x); x', f(x'))u(x' + nP) \\ &= e^{ik_{ix}nP} \int_{-\infty}^{\infty} dx' G(x, f(x); x', f(x'))u(x') \end{aligned} \tag{3.1.10}$$

The second equality in (3.1.10) is a result of translational invariant property of G. Thus we have

$$u(x' + nP) = e^{ik_{ix}nP} u(x') \tag{3.1.11}$$

Equation (3.1.11) is known as the Floquet's theorem and is generally true for periodic structures. Thus one write

$$u(x) = e^{ik_{ix}x} w(x) \tag{3.1.12}$$

where $w(x)$ is a periodic function with period P. From (3.1.12), we have the Bloch condition

$$u(x + P) = e^{ik_{ix}P} u(x) \tag{3.1.13}$$

The surface integral in (3.1.8) is over infinite domain. However, it can be condensed into a single period. Let the center period be from $-P/2$ to $P/2$. Thus, using (3.1.8), for $-P/2 \leq x \leq P/2$ we have

$$\begin{aligned} e^{ik_{ix}x - ik_{iz}f(x)} &= \sum_{m=-\infty}^{\infty} \int_{-P/2+mP}^{P/2+mP} dx' G(x, f(x); x', f(x')) u(x') \\ &= \int_{-P/2}^{P/2} dx' \sum_{m=-\infty}^{\infty} G(x, f(x); x' + mP, f(x')) e^{ik_{ix}mP} u(x') \end{aligned} \tag{3.1.14}$$

The second equality in (3.1.14) is a result of changing dummy variables of integration from x' to $x' + mP$ and using Floquet's theorem for $u(x)$.

1.3 2-D Green's Function in 1-D Lattice

From (3.1.14), one can define the periodic Green's function

$$G_p(x, z; x', z') = \sum_{m=-\infty}^{\infty} G(x, z; x' + mP, z') e^{ik_{ix}mP} \tag{3.1.15}$$

Thus for $-P/2 \leq x \leq P/2$ the integral equation of (3.1.14) becomes

$$e^{ik_{ix}x - ik_{iz}f(x)} = \int_{-P/2}^{P/2} dx' G_p(x, f(x); x', f(x')) u(x') \tag{3.1.16}$$

Thus the advantage of (3.1.16) is that the integral equation is reduced to matching the left- and right-hand sides over only one period of the periodic medium instead of over an infinite domain. However, instead of the free space Green's function G, we have to compute the periodic Green's function G_p which is an infinite series as represented by (3.1.15).

From (3.1.3) and (3.1.15) we obtain

$$G_p(x,z;x',z') = \frac{i}{4}\sum_{m=-\infty}^{\infty} e^{ik_{ix}mP} H_0^{(1)}(k\sqrt{(x-x'-mP)^2+(z-z')^2}) \tag{3.1.17}$$

Equation (3.1.17) is the periodic Green's function in spatial domain. As $m \to \infty$, the terms inside the summation decay as $1/m$. One can express the result in the spectral domain by making use of the spectral representation of the free space Green's function

$$G_p(x,z;x',z') = \frac{i}{4\pi}\sum_{m=-\infty}^{\infty}\int_{-\infty}^{\infty} dk_x \frac{1}{k_z}\exp(ik_x(x-x'-mP)+ik_z|z-z'|)e^{ik_{ix}mP} \tag{3.1.18}$$

We next make use of the property of Fourier series that

$$\sum_{m=-\infty}^{\infty} e^{i\alpha mP} = \frac{2\pi}{P}\sum_{m=-\infty}^{\infty}\delta\left(-\alpha+\frac{2\pi m}{P}\right) \tag{3.1.19}$$

where δ is Dirac delta function. Equation (3.1.19) simply states that a periodic train of impulses with period $2\pi/P$ can be represented by a Fourier series with the Fourier coefficient equal to $P/2\pi$. Substituting (3.1.19) in (3.1.18) gives the periodic Green's function in the spectral domain.

$$G_p(x,z;x',z') = \frac{i}{2P}\sum_{m=-\infty}^{\infty}\frac{1}{k_{zm}}\exp(ik_{xm}(x-x')+ik_{zm}|z-z'|) \tag{3.1.20}$$

where $k_{xm} = k_{ix} + 2\pi m/P$ and $k_{zm} = (k^2 - k_{xm}^2)^{1/2}$. For $k_{xm} < k$, we have propagating Floquet modes. For $k_{xm} > k$, k_{zm} becomes imaginary giving rise to evanescent Floquet modes. The spectral form of (3.1.20) does converge rapidly for large values of $|z-z'|$ due to exponential decay, but we often need to evaluate this function for small or zero values of $|z-z'|$ in which case the summation is slowly convergent. Here, we present a general transformation to speed up the convergence of summation that is due to Veysoglu et al. [1991]. We start from the expansion

$$\frac{1}{e^v - e^{it}} = e^{-it}\sum_{m=1}^{\infty} e^{imt}e^{-mv} \tag{3.1.21}$$

Multiply both sides by $q(v)$ and integrate from 0 to ∞. This gives

$$\sum_{m=1}^{\infty} e^{imt}Q(m) = e^{it}\int_0^{\infty} dv \frac{q(v)}{e^v - e^{it}} \tag{3.1.22}$$

where

$$Q(m) = \int_0^\infty dv\, e^{-mv} q(v) \tag{3.1.23}$$

is the Laplace transform of $q(v)$.

Next, we make use of the following Laplace transform integral using Bessel function of order 1/2 [Oberhettinger and Badii, 1973].

$$e^{-is} H_0^{(1)}\left((s^2 + a^2)^{1/2}\right) = -\frac{i2}{\pi} \int_0^\infty dy\, e^{-sy} \frac{\cos\left(a(y^2 - 2iy)^{1/2}\right)}{(y^2 - 2iy)^{1/2}} \tag{3.1.24}$$

Let

$$s = -k(x - x' - mP) \tag{3.1.25}$$

$$a = k(z - z') \tag{3.1.26}$$

in (3.1.24). This gives

$$e^{-ikmP} H_0^{(1)}(k\sqrt{(x - x' - mP)^2 + (z - z')^2})$$
$$= -\frac{i2}{\pi} e^{-ik(x-x')} \int_0^\infty dy\, e^{k(x-x'-mP)y} \frac{\cos\left(a(y^2 - 2iy)^{1/2}\right)}{(y^2 - 2iy)^{1/2}} \tag{3.1.27}$$

Changing dummy integration variable from y to $v = kPy$ gives

$$e^{-ikmP} H_0^{(1)}(k\sqrt{(x - x' - mP)^2 + (z - z')^2})$$
$$= \int_0^\infty dv\, e^{-mv} \left[-\frac{2i}{\pi k P} e^{-ik(x-x')} e^{(x-x')v/P} \frac{\cos\left(a\left(\frac{v^2}{k^2P^2} - \frac{2iv}{kP}\right)^{1/2}\right)}{\left(\frac{v^2}{k^2P^2} - \frac{2iv}{kP}\right)^{1/2}} \right] \tag{3.1.28}$$

Comparing (3.1.28) with (3.1.23) shows that the left-hand side of (3.1.28) is $Q(m)$ while $q(v)$ is the square bracket on the right-hand side of (3.1.28). Use $Q(m)$ and $q(v)$ from (3.1.28) and substitute into (3.1.22). We also let

$$t = (k_{ix} + k)P \tag{3.1.29}$$

These give

$$\sum_{m=1}^\infty e^{ik_{ix}mP} H_0^{(1)}(k\sqrt{(x - x' - mP)^2 + (z - z')^2})$$
$$= e^{it} \int_0^\infty dv \frac{1}{e^v - e^{it}} \left[-\frac{2i}{\pi k P} e^{-ik(x-x')} e^{(x-x')v/P} \frac{\cos\left(a\left(\frac{v^2}{k^2P^2} - \frac{2iv}{kP}\right)^{1/2}\right)}{\left(\frac{v^2}{k^2P^2} - \frac{2iv}{kP}\right)^{1/2}} \right] \tag{3.1.30}$$

Transforming dummy variable from v to $u^2 = v/(kP)$ gives

$$G_p(x,z;x',z') = \frac{i}{4}\sum_{m=1}^{\infty} e^{ik_{ix}mP} H_0^{(1)}(k\sqrt{(x-x'-mP)^2+(z-z')^2})$$
$$= \frac{e^{i(k_{ix}+k)P}e^{-ik(x-x')}}{\pi}\times$$
$$\int_0^{\infty} du \frac{e^{-u^2kP+k(x-x')u^2}}{1-e^{-u^2kP+i(k_{ix}+k)P}} \frac{\cos\left(k(z-z')u(u^2-2i)^{1/2}\right)}{(u^2-2i)^{1/2}} \tag{3.1.31}$$

This integral is rapidly convergent due to exponential decay and can easily be evaluated by using Romberg integration. Having the formula for the periodic Green's function, one can use the method of moments to solve (3.1.16) for the unknown surface variable by matching over one period of rough surface $-P/2 \le x \le P/2$.

1.4 Bistatic Scattering Coefficients

Once the surface field $u(x)$ is calculated, the scattered field for $z > f(x)$ is, on using (3.1.4) and (3.1.6)–(3.1.8),

$$E_y^s(\bar{r}) = -\int_{-\infty}^{\infty} dx'\, G(x,z;x',f(x'))u(x') \tag{3.1.32}$$

The integral in (3.1.32) can again be condensed into one period

$$E_y^s(\bar{r}) = -\int_{-P/2}^{P/2} dx'\, G_p(x,z;x',f(x'))u(x') \tag{3.1.33}$$

We use (3.1.20) in (3.1.33) to get

$$E_y^s(\bar{r}) = \sum_{m=-\infty}^{\infty} e^{ik_{xm}x}e^{ik_{zm}z}B_m \tag{3.1.34}$$

where

$$B_m = -\frac{i}{2Pk_{zm}}\int_{-P/2}^{P/2} dx'\, e^{-ik_{xm}x'-ik_{zm}f(x')}u(x') \tag{3.1.35}$$

Note that only propagating Floquet modes carry time-averaged radiation power. The incident power on the rough surface over a period P is $Pw/2\eta$, where w is the width in the $\hat{y}$-direction. The time-averaged power contained in scattered wave is equal to $-E_y^s/(2i\omega\mu)\,(\partial E_y^{s*}/\partial z)\,Pw$. Thus the fractional power is

$$P_m = |B_m|^2\frac{k_{zm}}{k} \tag{3.1.36}$$

for the mth propagating Floquet mode. Conservation of energy requires

$$\sum_{\substack{m=\text{propagat.}\\ \text{modes}}} P_m = 1 \tag{3.1.37}$$

For the case of large P with P much bigger than wavelength, the propagating Floquet modes can approximate a continuum of scattered angles. Let

$$k_{xm} = k_{ix} + \frac{2\pi m}{P} = k \sin\theta_s \tag{3.1.38}$$

represent the transformation between k_{xm}, m, and scattered angle θ_s. Thus

$$\Delta k_{xm} = \frac{2\pi \Delta m}{P} \tag{3.1.39}$$

Since $\Delta m = 1$,

$$\sum_m \Delta m(\,) = \frac{P}{2\pi}\int dk_{xm}\,(\,) = \frac{P}{2\pi}\int_{-\pi/2}^{\pi/2} d\theta_s k\cos\theta_s\,(\,) \tag{3.1.40}$$

where ( ) stands for a mathematical expression. Such transformation between discrete and continuum as represented in (3.1.40) is customarily done in solid state physics when a periodic boundary condition is applied to truncate the domain to a finite size. Thus

$$\sum_m P_m = \frac{P}{2\pi}\int_{-\pi/2}^{\pi/2} d\theta_s k\cos\theta_s\,|B_m|^2 \frac{k_{zm}}{k} \tag{3.1.41}$$

If we let

$$\sum_m P_m = \int_{-\pi/2}^{\pi/2} d\theta_s \sigma(\theta_s) \tag{3.1.42}$$

where $\sigma(\theta_s)$ is the bistatic scattering coefficient, we have

$$\sigma(\theta_s) = \frac{Pk}{2\pi}\cos^2\theta_s\,|B_m|^2 \tag{3.1.43}$$

for the propagating mth Floquet mode (i.e., $|k_{xm}| \leq k$). Equation (3.1.43) expresses the bistatic scattering coefficient in terms of Floquet mode amplitudes.

2 Dielectric Periodic Surface: T-Matrix Method

In this section the scattering of electromagnetic waves from dielectric periodic rough surface is studied. We consider the case of oblique incidence and also at an arbitrary azimuthal angle with respect to the row direction. In the formulation, the components of the electric and magnetic fields along the

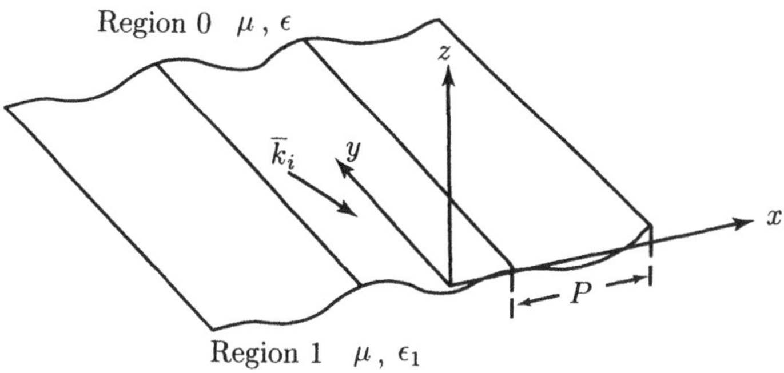

Figure 3.2.1 Geometrical configuration of the problem.

row direction are used as unknown scalar functions to reduce the vector nature of the problem to a scalar one. The rough surface is invariant along the row direction. Then, the extended boundary condition (EBC) approach with Fourier series expansion for the surface fields is used to obtain the matrix equations governing the scattered field amplitudes. In general, the E-waves, which are characterized by the components of the electric fields along the row direction, and the H-waves, which are characterized by the components of the magnetic fields along the row direction, are coupled together. Results are illustrated with sinusoidal profiles. The scattered power calculated is shown to satisfy reciprocity and energy conservation. The emissivity of a periodic rough surface is calculated from one minus the reflectivity. We also show good comparison with experimental data.

2.1 Formulation in Longitudinal Field Components

Consider a plane wave incident upon a periodic surface described by $f(x) = f(x + P)$, with P denoting the period of the surface in the $\hat{x}$-direction (Fig. 3.2.1). The electric field of the incident wave is given by

$$\overline{E}_i = \hat{e}_i E_o e^{i\overline{k}_i \cdot \overline{r}} \tag{3.2.1}$$

where $\overline{k}_i$ denotes the incident wave vector and is equal to $\hat{x}k_{ix} + \hat{y}k_{iy} - \hat{z}k_{iz}$ and $\hat{e}_i$ is the polarization of the electric field vector.

Since the structure is uniform in the $\hat{y}$-direction, all the field components in both region 0 and region 1 will have the same $\exp(ik_{iy}y)$ dependence. With this dependence, we can replace $\partial/\partial y$ in Maxwell's equation by ik_{iy}. It is possible to express all field components transverse to $\hat{y}$ in terms of longitudinal field components in the $\hat{y}$-direction. Unless otherwise specified, we will suppress the $\exp(ik_{iy}y)$ dependence.

In terms of the longitudinal components, the transverse components are

$$\overline{E}_{js}(\overline{r}) = \frac{i}{k_j^2 - k_{iy}^2}\left[k_{iy}\nabla_s E_{jy}(\overline{r}) + \omega\mu\nabla_s \times \overline{H}_{jy}(\overline{r})\right] \tag{3.2.2a}$$

$$\overline{H}_{js}(\overline{r}) = \frac{i}{k_j^2 - k_{iy}^2}\left[k_{iy}\nabla_s H_{jy}(\overline{r}) - \omega\epsilon_j\nabla_s \times \overline{E}_{jy}(\overline{r})\right] \tag{3.2.2b}$$

where $j = 0, 1$ signify regions 0 and 1, respectively, ∇_s is the transverse gradient operator that is transverse to $\hat{y}$-direction

$$\nabla_s = \hat{x}\frac{\partial}{\partial x} + \hat{z}\frac{\partial}{\partial z} \tag{3.2.3}$$

and $\overline{E}_{js}$ and $\overline{H}_{js}$ denote the transverse components of the electric and magnetic fields for region j. The subscript $j = 0$ is suppressed. The E-waves are described by $H_{jy} = 0$ and H-waves are described by $E_{jy} = 0$. The longitudinal components E_{jy} and H_{jy} satisfy the equations

$$\left(\nabla_s^2 + k_j^2 - k_{iy}^2\right)\left\{\begin{matrix} E_{jy} \\ H_{jy} \end{matrix}\right\} = 0 \tag{3.2.4}$$

We let

$$\overline{\rho}_s = \hat{x}x + \hat{z}z \tag{3.2.5}$$

be the position vector that is transverse to the $\hat{y}$-direction. Since E_{jy} and H_{jy} satisfy a two-dimensional wave equation, we shall use a two-dimensional Green's function. The Green's function is

$$G_j(\overline{\rho}_s, \overline{\rho}_s') = \frac{i}{4}H_0^{(1)}\left(k_{js}\left|\overline{\rho}_s - \overline{\rho}_s'\right|\right) \tag{3.2.6}$$

where $j = 0, 1$ and

$$k_{js} = \left(k_j^2 - k_{iy}^2\right)^{1/2} \tag{3.2.7}$$

Note that the Green's function is similar to (3.1.3), except that k has been replaced by k_{js} of (3.2.7). Integral equation can be formed by applying extinction theorem to E_y and H_y separately. We also make use of Floquet's theorem to condense the integral equation for one period using a procedure similar to that in Section 1. We have

$$E_{yi}(\overline{\rho}_s) - \int_P d\sigma'\left\{G_P(\overline{\rho}_s, \overline{\rho}_s')\,\hat{n}\cdot\nabla_s' E_y(\overline{\rho}_s') - E_y(\overline{\rho}_s')\,\hat{n}\cdot\nabla_s' G_P(\overline{\rho}_s, \overline{\rho}_s')\right\}$$

$$= \begin{cases} E_y(\overline{\rho}_s) & z > f(x) \quad (3.2.8a) \\ 0 & z < f(x) \quad (3.2.8b) \end{cases}$$

where the integration $d\sigma'$ is over one period P, and similarly

$$H_{yi}(\overline{\rho}_s) - \int_P d\sigma' \Big\{ G_P(\overline{\rho}_s, \overline{\rho}_s')\, \hat{n}\cdot\nabla_s' H_y(\overline{\rho}_s') - H_y(\overline{\rho}_s')\, \hat{n}\cdot\nabla_s' G_P(\overline{\rho}_s, \overline{\rho}_s') \Big\}$$

$$= \begin{cases} H_y(\overline{\rho}_s) & z > f(x) \\ 0 & z < f(x) \end{cases} \quad \begin{matrix}(3.2.9a)\\(3.2.9b)\end{matrix}$$

where

$$d\sigma'\hat{n} = \left[\hat{z} - \hat{x}\,\frac{df(x')}{dx'}\right] dx' \tag{3.2.10}$$

and

$$G_P(\overline{\rho}_s, \overline{\rho}_s') = \frac{-1}{2iP}\sum_n \frac{1}{k_{zn}} \exp[ik_{xn}(x - x') + ik_{zn}|z - z'|] \tag{3.2.11}$$

is the periodic Green's function of region 0. In (3.2.11)

$$k_{xn} = k_{ix} + \frac{2\pi n}{P} = k_s\alpha_n \tag{3.2.12}$$

$$k_{zn} = \sqrt{k_s^2 - k_{xn}^2} = \sqrt{k^2 - k_{iy}^2 - k_{xn}^2} = k_s\beta_n \tag{3.2.13}$$

Making use of periodic Green's function of region 1, we have

$$\int_P d\sigma' \Big\{ G_{1P}(\overline{\rho}_s, \overline{\rho}_s')\, \hat{n}\cdot\nabla_s' E_{1y}(\overline{\rho}_s') - E_{1y}(\overline{\rho}_s')\, \hat{n}\cdot\nabla_s' G_{1P}(\overline{\rho}_s, \overline{\rho}_s') \Big\}$$

$$= \begin{cases} 0 & z > f(x) \\ E_{1y}(\overline{\rho}_s) & z < f(x) \end{cases} \quad \begin{matrix}(3.2.14a)\\(3.2.14b)\end{matrix}$$

$$\int_P d\sigma' \Big\{ G_{1P}(\overline{\rho}_s, \overline{\rho}_s')\, \hat{n}\cdot\nabla_s' H_{1y}(\overline{\rho}_s') - H_{1y}(\overline{\rho}_s')\, \hat{n}\cdot\nabla_s' G_{1P}(\overline{\rho}_s, \overline{\rho}_s') \Big\}$$

$$= \begin{cases} 0 & z > f(x) \\ H_{1y}(\overline{\rho}_s) & z < f(x) \end{cases} \quad \begin{matrix}(3.2.15a)\\(3.2.15b)\end{matrix}$$

where

$$G_{1P}(\overline{\rho}_s, \overline{\rho}_s') = \frac{-1}{2iP}\sum_n \frac{1}{k_{1zn}} \exp[ik_{xn}(x - x') + ik_{1zn}|z - z'|] \tag{3.2.16}$$

is the periodic Green's function of medium 1, and

$$k_{1zn} = (k_1^2 - k_{iy}^2 - k_{xn}^2)^{1/2} = (k_{1s}^2 - k_{xn}^2)^{1/2} = k_{1s}\beta_n' \tag{3.2.17a}$$

$$k_{xn} = k_{ix} + \frac{2n\pi}{P} = k_{1s}\alpha_n' \tag{3.2.17b}$$

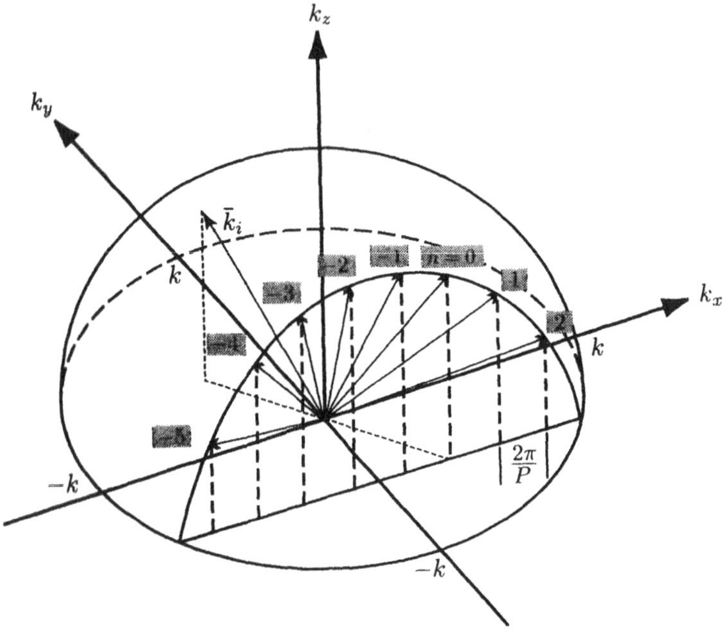

Figure 3.2.2 Conical diffraction of the reflected wave in region 0.

We see that the waves are propagating in discrete Floquet modes forming a cone shape when the observation point is either above the highest point or below the lowest point of the surface (Fig. 3.2.2).

Let f_{max} and f_{min} be respectively the maximum and minimum values of the surface profile $f(x)$. Then

$$E_y(\overline{\rho}_s) = E_{yi}(\overline{\rho}_s) + \sum_n b_n \frac{e^{i\overline{k}_n^+ \cdot \overline{\rho}_s}}{\sqrt{\beta_n}} \qquad z > f_{max} \tag{3.2.18a}$$

$$0 = E_{yi}(\overline{\rho}_s) - \sum_n a_n \frac{e^{i\overline{k}_n^- \cdot \overline{\rho}_s}}{\sqrt{\beta_n}} \qquad z < f_{min} \tag{3.2.18b}$$

$$H_y(\overline{\rho}_s) = H_{yi}(\overline{\rho}_s) + \sum_n b_n^{(h)} \frac{e^{i\overline{k}_n^+ \cdot \overline{\rho}_s}}{\sqrt{\beta_n}} \qquad z > f_{max} \tag{3.2.19a}$$

$$0 = H_{yi}(\overline{\rho}_s) - \sum_n a_n^{(h)} \frac{e^{i\overline{k}_n^- \cdot \overline{\rho}_s}}{\sqrt{\beta_n}} \qquad z < f_{min} \tag{3.2.19b}$$

where

$$\overline{k}_n^\pm = \hat{x} k_{xn} \pm \hat{z} k_{zn} \tag{3.2.20}$$

denote the propagation vectors of Floquet modes. We recognize that b_n and $b_n^{(h)}$ are scattered field amplitudes. The coefficients a_n, b_n, $a_n^{(h)}$ and $b_n^{(h)}$ are related to the surface fields by the following integrals,

$$b_n = \frac{1}{2ik_s P}\int_P d\sigma' \left\{ \frac{e^{-i\overline{k}_n^+\cdot\overline{\rho}_s(x')}}{\sqrt{\beta_n}}\hat{n}'\cdot\nabla'_s E_y(\overline{\rho}'_s) - E_y(\overline{\rho}'_s)\,\hat{n}'\cdot\nabla'_s\frac{e^{-i\overline{k}_n^+\cdot\overline{\rho}_s(x')}}{\sqrt{\beta_n}}\right\} \tag{3.2.21}$$

and $b_n^{(h)}$ is the same expression as b_n with E_y replaced by H_y.

$$a_n = \frac{-1}{2ik_s P}\int_P d\sigma' \left\{ \frac{e^{-i\overline{k}_n^-\cdot\overline{\rho}_s(x')}}{\sqrt{\beta_n}}\hat{n}'\cdot\nabla'_s E_y(\overline{\rho}'_s) - E_y(\overline{\rho}'_s)\,\hat{n}'\cdot\nabla'_s\frac{e^{-i\overline{k}_n^-\cdot\overline{\rho}_s(x')}}{\sqrt{\beta_n}}\right\} \tag{3.2.22}$$

and $a_n^{(h)}$ is the same expression as a_n with E_y replaced by H_y. In (3.2.21) and (3.2.22)

$$\overline{\rho}_s(x') = \hat{x}x + \hat{z}f(x') \tag{3.2.23}$$

is a point on the periodic surface. Similarly, making use of (3.2.14) and (3.2.15), we obtain

$$0 = -\sum_n B_n \frac{e^{i\overline{k}_{1n}^+\cdot\overline{\rho}_s}}{\sqrt{\beta'_n}} \qquad z > f_{max} \tag{3.2.24a}$$

$$E_{1y}(\overline{\rho}_s) = \sum_n A_n \frac{e^{i\overline{k}_{1n}^-\cdot\overline{\rho}_s}}{\sqrt{\beta'_n}} \qquad z < f_{min} \tag{3.2.24b}$$

$$0 = -\sum_n B_n^{(h)} \frac{e^{i\overline{k}_{1n}^+\cdot\overline{\rho}_s}}{\sqrt{\beta'_n}} \qquad z > f_{max} \tag{3.2.25a}$$

$$H_{1y}(\overline{\rho}_s) = \sum_n A_n^{(h)} \frac{e^{i\overline{k}_{1n}^-\cdot\overline{\rho}_s}}{\sqrt{\beta'_n}} \qquad z < f_{min} \tag{3.2.25b}$$

where

$$B_n = \frac{1}{2ik_{1s}P}\int_P d\sigma' \left\{ \frac{e^{-i\overline{k}_{1n}^+\cdot\overline{\rho}_s(x')}}{\sqrt{\beta'_n}}\hat{n}'\cdot\nabla'_s E_{1y}(\overline{\rho}'_s) - E_{1y}(\overline{\rho}'_s)\,\hat{n}'\cdot\nabla'_s\frac{e^{-i\overline{k}_{1n}^+\cdot\overline{\rho}_s(x')}}{\sqrt{\beta'_n}}\right\} \tag{3.2.26}$$

$B_n^{(h)}$ is the same expression as B_n with E_{1y} replaced by H_{1y},

$$A_n = \frac{-1}{2ik_{1s}P}\int_P d\sigma' \left\{ \frac{e^{-i\overline{k}_{1n}^{-}\cdot\overline{\rho}_s(x')}}{\sqrt{\beta_n'}}\,\hat{n}'\cdot\nabla_s' E_{1y}(\overline{\rho}_s') - E_{1y}(\overline{\rho}_s')\,\hat{n}'\cdot\nabla_s' \frac{e^{-i\overline{k}_{1n}^{-}\cdot\overline{\rho}_s(x')}}{\sqrt{\beta_n'}} \right\} \tag{3.2.27}$$

$A_n^{(h)}$ is the same expression as A_n with E_{1y} replaced by H_{1y}, and

$$\overline{k}_{1n}^{\pm} = \hat{x}k_{1xn} \pm \hat{z}k_{1zn} \tag{3.2.28}$$

are the propagation vectors of the Floquet modes in region 1. By using (3.2.18b) and (3.2.19b), we can readily solve for a_n and $a_n^{(h)}$ in terms of incident wave amplitudes.

$$a_n = \delta_{n0}\sqrt{\beta_0}\,E_o\,(\hat{e}_i\cdot\hat{y}) \tag{3.2.29}$$

$$a_n^{(h)} = \delta_{n0}\sqrt{\beta_0}\,\frac{E_o}{\omega\mu}\,(\overline{k}_i\times\hat{e}_i\cdot\hat{y}) \tag{3.2.30}$$

Also, from (3.2.24) and (3.2.25) we obtain

$$B_n = B_n^{(h)} \equiv 0 \quad \text{for all } n \tag{3.2.31}$$

Given a_n, $a_n^{(h)}$, B_n, and $B_n^{(h)}$ in (3.2.29)–(3.2.31), we next formulate matrix equations to solve for the surface fields. After the surface fields are solved, then the scattered field amplitudes b_n, $b_n^{(h)}$, A_n, and $A_n^{(h)}$ can be obtained using (3.2.21) and (3.2.27).

2.2 Surface Field Integral Equations and Coupled Matrix Equations

The surface integral equations are, from (3.2.22), (3.2.29), (3.2.30), and (3.2.26),

$$a_n = \delta_{n0}\sqrt{\beta_0}E_o(\hat{e}_i\cdot\hat{y}) = \frac{-1}{2ik_sP}\int_P d\sigma' \left\{ \frac{e^{-i\overline{k}_n^{-}\cdot\overline{\rho}_s(x')}}{\sqrt{\beta_n}}\,\hat{n}'\cdot\nabla_s' E_y(\overline{\rho}_s') - E_y(\overline{\rho}_s')\,\hat{n}'\cdot\nabla_s' \frac{e^{-i\overline{k}_n^{-}\cdot\overline{\rho}_s(x')}}{\sqrt{\beta_n}} \right\} \tag{3.2.32a}$$

$$a_n^{(h)} = \delta_{n0}\sqrt{\beta_0}\frac{E_o}{\omega\mu}(\overline{k}_i \times \hat{e}_i \cdot \hat{y})$$

$$= \frac{-1}{2ik_sP}\int_P d\sigma' \left\{ \frac{e^{-i\overline{k}_n^- \cdot \overline{\rho}_s(x')}}{\sqrt{\beta_n}} \hat{n}' \cdot \nabla_s' H_y(\overline{\rho}_s') - H_y(\overline{\rho}_s')\,\hat{n}' \cdot \nabla_s' \frac{e^{-i\overline{k}_n^- \cdot \overline{\rho}_s(x')}}{\sqrt{\beta_n}} \right\} \quad (3.2.32b)$$

$$B_n = 0 = \frac{1}{2ik_{1s}P}\int_P d\sigma' \left\{ \frac{e^{-i\overline{k}_{1n}^+ \cdot \overline{\rho}_s(x')}}{\sqrt{\beta_n'}} \hat{n}' \cdot \nabla_s' E_{1y}(\overline{\rho}_s') - E_{1y}(\overline{\rho}_s')\,\hat{n}' \cdot \nabla_s' \frac{e^{-i\overline{k}_{1n}^+ \cdot \overline{\rho}_s(x')}}{\sqrt{\beta_n'}} \right\} \quad (3.2.32c)$$

$$B_n^{(h)} = 0 = \frac{1}{2ik_{1s}P}\int_P d\sigma' \left\{ \frac{e^{-i\overline{k}_{1n}^+ \cdot \overline{\rho}_s(x')}}{\sqrt{\beta_n'}} \hat{n}' \cdot \nabla_s' H_{1y}(\overline{\rho}_s') - H_{1y}(\overline{\rho}_s')\,\hat{n}' \cdot \nabla_s' \frac{e^{-i\overline{k}_{1n}^+ \cdot \overline{\rho}_s(x')}}{\sqrt{\beta_n'}} \right\} \quad (3.2.32d)$$

However, in the four equations of (3.2.32*a*)–(3.2.32*d*), there are eight unknowns E_y, H_y, $\hat{n} \cdot \nabla_s E_y$, $\hat{n} \cdot \nabla_s H_y$, E_{1y}, H_{1y}, $\hat{n} \cdot \nabla_s E_{1y}$, and $\hat{n} \cdot \nabla_s H_{1y}$. We need to impose four boundary conditions to obtain a total of eight equations for the eight unknowns.

Applying the boundary conditions for the tangential electromagnetic fields on the periodic surface S, we have

$$E_y = E_{1y} \quad (3.2.33a)$$

$$H_y = H_{1y} \quad (3.2.33b)$$

$$\hat{n} \times \overline{E}_s = \hat{n} \times \overline{E}_{1s} \quad (3.2.33c)$$

$$\hat{n} \times \overline{H}_s = \hat{n} \times \overline{H}_{1s} \quad (3.2.33d)$$

where $\overline{E}_s$, $\overline{H}_s$, $\overline{E}_{1s}$, and $\overline{H}_{1s}$ are related to $\overline{E}_y$, $\overline{H}_y$, $\overline{E}_{1y}$, and $\overline{H}_{1y}$ by (3.2.2). Equations (3.2.33*a*) and (3.2.33*b*) relate the four unknowns. Next we need to put (3.2.33*c*) and (3.2.33*d*) as conditions on the eight unknowns.

From $\hat{n} \times \overline{E}_s = \hat{n} \times \overline{E}_{1s}$, (3.2.2*a*) and (3.2.2*b*),

$$\frac{k_{iy}}{k_s^2}\hat{n} \times \nabla_s E_y + \frac{\omega\mu}{k_s^2}\hat{n} \times (\nabla_s \times \overline{H}_y) = \frac{k_{iy}}{k_{1s}^2}\hat{n} \times \nabla_s E_{1y} + \frac{\omega\mu_1}{k_{1s}^2}\hat{n} \times (\nabla_s \times \overline{H}_{1y}) \quad (3.2.34)$$

Note that

$$\hat{n} \times (\nabla_s \times \overline{A}_y) = -\hat{y}\,\hat{n} \cdot \nabla_s A_y \tag{3.2.35}$$

where $\overline{A}_y = \hat{y} A_y(\overline{\rho}_s)$. Thus,

$$k_{iy}\hat{n}\times\nabla_s E_y - \omega\mu\hat{y}(\hat{n}\cdot\nabla_s H_y) = \frac{k_s^2}{k_{1s}^2}\left[k_{iy}\hat{n}\times\nabla_s E_{1y} - \omega\mu_1\hat{y}(\hat{n}\cdot\nabla_s H_{1y})\right] \tag{3.2.36}$$

From (3.2.33*a*) and (3.2.33*b*) we have two relations

$$E_y[x, z = f(x)] = E_{1y}[x, z = f(x)] \tag{3.2.37a}$$

$$H_y[x, z = f(x)] = H_{1y}[x, z = f(x)] \tag{3.2.37b}$$

Note that the surface fields (with $z = f(x)$) E_y, H_y, E_{1y}, and H_{1y} are functions of x only. If we take total derivative of (3.2.37*a*) with respect to x, we have

$$\frac{dE_y}{dx} = \frac{\partial E_y}{\partial x} + \frac{\partial E_y}{\partial z}\frac{df(x)}{dx} = \frac{\partial E_{1y}}{\partial x} + \frac{\partial E_{1y}}{\partial z}\frac{df(x)}{dx} = \frac{dE_{1y}}{dx} \tag{3.2.38}$$

with $z = f(x)$. Since

$$\hat{n} = \left[-\frac{df}{dx}\hat{x} + \hat{z}\right] \cdot \left[1 + \left(\frac{df}{dx}\right)^2\right]^{-\frac{1}{2}}$$

we get

$$\begin{aligned} \hat{n} \times \nabla_s E_y &= \left[1 + \left(\frac{df}{dx}\right)^2\right]^{-\frac{1}{2}} \left[\frac{\partial E_y}{\partial x} + \frac{df}{dx}\frac{\partial E_y}{\partial z}\right]_{z=f(x)} \\ &= \left[1 + \left(\frac{df}{dx}\right)^2\right]^{-\frac{1}{2}} \frac{dE_y}{dx} \end{aligned} \tag{3.2.39}$$

From (3.2.38) and (3.2.39), it follows that

$$\hat{n} \times \nabla_s E_y = \hat{n} \times \nabla_s E_{1y} \tag{3.2.40a}$$

Similarly, $\dfrac{dH_y}{dx} = \dfrac{dH_{1y}}{dx}$, so that

$$\hat{n} \times \nabla_s H_y = \hat{n} \times \nabla_s H_{1y} \tag{3.2.40b}$$

Putting (3.2.40*a*) and (3.2.40*b*) in (3.2.36), we have

$$\hat{y}(\hat{n} \cdot \nabla_s H_y) = -d_0\hat{n} \times \nabla_s E_{1y} + d_2\hat{y}(\hat{n} \cdot \nabla_s H_{1y}) \tag{3.2.41}$$

where

$$d_0 = \left[\frac{k_s^2}{k_{1s}^2} - 1\right]\frac{k_{iy}}{\omega\mu} \tag{3.2.42a}$$

$$d_2 = \frac{\mu_1}{\mu}\frac{k_s^2}{k_{1s}^2} \tag{3.2.42b}$$

Equation (3.2.41) is a result of applying $\hat{n} \times \overline{E}_s = \hat{n} \times \overline{E}_{1s}$. If we apply the same procedure to $\hat{n} \times \overline{H}_s = \hat{n} \times \overline{H}_{1s}$, we get

$$\hat{y}(\hat{n} \cdot \nabla_s E_y) = c_0 \hat{n} \times \nabla_s H_{1y} + c_2 \hat{y}(\hat{n} \cdot \nabla_s E_{1y}) \tag{3.2.43}$$

where

$$c_0 = \left[\frac{k_s^2}{k_{1s}^2} - 1\right] \frac{k_{iy}}{\omega \epsilon} \tag{3.2.44a}$$

$$c_2 = \frac{\epsilon_1}{\epsilon} \frac{k_s^2}{k_{1s}^2} \tag{3.2.44b}$$

Using (3.2.39) and the like in (3.2.41) and (3.2.43), we get respectively

$$\hat{n} \cdot (\nabla_s H_y) = -\frac{d_0 \dfrac{dE_{1y}}{dx}}{\left[1 + \left(\dfrac{df}{dx}\right)^2\right]^{1/2}} + d_2(\hat{n} \cdot \nabla_s H_{1y}) \tag{3.2.45a}$$

$$\hat{n} \cdot (\nabla_s E_y) = \frac{c_0 \dfrac{dH_{1y}}{dx}}{\left[1 + \left(\dfrac{df}{dx}\right)^2\right]^{1/2}} + c_2(\hat{n} \cdot \nabla_s E_{1y}) \tag{3.2.45b}$$

Equations (3.2.37*a*), (3.2.37*b*), (3.2.45*a*), and (3.2.45*b*) provide the four relations for the eight surface fields E_y, H_y, E_{1y}, H_{1y}, $\hat{n} \cdot \nabla_s E_y$, $\hat{n} \cdot \nabla_s H_y$, $\hat{n} \cdot \nabla_s E_{1y}$, and $\hat{n} \cdot \nabla_s H_{1y}$. The integral equations (3.2.32*a*–*d*) provide the additional four integral equations.

Surface Field Expansion

As indicated in Section 2.1, the surface fields obey Floquet's theorem; that is, $\psi(x) = \exp(ik_{ix}x)w(x)$, where $w(x)$ is a periodic function with period P. Thus one can expand $w(x)$ in a Fourier series. We use such expansion for the surface field components as follows. With $\overline{\rho}_s(x) = \hat{x}x + \hat{z}f(x)$, let

$$E_y[\overline{\rho}_s(x)] = E_{1y}[\overline{\rho}_s(x)] = \sum_n 2\alpha_n^s \exp\left[ik_{ix}x + in\frac{2\pi}{P}x\right] \tag{3.2.46a}$$

$$d\sigma \hat{n} \cdot \nabla_s E_{1y}[\overline{\rho}_s(x)] = ik_{1s}dx \sum_n 2\beta_n^s \exp\left[ik_{ix}x + in\frac{2\pi}{P}x\right] \tag{3.2.46b}$$

$$H_y[\overline{\rho}_s(x)] = H_{1y}[\overline{\rho}_s(x)] = \sum_n 2\gamma_n^s \exp\left[ik_{ix}x + in\frac{2\pi}{P}x\right] \tag{3.2.46c}$$

$$d\sigma\hat{n}\cdot\nabla_s H_{1y}[\overline{\rho}_s(x)] = ik_{1s}dx\sum_n 2\delta_n^s \exp\left[ik_{ix}x + in\frac{2\pi}{P}x\right] \qquad (3.2.46d)$$

From (3.2.45*a*), (3.2.45*b*), and (3.2.46*a*–*d*),

$$d\sigma\hat{n}\cdot\nabla_s\overline{H}_y = dx\sum_n\left[-d_0 2\alpha_n^s i\left(k_{ix}+\frac{n2\pi}{P}\right)+d_2 ik_{1s}2\delta_n^s\right] \times\exp\left[ik_{ix}x+in\frac{2\pi}{P}x\right] \qquad (3.2.46e)$$

$$d\sigma\hat{n}\cdot\nabla_s\overline{E}_y = dx\sum_n\left[c_0 2\gamma_n^s i\left(k_{ix}+\frac{n2\pi}{P}\right)+c_2 ik_{1s}2\beta_n^s\right] \times\exp\left[ik_{ix}x+in\frac{2\pi}{P}x\right] \qquad (3.2.46f)$$

Thus all eight surface field components are now expressed in terms of four sets of unknown coefficients α_n^s, β_n^s, γ_n^s, and δ_n^s. Substitute the surface field expressions of (3.2.46*a*–*d*) in the two surface integral equations of (3.2.32*c*) and (3.2.32*d*). Define $\overline{\overline{Q}}_{D_2}^{\pm}$, $\overline{\overline{Q}}_{N_2}^{\pm}$ as the Dirichlet and Neumann matrices with elements

$$\left[\overline{\overline{Q}}_{D_2}^{\pm}\right]_{mn} \equiv \frac{-1}{P}\int_P dx\frac{e^{-i\overline{k}_{1m}^{\pm}\cdot\overline{\rho}_s(x)}}{\sqrt{\beta'_m}}\exp\left[ik_{ix}x+in\frac{2\pi}{P}x\right] = \frac{-1}{P\sqrt{\beta'_m}}\int_P dx\exp\left[-i(m-n)\frac{2\pi}{P}x - ik_{1s}(\pm\beta'_m)f(x)\right] \qquad (3.2.47a)$$

$$\left[\overline{\overline{Q}}_{N_2}^{\pm}\right]_{mn} \equiv \frac{1}{ik_{1s}P}\int_P d\sigma\,\hat{n}\cdot\nabla_s\frac{e^{-i\overline{k}_{1m}^{\pm}\cdot\overline{\rho}_s(x)}}{\sqrt{\beta'_m}}\exp\left[ik_{ix}x+in\frac{2\pi}{P}x\right] = \frac{-1+\alpha'_m\alpha'_n}{\pm\beta'_m\sqrt{\beta'_m}P}\int_P dx\exp\left[-i(m-n)\frac{2\pi}{P}x - ik_{1s}(\pm\beta'_m)f(x)\right] \qquad (3.2.47b)$$

where the integrations are performed over one period. We obtain the two matrix equations

$$-\overline{\overline{Q}}_{D_2}^{+}\cdot\overline{\beta}^s - \overline{\overline{Q}}_{N_2}^{+}\cdot\overline{\alpha}^s = \overline{B} = 0 \qquad (3.2.48a)$$

$$-\overline{\overline{Q}}_{D_2}^{+}\cdot\overline{\delta}^s - \overline{\overline{Q}}_{N_2}^{+}\cdot\overline{\gamma}^s = \overline{B}^{(h)} = 0 \qquad (3.2.48b)$$

In deriving the second equality in (3.2.47*b*), we have performed an integration by parts. Here the vectors $\overline{B}$, $\overline{B}^{(h)}$, $\overline{\alpha}^s$, $\overline{\beta}^s$, $\overline{\gamma}^s$, and $\overline{\delta}^s$ contain the elements B_n, $B_n^{(h)}$, α_n^s, β_n^s, γ_n^s, δ_n^s, respectively. Similarly, substituting the surface field expressions of (3.2.46*a–d*) in the surface integral equations of (3.2.32*a*) and (3.2.32*b*), we get the equations in the following matrix forms

$$\overline{a} = c_0 \overline{\overline{Q}}^-_{hy1} \cdot \overline{\gamma}^s + c_2 \frac{k_{1s}}{k_s} \overline{\overline{Q}}^-_{D_1} \cdot \overline{\beta}^s + \overline{\overline{Q}}^-_{N_1} \cdot \overline{\alpha}^s \tag{3.2.48c}$$

$$\overline{a}^{(h)} = -d_0 \overline{\overline{Q}}^-_{hy1} \cdot \overline{\alpha}^s + d_2 \frac{k_{1s}}{k_s} \overline{\overline{Q}}^-_{D_1} \cdot \overline{\delta}^s + \overline{\overline{Q}}^-_{N_1} \cdot \overline{\gamma}^s \tag{3.2.48d}$$

where $\overline{a}$ and $\overline{a}^{(h)}$ are the column vectors containing the known coefficients a_n and $a_n^{(h)}$ respectively, while $\overline{\overline{Q}}^-_{hy1}$ denotes the hybrid matrix which couples $\overline{\alpha}^s$ and $\overline{\gamma}^s$ to $\overline{a}$ and $\overline{a}^{(h)}$, and

$$\left[\overline{\overline{Q}}^{\pm}_{D_1}\right]_{mn} = \frac{-1}{P\sqrt{\beta_m}} \int_P dx \exp\left[-i(m-n)\frac{2\pi}{P}x - ik_s(\pm\beta_m)f(x)\right] \tag{3.2.49a}$$

$$\left[\overline{\overline{Q}}^{\pm}_{N_1}\right]_{mn} = \frac{-1+\alpha_m\alpha_n}{\pm\beta_m\sqrt{\beta_m}P} \int_P dx \exp\left[-i(m-n)\frac{2\pi}{P}x - ik_s(\pm\beta_m)f(x)\right] \tag{3.2.49b}$$

$$\left[\overline{\overline{Q}}^{\pm}_{hy1}\right]_{mn} = -\alpha_n \left[\overline{\overline{Q}}^{\pm}_{D_1}\right]_{mn} \tag{3.2.50}$$

Equations (3.2.48*a–d*) can be put into the following matrix equation.

$$\begin{bmatrix} \overline{\overline{Q}}^-_{N_1} & c_2\frac{k_{1s}}{k_s}\overline{\overline{Q}}^-_{D_1} & c_0\overline{\overline{Q}}^-_{hy1} & 0 \\ -d_0\overline{\overline{Q}}^-_{hy1} & 0 & \overline{\overline{Q}}^-_{N_1} & d_2\frac{k_{1s}}{k_s}\overline{\overline{Q}}^-_{D_1} \\ \overline{\overline{Q}}^+_{N_2} & \overline{\overline{Q}}^+_{D_2} & 0 & 0 \\ 0 & 0 & \overline{\overline{Q}}^+_{N_2} & \overline{\overline{Q}}^+_{D_2} \end{bmatrix} \begin{bmatrix} \overline{\alpha}^s \\ \overline{\beta}^s \\ \overline{\gamma}^s \\ \overline{\delta}^s \end{bmatrix} = \begin{bmatrix} \overline{a} \\ \overline{a}^{(h)} \\ 0 \\ 0 \end{bmatrix} \tag{3.2.51}$$

The unknown vectors $\overline{\alpha}^s$, $\overline{\beta}^s$, $\overline{\gamma}^s$, and $\overline{\delta}^s$ for the surface field expansions are obtained by solving the above matrix equation. After the coefficients $\overline{\alpha}^s$, $\overline{\beta}^s$, $\overline{\gamma}^s$, and $\overline{\delta}^s$ are solved, the upward-going field amplitudes of (3.2.18*a*) and (3.2.19*a*) by using (3.2.21) for b_n and the like for $b_n^{(h)}$

$$\overline{b} = -c_0 \overline{\overline{Q}}^+_{hy1} \cdot \overline{\gamma}^s - c_2 \frac{k_{1s}}{k_s} \overline{\overline{Q}}^+_{D_1} \cdot \overline{\beta}^s - \overline{\overline{Q}}^+_{N_1} \cdot \overline{\alpha}^s \tag{3.2.52a}$$

$$\overline{b}^{(h)} = d_0 \overline{\overline{Q}}^+_{hy1} \cdot \overline{\alpha}^s - d_2 \frac{k_{1s}}{k_s} \overline{\overline{Q}}^+_{D_1} \cdot \overline{\delta}^s - \overline{\overline{Q}}^+_{N_1} \cdot \overline{\gamma}^s \tag{3.2.52b}$$

and the downward-going field amplitudes of (3.2.24*b*) and (3.2.25*b*) by using

(3.2.27) for A_n and the like for $A_n^{(h)}$ are

$$\overline{A} = \overline{\overline{Q}}_{D_2}^- \cdot \overline{\beta}^s + \overline{\overline{Q}}_{N_2}^- \cdot \overline{\alpha}^s \tag{3.2.53a}$$

$$\overline{A}^{(h)} = \overline{\overline{Q}}_{D_2}^- \cdot \overline{\delta}^s + \overline{\overline{Q}}_{N_2}^- \cdot \overline{\gamma}^s \tag{3.2.53b}$$

From the matrix equations we see that the E- and H-waves are generally coupled. An incident wave of E-type ($\overline{a}^{(h)} \equiv 0$) will be diffracted from the dielectric periodic surface into waves that have both E-wave and H-wave components. Both types of wave will coexist to satisfy the boundary conditions. When the incident wave vector is in the x-z plane ($k_{iy} = 0$), then $c_0 = d_0 = 0$ and we can easily see that the scattered waves arising from $\overline{E}_{yi}$ are decoupled from those arising from $\overline{H}_{yi}$.

We next apply to the case when the surface is a sinusoidal function

$$f(x) = -h \cos\left(\frac{2\pi x}{P}\right) \tag{3.2.54}$$

The $\overline{\overline{Q}}^{\pm}$ matrices can be calculated by carrying out the integrations in (3.2.47), (3.2.49), and (3.2.50), and expressed in terms of Bessel functions.

$$\left[\overline{\overline{Q}}_{D_1}^{\pm}\right]_{mn} = \frac{-1}{\sqrt{\beta_m}}(\pm i)^{|m-n|} J_{|m-n|}(k_s h \beta_m) \tag{3.2.55a}$$

$$\left[\overline{\overline{Q}}_{N_1}^{\pm}\right]_{mn} = \frac{-1 + \alpha_m \alpha_n}{\pm \beta_m \sqrt{\beta_m}}(\pm i)^{|m-n|} J_{|m-n|}(k_s h \beta_m) \tag{3.2.55b}$$

$$\left[\overline{\overline{Q}}_{D_2}^{\pm}\right]_{mn} = \frac{-1}{\sqrt{\beta'_m}}(\pm i)^{|m-n|} J_{|m-n|}(k_{1s} h \beta'_m) \tag{3.2.55c}$$

$$\left[\overline{\overline{Q}}_{N_2}^{\pm}\right]_{mn} = \frac{-1 + \alpha'_m \alpha'_n}{\pm \beta'_m \sqrt{\beta'_m}}(\pm i)^{|m-n|} J_{|m-n|}(k_{1s} h \beta'_m) \tag{3.2.55d}$$

$$\left[\overline{\overline{Q}}_{hy1}^{\pm}\right]_{mn} = \frac{\alpha_n}{\sqrt{\beta_m}}(\pm i)^{|m-n|} J_{|m-n|}(k_s h \beta_m) \tag{3.2.55e}$$

where $J_{|m-n|}$ denotes the Bessel function of order $|m-n|$.

The T-matrix approach, which makes use of Green's theorem to derive the extended boundary conditions, is exact. However, the matrices used may become ill-conditioned when the surface corrugation is deep or when the corrugation depth divided by the period is large. This limits the applicable regime of this method. The reason for ill-conditioning is that entire basis functions rather than subsectional basis functions are used in surface field expansions. In Section 3 we shall apply subsectional basis functions to this problem which can handle surfaces with deeper corrugations.

2.3 Emissivity and Comparison with Experiments

In this section a sinusoidal surface is used to model a row-structured plowed field, and the theoretical results of emissivity are illustrated and compared with the experimental data obtained from field measurements.

The reflected power P_r for one period is

$$P_r = \frac{1}{2P} \int_0^P \text{Re}[(\overline{E}_r \times \overline{H}_r^*) \cdot \hat{z}]\, dx \tag{3.2.56}$$

and incident power is

$$P_{inc} = \frac{1}{2P} \int_0^P \text{Re}[(\overline{E}_{inc} \times \overline{H}_{inc}^*) \cdot (-\hat{z})]\, dx \tag{3.2.57}$$

For the reflected E-modes, by using (3.2.2) and (3.2.18a) for $z > f_{max}$, we get

$$\overline{E}_r = \sum_n \frac{b_n}{\sqrt{\beta_n}} e^{i\overline{k}_n^+ \cdot \overline{\rho}_s} \begin{bmatrix} -k_{iy}k_{xn}/k_s^2 \\ 1 \\ -k_{iy}k_{zn}/k_s^2 \end{bmatrix} \tag{3.2.58}$$

$$\overline{H}_r = \sum_n \frac{\omega\epsilon b_n}{\sqrt{\beta_n}} e^{i\overline{k}_n^+ \cdot \overline{\rho}_s} \begin{bmatrix} -k_{zn}/k_s^2 \\ 0 \\ k_{xn}/k_s^2 \end{bmatrix} \tag{3.2.59}$$

By substituting (3.2.58) and (3.2.59) into (3.2.56) and integrating, the Floquet modes are orthogonal and for the E-modes we have

$$P_r = \sum_n \frac{1}{2} \text{Re} \frac{\omega k_{zn}\epsilon}{k_s^2} \frac{|b_n|^2}{|\beta_n|} \tag{3.2.60}$$

Similarly, the reflected power for the H-modes can be obtained. The total reflected power is

$$P_r = \frac{1}{2} \sum_n \text{Re} \frac{\omega k_{zn}}{k_s^2} \left(\epsilon \frac{|b_n|^2}{|\beta_n|} + \mu \frac{|b_n^{(h)}|^2}{|\beta_n|} \right) = \sum_n P_n^r \tag{3.2.61}$$

where P_n^r can be interpreted as the reflected power for the nth Floquet mode. When the mode is evanescent, k_{zn} is purely imaginary. The power $P_n^r = 0$ for the evanescent mode. The incident power is

$$P_{inc} = \frac{1}{2} \frac{\omega\epsilon}{k_s} \left(|a_0|^2 + \frac{\mu}{\epsilon} |a_0^{(h)}|^2 \right) \tag{3.2.62}$$

Thus, the reflectivity for a wave with horizontal or vertical polarization is

$$r_\alpha = \frac{1}{P_{inc}} \sum_n P_n^r = \sum_n \frac{|b_n|^2 + |\eta b_n^{(h)}|^2}{|a_0|^2 + |\eta a_0^{(h)}|^2} \tag{3.2.63}$$

where $\alpha = v, h$ incident polarization. The summation in (3.2.63) is over the propagating modes and a_0 and $a_0^{(h)}$ are obtained from (3.2.29) and (3.2.30), by setting

$$\hat{e}_i = \hat{\alpha}_i \tag{3.2.64}$$

The transmitted power per period of the nth mode passing through the surface defined by $z = z_{min}$ is

$$P_n^t = \frac{1}{2} \text{Re} \frac{\omega k_{1zn}}{k_{1s}^2} \left(\epsilon_1 \frac{|A_n|^2}{|\beta_n'|} + \mu_1 \frac{|A_n^{(h)}|^2}{|\beta_n'|} \right) e^{2\text{Im}[k_{1zn}]z_{min}} \tag{3.2.65}$$

for all modes, where $k_{1zn} = k_{1s}\beta_n'$, and $\text{Im}[k_{1zn}] > 0$ should be used. When medium 1 is lossless, we have (letting $\eta_1 = \sqrt{\mu_1/\epsilon_1}$)

$$P_n^t = \frac{\omega \epsilon_1}{2k_{1s}} \left(|A_n|^2 + |\eta_2 A_n^{(h)}|^2 \right) \tag{3.2.66}$$

for the nth propagating mode and $P_n^t = 0$ for the evanescent mode. The transmissivity is given by

$$t_\alpha = \sum_n \frac{\epsilon_1}{\epsilon} \frac{k_s}{k_{1s}} \frac{|A_n|^2 + |\eta_1 A_n^{(h)}|^2}{|a_0|^2 + |\eta a_0^{(h)}|^2} \qquad (\alpha = v, h) \tag{3.2.67}$$

For the lossless case, the power conservation relation is

$$r_\alpha + t_\alpha = 1 \tag{3.2.68}$$

For both lossless and lossy media, the emissivity is given by

$$e_\alpha = 1 - r_\alpha \tag{3.2.69}$$

The theoretical results are illustrated for a sinusoidal surface in Figs. 3.2.3, 3.2.4, and 3.2.5 at a frequency of 1.4 GHz. The effect of the row structure on the microwave emission from a bare agricultural field has been reported [Wang et al. 1980] together with the soil moisture contents for the measured data. The periodic surface has a height $h = 10$ cm and a period $P = 95$ cm and can be approximated by a sinusoidal function. For the upper medium, we let $\epsilon = \epsilon_o$ and $\mu = \mu_o$.

In Fig. 3.2.3 we illustrate the comparison between the theoretical results and the experimental data for both the vertical and horizontal polarizations when the radiometer observation angle is along the row direction ($\phi = 90°$). The reported soil moisture content varies from 26% by dry weight at top 0 to 1 cm to 21.4% at 9 to 15 cm. In the theoretical results, we take $\epsilon_1 = (5.5 + i1.2)\epsilon_o$ which corresponds to a soil moisture content of approximately 18%. In the same figure, we also show the theoretical curves for the flat surface case. It is seen that the brightness temperature for the periodic rough

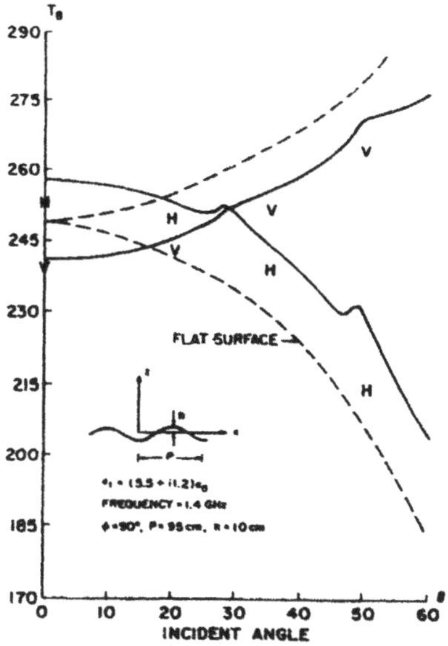

Figure 3.2.3 Brightness temperature as a function of viewing angle. Radiometer observation plane is parallel to the row direction ($\phi = 90°$). $\epsilon_1 = (5.5 + i1.2)\epsilon_o$

surfaces for the horizontal polarization is higher than that for the flat surface, whereas for the vertical polarization the brightness temperature is lower. For the flat surface, both polarizations have the same brightness temperature value when viewed from nadir, whereas for the periodic surface, the values for the horizontal polarization are higher than the vertical polarization at near-nadir angles and become lower at larger angles of observation.

In Fig. 3.2.4, the radiometer observation angles are perpendicular to the row direction ($\phi = 0°$). The soil moisture content is 29% by dry weight at the top 0 to 5 cm and becomes drier with depth. We use $\epsilon_1 = (10 + i2)\epsilon_o$. We see that at near-nadir angles, as compared with the flat surface cases, the brightness temperatures for the horizontal polarization are lower and for the vertical polarization are higher. The effect of the rough surface as compared with the flat surface appears to bring both the horizontal and vertical polarization results closer together at higher incident angles. The case of observation plane at $\phi = 30°$ is illustrated in Fig. 3.2.5.

In Figs. 3.2.3, 3.2.4, and 3.2.5, we observe that the T_B curves for the periodic rough surface are not smoothly varying. For instance, in Fig. 3.2.4, there are kinks appearing at observation angles θ near 6°, 19°, 34°, and 51°. The corresponding change in T_B may be as high as 10 K. Such a phenomenon can be explained by the appearance and disappearance of Floquet modes at various threshold angles. The kinks are caused by the redistribution of the

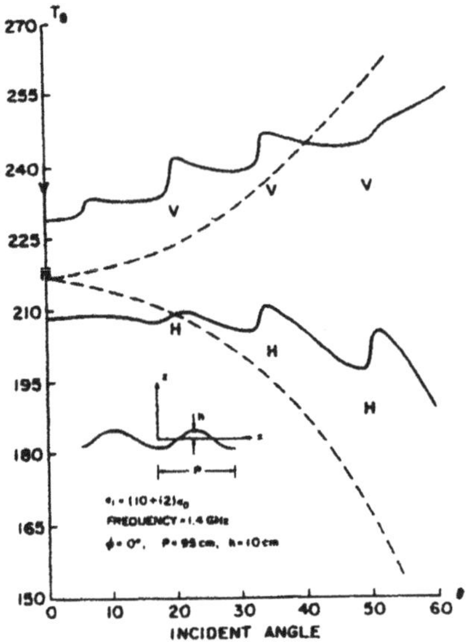

Figure 3.2.4 Brightness temperature as a function of viewing angle. Radiometer observation plane is perpendicular to the row direction ($\phi = 0°$). $\epsilon_1 = (10 + i2)\epsilon_o$.

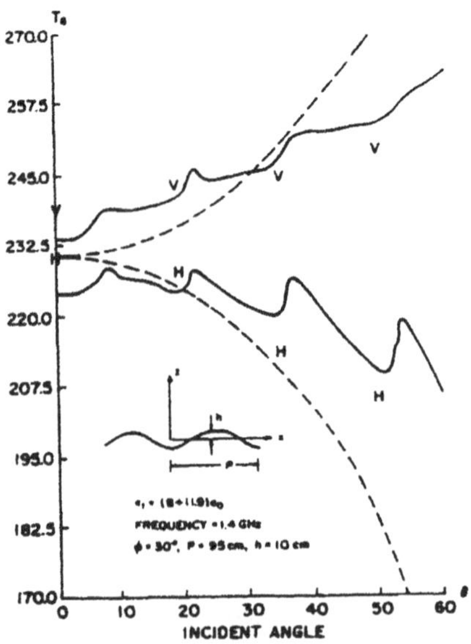

Figure 3.2.5 Brightness temperature as a function of viewing angle. Radiometer observation plane is slanted with respect to the row direction ($\phi = 30°$). $\epsilon_1 = (8 + i1.9)\epsilon_o$.

scattered power during the course of the disappearance and appearance of the propagating Floquet modes. For real soil surfaces, such kinks will not appear because real soil surfaces are not periodic. They have randomness superimposed on the periodic structure.

3 Scattering of Waves Obliquely Incident on Periodic Rough Surfaces: Integral Equation Approach

In this section, we use subsectional basis functions to solve the integral equations. The speed-up computation of the periodic Green's functions as discussed in Section 1 will be used.

3.1 Formulation

We use the integral equations developed in Section 2.1. From (3.2.8), (3.2.9), (3.2.14), and (3.2.15), we have

$$E_{yi}(\overline{\rho}_s) - \int_P d\sigma' \left\{ G_P(\overline{\rho}_s, \overline{\rho}'_s) \hat{n}' \cdot \nabla'_s E_y(\overline{\rho}'_s) - E_y(\overline{\rho}'_s) \hat{n}' \cdot \nabla'_s G_P(\overline{\rho}_s, \overline{\rho}'_s) \right\} = \begin{cases} E_y(\overline{\rho}_s) & \text{for } \overline{\rho}_s = \overline{\rho}_s^+ \\ 0 & \text{for } \overline{\rho}_s = \overline{\rho}_s^- \end{cases} \tag{3.3.1}$$

$$H_{yi}(\overline{\rho}_s) - \int_P d\sigma' \left\{ G_P(\overline{\rho}_s, \overline{\rho}'_s) \, \hat{n}' \cdot \nabla'_s H_y(\overline{\rho}'_s) - H_y(\overline{\rho}'_s) \, \hat{n}' \cdot \nabla'_s G_P(\overline{\rho}_s, \overline{\rho}'_s) \right\} = \begin{cases} H_y(\overline{\rho}_s) & \text{for } \overline{\rho}_s = \overline{\rho}_s^+ \\ 0 & \text{for } \overline{\rho}_s = \overline{\rho}_s^- \end{cases} \tag{3.3.2}$$

$$\int_P d\sigma' \left\{ G_{1P}(\overline{\rho}_s, \overline{\rho}'_s) \, \hat{n}' \cdot \nabla'_s E_{1y}(\overline{\rho}'_s) - E_{1y}(\overline{\rho}'_s) \, \hat{n}' \cdot \nabla'_s G_{1P}(\overline{\rho}_s, \overline{\rho}'_s) \right\} = \begin{cases} 0 & \text{for } \overline{\rho}_s = \overline{\rho}_s^+ \\ E_{1y}(\overline{\rho}_s) & \text{for } \overline{\rho}_s = \overline{\rho}_s^- \end{cases} \tag{3.3.3}$$

$$\int_P d\sigma' \left\{ G_{1P}(\overline{\rho}_s, \overline{\rho}'_s) \, \hat{n}' \cdot \nabla'_s H_{1y}(\overline{\rho}'_s) - H_{1y}(\overline{\rho}'_s) \, \hat{n}' \cdot \nabla'_s G_{1P}(\overline{\rho}_s, \overline{\rho}'_s) \right\} = \begin{cases} 0 & \text{for } \overline{\rho}_s = \overline{\rho}_s^+ \\ H_{1y}(\overline{\rho}_s) & \text{for } \overline{\rho}_s = \overline{\rho}_s^- \end{cases} \tag{3.3.4}$$

In (3.3.1)–(3.3.4),

$$\overline{\rho}'_s = x'\hat{x} + f(x')\hat{z} \tag{3.3.5}$$

$$\overline{\rho}_s^+ = x\hat{x} + f^+(x)\hat{z} \tag{3.3.6a}$$

$$\overline{\rho}_s^- = x\hat{x} + f^-(x)\hat{z} \tag{3.3.6b}$$

where $f^+(x)$ means infinitesimally larger than $f(x)$ and $f^-(x)$ means infinitesimally smaller than $f(x)$. The discontinuity is due to the fact that $\hat{n} \cdot \nabla G_P$ and $\hat{n} \cdot \nabla G_{1P}$ have a nonintegrable singularity when $\overline{\rho}_s$ and $\overline{\rho}'_s$ coincide. In the spatial domain, the periodic Green's functions are

$$G_{jP}(x, z; x', z') = \frac{i}{4} \sum_{m=-\infty}^{\infty} e^{ik_{ix}mP} H_0^{(1)}(k_{js}\sqrt{(x - x' - mP)^2 + (z - z')^2}) \tag{3.3.7}$$

where $j = 0, 1$. Computation of the periodic Green's function can be speeded up by using transformation as described in Section 1. We have, from (3.1.31),

$$\begin{aligned} &\frac{i}{4} \sum_{m=1}^{\infty} e^{ik_{ix}mP} H_0^{(1)}(k_{js}\sqrt{(x - x' - mP)^2 + (z - z')^2}) \\ &= \frac{e^{i(k_{ix}+k_{js})P} e^{-ik_{js}(x-x')}}{\pi} \times \\ &\int_0^{\infty} du \frac{e^{-u^2 k_{js} P + k_{js}(x-x')u^2}}{1 - e^{-u^2 k_{js} P + i(k_{ix}+k_{js})P}} \frac{\cos\left(k_{js}(z - z')u(u^2 - 2i)^{1/2}\right)}{(u^2 - 2i)^{1/2}} \end{aligned} \tag{3.3.8}$$

where $j = 0, 1$.

Equations (3.3.1)–(3.3.4) are four integral equations with eight unknowns, namely, E_y, H_y, E_{1y}, H_{1y}, $\hat{n} \cdot \nabla_s E_y$, $\hat{n} \cdot \nabla_s H_y$, $\hat{n} \cdot \nabla_s E_{1y}$, and $\hat{n} \cdot \nabla_s H_{1y}$. Thus as in Section 2, we have the four boundary conditions of (3.2.37*a*), (3.2.37*b*), (3.2.45*a*), and (3.2.45*b*) so that, when combined with (3.3.1)–(3.3.4), we have eight equations for the eight unknowns.

Next we apply MoM with pulse basis functions and point matching at the midpoint. The midpoint between x_{n-1} and x_n designated as $\overline{\rho}_n$ is the testing point of the point matching. For the nth internal between x_{n-1} and x_n,

$$E_y = E_{1y} = \gamma_n \tag{3.3.9a}$$

$$\hat{n} \cdot \nabla_s E_{1y} = \delta_n \tag{3.3.9b}$$

$$H_y = H_{1y} = \zeta_n \tag{3.3.9c}$$

$$\hat{n} \cdot \nabla_s H_{1y} = \xi_n \tag{3.3.9d}$$

Let us consider (3.3.3). The discontinuity is due to the singularity of $\hat{n}' \cdot \nabla'_s G_{1P}$ for $\overline{\rho}_s$ and $\overline{\rho}'_s$ on the same patch (self patch). Thus (3.3.3) can be written, assuming $\overline{\rho}_s$ to be on the same patch so that $\overline{\rho}_s = \overline{\rho}_m$,

$$\begin{aligned} &\sum_n \delta_n \int_{x_{n-1}}^{x_n} d\sigma' G_{1P}(\overline{\rho}_m, \overline{\rho}'_s) - \sum_{n \neq m} \gamma_n \int_{x_{n-1}}^{x_n} d\sigma' \hat{n}' \cdot \nabla'_s G_{1P}(\overline{\rho}_m, \overline{\rho}'_s) \\ &- \gamma_m \int_{x_{m-1}}^{x_m} d\sigma' \hat{n}' \cdot \nabla'_s G_{1P}(\overline{\rho}_m^{\pm}, \overline{\rho}'_s) = \begin{cases} 0 & \text{for } \overline{\rho}_m^{+} \\ \gamma_m & \text{for } \overline{\rho}_m^{-} \end{cases} \end{aligned} \tag{3.3.10}$$

Hence, if we take the difference between upper half and lower half of (3.3.10), we have

$$\gamma_m = -\gamma_m \int_{x_{m-1}}^{x_m} d\sigma' \hat{n}' \cdot \nabla_s' G_{1P}(\overline{\rho}_m^-, \overline{\rho}_s') + \gamma_m \int_{x_{m-1}}^{x_m} d\sigma' \hat{n}' \cdot \nabla_s' G_{1P}(\overline{\rho}_m^+, \overline{\rho}_s') \tag{3.3.11}$$

If the patch is approximated by a straight segment, then the two parts of (3.3.11) should be equal to and opposite each other. Thus

$$\int_{x_{m-1}}^{x_m} d\sigma' \hat{n}' \cdot \nabla_s' G_{1P}(\overline{\rho}_m^+, \overline{\rho}_s') = -\int_{x_{m-1}}^{x_m} d\sigma' \hat{n}' \cdot \nabla_s' G_{1P}(\overline{\rho}_m^-, \overline{\rho}_s') = \frac{1}{2} \tag{3.3.12}$$

Using (3.3.12), (3.3.10) can be written as

$$\sum_n \delta_n E_{mn} + \sum_n \gamma_n D_{mn} = 0 \tag{3.3.13}$$

where

$$E_{mn} = -\int_{x_{n-1}}^{x_n} d\sigma' \, G_{1P}(\overline{\rho}_m, \overline{\rho}_s') \tag{3.3.14}$$

$$D_{mn} = \begin{cases} \dfrac{1}{2} & \text{if } m = n \\ \displaystyle\int_{x_{n-1}}^{x_n} d\sigma' \hat{n}' \cdot \nabla_s' G_{1P}(\overline{\rho}_m, \overline{\rho}_s') & \text{if } m \neq n \end{cases} \tag{3.3.15}$$

Equation (3.3.4) becomes, in a similar manner,

$$\sum_n E_{mn} \xi_n + \sum_n D_{mn} \zeta_n = 0 \tag{3.3.16}$$

Next we consider (3.3.1). Note that at the boundary, $E_y = E_{1y}$, $H_y = H_{1y}$, $\dfrac{dE_y}{dx} = \dfrac{dE_{1y}}{dx}$, and $\dfrac{dH_y}{dx} = \dfrac{dH_{1y}}{dx}$. Consider the term $\int_0^P d\sigma' G_P(\overline{\rho}_m, \overline{\rho}_s')$ $\hat{n}' \cdot \nabla_s' E_y$ and use boundary condition (3.2.45*b*),

$$\int_0^P d\sigma' G_P(\overline{\rho}_m, \overline{\rho}_s') \hat{n}' \cdot \nabla_s' E_y$$

$$= \int_0^P d\sigma' G_P(\overline{\rho}_m, \overline{\rho}_s') \left\{ \frac{c_0 \dfrac{dH_{1y}}{dx'}}{\left[1 + \left(\dfrac{df(x')}{dx'}\right)^2\right]^{1/2}} + c_2 (\hat{n}' \cdot \nabla_s' E_{1y}) \right\}$$

$$= \left[G_P(\overline{\rho}_m, \overline{\rho}'_s) c_0 H_y(\overline{\rho}'_s) \right]_{x'=0}^{x'=P} - \int_0^P dx' \left[\frac{d}{dx'} G_P(\overline{\rho}_m, \overline{\rho}'_s) \right] c_0 H_y(\overline{\rho}'_s)$$

$$+ c_2 \int_0^P d\sigma' G_P(\overline{\rho}_m, \overline{\rho}'_s) \hat{n}' \cdot \nabla'_s E_{1y} \tag{3.3.17}$$

The second equality is a result of integration by parts. Since

$$\left[G_P(\overline{\rho}_m, \overline{\rho}'_s) \right]_{x'=P} = e^{-ik_{ix}P} \left[G_P(\overline{\rho}_m, \overline{\rho}'_s) \right]_{x'=0} \tag{3.3.18}$$

and

$$\left[H_y(\overline{\rho}'_s) \right]_{x'=P} = e^{ik_{ix}P} \left[H_y(\overline{\rho}'_s) \right]_{x'=0} \tag{3.3.19}$$

the first term of (3.3.17) vanishes. Thus

$$\int_0^P d\sigma' G_P(\overline{\rho}_m, \overline{\rho}'_s) \hat{n}' \cdot \nabla'_s E_y = -c_0 \sum_n C_{mn} \zeta_n + c_2 \sum_n B_{mn} \delta_n \tag{3.3.20}$$

where

$$B_{mn} = \int_{x_{n-1}}^{x_n} d\sigma' \, G_P(\overline{\rho}_m, \overline{\rho}'_s) \tag{3.3.21}$$

$$C_{mn} = G_P(\overline{\rho}_m, \overline{\rho}_n) - G_P(\overline{\rho}_m, \overline{\rho}_{n-1}) \tag{3.3.22}$$

with $\overline{\rho}_n = (x_n, f(x_n))$. Also let

$$A_{mn} = \begin{cases} \frac{1}{2} & \text{if } m = n \\ -\displaystyle\int_{x_{n-1}}^{x_n} d\sigma' \hat{n}' \cdot \nabla'_s G_P(\overline{\rho}_m, \overline{\rho}'_s) & \text{if } m \neq n \end{cases} \tag{3.3.23}$$

Integral equation (3.3.1) becomes, using (3.3.23) and (3.3.20),

$$E_{yi}(\overline{\rho}_m) - c_2 \sum_n B_{mn} \delta_n + c_0 \sum_n C_{mn} \zeta_n = \sum_n A_{mn} \gamma_n \tag{3.3.24}$$

Similarly, from (3.3.2) we have, using boundary condition of (3.2.45a),

$$H_{yi}(\overline{\rho}_m) - d_2 \sum_n B_{mn} \xi_n - d_0 \sum_n C_{mn} \gamma_n = \sum_n A_{mn} \zeta_n \tag{3.3.25}$$

where c_0, c_2, d_0, and d_2 are given in (3.2.42a–b) and (3.2.44a–b).

Equations (3.3.13), (3.3.16), and (3.3.24)–(3.3.25) are the equations governing the unknowns γ_n, δ_n, ζ_n and ξ_n. In matrix notation, they can be put in the form

$$\begin{bmatrix} A & c_2 B & -c_0 C & 0 \\ D & E & 0 & 0 \\ d_0 C & 0 & A & d_2 B \\ 0 & 0 & D & E \end{bmatrix} \begin{bmatrix} \gamma \\ \delta \\ \zeta \\ \xi \end{bmatrix} = \begin{bmatrix} E_{yi} \\ 0 \\ H_{yi} \\ 0 \end{bmatrix} \tag{3.3.26}$$

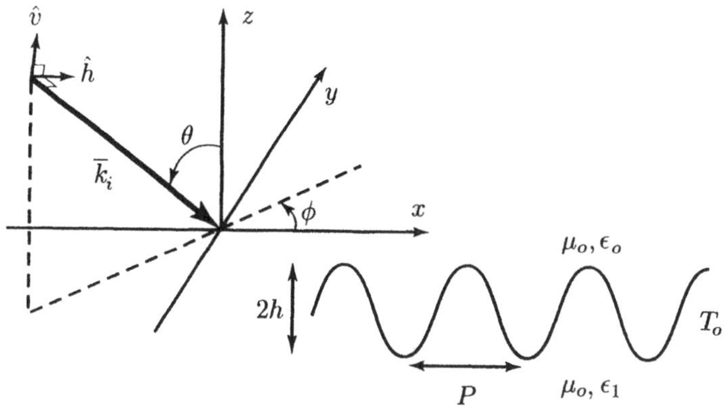

Figure 3.3.1 Polarimetric emission from a sinusoidal surface at temperature T_o.

Solving (3.3.26) numerically gives the surface field unknowns.

After the surface fields are determined, the coefficients of the reflected Floquet modes can be calculated. The reflected fields for $z > f(x)$ can be written as

$$E_y^r = \sum_n b_n e^{i\overline{k}_n^+ \cdot \overline{\rho}_s} \tag{3.3.27a}$$

$$H_y^r = \sum_n b_n^{(h)} e^{i\overline{k}_n^+ \cdot \overline{\rho}_s} \tag{3.3.27b}$$

Using (3.2.21) for b_n and the like for $b_n^{(h)}$, we have

$$b_n = \frac{1}{2ik_s P}\int_0^P d\sigma' \left[\frac{e^{-i\overline{k}_n^+ \cdot \overline{\rho}_s'}}{\beta_n}\hat{n}' \cdot \nabla_s' E_y(\overline{\rho}_s') - E_y(\overline{\rho}_s')\hat{n}' \cdot \nabla_s' \frac{e^{-i\overline{k}_n^+ \cdot \overline{\rho}_s'}}{\beta_n}\right] \tag{3.3.28a}$$

$$b_n^{(h)} = \frac{1}{2ik_s P}\int_0^P d\sigma' \left[\frac{e^{-i\overline{k}_n^+ \cdot \overline{\rho}_s'}}{\beta_n}\hat{n}' \cdot \nabla_s' H_y(\overline{\rho}_s') - H_y(\overline{\rho}_s')\hat{n}' \cdot \nabla_s' \frac{e^{-i\overline{k}_n^+ \cdot \overline{\rho}_s'}}{\beta_n}\right] \tag{3.3.28b}$$

3.2 Polarimetric Brightness Temperatures

In this section, polarimetric brightness temperatures are illustrated for the emission from a sinusoidal surface (Fig. 3.3.1). The frequency is fixed at 1 GHz and the period is fixed at 50 cm. The first region is free space and the physical temperature of the lossy dielectric, T_o, is fixed at 300 K. The expressions for emissivity of the four Stokes parameters can be found in Chapter 3, Section 5.4 of Volume I.

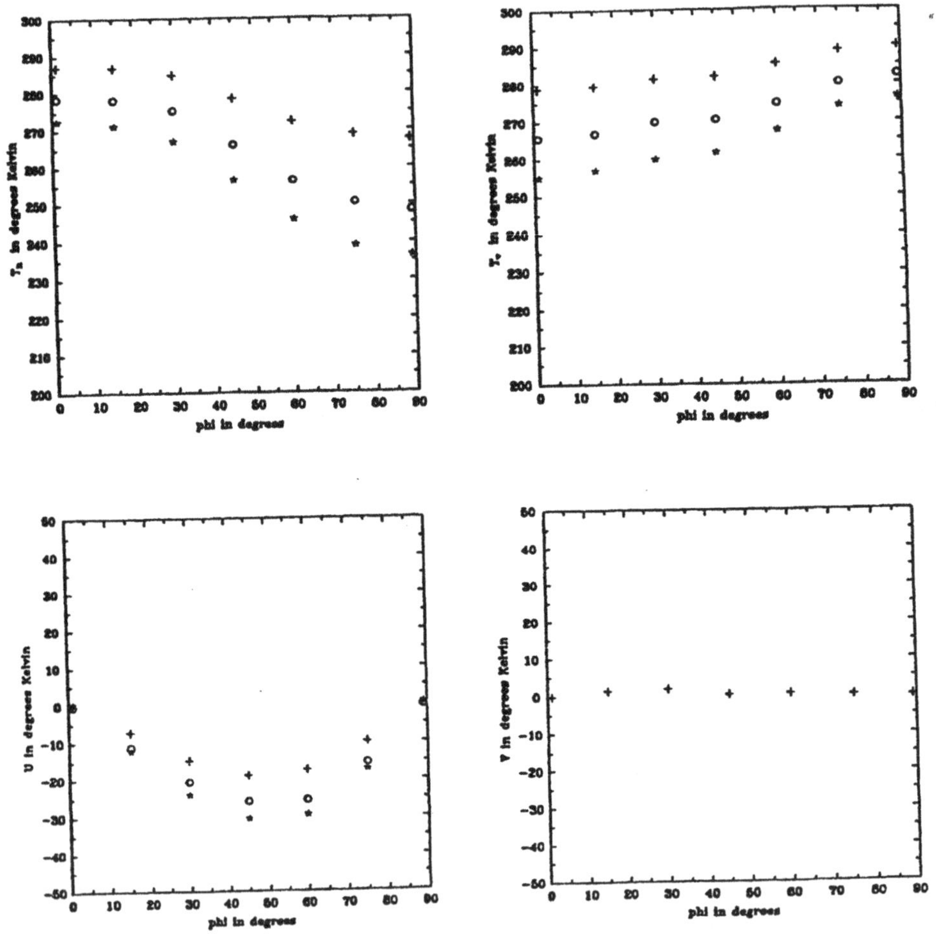

Figure 3.3.2 Brightness temperature versus ϕ. $+$: $\epsilon_{r1} = 6$, $\circ$: $\epsilon_{r1} = 9$, $*$: $\epsilon_{r1} = 12$.

Figure 3.3.2 plots the brightness temperatures versus the azimuth angle ϕ. For these plots, the height is 15 cm and the incident angle θ is 20°. Calculations are made for three different values of $\epsilon_{r1} = \epsilon_1/\epsilon_o$. It is seen that T_h and T_v tend to be higher for smaller ϵ_{r1}, while $|U|$ increases with increasing ϵ_{r1}. No significant change is observed in V. With increasing ϕ, T_h decreases and T_v increases. The brightness temperature of U is zero when $\phi = 0°$ or $\phi = 90°$ as expected. In between, it decreases to about -30 K. Values of V are small for the whole range of ϕ.

Figure 3.3.3 shows the variation with respect to θ with ϕ fixed at 45° and height kept at 15 cm. Again, the plots are given for three different values of

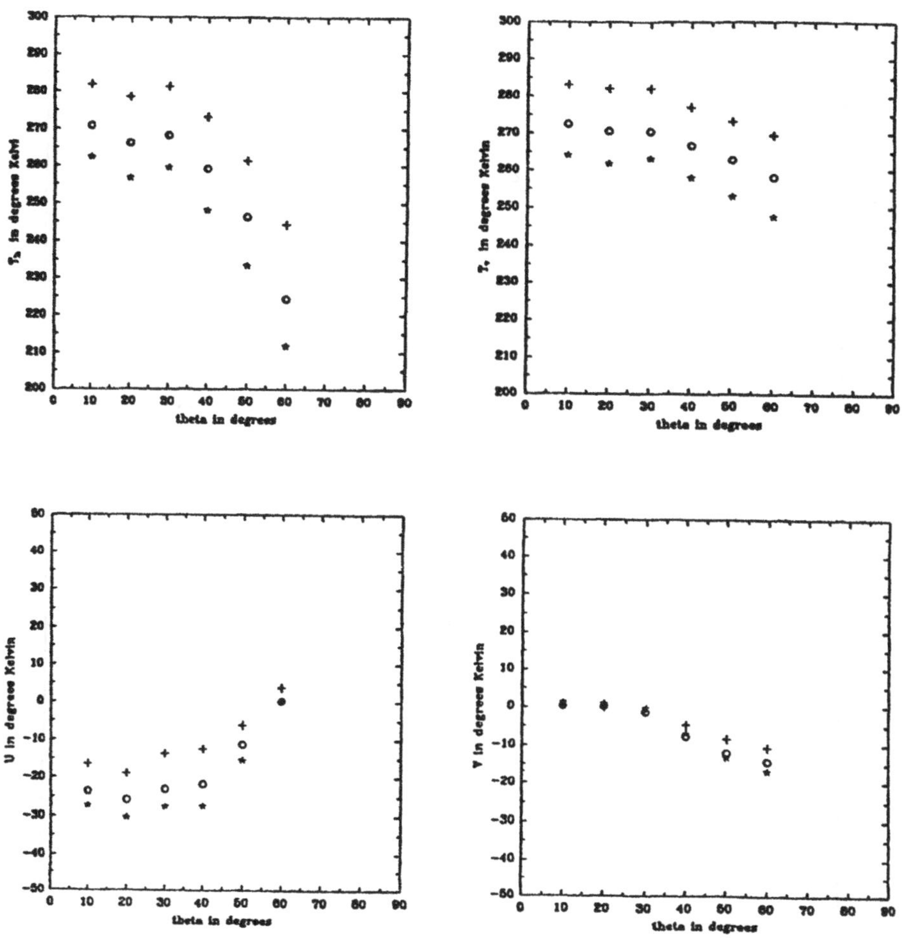

Figure 3.3.3 Brightness temperature versus θ. $+$: $\epsilon_{r1} = 6$, $\circ$: $\epsilon_{r1} = 9$, $*$: $\epsilon_{r1} = 12$.

ϵ_{r1} and we observe the same behavior. There is a more than 30 K difference in T_h for different ϵ_{r1} values at $\theta = 60°$. We also observe a change in V with increasing θ.

Dependence of brightness temperatures on the surface height is given in Fig. 3.3.4. For this case, $\epsilon_{r1} = 12$, $\theta = 20°$, and $\phi = 30°$. We see a general tendency of increasing T_h and T_v. V does not change much, but U varies significantly with height. We note that when the height is 10 cm, $|U|$ is as high as 48 K.

Finally, we analyze the effect of complex ϵ_{r1} on the brightness temperatures in Fig. 3.3.5, when θ is 20°, ϕ is 45°, and height is 15 cm. The real

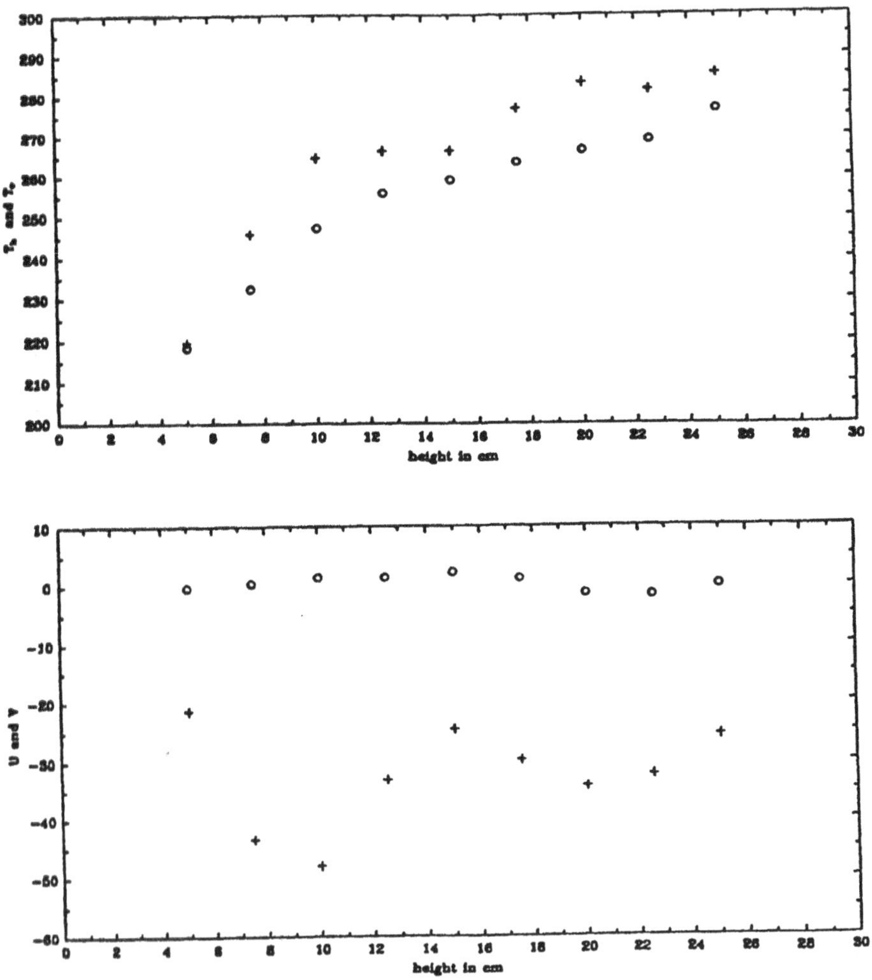

Figure 3.3.4 Brightness temperature vs. height. $+ : T_h(U)$, $\circ : T_v(V)$.

part of ϵ_{r1} is fixed at 12 and the imaginary part is varied from 0 to 8. Real ϵ_{r1} is taken as a base and the difference in temperatures is plotted. It is seen that variations in T_h and T_v are almost identical while the decrease in U is less. There is a slight increase in V.

Note also that the plots are not given as continuous curves. We merely show the results of some calculations, and linear interpolation between these may be misleading. This is due to the presence of kinks in actual continuous curves as shown in Section 2.

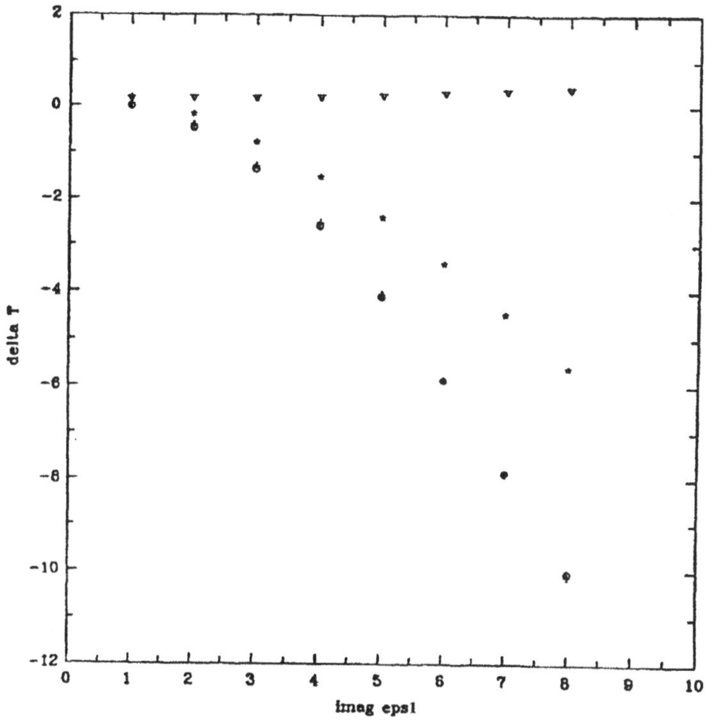

Figure 3.3.5 Brightness temperature vs. ϵ_{r1}. $+ : T_h$, $\circ : T_v$, $* : U$, $v : V$

4 Ewald's Method

A method that is used in computing periodic Green's function is the Ewald's method, which we will discuss in this section.

4.1 Preliminaries

We first describe several properties that are used in the derivation of Ewald's method. The first is an integral identity of the Hankel function.

From Abramowitz and Stegun [1965], the integral representation of the Hankel function of order ν is

$$H_\nu^{(1)}(z) = \frac{1}{\pi i} \int_{-\infty, C'}^{\infty + \pi i} dt\, e^{z \sinh t - \nu t} \qquad \text{with } |\arg z| < \frac{\pi}{2} \tag{3.4.1}$$

where the integral contour in the complex t plane is labeled by C'. Next we make the transformation $u = e^t$. Then $\sinh t = \frac{1}{2}\left(u - \frac{1}{u}\right); dt = du/u$. The integral becomes

$$H_\nu^{(1)}(z) = \frac{1}{\pi i}\int_{C'} du\ \exp\left[\frac{z}{2}\left(u - \frac{1}{u}\right)\right] u^{-\nu-1} \tag{3.4.2}$$

The contour C' in the complex u plane is from 0 to $-\infty$ along some suitable chosen path.

The spherical Hankel function of order 0 is, letting $z = k_o r$,

$$-\frac{ie^{ik_o r}}{k_o r} = h_0^{(1)}(k_o r) = \sqrt{\frac{\pi}{2k_o r}}H_{\frac{1}{2}}^{(1)}(k_o r)$$

$$= \sqrt{\frac{\pi}{2k_o r}}\frac{1}{\pi i}\int_{C'} du\ \exp\left[\frac{k_o r}{2}\left(u - \frac{1}{u}\right)\right] u^{-\frac{3}{2}} \tag{3.4.3}$$

Next we make transformation of variable to s. Let $u = \frac{k_o}{2rs^2}$, $du = -\frac{k_o}{rs^3}ds$, $du\, u^{-\frac{3}{2}} = -\sqrt{\frac{8r}{k_o}}ds$. Then

$$h_0^{(1)}(k_o r) = \frac{i2}{\sqrt{\pi}k_o}\int_{C'} ds\ \exp\left[\frac{k_o^2}{4s^2} - r^2 s^2\right] \tag{3.4.4}$$

The contour C' in the complex s plane is from ∞ to 0 [Watson, 1966]. We define the contour C to be the reverse of C' so that C is from $s = 0$ to $s = \infty$.

$$h_0^{(1)}(k_o r) = \frac{2}{i\sqrt{\pi}k_o}\int_C ds \exp\left[\frac{k_o^2}{4s^2} - r^2 s^2\right] \tag{3.4.5}$$

To choose the proper contour. Note that as $s \to \infty$, the convergence of the integral is dictated by $\exp(-r^2 s^2)$ and we require $\mathrm{Re}(s^2) > 0$. Thus

$$\text{as } s \to \infty,\ |\arg s| < \frac{\pi}{4} \tag{3.4.6}$$

On the other hand, as $s \to 0$, the convergence of the integral is dictated by $\exp\left(\frac{k_o^2}{4s^2}\right)$ and we require $\mathrm{Re}\left(\frac{k_o^2}{4s^2}\right) < 0$ which means $\frac{\pi}{2} < \arg\left(k_o^2\right) - 2\arg s < \frac{3\pi}{2}$. Thus

$$\text{as } s \to 0,\ -\frac{3\pi}{4} + \beta < \arg s < -\frac{\pi}{4} + \beta \tag{3.4.7}$$

where $\beta = \frac{1}{2}\arg\left(k_o^2\right) = \arg(k_o)$.

The contour C is as shown in Fig. 3.4.1. For $0 \le \beta \le \frac{\pi}{2}$, the intersection of these two regions is [Jordan et al. 1986]

$$-\frac{\pi}{4} \le \arg s \le -\frac{\pi}{4} + \beta \tag{3.4.8}$$

The second relation is the definition of the lattice vectors and reciprocal lattice vectors. Consider a periodic lattice that is three-dimensional. Let

$$\overline{R} = n_1\overline{a}_1 + n_2\overline{a}_2 + n_3\overline{a}_3 \tag{3.4.9}$$

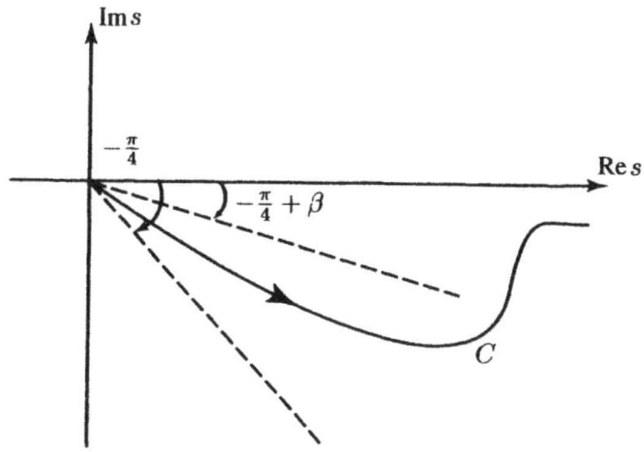

Figure 3.4.1 Contour C with $\beta = \arg k_o$.

where $\overline{a}_1$, $\overline{a}_2$, and $\overline{a}_3$ are the basis lattice vectors and n_1, n_2, and n_3 are integers.

By a periodic medium, we mean that the wave function ψ obeys a wave equation with periodic potential $V(\overline{r})$. The periodic potential obeys the condition

$$V(\overline{r}+\overline{R}) = V(\overline{r}) \tag{3.4.10}$$

where $\overline{R}$ is as given in (3.4.9). Note that

$$\sum_{\overline{R}} = \sum_{n_1=-\infty}^{\infty}\sum_{n_2=-\infty}^{\infty}\sum_{n_3=-\infty}^{\infty} \tag{3.4.11}$$

The reciprocal space is defined by

$$\overline{K} = l_1\overline{b}_1 + l_2\overline{b}_2 + l_3\overline{b}_3 \tag{3.4.12}$$

and

$$\Omega = \overline{a}_1 \cdot \overline{a}_2 \times \overline{a}_3 \tag{3.4.13}$$

is the cell volume. The vectors $\overline{b}_1$, $\overline{b}_2$, and $\overline{b}_3$ are the basis in the reciprocal vector space

$$\overline{b}_1 = \frac{2\pi}{\Omega}\overline{a}_2 \times \overline{a}_3 \tag{3.4.14}$$

$$\overline{b}_2 = \frac{2\pi}{\Omega}\overline{a}_3 \times \overline{a}_1 \tag{3.4.15}$$

$$\overline{b}_3 = \frac{2\pi}{\Omega}\overline{a}_1 \times \overline{a}_2 \tag{3.4.16}$$

Thus

$$\overline{b}_i \cdot \overline{a}_j = 2\pi\delta_{ij} \tag{3.4.17}$$

The sum $\sum_{\overline{K}}$ is used to denote summation over the reciprocal lattice space

$$\sum_{\overline{K}} = \sum_{l_1=-\infty}^{\infty} \sum_{l_2=-\infty}^{\infty} \sum_{l_3=-\infty}^{\infty} \tag{3.4.18}$$

Note that $\exp(i\overline{K} \cdot \overline{R}) = 1$.

The third relation is the Poisson's summation which replaces summation over lattice vectors by summation over reciprocal lattice vectors. Consider any function $f(\overline{r})$ and define $q(\overline{r})$ by

$$q(\overline{r}) = \sum_{\overline{R}} e^{i\overline{k}\cdot\overline{R}} f(\overline{r} - \overline{R}) \tag{3.4.19}$$

Consider

$$q(\overline{r} + \overline{R}') = \sum_{\overline{R}} e^{i\overline{k}\cdot\overline{R}} f(\overline{r} + \overline{R}' - \overline{R}) \tag{3.4.20}$$

Then

$$q(\overline{r} + \overline{R}') = \sum_{\overline{R}''} e^{i\overline{k}\cdot(\overline{R}'+\overline{R}'')} f(\overline{r} - \overline{R}'') = e^{i\overline{k}\cdot\overline{R}'} q(\overline{r}) \tag{3.4.21}$$

Thus $q(\overline{r})$ obeys Bloch's condition. We then have

$$q(\overline{r}) = e^{i\overline{k}\cdot\overline{r}} w(\overline{r}) = \sum_{\overline{R}} e^{i\overline{k}\cdot\overline{R}} f(\overline{r} - \overline{R}) \tag{3.4.22}$$

where $w(\overline{r})$ is periodic in $\overline{R}$. The periodic function $w(\overline{r})$ can be represented by a Fourier series in spectral domain

$$w(\overline{r}) = \sum_{\overline{K}} w_{\overline{K}} e^{i\overline{K}\cdot\overline{r}} = e^{-i\overline{k}\cdot\overline{r}} \sum_{\overline{R}} e^{i\overline{k}\cdot\overline{R}} f(\overline{r} - \overline{R}) \tag{3.4.23}$$

The Fourier coefficient is given by

$$w_{\overline{K}} = \frac{1}{\Omega} \int_{\Omega_0} d\overline{r} w(\overline{r}) e^{-i\overline{K}\cdot\overline{r}} \tag{3.4.24}$$

where Ω is the area of the unit cell and Ω_0 is the domain of cell 0. Hence

$$\begin{aligned} w_{\overline{K}} &= \frac{1}{\Omega} \int_{\Omega_0} d\overline{r} e^{-i\overline{K}\cdot\overline{r}} e^{-i\overline{k}\cdot\overline{r}} \sum_{\overline{R}} e^{i\overline{k}\cdot\overline{R}} f(\overline{r} - \overline{R}) \\ &= \frac{1}{\Omega} \sum_{\overline{R}} \int_{\Omega_0} d\overline{r} e^{-i(\overline{K}+\overline{k})\cdot(\overline{r}-\overline{R})} f(\overline{r} - \overline{R}) \end{aligned} \tag{3.4.25}$$

Let $F(\overline{k})$ be the Fourier transform of $f(\overline{r})$

$$F(\overline{k}) = \int d\overline{r} e^{-i\overline{k}\cdot\overline{r}} f(\overline{r}) \tag{3.4.26}$$

The integration is over all space. We can divide the infinite space into periodic array of cells

$$F(\overline{k}) = \sum_{\overline{R}} \int_{\Omega_{-\overline{R}}} d\overline{r} e^{-i\overline{k}\cdot\overline{r}} f(\overline{r}) \tag{3.4.27}$$

where $\Omega_{-\overline{R}}$ is the cell with center at $-\overline{R}$. Let $\overline{r} = \overline{r}' - \overline{R}$. Then

$$F(\overline{k}) = \sum_{\overline{R}} \int_{\Omega_0} d\overline{r}' e^{-i\overline{k}\cdot(\overline{r}'-\overline{R})} f(\overline{r}' - \overline{R}) \tag{3.4.28}$$

Using (3.4.25) and (3.4.28),

$$w_{\overline{K}} = \frac{1}{\Omega} F(\overline{k} + \overline{K}) \tag{3.4.29}$$

Putting (3.4.29) in (3.4.22), we establish Poisson's summation formula

$$q(\overline{r}) = \sum_{\overline{R}} e^{i\overline{k}\cdot\overline{R}} f\left(\overline{r} - \overline{R}\right) = \sum_{\overline{K}} \frac{1}{\Omega} F(\overline{k} + \overline{K}) e^{i(\overline{k}+\overline{K})\cdot\overline{r}} \tag{3.4.30}$$

In the above derivation, $\overline{r}$, $\overline{R}$, $\overline{K}$, and $\overline{k}$ are 3-D vectors. For the case when $\overline{r} = \overline{\rho} + z\hat{z}$ are 3-D while $\overline{\rho}$, $\overline{R}$, $\overline{K}$, and $\overline{k}$ are 2-D vectors, the corresponding results are as follows.

Let

$$\overline{R} = n_1 \overline{a}_1 + n_2 \overline{a}_2 \tag{3.4.31}$$

where n_1 and n_2 are integers and $\overline{a}_1$ and $\overline{a}_2$ are two-dimensional lattice vectors in the x-y plane. Let $\overline{K}$ is the reciprocal lattice vector

$$\overline{K} = l_1 \overline{b}_1 + l_2 \overline{b}_2 \tag{3.4.32}$$

$$\overline{b}_i \cdot \overline{a}_j = 2\pi \delta_{ij} \tag{3.4.33}$$

$$\sum_{\overline{K}} = \sum_{l_1=-\infty}^{\infty} \sum_{l_2=-\infty}^{\infty} \tag{3.4.34}$$

Also $\overline{b}_1 \times \overline{b}_2$ and $\overline{a}_1 \times \overline{a}_2$ are both in positive $\hat{z}$-direction, and $\exp(i\overline{K}\cdot\overline{R}) = 1$.

Let $\overline{k}$ be a 2-D vector and

$$q(\overline{r}) = \sum_{\overline{R}} e^{i\overline{k}\cdot\overline{R}} f(\overline{r} - \overline{R}) = \sum_{\overline{R}} e^{i\overline{k}\cdot\overline{R}} f(\overline{\rho} - \overline{R} + z\hat{z}) \tag{3.4.35}$$

The Poisson's summation formula is

$$q(\overline{r}) = \sum_{\overline{R}} e^{i\overline{k}\cdot\overline{R}} f\left(\overline{r} - \overline{R}\right) = \sum_{\overline{K}} \frac{1}{\Omega} F(\overline{k} + \overline{K}) e^{i(\overline{k}+\overline{K})\cdot\overline{\rho}} \tag{3.4.36}$$

where $F(\overline{k})$ is the 2-D Fourier transform of $f(\overline{r})$:

$$F(\overline{k}) = \int d\overline{\rho}\, e^{-i\overline{k}\cdot\overline{\rho}} f(\overline{\rho} + z\hat{z}) \tag{3.4.37}$$

$$f(\overline{\rho} + z\hat{z}) = \frac{1}{4\pi^2} \int d\overline{k}\, e^{i\overline{k}\cdot\overline{\rho}} F(\overline{k}) \tag{3.4.38}$$

4.2 3-D Green's Function in 3-D Lattices

We first consider the case where $\overline{r}$ and $\overline{R}$ are three-dimensional. A periodic Green's function involves summation of radiation from sources at all the $\overline{R}$ in the periodic lattice. Let

$$G_{\overline{k}}(\overline{r}, 0) = \sum_{\overline{R}} \exp(i\overline{k}\cdot\overline{R}) \frac{e^{ik_o|\overline{r}-\overline{R}|}}{4\pi|\overline{r}-\overline{R}|} \tag{3.4.39}$$

where $k_o = \omega\sqrt{\mu\epsilon}$ is the wavenumber, $\overline{k}$ is a wavevector, and $\sum_{\overline{R}}$ is a threefold summation over the lattice. In solid state theory, $\overline{k}$ is a vector in the first Brillouin zone. More generally, we need to evaluate

$$G_{\overline{k}}(\overline{r}, \overline{r}') = \sum_{\overline{R}} \exp(i\overline{k}\cdot\overline{R}) \frac{e^{ik_o|\overline{r}-\overline{r}'-\overline{R}|}}{4\pi|\overline{r}-\overline{r}'-\overline{R}|} \tag{3.4.40}$$

where $\overline{r}$ and $\overline{r}'$ are both in the unit cell centered at the origin. The unit cell centered at the origin has $\overline{R} = 0$. Direct summation of (3.4.40) converges slowly because the decay for large indices is only 1/(distance).

Without loss of generality, let $r > r'$. Note that

$$\nabla^2 G_{\overline{k}}(\overline{r}, \overline{r}') + k_o^2 G_{\overline{k}}(\overline{r}, \overline{r}') = -\delta(\overline{r} - \overline{r}') \tag{3.4.41}$$

We use the addition theorem for $r > r'$,

$$\frac{e^{ik_o|\overline{r}-\overline{r}'|}}{4\pi|\overline{r}-\overline{r}'|} = \sum_{l=0}^{\infty} \sum_{m=-l}^{l} ik_o h_l(k_o r) Y_{lm}(\hat{r}) j_l(k_o r') Y_{lm}^*(\hat{r}') \tag{3.4.42}$$

Then we define $\mathcal{D}_{lm}(\overline{r})$ such that

$$G_{\overline{k}}(\overline{r}, \overline{r}') = \sum_{l=0}^{\infty} \sum_{m=-l}^{l} (-i)^l \mathcal{D}_{lm}(\overline{r}) j_l(k_o r') Y_{lm}^*(\hat{r}') \tag{3.4.43}$$

In (3.4.43),

$$\hat{r}' = (\theta', \phi') \tag{3.4.44}$$

is the direction vector with angular variables (θ', ϕ'). and $Y_{lm}(\theta', \phi')$ are the spherical harmonics. In this section, unlike in other chapters, we shall define them as

$$Y_{lm}(\theta, \phi) = \left[\frac{(2l+1)}{4\pi} \frac{(l-m)!}{(l+m)!} \right]^{\frac{1}{2}} P_l^m(\cos\theta) e^{im\phi} \tag{3.4.45}$$

where $P_l^m(\cos\theta)$ is the associated Legendre polynomial as defined in (1.4.37) of Volume I. The definition in (3.4.45) differs from (1.4.45) of Volume I by a scale factor. The orthonormality relation for the spherical harmonic is

$$\int_0^{2\pi} d\phi \int_0^{\pi} d\theta \, \sin\theta Y_{lm}^*(\theta, \phi) Y_{l'm'}(\theta, \phi) = \delta_{ll'} \delta_{mm'} \tag{3.4.46}$$

From (3.4.43) and (3.4.46)

$$\mathcal{D}_{lm}(\overline{r}) = ik_o \sum_{\overline{R}} e^{i\overline{k}\cdot\overline{R}} h_l(k_o|\overline{r} - \overline{R}|) i^l Y_{lm}(\widehat{\overline{r} - \overline{R}}) \tag{3.4.47}$$

where $\widehat{\overline{r} - \overline{R}}$ refers to the unit vector point from $\overline{R}$ to $\overline{r}$ and is in the direction of the spherical coordinate angles, Θ and Φ. Note that $G_{\overline{k}}(\overline{r}, 0)$ is a special case of (3.4.47) with

$$G_{\overline{k}}(\overline{r}, 0) = \sqrt{\frac{1}{4\pi}} \mathcal{D}_{00}(\overline{r}) \tag{3.4.48}$$

From (3.4.47), we can interpret $\mathcal{D}_{lm}(\overline{r})$ as multipole radiation from the lattice points.

Next, use the integral identity

$$h_0(k_o r) = \frac{1}{ik_o} \frac{2}{\sqrt{\pi}} \int_{\substack{0 \\ C}}^{\infty} ds \, \exp\left[-r^2 s^2 + \frac{k_o^2}{4s^2} \right] \tag{3.4.49}$$

where C is a contour (Fig. 3.4.1) that ensures the convergence of the integral. For C, as indicated in (3.4.49), the arg s of C obeys the condition,

$$-\frac{\pi}{4} \leq \arg s \leq \beta - \frac{\pi}{4} \tag{3.4.50}$$

where $\beta = \arg k_o$.

By using (3.4.49), $h_1^{(1)}(k_o r) = -h_0^{(1)'}(k_o r)$, the recurrence relation

$$h_{l+1}(k_o r) = \frac{(2l+1)}{k_o r} h_l(k_o r) - h_{l-1}(k_o r) \tag{3.4.51}$$

and mathematical induction, it can be shown that

$$h_l(k_o r) = \frac{2^l r^l}{ik_o^{l+1}} \frac{2}{\sqrt{\pi}} \int_{\substack{0 \\ C}}^{\infty} ds \, s^{2l} \exp\left[-r^2 s^2 + \frac{k_o^2}{4s^2} \right] \tag{3.4.52}$$

Next, we split the integral of s into two parts from 0 to E and from E to ∞, where E is known as the splitting parameter. Thus

$$\mathcal{D}_{lm}(\overline{r}) = \mathcal{D}_{lm}^{(1)}(\overline{r}) + \mathcal{D}_{lm}^{(2)}(\overline{r}) \tag{3.4.53}$$

where

$$\mathcal{D}_{lm}^{(1)}(\overline{r}) = \frac{2^l}{k_o^l} \sum_{\overline{R}} e^{i\overline{k}\cdot\overline{R}} |\overline{r} - \overline{R}|^l i^l Y_{lm}(\widehat{\overline{r} - \overline{R}}) \frac{2}{\sqrt{\pi}} \int_0^E ds\ s^{2l} \exp\left[-|\overline{r} - \overline{R}|^2 s^2 + \frac{k_o^2}{4s^2}\right] \tag{3.4.54a}$$

$$\mathcal{D}_{lm}^{(2)}(\overline{r}) = \frac{2^l}{k_o^l} \sum_{\overline{R}} e^{i\overline{k}\cdot\overline{R}} |\overline{r} - \overline{R}|^l i^l Y_{lm}(\widehat{\overline{r} - \overline{R}}) \frac{2}{\sqrt{\pi}} \int_E^\infty ds\ s^{2l} \exp\left[-|\overline{r} - \overline{R}|^2 s^2 + \frac{k_o^2}{4s^2}\right] \tag{3.4.54b}$$

The quantity $\mathcal{D}_{lm}^{(2)}(\overline{r})$ of (3.4.54b) can be calculated by direct numerical integration since an exponential decay is endowed in the integral, particularly for large $\overline{R}$. For $l = 0$, it can be evaluated exactly as will be shown in Section 4.3.

For the calculation of $\mathcal{D}_{lm}^{(1)}(\overline{r})$, we let

$$f(\overline{r} - \overline{R}) = i^l Y_{lm}(\widehat{\overline{r} - \overline{R}}) |\overline{r} - \overline{R}|^l \exp(-|\overline{r} - \overline{R}|^2 s^2) \tag{3.4.55a}$$

$$\sum_{\overline{R}} e^{i\overline{k}\cdot\overline{R}} f(\overline{r} - \overline{R}) = u(\overline{r}) \tag{3.4.55b}$$

Then

$$\mathcal{D}_{lm}^{(1)}(\overline{r}) = \frac{2^l}{k_o^l} \frac{2}{\sqrt{\pi}} \int_0^E ds\ s^{2l} \exp\left(\frac{k_o^2}{4s^2}\right) u(\overline{r}) \tag{3.4.56}$$

Using Poisson's summation formula (3.4.30) from Section 4.1 and (3.4.55b),

$$u(\overline{r}) = \frac{1}{\Omega} \sum_{\overline{K}} e^{i(\overline{k}+\overline{K})\cdot\overline{r}} F(\overline{k} + \overline{K}) \tag{3.4.57}$$

where $F(\overline{k})$ is the Fourier transform of $f(\overline{r})$. From (3.4.55a) and the property of Fourier transform

$$F(\overline{k}) = \int_{-\infty}^{\infty} d\overline{r}\ i^l Y_{lm}(\hat{r}) r^l \exp(-r^2 s^2) e^{-i\overline{k}\cdot\overline{r}} \tag{3.4.58}$$

We use the spherical wave expansion of

$$e^{-i\overline{k}\cdot\overline{r}} = 4\pi \sum_{lm} (-i)^l j_l(kr) Y_{lm}^*(\hat{r}) Y_{lm}(\hat{k}) \tag{3.4.59}$$

Then using the orthogonality relation of (3.4.46), we obtain

$$F(\bar{k}) = \int_0^\infty dr\; r^2 \int_0^\pi d\theta\; \sin\theta \int_0^{2\pi} d\phi i^l Y_{lm}(\theta,\phi) r^l \exp(-r^2 s^2) \cdot 4\pi \sum_{l'm'} (-i)^{l'} j_{l'}(kr) Y^*_{l'm'}(\hat{r}) Y_{l'm'}(\hat{k}) = 4\pi\sqrt{\frac{\pi}{2}} Y_{lm}(\hat{k}) \frac{k^l}{(2s^2)^{l+\frac{3}{2}}} \exp\left(-\frac{k^2}{4s^2}\right) \tag{3.4.60}$$

Putting (3.4.60) and (3.4.57) in (3.4.56), we have

$$\mathcal{D}^{(1)}_{lm}(\bar{r}) = \frac{1}{k_o^l}\frac{2}{\sqrt{\pi}} \int_0^E ds\; \exp\left(\frac{k_o^2}{4s^2}\right) \frac{1}{\Omega} \sum_{\overline{K}} e^{i(\bar{k}+\overline{K})\cdot\bar{r}} \cdot 4\pi\sqrt{\frac{\pi}{2}} Y_{lm}(\widehat{\bar{k}+\overline{K}}) \frac{\left|\bar{k}+\overline{K}\right|^l}{2^{\frac{3}{2}} s^3} \exp\left[-\frac{\left|\bar{k}+\overline{K}\right|^2}{4s^2}\right] \tag{3.4.61}$$

The ds integration in (3.4.61) can be performed. We get

$$\mathcal{D}^{(1)}_{lm}(\bar{r}) = \frac{4\pi}{k_o^l \Omega} \sum_{\overline{K}} Y_{lm}(\widehat{\bar{k}+\overline{K}}) e^{i(\bar{k}+\overline{K})\cdot\bar{r}} \frac{\left|\bar{k}+\overline{K}\right|^l}{\left|\bar{k}+\overline{K}\right|^2 - k_o^2} \exp\left[\frac{k_o^2 - \left|\bar{k}+\overline{K}\right|^2}{4E^2}\right] \tag{3.4.62}$$

This converges rapidly in $\overline{K}$ because of the exponential decay in $\overline{K}$. To summarize, using (3.4.62), (3.4.53) and (3.4.54b), we have

$$\begin{aligned} \mathcal{D}_{lm}(\bar{r}) &= \mathcal{D}^{(1)}_{lm}(\bar{r}) + \mathcal{D}^{(2)}_{lm}(\bar{r}) \\ &= \frac{1}{k_o^l}\left\{ \frac{4\pi}{\Omega} \sum_{\overline{K}} Y_{lm}(\widehat{\bar{k}+\overline{K}}) e^{i(\bar{k}+\overline{K})\cdot\bar{r}} \frac{\left|\bar{k}+\overline{K}\right|^l \exp\left(\frac{k_o^2-\left|\bar{k}+\overline{K}\right|^2}{4E^2}\right)}{\left|\bar{k}+\overline{K}\right|^2 - k_o^2} \right. \\ &\quad + 2^l \sum_{\overline{R}} e^{i\bar{k}\cdot\overline{R}} |\bar{r}-\overline{R}|^l i^l Y_{lm}(\widehat{\bar{r}-\overline{R}}) \frac{2}{\sqrt{\pi}} \\ &\quad \left. \cdot \int_E^\infty ds\; s^{2l} \exp\left[-|\bar{r}-\overline{R}|^2 s^2 + \frac{k_o^2}{4s^2}\right]\right\} \end{aligned} \tag{3.4.63}$$

In numerical implementation, the parameter E in (3.4.63) has to be chosen. The optimum choice of the splitting parameter E is when $\mathcal{D}^{(1)}_{lm}(\bar{r})$ and $\mathcal{D}^{(2)}_{lm}(\bar{r})$ do not differ by more than several orders of magnitude. For the special case of $l = m = 0$,

$$G_{\overline{k}}(\overline{r},0) = \sqrt{\frac{1}{4\pi}}\mathcal{D}_{00}(\overline{r}) = \frac{1}{4\pi}\left\{\frac{4\pi}{\Omega}\sum_{\overline{K}} e^{i(\overline{k}+\overline{K})\cdot\overline{r}}\,\frac{\exp\left(\frac{k_o^2-\left|\overline{k}+\overline{K}\right|^2}{4E^2}\right)}{\left|\overline{k}+\overline{K}\right|^2-k_o^2}\right.$$
$$\left.+\sum_{\overline{R}} e^{i\overline{k}\cdot\overline{R}}\frac{2}{\sqrt{\pi}}\int_E^{\infty} ds\,\exp\left[-|\overline{r}-\overline{R}|^2 s^2+\frac{k_o^2}{4s^2}\right]\right\} \tag{3.4.64}$$

4.3 3-D Green's Function in 2-D Lattices

We next consider 3-D Green's function in 2-D lattices. In this case the space is three-dimensional so that the field point $\overline{r}$ and source point $\overline{r}'$ are in three-dimensional space. However, the periodic lattice is two-dimensional. This problem has applications in scattering by frequency selective surfaces, periodic surfaces, and random rough surfaces with periodic boundary conditions. Let $\overline{k}$ be a 2-D vector and the Green's function be

$$G(\overline{r}) = \sum_{\overline{R}} \exp(i\overline{k}\cdot\overline{R})\frac{e^{ik_o|\overline{r}-\overline{R}|}}{4\pi|\overline{r}-\overline{R}|} = \sum_{\overline{R}} e^{i\overline{k}\cdot\overline{R}}\frac{ik_o h_0(k_o|\overline{r}-\overline{R}|)}{4\pi} \tag{3.4.65}$$

where $\sum_{\overline{R}} = \sum_{n_1=-\infty}^{\infty}\sum_{n_2=-\infty}^{\infty}$. To put (3.4.65) in spectral domain, we let $\overline{k} = k_x\hat{x} + k_y\hat{y}$ and

$$f(\overline{r}) = \frac{e^{ik_o r}}{4\pi r} = \frac{i}{4\pi^2}\int d\overline{k}\,\frac{e^{i\overline{k}\cdot\overline{\rho}+ik_z|z|}}{2k_z} \tag{3.4.66}$$

Then

$$F(\overline{k}) = \frac{ie^{i\sqrt{k_o^2-k^2}|z|}}{2\sqrt{k_o^2-k^2}} \tag{3.4.67}$$

Using (3.4.67) in the Poisson summation formulation, the reciprocal lattice domain solution is

$$G(\overline{r}) = \frac{i}{2\Omega}\sum_{\overline{K}}\frac{e^{ik_z|z|}}{k_z}e^{i(\overline{k}+\overline{K})\cdot\overline{\rho}} \tag{3.4.68}$$

where

$$k_z = \sqrt{k_o^2 - |\overline{k}+\overline{K}|^2}, \quad \mathrm{Im} k_z \ge 0 \tag{3.4.69}$$

We make use of the integral identity of (3.4.49) for $h_0(k|\overline{r}-\overline{R}|)$ and a splitting parameter of E. Thus

$$G(\overline{r}) = G_1(\overline{r}) + G_2(\overline{r}) \tag{3.4.70}$$

where

$$G_1(\overline{r}) = \frac{1}{4\pi}\sum_{\overline{R}} \exp(i\overline{k}\cdot\overline{R})\frac{2}{\sqrt{\pi}}\int_{\substack{0\\C}}^{E} ds \exp\left[-|\overline{r}-\overline{R}|^2 s^2 + \frac{k_o^2}{4s^2}\right] \qquad (3.4.71a)$$

$$G_2(\overline{r}) = \frac{1}{4\pi}\sum_{\overline{R}} \exp(i\overline{k}\cdot\overline{R})\frac{2}{\sqrt{\pi}}\int_{E}^{\infty} ds \exp\left[-|\overline{r}-\overline{R}|^2 s^2 + \frac{k_o^2}{4s^2}\right] \qquad (3.4.71b)$$

The integral of $G_2(\overline{r})$ in (3.4.71b) can be calculated as follows

$$\begin{aligned}
&\frac{2}{\sqrt{\pi}}\int_E^{\infty} ds \exp\left[-|\overline{r}-\overline{R}|^2 s^2 + \frac{k_o^2}{4s^2}\right] \\
&= \frac{1}{\sqrt{\pi}}\frac{1}{|\overline{r}-\overline{R}|}\left\{\int_E^{\infty} ds\left[|\overline{r}-\overline{R}| - \frac{ik_o}{2s^2}\right]\exp\left[-|\overline{r}-\overline{R}|^2 s^2 + \frac{k_o^2}{4s^2}\right]\right. \\
&\quad \left. + \int_E^{\infty} ds\left[|\overline{r}-\overline{R}| + \frac{ik_o}{2s^2}\right]\exp\left[-|\overline{r}-\overline{R}|^2 s^2 + \frac{k_o^2}{4s^2}\right]\right\} \\
&= \frac{1}{\sqrt{\pi}}\frac{1}{|\overline{r}-\overline{R}|}\left\{\exp(ik_o|\overline{r}-\overline{R}|)\right. \\
&\quad \cdot\int_E^{\infty} ds\left(|\overline{r}-\overline{R}| - \frac{ik_o}{2s^2}\right)\exp\left[-\left(|\overline{r}-\overline{R}|s + \frac{ik_o}{2s}\right)^2\right] \\
&\quad \left. + \exp(-ik_o|\overline{r}-\overline{R}|)\int_E^{\infty} ds\left(|\overline{r}-\overline{R}| + \frac{ik_o}{2s^2}\right)\exp\left[-\left(|\overline{r}-\overline{R}|s - \frac{ik_o}{2s}\right)^2\right]\right\}
\end{aligned} \qquad (3.4.72)$$

The complementary error function is

$$\operatorname{erfc}(z) = \frac{2}{\sqrt{\pi}}\int_z^{\infty} dw\, e^{-w^2} \qquad (3.4.73)$$

Then

$$\begin{aligned}
&\frac{2}{\sqrt{\pi}}\int_E^{\infty} ds \exp\left(-|\overline{r}-\overline{R}|^2 s^2 + \frac{k_o^2}{4s^2}\right) \\
&= \frac{1}{2|\overline{r}-\overline{R}|}\left\{\exp(ik_o|\overline{r}-\overline{R}|)\operatorname{erfc}\left(|\overline{r}-\overline{R}|E + \frac{ik_o}{2E}\right)\right. \\
&\quad \left. + \exp(-ik_o|\overline{r}-\overline{R}|)\operatorname{erfc}\left(|\overline{r}-\overline{R}|E - \frac{ik_o}{2E}\right)\right\}
\end{aligned} \qquad (3.4.74)$$

Putting in (3.4.71b) gives the expression of G_2 that is written in (3.4.83b).

For $G_1(\overline{r})$, we note that since $\overline{R}$ is in the x-y plane and $\overline{r} = (x, y, z)$,

$$\exp(-|\overline{r}-\overline{R}|^2 s^2) = \exp(-z^2 s^2)\exp(-|\overline{\rho}-\overline{R}|^2 s^2) \qquad (3.4.75)$$

where $\overline{\rho} = x\hat{x} + y\hat{y}$ is the two-dimensional position vector. Let

$$f(\overline{\rho}-\overline{R}) = \exp(-|\overline{\rho}-\overline{R}|^2 s^2) \tag{3.4.76a}$$

$$q(\overline{\rho}) = \sum_{\overline{R}} \exp(i\overline{k}\cdot\overline{R}) f(\overline{\rho}-\overline{R}) \tag{3.4.76b}$$

From (3.4.36) of Section 4.1,

$$q(\overline{\rho}) = \sum_{\overline{R}} \exp(i\overline{k}\cdot\overline{R}) \exp(-|\overline{\rho}-\overline{R}|^2 s^2) = \sum_{\overline{K}} \frac{1}{\Omega} F(\overline{k}+\overline{K}) e^{i(\overline{k}+\overline{K})\cdot\overline{\rho}} \tag{3.4.77}$$

where $F(\overline{k})$ is the 2-D Fourier transform of $f(\overline{\rho})$

$$F(\overline{k}) = \iint_{-\infty}^{\infty} d\overline{\rho}\, e^{-i\overline{k}\cdot\overline{\rho}} \exp(-\rho^2 s^2) = \frac{\pi}{s^2} \exp\left(-\frac{k^2}{4s^2}\right) \tag{3.4.78}$$

Thus we have

$$q(\overline{\rho}) = e^{i\overline{k}\cdot\overline{\rho}} \sum_{\overline{K}} e^{i\overline{K}\cdot\overline{\rho}} \frac{\pi}{\Omega s^2} \exp\left[-\frac{|\overline{k}+\overline{K}|^2}{4s^2}\right] \tag{3.4.79}$$

Putting (3.4.79) into (3.4.71a) gives

$$\begin{aligned} G_1(\overline{r}) &= \frac{1}{4\pi}\frac{2}{\sqrt{\pi}} \int_0^E ds \exp\left(\frac{k_o^2}{4s^2}\right) u(\overline{\rho}) \exp\left(-z^2 s^2\right) \\ &= \frac{1}{4\pi}\frac{2}{\sqrt{\pi}} \sum_{\overline{K}} e^{i(\overline{k}+\overline{K})\cdot\overline{\rho}} \frac{\pi}{\Omega} \int_{\substack{0\\C}}^{E} \frac{ds}{s^2} \exp\left[-\frac{|\overline{k}+\overline{K}|^2}{4s^2} - z^2 s^2 + \frac{k_o^2}{4s^2}\right] \end{aligned} \tag{3.4.80}$$

Letting $s \to 1/s$ in (3.4.80), we have

$$G_1(\overline{r}) = \frac{1}{2\Omega\sqrt{\pi}} \sum_{\overline{K}} e^{i(\overline{k}+\overline{K})\cdot\overline{\rho}} \int_{\substack{1/E\\C}}^{\infty} ds \exp\left[-\left(\frac{|\overline{k}+\overline{K}|^2}{4} - \frac{k_o^2}{4}\right) s^2 - \frac{z^2}{s^2}\right] \tag{3.4.81}$$

where C is a contour that ensures convergence at ∞. Equation (3.4.81) is now in similar now to (3.4.71b) and can be manipulated in a similar manner to be in the form of complementary error function.

To summarize, we have

$$G(\overline{r}) = G_1(\overline{r}) + G_2(\overline{r}) \tag{3.4.82}$$

where

$$\begin{aligned} G_1(\overline{r}) = \frac{i}{4\Omega} \sum_{\overline{K}} \frac{e^{i(\overline{k}+\overline{K})\cdot\overline{\rho}}}{k_z} \Bigg\{ &\exp(ik_z z) \operatorname{erfc}\left(-\frac{ik_z}{2E} - E_z\right) \\ &+ \exp(-ik_z z) \operatorname{erfc}\left(-\frac{ik_z}{2E} + E_z\right) \Bigg\} \end{aligned} \tag{3.4.83a}$$

with $k_z = \sqrt{k_o^2 - |\overline{k} + \overline{K}|^2}$, $\mathrm{Im}k_z \geq 0$, and

$$G_2(\overline{r}) = \frac{1}{8\pi} \sum_{\overline{R}} \exp(i\overline{k} \cdot \overline{R}) \frac{1}{|\overline{r} - \overline{R}|} \left\{ \exp(ik_o|\overline{r} - \overline{R}|) \,\mathrm{erfc}\left(|\overline{r} - \overline{R}|E + \frac{ik_o}{2E} \right) \right.$$

$$\left. + \exp(-ik_o|\overline{r} - \overline{R}|) \,\mathrm{erfc}\left(|\overline{r} - \overline{R}|E - \frac{ik_o}{2E} \right) \right\} \quad (3.4.83b)$$

The splitting parameter E is optimally chosen such that $G_1(\overline{r})$ and $G_2(\overline{r})$ do not differ by more than several orders of magnitude.

4.4 Numerical Results

We illustrate the results for the case of 3-D Greens function in a 2-D square lattice. The case that the Ewald's method gains is when the medium is lossless and $z = 0$. Let the lattice be $\overline{a}_1 = a_x\hat{x}$, $\overline{a}_2 = a_y\hat{y}$. Then $\overline{b}_1 = \frac{2\pi}{a_x}\hat{x}$, $\overline{b}_2 = \frac{2\pi}{a_y}\hat{y}$. The lattice vectors and the reciprocal lattice vectors are respectively

$$\overline{R} = n_1 a_x \hat{x} + n_2 a_y \hat{y} \quad (3.4.84)$$

$$\overline{K} = \frac{2\pi l_1}{a_x}\hat{x} + \frac{2\pi l_2}{a_y}\hat{y} \quad (3.4.85)$$

Also $\Omega = a_x a_y$. We consider the case that $\overline{k} = k_{ix}\hat{x} + k_{iy}\hat{y}$. The spatial domain solution is

$$G(x,y,z) = \sum_{n_1=-N_S}^{N_S} \sum_{n_2=-N_S}^{N_S} e^{i(k_{ix}n_1a_x + k_{iy}n_2a_y)} \frac{e^{ik_0 R_{n_1n_2}}}{4\pi R_{n_1n_2}} \quad (3.4.86)$$

where

$$R_{n_1n_2} = \sqrt{(x - n_1a_x)^2 + (y - n_2a_y)^2 + z^2} \quad (3.4.87)$$

and we also truncate at N_s.

The reciprocal lattice domain, or the spectral domain solution is

$$G(x,y,z) = \frac{i}{2\Omega} \sum_{l_1=-N_r}^{N_r} \sum_{l_2=-N_r}^{N_r} \frac{e^{i\left[\left(k_{ix}+\frac{2\pi l_1}{a_x}\right)x + \left(k_{iy}+\frac{2\pi l_2}{a_y}\right)y\right]} e^{ik_{zl_1l_2}|z|}}{k_{zl_1l_2}} \quad (3.4.88)$$

where

$$k_{zl_1l_2} = \sqrt{k_o^2 - \left[\left(k_{ix} + \frac{2\pi l_1}{a_x}\right)^2 + \left(k_{iy} + \frac{2\pi l_2}{a_y}\right)^2 \right]} \quad (3.4.89)$$

and truncation is at N_r. Also $\mathrm{Im}(k_{zl_1l_2}) \geq 0$.

On the other hand, in Ewald's method,

$$G(x,y,z) = G_1(x,y,z) + G_2(x,y,z) \tag{3.4.90}$$

where

$$G_1(x,y,z) = \frac{i}{4\Omega} \sum_{l_1=-N_1}^{N_1} \sum_{l_2=-N_1}^{N_1} \frac{e^{i\left[\left(k_{ix}+\frac{2\pi l_1}{a_x}\right)x+\left(k_{iy}+\frac{2\pi l_2}{a_y}\right)y\right]}}{k_{zl_1l_2}}$$
$$\left\{e^{ik_{zl_1l_2}z}\operatorname{erfc}\left(-i\frac{k_{zl_1l_2}}{2E} - Ez\right) + e^{-ik_{zl_1l_2}z}\operatorname{erfc}\left(-i\frac{k_{zl_1l_2}}{2E} + Ez\right)\right\} \tag{3.4.91}$$

and

$$G_2(x,y,z) = \sum_{n_1=-N_2}^{N_2} \sum_{n_2=-N_2}^{N_2} \frac{e^{i(k_{ix}n_1a_x+k_{iy}n_2a_y)}e^{ik_oR_{n_1n_2}}}{8\pi R_{n_1n_2}} \operatorname{erfc}\left(R_{n_1n_2}E + \frac{ik_o}{2E}\right)$$
$$+ \sum_{n_1=-N_2}^{N_2} \sum_{n_2=-N_2}^{N_2} \frac{e^{i(k_{ix}n_1a_x+k_{iy}n_2a_y)}e^{-ik_oR_{n_1n_2}}}{8\pi R_{n_1n_2}} \operatorname{erfc}\left(R_{n_1n_2}E - \frac{ik_o}{2E}\right) \tag{3.4.92}$$

where truncations of G_1 and G_2 are done respectively at N_1 and N_2.

Asymptotically,

$$\operatorname{erfc}(w) \to \frac{e^{-w^2}}{\sqrt{\pi}w} \tag{3.4.93}$$

Thus both series of $G_1(x,y)$ and $G_2(x,y)$ have exponential decay:

$$G_1(x,y) \to \exp\left[-\frac{\left(\frac{2\pi l_1}{a_x}\right)^2 + \left(\frac{2\pi l_2}{a_y}\right)^2}{4E^2}\right] \tag{3.4.94}$$

$$G_2(x,y) \to \exp\left[-\left((n_1a_x)^2 + (n_2a_y)^2\right)E^2\right] \tag{3.4.95}$$

For the 1-D case, the two series are respectively $e^{-\left(\frac{\pi l_1}{a_xE}\right)^2}$ and $e^{-[(n_1a_x)^2]E^2}$. The two series have the same exponential decay rate if $E = \sqrt{\pi}/a_x$.

For the 2-D case, we choose splitting parameter E such that

$$E = \sqrt{\frac{\pi}{a_xa_y}} \tag{3.4.96}$$

In the numerical simulations, we use the following parameters: $\lambda = 1$, $a_x = 0.95\lambda$, $a_y = 0.95\lambda$, $k_{ix} = \operatorname{Re}(k_o)\sin\theta_i\cos\phi_i$, $k_{iy} = \operatorname{Re}(k_o)\sin\theta_i\sin\phi_i$, $N_1 = N_2 = 2$, $E = \sqrt{\frac{\pi}{a_xa_y}} = 1.866\lambda^{-1}$.

Case	N_s	Spatial	N_r	Spectral	G_1	G_2	Ewald
(a)	300	$0.202 + i0.317$	250	$0.202 + i0.317$	$0.356 + i0.334$	$-0.153 - i0.0170$	$0.202 + i0.317$
(b)	1000	$-0.164 + i0.069$	10	$-.0.166 + i0.071$	$-0.623 + i0.071$	0.457	$-0.165 + i0.071$
(c)	1000	$0.0566 + i0.473$	10	$0.0163 + i0.457$	$0.162 + i0.468$	$-0.146 - i0.010$	$0.0162 + i0.457$
(d)	1000	$0.0609 + i0.465$	300	$0.0216 + i0.450$	$0.175 + i0.460$	$-0.154 - i0.0106$	$0.0206 + i0.449$

Table 3.4.1 Computation of the periodic Green's function using Ewald's method.

We use very few terms in the Ewald summation. The slight difference in the results could be due to the accuracy in computing the complementary error function of complex arguments.

The results for the four cases considered below are tabulated in Table 3.4.1.

Case (a) Lossy medium: $k_o = \frac{2\pi}{\lambda}(1 + i0.01)$, $\theta_i = 45°$, $\phi_i = 25°$, $z = 0$, $x = 0.48\lambda$, $y = -0.91\lambda$. In Fig. 3.4.2, we plot the convergence tests of the spatial solution (dotted line) and the spectral solution (solid line) for the real part of the Green's function as a function of N_s and N_r respectively. There are good convergence for both spatial and spectral solutions.

Case (b) Normal incidence: $k_o = \frac{2\pi}{\lambda}$, $\theta_i = 0°$, , $\phi_i = 25°$, $z = 0.1\lambda$, $x = 0.48\lambda$, $y = -0.91\lambda$. In Fig. 3.4.3, we plot the convergence tests of the spatial solution (dotted line) and the spectral solution (solid line) for the real part of the Green's function as a function of N_s and N_r respectively. There is good convergence for the spectral solution. Because z is not equal to zero, there is good exponential decay for the evanescent Floquet modes in the spectral solution. On the other hand, the spatial solution needs many more terms.

Case (c) Oblique incidence: $z \neq 0$, $k_o = \frac{2\pi}{\lambda}$, $\theta_i = 45°$, $\phi_i = 25°$, $z = 0.1\lambda$, $x = 0.48\lambda$, $y = -0.91\lambda$. In Fig. 3.4.4, we plot the convergence tests. There is good convergence for the spectral solution. On the other hand, the spatial solution does not converge even for $N_s = 1000$.

Case (d) Oblique incidence: $z = 0$, $k_o = \frac{2\pi}{\lambda}$, $\theta_i = 45°$, $\phi_i = 25°$, $z = 0$, $x = 0.48\lambda$, $y = -0.91\lambda$. In Fig. 3.4.5, we plot the convergence tests. Neither spatial solution nor spectral solution converge well. However, the spectral solution shows a better convergence.

It is important to emphasize that Ewald's method requires very few terms for all the four cases considered. As shown in Table 3.4.1, the results of Ewald's method are also accurate.

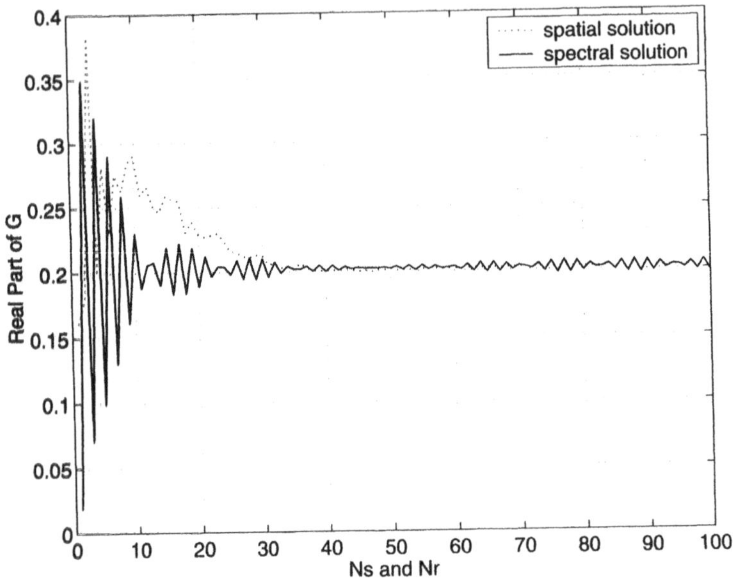

Figure 3.4.2 Convergence of the spatial solution (dotted line) and spectral solution (solid line) for case (a) — Lossy medium: $k_o = \frac{2\pi}{\lambda}(1 + i0.01)$, $\theta_i = 45°$, $\phi_i = 25°$, $z = 0$, $x = 0.48\lambda$, $y = -0.91\lambda$.

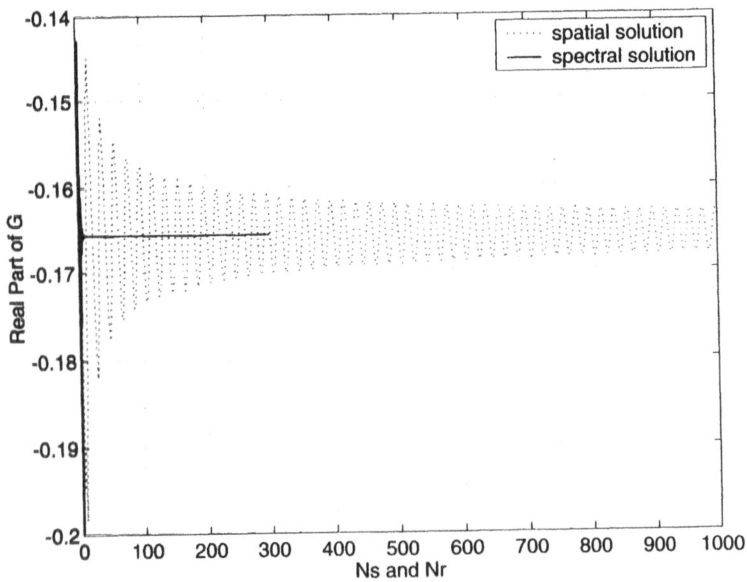

Figure 3.4.3 Convergence of the spatial solution (dotted line) and spectral solution (solid line) for case (b) — Normal incidence: $k_o = \frac{2\pi}{\lambda}$, $\theta_i = 0°$, $\phi_i = 25°$, $z = 0.1\lambda$, $x = 0.48\lambda$, $y = -0.91\lambda$.

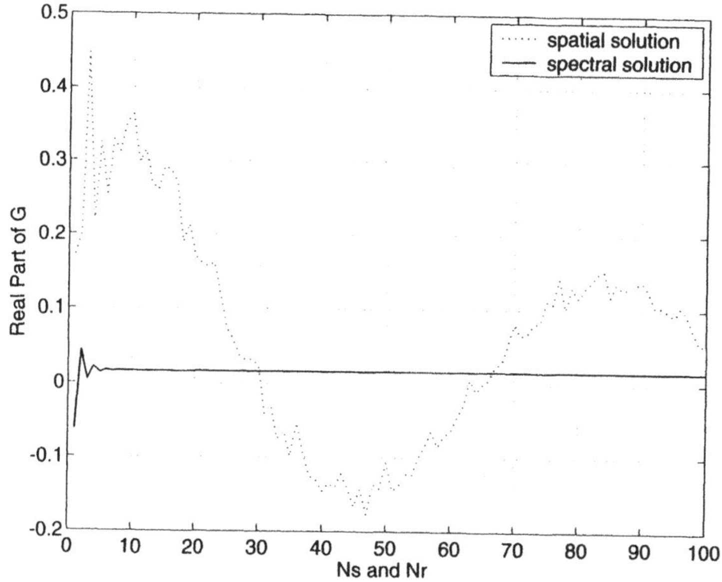

Figure 3.4.4 Convergence of the spatial solution (dotted line) and spectral solution (solid line) for case (c) — Oblique incidence: $z \neq 0$, $k_o = \frac{2\pi}{\lambda}$, $\theta_i = 45°$, $\phi_i = 25°$, $z = 0.1\lambda$, $x = 0.48\lambda$, $y = -0.91\lambda$.

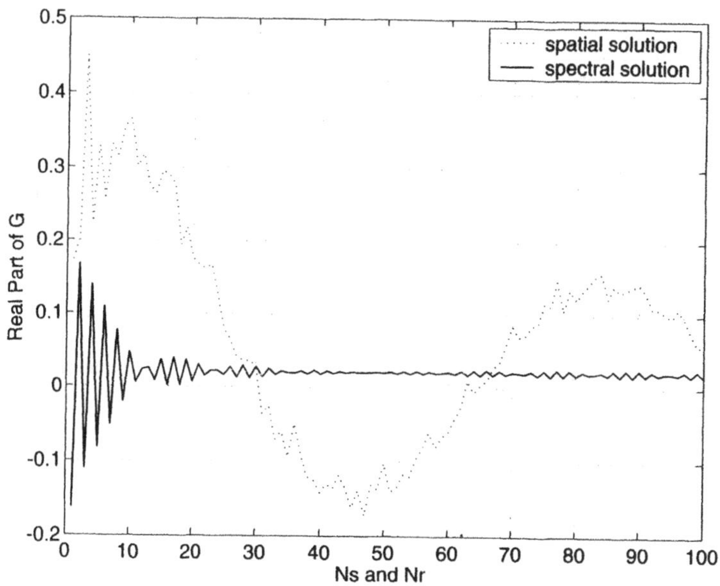

Figure 3.4.5 Convergence of the spatial solution (dotted line) and spectral solution (solid line) for case (d) — Oblique incidence: $z = 0$, $k_o = \frac{2\pi}{\lambda}$, $\theta_i = 45°$, $\phi_i = 25°$, $z = 0$, $x = 0.48\lambda$, $y = -0.91\lambda$.

REFERENCES AND ADDITIONAL READINGS

Abramowitz, M. and J. A. Stegun (1965), *Handbook of Mathematical Functions*, Dover Publications, New York.

Chan, C. H. (1995), Analysis of frequency selective surfaces, in *Frequency Selective Surface and Grid Array*, edited by T. K. Wu, Chapter 2, 27–85, Wiley-Interscience, New York.

Cohen, E. (1995), An Ewald transformation of frequency domain integral formulations, *Electromagnetics*, **15**, 427–439.

Ewald, P. P. (1921), Die berechnug optischer und elekrostatischen gitterpotential, *Ann. Phys.*, **64**, 253–268.

Kipp, R. A. and C. H. Chan (1994), A numerically efficient technique for the method of moments solution to planar periodic structures in a layered media, *IEEE Trans. Microwave Theory Tech.*, **42(4)**, 635–643.

Joannopoulos, J. D., R. D. Meade, and J. M. Winn (1995), *Photonic Crystals: Molding the Flow of Light*, Princeton University Press, Princeton.

Johnson, J. T., J. A. Kong, R. T. Shin, D. H. Staelin, K. O'Neill, and A. W. Lohanick (1993), Third Stokes parameter emission from a periodic water surface, *IEEE Trans. Geosci. Remote Sens.*, **31(5)**, 1066–1080.

Jordan, K. E., G. R. Richter, and P. Sheng (1986), An efficient numerical evaluation of the Green's function for the Helmholtz operator on periodic structures, *J. of Comp. Phys.*, **63**, 222–235.

Kittel, C. (1996), *Introduction to Solid State Physics*, 7th edition, Wiley, New York.

Mathis, A. W. and A. F. Peterson (1996), A comparison of acceleration procedures for the two-dimensional periodic Green's function, *IEEE Trans. Antennas Propagat.*, **44(1)**, 567–571.

Mathis, A. W. and A. F. Peterson (1998), Efficient electromagnetic analysis of a doubly infinite array of rectangular apertures, *IEEE Trans. Microwave Theory Tech.*, **46(1)**, 46–54.

Munk, B. A. (2000), *Frequency Selective Surfaces: Theory and Design*, Wiley-Interscience, New York.

Nghiem, S. V., M. E. Veysoglu, J. A. Kong, R. T. Shin, K. O'Neill, and A. W. Lohanick (1991), Polarimetric passive remote sensing of a periodic soil surface: Microwave measurements and analysis, *J. Electromag. Waves and Appl.*, **5**, 997–1005.

Oberhettinger, F. and L. Badii (1973), *Tables of Laplace Transforms*, Springer-Verlag, Berlin.

Radisic, V., Y. Qian, and T. Itoh (1998), Broadband power amplifier using dielectric photonic bandgap structures, *IEEE Microwave Guided Wave Lett.*, **8**, 13–14.

Veysoglu, M. E., S. H. Yueh, R. T. Shin, and J. A. Kong (1991), Polarimetric passive remote sensing of periodic surfaces, *J. Electromag. Waves and Appl.*, **5**, 267–280.

Wang, J. R., R. W. Newton, and J. W. Rouse (1980), Passive microwave remote sensing of soil moisture: The effect of tilled row structure, *IEEE Trans. Geosci. Remote Sens.*, **18**, 296–302.

Watson, G. N. (1966), *A Treatise on the Theory of Bessel Functions*, 2nd edition, Cambridge University Press, Cambridge.

Yablonovitch, E. (1987), Inhibited spontaneous emission is solid state physics and electronics, *Phys. Rev. Lett.*, **58(20)**, 2059–2062.

Yu, Y. X. and C. H. Chan (1998), On the extension of Ewald's method to periodic structures, *Microwave Opt. Technol. Lett.*, **19(2)**, 125–131.

Chapter 4

RANDOM ROUGH SURFACE SIMULATIONS

In this chapter we study random rough surface simulations of one-dimensional surface for two-dimensional scattering problem. The simulations of rough surface scattering started in the late 1970's and continue to the present day [Axline and Fung, 1978; Thorsos, 1988; Thorsos and Jackson, 1991; Maystre et al. 1991; Devayya and Wingham, 1992; Thorsos and Jackson, 1989; Thorsos and Broschat, 1995; Maradudin et al. 1990; Michel and O'Donnell, 1992; McGurn and Maradudin, 1993; Chan et al. 1991; Nieto-Vesperinas and Soto-Crespo, 1987]. The main purposes of the early simulations were to validate analytic scattering theory and to investigate backscattering enhancement.

The numerical method in this chapter is based on the formulation of integral equations and converting the integral equations into matrix equations using the method of moments. We discuss the Dirichlet problem and Neumann problem and illustrate the results using Gaussian surfaces, surfaces with ocean spectrum, and fractal surfaces. Next, we discuss dielectric surface and the calculation of emissivity for applications in passive remote sensing. In particular, we address the accuracy issue in the calculation of emissivity. The accurate calculation of emissivity distinguishes the emphasis of rough surface simulations in this book. Most researchers on rough surface simulations emphasize on bistatic scattering and backscattering, which are usually measured and plotted in dB scale. For such simulations, an accuracy of 25% or 1 dB is acceptable. However, for passive remote sensing calculations, the physics is based on energy conservation. The key result in passive remote sensing is the difference of emissivity between a rough surface and a flat surface. The difference is small and can be a few percent to less than 1%. For ocean remote sensing, that difference is particularly small, e.g., 0.003 or 0.3%. This corresponds to a brightness temperature difference of less than a Kelvin between a rough surface and a flat surface. The ability to distinguish that small difference actually forms the basis of passive remote sensing of ocean wind. This means that for passive remote sensing numerical simulations, energy conservation has to be within 0.3%. Such a stringent requirement is not needed in active remote sensing simulations, where an energy conservation of 96% is deemed to be good. Thus numerical methods of simulations for active remote sensing can be different from passive remote sensing because of the large difference in accuracy requirements. One key implication for calculations in passive remote sensing is that the rough surface needs to have a fine discretization. In this chapter, we will also introduce examples of real life surface profiles measured for rocky surfaces, soil surfaces, and snow surfaces.

1 Perfect Electric Conductor (Non-Penetrable Surface)

1.1 Integral Equation

Consider an incident wave $\psi_{inc}(\overline{r})$ impinging upon a random surface (Fig. 4.1.1) with height profile $z = f(x)$. In two-dimensional scattering problem $\overline{r} = x\hat{x} + z\hat{z}$, the wavefunction $\psi(\overline{r})$ is

$$\psi(\overline{r}) = \psi_{inc}(\overline{r}) + \psi_s(\overline{r}) \tag{4.1.1}$$

where $\psi_s(\overline{r})$ is the scattered wave. The wavefunction obeys the equation

$$\left(\nabla^2 + k^2\right)\psi = 0 \tag{4.1.2}$$

The two-dimensional Green's function obeys the equation

$$\left(\nabla^2 + k^2\right) g(\overline{r}, \overline{r}') = -\delta(\overline{r} - \overline{r}') \tag{4.1.3}$$

and

$$g(\overline{r}, \overline{r}') = \frac{i}{4} H_0^{(1)}(k|\overline{r} - \overline{r}'|) \tag{4.1.4}$$

Let the spaces above and below the rough surface be denoted by region 0 (V_0) and region 1 (V_1). We use the Green's theorem to get

$$\begin{aligned} &\iint_{V_0} d\overline{r}\left[\psi(\overline{r})\nabla^2 g(\overline{r}, \overline{r}') - g(\overline{r}, \overline{r}')\nabla^2\psi(\overline{r})\right] \\ &\quad = -\int_S ds\, \hat{n} \cdot \left[\psi(\overline{r})\nabla g(\overline{r}, \overline{r}') - g(\overline{r}, \overline{r}')\nabla\psi(\overline{r})\right] \\ &\qquad + \int_{S_\infty} ds\, \hat{n} \cdot \left[\psi(\overline{r})\nabla g(\overline{r}, \overline{r}') - g(\overline{r}, \overline{r}')\nabla\psi(\overline{r})\right] \end{aligned} \tag{4.1.5}$$

where S_∞ is the surface at infinity. Using (4.1.2) and (4.1.3) in the left-hand side of (4.1.5), we have

$$\begin{aligned} &\iint_{V_0} d\overline{r}\left[\psi(\overline{r})\left(-k^2 g(\overline{r}, \overline{r}') - \delta(\overline{r} - \overline{r}')\right) + g(\overline{r}, \overline{r}')k^2\psi(\overline{r})\right] \\ &\quad = -\iint_{V_0} d\overline{r}\, \delta(\overline{r} - \overline{r}')\, \psi(\overline{r}) \end{aligned} \tag{4.1.6}$$

To evaluate (4.1.6), one needs to define where $\overline{r}'$ is. One can have $\overline{r}'$ above the rough surface or $\overline{r}'$ below the rough surface. Note that $\overline{r}'$ can be infinitesimally close to the surface. If $\overline{r}'$ is infinitesimally close but above the rough surface, we denote it by $\overline{r}'_+$. If it is infinitesimally close but below the rough surface, we denote it by $\overline{r}'_-$. Also

$$-\iint_{V_0} d\overline{r}\, \delta(\overline{r} - \overline{r}')\, \psi(\overline{r}) = \begin{cases} -\psi(\overline{r}') & \text{if } \overline{r}' \text{ in region } V_0 \\ 0 & \text{if } \overline{r}' \text{ in region } V_1 \end{cases} \tag{4.1.7}$$

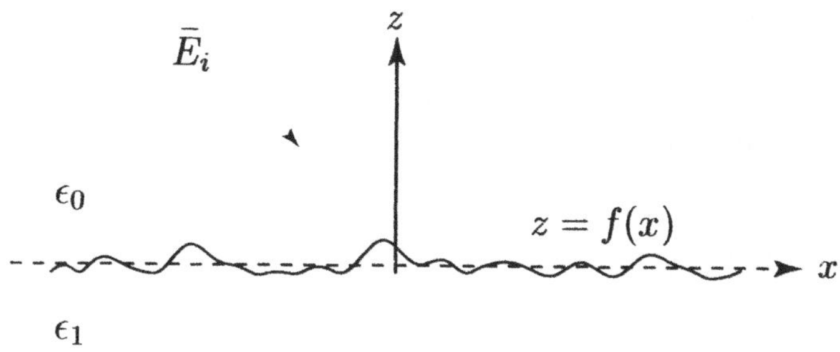

Figure 4.1.1 Wave impinging upon a dielectric surface.

The surface integral at infinity in (4.1.5) gives the incident wave. Thus

$$\psi_{inc}(\bar{r}') + \int_S ds\, \hat{n} \cdot \left[\psi(\bar{r})\nabla g(\bar{r}, \bar{r}') - g(\bar{r}, \bar{r}')\nabla\psi(\bar{r})\right]$$

$$= \begin{cases} \psi(\bar{r}') & \bar{r}' \in V_0 \qquad (4.1.8a) \\ 0 & \bar{r}' \in V_1 \qquad (4.1.8b) \end{cases}$$

The zero in (4.1.8b) corresponds to the extinction theorem. Note that in (4.1.8a) and (4.1.8b), $\bar{r}$ is on surface S while $\bar{r}'$ is in region V_0 or V_1. To obtain integral equations, we consider the following two cases.

A. Dirichlet Boundary Condition

Dirichlet boundary condition represents the TE case of electromagnetic scattering with $\overline{E} = \hat{y}E$ and the surface is perfectly conducting.

The Dirichlet boundary condition is

$$\psi(\bar{r}) = 0 \tag{4.1.9}$$

for $\bar{r}$ on S. Then

$$\psi_{inc}(\bar{r}') - \int_S ds\, g(\bar{r}, \bar{r}')\hat{n} \cdot \nabla\psi(\bar{r}) = \begin{cases} \psi(\bar{r}') & \bar{r}' \in V_0 \qquad (4.1.10a) \\ 0 & \bar{r}' \in V_1 \qquad (4.1.10b) \end{cases}$$

If we let $\bar{r}'$ approach the surface, we note that both (4.1.10a) and (4.1.10b) will approach zero. Thus one does not have to distinguish between $\bar{r}'_+$ and $\bar{r}'_-$. For $\bar{r}'$ on S, (4.1.10) takes the form of a surface integral equation for the surface unknown $\hat{n} \cdot \nabla\psi(\bar{r})$

$$\psi_{inc}(\bar{r}') = \int_S ds\, g(\bar{r}, \bar{r}')\hat{n} \cdot \nabla\psi(\bar{r}) \tag{4.1.11}$$

Note that both $\bar{r}$ and $\bar{r}'$ are on S in (4.1.11).

We also note that as $\bar{r} \to \bar{r}'$, $g(\bar{r}, \bar{r}')$ has an integrable singularity for a one-dimensional integration. Equation (4.1.11) is also known as the EFIE (electric field integral equation) for TE case because ψ represents the electric field which points in the horizontal direction.

B. Neumann Boundary Condition

This represents a TM case when the magnetic field is $\overline{H} = \hat{y}H$ and the surface is perfectly conducting. Let ψ represent H. Then the boundary condition is

$$\hat{n} \cdot \nabla\psi = 0 \tag{4.1.12}$$

From (4.1.8), we have

$$\psi_{inc}(\overline{r}') + \int_S ds\ \psi(\overline{r})\hat{n} \cdot \nabla g(\overline{r}, \overline{r}') = \begin{cases} \psi(\overline{r}') & \overline{r}' \in V_0 \quad (4.1.13a) \\ 0 & \overline{r}' \in V_1 \quad (4.1.13b) \end{cases}$$

Next, we let $\overline{r}'$ approach the surface. In this case, it makes a difference whether one approaches the surface from above which gives $\psi(\overline{r}'_+)$ or from below which gives zero by (4.1.13b). Thus, we have

$$\psi_{inc}(\overline{r}'_+) + \int_S ds\ \psi(\overline{r})\hat{n} \cdot \nabla g(\overline{r}, \overline{r}'_+) = \psi(\overline{r}'_+) \tag{4.1.14a}$$

$$\psi_{inc}(\overline{r}'_-) + \int_S ds\ \psi(\overline{r})\hat{n} \cdot \nabla g(\overline{r}, \overline{r}'_-) = 0 \tag{4.1.14b}$$

In Section 1.5, we shall show that because of the singularity of Green's function, (4.1.14a) and (4.1.14b) are consistent with each other. One can use (4.1.14a) or (4.1.14b) to solve for the surface unknown $\psi(\overline{r})$. The discontinuity of $\psi(\overline{r}'_+)$ and $\psi(\overline{r}'_-)$ can be accounted for by the singularity of the normal derivative of the Green's function. As $\overline{r}' \to \overline{r}$, $\hat{n} \cdot \nabla g(\overline{r}, \overline{r}')$ is more singular than $g(\overline{r}, \overline{r}')$.

1.2 Matrix Equation: Dirichlet Boundary Condition (EFIE for TE Case)

Integral equations can be readily converted to matrix equations. The surface integral equation for Dirichlet boundary condition is, for $\overline{r}'$ on S,

$$\psi_{inc}(\overline{r}') = \int_S ds g(\overline{r}, \overline{r}')\hat{n} \cdot \nabla\psi(\overline{r}) \tag{4.1.15}$$

Note that both $\overline{r}$ and $\overline{r}'$ are on S. On S, $z = f(x)$ and $z' = f(x')$. Hence (4.1.15) becomes

$$\psi_{inc}(x', f(x')) = \int_{-\frac{L}{2}}^{\frac{L}{2}} dx \sqrt{1 + \left(\frac{df}{dx}\right)^2} g(x, f(x); x', f(x')) \left(\hat{n} \cdot \nabla\psi(\overline{r})\right)_{z=f(x)} \tag{4.1.16}$$

where we have limited the surface to between $-L/2$ to $L/2$. The quantity $\psi_{inc}(x', f(x'))$ is only a function of x'. Let

$$b(x') = \psi_{inc}(x', f(x')) \tag{4.1.17}$$

Also let

$$\sqrt{1 + \left(\frac{df}{dx}\right)^2} \, (\hat{n} \cdot \nabla\psi(\overline{r}))_{z=f(x)} = U(x) \tag{4.1.18}$$

be the surface unknown. The kernel of the integral equation is

$$K(x', x) = g(x, f(x); x', f(x')) = g(x', f(x'); x, f(x)) \tag{4.1.19}$$

Putting (4.1.17), (4.1.18), and (4.1.19) into (4.1.16), we have

$$\int_{-\frac{L}{2}}^{\frac{L}{2}} dx \; K(x', x) U(x) = b(x') \tag{4.1.20}$$

Next we convert (4.1.20) into a matrix equation using MoM. The domain $-L/2 \le x \le L/2$ is divided into N intervals, each of width $\Delta = L/N$. The intervals are centered at x_m, $m = 1, 2, \ldots, N$. Thus $U(x) = U_n$ in the nth interval and we point match the integral equation at $x' = x_m$.

First we set $x' = x_m$ in (4.1.20) (point matching):

$$\int_{-\frac{L}{2}}^{\frac{L}{2}} dx \; K(x_m, x) U(x) = b(x_m) \tag{4.1.21}$$

$m = 1, 2, \ldots, N$. Next the integral (4.1.21) can be replaced by a summation, assuming that $U(x)$ is constant in each interval (pulse basis functions):

$$\Delta x \sum_{\substack{n=1 \\ n \neq m}}^{N} K(x_m, x_n) U(x_n) + \left(\int_m dx \; K(x_m, x) \right) U(x_m) = b(x_m) \tag{4.1.22}$$

where $m = 1, 2, \ldots, N$ and $\int_m$ implies integration over the mth interval. We have to single out the mth interval because $K(x_m, x)$ is singular at $x = x_m$. The second term in (4.1.22) is known as the self-patch contribution. For x and x_m close to each other, the argument of the Green's function is small. For small argument ω,

$$H_0^{(1)}(\omega) = 1 + i\frac{2}{\pi} \ln\left(\frac{\gamma\omega}{2}\right) \tag{4.1.23}$$

where $\gamma = 1.78107$. We further approximate

$$-f(x_m) + f(x) \simeq f'(x_m)(x - x_m) \tag{4.1.24}$$

Thus

$$
\begin{aligned}
\int_m dx\ K(x_m, x) &= 2\int_{x_m}^{x_m+\frac{\Delta x}{2}} dx\ K(x_m, x) \\
&\simeq \frac{i}{2}\int_0^{\frac{\Delta x}{2}} dx\left[1 + i\frac{2}{\pi}\ln\left(\frac{\gamma}{2}kx\sqrt{1+(f'(x_m))^2}\right)\right] \\
&= \frac{i\Delta x}{4}\left\{1 + i\frac{2}{\pi}\ln\left[\frac{\gamma k}{4e}\Delta x\sqrt{1+(f'(x_m))^2}\right]\right\} \\
&= \frac{i\Delta x}{4}\left\{1 + i\frac{2}{\pi}\ln\left(\frac{\gamma k}{4e}\Delta l_m\right)\right\}
\end{aligned}
\tag{4.1.25}
$$

where $\Delta l_m = \Delta x\sqrt{1+(f'(x_m))^2}$ is the length of the segment of the surface that is centered at x_m.

Let

$$U(x_n) = U_n \tag{4.1.26}$$

$$b(x_m) = b_m \tag{4.1.27}$$

$$
A_{mn} = \begin{cases} \Delta x\, K(x_m, x_n) & \text{for } m \neq n \\ \dfrac{i\Delta x}{4}\left[1 + i\dfrac{2}{\pi}\ln\left(\dfrac{\gamma k}{4e}\Delta l_m\right)\right] & \text{for } m = n \end{cases}
\tag{4.1.28}
$$

In Section 2.3, we will describe numerical integration to obtain more accurate matrix elements. Putting (4.1.26)–(4.1.28) into (4.1.22), we get the equation

$$\sum_{n=1}^{N} A_{mn} U_n = b_m \tag{4.1.29}$$

In matrix notation, we have

$$\overline{\overline{A}}\,\overline{U} = \overline{b} \tag{4.1.30}$$

1.3 Tapering of Incident Waves and Calculation of Scattered Waves

In numerical simulations, the rough surface is truncated at $x = \pm L/2$. This means that the surface current is forced to be zero for $|x| > L/2$. If there is an abrupt change of surface current from nonzero to zero, artificial reflection from the two endpoints will occur. To avoid these problems, one way is to taper the incident wave so that the incident wave decays to zero in a Gaussian manner for large x.

A tapered incident wave is [Thorsos, 1988]

$$\psi_{inc}(\overline{r}) = \exp\left(ik(x\sin\theta_i - z\cos\theta_i)(1+w(\overline{r}))\right)\exp\left(-\frac{(x+z\tan\theta_i)^2}{g^2}\right) \tag{4.1.31}$$

where g is the tapering parameter, and the incident wave vector is

$$\overline{k}_i = k(\hat{x}\sin\theta_i - \hat{z}\cos\theta_i) \tag{4.1.32}$$

The choice of decay factor is dictated by the fact that the direction of constant phase and the direction of constant amplitude are perpendicular. We have

$$\nabla(x\sin\theta_i - z\cos\theta_i)\cdot\nabla\left((x+z\tan\theta_i)^2\right) = 0 \tag{4.1.33}$$

Note also that at $z = 0$,

$$|\psi_{inc}(\overline{r})|_{z=0} = \exp\left(-\frac{x^2}{g^2}\right) \tag{4.1.34}$$

so that it is Gaussian in the plane $z = 0$ and decays rapidly for $|x| > g$.

The additional factor in the phase, $w(\overline{r})$ is inserted such that ψ_{inc} obeys the wave equation to a higher order. The choice of $w(\overline{r})$ is

$$w(\overline{r}) = \frac{\left[2\dfrac{(x+z\tan\theta_i)^2}{g^2} - 1\right]}{(kg\cos\theta_i)^2} \tag{4.1.35}$$

By straightforward differentiation, it follows that

$$\begin{aligned}
&\frac{\partial^2\psi_{inc}}{\partial x^2} + \frac{\partial^2\psi_{inc}}{\partial z^2} + k^2\psi_{inc} \\
&= k^2\psi_{inc}\left\{-w^2 - 16\frac{(x\sin\theta_i - z\cos\theta_i)^2(x+z\tan\theta_i)^2}{k^4g^8\cos^6\theta_i}\right. \\
&\quad \left. + \frac{4ik(x\sin\theta_i - z\cos\theta_i)}{k^4g^4\cos^4\theta_i}\left[1 - \frac{4(x+z\tan\theta_i)^2}{g^2}\right]\right\}
\end{aligned} \tag{4.1.36}$$

The right-hand side of (4.1.36) is much smaller than $|k^2\psi_{inc}|$. The curly bracket in (4.1.36) is of the order of $\mathcal{O}(\frac{1}{k^3g^3\cos^3\theta_i})$ and is usually small for large g and θ_i not close to grazing. In numerical simulations, g is usually chosen to be somewhere between $\frac{L}{4}$ to $\frac{L}{10}$, depending on the incident angle. The advantage of the analytical expression of (4.1.31) is that ψ_{inc} can be evaluated readily for any x and z. However, the right-hand side of (4.1.36) can grow as $\frac{1}{\cos\theta_i}$ as $\theta_i \to 90°$. Thus ψ_{inc} of (4.1.31) should not be used for problems of low grazing angle (LGA) incidence where $\theta_i \to 90°$. To calculate the Poynting's vector of the incident wave, note that

$$\overline{S} = -\frac{1}{2\eta k}\mathrm{Im}(\psi\nabla\psi^*) \tag{4.1.37}$$

At $z = 0$, by using (4.1.37) and (4.1.31), we have

$$(\hat{z}\cdot\overline{S}_{inc})_{z=0} = -\frac{1}{2\eta k}\mathrm{Im}\left(\psi_{inc}\frac{\partial\psi^*_{inc}}{\partial z}\right)_{z=0} \tag{4.1.38}$$

$$= -\frac{1}{2\eta k}\left\{k\cos\theta_i\left[1+\frac{\frac{2x^2}{g^2}-1}{(kg\cos\theta_i)^2}\right] - \frac{4k\sin\theta_i\tan\theta_i x^2}{k^2g^4\cos^2\theta_i}\right\}e^{-\frac{2x^2}{g^2}} \tag{4.1.39}$$

The power received by the rough surface is obtained by integrating over x from $-\infty$ to ∞

$$P_{inc} = -\int_{-\infty}^{\infty} dx\ (\overline{S}\cdot\hat{z})_{z=0} \tag{4.1.40}$$

On integration,

$$P_{inc} = \frac{\cos\theta_i}{2\eta}g\sqrt{\frac{\pi}{2}}\left\{1-\frac{1+2\tan^2\theta_i}{2k^2g^2\cos^2\theta_i}\right\} \tag{4.1.41}$$

A second way of tapering is to taper the incident wave in the spectral domain. Let

$$\psi_{inc}(x,z) = \frac{g}{2\sqrt{\pi}}\int_{-\infty}^{\infty} dk_x\ e^{ik_x x - ik_z z}\exp\left[-\frac{(k_x-k_{ix})^2g^2}{4}\right] \tag{4.1.42}$$

The advantage of using (4.1.42) is that it obeys the wave equation exactly since it is a spectrum of plane waves. Note that $|x|$ can be very large while $|z|$ is moderate. The disadvantage of using (4.1.42) is that the $\psi_{inc}(x, f(x))$ has to be numerically evaluated by performing the integration of (4.1.42). The integrand in (4.1.42) can be highly oscillatory for large $|x|$. To avoid the oscillatory integrand, one can carry out a numerical contour integration as shown in Fig. 4.1.2. Let $x > 0$ and consider general z. Consider the dominant exponential factor $\exp(ik_x x - \frac{k_x^2 g^2}{4})$. Let $k_x = k_x' + ik_x''$. The dominant exponent term for large value of x is,

$$\text{exponent} = ik_x x - \frac{(k_x-k_{ix})^2g^2}{4}$$

$$= i\left[k_x'x - \frac{k_x''g^2}{2}(k_x'-k_{ix})\right] - \left[k_x''x + \frac{[(k_x'-k_{ix})^2-k_x''^2]g^2}{4}\right]$$

If we let

$$k_x'' = \frac{2x}{g^2}$$

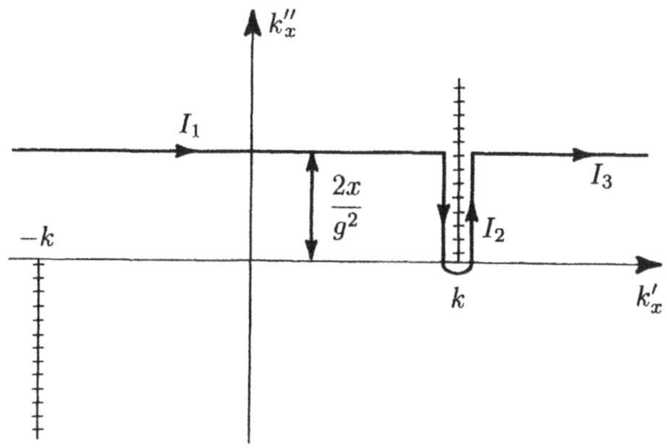

Figure 4.1.2 Contour in complex plane of k_x.

then

$$\text{exponent} = ik_{ix}x - \left[\frac{x^2}{g^2} + \frac{(k_x' - k_{ix})^2 g^2}{4}\right]$$

The exponential has constant phase and the real part decays rapidly from $k_x' = k_{ix}$.

The contour C consists of I_1, I_2, and I_3 in the complex k_x-plane. Contour I_1 is parallel to real k_x-axis at a distance $2x/g^2$ above it, and $\text{Re}(k_x)$ runs from $-\infty$ to k. Contour I_2 goes down on the left-hand side of the vertical branch cut at k and goes up on the right-hand side of the branch cut. Contour I_3 is parallel to the $\text{Re}(k_x)$-axis and at a distance of $2x/g^2$ above it and runs from k to ∞.

$$\begin{aligned}\psi_{inc}(x,z) = {} & \frac{g}{2\sqrt{\pi}} \int_{-\infty(I_1)}^{k} dk_x' \left[e^{-ik_z z}\right]_{k_x = k_x' + i2x/g^2} \\ & \exp\left[ik_{ix}x - \frac{(k_x' - k_{ix})^2 g^2}{4} - \frac{x^2}{g^2}\right] \\ & + \frac{g}{2\sqrt{\pi}} \int_{k(I_3)}^{\infty} dk_x' \left[e^{-ik_z z}\right]_{k_x = k_x' + i2x/g^2} \\ & \exp\left[ik_{ix}x - \frac{(k_x' - k_{ix})^2 g^2}{4} - \frac{x^2}{g^2}\right] \\ & + \frac{g}{2\sqrt{\pi}} \int_{(I_2)} dk_x \, e^{ik_x x - ik_z z} \exp\left[-\frac{(k_x - k_{ix})^2 g^2}{4}\right]\end{aligned} \quad (4.1.43)$$

In numerical integration, (4.1.43) converges much faster. For $x < 0$, similar formula can be established. Note that I_2 is a short contour as the "length" of the contour normalized by k is $2|x|/(kg^2)$, which is much less than 1 even when $|x| = L$.

To calculate the power impinging upon the surface, we have, from (4.1.38) and (4.1.42)

$$\overline{S}_{inc} \cdot \hat{z} = -\frac{1}{2\eta k} \mathrm{Im} \left\{ \frac{g^2}{4\pi} \int_{-\infty}^{\infty} dk'_x \, e^{ik'_x x - i\sqrt{k^2 - k'^2_x} z} \exp\left[-\frac{(k'_x - k_{ix})^2 g^2}{4} \right] \right.$$
$$\left. \cdot \int_{-\infty}^{\infty} dk_x \, e^{-ik_x x + ik_z^* z} ik_z^* \exp\left[-\frac{(k_x - k_{ix})^2 g^2}{4} \right] \right\} \tag{4.1.44}$$

On substituting (4.1.44) into (4.1.40) and integrating over dx, we get a Dirac delta function so that $k'_x = k_x$. Because the imaginary part is taken in (4.1.44), only propagating waves contribute to power. Thus

$$P_{inc} = \frac{g^2}{4\eta k} \int_{-k}^{k} dk_x \, k_z \exp\left[-\frac{(k_x - k_{ix})^2 g^2}{2} \right] \tag{4.1.45}$$

Scattered Wave

After the surface fields $\psi(\overline{r})$ and $\hat{n} \cdot \nabla\psi(\overline{r})$ are calculated by numerical methods, the scattered wave can be calculated by using Huygen's principle. We describe general nonzero surface fields of $\psi(\overline{r})$ and $\hat{n} \cdot \nabla\psi(\overline{r})$ so that the results of the scattered wave are also applicable to the two media problem to be treated in Section 2. From (4.1.8a), the scattered wave is

$$\psi_s(\overline{r}') = -\int_S ds \, \left[\psi(\overline{r}) \hat{n} \cdot \nabla g(\overline{r}, \overline{r}') - g(\overline{r}, \overline{r}') \hat{n} \cdot \nabla\psi(\overline{r}) \right] \tag{4.1.46}$$

Given $\psi(\overline{r})$ and $\hat{n} \cdot \nabla\psi(\overline{r})$ on the surface, $\psi_s(\overline{r}')$ can be calculated by carrying out the integration in (4.1.46).

To calculate bistatic scattering coefficients, we put $\overline{r}'$ in the far field. For observation in the $\hat{k}_s = \sin\theta_s \hat{x} + \cos\theta_s \hat{z}$ direction,

$$g(\overline{r}, \overline{r}') = \frac{i}{4} \sqrt{\frac{2}{\pi k r'}} e^{-i\frac{\pi}{4}} e^{ikr'} e^{-ik(\sin\theta_s x + \cos\theta_s z)} \tag{4.1.47}$$

Then

$$\left(\hat{n} \cdot \nabla g(\overline{r}', \overline{r}) \right)_{z=f(x)} \sqrt{1 + \left(\frac{df}{dx} \right)^2} = \frac{i}{4} \sqrt{\frac{2}{\pi k r'}} e^{-i\frac{\pi}{4}} e^{ikr'} \cdot$$
$$\left[\frac{df(x)}{dx} (ik \sin\theta_s) - ik \cos\theta_s \right] e^{-ik(\sin\theta_s x + \cos\theta_s f(x))} \tag{4.1.48}$$

Putting (4.1.47) and (4.1.48) in (4.1.46), we have

$$\psi_s(\overline{r}') = \frac{i}{4}\sqrt{\frac{2}{\pi k r'}}e^{-i\frac{\pi}{4}}e^{ikr'}\psi_s^{(N)}(\theta_s) \tag{4.1.49}$$

where

$$\psi_s^{(N)}(\theta_s) = -\int_{-\infty}^{\infty} dx \left\{ -U(x) + \psi(x)ik\left[\frac{df}{dx}\sin\theta_s - \cos\theta_s\right]\right\} \cdot e^{-ik(\sin\theta_s x + f(x)\cos\theta_s)} \tag{4.1.50}$$

and $\psi(x) = \psi(x, z = f(x))$ is the surface field, and $U(x) = (\hat{n}\cdot\nabla\psi)_{z=f(x)}$ $\sqrt{1+(\frac{df}{dx})^2}$ is proportional to the normal derivative of the surface field. The Poynting's vector in direction $\hat{k}_s$ is

$$\overline{S}_s(\overline{r}') = -\frac{1}{2\eta k}\text{Im}\left(\psi_s(\overline{r}')\nabla\psi_s^*(\overline{r}')\right) \tag{4.1.51}$$

In the far field, this becomes

$$\overline{S}_s(\overline{r}') = \frac{\hat{r}'}{2\eta}\left(\frac{1}{8\pi k r'}\right)\left|\psi_s^{(N)}(\theta_s)\right|^2 \tag{4.1.52}$$

The total power scattered is

$$P_s = \int_{-\frac{\pi}{2}}^{\frac{\pi}{2}} d\theta_s\, r' S_s(\overline{r}') = \int_{-\frac{\pi}{2}}^{\frac{\pi}{2}} d\theta_s \frac{1}{2\eta}\left(\frac{1}{8\pi k}\right)\left|\psi_s^{(N)}(\theta_s)\right|^2 \tag{4.1.53}$$

The bistatic scattering coefficient $\sigma(\theta_s)$ is defined so that

$$\frac{P_s}{P_{inc}} = \int_{-\frac{\pi}{2}}^{\frac{\pi}{2}} d\theta_s\, \sigma(\theta_s) \tag{4.1.54}$$

giving

$$\sigma(\theta_s) = \frac{\frac{1}{2\eta}\cdot\frac{1}{8\pi k}\left|\psi_s^{(N)}(\theta_s)\right|^2}{P_{inc}} \tag{4.1.55}$$

The definition of $\sigma(\theta_s)$ is such that $\int_{-\frac{\pi}{2}}^{\frac{\pi}{2}} d\theta_s \sigma(\theta_s) = 1$ for non-penetrable rough surface. For the spatial domain tapered incident wave, with P_{inc} given by (4.1.41), we have

$$\sigma(\theta_s) = \frac{\left|\psi_S^{(N)}(\theta_s)\right|^2}{8\pi k g\sqrt{\frac{\pi}{2}}\cos\theta_i\left[1 - \frac{1+2\tan^2\theta_i}{2k^2g^2\cos^2\theta_i}\right]} \tag{4.1.56}$$

For the spectral domain tapered incident wave of (4.1.42) and (4.1.45) we obtain

$$\sigma(\theta_s) = \frac{\left|\psi_s^{(N)}(\theta_s)\right|^2}{4\pi g^2 \int_{-k}^{k} dk_x\, k_z \exp\left[-\frac{(k_x - k_{ix})^2 g^2}{2}\right]} \tag{4.1.57}$$

1.4 Random Rough Surface Generation

In this section, we describe how to generate realizations of random rough surface. We generate Gaussian rough surfaces with Gaussian correlation function, Gaussian rough surface with band-limited ocean spectrum, and fractal rough surface.

1.4.1 Gaussian Rough Surface

A process $f(x)$ is Gaussian if the random variables $f(x_1)$, $f(x_2), \ldots, f(x_n)$ are jointly Gaussian for any n, x_1, x_2, $\ldots$, x_n [Papoulis, 1984]. The Gaussian process is completely characterized by the correlation function $\langle f(x_1)f(x_2)\rangle = h^2C(x_1, x_2)$. If the rough surface $f(x)$ is statistically translational invariant, then $C(x_1, x_2) = C(x_1 - x_2)$. The Fourier transform of $h^2C(x)$ is the spectral density $W(k_x)$.

To generate Gaussian random rough surfaces, we use the results from Chapter 9, Section 2 of Volume I. We first note that from Eq. (9.2.10) of Volume I,

$$\langle F(k_x)F^*(k'_x)\rangle = 0 \tag{4.1.58}$$

for $k_x \neq k'_x$. Thus

$$\langle F(k_x)F^*(k'_x)\rangle = \langle F(k_x)\rangle\langle F^*(k'_x)\rangle = 0 \tag{4.1.59}$$

for $k_x \neq k'_x$. This means that $F(k_x)$ and $F(k'_x)$ are independent random variables for $k_x \neq k'_x$.

A surface of finite length L is to be generated. We make $f(x)$ periodic outside L, i.e., $f(x) = f(x + L)$. A Fourier series is used to represent $f(x)$

$$f(x) = \frac{1}{L}\sum_{n=-\infty}^{\infty} b_n e^{\frac{i2\pi nx}{L}} \tag{4.1.60}$$

where b_n is a Gaussian random variable. From (4.1.60) we have

$$\langle f(x_1)f(x_2)\rangle = \frac{1}{L^2}\sum_{n=-\infty}^{\infty}\sum_{m=-\infty}^{\infty} \langle b_n b_m^*\rangle e^{\frac{i2\pi nx_1}{L}} e^{\frac{-i2\pi mx_2}{L}} \tag{4.1.61}$$

From (9.2.5) and (9.2.12) of Volume I,

$$\langle f(x_1)f(x_2)\rangle = h^2C(x_1 - x_2) = \int_{-\infty}^{\infty} dk_x e^{ik_x(x_1-x_2)}W(k_x) \tag{4.1.62}$$

Comparing (4.1.61) and (4.1.62),

$$\langle b_n b_m^*\rangle = \delta_{nm}B_n \tag{4.1.63}$$

$$\int_{-\infty}^{\infty} dk_x e^{ik_x(x_1-x_2)}W(k_x) = \frac{1}{L^2}\sum_{n=-\infty}^{\infty} B_n \exp\left(i\frac{2\pi n}{L}(x_1 - x_2)\right) \tag{4.1.64}$$

Let

$$\Delta k_x = \frac{2\pi}{L} \tag{4.1.65}$$

$$K_n = \frac{2\pi n}{L} = n\Delta k_x \tag{4.1.66}$$

we sample k_x at K_n. Then

$$\frac{2\pi}{L}\sum_{n=-\infty}^{\infty} e^{iK_n(x_1-x_2)}W(K_n) = \frac{1}{L^2}\sum_{n=-\infty}^{\infty} B_n e^{iK_n(x_1-x_2)} \tag{4.1.67}$$

giving

$$B_n = 2\pi LW(K_n) \tag{4.1.68}$$

From (4.1.63) and (4.1.68)

$$\langle | b_n |^2\rangle = 2\pi LW(K_n) \tag{4.1.69}$$

From (9.2.11) of Volume I,

$$F(K_n) = F^*(-K_n) \tag{4.1.70}$$

Since b_n is proportional to $F(K_n)$,

$$b_n = b_{-n}^* \tag{4.1.71}$$

Let $m = -n$ in (4.1.63). Then

$$\langle b_n b_{-n}^*\rangle = 0 \tag{4.1.72}$$

and from (4.1.71) and (4.1.72)

$$\langle b_n b_n\rangle = 0 \tag{4.1.73}$$

Let

$$b_n = \text{Re}b_n + i\text{Im}b_n \tag{4.1.74}$$

Then (4.1.73) gives

$$\langle(\text{Re}b_n)^2\rangle = \langle(\text{Im}b_n)^2\rangle \tag{4.1.75}$$

$$\langle\text{Re}b_n\rangle\langle\text{Im}b_n\rangle = 0 \tag{4.1.76}$$

Thus Reb_n and Imb_n are independent Gaussian random variables with variance equal to half of that of $\langle |b_n|^2 \rangle$.

We further use a DFT (discretized Fourier transform) version of (4.1.60). Let there be N points in both space and spectral domains

$$\Delta x = \frac{L}{N} \tag{4.1.77}$$

and

$$x_m = m\Delta x \tag{4.1.78}$$

for

$$m = -\frac{N}{2} + 1, ..., 0, 1, ..., \frac{N}{2}$$

$$f(x_m) \equiv f_m \tag{4.1.79}$$

Then

$$f_m = \frac{1}{L} \sum_{n=-\frac{N}{2}+1}^{\frac{N}{2}} b_n \exp\left(i\frac{2\pi nm}{N}\right) \tag{4.1.80}$$

The inverse DFT is

$$b_n = \frac{L}{N} \sum_{m=-\frac{N}{2}+1}^{\frac{N}{2}} f_m e^{-i\frac{2\pi nm}{N}} \tag{4.1.81}$$

Equations (4.1.80) and (4.1.81) can readily be computed from FFT. Both f_m and b_n are periodic sequences with period N. That is,

$$b_{n+N} = b_n \tag{4.1.82}$$

$$f_{m+N} = f_m \tag{4.1.83}$$

Hence

$$b_{-\frac{N}{2}} = b_{\frac{N}{2}} \tag{4.1.84}$$

However from (4.1.71)

$$b_{-\frac{N}{2}} = b_{\frac{N}{2}} = b^*_{-\frac{N}{2}} \tag{4.1.85}$$

Then

$$b_{+\frac{N}{2}} \quad \text{is real} \tag{4.1.86}$$

Also from (4.1.71)

$$b_0 \quad \text{is real} \tag{4.1.87}$$

To summarize, we have b_0 and $b_{+\frac{N}{2}}$, which account for two Gaussian random numbers. Also $b_{-\frac{N}{2}+1}, b_{-\frac{N}{2}+2}, ..., b_{-2}, b_{-1}$ are complex with real and imaginary parts and account for $2(\frac{N}{2}-1)=2N-2$ Gaussian random numbers. The rest of the b_n can be computed as follows. The quantities b_n for $n=1,2,...,\frac{N}{2}-1$ can be calculated by using the condition $b_n=b^*_{-n}$. Other values obey the periodic relation of (4.1.82). This is how the N independent Gaussian random numbers are distributed to the b_n's. The algorithm is as follows.

(i) With a given seed, get N Gaussian distributed random numbers that have zero mean and unit variance. The IMSL subroutine is **rnnor** and the MATLAB function is **randn**. Note that these N numbers are independent and they need not be grouped or arranged in any order. Let the numbers be labeled $r_1, r_2, ..., r_N$.

(ii) Calculate

$$b_0 = \sqrt{2\pi L W(0)}\, r_\alpha \tag{4.1.88}$$

$$b_{+\frac{N}{2}} = \sqrt{2\pi L W\left(\frac{\pi N}{L}\right)}\, r_\beta \tag{4.1.89}$$

where $\alpha \neq \beta$ and α and β assume one of the values of $1, 2, ..., N$. These use up two of the random numbers $r_1, r_2, \ldots, r_N$.

(iii) Calculate

$$b_n = \sqrt{2\pi L W(|K_n|)} \left\{ \frac{1}{\sqrt{2}} (r_\sigma + i r_\xi) \right\} \tag{4.1.90}$$

for $n=-\frac{N}{2}+1, ..., -2, -1$ where σ, ξ are distinct indices of $1, 2, ..., N$. Thus (4.1.90) will use up the remaining $N-2$ of the random numbers $r_1, r_2, \ldots, r_N$.

(iv) Calculate

$$b_n = b^*_{-n} \tag{4.1.91}$$

for $n = 1, 2, ..., \frac{N}{2}-1$ by using (4.1.90).

For general DFT relations

$$X(k) = \sum_{j=-\frac{N}{2}+1}^{\frac{N}{2}} x(j) e^{-\frac{2\pi i}{N} jk} \tag{4.1.92}$$

$$x(j) = \frac{1}{N} \sum_{k=-\frac{N}{2}+1}^{\frac{N}{2}} X(k) e^{\frac{2\pi i}{N} jk} \tag{4.1.93}$$

where $X(k)$ and $x(j)$ are periodic:

$$x(j+N) = x(j) \tag{4.1.94}$$

$$X(k+N) = X(k) \tag{4.1.95}$$

An alternative way of writing (4.1.92)–(4.1.93)

$$X(k) = \sum_{j=0}^{N-1} x(j) e^{\frac{-2\pi i}{N} jk} \tag{4.1.96}$$

$$x(j) = \frac{1}{N} \sum_{k=0}^{N-1} X(k) e^{\frac{2\pi i}{N} jk} \tag{4.1.97}$$

Define

$$\tilde{x}(j+1) = x(j) \tag{4.1.98}$$

$$\widetilde{X}(k+1) = X(k) \tag{4.1.99}$$

Both $\tilde{x}(j)$ and $\widetilde{X}(k)$ are also periodic sequences. Then

$$\widetilde{X}(k) = \sum_{j=1}^{N} \tilde{x}(j) e^{\frac{-2\pi i}{N}(j-1)(k-1)} \tag{4.1.100}$$

$$\tilde{x}(j) = \frac{1}{N} \sum_{k=1}^{N} \widetilde{X}(k) e^{\frac{2\pi i}{N}(j-1)(k-1)} \tag{4.1.101}$$

Equations (4.1.96) and (4.1.97) and (4.1.100) and (4.1.101) are common forms of DFT subroutines. [MATLAB utilizes (4.1.100) and (4.1.101).] We have

$$X(k) = X^*(-k) \tag{4.1.102}$$

Then

$$\widetilde{X}(k+1) = \widetilde{X}^*(-k+1) \tag{4.1.103}$$

We first periodically extend b_n

$$b_{l+\frac{N}{2}} = b_{l-\frac{N}{2}} = b^*_{\frac{N}{2}-l} \tag{4.1.104}$$

for $l = 1, 2, \ldots, \frac{N}{2} - 1$. First let

$$X(n) = b_n \tag{4.1.105}$$

We can then obtain $\widetilde{X}(n)$ from $X(n)$ by using (4.1.99). From $\widetilde{X}(n)$ we calculate $\tilde{x}(n)$, $n = 1, 2, \ldots, N$ by FFT of (4.1.101). Then we obtain $x(n)$, $n = 0, 1, 2, \ldots, N-1$ from $\tilde{x}(n)$ by (4.1.98). We next periodically extend $x(n)$. Finally, we obtain the rough surface height profiles using

$$f_m = \frac{N}{L} x(m) \tag{4.1.106}$$

for $m = -\frac{N}{2}+1, \ldots, \frac{N}{2}$.

To evaluate the derivative and higher order derivatives of the rough surface profile, one method is to evaluate by means of finite difference

$$f'(x_m) = \frac{f(x_{m+1}) - f(x_{m-1})}{2\Delta x} \tag{4.1.107}$$

For the two endpoints $m = -\frac{N}{2}+1$ and $m = \frac{N}{2}$, one can use the periodic condition of DFT to get $f(x_m)$ for $m = -\frac{N}{2}$ and $m = \frac{N}{2}+1$. Another method of calculating derivatives is to differentiate (4.1.60) directly.

$$f'(x) = \frac{1}{L}\sum_{n=-\infty}^{\infty} b_n \frac{i2\pi n}{L} e^{\frac{i2\pi n x}{L}} \tag{4.1.108}$$

In terms of DFT, this becomes

$$f'(x_m) = \frac{1}{L}\sum_{n=-\frac{N}{2}+1}^{\frac{N}{2}} b_n \frac{i2\pi n}{L} e^{\frac{i2\pi n x_m}{L}} \tag{4.1.109}$$

Also, the second derivative is

$$f''(x_m) = -\frac{1}{L}\sum_{n=-\frac{N}{2}+1}^{\frac{N}{2}} b_n \left(\frac{2\pi n}{L}\right)^2 e^{\frac{i2\pi n x_m}{L}} \tag{4.1.110}$$

FFT can be performed in ways similar to that of calculating f_m.

Gaussian Spectrum

For Gaussian rough surface with Gaussian spectrum

$$\langle f(x_1) f(x_2)\rangle = h^2 C(x_1 - x_2) \tag{4.1.111}$$

$$= h^2 e^{-\frac{(x_1-x_2)^2}{l^2}} \tag{4.1.112}$$

The correlation length is related to the rms slope s by

$$l = \sqrt{2}\frac{h}{s} \tag{4.1.113}$$

The radius of curvature is defined by (9.4.16) of Volume I. The radius of curvature ρ can be estimated by calculating the fourth derivative of the correlation function if we let (rms of ρ) $\approx \dfrac{1}{\text{rms of } |f''(x)|} = \dfrac{l^2}{2\sqrt{3}h}$.

The Gaussian spectrum is

$$W(k_x) = \frac{h^2 l}{2\sqrt{\pi}} e^{-\frac{k_x^2 l^2}{4}} \tag{4.1.114}$$

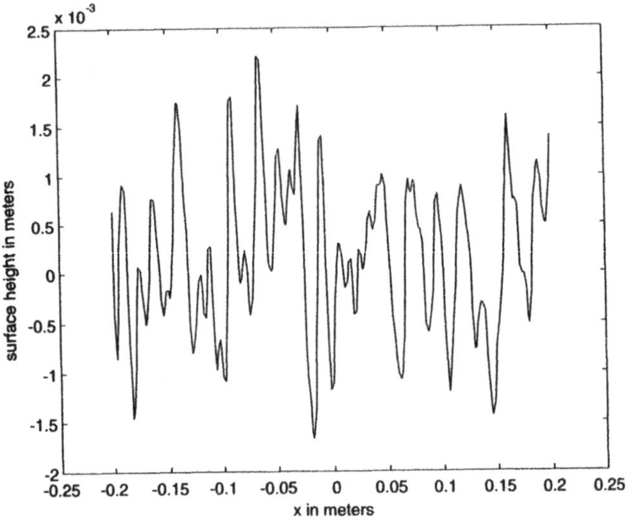

Figure 4.1.3 Gaussian rough surface.

Note that the Gaussian spectrum is for $0 \leq |k_x| \leq \infty$. In Fig. 4.1.3, we plot the height profile with a Gaussian spectrum that has the same rms height and rms slope of the band-limited ocean spectrum to be discussed below. For this case $h = 8.1202 \times 10^{-4}$ meters, $s = 0.2252$ and $l = 5.0993 \times 10^{-3}$ meter.

Bandlimited Ocean Spectrum

The ocean spectrum has been discussed in Chapter 4, Section 8 of Volume I. Note that the ocean spectrum case is still a Gaussian process with the ocean spectral density. For 2-D spectrum $W_2(k_x, k_y)$, we have

$$h^2 C(x,y) = \int_{-\infty}^{\infty} dk_x \int_{-\infty}^{\infty} dk_y e^{ik_x x + ik_y y} W_2(k_x, k_y) \tag{4.1.115}$$

$$h^2 = \int_0^{2\pi} d\phi_k \int_0^{\infty} dk_\rho k_\rho W_2(k_\rho) = 2\pi \int_0^{\infty} dk_\rho k_\rho W_2(k_\rho) \tag{4.1.116}$$

On the other hand for 1-D spectrum,

$$h^2 = 2 \int_0^{\infty} dk_x W(k_x) \tag{4.1.117}$$

Also

$$W_2(k_\rho) = \frac{1}{2k_\rho \pi} S(k_\rho) \tag{4.1.118}$$

To set up a correspondence between 2-D spectrum $W_2(k_\rho)$ and 1-D spectrum

$W(k_x)$, we compare (4.1.116) and (4.1.117). Then

$$W(k_x) = W(|k_x|) = \pi k_\rho W_2(k_\rho) \tag{4.1.119}$$

$$W(|k_x|) = \frac{S(|k_\rho|)}{2} \tag{4.1.120}$$

For ocean spectrum, letting $W_2(k_\rho) = W_{DV}(k_\rho)$ of (4.8.13) in Volume I (*DV* stands for Durden-Vesecky) and ignoring anisotropy, $S(k_\rho)$ is found to be

$$S(k_\rho) = \begin{cases} \dfrac{a_0}{k_\rho^3}\left(\dfrac{bk_\rho u_*^2}{g_*}\right)^{a\log_{10}\left(\frac{k_\rho}{k_j}\right)} & k_\rho > k_j \\ \dfrac{b_0}{k_\rho^3}\exp\left[-0.74\left(\dfrac{k_c}{k_\rho}\right)^2\right] & k_\rho \le k_j \end{cases} \tag{4.1.121}$$

where

$$g_* = g + \gamma k_\rho^2 \tag{4.1.122}$$

$$\gamma = 7.25 \times 10^{-5} \tag{4.1.123}$$

$$g = 9.81 \tag{4.1.124}$$

$$a = 0.225 \tag{4.1.125}$$

$$b = 1.25 \tag{4.1.126}$$

$$a_0 = 0.008 \tag{4.1.127}$$

$$k_j = 2m^{-1} \tag{4.1.128}$$

$$k_c = \frac{g}{(U(19.5))^2} \tag{4.1.129}$$

$$U(z) = \frac{u_*}{0.4}\ln\left(\frac{z}{z_0}\right) \quad \text{(ocean wind speed at elevation } z \text{ in meters)} \tag{4.1.130}$$

$$z_0 = \frac{0.0000684}{u_*} + 0.00428u_*^2 - 0.000443 \tag{4.1.131}$$

Suppose the wind speed is 10 m/s at elevation 5 m, then

$$U(5) = 10 = \frac{u_*}{0.4}\ln\left(\frac{5}{z_0}\right) \tag{4.1.132}$$

We solve for u_* by using (4.1.131) and (4.1.132). (Use `fzero` in MATLAB to find u_*.)

To perform numerical simulations, it is customary to bandlimit the ocean spectrum. For ocean spectrum, let $W_{HWO}(k_\rho)$ be bandlimited between k_L and k_U. Then

$$W_{HWO}(k_\rho) = \begin{cases} W(k_\rho), & k_L < k_\rho < k_U \\ 0, & \text{otherwise} \end{cases} \tag{4.1.133}$$

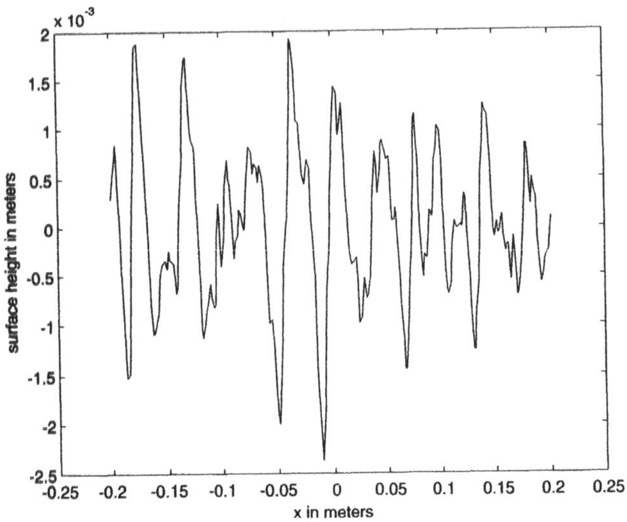

Figure 4.1.4 Surface profile with ocean spectrum.

For the bandlimited spectrum between k_L and k_U, we can calculate the rms height h and the rms slope s

$$h^2 = 2\int_0^{\infty} dk_x W(k_x) = 2\int_{k_L}^{k_U} dk_x W_{HWO}(k_x) \tag{4.1.134}$$

$$s^2 = 2\int_{k_L}^{k_U} dk_x k_x^2 W_{HWO}(k_x) \tag{4.1.135}$$

For example, for microwave scattering at 19 GHz, we can choose $k_L = 100\,\mathrm{m}^{-1}$ and $k_U = 4000\,\mathrm{m}^{-1}$. Then for wind speed at 10 m/s, we have $h = 8.1202 \times 10^{-4}\,\mathrm{m}$ and $s = 0.2252$.

In Fig. 4.1.4, we show a surface height profile created for a Gaussian range surface with the bandlimited ocean spectrum ($N = 256$; $\Delta x = \lambda/10$). Note that Figs. 4.1.3 and 4.1.4 have the same rms height and rms slope. We can see that the bandlimited ocean surface of Fig. 4.1.4 is more spiky than the Gaussian spectrum surface of Fig. 4.1.3.

1.4.2 Fractal Rough Surface

For fractal surface, we can use the Weierstrass-Mandelbrot function [Berizzi et al. 1999]. Unlike the previous two cases, the fractal surface is not a Guas-

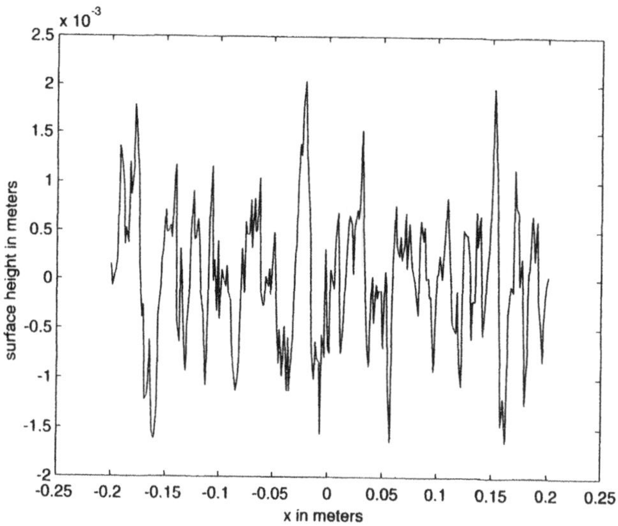

Figure 4.1.5 Surface profile for fractal surface.

sian process. The Weierstrass-Mandelbrot function is

$$f(x) = hC_N \sum_{n=0}^{N_f-1} b^{(S-2)n} \sin(K_0 b^n x + \Phi_n) \tag{4.1.136}$$

where

$$h = \text{rms height}$$

$$N_f = \ \# \text{ of tones} \quad (\text{e.g., } 100) \tag{4.1.137}$$

$$s = \text{fractal dimension} \quad (1 \le s < 2)) \tag{4.1.138}$$

$$C_N = \sqrt{\frac{2\left(1 - b^{2(S-2)}\right)}{1 - b^{2(S-2)N_f}}} = \text{normalization constant} \tag{4.1.139}$$

$$\Phi_n = \text{phase and is a Gaussian random variable uniformly distributed between 0 and } 2\pi \text{ (in MATLAB use } 2\pi * \texttt{rand}(N_f, 1)) \tag{4.1.140}$$

We also bandlimit the fractal surface between k_L and k_U. This means

$$K_0 = k_L \tag{4.1.141}$$

$$K_0 b^{N_f - 1} = k_U \tag{4.1.142}$$

so that

$$b = e^{\frac{1}{N_f - 1} \ln \frac{k_U}{k_L}} = \left(\frac{k_U}{k_L}\right)^{\frac{1}{N_f - 1}} \tag{4.1.143}$$

In Fig. 4.1.5, we plot one height profile for $k_L = 100\,\mathrm{m}^{-1}$ and $k_U = 4000\,\mathrm{m}^{-1}$. For this case rms height $h = 8.1202 \times 10^{-4}\,\mathrm{m}$, and fractal dimension is $s = 1.5$. We compare between Figs. 4.1.4 and 4.1.5. Both have the same rms height h and the same k_L and k_U. We note that Fig. 4.1.5 of the fractal surface is more spiky.

1.5 Neumann Boundary Condition (MFIE for TM Case)

From (4.1.13a) and (4.1.13b), we have the integral equations for the Neumann boundary condition. We let $\overline{r}'$ approach the boundary S. If we zoom in at the point $\overline{r}'$, we can have a small piece of surface of length $2a$ that is symmetrical about the point $\overline{r}'$. This piece can be regarded as a straight line for the limit of vanishing a (Fig. 4.1.6). Then

$$\int_S ds\psi(\overline{r})\hat{n}\cdot\nabla g(\overline{r},\overline{r}') = ⨍_S ds\psi(\overline{r})\hat{n}\cdot\nabla g(\overline{r},\overline{r}') + \int_{\text{piece}} \psi(\overline{r})\hat{n}\cdot\nabla g(\overline{r},\overline{r}'). \tag{4.1.144}$$

where $⨍$ is the principal value integral with a vanishingly small piece subtracted out from the domain of the integration. For the small piece, we use the coordinate system X and Z that are tangential and normal to the straight line respectively. Thus

$$ds = dX, \quad \hat{n} = \hat{Z}, \quad X' = 0$$
$$Z = 0, \quad Z' = \text{infinitesimal}$$

For Z' positive, $\overline{r}'$ is above the surface. For Z' negative, $\overline{r}'$ is below the surface.

$$\int_{\text{piece}} \psi(\overline{r})\hat{n}\cdot\nabla g(\overline{r},\overline{r}') = \lim_{a\to 0}\lim_{|Z'|\to 0} \psi(\overline{r}')\int_{-a}^{a} dX \frac{\partial}{\partial Z} g \tag{4.1.145}$$

Note that the order of the two limits cannot be interchanged. In the small piece,

$$|\overline{r}-\overline{r}'| = \sqrt{X^2+(Z-Z')^2} \tag{4.1.146}$$

$$g = \frac{i}{4}H_0^{(1)}(k|\overline{r}-\overline{r}'|) \to -\frac{1}{2\pi}\ln\frac{|\overline{r}-\overline{r}'|}{2} \tag{4.1.147}$$

$$\left(\frac{\partial g}{\partial Z}\right)_{Z=0} = \frac{Z'}{2\pi(X^2+(Z')^2)} \tag{4.1.148}$$

Thus

$$\int_{\text{piece}} \psi(\overline{r})\hat{n}\cdot\nabla g(\overline{r},\overline{r}') = \lim_{a\to 0}\lim_{|Z'|\to 0} \psi(\overline{r}')\int_{-a}^{a} dX \frac{Z'}{2\pi(X^2+(Z')^2)}$$

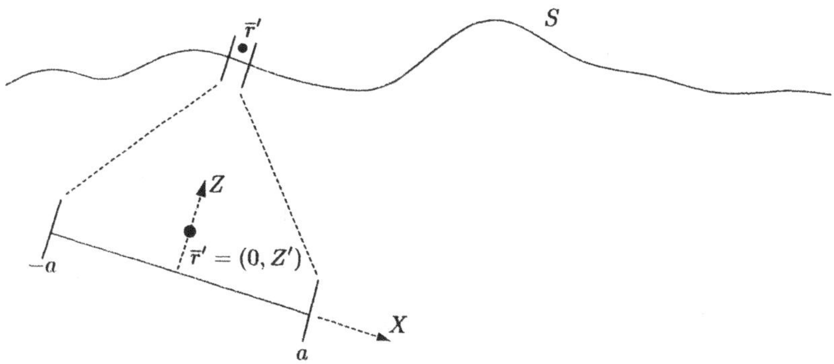

Figure 4.1.6 Principal value integral obtained by subtracting a small piece around x'.

$$= \lim_{a\to 0}\lim_{|Z'|\to 0}\frac{\psi(\bar{r}')}{2\pi}\left[\tan^{-1}\frac{X}{Z'}\right]_{-a}^{a} = \lim_{a\to 0}\lim_{|Z'|\to 0}\frac{\psi(\bar{r}')}{\pi}\tan^{-1}\frac{a}{Z'}$$

$$= \begin{cases} \frac{1}{2}\psi(\bar{r}') & \text{for } Z' > 0 \\ -\frac{1}{2}\psi(\bar{r}') & \text{for } Z' < 0 \end{cases} \tag{4.1.149}$$

The integral equation of (4.1.13*a*) for $\bar{r}'$ approaching the surface from region 0 then becomes

$$\psi_{inc}(\bar{r}') + \fint_S ds\psi(\bar{r})\hat{n}\cdot\nabla g(\bar{r},\bar{r}') + \frac{1}{2}\psi(\bar{r}') = \psi(\bar{r}') \tag{4.1.150}$$

while the integral equation of (4.1.13*b*) for $\bar{r}'$ approaching the surface from region 1 becomes

$$\psi_{inc}(\bar{r}') + \fint_S ds\psi(\bar{r})\hat{n}\cdot\nabla g(\bar{r},\bar{r}') - \frac{1}{2}\psi(\bar{r}') = 0 \tag{4.1.151}$$

These two equations of (4.1.150) and (4.1.151) become identical, which mean that Eqs. (4.1.13*a*) and (4.1.13*b*) are well explained by the singularity of the normal derivative of the Green's function. Equation (4.1.151) is known as MFIE (magnetic field integral equation) because ψ represents the magnetic field which is in the horizontal direction for the TM case.

Then if we let the kernel be

$$K_N(x', x) = \left[\sqrt{1+\left(\frac{df}{dx}\right)^2}\,\hat{n}\cdot\nabla g(\bar{r},\bar{r}')\right]_{z=f(x),\ z'=f(x')}$$

$$= \sqrt{1+\left(\frac{df}{dx}\right)^2}\,\hat{n}\cdot\hat{R}\frac{\partial g}{\partial R} \tag{4.1.152}$$

the surface integral equation is

$$\psi_{inc}(x') + \fint_{-\infty}^{\infty} dx\ \psi(x) K_N(x', x) = \frac{1}{2}\psi(x') \tag{4.1.153}$$

where $\psi_{inc}(x') = \psi_{inc}(x', z' = f(x'))$ and $\psi(x) = \psi(x, z = f(x))$.

The integral equation is solved using the method of moment with pulse basis functions and point matching. Let there be N intervals with length Δx for each interval. The simplest approximation of $\fint dx$ is to exclude the interval about x'. Then the diagonal matrix element is 1/2. Let

$$\psi_{inc}(x_m) = b(x_m) = b_m \tag{4.1.154}$$

$$\psi(x_m) = \psi_m \tag{4.1.155}$$

$$B_{mn} = \begin{cases} -\Delta x\ K_N(x_m, x_n) & \text{for } m \neq n \\ \dfrac{1}{2} & \text{for } m = n \end{cases} \tag{4.1.156}$$

Then the matrix equation is

$$\sum_n B_{mn}\psi_n = b_m \tag{4.1.157}$$

Equations (4.1.154)–(4.1.157) are of sufficient accuracy for many cases.

Note that

$$K_N(x', x) = \frac{ik}{4} \frac{H_1^{(1)}\left(k\sqrt{(x'-x)^2 + (f(x') - f(x))^2}\right)}{\sqrt{(x'-x)^2 + (f(x') - f(x))^2}} \cdot \{f'(x)(x - x') - (f(x) - f(x'))\} \tag{4.1.158}$$

To have a more accurate evaluation of the self patch of $m = n$, we need to calculate

$$\begin{aligned} I &= \fint_{x_m - \frac{\Delta x}{2}}^{x_m + \frac{\Delta x}{2}} dx\ K_N(x_m, x) \\ &= \frac{ik}{4} \fint_{x_m - \frac{\Delta x}{2}}^{x_m + \frac{\Delta x}{2}} dx \frac{H_1^{(1)}\left(k\sqrt{(x_m - x)^2 + (f(x_m) - f(x))^2}\right)}{\sqrt{(x_m - x)^2 + (f(x_m) - f(x))^2}} \\ &\quad \cdot \left\{\frac{df}{dx}(x - x_m) - (f(x) - f(x_m))\right\} \end{aligned} \tag{4.1.159}$$

We next approximate $f(x)$ and $\frac{df(x)}{dx}$ by Taylor expansions about x_m

$$f(x) = f(x_m) + f'(x_m)(x - x_m) + \frac{f''(x_m)}{2}(x - x_m)^2 \tag{4.1.160}$$

$$\frac{df(x)}{dx} = f'(x_m) + f''(x_m)(x - x_m) \tag{4.1.161}$$

Then

$$\sqrt{(x_m - x)^2 + (f(x_m) - f(x))^2} \simeq \sqrt{(x_m - x)^2 + (f'(x_m)(x_m - x))^2} \tag{4.1.162}$$

$$\frac{df}{dx}(x - x_m) - (f(x) - f(x_m)) = \frac{f''(x_m)}{2}(x_m - x)^2 \tag{4.1.163}$$

For small argument w

$$H_1^{(1)}(w) = -\frac{2i}{\pi w} \tag{4.1.164}$$

Thus

$$\begin{aligned} I &= \frac{1}{2\pi} \fint_{x_m - \frac{\Delta x}{2}}^{x_m + \frac{\Delta x}{2}} dx \, \frac{\frac{f''(x_m)}{2}(x - x_m)^2}{(x - x_m)^2 + (f'(x_m)(x_m - x))^2} \\ &= \frac{1}{2\pi} \fint_{x_m - \frac{\Delta x}{2}}^{x_m + \frac{\Delta x}{2}} dx \, \frac{f''(x_m)}{2\left[1 + (f'(x_m))^2\right]} \end{aligned} \tag{4.1.165}$$

Since the integrand is now well-behaved, subtracting out an infinitesimally small piece does not make any difference. Thus

$$I = \frac{\Delta x}{4\pi} \frac{f''(x_m)}{1 + (f'(x_m))^2} \tag{4.1.166}$$

Thus the matrix elements become, from (4.1.156) and (4.1.166),

$$B_{mn} = \begin{cases} -\Delta x \ K_N(x_m, x_n) & \text{for } \ m \neq n \\ \dfrac{1}{2} - \dfrac{f''(x_m)}{4\pi} \dfrac{\Delta x}{1 + (f'(x_m))^2} & \text{for } \ m = n \end{cases} \tag{4.1.167}$$

In Section 2.3, we will include numerical integration to further improve the accuracy of the matrix elements.

2 Two-Media Problem

For the case of a two media problem where the lower medium has permittivity ϵ_1 (Fig. 4.1.1), we need dual integral equations. The first one is given by (4.1.8). We also use the relation given in (4.1.149). Then we have, for $\overline{r}$ and $\overline{r}'$ on S,

$$\psi_{inc}(\overline{r}') = \frac{1}{2}\psi(\overline{r}') - \fint ds \psi(\overline{r}) \hat{n} \cdot \nabla g(\overline{r}, \overline{r}') + \int_S ds g(\overline{r}, \overline{r}') \hat{n} \cdot \nabla \psi(\overline{r}) \tag{4.2.1}$$

We next apply Green's theorem to the lower medium.

$$(\nabla^2 + k_1^2)\psi_1 = 0 \tag{4.2.2}$$

where $k_1 = \omega\sqrt{\mu\epsilon_1}$. The Green's function of the lower half-space $g_1(\overline{r}, \overline{r}')$ is

$$g_1(\overline{r}, \overline{r}') = \frac{i}{4} H_0^{(1)}(k_1|\overline{r} - \overline{r}'|) \tag{4.2.3}$$

$$(\nabla^2 + k_1^2) g_1(\overline{r}, \overline{r}') = -\delta(\overline{r} - \overline{r}') \tag{4.2.4}$$

Applying Green's theorem,

$$\begin{aligned}\iint_{V_1} d\overline{r} \left[\psi_1(\overline{r})\nabla^2 g_1(\overline{r}, \overline{r}') - g_1(\overline{r}, \overline{r}')\nabla^2\psi_1(\overline{r})\right] \\ = \int_{S_{-\infty}} ds\hat{n} \cdot \left[\psi_1(\overline{r})\nabla g_1(\overline{r}, \overline{r}') - g_1(\overline{r}, \overline{r}')\nabla\psi_1(\overline{r})\right] \\ + \int_S ds\hat{n} \cdot \left[\psi_1(\overline{r})\nabla g_1(\overline{r}, \overline{r}') - g_1(\overline{r}, \overline{r}')\nabla\psi_1(\overline{r})\right]\end{aligned} \tag{4.2.5}$$

where $\int_{S_{-\infty}}$ is the surface integral at infinity of the lower half-space and contributes zero in this case.

We use (4.2.2) and (4.2.4) in the left-hand side of (4.2.5):

$$\begin{aligned}\iint_{V_1} d\overline{r} \left[\psi_1(\overline{r})\nabla^2 g_1(\overline{r}, \overline{r}') - g_1(\overline{r}, \overline{r}')\nabla^2\psi_1(\overline{r})\right] \\ = -\iint_{V_1} d\overline{r}\ \psi_1(\overline{r})\ \delta(\overline{r} - \overline{r}') = \begin{cases} 0 & \text{if } \overline{r}' \in V_0 \\ -\psi_1(\overline{r}') & \text{if } \overline{r}' \in V_1 \end{cases}\end{aligned} \tag{4.2.6}$$

Thus, we have

$$\int_S ds \left[\psi_1(\overline{r})\hat{n} \cdot \nabla g_1(\overline{r}, \overline{r}') - g_1(\overline{r}, \overline{r}')\hat{n} \cdot \nabla\psi_1(\overline{r})\right] = \begin{cases} 0 & \text{if } \overline{r}' \in V_0 \\ -\psi_1(\overline{r}') & \text{if } \overline{r}' \in V_1 \end{cases} \tag{4.2.7}$$

We have a similar relation to (4.1.149). Let δ be an infinitesimal positive number,

$$\begin{aligned}\left[\int_S ds\psi_1(\overline{r})\hat{n} \cdot \nabla g_1(\overline{r}, \overline{r}')\right]_{r'=(x', f(x')\pm\delta)} \\ = \pm\frac{1}{2}\psi_1(\overline{r}') + ⨍ ds\psi_1(\overline{r})\hat{n} \cdot \nabla g_1(\overline{r}, \overline{r}')\end{aligned} \tag{4.2.8}$$

Using this in (4.2.7) gives, for $\overline{r}$ and $\overline{r}'$ on S,

$$\frac{1}{2}\psi_1(\overline{r}') + ⨍ ds\psi_1(\overline{r})\hat{n} \cdot \nabla g_1(\overline{r}, \overline{r}') - \int_S ds g_1(\overline{r}, \overline{r}')\hat{n} \cdot \nabla\psi_1(\overline{r}) = 0 \tag{4.2.9}$$

Equations (4.2.1) and (4.2.9) are the dual integral equations for the two-media problem.

2.1 TE and TM Waves

Surface integral equations are also conveniently expressed in terms of surface electric current $\overline{J}_s$ and surface magnetic current $\overline{M}_s$. For $\overline{r}$ on S,

$$\overline{J}_s = \hat{n} \times \overline{H} \tag{4.2.10}$$

$$\overline{M}_s = \overline{E} \times \hat{n} \tag{4.2.11}$$

(A) TE case: ψ represents the electric field

$$\overline{E} = \hat{y}\psi \tag{4.2.12}$$

$$\overline{H} = -\frac{1}{i\omega\mu}\hat{y} \times \nabla\psi \tag{4.2.13}$$

$$\hat{n} \times \overline{H} = -\frac{\hat{y}}{i\omega\mu}(\hat{n} \cdot \nabla\psi) \tag{4.2.14}$$

Similar relations apply to $\overline{E}_1$, $\overline{H}_1$, and ψ_1. Thus for $\overline{r}$ on S, the boundary conditions of continuity of the tangential electric and magnetic fields become respectively,

$$\psi(\overline{r}) = \psi_1(\overline{r}) \tag{4.2.15}$$

$$\hat{n} \cdot \nabla\psi(\overline{r}) = \hat{n} \cdot \nabla\psi_1(\overline{r}) \tag{4.2.16}$$

(B) TM case: ψ represents the magnetic field

$$\overline{H} = \hat{y}\psi \tag{4.2.17}$$

$$\overline{E} = \frac{1}{i\omega\epsilon}\hat{y} \times \nabla\psi \tag{4.2.18}$$

$$\overline{E} \times \hat{n} = -\frac{\hat{y}}{i\omega\epsilon}(\hat{n} \cdot \nabla\psi) \tag{4.2.19}$$

$$\overline{H}_1 = \hat{y}\psi_1 \tag{4.2.20}$$

$$\overline{E}_1 \times \hat{n} = -\frac{\hat{y}}{i\omega\epsilon_1}(\hat{n} \cdot \nabla\psi_1) \tag{4.2.21}$$

For $\overline{r}$ on S, the boundary conditions of continuity of tangential electric and magnetic field become respectively,

$$\hat{n} \cdot \nabla\psi(\overline{r}) = \frac{\epsilon}{\epsilon_1}\hat{n} \cdot \nabla\psi_1(\overline{r}) \tag{4.2.22}$$

$$\psi(\overline{r}) = \psi_1(\overline{r}) \tag{4.2.23}$$

The TE and TM cases when put into (4.2.9) become

$$\frac{1}{2}\psi(\overline{r}') + \fint ds\psi(\overline{r})\hat{n} \cdot \nabla g_1(\overline{r}, \overline{r}') - \int_S dsg_1(\overline{r}, \overline{r}')\rho\hat{n} \cdot \nabla\psi(\overline{r}) = 0 \tag{4.2.24}$$

where $\rho = 1$ for TE wave and $\rho = \epsilon_1/\epsilon$ for TM wave.

Let

$$u(x) = \sqrt{1 + \left(\frac{df}{dx}\right)^2} \left(\hat{n} \cdot \nabla\psi(\bar{r})\right)_{z=f(x)} \tag{4.2.25}$$

Then the dual integral equations of (4.2.1) and (4.2.24) become

$$\psi_{inc}(x') + \int_{-\infty}^{\infty} dx \; \left[\psi(x)K_N(x',x) - K(x',x)u(x)\right] = \frac{1}{2}\psi(x') \tag{4.2.26}$$

$$\frac{1}{2}\psi(x') + \int_{-\infty}^{\infty} dx \left[\psi(x)K_{1N}(x',x) - K_1(x',x)\rho u(x)\right] = 0 \tag{4.2.27}$$

In (4.2.27), the kernels $K_{1N}(x',x)$ and $K_1(x',x)$ are that of using Green's function g_1 of the lower medium that has wavenumber k_1

$$\begin{aligned} K_1(x',x) &= g_1(x', f(x'); x, f(x)) \\ &= \frac{i}{4} H_0^{(1)}\left(k_1\sqrt{(x'-x)^2 + (f(x') - f(x))^2}\right) \end{aligned} \tag{4.2.28}$$

$$\begin{aligned} K_{1N}(x',x) = \frac{ik_1}{4} & \frac{H_1^{(1)}\left(k_1\sqrt{(x'-x)^2 + (f(x') - f(x))^2}\right)}{\sqrt{(x'-x)^2 + (f(x') - f(x))^2}} \\ & \cdot \left\{f'(x)(x - x') - \left(f(x) - f(x')\right)\right\} \end{aligned} \tag{4.2.29}$$

We obtain the matrix equation from (4.2.26)

$$\sum_{n=1}^{N} A_{mn}u_n + \sum_{n=1}^{N} B_{mn}\psi_n = b_m \tag{4.2.30}$$

where A_{mn} is given by (4.1.28) of the Dirichlet case and B_{mn} is given by (4.1.156) or (4.1.167) of the Neumann case.

For the surface integral equation of (4.2.27)

$$\sum_{n=1}^{N} A_{mn}^{(1)} \; \rho \; u_n + \sum_{n=1}^{N} B_{mn}^{(1)}\psi_n = 0 \tag{4.2.31}$$

where

$$A_{mn}^{(1)} = \begin{cases} \Delta x \, K_1(x_m, x_n) & \text{for } m \neq n \\ \dfrac{i\Delta x}{4}\left[1 + \dfrac{i2}{\pi}\ln(\dfrac{\gamma k_1}{4e}\Delta l_m)\right] & \text{for } m = n \end{cases} \tag{4.2.32}$$

$$B_{mn}^{(1)} = \begin{cases} -\Delta x \, K_{1N}(x_m, x_n; 0) & \text{for } m \neq n \\ -\dfrac{1}{2} - \dfrac{f''(x_m)}{4\pi}\dfrac{\Delta x}{1 + (f'(x_m))^2} & \text{for } m = n \end{cases} \tag{4.2.33}$$

Note that there is a sign difference in the $\frac{1}{2}$ term between B_{mn} and $B_{mn}^{(1)}$. Equations (4.2.30) and (4.2.31) are for $m = 1, 2, \ldots, N$. They provide $2N$ equations for the $2N$ unknowns of u_n, ψ_n, $n = 1, 2, \ldots, N$. In Section 2.3, we discuss numerical integration to obtain more accurate matrix elements.

2.2 Absorptivity, Emissivity and Reflectivity

Consider the incident wave impinging on the dielectric medium at an incident angle of θ_i. We can calculate $\sigma(\theta_s, \theta_i)$, the bistatic scattering coefficient from incident direction θ_i into scattered direction θ_s. The reflectivity is the integration of the bistatic scattering coefficient

$$r(\theta_i) = \int_{-\frac{\pi}{2}}^{\frac{\pi}{2}} d\theta_s \sigma(\theta_s) \tag{4.2.34}$$

This is a way of calculating reflectivity. For perfect conductor the reflectivity should be equal to unity. This provides a useful check on the numerical accuracy of the bistatic scattering coefficient. For penetrable medium, the absorptivity is

$$a(\theta_i) = 1 - r(\theta_i) \tag{4.2.35}$$

The emissivity is equal to the absorptivity

$$e(\theta_i) = a(\theta_i) = 1 - r(\theta_i) \tag{4.2.36}$$

and the brightness temperature is equal to

$$T_B(\theta_i) = e(\theta_i) T \tag{4.2.37}$$

where T is the physical temperature. We can use (4.2.34) and (4.2.36) to calculate emissivity.

In Monte Carlo simulations, we can calculate the surface fields for each realization of the rough surface. We can also use the surface fields to calculate reflectivity and absorptivity. The power absorbed in medium 1 is

$$P_a = \frac{1}{2} \int_{V_1} d\overline{r}\, \omega \epsilon_1'' |\overline{E}_1|^2 \tag{4.2.38}$$

Using Maxwell's equations, it follows that

$$P_a = -\frac{1}{2} \mathrm{Re} \int_{V_1} d\overline{r}\, \nabla \cdot \left(\overline{E}_1 \times \overline{H}_1^* \right)$$

Using the divergence theorem and the continuity of tangential fields,

$$P_a = -\frac{1}{2} \mathrm{Re} \int_S ds \hat{n} \cdot \left(\overline{E}_1 \times \overline{H}_1^* \right)$$

$$= -\frac{1}{2}\mathrm{Re}\int_S ds\hat{n}\cdot\left(\overline{E}\times\overline{H}^*\right)$$
$$= -\int_S ds\hat{n}\cdot\overline{S} \tag{4.2.39}$$

In terms of the wavefunction, we then have

$$P_a = \frac{1}{2\eta k}\mathrm{Im}\int_S ds\hat{n}\cdot(\psi\nabla\psi^*)$$
$$= \frac{1}{2\eta k}\mathrm{Im}\int dx\sqrt{1+\left(\frac{df}{dx}\right)^2}\,\hat{n}\cdot(\psi\nabla\psi^*)$$
$$= \frac{1}{2\eta k}\mathrm{Im}\int dx\psi(x)\,u^*(x) \tag{4.2.40}$$

where $u(x) = \sqrt{1+\left(\frac{df}{dx}\right)^2}(\hat{n}\cdot\nabla\psi)_{z=f(x)}$.

The absorptivity, which is equal to emissivity, is

$$a(\theta_i) = e(\theta_i) = \frac{P_a}{P_{inc}} \tag{4.2.41}$$

For the case of tapered incident wave,

$$a(\theta_i) = e(\theta_i) = \frac{\mathrm{Im}\displaystyle\int dx\,\psi(x)u^*(x)}{k\cos\theta_i g\sqrt{\frac{\pi}{2}}\left\{1-\frac{1+2\tan^2\theta_i}{2k^2g^2\cos^2\theta_i}\right\}} \tag{4.2.42}$$

This formula allows the calculation of absorptivity and emissivity in terms of surface fields. The reflectivity can also be calculated using surface fields. The scattered power is

$$P_s = \frac{1}{2}\mathrm{Re}\int_{S_\infty} ds\,\hat{n}\cdot\left(\overline{E}_s\times\overline{H}_s^*\right) = \frac{1}{2}\mathrm{Re}\int_S ds\,\hat{n}\cdot\left(\overline{E}_s\times\overline{H}_s^*\right) \tag{4.2.43}$$

Thus the reflectivity is

$$r(\theta_i) = \frac{P_s}{P_{inc}} \tag{4.2.44}$$

and for the case of tapered wave, we have

$$r(\theta_i) = \frac{\mathrm{Im}\displaystyle\int dx\,\psi_s(x)\,u_s^*(x)}{k\cos\theta_i g\sqrt{\frac{\pi}{2}}\left\{1-\frac{1+2\tan^2\theta_i}{2k^2g^2\cos^2\theta_i}\right\}} \tag{4.2.45}$$

where the surface scattered fields are

$$\psi_s(x) = \psi(x) - \psi_{inc}(x) \tag{4.2.46}$$

$$u_s(x) = u(x) - u_{inc}(x) \tag{4.2.47}$$

On the rough surface, we have, for the case of a tapered incident wave,

$$\psi_{inc}(x) = \psi_{inc}(x, z = f(x)) \tag{4.2.48}$$

$$\begin{aligned} u_{inc}(x) &= \sqrt{1 + \left(\frac{df}{dx}\right)^2}\,[\hat{n} \cdot \nabla\psi_{inc}]_{z=f(x)} \\ &= \psi_{inc}(x)\left\{ik(1+w)\left(-\frac{df}{dx}\sin\theta_i - \cos\theta_i\right)\right. \\ &\quad + 4ik(x\sin\theta_i - z\cos\theta_i)\frac{(x + z\tan\theta_i)}{g^2(kg\cos\theta_i)^2}\left(\tan\theta_i - \frac{df}{dx}\right) \\ &\quad \left. + 2\frac{(x + z\tan\theta_i)}{g^2}\left(\frac{df}{dx} - \tan\theta_i\right)\right\}_{z=f(x)} \end{aligned} \tag{4.2.49}$$

2.3 Impedance Matrix Elements: Numerical Integrations

The rough surface can have fine scale structures. By fine scale structures, we mean irregularities that are smaller than a wavelength and yet have large slopes. They can also have radius of curvature smaller or much smaller than a wavelength. It can become necessary to perform numerical integrations to calculate the impedance matrix elements accurately. The matrix equations are

$$b_m - \sum_{n=1}^{N} A_{mn}u_n - \sum_{n=1}^{N} B_{mn}\psi_n = 0 \tag{4.2.50}$$

$$\sum_{n=1}^{N} A_{mn}^{(1)}u_n + \sum_{n=1}^{N} B_{mn}^{(1)}\psi_n = 0 \tag{4.2.51}$$

using the pulse basis functions and point matching.

For non-self patches, we can perform integration on the integrals as follows. For $m \neq n$,

$$A_{mn} = \int_{x_n - \frac{\Delta x}{2}}^{x_n + \frac{\Delta x}{2}} dx K(x_m, x) \tag{4.2.52}$$

$$A_{mn}^{(1)} = \int_{x_n - \frac{\Delta x}{2}}^{x_n + \frac{\Delta x}{2}} dx K_1(x_m, x) \tag{4.2.53}$$

$$B_{mn} = -\int_{x_n-\frac{\Delta x}{2}}^{x_n+\frac{\Delta x}{2}} dx K_N(x_m, x) \tag{4.2.54}$$

$$B^{(1)}_{mn} = -\int_{x_n-\frac{\Delta x}{2}}^{x_n+\frac{\Delta x}{2}} dx K_{1N}(x_m, x) \tag{4.2.55}$$

For self patches, the formulas for numerical integrations are

$$B_{mm} = \widetilde{B}_{mm} + \frac{1}{2} \tag{4.2.56}$$

$$B^{(1)}_{mm} = \widetilde{B}^{(1)}_{mm} - \frac{1}{2} \tag{4.2.57}$$

where

$$\widetilde{B}_{mm} = -\!\!\!\!\!\!\int_{x_m-\frac{\Delta x}{2}}^{x_m+\frac{\Delta x}{2}} dx \left\{ \frac{ik}{4} \frac{H_1^{(1)}\left(k\sqrt{(x_m-x)^2+(f(x_m)-f(x))^2}\right)}{\sqrt{(x_m-x)^2+(f(x_m)-f(x))^2}} \cdot \left[\frac{df(x)}{dx}(x-x_m)-(f(x)-f(x_m))\right]\right\} \tag{4.2.58}$$

$$\widetilde{B}^{(1)}_{mm} = -\!\!\!\!\!\!\int_{x_m-\frac{\Delta x}{2}}^{x_m+\frac{\Delta x}{2}} dx \left\{ \frac{ik_1}{4} \frac{H_1^{(1)}\left(k_1\sqrt{(x_m-x)^2+(f(x_m)-f(x))^2}\right)}{\sqrt{(x_m-x)^2+(f(x_m)-f(x))^2}} \cdot \left[\frac{df(x)}{dx}(x-x_m)-(f(x)-f(x_m))\right]\right\} \tag{4.2.59}$$

$$A_{mm} = \widetilde{A}_{mm} + \frac{i\Delta x}{4}\left[1+\frac{i2}{\pi}\ln\left(\frac{\gamma k}{4e}\Delta l_m\right)\right] \tag{4.2.60}$$

$$\widetilde{A}_{mm} = \int_{x_m-\frac{\Delta x}{2}}^{x_m+\frac{\Delta x}{2}} dx \left\{ \frac{i}{4} H_0^{(1)}\left(k\sqrt{(x-x_m)^2+(f(x)-f(x_m))^2}\right) - \frac{i}{4}\left[1+i\frac{2}{\pi}\ln\left(\frac{\gamma}{2}k\sqrt{(x-x_m)^2+(f(x)-f(x_m))^2}\right)\right]\right\} \tag{4.2.61}$$

$$A^{(1)}_{mm} = \widetilde{A}^{(1)}_{mm} + \frac{i\Delta x}{4}\left[1+\frac{i2}{\pi}\ln\left(\frac{\gamma k_1}{4e}\Delta l_m\right)\right] \tag{4.2.62}$$

$$\widetilde{A}^{(1)}_{mm} = \int_{x_m-\frac{\Delta x}{2}}^{x_m+\frac{\Delta x}{2}} dx \left\{ \frac{i}{4} H_0^{(1)}\left(k_1\sqrt{(x-x_m)^2+(f(x)-f(x_m))^2}\right) - \frac{i}{4}\left[1+i\frac{2}{\pi}\ln\left(\frac{\gamma}{2}k_1\sqrt{(x-x_m)^2+(f(x)-f(x_m))^2}\right)\right]\right\} \tag{4.2.63}$$

where

$$\Delta l_m = \Delta x \sqrt{1 + (f(x_m))^2} \tag{4.2.64}$$

and $⨍$ denotes principal value, i.e., with a section of very small radius a subtracted from the interval of integration. Numerical integrations are performed over the integrals in (4.2.52)–(4.2.55) and (4.2.58)–(4.2.63) to improve the accuracy of the impedance matrix elements.

2.4 Simulation Results

The MATLAB programs used to generate the simulation results in this section can be found on the World Wide Web at `http://www.emwave.com`.

2.4.1 Gaussian Surface and Comparisons with Analytical Methods

In this section, we simulate scattering from Gaussian random rough surface profiles with rms height h for the Dirichlet problem. The correlation function is also Gaussian with

$$C(x) = \exp\left(-\frac{x^2}{l^2}\right) \tag{4.2.65}$$

In the simulation, $N = 256$ points are used. All distance units are in terms of free space wavelength.

For Gaussian random rough surface and Gaussian correlation function, the power spectrum is, from (9.2.14) of Volume I,

$$W(k_x) = \frac{h^2 l}{2\sqrt{\pi}} \exp\left(-\frac{k_x^2 l^2}{4}\right) \tag{4.2.66}$$

The bistatic scattering coefficient for the incoherent wave from the small perturbation method (SPM) is, from (9.3.40) of Volume I,

$$\sigma(\theta_s) = 4k^3 \cos^2\theta_s \cos\theta_i \frac{h^2 l}{2\sqrt{\pi}} \exp\left[-\frac{(\sin\theta_s - \sin\theta_i)^2 k^2 l^2}{4}\right] \tag{4.2.67}$$

For the Kirchhoff approximation (KA), we first calculate, from (9.4.45) of Volume I,

$$W^{(m)}(k_x) = \frac{1}{2\pi}\int_{-\infty}^{\infty} dx\, e^{-ik_x x} h^{2m} C^m(x) = \frac{h^{2m} l}{2\sqrt{m\pi}} \exp\left[-\frac{k_x^2 l^2}{4m}\right] \tag{4.2.68}$$

Then the bistatic scattering coefficient for the incoherent wave from the Kirchhoff approximation follows from (9.4.47) of Volume I,

$$\sigma(\theta_s) = |1+\cos(\theta_i+\theta_s)|^2 \frac{k^3}{\cos\theta_i} \exp\left[-k^2h^2(\cos\theta_i+\cos\theta_s)^2\right]$$
$$\sum_{m=1}^{\infty} \frac{(k(\cos\theta_i+\cos\theta_s))^{2(m-1)}}{m!} \frac{h^{2m}l}{2\sqrt{m\pi}} \exp\left[-\frac{k^2l^2}{4m}(\sin\theta_s-\sin\theta_i)^2\right] \quad (4.2.69)$$

In Fig. 4.2.1, we show a single realization of rough surface profile with $h = 0.05$, $l = 0.35$, $L = 25.6$, and $\Delta x = 0.1$. Since $l \ll L$, we see many peaks and valleys in the realization. Note that the vertical and horizontal scales are different so that the slopes are exaggerated in the figure. Although the rms height is $h = 0.05$, the maximum and minimum $f(x)$ can go up to 0.1 and -0.1 respectively. Thus the maximum peak to valley separation can be as large as 0.2 for an rms height of 0.05.

We use the spatial tapered wave of (4.1.31). We use an incident angle of $\theta_i = 30°$, and $g = \frac{L}{4}$. We calculate $\sigma(\theta_s)$ for each realization. The average bistatic scattering coefficient $\langle\sigma(\theta_s)\rangle$ is $\sigma(\theta_s)$ averaged over 100 realizations.

Four cases are considered: (a) $h = 0.005$, $l = 0.35$; (b) $h = 0.005$, $l = 0.20$; (c) $h = 0.2$, $l = 1.5$; (d) $h = 0.2$, $l = 0.6$.

In Fig. 4.2.2, a single realization of $\sigma(\theta_s)$ for case (a) is shown. Because of the small rms height, we see a distinct coherent wave peak in the specular direction of $\theta_s = \theta_i$. Since the result is only for one realization, there are angular fluctuations as a result of constructive and destructive interferences as a function of θ_s.

In Figs. 4.2.3–4.2.6, we show the results of $\langle\sigma(\theta_s)\rangle$ for cases (a), (b), (c), and (d), respectively, averaged over 100 realizations. The results of $\langle\sigma(\theta_s)\rangle$ for the incoherent wave based on the small perturbation method and the Kirchhoff approximation are also shown for comparison. We see that after averaging, the plot of $\langle\sigma(\theta_s)\rangle$ is much smoother than that of $\sigma(\theta_s)$ of one realization.

In Fig. 4.2.3, the parameters are $h = 0.05$ and $l = 0.35$. Because of the small height, we see a distinct angular peak in the specular direction of $\theta_s = \theta_i = 30°$. Because of the small slope, $\langle\sigma(\theta_s)\rangle$ decreases rapidly away from the specular direction. In Fig. 4.2.4, the results are for the same h and a smaller l of 0.2. We see that $\langle\sigma(\theta_s)\rangle$ decreases less rapidly with θ_s when compared with Fig. 4.2.3. Note that the results of $\langle\sigma(\theta_s)\rangle$ for SPM and KA are for the incoherent wave only. The results of SPM are in excellent agreement with numerical simulations. However, the results of KA only agree in the vicinity of the specular direction. The results of KA are poor for large θ_s because shadowing effects have been neglected.

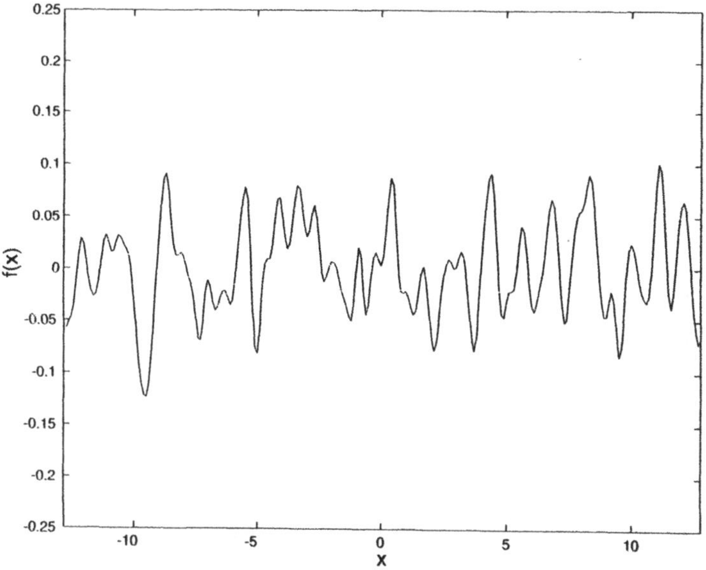

Figure 4.2.1 A single realization of a Gaussian rough surface profile with $h = 0.05$, $l = 0.35$, $L = 25.6$, $N = 256$.

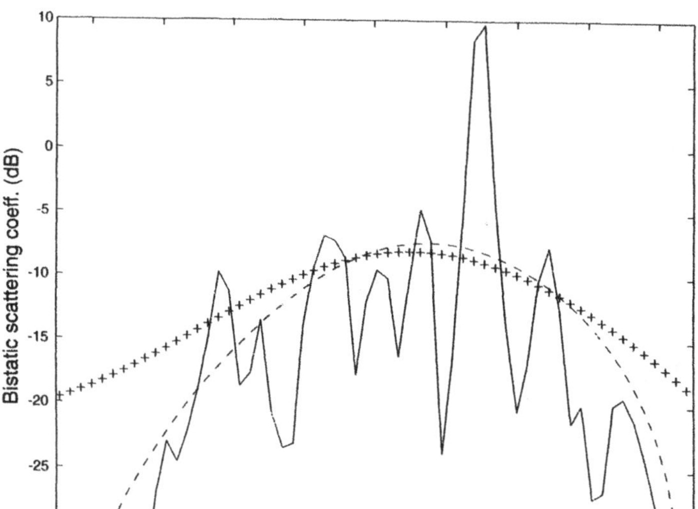

Figure 4.2.2 $N = 256$, $\Delta x = 0.1$, $L = 25.6$, $\theta_i = 30°$, $h = 0.05$, $l = 0.35$, $g = \frac{L}{4}$, and one realization. Solid curve: MoM; dashed curve: SPM; +: KA.

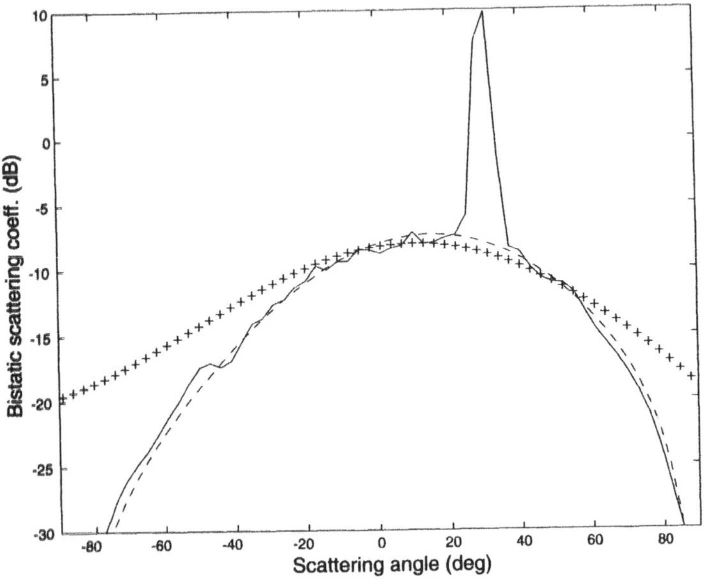

Figure 4.2.3 $N = 256$, $\Delta x = 0.1$, $L = 25.6$, $\theta_i = 30°$, $h = 0.05$, $l = 0.35$, $g = \frac{L}{4}$, and 100 realizations. Solid curve: MoM; dashed curve: SPM; +: KA.

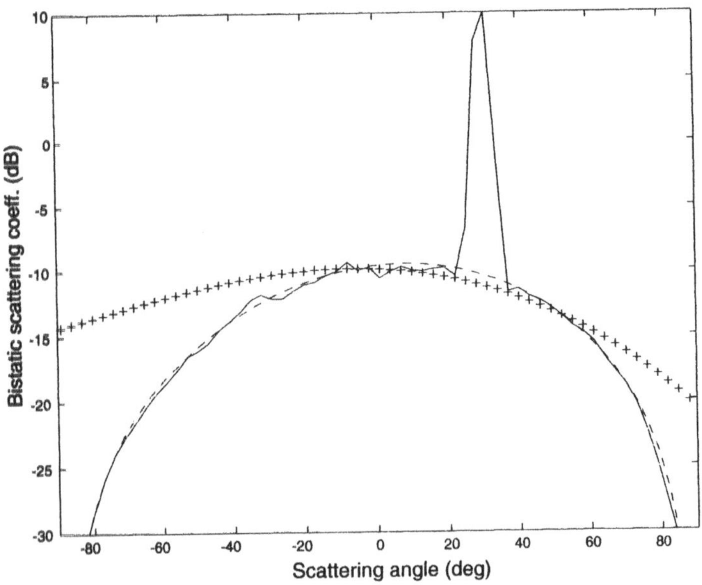

Figure 4.2.4 $N = 256$, $\Delta x = 0.1$, $L = 25.6$, $\theta_i = 30°$, $h = 0.05$, $l = 0.2$, $g = \frac{L}{4}$, and 100 realizations. Solid curve: MoM; dashed curve: SPM; +: KA.

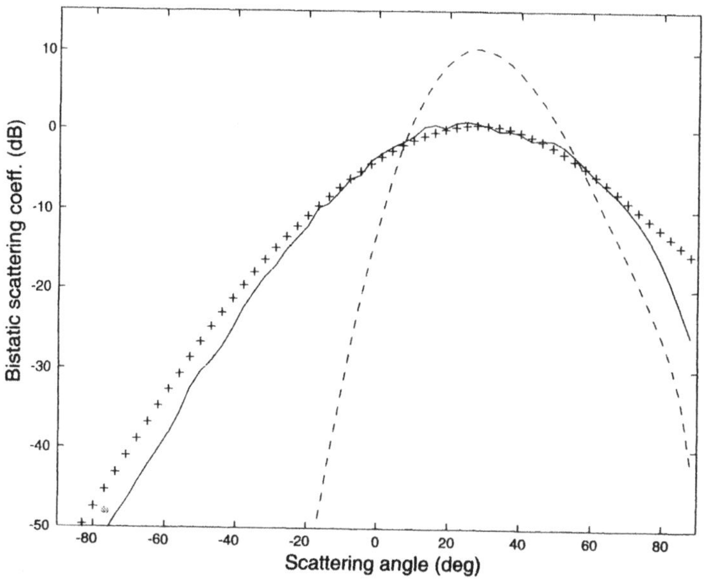

Figure 4.2.5 $N = 256$, $L = 25.6$, $\theta_i = 30°$, $h = 0.2$, $l = 1.5$, $g = \frac{L}{4}$, and 100 realizations. Solid curve: MoM; dashed curve: SPM; +: KA.

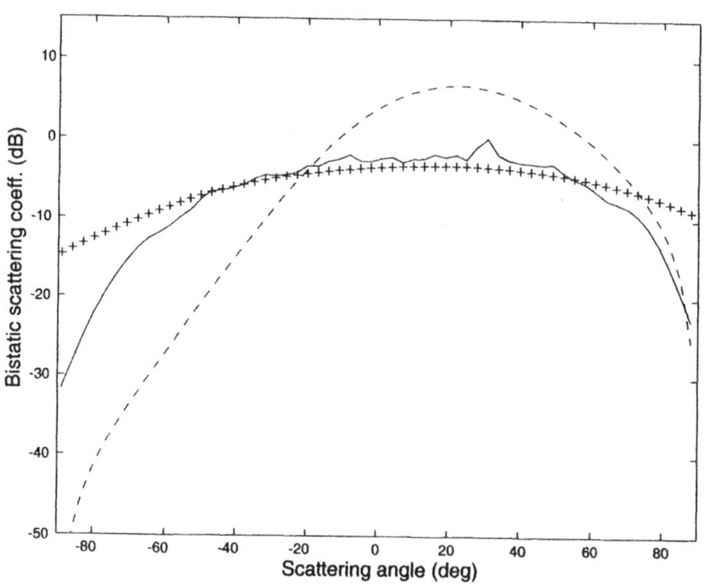

Figure 4.2.6 $N = 256$, $L = 25.6$, $\theta_i = 30°$, $h = 0.2$, $l = 0.6$, $g = \frac{L}{4}$, and 100 realizations. Solid curve: MoM; dashed curve: SPM; +: KA.

In Figs. 4.2.5 and 4.2.6, we show the results of a larger rms height of $h = 0.2$. The coherent specular peak disappears in the numerical simulation. For Fig. 4.2.5, with a large correlation length of $l = 1.5$, there is good agreement between KA and numerical simulations. The agreement is particularly good in the vicinity of the specular direction. This case has a large radius of curvature that accounts for the agreement. However, for this case the fact that $kh = 1.26$ is larger than unity accounts for the poor agreement of SPM with numerical simulations. For Fig. 4.2.6, we use a smaller correlation length of $l = 0.6$. Because of the larger slope, we have larger bistatic scattering return. SPM results are incorrect. KA also gives poor results for larger θ_s because of the neglect of multiple scattering and shadowing.

2.4.2 Dirichlet Case of Gaussian Surface with Ocean Spectrum and Fractal Surface

We next illustrate rough surface scattering from Gaussian surfaces with ocean spectrum and from fractal surfaces. To compare scattering from different surfaces, the equivalent rms height and correlation length are first calculated from the ocean spectrum.

Assuming the wind speed at $z = 5$ is 10 m/s. We can first solve for u_* ($u_* = 0.437198$). The lower and upper limits of the ocean spectrum chosen are $k_L = 100\,\text{m}^{-1}$ and $k_U = 4000\,\text{m}^{-1}$. We then calculate the equivalent rms height as $h = 8.1202 \times 10^{-4}\,\text{m}$ from (4.1.134). Then the rms slope s can be calculated from (4.1.135). The equivalent correlation length is $l = \sqrt{2}h/s$. For this case, the equivalent correlation length is calculated to be $l = 5.0993 \times 10^{-3}\,\text{m}$. Frequency is 19 GHz. The surface length is $L = 25.6\lambda$ and 10 points per wavelength are taken on the rough surface ($N = 256$). For fractal surface, the number of tones (N_f) is 100 and the fractal dimension s is 1.5. Numerical integration is not performed to calculate the impedance matrix elements. The formulas of (4.1.27) and (4.1.28) are used.

For the incident field, we use tapered plane wave at $\theta_i = 50°$ with $g = L/4$. The bistatic scattering coefficients averaged over 100 realizations are plotted as a function of observation angle. In Fig. 4.2.7, we show, for reference, the results for Gaussian surfaces with Gaussian spectrum. Figures. 4.2.8 and 4.2.9 show the cases of Gaussian surfaces with ocean spectrum and of fractal surfaces, respectively. Note that a forward scattering peak appears at $\theta_s = 50°$ (the specular direction). In comparison with Gaussian surfaces with Gaussian correlation function, the bistatic scattering coefficients for the ocean surfaces of Fig. 4.2.8 and the fractal surfaces of Fig. 4.2.9 both show a "dip" near the specular direction. This is because for both cases, the

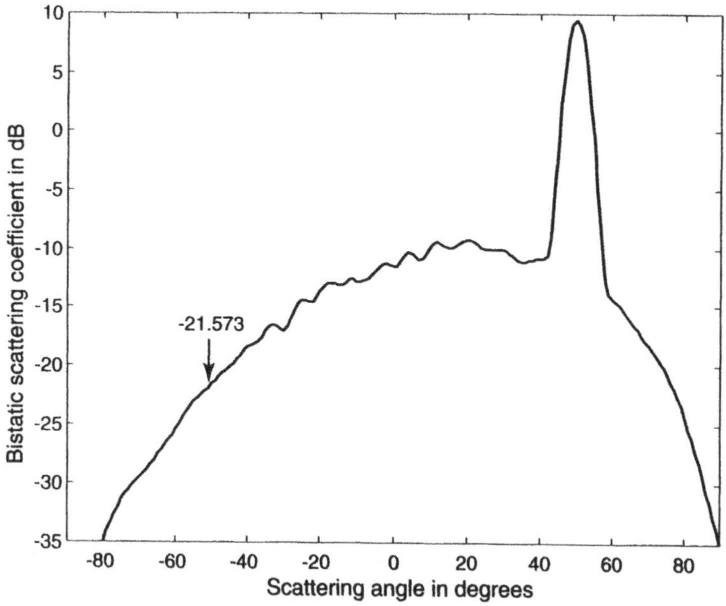

Figure 4.2.7 Bistatic scattering coefficient for Gaussian surfaces with Gaussian spectrum.

rough surface spectrum is bandlimited. The bandlimited spectrum means that there is no spectral component less than k_L to contribute to scattering near the specular direction. The backscattering coefficients for random rough surfaces are at $\theta_s = -50°$. We found that fractal surfaces have the largest backscattering.

2.4.3 Bistatic Scattering for Two Media Problem with Ocean Spectrum

In the two media problem, we use a relative permittivity of ocean ϵ_r of $28.9541 + i36.8430$ for 19 GHz and $13.2444 + i24.5221$ for 37 GHz for the lower medium. After solving u_n and ψ_n, the bistatic scattering coefficient can be calculated by integrating the fields over the rough surface using (4.1.50). Parameters for the incident field and for ocean spectrum are the same as Section 2.4.2. The surface length is $L = 20\lambda$. Results are shown for impedance matrix elements with and without numerical integrations for TE and TM. The numerical integration is done for both self-patch term and near field terms ($(m - n)\,\Delta x \leq \lambda$). For bistatic scattering coefficients, numerical integrations do not seem to result in significant differences.

For the simulations, a discretization of 40 points per wavelength is used for this simulation. The bistatic scattering coefficients averaged over 50 real-

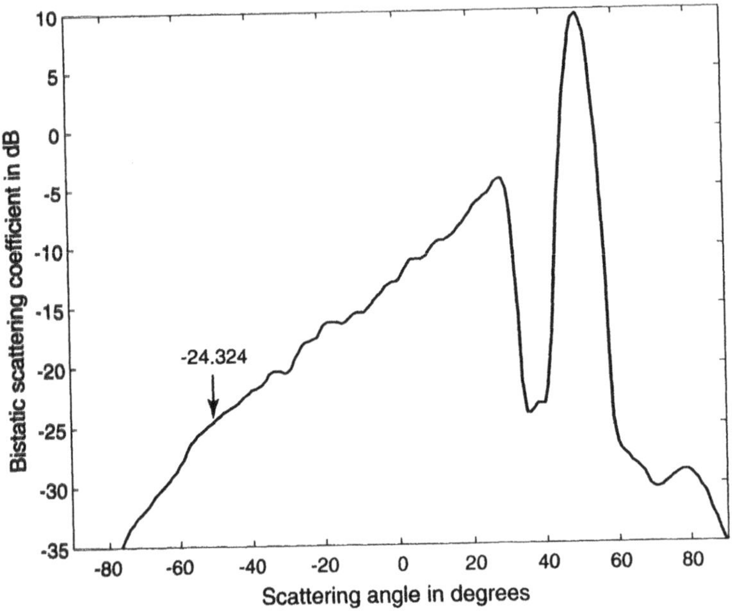

Figure 4.2.8 Bistatic scattering coefficient for Gaussian surfaces with ocean spectrum.

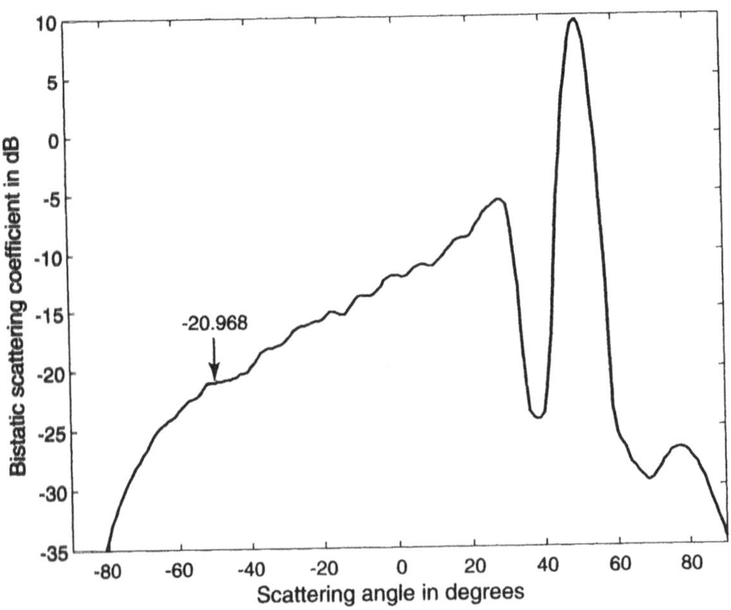

Figure 4.2.9 Bistatic scattering coefficient for fractal surfaces.

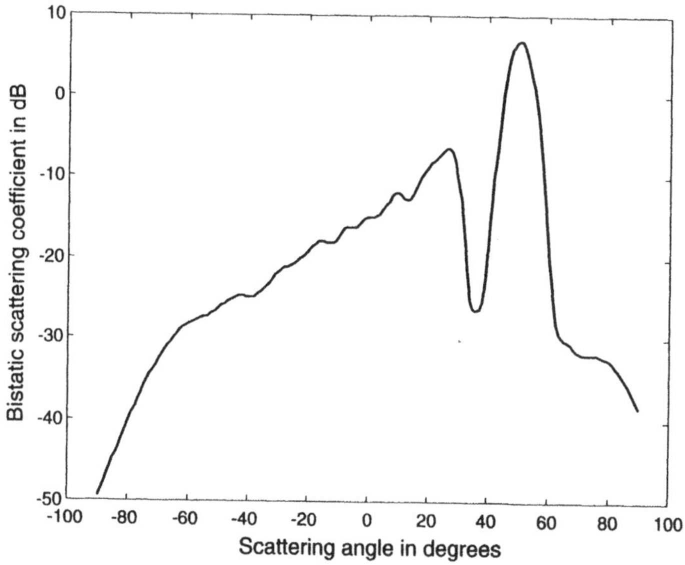

Figure 4.2.10 Bistatic scattering coefficient from ocean surface. TE polarization. Without numerical integration of impedance matrix elements: $\sigma(-50°) = -26.316\,\text{dB}$ and $\sigma(50°) = 7.028\,\text{dB}$. With integration: $\sigma(-50°) = -26.283\,\text{dB}$ and $\sigma(-50°) = 7.043\,\text{dB}$.

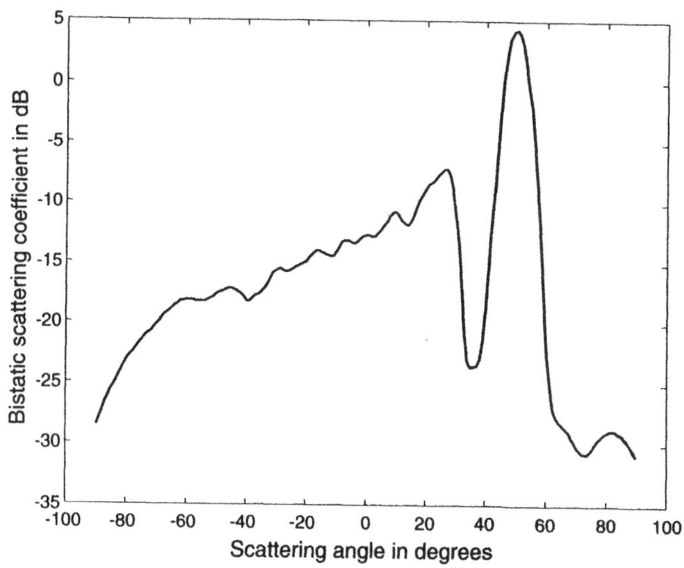

Figure 4.2.11 Bistatic scattering coefficient from ocean surface. TM polarization. Without numerical integration of impedance matrix elements: $\sigma(-50°) = -17.697\,\text{dB}$ and $\sigma(50°) = 4.378\,\text{dB}$. With integration: $\sigma(-50°) = -17.693\,\text{dB}$ and $\sigma(-50°) = 4.349\,\text{dB}$.

izations are plotted as a function of observation angle. Results with and without numerical integrations to obtain matrix elements are plotted in Figs. 4.2.10 and 4.2.11 respectively for TE and TM waves. There is no appreciable difference between the results with and without numerical integrations. Results of backscattering coefficients can have 1.3 dB difference between using 30 or 40 points per wavelength.

3 Topics of Numerical Simulations

In this section, we discuss several aspects in the numerical simulations of rough surface scattering. The topics include periodic boundary condition, MFIE for TE case, and the impedance boundary condition.

3.1 Periodic Boundary Condition

To prevent current discontinuity at $x = \pm L/2$, a tapered incident wave was introduced either in the spatial domain or in the spectral domain. An alternative method is to use the periodic boundary condition. In this case, an incident plane wave is used:

$$\psi_{inc}(\overline{r}) = e^{ik_{ix}x - ik_{iz}z} \tag{4.3.1}$$

To prevent current discontinuity, the surface is assumed to be infinite, that is, x is from $-\infty$ to ∞. Obviously, in numerical methods, we cannot deal with a rough surface of infinite length. To circumvent this problem, a periodic surface with period P is assumed. Thus

$$f(x + P) = f(x) \tag{4.3.2}$$

for the height profile. However, the period P is quite large, $P \gg l$, where l is the correlation length, so that there are many peaks and valleys of the height profile within the period P.

The integral equation is given by (4.1.15) for the Dirichlet boundary condition:

$$\psi_{inc}(\overline{r}') = \int_S ds\ g(\overline{r}, \overline{r}')\hat{n} \cdot \nabla\psi(\overline{r}) \tag{4.3.3}$$

for $\overline{r}$ and $\overline{r}'$ on S. This means that

$$\psi_{inc}(x') = e^{ik_{ix}x' - ik_{iz}z'} = \int_{-\infty}^{\infty} dx\ g(x', z'; x, f(x))u(x) \tag{4.3.4}$$

where

$$u(x) = \sqrt{1 + \left(\frac{df}{dx}\right)^2}\ (\hat{n} \cdot \nabla\psi(\overline{r}))_{z=f(x)} \tag{4.3.5}$$

is the surface unknown.

To solve (4.3.4), we use the Bloch condition as in Chapter 3. Let

$$u(x) = e^{ik_{ix}x} w(x) \tag{4.3.6}$$

and $w(x)$ is a periodic function obeying

$$w(x + P) = w(x) \tag{4.3.7}$$

We next convert the integral equation to be with one period from $-P/2$ to $P/2$. We follow Section 1 of Chapter 3. Let the periodic Green's function be defined by

$$g_p(x', z'; x, z) = \sum_{m=-\infty}^{\infty} g(x', z'; x + mL, z) e^{ik_{ix}mL} \tag{4.3.8}$$

Then the integral equation (4.3.4) becomes

$$e^{ik_{ix}x' - ik_{iz}z'} = \int_{-\frac{P}{2}}^{\frac{P}{2}} dx\; g_p(x', z'; x, f(x)) e^{ik_{ix}x} w(x) \tag{4.3.9}$$

The integral equation in (4.3.9) needs to be solved only over one period. Solution methods were discussed in Chapter 3. The periodic Green's function can be presented in the spectral domain as in Section 1 of Chapter 3. Equation (4.3.9) can be solved using the T-matrix method as in Section 2 of Chapter 3.

Let

$$k_{xm} = k_{ix} + \frac{2\pi m}{P} \tag{4.3.10}$$

$$k_{zm} = \sqrt{k^2 - k_{xm}^2} \tag{4.3.11}$$

be the wave vector components of the mth Floquet mode.

The integral equation becomes

$$\delta_{m0} = \frac{i}{2P} \int_{-\frac{P}{2}}^{\frac{P}{2}} dx\; \frac{e^{-ik_{xm}x + ik_{zm}f(x)}}{k_{zm}} u(x) \tag{4.3.12}$$

where $u(x)$ is as given by (4.3.6). To solve (4.3.12), a Fourier series expansion of $w(x)$ can be made

$$w(x) = \sum_{n=-\infty}^{\infty} w_n e^{\frac{i2n\pi}{P} x} \tag{4.3.13}$$

Let

$$A_{mn} = \frac{i}{2P} \int_{-\frac{P}{2}}^{\frac{P}{2}} dx\; \frac{e^{-ik_{xm}x + ik_{zm}f(x) + ik_{xn}x}}{k_{zm}} \tag{4.3.14}$$

to denote the coupling between the mth and nth Floquet modes, then the matrix equation is

$$\sum_{n=-\infty}^{\infty} A_{mn} w_n = \delta_{m0} \tag{4.3.15}$$

In actual numerical implementation, the number of modes are truncated by keeping all the propagating modes $|k_{xm}| < k$ and a reasonable number of evanescent modes. Suppose we keep modes from $-N$ to M so that we have a total of $N + M + 1$ modes, we then have the standard matrix equation of dimension $M + N + 1$.

$$\sum_{n=-N}^{M} A_{mn} w_n = \delta_{m0} \tag{4.3.16}$$

$m = -N, -N+1, \ldots, -1, 0, 1, \ldots, M$.

Suppose $P = 25.6$ wavelengths. Then

$$\frac{2\pi}{P} = \frac{2\pi}{(25.6)\lambda} = \frac{2\pi}{(25.6)\lambda} = \frac{k}{25.6}$$

Thus

$$k_{xm} = k_{ix} + \frac{mk}{25.6}$$

To illustrate, let $k_{ix} = 0$ (normal incidence). Then

$$k_{xm} = \frac{mk}{25.6}$$

The propagating modes are $m = -25, -24, \ldots, -1, 0, 1, \ldots, 25$, a total of 51 propagating modes. Suppose we put 5 evanescent modes on both sides giving us 61 modes. Thus we have p_n, $n = -30, -29, -28, -27, \ldots, -1, 0, 1, \ldots, 30$. The larger the value of P, the smaller is the mode spacing, and the more modes we have. For example, for $P = 25.6$, we have $60/25.6 = 2.3$ modes per wavelength. This method of mode expansion has the advantage over the method of subsectional basis function, which needs 10 points (i.e., 10 subsectional basis functions) per wavelength. The method of subsectional basis function needs 256 unknowns for the case $P = 25.6\lambda$. On the other hand, the periodic boundary condition method uses 51 modes. However, the disadvantage of the mode expansion method is that the diagonal elements of the matrix equation may not be dominant. As can be seen from (4.3.14), A_{mn} depends on $f(x)$. If $f(x)$ is small, then A_{mn} has large diagonal elements. For the special case when $f(x) = 0$, we have $A_{mn} = \frac{i}{2}\delta_{mn}$ and all off-diagonal elements are zero. However, for large values of $f(x)$, the diagonal elements may not be large compared with the off-diagonal elements. Also for k_{zm} imaginary, depending on how large is $f(x)$, the integrand in (4.3.14) can be exponentially large or exponentially small. Thus some off-diagonal elements can be exponentially large or exponential small, giving an ill-conditioned matrix. Nevertheless, the periodic boundary condition method is quite useful in the small height limit.

After the w_n's are solved, we can calculate the scattered field as follows:

$$\psi_s(x', z') = \sum_{m=-\infty}^{\infty} b_m e^{ik_{xm}x' + ik_{zm}z'} \tag{4.3.17}$$

where

$$b_m = \sum_{n=-\infty}^{\infty} B_{mn} w_n \tag{4.3.18}$$

$$B_{mn} = -\frac{i}{2P} \int_{-\frac{P}{2}}^{\frac{P}{2}} dx \, \frac{e^{-ik_{xm}x - ik_{zm}f(x)}}{k_{zm}} e^{ik_{xn}x} \tag{4.3.19}$$

The incident power per unit area is

$$\overline{S}_{inc} \cdot \hat{z} = -\frac{1}{2\eta} \cos\theta_i \tag{4.3.20}$$

The power per unit area outflowing from the surface is

$$\langle \overline{S}_s \cdot \hat{z} \rangle = \text{Re}\left(\frac{i}{2\omega\mu} \psi_s \frac{\partial \psi_s^*}{\partial z} \right) \tag{4.3.21}$$

If we integrate over a period, then

$$\frac{1}{P} \int_{-\frac{P}{2}}^{\frac{P}{2}} dx \, \langle \overline{S}_s \cdot \hat{z} \rangle = \text{Re}\left(\sum_{m=-\infty}^{\infty} |b_m|^2 \frac{k_{zm}}{2\omega\mu} \right) = \sum_{\substack{m,\ \text{propagating} \\ \text{modes only}}} |b_m|^2 \frac{k_{zm}}{2k\eta} \tag{4.3.22}$$

We only have propagating modes in (4.3.22) because k_{zm} of evanescent modes are imaginary. Next we need to convert the result to bistatic scattering coefficients as a function of scattered angle. Let

$$k_{xm} = k \sin\theta_s \tag{4.3.23}$$

For each discrete m, there corresponds a scattered angle θ_s. Thus we have discrete scattered angles. The spacing of k_{xm} is $2\pi/P$, so that in the limit of large P the discrete angles almost form a continuum. To convert the summation to integration, we note that since the k_{xm}'s are spaced $2\pi/P$ apart we have

$$\Delta k_{xm} = \frac{2\pi}{P} \tag{4.3.24}$$

Thus

$$\sum_{\substack{m \\ \text{propagat.}}} = \frac{P}{2\pi} \sum_{\substack{m \\ \text{propagat.}}} \Delta k_{xm} = \frac{P}{2\pi} \int_{-k}^{k} dk_{xm} = \frac{P}{2\pi} \int_{-\frac{\pi}{2}}^{\frac{\pi}{2}} d\theta_s \; k \cos\theta_s \tag{4.3.25}$$

From (4.3.22) and (4.3.25) and using $k_{zm} = k\cos\theta_s$ we obtain

$$\frac{1}{P} \int_{-\frac{P}{2}}^{\frac{P}{2}} dx \; \langle \overline{S}_s \cdot \hat{z} \rangle = \frac{P}{2\pi} \frac{1}{2\eta} \int_{-\frac{\pi}{2}}^{\frac{\pi}{2}} d\theta_s \; k \cos^2\theta_s |b_m|^2 \tag{4.3.26}$$

Since the incident power per unit area is $\frac{\cos\theta_i}{2\eta}$, the bistatic scattering coefficient is

$$\sigma(\theta_s) = \frac{Pk}{2\pi\cos\theta_i} \cos^2\theta_s |b_m|^2 \tag{4.3.27}$$

In Chapter 5, we shall compare the numerical results of using periodic boundary condition with that of using a large surface length L of many wavelengths.

3.2 MFIE for TE Case of PEC

For scattering by PEC, we have used EFIE as given by (4.1.11) for TE case and MFIE as given by (4.1.151) for TM case. For EFIE of the TE case, the self patch is of order $\mathcal{O}(\Delta x \ln \Delta x)$ as given by (4.1.28) and the non-self patch is of order $\mathcal{O}(\Delta x)$. On the other hand, for MFIE of the TM case, the self patch is equal to 1/2 as represented by (4.1.156). Non-self patch is of order $\mathcal{O}(\Delta x)$. Thus MFIE gives larger diagonal elements for the impedance matrix and that gives a better condition number. In this section, the MFIE for the TE case is derived which results in larger diagonal elements than the EFIE for TE case.

We use (4.1.10) and take normal derivative with respect to $\overline{r}'$ and let $\overline{r}'$ approach the surface

$$\hat{n}' \cdot \nabla' \psi_{inc}(\overline{r}') - \int_S ds [\hat{n}' \cdot \nabla' g(\overline{r}, \overline{r}')][\hat{n} \cdot \nabla\psi(\overline{r})] = \begin{cases} \hat{n}' \cdot \nabla'\psi(\overline{r}') & \overline{r}' \in V_0 \\ 0 & \overline{r}' \in V_1 \end{cases} \tag{4.3.28}$$

The self patch analysis can be performed as in Section 1.5 for the Neumann case. The integral equation becomes, for $\overline{r}$ and $\overline{r}'$ on surface S,

$$\begin{aligned} \hat{n}' \cdot \nabla' \psi_{inc}(\overline{r}') &= \frac{1}{2} \hat{n}' \cdot \nabla'\psi(\overline{r}') + \fint_S ds [\hat{n}' \cdot \nabla' g(\overline{r}, \overline{r}')][\hat{n} \cdot \nabla\psi(\overline{r})] \\ &= \frac{1}{2} \hat{n}' \cdot \nabla'\psi(\overline{r}') + \fint_S ds \left[\hat{n}' \cdot \hat{R}' \frac{\partial g}{\partial R} \right] [\hat{n} \cdot \nabla\psi(\overline{r})] \end{aligned} \tag{4.3.29}$$

where $\displaystyle -\!\!\!\!\!\!\int_S ds$ is the principal value integral which is integration by subtracting out a very small section of radius a about $(x', f(x'))$. Let the surface unknown be

$$V(x) = [\hat{n} \cdot \nabla\psi(\overline{r})]_{z=f(x)} \tag{4.3.30}$$

and

$$V_{inc}(x) = [\hat{n} \cdot \nabla\psi_{inc}(\overline{r})]_{z=f(x)} \tag{4.3.31}$$

Note that the self patch is of order unity. The kernel is also different from the Neumann case because the surface normal $\hat{n}'$ is at the field point $\overline{r}'$ rather than at the source point $\overline{r}$. The integral equation is

$$V_{inc}(x') = \frac{1}{2}V(x') + -\!\!\!\!\!\!\int dx K_D^{(M)}(x', x)V(x) \tag{4.3.32}$$

where

$$\begin{aligned} K_D^{(M)}(x', x) &= \sqrt{1 + (f'(x))^2}\, \hat{n}' \cdot \hat{R}' \frac{\partial g}{\partial R} \\ &= -\frac{ik}{4} \frac{H_1^{(1)}(k\sqrt{(x'-x)^2 + (f(x') - f(x))^2})}{\sqrt{(x'-x)^2 + (f(x') - f(x))^2}} \\ &\quad \sqrt{\frac{1 + (f'(x))^2}{1 + (f'(x'))^2}} \left[-f'(x')(x'-x) + (f(x') - f(x))\right] \end{aligned} \tag{4.3.33}$$

To compute numerical results, we use MoM with pulse basis functions and point matching. To obtain accurate results, particularly for energy conservation, we use numerical integration to calculate near-field impedance matrix elements including the self patch.

The matrix equation is

$$\sum_{n=1}^{N} A_{mn}^{(M)} V_n = b_m \tag{4.3.34}$$

where

$$b_m = V_{inc}(x_m) \tag{4.3.35}$$

$$V_n = V(x_n) \tag{4.3.36}$$

Without numerical integration, the matrix elements are

$$A_{mn}^{(M)} = \begin{cases} \dfrac{1}{2} & \text{for } m = n \\ K_D^{(M)}(x_m, x_n)\Delta x & \text{for } m \neq n \end{cases} \tag{4.3.37a}$$

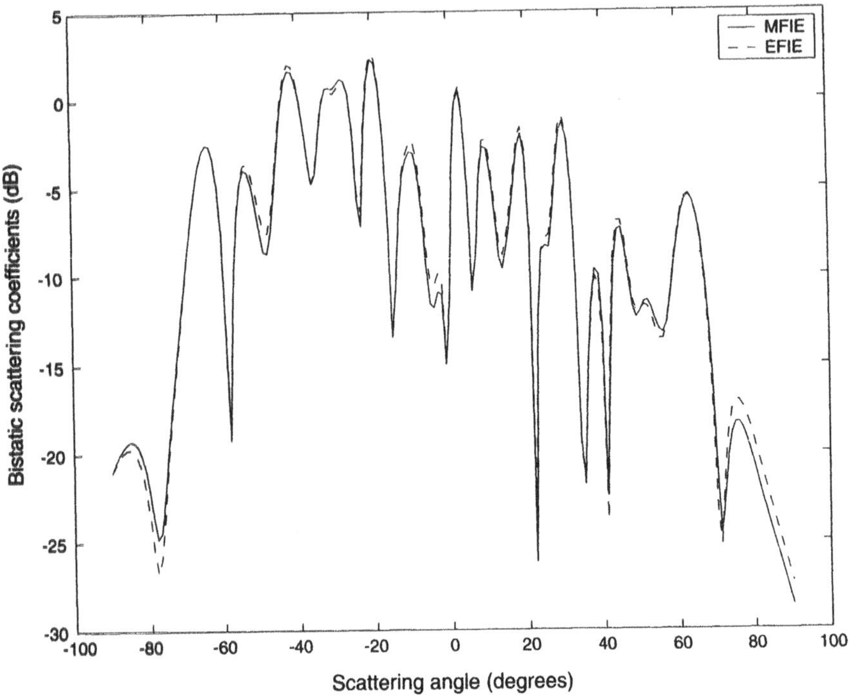

Figure 4.3.1 Comparison of the bistatic scattering coefficient of 1 realization for TE case using MFIE and EFIE at $\theta_i = 30°$ with $h = 0.4\lambda$, $l = 0.2\lambda$, and $L = 25.6\lambda$. 80 points/λ discretization and near-field integration are used in the simulation.

With numerical integration, the matrix elements are

$$A_{mn}^{(M)} = \begin{cases} \dfrac{1}{2} + \displaystyle\fint_{x_m - \frac{\Delta x}{2}}^{x_m + \frac{\Delta x}{2}} dx\ K_D^{(M)}(x_m, x) & \text{for } m = n \\ \displaystyle\int_{x_n - \frac{\Delta x}{2}}^{x_n + \frac{\Delta x}{2}} dx\ K_D^{(M)}(x_m, x) & \text{for } m \neq n \end{cases} \tag{4.3.37b}$$

where $\fint$ denotes principal value integration. In Fig. 4.3.1, we compare the bistatic scattering coefficients for one realization using MFIE and EFIE for TE wave with $\theta_i = 30°$, $h = 0.4\lambda$, $l = 0.2\lambda$, and $L = 25.6\lambda$. Because the correlation length is small, we use 80 points per wavelength. The results for bistatic scattering coefficients are almost indistinguishable in dB scale. Nevertheless there are major differences in energy conservation. The test is to compute

	MFIE		EFIE	
No of points/λ	No. num. int.	With num. int.	No. num. int.	With num. int.
40	0.83603	0.9158	1.0021	1.0023
80	0.9036	0.9530	0.99955	0.9996

Table 4.3.1 Comparison of the energy conservation for TE case using MFIE at $\theta_i = 30°$ with $h = 0.4\,\lambda$, $l = 0.2\,\lambda$, and $L = 25.6\lambda$.

$$P = \int_{-\frac{\pi}{2}}^{\frac{\pi}{2}} d\theta_s\, \sigma(\theta_s)$$

which should be equal to unity. The results are tabulated in Table 4.3.1 for 40 points per wavelength and 80 points per wavelength. We note that energy conservation for MFIE is inadequate while EFIE has good energy conservation. EFIE also obeys energy conservation even for 40 points per wavelength.

3.3 Impedance Boundary Condition

The impedance boundary condition is a common approximation to the two media problem when the lower medium is lossy. Its approximation is similar to that of the transmission line concept when the voltage is equal to the product of the impedance and the current. In terms of wave reflection by a lossy medium, the impedance boundary condition is

$$\overline{E}_{\tan} = \eta_1 \overline{J}_s = \eta_1 \hat{n} \times \overline{H} \tag{4.3.38}$$

where $\overline{E}_{\tan}$ is the tangential electric field and $\overline{J}_s$ is the surface electric current, and $\eta_1 = \sqrt{\mu/\epsilon_1}$ is the wave impedance of the lower medium.

The surface integral equation is, from (4.2.1)

$$\psi_{inc}(\overline{r}') = \frac{1}{2}\psi(\overline{r}') - \fint_S ds \psi(\overline{r})\hat{n}\cdot\nabla g(\overline{r},\overline{r}') + \int_S ds g(\overline{r},\overline{r}')\hat{n}\cdot\nabla\psi(\overline{r}) \tag{4.3.39}$$

The impedance boundary condition can be used to establish a relation between ψ and $\hat{n}\cdot\nabla\psi$.

For TM case, taking the cross product of both sides of (4.3.38) with the surface normal gives

$$-\hat{n} \times \overline{E}_{\tan} = \overline{M}_s = -\eta_1 \hat{n} \times \overline{J}_s \tag{4.3.40}$$

where $\overline{M}_s$ is the surface magnetic current. Using (4.2.19) in (4.3.40), we have

$$-\frac{\hat{y}}{i\omega\epsilon}(\hat{n}\cdot\nabla\psi) = -\eta_1\hat{n}\times(\hat{n}\times\hat{y}\psi) = \eta_1\hat{y}\psi \tag{4.3.41}$$

Hence

$$\hat{n} \cdot \nabla \psi(\bar{r}) = -i \frac{k^2}{k_1} \psi(\bar{r}) \qquad (4.3.42)$$

Then the surface integral equation (4.3.39) becomes

$$\psi_{inc}(\bar{r}') = \frac{1}{2}\psi(\bar{r}') - \int_S ds \psi(\bar{r}) \hat{n} \cdot \nabla g(\bar{r}, \bar{r}') - i \frac{k^2}{k_1} \int_S ds g(\bar{r}, \bar{r}') \psi(\bar{r}) \qquad (4.3.43)$$

For TE case, from (4.2.12), (4.2.13), and (4.3.38),

$$\hat{y}\psi = \eta_1 \left[\frac{-\hat{y}}{i\omega\mu} (\hat{n} \cdot \nabla \psi) \right] \qquad (4.3.44)$$

Thus the impedance boundary condition becomes

$$\frac{1}{ik_1} \hat{n} \cdot \nabla \psi(\bar{r}) = -\psi(\bar{r}) \qquad (4.3.45)$$

which results in the integral equation

$$\psi_{inc}(\bar{r}') = -\frac{1}{2ik_1} [\hat{n}' \cdot \nabla' \psi(\bar{r}')] + \frac{1}{ik_1} \int_S ds [\hat{n} \cdot \nabla g(\bar{r}, \bar{r}')][\hat{n} \cdot \nabla \psi(\bar{r})]$$
$$+ \int_S ds g(\bar{r}, \bar{r}') [\hat{n} \cdot \nabla \psi(\bar{r})] \qquad (4.3.46)$$

When applying MoM to solve these integral equations, matrix elements including numerical integrations can be derived using procedures described in previous sections.

For the case of flat surface, the Fresnel reflection coefficients for TE and TM waves for impedance boundary conditions are respectively

$$R^{\mathrm{TE}} = \frac{k_{iz} - k_1}{k_{iz} + k_1} \qquad (4.3.47)$$

$$R^{\mathrm{TM}} = \frac{k_{iz} - \dfrac{k^2}{k_1}}{k_{iz} + \dfrac{k^2}{k_1}} \qquad (4.3.48)$$

The emissivities are

$$e_{\mathrm{TE}} = 1 - |R^{\mathrm{TE}}|^2 \qquad (4.3.49)$$

$$e_{\mathrm{TM}} = 1 - |R^{\mathrm{TM}}|^2 \qquad (4.3.50)$$

In Section 4, we compare emissivities of rough surface computed using the impedance boundary condition with the dual integral equation method.

4 Microwave Emission of Rough Ocean Surfaces

Numerical simulation of passive microwave remote sensing of ocean surfaces has a strict accuracy requirement. This is because the key output of the simulations is the difference of brightness temperature between a rough surface and a flat surface. Since the difference can be as small as 0.5 K, it is important to be able to simulate the scattering and emission accurately. In this section, we perform accurate simulations of TE and TM waves for ocean surfaces with relative permittivity $28.9541 + i36.8430$ at 19 GHz. Because the ocean permittivity is large, we use up to 80 points per free space wavelength. To ensure accuracy, matrix equation is solved by direct inversion. Conservation of energy is within 0.001 in the simulations. Numerical results are illustrated for rough surfaces with Gaussian spectrum, bandlimited ocean spectrum, and bandlimited fractal surfaces. We show convergence with respect to the density of sampling points and to the upper limit of the bandlimited ocean spectrum. Comparisons are also made with the results of impedance boundary condition. Numerical results indicate that fine discretization is required for ocean surfaces with fine scale roughness.

The rough surfaces are generated as described in Section 1.4. The ocean surfaces are generated using 240 points per wavelength. In the numerical results, an incident angle of 50° is used, and the physical temperature is 283 K. Energy conservation check is done by using $e(\theta_i) + r(\theta_i)$, where $e(\theta_i)$ and $r(\theta_i)$ are given by (4.2.42) and (4.2.45), respectively. Table 4.4.1 presents the emissivity for both TE and TM waves with various surface lengths. Based on energy conservation check, a surface length of 8λ is not large enough to give the correct results. Hence, a surface length of 20λ is used in all subsequent simulations.

To demonstrate the significance of the numerical integrations taken between near field interaction, we show in Table 4.4.2 the significant improvement for TM waves after near field integration was performed for the matrix elements.

In Table 4.4.3, the simulations are performed for one realization of the same profile with a sampling density that various from 10 to 80 points per wavelength. It is shown that the results converge and accuracy requirement is satisfied with 40 points per wavelength discretization for TE polarization and 80 points per wavelength discretization for TM polarization.

In Table 4.4.4, k_U is chosen to be 400, 1000, 4000, and 6000 rads/m, while k_L is fixed at 100 rads/m. The rms heights for these four cases correspond to $kh = 0.31130, 0.32257, 0.32404, 0.32406$, respectively. As k_U increases, there are more fine scale structures on the ocean spectrum. Results show

Polar	Surface length(λ)	Emission	Energy cons. check	Emissivity of flat surface	$T_B(K)$	ΔT_B(rough – flat surface)
TE	8	0.28914	0.99384	0.28728	81.83	0.53
TE	16	0.29738	0.99974	0.28728	84.16	2.86
TE	20	0.29695*	1.00016	0.28728	84.04	2.74*
TM	8	0.57340	1.0006	0.55927	162.27	4.00
TM	16	0.56778	1.00084	0.55927	160.68	2.41
TM	20	0.56823*	1.00096	0.55927	160.81	2.53*

Table 4.4.1 Comparison of emissivities with various surface length for TE and TM polarizations. $k_L = 100\,\text{rads/m}$, $k_U = 4000\,\text{rads/m}$, and 60 points/$\lambda$ discretization. (* represents the correct result.)

Codes with near field int.	Polar	Emissiity	Energy cons. check	Emissivity of flat surface	$T_B(K)$	ΔT_B(rough – flat surface)
YES	TE	0.29695*	1.00016	0.28728	84.04	2.74*
NO	TE	0.29696	0.9997	0.28728	84.04	2.74
YES	TM	0.56823*	1.00096	0.55927	160.81	2.54*
NO	TM	0.56124	0.9968	0.55927	158.83	0.56

Table 4.4.2 Comparison of emissivities with and without near field integration for TE and TM waves. $k_L = 100\,\text{rads/m}$, $k_U = 4000\,\text{rads/m}$, and 60 points/$\lambda$ discretization. (* represents the correct result.)

No. of points/λ	Polar	Emissiity	Energy cons. check	Emissivity of flat surface	$T_B(K)$	ΔT_B(rough – flat surface)
10	TE	0.30400	1.0189	0.28728	86.03	4.73
40	TE	0.29710	1.00050	0.28728	84.08	2.78
60	TE	0.29695	1.00016	0.28728	84.04	2.74
80	TE	0.29686	0.999968	0.28728	84.01	2.71
10	TM	0.56097	0.99511	0.55927	158.75	0.48
40	TM	0.56894	1.00155	0.55927	161.01	2.74
60	TM	0.56849	1.00121	0.55927	160.88	2.61
80	TM	0.56823	1.00096	0.55927	160.81	2.54

Table 4.4.3 Emissivity with various sampling density (# of pts/wavelength) number of points/λ. $k_L = 100\,\text{rads/m}$, and $k_U = 4000\,\text{rads/m}$.

Polar	k_U (rads/m)	Emission	Energy cons. check	Emissivity of flat surface	$T_B(K)$	ΔT_B(rough – flat surface)
TE	400	0.29358	1.00010	0.28728	83.08	1.78
TE	1000	0.29507	1.00006	0.28728	83.50	2.20
TE	4000	0.29695	1.00016	0.28728	84.04	2.74
TE	6000	0.29709	1.00011	0.28728	84.08	2.78
TM	400	0.56250	1.00010	0.55927	159.19	0.91
TM	1000	0.56513	1.00024	0.55927	159.93	1.66
TM	4000	0.56823	1.00096	0.55927	160.81	2.54
TM	6000	0.56862	1.00110	0.55927	160.92	2.65

Table 4.4.4 Emissivity and brightness temperature with various value of k_U. k_L = 100 rads/m.

Polar	k_U (rads/m)	Emission	Energy cons. check	Emissivity of flat surface	$T_B(K)$	ΔT_B(rough – flat surface)
TE	400	0.29299	1.0001	0.28684	82.92	1.74
TE	1000	0.29403	1.0001	0.28684	83.21	2.03
TE	4000	0.29403	1.0001	0.28684	83.21	2.03
TE	6000	0.29386	1.0002	0.28684	83.16	1.99
TM	400	0.56286	1.0001	0.55984	159.29	0.85
TM	1000	0.56557	1.0004	0.55984	160.06	1.62
TM	4000	0.56879	1.0016	0.55984	160.97	2.53
TM	6000	0.56935	1.0018	0.55984	161.13	2.69

Table 4.4.5 Emissivity and brightness temperature with various value of k_U using *impedance boundary conditions*. The parameters used are the same as in Table 4.4.4.

the importance of fine scale structures on the observed brightness temperatures and emissivities. We also list results calculated using the impedance boundary condition in Table 4.4.5. Comparing the results in Table 4.4.5 and Table 4.4.4, we find that good agreement between the results using the impedance boundary condition and the dual integral equations when k_U is small. However, as k_U increases, the differences between the two methods become larger. Since a larger k_U represents a finer scale structure, this indicates that the impedance boundary condition is applicable only when the radius of curvature is not much less than a wavelength.

Tables 4.4.6 and 4.4.7 show the emissivities and brightness temperatures for different kinds of rough surfaces, using both the dual integral equations and the impedance boundary condition. In Table 4.4.6, a Gaussian spectrum

Method	Polar	Emission	Energy cons. check	Emissivity of flat surface	$T_B(K)$	ΔT_B(rough – flat surface)
Dual	TE	0.42639	1.00067	0.28728	120.67	39.37
Dual	TM	0.63463	1.0084	0.55927	179.60	21.33
Impedance	TE	0.40160	1.0007	0.28684	113.65	32.48
Impedance	TM	0.63428	1.011	0.55984	179.50	21.07

Table 4.4.6 Emissivity and brightness temperature for Gaussian rough surface with rms height $h = 0.2\,\lambda$ and correlation length $l = 0.2\,\lambda$.

Method	Polar	Emission	Energy cons. check	Emissivity of flat surface	$T_B(K)$	ΔT_B(rough – flat surface)
Dual	TE	0.32536	0.99974	0.28728	92.08	10.78
Dual	TM	0.60870	1.0147	0.55927	172.26	13.99
Impedance	TE	0.29923	1.0010	0.28684	84.68	3.51
Impedance	TM	0.60536	1.0149	0.55984	171.32	12.88

Table 4.4.7 Emissivity and brightness temperature for fractal surface. $K_0 = k_L = 100$, and $K_0 b^{N_f - 1} = k_U = 4000$.

with rms height $h = 0.2\lambda$ and correlation length $l = 0.2\lambda$ is used the simulations. In Table 4.4.7, bandlimited fractal surfaces are used with the number of tones $N_f = 100$ and fractal dimension $s = 1.5$. The equivalent rms height, $h = 0.051572\lambda$ is the same as chosen in the ocean spectrum with bandlimits $k_L = 100$ rads/m and $k_U = 4000$ rads/m.

5 Waves Scattering from Real-Life Rough Surface Profiles

5.1 Introduction

In modeling random rough surface scattering, the rough surface is usually modeled as a random process $f(x)$. It is characterized by a height probability density function (PDF) and its surface power spectral density. In real life, one is often faced with a problem of estimating the average surface spectrum and PDF from a limited amount of surface profile data. A common procedure is to use a surface spectrum with a power law or a correlation function with an exponential function or a Gaussian correlation function. Thus, a best-fit spectrum has been used in scattering models. However, scattering solutions based on an average or a best-fit surface spectrum has not been rigorously verified. Also, in fitting a correlation function or a spectrum, it is unclear which portion of the correlation function or spectrum in the spatial and spectral domain, respectively, should be fitted to produce the best wave scattering results. Thus, depending on how the correlation function is fitted, several correlation lengths can be obtained. This complicates the matter because scattering results are strongly dependent on correlation lengths. Furthermore, a particular fit may produce a good scattering agreement at one frequency and incident angle, but not at other frequencies and incident angles.

In this section, we illustrate rough surface simulations on real life measured profiles. Three methods are used. In the first method, we use a set of field-measured profiles of a natural surface to calculate the scattering directly for each profile. This will be called the correct result. In the second method, an average surface spectrum is found from the same field-measured profiles of the first method. Because of the limited number of real-life profiles that are measured, the average spectrum still oscillates. Based on this average spectrum that still contains oscillations, a large number of realizations of synthetic surfaces are numerically generated by the spectral method. Scattering is calculated for these synthetic profiles. Then, an average scattering cross section of the generated profiles is found. In the third method, which is a commonly adopted procedure, the average surface spectrum found in the second method is fitted with a power-law function. From the best-fit spectrum, many independent realizations of surfaces are numerically generated. Then, the average scattering cross section is calculated for these realizations. The third method is the most common because an analytic expression is obtained for the power spectrum. Based on this analytic expression, analytical scattering results and numerical simulated scattering results can be calculated. However, we shall show in this section that this method can give poor results.

5.2 Rough Surface Generated by Three Methods

The surfaces generated by methods 1–3 shall be labeled S_1, S_2, and S_3, respectively. The profiles S_1 are the actual measured profiles.

Method 1

In Method 1 we use a set of measured profiles of soil, snow, and rock surfaces. The measured profiles are called S_1. Based on the measured profile $z = f_t(x_n)$, where t denotes true profiles, we compute the average bistatic scattering coefficient. From Table 4.5.1 the number of profiles used for rock, snow, and soil are 20, 35, and 14, respectively. We denote the number of

	Rock	Snow	Soil
Number of samples (N_t)	20	35	14
Surface length (λ)	12.5	34.13	9.6
Number of points N	300	512	96
rms height(λ)	0.4	0.156	0.1

Table 4.5.1 Real-life surface profile data.

measured samples by N_t. Note that because the profiles are actually measured in the field, there can only be a moderate number of profiles for each case. Surface length, rms height, and total number of points are also listed in Table 4.5.1. The sampling rate is at least 10 points per wavelength.

Methods 2 and 3 of the scattering study use numerically generated surface profiles based on the calculated average spectrum.

Synthetic Surfaces

In Methods 2 and 3, realizations of height profiles are generated from spectral density $W(K_n)$ with $K_n = 2\pi n/L$. Methods 2 and 3 use different $W(K_n)$. Using $W(K_n)$, we generate a realization of $F(K_n)$ by the method described in Section 1.4,

$$F(K_n) = \sqrt{2\pi L W(K_n)} \begin{cases} \dfrac{1}{\sqrt{2}}[N(0,1) + iN(0,1)] & n \neq 0, \dfrac{N}{2} \\ N(0,1) & n = 0, \dfrac{N}{2} \end{cases} \tag{4.5.1}$$

where $N(0,1)$ is a sequence of normally distributed random numbers in $[0,1]$ with zero mean and unity standard deviation and N is the total number of sampling points.. From $F(K_n)$, the rough surface profile is given by

$$f(x) = \frac{1}{L} \sum_{n=-N/2}^{N/2-1} F(K_n) \exp(iK_n x) \tag{4.5.2}$$

Method 2: S_2 Surfaces

We compute the spectral density from the true measured profiles $f_t(x_n)$ by taking averages of the N_t profiles as follows:

$$\begin{aligned} \widehat{W}(K_n) &= \frac{L}{2\pi U} \left\langle \left| \frac{1}{N} \sum_{n=0}^{N-1} f_t(x_n) g(x_n) \exp(-j2\pi n/L) \right|^2 \right\rangle \\ &= \frac{L}{2\pi U N_t} \sum_{t=1}^{N_t} \left| \frac{1}{N} \sum_{n=0}^{N-1} f_t(x_n) g(x_n) \exp(-j2\pi n/L) \right|^2 \end{aligned} \tag{4.5.3}$$

where $g(x_n)$ is a space-domain window function, the angular brackets denote the ensemble average, and U is a normalization constant. We then set $W(K_n) = \widehat{W}(K_n)$ in Method 2. Because of the limited number of samples in the true profiles, there are still oscillations in $\widehat{W}(K)$. We then use this oscillatory $W(K) = \widehat{W}(K)$ to generate many profiles to compute the scattering results. However, in spite of the presence of oscillations in $\widehat{W}(K_n)$ we find that the scattering results computed with this method are in good agreement with Method 1.

Method 3: S_3 **Surfaces**

This is a common procedure. The averaged power spectrum $\widehat{W}(K)$ found in (4.5.3) is fitted with a function $W_p(K)$. A power-law spectrum is frequently used. This can be a controversial procedure. For example, different correlation lengths can be obtained, depending on the fitting procedure. In the following we use a weighted least-squares method to fit a calculated spectrum of (4.5.3) with a power-law spectrum

$$W_p(K) = C|K|^{-\alpha} \tag{4.5.4}$$

where C and α are parameters to be determined. They are tabulated in Table 4.5.2. When the power-law spectrum is plotted on a log-log scale, we have a linear function given by

$$\log(W(K)) = -\alpha \log(|K|) + \log(C) \tag{4.5.5}$$

Then we let $W(K) = W_p(K)$ in (4.5.3) to generate independent realizations of synthetic surfaces in Method 3.

5.3 Numerical Results of the Three Methods

In Fig. 4.5.1 a real-life rock surface profile (S_1), a synthetically generated rock surface profile based on the calculated average spectrum (S_2), and a surface based on the best-fit spectrum (S_3) are illustrated. In Figs. 4.5.2 and 4.5.3, we illustrate S_1, S_2, and S_3 for the cases of snow and soil surfaces, respectively. The profiles are shifted vertically for comparison. It can be observed that none of the S_3 surfaces have the appearance of the measured profile. In fact, the S_3 surface of snow has a closer resemblance to a measured soil profile than to the snow profile. The S_2 surfaces of snow and soil have the best resemblance to the real-life profiles. For all three surfaces, S_2 surfaces resemble the real-life surface profiles better than the S_3 surface profiles. Thus, one can foresee (and it will be demonstrated) that the scattering from the S_2 surfaces will have better agreement with the correct scattering results of

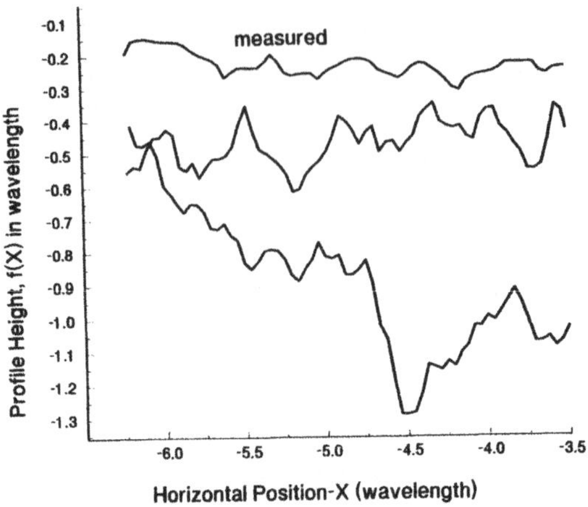

Figure 4.5.1 Comparison of three profiles of rock surfaces: real-life S_1 (top), synthetic surface based on the calculated average spectrum S_2 (center), synthetic surface based on the power-law spectrum S_3 (bottom).

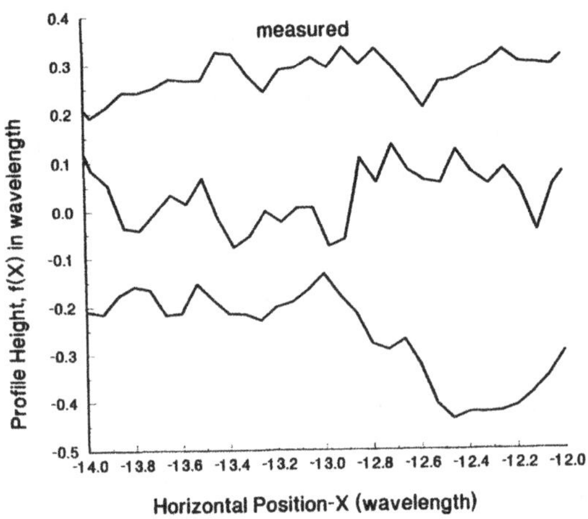

Figure 4.5.2 Comparison of three profiles of snow surfaces: real-life S_1 (top), S_2 (center), S_3 (bottom).

S_1 surfaces. In Table 4.5.2, the rms heights of three type of surfaces S_1, S_2, and S_3 for rock, snow, and soil are compared.

In Fig. 4.5.4 the power spectra of S_2 and S_3 surfaces for rock, soil, and snow are illustrated on a log-log plot. Note that the power spectra of S_1 and S_2 are the same. We use a weighted least-squares best fit to find the parameters C and α in (4.5.5). It should be noted that in fitting the calcu-

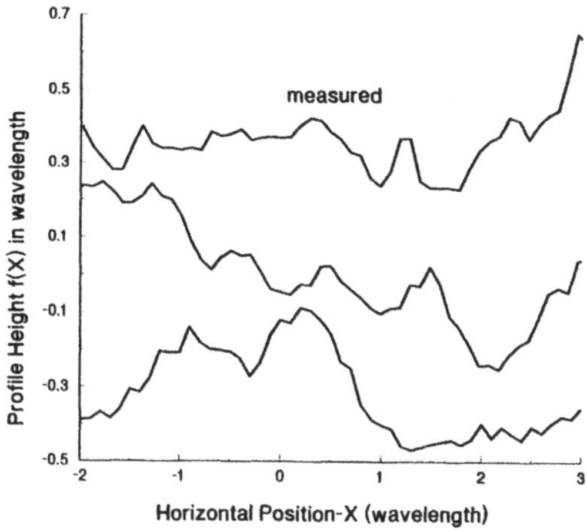

Figure 4.5.3 Comparison of three profiles of soil surfaces: real-life S_1 (top), S_2 (center), S_3 (bottom).

	Rocks (20)	Snow (35)	Soil (14)
S_1(rms)	0.4	0.156	0.10
S_2(rms)	0.3	0.143	0.106
S_3(rms)	0.55	0.145	0.102
α	2.25	1.8	2.5
C	$10^{-2.4}$	$10^{-2.85}$	$10^{-1.4}$

Table 4.5.2 Comparisons of rms heights.

lated power spectrum with a power law, we have disregarded regions that do not have a linear behavior on a log-log plot. In Table 4.5.2, we list the parameters C and α of the power law. Because the power-law function blows up for $K = 0$, we have assigned $W(0) = 0$ in the surface generation. An alternative approach is to taper the power-law spectrum by an exponential function for small values of K_n. However, we find that the alternative approach of an exponential tapering did not produce a noticeable difference in the surface profile or the scattering result. It can be seen from Fig. 4.5.4 that the power law gives a good fit for the cases of rock and soil surface spectra. For a rock surface, a power-law fit overestimates the calculated spectrum for spatial frequencies lower than $3/\lambda$. Therefore, a rock surface generated with the fitted spectrum of Fig. 4.5.4a generates surfaces with larger rms surface heights. From Table 4.5.2, we see that the rms height difference between the real-life surfaces and those generated with a power-law spectrum is 38%. For a soil-surface spectrum, the agreement between the calculated spectrum and the best-fit spectrum is good over a wider range than the rock-surface case.

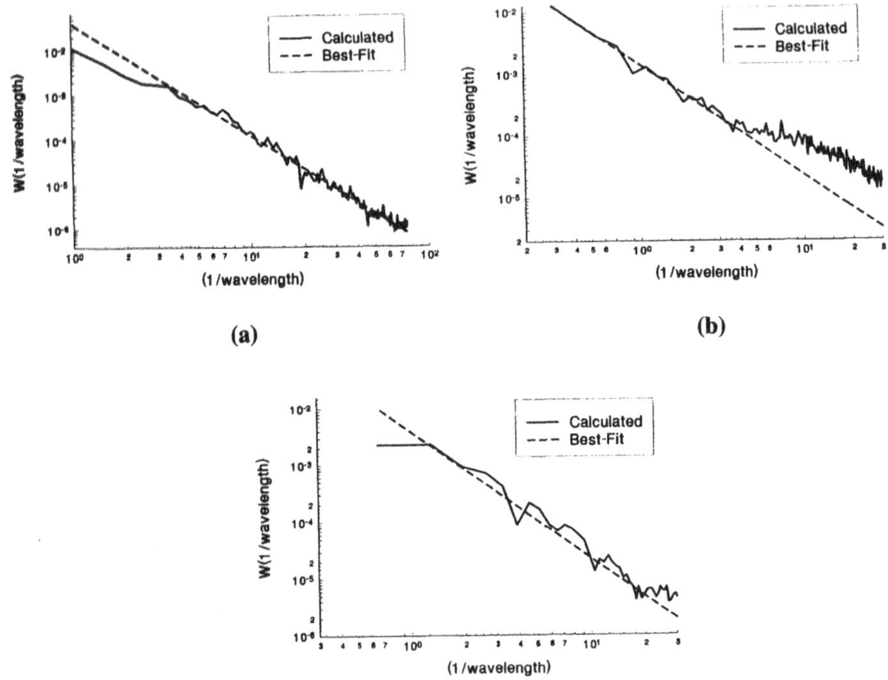

Figure 4.5.4 Comparison of calculated average spectrum and a best-fit spectrum for (a) rock ($\alpha = 2.5$, $C = 10^{-1.4}$), (b) snow ($\alpha = 1.8$, $C = 10^{-2.85}$), and (c) soil ($\alpha = 2.25$, $C = 10^{-2.4}$).

	Rocks (dB)	Snow (dB)	Soil (dB)
S_1	−9.6	−16.0	−15.1
S_2	−9.1	−14.9	−14.8
S_3	−4.0	−19.8	−19.4

Table 4.5.3 Backscattering level comparison −45°.

This is further evident in the rms height comparison, where the agreement is within 1%.

Next, bistatic scattering cross sections from the surfaces of Figs. 4.5.1–4.5.3 are presented in Figs. 4.5.5a–4.5.5c, respectively. The incident angle is 45° for all simulations. Since the study is a profile testing, all scattering simulations are based on perfect electric conductors. In Table 4.5.3, we list the backscattering ($\theta_s = -\theta_i$) levels in decibel scale; the number of surface realizations is found in Table 4.5.2. All simulations satisfy the power con-

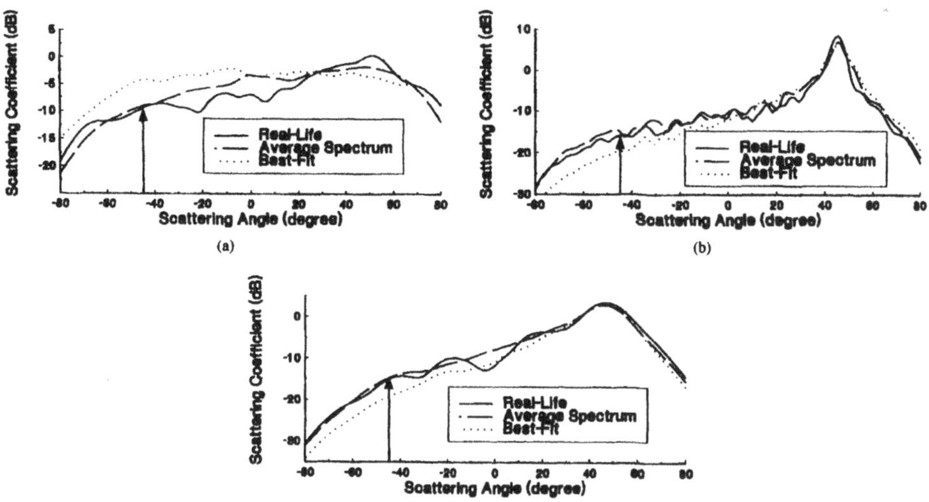

Figure 4.5.5 Comparison of bistatic scattering coefficient from three types of surfaces: (a) rock surfaces of Fig. 4.5.1, (b) snow surfaces of Fig. 4.5.2, and (c) soil surfaces of Fig. 4.5.3.

servation check to less than 1%. From Figs. 4.5.5a–4.5.5c and Table 4.5.3, it is seen that the backscattering from the calculated spectra (S_2 surfaces) are within 1.1 dB of real-life surfaces (S_1) for all cases. Note that because of the limited number of real-life profiles, the calculated power spectrum $\widehat{W}(K_n)$ from S_1 still oscillates (Fig. 4.5.4). This $\widehat{W}(K_n)$ is also used to generate S_2 profiles. Nevertheless, the backscattering from S_2 surfaces all converge to within 1.1 dB of the correct results.

From Table 4.5.3, we can see that the backscattering level of the power-law spectrum that are used to fit the measured spectrum and the real-life surfaces are not in good agreement. For rocks (Fig. 4.5.5a) the best-fit spectrum model scattering result is 5.6 dB above that of the real-life surface. For snow surfaces, the scattering solutions between the real-life and best-fit spectra have a 4 dB difference. The power-law spectrum of soil provides a better fit of the spectrum than snow or rocks over a wider range of spatial frequencies. Based on this observation, it would seem that the soil-surface case will likely give the best scattering agreement. However, it gave a 4.3-dB difference in backscattering. If we take the point of view of the first-order small perturbation model, the average backscattering is proportional to the spatial frequency component $2k \sin\theta_{inc}$ of the spectral density, i.e., $W(2k \sin\theta_{inc})$. Thus SPM says that only the spectral component $2k \sin\theta_{inc}$ in the spectrum matters in backscattering. For the cases considered, $2k \sin\theta_{inc}$ is approximatêly equal to $9/\lambda$ for the calculation. In fitting the spectrum, we paid particular attention to matching up the spectrum at $2k \sin \theta_{inc}$. At this spatial frequency the calculated power spectrum and the power-law best-fit spectrum are within 1 dB for rocks (Fig. 4.5.4a) and soil surfaces (Fig. 4.5.4c). However, although the spectral are matched at $2k \sin \theta_{inc}$, the differences in the backscattering levels were 5.6 and 4.3 dB, respectively. For a snow surface the calculated power spectrum and the power-law best-fit spectrum have a 6-dB difference (Fig. 4.5.4), but they have a 4.0-dB difference in backscattering. These observations suggest that even though the power spectrum can be fitted to have good agreement at $W(2k \sin \theta_{inc})$, the backscattering levels can be significantly different. It is also interesting that S_1 and S_2 backscattering are larger than S_3 backscattering for snow and soil while the opposite is true for rock surfaces.

In the past, Methods 1 and 2 are not adopted because (i) there are only a limited number of surface profiles and (ii) the calculated average surface spectrum based on the limited number of samples still have oscillations. However, we show that even with this limited number of profiles, Methods 1 and 2 give results that are within 1.1 dB of each other. On the other hand, the commonly adopted method of fitting a power spectrum is a controversial procedure because it is unclear which part of the spectrum is to be fitted. Furthermore, the generated surface profiles S_3 produce backscattering results of up to a 6-dB difference from Method 1. Results from this section indicate that further studies of surface profiles and rough surface characterization are needed.

REFERENCES AND ADDITIONAL READINGS

Agnon, Y. and M. Stiassnie (1991), Remote sensing of the roughness of a fractal sea surface, *J. Geophys. Res.*, **96(C7)**, 12773–12779.

Austin, T. R., A. W. England, and G. H. Wakefield (1994), Special problems in the estimation of power-law spectra as applied to topographical modeling, *IEEE Trans. Geosci. Remote Sens.*, **32**, 928–939.

Axline, R. M. and A. K. Fung (1978), Numerical computation of scattering from a perfectly conducting random surface, *IEEE Trans. Antennas Propagat.*, **26(3)**, 482–488.

Berizzi, F., E. D. Dalle Mese, and G. Pinelli (1999), One dimensional fractal model of the sea surface, *IEE Proc.-Radar, Sonar, and Navig.*, **146**, 55–64.

Brown, S. R. and C. H. Scholz (1985), Broad bandwidth study of the topography of natural rock surfaces, *J. Geophys. Res.*, **90(B14)**, 12575–12582.

Chan, C. H., S. H. Lou, L. Tsang, and J. A. Kong (1991), Electromagnetic scattering of waves by random rough surface: a finite-difference time-domain approach, *Microwave Opt. Technol. Lett.*, **4**, 355–359.

Chen, J., T. K. L. Lo, H. Leung, and J. Litva (1996), The use of fractals for modeling EM waves scattering from rough sea surface, *IEEE Trans. Geosci. Remote Sens.*, **34**, 966–972.

Chen, M. F. and A. K. Fung (1988), A numerical study of the regions of validity of the Kirchhoff and small-perturbation rough surface scattering models, *Radio Sci.*, **23**, 163–170.

Devayya, R. and D. J. Wingham (1992), The numerical calculation of rough surface scattering by the conjugate gradient method, *IEEE Trans. Geosci. Remote Sens.*, **30(3)**, 645–648.

Franceschetti, G., A. Iodice, M. Migliaccio, and D. Riccio (1999), Scattering from natural rough surfaces modeled by fractional browning motion two-dimensional process, *IEEE Trans. Antennas Propagat.*, **47(9)**, 1405–1415.

Franceschetti, G., A. Iodice, and D. Riccio (2000), Scattering from dielectric random fractal surfaces via method of moments, *IEEE Trans. Geosci. Remote Sens.*, **38(4)**, 1644–1655.

Fung, A. K. and M. F. Chen (1985), Numerical simulation of scattering from simple and composite random surfaces, *J. Opt. Soc. Am. A*, **2**, 2274–2284.

Garcia, N. and E. Stoll (1984), Monte Carlo calculation for electromagnetic-wave scattering from random rough surfaces, *Phys. Rev. Lett.*, **52**, 1798–1801.

Irisov, V. G. (2000), Azimuthal variations of the microwave radiation from a slightly non-Gaussian sea surface, *Radio Sci.*, **35(1)**, 65–82.

Ishimaru, A. (1978), *Wave Propagation and Scattering in Random Media*, **1** and **2**, Academic Press, New York.

Jaggard, D. L. (1990), On fractal electromagnetics, *Recent Advances in Electromagnetic Theory*, H. N. Kritios and D. L. Jaggard, Eds., Springer-Verlag, Berlin, Germany, 183–223.

Jaggard, D. L. and X. Sun (1990), Scattering from fractally corrugated surfaces, *J. Opt. Soc. Am. A*, **7**, 1131–1139.

Johnson, J. T., R. T. Shin, J. A. Kong, and L. Tsang (1999), A numerical study of ocean polarimetric thermal emission, *IEEE Trans. Geosci. Remote Sens.*, **37(1)**, 8–20.

Lou, S. H. (1991), Application of numerical methods to Monte Carlo simulations of scattering of waves by random rough surfaces, University of Washington, Seattle.

Lou, S. H., L. Tsang, and C. H. Chan (1991), Application of the finite element method to Monte Carlo simulations of scattering of waves by random rough surface: penetrable case, *Waves in Random Media*, **1**, 287–307.

Lou, S. H., L. Tsang, C. H. Chan, and A. Ishimaru (1991), Application of the finite element method to Monte Carlo simulations of scattering of waves by random rough surface with the periodic boundary conditions, *J. Electromag. Waves and Appl.*, **5**, 835–855.

Mandelbrot, B. B. (1983), *The Fractal Geometry of Nature*, W. H. Freeman & Co., New York.

Maradudin, A. A., E. R. Mendez, and T. Michel (1990), Backscattering effects in the elastic scattering of p-polarization light from a large amplitude random grating, *Scattering in Volumes and Surfaces*, M. Nieto-Vesperians and J. C. Dainty, Eds., Elsevier Science Publishers, North-Holland.

Maradudin, A. A., T. Michel, A. R. McGurn, and E. R. Mendez (1990), Enhanced backscattering of light from a random grating, *Ann. Phys.*, **203(2)**, 255–307.

Maystre, D., M. Saillard, and J. Ingers (1991), Scattering by one- or two-dimensional randomly rough surfaces, *Waves in Random Media*, **1**, 143–155.

McGurn, A. R. and A. A. Maradudin (1993), Weak transverse localization of the light scattered incoherently from a one-dimensional random metal surfaces, *J. Optical Soc. Am. B*, **10(3)**, 539–545.

Michel, T. R. and K. A. O'Donnell (1992), Angular correlation functions of amplitudes scattered from a one-dimensional, perfectly conducting rough surface, *J. Opt. Soc. Am. A*, **9(8)**, 1374–1384.

Nieto-Vesperinas, M. and J. M. Soto-Crespo (1987), Monte-Carlo simulations for scattering of electromagnetic waves from perfectly conducting random rough surfaces, *Optics Lett.*, **12**, 979–981.

Papoulis, A. (1984), *Probability, Random Variables, and Stochastic Processes*, McGraw Hill, New York.

Rino, C. L., T. L. Crystal, A. K. Koide, H. D. Ngo, and H. Guthart (1991), Numerical simulations of backscatterer from linear and nonlinear ocean surface realizations, *Radio Sci.*, **26**, 51–71.

Thorsos, E. I. (1988), The validity of the Kirchhoff approximation for rough surface scattering using a Gaussian roughness spectrum, *J. Acous. Soc. Am.*, **83(1)**, 78–92.

Thorsos, E. I. and S. L. Broschat (1995), An investigation of the small slope approximation for scattering from rough surfaces. Part I. Theory, *J. Acous. Soc. Am.*, **97(4)**, 2082–2093.

Thorsos, E. I. and D. R. Jackson (1989), The validity of the perturbation approximation for rough surface scattering using a Gaussian roughness spectrum, *J. Acous. Soc. Am.*, **86(1)**, 261–272.

Thorsos, E. I. and D. R. Jackson (1991), Studies of scattering theory using numerical methods, *Waves in Random Media*, **1(3)**, S165–S190.

Toporkov, J. V., R. S. Awadallah, and G. S. Brown (1999), Issues related to the use of a Gaussian-like incident field for low-grazing-angle scattering, *J. Opt. Soc. Am. A*, **16**, 176–187.

West, J. C. (2000), Integral equation formulation for iterative calculation of scattering from lossy rough surfaces, *IEEE Trans. Geosci. Remote Sens.*, **38(4)**, 1609–1615.

Yueh, S. H. (1997), Modeling of wind direction signal in polarimetric sea surface brightness temperatures, *IEEE Trans. Geosci. Remote Sens.*, **35(6)**, 1400–1418.

Chapter 5

FAST COMPUTATIONAL METHODS FOR SOLVING ROUGH SURFACE SCATTERING PROBLEMS

In this chapter we study the fast computational methods that can be used to numerically simulate random rough surface scattering. Many methods have been used for rough surface simulations. These include the finite difference time domain (FDTD) [Chan et al. 1991] method, finite element method (FEM) [Lou et al. 1991], wavelet method [Lin et al. 1999], multigrid method [Donohue et al. 1998], method of multiple interactions (MOMI) [Kapp and Brown, 1996], recursive T-matrix [Michelssen et al. 1996], and the fast multipole method (FMM) [Jandhyala et al. 1998a,b]. In this book, we choose to illustrate the sparse matrix canonical grid (SMCG) and the physics-based two-grid (PBTG) method in details. These two methods have been applied extensively to 3-D simulations for dielectric surfaces and lossy dielectric surfaces which are practical problems in active and passive microwave remote sensing. In this chapter, we illustrate the SMCG and PBTG for 2-D problems. The 3-D cases will be discussed in Chapter 6. We also discuss the steepest descent fast multipole method (SDFMM), MOMI, and the combination of PBTG and SDFMM. Numerical results presented in this chapter are limited to cases when the results were first published using computer resources available at the time.

1 Banded Matrix Canonical Grid Method for Two-Dimensional Scattering for PEC Case

1.1 Introduction

We have illustrated the surface integral equation method. The integral equation is converted to a matrix equation by the method of moments, and the resulting equation is solved with a full matrix inversion. Surface lengths of the order of 40 to 100 wavelengths have been used in Chapter 4. Many practical problems such as near-grazing incidence or rough surfaces with large correlation lengths are considered large-scale rough surface problems. For such large-scale rough surface problems, a much larger surface length is required and a more efficient method is needed.

A banded matrix iterative approach (BMIA) has been applied to rough surface scattering. In the BMIA, the original full matrix equation is decomposed into (a) a banded matrix which represents strong interaction and (b) the remainder of the full matrix which represents the weak interaction part. An iterative approach is then adopted. The size of the neighborhood varies depending on the rough surface statistics and the incident angle. It is always much less than the required surface length and is also much less than the tapering parameter g.

Improvements to the BMIA method are further made by using the flat surface as a canonical grid (CAG or CG). We call the method BMIA/CAG. For two points outside the neighborhood, the Green's function connecting the two points on the rough surface is close to that of a flat surface. Thus in this case, the flat surface provides the CAG for distances larger than neighborhood distance. A Taylor expansion is carried out for the Green's function outside the neighborhood distance. The impedance matrix from the Green's function is decomposed into a sum of the banded matrix and the Taylor expanded flat surface impedance matrix that replaces the weak matrix from the BMIA method. The advantage of the canonical grid (flat surface in this case) is that the product of the Taylor expanded flat surface matrix and the surface current column vector can be computed by the fast Fourier transform (FFT). The BMIA/CAG is much faster than the customary full matrix inversion. When the method was first used in 1994 [Tsang et al. 1994, 1995], we studied rough surfaces with lengths up to 2500 wavelengths with 25,000 surface unknowns. The method was also applied to the case of near grazing incidence. We also make comparisons between using a large surface length and the method of using a periodic boundary condition which emulates a large surface length.

Other methods known as the adaptive integral method [Belszynski et al. 1994; Anastassiu et al. 1998] and the pre-corrected FFT method [Phillips and White, 1997] are similar in concept to the BMIA/CAG.

1.2 Formulation and Computational Procedure

Consider a tapered plane wave $\psi_{inc}(x, z)$ impinging upon a one-dimensional (1-D) rough surface with a random height profile $z = f(x)$.

$$\psi_{inc}(\overline{r}) = e^{i\overline{k}\cdot\overline{r}[1+w(\overline{r})]}e^{-(x+z\tan\theta_{inc})^2/g^2} \tag{5.1.1}$$

where w and g are as given in Chapter 4. Applying the boundary condition that $\psi(\overline{r}) = 0$ on the rough surface, the Fredholm integral equation of the first kind can be formed. For a point $\overline{r} = x\hat{x} + f(x)\hat{z}$ on the rough surface, the integral equation for the Dirichlet boundary condition is

$$0 = \psi_{inc}(\overline{r}) - \int_{-\infty}^{\infty} dx' G(\overline{r}, \overline{r}') u(x') \tag{5.1.2}$$

where

$$u(x') = \sqrt{1 + (df(x')/dx')^2}\frac{\partial\psi(\overline{r}')}{\partial n'}$$

is the surface unknown to be calculated. Let $x_d = |x - x'|$ be the separation distance between the field point and the source point on the rough surface.

We select a neighborhood distance within which there is a strong interaction. In numerical simulations, r_d is an adjustable parameter and $r_d \gg h$. Then, (5.1.2) can be put into the form

$$\int_{-\infty}^{+\infty} dx' \frac{i}{4} H_0^{(1)} \left(k\sqrt{x_d^2 + (f(x) - f(x'))^2} \right) u(x') U(r_d - x_d)$$
$$+ \int_{-\infty}^{+\infty} dx' \frac{i}{4} H_0^{(1)} \left(k\sqrt{x_d^2 + (f(x) - f(x'))^2} \right) u(x') U(x_d - r_d)$$
$$= \psi_{inc}(x, f(x)) = b(x) \tag{5.1.3}$$

with U as the Heaviside step function. Thus the first term in (5.1.3) represents strong interaction, and the second term represents weak interaction. The strong interaction term corresponds to a banded matrix (a sparse matrix for the 3-D scattering case), and the weak interaction term is a dense matrix. The weak interaction matrix is the computationally intensive part because of the large number of pairs of points that are not within the neighborhood of each other. We then rewrite the weak interaction matrix by using Taylor series expansion to translate it to a flat surface which acts as the canonical grid. Since in the weak interaction term, we have $x_d > r_d \gg h$, we obtain $x_d \gg |f(x) - f(x')|$. Thus a Taylor expansion can be made on Green's function. Let $z_d = |f(x) - f(x')|$.

From Chapter 4, the impedance matrix is $Z_{mn} = A_{mn}$, where

$$A_{mn} = \int_{x_n - \frac{\Delta x}{2}}^{x_n + \frac{\Delta x}{2}} dx' \frac{i}{4} H_0^{(1)} \left(k\sqrt{(x_m - x')^2 + (f(x_m) - f(x'))^2} \right) \tag{5.1.4}$$

$$\sum_{n=1}^{N} Z_{mn} u_n = b_m \tag{5.1.5}$$

Decompose into strong part and weak part.

$$\sum_{n=1}^{N} Z_{mn}^{(s)} u_n + \sum_{n=1}^{N} Z_{mn}^{(w)} u_n = b_m \tag{5.1.6}$$

where

$$Z_{mn}^{(s)} = \begin{cases} A_{mn} & \text{for } |m-n| \le b_w \\ 0 & \text{for } |m-n| \ge b_w \end{cases} \tag{5.1.7}$$

$$Z_{mn}^{(w)} = \begin{cases} 0 & \text{for } |m-n| \le b_w \\ A_{mn} & \text{for } |m-n| \ge b_w \end{cases} \tag{5.1.8}$$

where $b_w = r_d/\Delta x$. Without loss of generality, we can make b_w an integer, e.g., if $r_d = 2\lambda$, $\Delta x = \lambda/10$, then $b_w = 20$. Since $Z_{mn}^{(w)}$ is in the non-near field, we can calculate the matrix elements without numerical integration.

$$Z_{mn}^{(w)} = \begin{cases} 0 & \text{for } |m-n| \le b_w \\ \frac{i}{4} H_0^{(1)} \left(k\sqrt{(x_m - x_n)^2 + (f(x_m) - f(x_n))^2} \right) \Delta x & \text{for } |m-n| \ge b_w \end{cases} \tag{5.1.9}$$

Let

$$Z_{mn}^{(w)} = Z_{mn}^{(w)(0)} + \left(Z_{mn}^{(w)} - Z_{mn}^{(w)(0)} \right) \tag{5.1.10}$$

$$Z_{mn}^{(w)(0)} = \begin{cases} 0 & \text{for } |m-n| \le b_w \\ \frac{i}{4} H_0^{(1)} \left(k\sqrt{x_d^2} \right) \Delta x & \text{for } |m-n| \ge b_w \end{cases} \tag{5.1.11}$$

where

$$x_d = |m-n|\Delta x \tag{5.1.12}$$

Note that $Z_{mn}^{(w)(0)}$ is translational invariant.

The matrix equation is

$$\overline{\overline{Z}}\,\overline{u} = \overline{b} \tag{5.1.13}$$

so that

$$\left[\overline{\overline{Z}}^{(s)} + \overline{\overline{Z}}^{(w)} \right] \overline{u} = \overline{b} \tag{5.1.14}$$

We rewrite (5.1.14) as

$$\left[\overline{\overline{Z}}^{(s)} + \overline{\overline{Z}}^{(w)(0)} \right] \overline{u} = \overline{b} - \left(\overline{\overline{Z}}^{(w)} - \overline{\overline{Z}}^{(w)(0)} \right) \overline{u} \tag{5.1.15}$$

Note that the sparse matrix and the flat surface impedance matrix are kept on the left-hand side of (5.1.15). We discuss three ways of iteration listed below as (A), (B), and (C).

(A) Iteration Based on Strong Matrix and Zeroth Order Weak Matrix

The zeroth order solution $\overline{u}^{(0)}$ obeys the equation

$$\left[\overline{\overline{Z}}^{(s)} + \overline{\overline{Z}}^{(w)(0)} \right] \overline{u}^{(0)} = \overline{b} \tag{5.1.16}$$

$$\overline{\overline{Z}}^{(0)} \overline{u}^{(0)} = \overline{b} \tag{5.1.17}$$

where the zeroth order impedance matrix $\overline{\overline{Z}}^{(0)}$ is defined as

$$\overline{\overline{Z}}^{(0)} = \overline{\overline{Z}}^{(s)} + \overline{\overline{Z}}^{(w)(0)} \tag{5.1.18}$$

We can apply the conjugate gradient method to solve $\overline{\overline{Z}}^{(0)}\overline{u}^{(0)} = \overline{b}$. Note that $\overline{\overline{Z}}^{(s)}$ is a banded matrix. It only has values near the diagonal. The product of matrix and column vector $\overline{\overline{Z}}^{(s)}\overline{u}$ requires $\mathcal{O}(Nb_w)$ iterations. The product of matrix and column vector $\overline{\overline{Z}}^{(w)(0)\dagger}\overline{u}$ requires $\mathcal{O}(N\log_2 N)$ iterations. Thus in implementing conjugate gradient method, the number of operations is $\mathcal{O}(Nb_w) + \mathcal{O}(N\log_2 N)$ per iteration. Using (5.1.15), higher order solutions can be obtained.

The nth order solution $\overline{u}^{(n)}$ is given by

$$\overline{\overline{Z}}^{(0)}\overline{u}^{(n)} = \overline{b}^{(n)} \tag{5.1.19}$$

where the updating of the right hand side is

$$\overline{b}^{(n)} = \overline{b} - \left(\overline{\overline{Z}}^{(w)} - \overline{\overline{Z}}^{(w)(0)}\right)\overline{u}^{(n-1)} \tag{5.1.20}$$

In (5.1.19) and (5.1.20), $n \geq 1$. We then rewrite the weak interaction matrix by using Taylor series expansion to translate to a flat surface which acts as the canonical grid. Let N_T be the number of terms in the Taylor series expansion. Let $z_d = |f(x) - f(x')|$. Define

$$y_m = \left\{\left(\overline{\overline{Z}}^{(w)} - \overline{\overline{Z}}^{(w)(0)}\right)\overline{u}\right\}_m \tag{5.1.21}$$

Then

$$\begin{aligned} y_m &= \sum_{n=1}^{N}\left[\frac{i}{4}H_0^{(1)}\left(k\sqrt{x_d^2 + z_d^2}\right) - \frac{i}{4}H_0^{(1)}(kx_d)\right]u_n \\ &= \sum_{n=1}^{N}\sum_{l=1}^{N_T} a_l(x_d)\left(\frac{z_d^2}{x_d^2}\right)^l u_n \end{aligned} \tag{5.1.22}$$

where $a_l(x_d)$ represents the lth order term in Taylor series expansion and

$$x_d = |m - n|\Delta x \tag{5.1.23}$$

$$z_d = f(x_m) - f(x_n) \tag{5.1.24}$$

Then

$$\frac{i}{4}H_0^{(1)}\left(k\sqrt{x_d^2 + z_d^2}\right) = \sum_{n=1}^{N_T} a_l(x_d)\left(\frac{z_d^2}{x_d^2}\right)^l \tag{5.1.25}$$

For the first three terms,

$$a_1(x_d) = -\frac{i}{4}H_1^{(1)}(kx_d)\frac{kx_d}{2} \tag{5.1.26}$$

$$a_2(x_d) = -\frac{i}{4}\frac{k^2x_d^2}{8}H_0^{(1)}(kx_d) + \frac{i}{4}\frac{kx_d}{4}H_1^{(1)}(kx_d) \tag{5.1.27}$$

$$a_3(x_d) = -\frac{i}{4}\frac{k^2x_d^2}{12}H_0^{(1)}(kx_d) + \frac{i}{4}\frac{k^3x_d^3}{48}H_1^{(1)}(kx_d) - \frac{i}{4}\frac{kx_d}{6}H_1^{(1)}(kx_d) \tag{5.1.28}$$

If we retain three terms in the Taylor expansion ($N_T = 3$),

$$\begin{aligned}
y_m &= \sum_{n=1}^{N}\left(Z_{mn}^{(w)} - Z_{mn}^{(w)(0)}\right)u_n \\
&= \sum_{n=1}^{N} a_1(x_d)\frac{z_d^2}{x_d^2}u_n + \sum_{n=1}^{N} a_2(x_d)\frac{z_d^4}{x_d^4}u_n + \sum_{n=1}^{N} a_3(x_d)\frac{z_d^6}{x_d^6}u_n \\
&= (f(x_m))^2\sum_{n=1}^{N}\frac{a_1(x_d)}{x_d^2}u_n - 2f(x_m)\sum_{n=1}^{N}\frac{a_1(x_d)}{x_d^2}f(x_n)u_n \\
&\quad + \sum_{n=1}^{N}\frac{a_1(x_d)}{x_d^2}(f(x_n))^2u_n \\
&\quad + (f(x_m))^4\sum_{n=1}^{N}\frac{a_2(x_d)}{x_d^4}u_n - 4(f(x_m))^3\sum_{n=1}^{N}\frac{a_2(x_d)}{x_d^4}f(x_n)u_n \\
&\quad + 6(f(x_m))^2\sum_{n=1}^{N}\frac{a_2(x_d)}{x_d^4}f(x_n)^2u_n - 4f(x_m)\sum_{n=1}^{N}\frac{a_2(x_d)}{x_d^4}(f(x_n))^3u_n \\
&\quad + \sum_{n=1}^{N}\frac{a_2(x_d)}{x_d^4}(f(x_n))^4u_n \\
&\quad + (f(x_m))^6\sum_{n=1}^{N}\frac{a_3(x_d)}{x_d^6}u_n - 6(f(x_m))^5\sum_{n=1}^{N}\frac{a_3(x_d)}{x_d^6}f(x_n)u_n \\
&\quad + 15(f(x_m))^4\sum_{n=1}^{N}\frac{a_3(x_d)}{x_d^6}(f(x_n))^2u_n \\
&\quad - 20(f(x_m))^3\sum_{n=1}^{N}\frac{a_3(x_d)}{x_d^6}(f(x_n))^3u_n \\
&\quad + 15(f(x_m))^2\sum_{n=1}^{N}\frac{a_3(x_d)}{x_d^6}(f(x_n))^4u_n - 6f(x_m)\sum_{n=1}^{N}\frac{a_3(x_d)}{x_d^6}(f(x_n))^5u_n
\end{aligned} \tag{5.1.29}$$

$$+\sum_{n=1}^{n}\frac{a_3(x_d)}{x_d^6}(f(x_n))^6u_n \tag{5.1.30}$$

We make use of FFT in the calculation of y_m in (5.1.30). For example, for the second term in (5.1.30), $-2(f(x_m))\sum_{n=1}^{N}\frac{a_1(x_d)}{x_d^2}f(x_n)u_n$, we calculate in the following manner:

(1) pre-multiply u_n by $f(x_n)$ to get $f(x_n)u_n = w_n$

(2) calculate $\displaystyle\sum_{n=1}^{N}\frac{a_1(x_d)}{x_d^2}w_n$ by FFT

(3) post-multiply by $-2f(x_m)$

In the BMIA/CAG method, the $\overline{\overline{Z}}^{(w)}$ is decomposed into a sum represented by the Taylor series expansion. Thus

$$\overline{\overline{Z}}^{(w)} = \sum_{m=0}^{M}\overline{\overline{Z}}_m^{(w)} \tag{5.1.31}$$

The $m = 0$ term corresponds to that of a flat surface. The form of $\overline{\overline{Z}}_m^{(w)}$ is such that it consists of terms that are products of a diagonal matrix $\overline{\overline{T}}_r$, a translationally invariant matrix $\overline{\overline{Z}}_d$, and a diagonal matrix $\overline{\overline{T}}_s$:

$$\overline{\overline{T}}_r\overline{\overline{Z}}_d\overline{\overline{T}}_s \tag{5.1.32}$$

where $\overline{\overline{T}}_s$ is a function of the coordinates x' of the scattering source, while $\overline{\overline{T}}_r$ is a function of the coordinates x of the field.

(B) Iteration Based on Updating the Right-Hand Side

Let $\overline{X}^{(0)}$ and $\overline{X}^{(n)}$ represent the zeroth-order solution and the nth-order solution, respectively. They obey the equations

$$\overline{\overline{Z}}^{(s)}\overline{X}^{(0)} = \overline{C} \tag{5.1.33}$$

$$\overline{\overline{Z}}^{(s)}\overline{X}^{(n+1)} = \overline{C}^{(n)} \tag{5.1.34}$$

where $\overline{C}^{(n)}$ represents the updated right-hand side with

$$\overline{C}^{(0)} = \overline{C}$$

$$\overline{C}^{(n)} = \overline{C} - \overline{\overline{Z}}^{(w)}\overline{X}^{(n)} = \overline{C} - \sum_{m=0}^{M}\overline{\overline{Z}}_m^{(w)}\overline{X}^{(n)} \tag{5.1.35}$$

Note that for this case of iteration, only $\overline{\overline{Z}}^{(s)}$ is kept in the left-hand side of (5.1.34).

A residual $\overline{R}^{(n)}$ can be defined as follows such that its norm provides the stopping criterion for the iterative procedure:

$$\overline{R}^{(n)} = -\overline{\overline{Z}}\,\overline{X}^{(n)} + \overline{C} = -(\overline{\overline{Z}}^{(s)} + \overline{\overline{Z}}^{(w)})\overline{X}^{(n)} + \overline{C} \tag{5.1.36}$$

where the normalized L-2 norm is defined as $\sqrt{\|\overline{R}^{(n)}\|/\|\overline{C}\|} \times 100\%$. From (5.1.33)–(5.1.36), it follows that

$$\overline{R}^{(0)} = -\overline{\overline{Z}}^{(s)}\overline{X}^{(0)} - \overline{\overline{Z}}^{(w)}\overline{X}^{(0)} + \overline{C} = -\overline{\overline{Z}}^{(w)}\overline{X}^{(0)} = -\sum_{m=0}^{M} \overline{\overline{Z}}_m^{(w)}\overline{X}^{(0)} \tag{5.1.37}$$

and

$$\overline{R}^{(n+1)} = -\overline{\overline{Z}}^{(s)}\overline{X}^{(n+1)} - \overline{\overline{Z}}^{(w)}\overline{X}^{(n+1)} + \overline{C} = \overline{C}^{(n+1)} - \overline{C}^{(n)} \tag{5.1.38}$$

Thus the residual vector can be computed readily from the updated right-hand sides. In the numerical results illustrated in this section, the stopping criterion of the iterative solution is set at 0.1%.

Computational Complexity

For the TE case the matrix is symmetric. The bandwidth b is usually much smaller than the order of the matrix N. To take full advantage of the banded matrix $\overline{\overline{Z}}^{(s)}$, a direct banded matrix solver is used to solve (5.1.33) and (5.1.34). The LU decomposition requires $\mathcal{O}(b^2N/2)$ operations, while the backsubstitution only requires $\mathcal{O}(2bN)$ operations. The $\overline{\overline{Z}}^{(w)}\overline{X}$ product is computed by the FFT. Therefore, we can evaluate $\overline{\overline{Z}}^{(w)}\overline{X}$ in $rN(\log N) + sN$ operations (where r accounts for the number of FFTs and s accounts for the number of pre- and post-multiplications before the FFT). The computational complexity up to the nth-order solutions are $\mathcal{O}(nb^2N) + \mathcal{O}(nrN\log N + nsN)$.

(C) Solution Based on Complete Impedance Matrix and Conjugate Gradient Method (CGM)

Another iteration approach is to keep the entire impedance matrix on the left-hand side. Then we apply a conjugate gradient method (CGM) to the matrix equation with the matrix decomposition.

$$\left(\overline{\overline{Z}}^{(s)} + \sum_{m=0}^{M} \overline{\overline{Z}}_m^{(w)}\right)\overline{X} = \overline{C} \tag{5.1.39}$$

For the CGM version, an initial guess of $\overline{X}^{(0)} = 0$ is chosen. Let N_c be the number of CGM iterations. By decomposing into $\overline{\overline{Z}}^{(s)}$ and $\overline{\overline{Z}}_m^{(w)}$ and the use of FFT in conjunction with CGM, the approach requires $\mathcal{O}(N_c(bN + rN\log N + sN))$.

Memory Requirements

The memory requirement of the strong matrix $\overline{\overline{Z}}^{(s)}$ is $\mathcal{O}(bN)$. The coefficients $a_m(x_d)$ in the Taylor expansions are translationally invariant. The storage requirement for $\overline{\overline{Z}}_m^{(w)}$, $m = 0, 1, 2, \ldots, M$, is $\mathcal{O}((M+1)N)$. The total memory requirement for the algorithm is $\mathcal{O}(bN + (M+1)N)$.

In the simulations, the bandwidth b is an adjustable parameter. In the updated right-hand-side approach there is a minimum bandwidth b_{min} for which the iteration process works. It requires many more iteration steps to converge at $b = b_{min}$ than at a larger bandwidth. Therefore, in the simulations, b is chosen to be greater than the b_{min} so as to reduce the number of iteration steps. For the CGM iterative approach discussed above, the bandwidth can be smaller than the one used in the updated right-hand-side approach. This is because the bandwidth in this case depends on the accuracy of the Taylor series expansion. As a result, this approach requires less computer memory, and therefore it is ideal for very large surface lengths. However, it usually takes more iteration steps to converge. In Section 1.5, only the 2500 wavelength surface examples are performed by applying the CGM iterative approach.

1.3 Product of a Weak Matrix and a Surface Unknown Column Vector

From (5.1.32), the product of a matrix term in the Taylor expanded impedance matrix with column vector can be expressed in the form

$$\overline{v} = \overline{\overline{T}}_r \overline{\overline{Z}}_d \overline{\overline{T}}_s \overline{X} \tag{5.1.40}$$

The calculation procedure of equation (5.1.40) is as follows. First, the product $\overline{\overline{T}}_s \overline{X}$ is performed (premultiplication). Next, the product $\overline{\overline{Z}}_d$ and $\overline{\overline{T}}_s \overline{X}$ is done by the FFT. Then the product of $\overline{\overline{T}}_r$ and $\overline{\overline{Z}}_d \overline{\overline{T}}_s \overline{X}$ is performed (postmultiplication).

In Fig. 5.1.1, we illustrate the physical interpretation of the two ways of calculating radiation from N source points to N field points on the rough surface. Direct calculation is from source S to field point R. Indirect calculation consists of first translation from source point to a point S_f on the flat

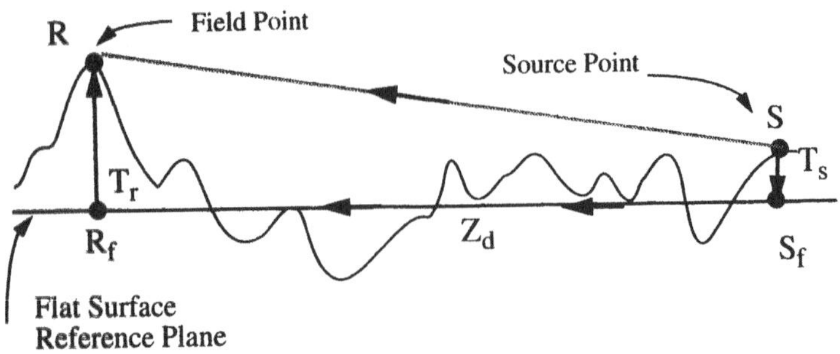

Figure 5.1.1 Direct (dotted line) and indirect (solid line) calculation of radiation from a source point S to a field point R. Points S_f and R_f are projections respectively of points S and R on flat surface. Indirect calculation procedure: (1) Translate from a source point to a point S_f on the flat surface via $\overline{\overline{T}}_s$, (2) translate from S_f to R_f on flat flat surface using $\overline{\overline{Z}}_d$, and (3) translate from R_f to R via $\overline{\overline{T}}_r$.

surface via $\overline{\overline{T}}_s$. Next, we calculate radiation from S_f to R_f on flat surface using $\overline{\overline{Z}}_d$. Finally, we calculate translation from R_f to R via $\overline{\overline{T}}_r$. For a single source point S to a single field point R, direct calculation is faster. However, for calculations of N source points to N field points, indirect calculation is faster. Direct calculation requires N^2 operations. Indirect calculation requires N operations for $\overline{\overline{T}}_r$ times a column vector, $N \log N$ for $\overline{\overline{Z}}_d$ times a column vector, and N for $\overline{\overline{T}}_s$ times a column vector. Indirect calculation is faster because of the use of FFT.

1.4 Convergence and Neighborhood Distance

In actual numerical computation, the method works for a neighborhood distance r_d which is much less than the surface length L. The kernel Green's function can be decomposed into two parts representing strong and weak interactions as limited by the neighborhood distance r_d.

$$G(\overline{r}, \overline{r}') = G(\overline{r}, \overline{r}')U(r_d - |x - x'|) + G(\overline{r}, \overline{r}')U(|x - x'| - r_d) \qquad (5.1.41)$$

where U is a Heaviside step function. Then the integral equation of (5.1.2) can be put into the following form:

$$\int_{x-r_d}^{x+r_d} dx' \left\{1 + \left[\frac{df(x')}{dx'}\right]^2\right\}^{\frac{1}{2}} G(\overline{r}, \overline{r}') \frac{\partial \psi(\overline{r}')}{\partial n'}$$

$$= \psi_{inc}(\overline{r}) - \int_{|x-x'|>r_d} dx' \left\{1 + \left[\frac{df(x')}{dx'}\right]^2\right\}^{\frac{1}{2}} G(\overline{r}, \overline{r}') \frac{\partial \psi(\overline{r}')}{\partial n'} \qquad (5.1.42)$$

The left-hand side represents the strong interaction $\overline{\overline{Z}}^{(s)}\overline{X}$, and the second term on the right-hand side represents the weak interaction $\overline{\overline{Z}}^{(w)}\overline{X}$, the magnitude of which is estimated as follows.

Because r_d is chosen to be at least several wavelengths, an asymptotic approximation can be made on the two-dimensional Green's function of the Hankel function. Thus

$$\mathcal{O}(\overline{\overline{Z}}^{(w)}\overline{X}) = \mathcal{O}\left(\int_{|x-x'|>r_d} dx' \left\{1 + \left[\frac{df(x')}{dx'}\right]^2\right\}^{\frac{1}{2}} \frac{\exp(ik|\overline{r}-\overline{r}'|)}{(k|\overline{r}-\overline{r}'|)^{\frac{1}{2}}} \frac{\partial\psi(\overline{r}')}{\partial n'}\right) \tag{5.1.43}$$

where $\mathcal{O}(\cdot)$ represents the order of magnitude. We note that ψ_{inc} is of order $\mathcal{O}(1)$ and that $\frac{\partial\psi}{\partial n}$ is of order $k\mathcal{O}(1)$, assuming that the order of magnitude of the normal derivative of the wavefunction is about the same as that of the incident wave. Thus, if $r_d \gg h$, where h is the rms height, then

$$\begin{aligned}\mathcal{O}(\overline{\overline{Z}}^{(w)}\overline{X}) = &\int_{x'=x+r_d}^{\infty} dx' \left\{1 + \left[\frac{df(x')}{dx'}\right]^2\right\}^{\frac{1}{2}} \frac{\exp[ik(x'-x)]}{[k(x'-x)]^{\frac{1}{2}}} \frac{\partial\psi(\overline{r}')}{\partial n'} \\ &+ \int_{-\infty}^{x-r_d} dx' \left\{1 + \left[\frac{df(x')}{dx'}\right]^2\right\}^{\frac{1}{2}} \frac{\exp[ik(x-x')]}{[k(x'-x)]^{\frac{1}{2}}} \frac{\partial\psi(\overline{r}')}{\partial n'}\end{aligned} \tag{5.1.44}$$

An asymptotic evaluation of (5.1.44) can be performed by integration by parts. This gives

$$\mathcal{O}(\overline{\overline{Z}}^{(w)}\overline{X}) = \mathcal{O}\left(\frac{1}{\sqrt{kr_d}}\right) \tag{5.1.45}$$

Thus the weak part $\overline{\overline{Z}}^{(w)}\overline{X}$ is of order $\mathcal{O}(1/\sqrt{kr_d})$ and is smaller than ψ_{inc}. Hence $\overline{\overline{Z}}^{(w)}\overline{X}$ is smaller than $\overline{\overline{Z}}^{(s)}\overline{X}$. The left-hand side of (5.1.42) balances most of ψ_{inc}. However, for the case of 2-D surface (3-D scattering problem), the result for the estimate of $\mathcal{O}(\overline{\overline{Z}}^{(w)}\overline{X})$ is different from that of the 1-D surface (2-D case), as shown Chapter 6. For iteration approach (A) with $\overline{\overline{Z}}^{(w)(0)}$ as well as $\overline{\overline{Z}}^{(s)}$ on the left-hand side of (5.1.15), the product $(\overline{\overline{Z}}^{(w)} - \overline{\overline{Z}}^{(w)(0)})\overline{u}$ will be smaller. This indicates that the order of magnitude of $\overline{\overline{Z}}^{(w)}\overline{X}$ is dimension-dependent.

1.5 Results of Composite Surfaces and Grazing Angle Problems

In this section, we present the results that were generated when the approach was initially proposed [Tsang et al. 1994, 1995]. All the computations were

Method	Speed	Memory
Gaussian elimination	10 minutes	1.30 MB
BMIA/CAG	10 seconds	0.20 MB

Table 5.1.1 $L = 40\lambda,\ N = 400$. Iteration for BMIA/CAG is based on Method B.

Method	Speed	Memory
Gaussian elimination	days	512 MB
BMIA/CAG	20 minutes	26 MB

Table 5.1.2 $L = 800\lambda,\ N = 8000$. Iteration for BMIA/CAG is based on Method B.

done on a SUN SPARC10 which has a speed of 60 MHz. We use 10 unknowns per wavelength. In the figures, the bistatic scattering coefficient $\sigma(\theta_s)$ as defined by (4.1.56) is plotted.

Example 1. Comparison of CPU and memory requirements

In Tables 5.1.1 and 5.1.2, we compare the CPU and memory requirements for $L = 40\lambda$ (400 unknowns) and $L = 800\lambda$ (8000 unknowns) for $h = 0.5\lambda$, correlation length $l = 1.0\lambda$, and incident angle of 10°. The BMIA/CAG is 60 times faster than the Gaussian elimination. The results of all these methods completely overlie each other. For the case of $L = 800\lambda$ (8000 unknowns), the BMIA/CAG requires only 20 minutes CPU for one realization and a memory of 26 MBytes. The CPU time for the method of Gaussian elimination consists of projected values only.

Example 2. Comparison of results for different surface lengths for one realization and for averages over realizations

In Fig. 5.1.2, we compare the results of surface length of (a) $L = 40\lambda$, (b) $L = 200\lambda$, and (c) $L = 800\lambda$ for one realization. The rms height is $h = 0.5\lambda$, correlation length $l = 1.0\lambda$, and incident angle $\theta_i = 10°$ with $g = L/4$. We note that the results of the three cases have different features. The largest surface length case of $L = 800\lambda$ has very rapid angular fluctuations of intensity, a feature that is very different from $L = 40\lambda$, which has much slower angular fluctuations of intensity. Instead of using a tapered wave, another method is the periodic boundary condition method (PBC). This method uses a plane wave and assumes that the random rough surface lengths re-

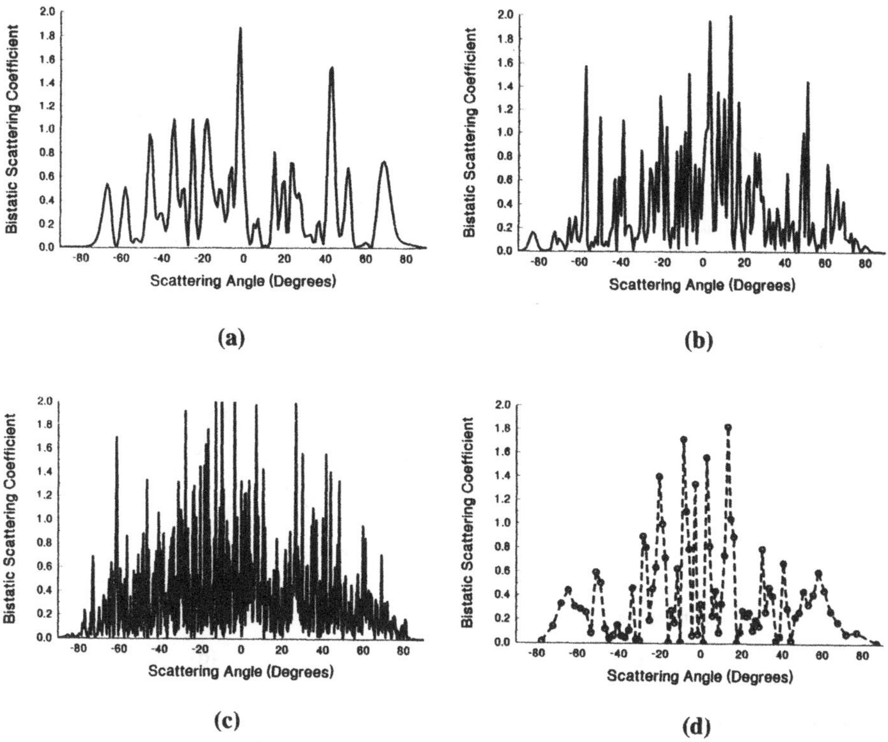

Figure 5.1.2 Comparison of various surface lengths of one realization with $h = 0.5\lambda$, $l = 1.0\lambda$, and $\theta_i = 10°$. The number of surface unknowns are varied from (a) $N = 400$ with $b = 40(4\lambda)$, (b) $N = 2000$ with $b = 200(20\lambda)$, (c) $N = 8000$ with $b = 400(40\lambda)$, and (d) periodic boundary condition (PBC) with $N = 400$.

peats itself for a given period P. Usually the period used is not very large. The PBC method was described in Chapter 4. The CPU associated with computing the periodic Green's function can be intensive. In Fig. 5.1.2d, we show the result of a periodic boundary condition with period $= 40\lambda$. Because of the periodic boundary condition, the scattered angles θ_{sm} are governed by the Floquet modes $k \sin\theta_{sm} = k \sin\theta_i + \frac{2\pi m}{P}$, $m = 0, \pm 1, \pm 2, \ldots$, which results in a finite number of angles. We note that the features of the periodic boundary condition are also different. Because of the finite number of discrete scattered angles, it does not have the rapid angular variations as in the 800λ surface length case. Its features are more like that of a surface length of 40λ of Fig. 5.1.2a. The imposition of the PBC does not imply that the rough surface is random from $-\infty$ to $+\infty$.

In Fig. 5.1.3, we compare the results when averaged over realizations.

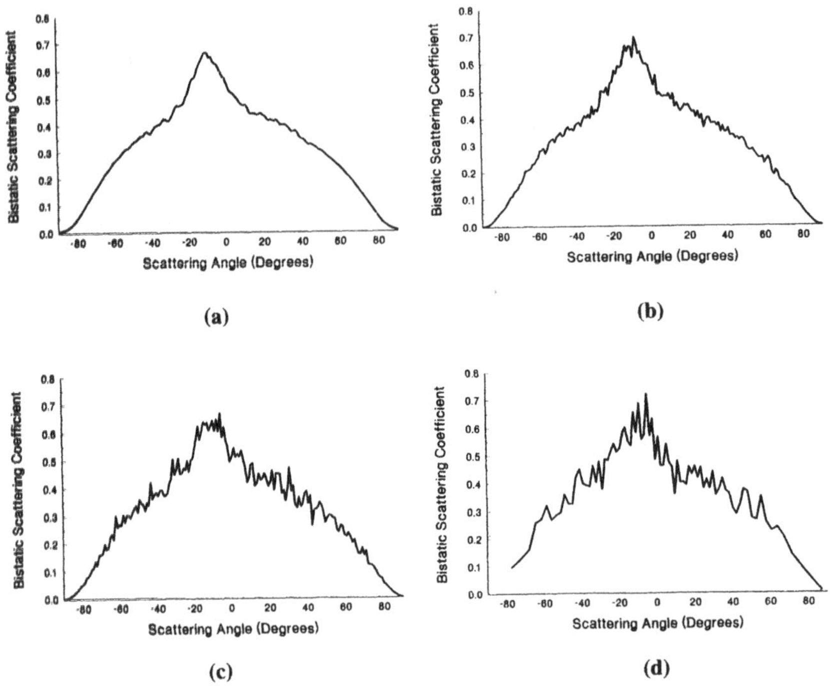

Figure 5.1.3 Comparison of Monte-Carlo simulations of small and large surface lengths with (a) $L = 40\lambda$ with 4000 realizations, (b) $L = 200\lambda$ with 800 realizations, (c) $L = 800\lambda$ with 200 realizations, and (d) periodic boundary condition method with $L = 40\lambda$ with 100 realizations. The parameters are $h = 0.5\lambda$, $l = 1.0\lambda$, and $\theta_i = 10°$.

The numbers of realizations are 4000 for $L = 40\lambda$, 800 for $L = 200\lambda$, 200 for $L = 800\lambda$, and 100 for the periodic case with $P = 40\lambda$. The results of all four cases are similar upon averaging over realizations. It is to be noted that in Fig. 5.1.3d of PBC, the data points for angles beyond $-78°$ are not available since the number of angular data points are limited by the Floquet modes.

Example 3. Close to grazing and moderate RMS height and correlation length

In Fig. 5.1.4a, we illustrate the results for an incidence angle $= 85°$, rms height $h = 0.5\lambda$ and $l = 1.0\lambda$ and averaged over 50 realizations. We compare three cases of ($L = 200\lambda$, $g = L/4$), ($L = 800\lambda$, $g = L/4$), and ($L = 800\lambda$, $g = L/8$). The results for the cases agree except for the forward specular direction (Fig. 5.1.4b) and the vicinity of the backscattering direction (Fig. 5.1.4c). The difference in the forward direction is due to the

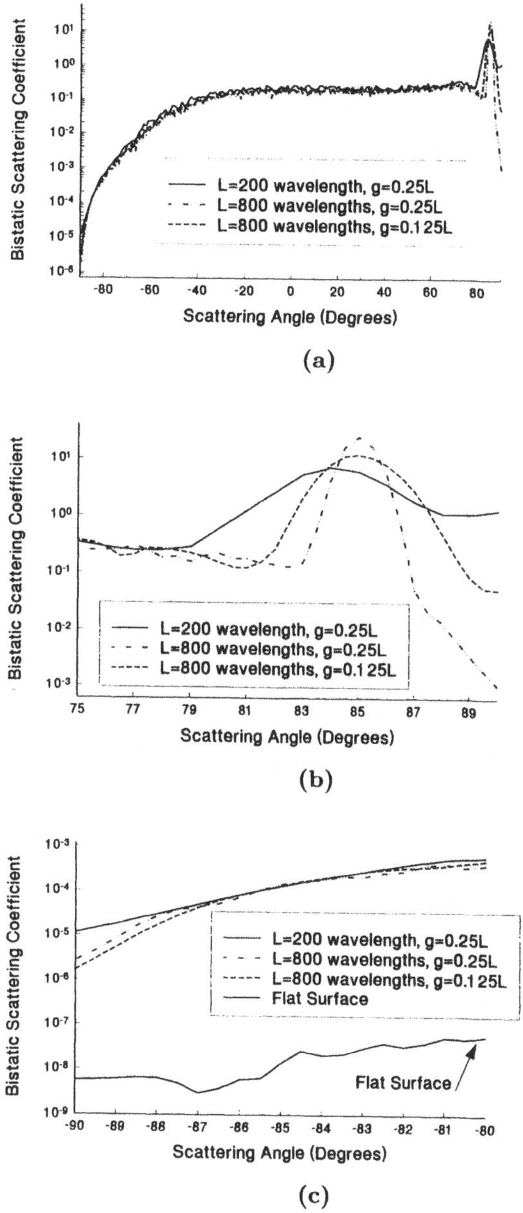

Figure 5.1.4 (a) Comparison of bistatic scattering coefficients for various surface lengths and g for 50 realizations with $h = 0.5\lambda$, $l = 1.0\lambda$, and $\theta_i = 85°$. For $L = 200\lambda$ the bandwidth is $b = 200(20\lambda)$, and for $L = 800\lambda$ it is $b = 400(40\lambda)$. (b) Comparison of bistatic scattering coefficients of (a) near-specular direction. (c) Comparison of bistatic scattering coefficients of (a) near backscattering direction. The flat surface result is for $L = 800\lambda$ and $g = 0.25L$.

fact that the forward scattering peak of the coherent wave depends strongly on the surface lengths. The bistatic scattering coefficient from the largest surface length and the smallest tapering gives the smallest scattering level for angles beyond $-85°$. In Fig. 5.1.4c, an additional simulation is performed with a flat surface to illustrate the edge diffraction contribution to backscattering. Note that the backscattering level is at least two orders of magnitude larger than the edge diffraction. For this particular example, we have achieved convergence up to $-87°$.

Example 4. Close to grazing angle and composite surface

Next we examine the case of scattering from a composite random rough surface at a near-grazing incidence angle of $\theta_i = 85°$. The composite surface has a small-scale roughness ($h_1 = 0.1\lambda$ and $l_1 = 0.3\lambda$) superimposed on a larger scale roughness ($h_2 = 0.5\lambda$ and $l_2 = 5\lambda$). Figure 5.1.5a shows the bistatic scattering coefficient for one realization for the surface length of 2500 wavelengths with 25,000 surface unknowns. From Table 5.1.3 we can see that in order to perform Monte Carlo simulations with the 25,000 unknowns case, other methods which require the storage of a full matrix would be impossible on a workstation. However, using the BMIA/CAG, we were able to compute the solution. For a problem of this size, it requires 6 CPU hours on a SPARC10 workstation with 75 Mbytes of memory. In Fig. 5.1.5b, the bistatic scattering coefficient for 50 realizations is shown.

In Fig. 5.1.5c, we compare the four cases of $L = 200\lambda$, $L = 400\lambda$, $L = 800\lambda$, and $L = 2500\lambda$ near the backscattering angle. Clearly, the $L = 200\lambda$ surface is not large enough for the parameters used for this example. On the other hand, the results for $L = 400\lambda$, $L = 800\lambda$, and $L = 2500\lambda$ agree near the backscattering angle of $-85°$ and up to $-88°$. This shows convergence of the bistatic coefficient with respect to the surface length for the scattering angles up to $-88°$. Next in Fig. 5.1.5d, the bistatic scattering coefficient from the 2500 wavelengths composite surface is compared with the result from the PBC with $P = 40\lambda$. Since the periodic surface has the periodicity of 40 wavelengths, the angular resolution is only 6° near the backscattering angle of $-85°$. This is illustrated in Fig. 5.1.6 for one realization with $P = 40\lambda$. In fact, for the 40λ surface, the maximum backscattering angle for PBC is $-78°$. If we use a period of 200λ for PBC, The matrix building time for this corresponding 2000×2000 matrix can be large.

Method	Speed	Memory
Gaussian elimination	weeks	5000 MB
BMIA/CAG	6 hours	75 MB

Table 5.1.3 $L = 2500\lambda$, $N = 25000$. Iteration for BMIA/CAG is based on conjugate gradient and matrix decomposition (Method C).

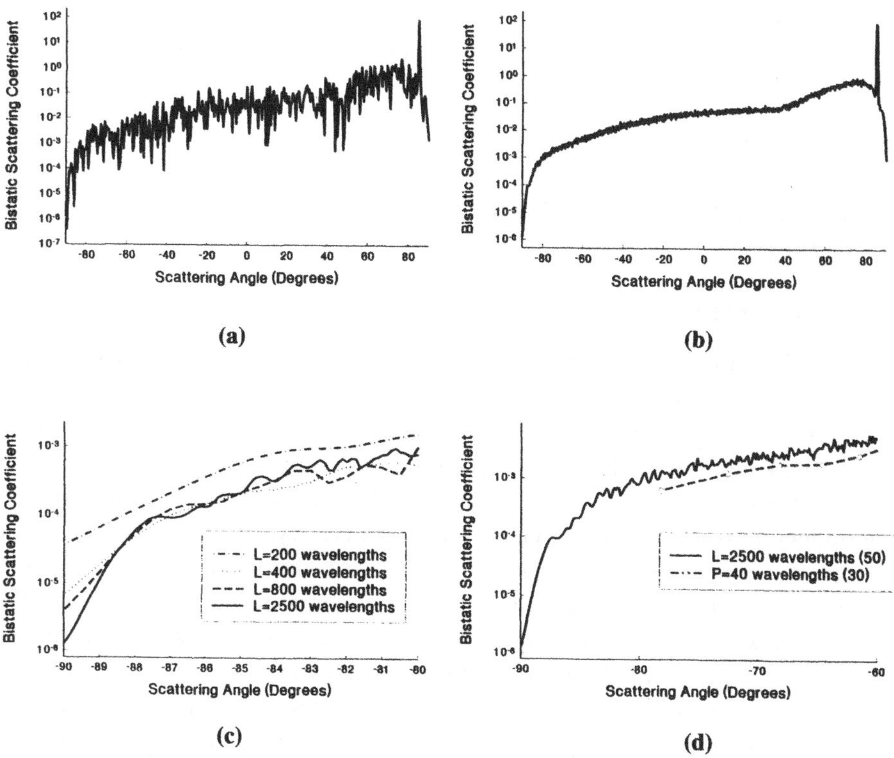

Figure 5.1.5 (a) Close to grazing angle and composite surface. Incident angle $\theta_i = 85°$, $g = 0.125L$, $b = 400$. $h_1 = 0.1\lambda$, and $l_1 = 0.3\lambda$, $h_2 = 0.5\lambda$, and $l_2 = 5\lambda$ for one realization. (b) Close to grazing angle and composite surface. Averaged over 50 realizations. (c) Comparison near-backscattering angle for various surface lengths. Parameters are those of (a). For 2500 wavelengths surface 50 realizations of Fig. 5.1.5b is used and others are averaged over 10 realizations. The bandwidths of $L = 200\lambda$ and $L = 800\lambda$ are the same as Fig. 5.1.4a, for $L = 400\lambda$, $b = 300(30\lambda)$. (d) Comparison near-backscattering angle with the periodic boundary condition method (PBC). For BMIA/CAG the result of (b) is used. The PBC is for the surface length of 40 wavelengths and averaged over 30 realizations. Parameters are those of (a).

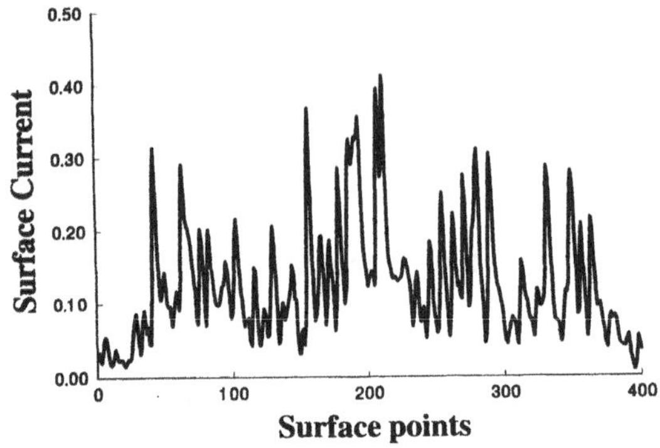

Figure 5.1.6 Magnitude of the surface current for one realization for the PBC method with $P = 40$ wavelengths. Parameters are those of Fig. 5.1.5a.

2 Physics-Based Two-Grid Method for Lossy Dielectric Surfaces

2.1 Introduction

In the application of the method of moments to the rough surface scattering problem, a common implementation is to use a grid of 10 points per free-space wavelength to discretize the surface. We shall call such a gridding a single coarse grid (SCG). However, in lossy dielectric surfaces, the wavelength in the dielectric medium is much shorter. Thus in scattering by lossy dielectric rough surfaces with high permittivity, there can be rapid spatial variation of surface fields. For microwave remote sensing applications, both wet soil surfaces and ocean surfaces can have large permittivity. Two alternatives were used. The first alternative is to use impedance boundary condition as shown in Chapter 4. The disadvantage of this alternative is that an approximation is used in the problem. The second alternative is to use a dense grid with a large number of points (say more than 30 points) per free-space wavelength. We shall call such a gridding a single dense grid (SDG). We have shown in Chapter 4 that dense sampling of points is a requirement for energy conservation and an accurate calculation of emissivity. The disadvantage of this second alternative is that there is a large increase in CPU and required memory. The physics-based two-grid (PBTG) method to be discussed in this section is an improvement over these two alternatives in that it has the same accuracy as the single dense grid and yet has a CPU comparable with that of the single coarse grid. To demonstrate the accuracy of the PBTG method, we use it to calculate the emissivity of a random rough surface. In PBTG, two grids were used: a dense grid and a sparse grid. The sparse grid is that of the usual 10 points per wavelength. The dense grid ranges from 20 to higher number points per wavelength, depending on the relative permittivity of the lossy dielectric medium. The method of PBTG is based on the following two observations: (1) Green's function of the lossy dielectric is attenuative, and (2) Green's function of free space is slowly varying on the dense grid. Because of the Kramer-Kronig relation, a large real part of dielectric constant is associated with a large imaginary part. The first property of the lossy dielectric gives a banded submatrix for the Green's function of the lossy dielectric. When the Green's functions act on the surface field on the dense grid, it corresponds to the product of a sparse matrix with that of a column vector. Thus the convolution of the lossy dielectric Green's function with surface fields is a spatial limited operation. The second property means that the convolution of the free space Green's functions with surface fields on the dense grid is a spatial frequency limited operation. This allows us, when using the free space Green's function to act on the surface fields of the dense grid, to first average the values of surface unknowns on the dense grid and then place them on the coarse grid. PBTG calculates surface field solutions on the dense grid. It needs to be mentioned that PBTG is different from multigrid method [Donohue et al. 1998, Briggs, 1987]. The multigrid method tries to facilitate the convergence of iterations in iterative techniques. It entails discretization of the structure into various grid sizes. The coarse grid corresponds to the low-frequency portion of the solution, while the fine grid corresponds to that of the high-frequency solution. An iterative solution is obtained for each level of discretization, and the solutions are interpolated from the coarse grid to the fine grid. The solution is first obtained in the coarse grid, and then one moves to the next level of fine grid. Once the iterative solution is obtained in the fine grid, one has to go back to the coarse grid to refine the solution. The present method of PBTG, on the other hand, is based on scattering physics. The purpose of PBTG is to speed up the matrix-vector product of two Green's functions convolving with the surface fields on the dense grid.

We use two grids in PBTG: a dense grid and a coarse grid. The interaction is divided into (1) a very near field of less than 1 wavelength, (2) a near field of between 1 wavelength and r_d wavelengths, and (3) a non-near-field beyond r_d wavelengths. In the numerical simulations performed in this section, r_d is fixed at 10 wavelengths. For very near-field interactions, we use a dense grid which is represented by four banded submatrices. For near-field and non-near-field interactions, the free-space Green's function is slowly varying on the dense grid. We average the fields on the dense grid to get fields on the sparse grid. For the non-near-field interactions, we further expand on a canonical grid of a horizontal surface so that the fast Fourier transform (FFT) can be applied. In the lower medium, non-near-field interactions were neglected because of lossy properties of the lower medium. The approach is denoted as PBTG-BMIA/CAG. The computational complexity and the memory requirements for the present algorithm are $\mathcal{O}(N\log(N))$ and $\mathcal{O}(N)$, respectively, where N is the number of surface unknowns on the coarse grid. Using this approach, we illustrate numerical results of TE and TM wave scattering up to surface length of 500 wavelengths and 30,000 surface unknowns. The salient features of the numerical results are as follows:

(1) A single coarse grid (SCG) has poorer accuracy for TM case than for TE case.
(2) PBTG-BMIA/CAG speeds up CPU and preserves the accuracy. It has accuracy comparable to that of a single dense grid and yet has a CPU comparable to that of a single coarse grid. It also gives surface fields on the dense grid and can give accurate results of the surface fields even when the surface fields have large spatial variations.
(3) PBTG-BMIA/CAG gives accurate results for emissivity calculations.

2.2 Formulation and Single-Grid Implementation

Consider a tapered plane wave, $\psi_{inc}(x,z)$, impinging upon a 1-D rough surface with a random height profile $z = f(x)$. It is tapered in the spectral domain so that the illuminated rough surface can be confined to surface length L. The incident wave is tapered in the spectral domain as given in (4.1.42),

$$\psi_{inc}(x, f(x)) = \frac{g}{2\sqrt{\pi}} \int_{-\infty}^{\infty} dk_x e^{i(k_x x - k_z z)} e^{-\frac{(k_x - k_{ix})^2 g^2}{4}} \tag{5.2.1}$$

Let ψ and ψ_1 denote, respectively, the wave functions for the upper medium and lower medium. From Chapter 4, Section 2, they satisfy the following surface integral equations

$$\frac{1}{2}\psi(\bar{r}') - \fint_S \left[\psi(\bar{r})\frac{\partial G(\bar{r},\bar{r}')}{\partial n} - G(\bar{r},\bar{r}')\frac{\psi(\bar{r}')}{\partial n}\right] ds = \psi_{inc}(\bar{r}') \tag{5.2.2}$$

$$\frac{1}{2}\psi(\bar{r}') + \fint_S \left[\psi_1(\bar{r})\frac{\partial G_1(\bar{r},\bar{r}')}{\partial n} - G_1(\bar{r},\bar{r}')\frac{\psi_1(\bar{r}')}{\partial n}\right] ds = 0 \tag{5.2.3}$$

where $\fint_S$ denotes a Cauchy principal value integral and G and G_1 are the 2-D Green's function of the upper and lower medium. The wave functions ψ and ψ_1 are related through the boundary conditions on the surface S as in

Chapter 4. The integral equation is next discretized using an evenly spaced single grid. The surface is discretized into a single grid of N points for x between $-L/2$ and $L/2$, and the x_m points are at $x_m = (m-0.5)\Delta x - \frac{L}{2}$, $m = 1, 2, \ldots, N$. The discretized equations become

$$\sum_{n=1}^{N} a_{mn} u(x_n) + \sum_{n=1}^{N} b_{mn} \psi(x_n) = \psi_{inc}(x_m) \tag{5.2.4}$$

$$\sum_{n=1}^{N} a_{mn}^{(1)} \rho u(x_n) + \sum_{n=1}^{N} b_{mn}^{(1)} \psi(x_n) = 0 \tag{5.2.5}$$

where $u(x) = \sqrt{1+[f'(x)]^2}\partial\psi/\partial n$, $\rho = \mu_1/\mu$, and ϵ_1/ϵ for TE and TM polarization, respectively. The matrix elements a_{mn}, b_{mn}, $a_{mn}^{(1)}$, and $b_{mn}^{(1)}$ are, to a first order approximation,

$$a_{mn} = \begin{cases} \Delta x \frac{i}{4} H_0^{(1)}(kr_{mn}) & m \neq n \\ \Delta x \frac{i}{4} H_0^{(1)}[k\Delta x \gamma_m/(2e)] & m = n \end{cases} \tag{5.2.6}$$

$$b_{mn} = \begin{cases} -\Delta x \frac{ik}{4} \frac{f'(x_n)(x_n - x_m) - [f(x_n) - f(x_m)]}{r_{mn}} H_1^{(1)}(kr_{mn}) & m \neq n \\ \frac{1}{2} - \frac{f''(x_m)}{4\pi} \frac{\Delta x}{\gamma_m^2} & m = n \end{cases} \tag{5.2.7}$$

$$a_{mn}^{(1)} = \begin{cases} -\Delta x \frac{i}{4} H_0^{(1)}(k_1 r_{mn}) & m \neq n \\ -\Delta x \frac{i}{4} H_0^{(1)}[k_1 \Delta x \gamma_m/(2e)] & m = n \end{cases} \tag{5.2.8}$$

$$b_{mn}^{(1)} = \begin{cases} \Delta x \frac{ik_1}{4} \frac{f'(x_n)(x_n - x_m) - [f(x_n) - f(x_m)]}{r_{mn}} H_1^{(1)}(k_1 r_{mn}) & m \neq n \\ \frac{1}{2} + \frac{f''(x_m)}{4\pi} \frac{\Delta x}{\gamma_m^2} & m = n \end{cases} \tag{5.2.9}$$

where $r_{mn} = \sqrt{(x_n - x_m)^2 + [f(x_n) - f(x_m)]^2}$ and $\gamma_m = \sqrt{1+[f'(x_m)]^2}$. The matrix equation in (5.2.4) and (5.2.5) is in the form of a single grid. Let $n = N/L$ be the number of points per wavelength. Usually a sample frequency of $n = 10$ is taken, meaning that we have 10 points per wavelength. We shall call such a sampling a single coarse grid (SCG). If the sampling frequency is two or more times denser than the coarse grid, we shall call it a single dense grid (SDG). The dense grid that we use ranges from $n = 20$ to $n = 30$ in this section.

2.3 Physics-Based Two-Grid Method Combined with Banded Matrix Iterative Approach/Canonical Grid Method

Assume that the upper medium is the free space and that the lower medium is lossy with the following relative permittivity:

$$\epsilon_1 = \epsilon_1'(1 + i \tan \delta) \tag{5.2.10}$$

where $\tan \delta$ stands for loss tangent. Let λ and λ_1 represent the wavelength of the wave in the free space and the lower medium, respectively, with $\lambda_1 = \lambda / \mathrm{Re}\sqrt{\epsilon_1'}$. The number of sampling points needed in the lower medium should be $\mathrm{Re}\sqrt{\epsilon_1'}$ times more than that in the free space.

In the physics-based two-grid method, we use two grids with samplings per wavelength of n_{cg} (coarse grid) and n_{dg} (dense grid), respectively. Let N_{dg} and N be, respectively, the total number of points on the dense grid and the coarse grid.

$$N_{dg} = n_{dg}\frac{L}{\lambda} \tag{5.2.11}$$

$$N = n_{cg}\frac{L}{\lambda} \tag{5.2.12}$$

Let $n_{dg}/n_{cg} = n_1$. For the sake of convenience, we choose n_1 to be an integer. We first rewrite (5.2.4) and (5.2.5) using the dense grid.

$$\sum_{n=1}^{N_{dg}} a_{mn} u(x_n) + \sum_{n=1}^{N_{dg}} b_{mn} \psi(x_n) = \psi_{inc}(x_m) \tag{5.2.13}$$

$$\sum_{n=1}^{N_{dg}} a_{mn}^{(1)} \rho u(x_n) + \sum_{n=1}^{N_{dg}} b_{mn}^{(1)} \psi(x_n) = 0 \tag{5.2.14}$$

The subscripts m, n denote indexing with the dense grid. In the method of PBTG, the surface fields on the dense grid are calculated. The matrix elements a_{mn} and b_{mn} represent Green's function of the upper medium, while $a_{mn}^{(1)}$ and $b_{mn}^{(1)}$ represent Green's function of the lower medium of the lossy dielectric. We make the following three observations:

(1) The Green's function in the lower region can have moderate to large attenuation. A medium with a large real part of dielectric constant is normally associated with a large imaginary part. Let k_1'' be the imaginary part of k_1. Let r_{mn} be the distance between the mth and the nth parts. If $k_1'' r_{mn} > C$, where C is a constant, the field interaction between the mth and the nth point is vanishingly small. We can define a distance limit as dictated by

attenuative loss:

$$r_l = \frac{C}{k_1''} \tag{5.2.15}$$

outside of which the lower medium Green's function can be set equal to zero. In the simulation of this section, we let C be fixed at 1.5.

Based on this observation, we calculate the left-hand sides of (5.2.14) as follows. We approximate

$$a_{mn}^{(1)} \approx \tilde{a}_{mn}^{(1)} = \begin{cases} a_{mn}^{(1)} & r_{mn} \le r_l \\ 0 & r_{mn} \ge r_l \end{cases} \tag{5.2.16}$$

$$b_{mn}^{(1)} \approx \tilde{b}_{mn}^{(1)} = \begin{cases} b_{mn}^{(1)} & r_{mn} \le r_l \\ 0 & r_{mn} \ge r_l \end{cases} \tag{5.2.17}$$

Thus $\tilde{a}_{mn}^{(1)}$ and $\tilde{b}_{mn}^{(1)}$ are banded matrices and (5.2.14) becomes

$$\sum_{n=1}^{N_{dg}} \tilde{a}_{mn}^{(1)} \rho u(x_n) + \sum_{n=1}^{N_{dg}} \tilde{b}_{mn}^{(1)} \psi(x_n) = 0 \tag{5.2.18}$$

(2) For non-near-field interaction, the free space Green's function is slowly varying (spatial frequency limited) on the dense grid. Thus when performing matrix and column vector multiplication on the dense grid as indicated in (5.2.13), the Green's function of the upper medium is essentially constant over an interval of n_1 points on the dense grid. Thus we can write

$$\sum_{l=1}^{n_1} a_{(m+l')(n+l)} u_{n+l} \approx a_{m_{mp} n_{mp}} \sum_{l=1}^{n_1} u_{n+l} = n_1 a_{m_{mp} n_{mp}} \left(\frac{1}{n_1} \sum_{l=1}^{n_1} u_{n+l} \right) \tag{5.2.19}$$

where $l' = 1, 2, \ldots, n_1$ and the points with indexes m_{mp} and n_{mp} are the middle points between the $(m+1)$th point and the $(m+n_1)$th point and between the $(n+1)$th point and the $(n+n_1)$th point, respectively. What is done in (5.2.19) is that the surface fields on the dense grid are first averaged before multiplied by the upper medium Green's function.

(3) The slowly varying nature of Green's function of the upper medium only applies to non-near-field interaction. For near-field interaction, Green's functions G and G_1 have roughly the same spatial rate of variation. Thus we need to separate out a distance, say 1λ, outside of which G_1 is much more rapidly varying than G. Within near field interaction for the upper medium Green's function, direct matrix and vector multiplication is performed.

Based on the observations above, we decompose the upper medium

Green's function into near-field and non-near-field interactions

$$\sum_{n=1}^{N_{dg}} a_{mn} u(x_n) = \sum_{n=1}^{N_{dg}} a_{mn}^s u_n + \sum_{n=1}^{N_{dg}} a_{mn}^{ns} u_n \tag{5.2.20}$$

$$\sum_{n=1}^{N_{dg}} b_{mn} \psi(x_n) = \sum_{n=1}^{N_{dg}} b_{mn}^s \psi_n + \sum_{n=1}^{N_{dg}} b_{mn}^{ns} \psi_n \tag{5.2.21}$$

where the superscript s represents near field interactions and superscript ns represents non-near field interactions. The matrices a_{mn}^s, b_{mn}^s, a_{mn}^{ns}, and b_{mn}^{ns} are determined by

$$a_{mn}^s = \begin{cases} a_{mn} & r_{mn} \le r_f \\ 0 & r_{mn} \ge r_f \end{cases} \tag{5.2.22}$$

$$b_{mn}^s = \begin{cases} b_{mn} & r_{mn} \le r_f \\ 0 & r_{mn} \ge r_f \end{cases} \tag{5.2.23}$$

$$a_{mn}^{ns} = \begin{cases} 0 & r_{mn} \le r_f \\ a_{mn} & r_{mn} \ge r_f \end{cases} \tag{5.2.24}$$

$$b_{mn}^{ns} = \begin{cases} 0 & r_{mn} \le r_f \\ b_{mn} & r_{mn} \ge r_f \end{cases} \tag{5.2.25}$$

Thus r_f is the distance outside which the Green's function of the lower medium is fast varying compared with that of free space Green's function.

We use the Greek indices to represent coarse grid and Roman indices for dense grid. The coarse grid has surface unknowns $\tilde{u}$ and $\tilde{\psi}$, which are averages of the dense grid surface unknowns. Thus if $\overline{r}_\beta$ is centered in the middle of the n_1 dense grid points of $n+1, n+2, \ldots, n+n_1$, we have

$$\tilde{u}_\beta = \frac{u_{n+1} + u_{n+2} + \cdots + u_{n+n_1}}{n_1} \tag{5.2.26}$$

$$\tilde{\psi}_\beta = \frac{\psi_{n+1} + \psi_{n+2} + \cdots + \psi_{n+n_1}}{n_1} \tag{5.2.27}$$

We calculate Green's function of the upper medium on the coarse grid. These are represented by $\tilde{a}_{\alpha\beta}$ and $\tilde{b}_{\alpha\beta}$.

$$\tilde{a}_{\alpha\beta} = \begin{cases} \Delta\tilde{x}\frac{i}{4}H_0^{(1)}(kr_{\alpha\beta}) & r_{\alpha\beta} > r_f \\ 0 & r_{\alpha\beta} \le r_f \end{cases} \tag{5.2.28}$$

$$\tilde{b}_{\alpha\beta} = \begin{cases} -\Delta\tilde{x}\frac{ik}{4}\frac{f'(x_\beta)(x_\beta - x_\alpha) - [f(x_\beta) - f(x_\alpha)]}{r_{\alpha\beta}} H_1^{(1)}(kr_{\alpha\beta}) & r_{\alpha\beta} > r_f \\ 0 & r_{\alpha\beta} \le r_f \end{cases} \tag{5.2.29}$$

where $\Delta\tilde{x}$ is the coarse grid sampling, $\Delta\tilde{x} = n_1 \Delta x$, and Δx is the dense grid sampling. Thus we use this averaging for the second terms of both (5.2.20) and (5.2.21). Equation (5.2.13) becomes

$$\left[\sum_{n=1}^{N_{dg}} a^s_{mn} u(x_n) + \sum_{n=1}^{N_{dg}} b^s_{mn} \psi(x_n)\right] + \left[\sum_{\beta=1}^{N} \tilde{a}^s_{\alpha\beta} \tilde{u}(x_\beta) + \sum_{\beta=1}^{N} \tilde{b}^s_{\alpha\beta} \tilde{\psi}(x_\beta)\right]_{\text{intp}} = \psi_{inc}(x_m) \tag{5.2.30}$$

In (5.2.30) we use subscript "intp" to represent linear interpolation. Note in (5.2.30) that $\sum_{n=1}^{N_{dg}} a^s_{mn} u(x_n)$ has N_{dg} values of $m = 1, 2, \ldots, N_{dg}$, while $\sum_{\beta=1}^{N} \tilde{a}^s_{\alpha\beta} \tilde{u}(x_\beta)$ only has N values of $\alpha = 1, 2, \ldots, N$. Thus we first compute

$$\sum_{\beta=1}^{N} \tilde{a}^s_{\alpha\beta} \tilde{u}(x_\beta) = d(x_\alpha) \tag{5.2.30a}$$

for $\alpha = 1, 2, \ldots, N$ on the coarse grid. To find $d(x_m)$ on the dense grid of x_m, $m = 1, 2, \ldots, N_{dg}$, we use linear interpolation of $d(x_\alpha)$'s to get $d(x_m)$'s.

We further use BMIA/CAG to solve matrix equation. We divide non-near-field interactions into two regions which are separated by r_d. We now have three distance ranges for the upper medium Green's function, $0 \le r \le r_f$, $r_f < r \le r_d$, and $r > r_d$ with different operations. For $0 \le r \le r_f$ known as very near-field region, we use direct matrix and column vector product on the dense grid. For $r_f < r \le r_d$ known as near-field region, we use direct matrix and column vector product on the coarse grid and interpolation as in (5.2.30). For $r > r_d$ known as non-near-field region, we expand $\tilde{a}_{\alpha\beta}$ and $\tilde{b}_{\alpha\beta}$ in Taylor series as in the BMIA/CAG so that the FFTs can be used to compute this part of the matrix-vector multiplication. The Taylor series expansion is as in Section 1

$$H\left(k\sqrt{x_d^2 + z_d^2}\right) = \sum_{m=0}^{\infty} a_m(x_d) \left(\frac{z_d}{x_d}\right)^{2m} \tag{5.2.31}$$

Here H represents both $H_0^{(1)}$ and $H_1^{(1)}$.

2.4 Bistatic Scattering Coefficient and Emissivity

After the matrix equation is solved, the surface field can be calculated. The bistatic scattering coefficient $\sigma(\theta_s, \theta_i)$ for the spectral domain tapered wave is given in (4.1.57).

We next illustrate the numerical simulation results of wave scattering from a rough lossy dielectric surface for both TE and TM waves [Tsang and

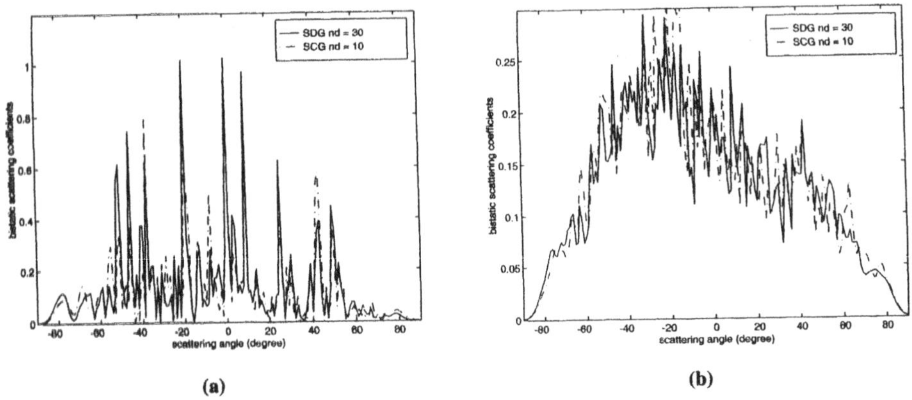

Figure 5.2.1 Comparison of the bistatic scattering coefficients between the single dense grid of 30 points per wavelength and the single coarse grid of 10 points per wavelength. TE wave, rms $h = 0.5\lambda$, correlation length of $l = 0.6\lambda$, dielectric constant of $\epsilon_r = 25 + i$, surface length of $L = 100\lambda$, and tapering parameter of $g = L/4$ at incidence angle of θ_i. (a) One realization, (b) 20 realizations.

Li, 1997; Li et al. 1999]. Simulations are based on Gaussian random rough surfaces with Gaussian correlation functions. First, we show the comparisons of bistatic scattering coefficients and surface fields based on a single dense grid (SDG) and a single coarse grid (SCG) with a complex dielectric constant of $25 + i$, surface length of 100 wavelengths, and at an incidence angle of 30°. The results show that the dense grid is required for the case with large dielectric constant. We shall regard the SDG results to be correct. Next, we compare the results based on PBTG-BMIA/CAG with that of SDG. Then we use the PBTG-BMIA/CAG method to calculate the cases with large surface length and compare with SDG.

The tapering parameter was taken to be $L/4$ for the case of surface length of 100 wavelengths and $L/8$ for the case of surface length of 500 wavelengths at near-grazing incidence. The critical distance r_f that defines the very near field is fixed at 1 wavelength. The cases with a surface length of 100 wavelengths were run on a SPARC 20 workstation, and the cases with a surface length of 500 wavelengths were run on a Pentium-Pro Personal Computer with a clock rate of 200 MHz.

a) Comparison Between a Single Dense Grid and a Single Coarse Grid

In Figs. 5.2.1a and 5.2.1b, we compare respectively the results of the bistatic scattering coefficients of a single realization of rough surface and averaged over 20 realizations for a TE wave, at incident angle $\theta_i = 30°$, and surface

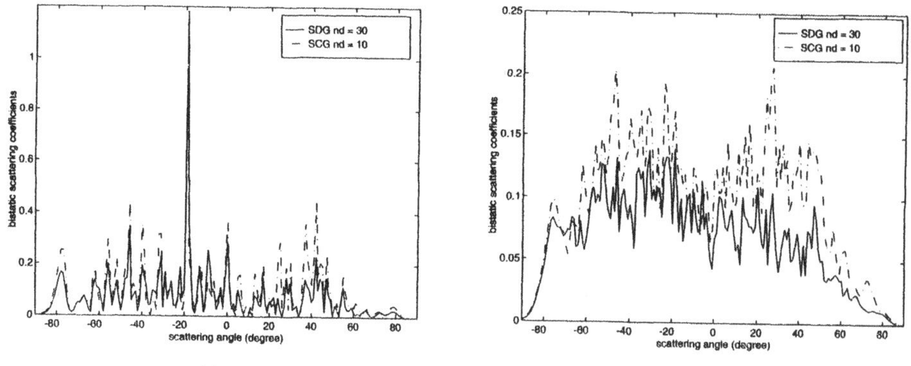

(a) (b)

Figure 5.2.2 Comparison of the bistatic scattering coefficients between the single dense grid of 30 points per wavelength and the single coarse grid of 10 points per wavelength. TM wave, rms $h = 0.5\lambda$, correlation length of $l = 0.6\lambda$, dielectric constant of $\epsilon_r = 25 + i$, surface length of $L = 100\lambda$, and tapering parameter of $g = L/4$ at incidence angle of $\theta_i = 30°$. (a) One realization, (b) 20 realizations.

length $L = 100\lambda$, where λ is the wavelength. The rms height and correlation length are 0.5 and 0.6 wavelength, respectively. We compare the cases of SCG of $n_{cg} = 10$ and SDG of $n_{dg} = 30$ points per wavelength. We note that the results of SCG and SDG are quite different both for one realization and for averages over 20 realizations. The results based on SCG are not accurate. The comparisons were also made in Figs. 5.2.2a and 5.2.2b for a TM wave with the same parameters. It is noted that the performance of SCG is poorer for the TM case. This is because more energy is transmitted into the lower medium for the TM case than for the TE case, and the lower medium dielectric requires a dense discretization. In Figs. 5.2.3a and 5.2.3b, we compare the surface electric fields between SDG and SCG for TE and TM waves, respectively. In Table 5.2.1, we compare the emissivities calculated for using one realization and 20 realizations for SCG and SDG for both TE and TM waves. We found that for SCG the emissivity is 0.614097, while for SDG the emissivity is 0.592344 for one realization of a TE wave. The difference of emissivities of 0.021753 can give a difference of 300 K $\times$ 0.021753 = 6.53 K in brightness temperature. Even after averaging over 20 realizations, the emissivity for SCG still has a -0.013456 difference from that of SDG. For a TM wave, the difference of emissivity is much larger. It is -0.061699 for one realization and -0.092587 for 20 realizations. This gives differences of 18.5 K and 27.78 K in brightness temperatures, respectively. Thus SCG is not accurate for problems of large dielectric constant and cannot be used to calculate the emissivities. In Table 5.2.2, we compare the CPU. We note that although SDG is accurate, it requires much more CPU than SCG.

Method	n_d	Polarization	Number of Realizations	Emissivity	Difference of Emissivity***
SCG*	10	TE	1	0.614097	+0.021753
SDG**	30	TE	1	0.592344	0
PBTG-BMIA/CAG	10 and 30	TE	1	0.602337	0.0099
SCG	10	TM	1	0.703530	−0.061699
SDG	30	TM	1	0.765229	0
PBTG-BMIA/CAG	10 and 30	TM	1	0.768926	0.003697
SCG	10	TE	20	0.592318	−0.013456
SDG	30	TE	20	0.605774	0
PBTG-BMIA/CAG	10 and 30	TE	20	0.613663	0.007889
SCG	10	TM	20	0.701566	−0.092587
SDG	30	TM	20	0.794153	0
PBTG-BMIA/CAG	10 and 30	TM	20	0.797086	−0.002933

*SCG = single coarse grid **SDG = single dense grid

***The difference of emissivity means the emissivity minus the emissivity of SDG.

Table 5.2.1 Comparison of emissivities based on PBTG-BMIA/CAG and single grid method ($L = 100$ wavelength).

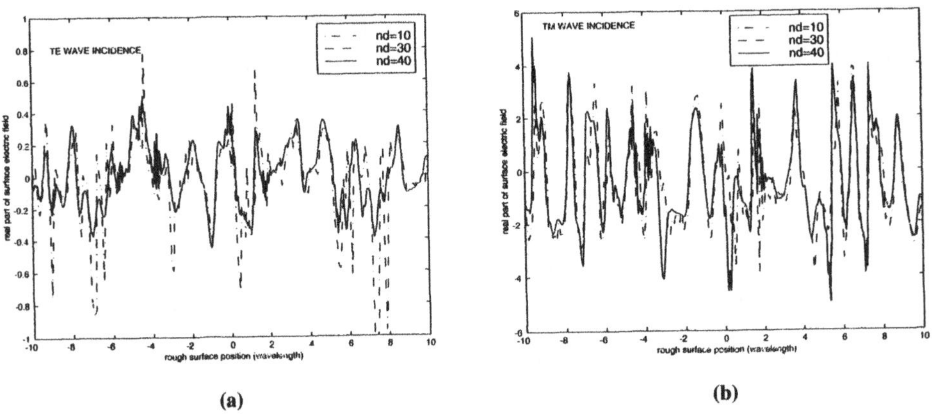

Figure 5.2.3 Comparison of the surface fields between the single dense grid of 30 points per wavelength and the single coarse grid of 10 points per wavelength. rms $h = 0.5\lambda$, correlation length of $l = 0.6\lambda$, dielectric constant of $\epsilon_r = 25 + i$, surface length of $L = 100\lambda$, and tapering parameter of $g = L/4$ at incidence angle of $\theta_i = 30°$. (a) TE wave (b) TM wave.

Method	n_d	No. of surface unknowns	Polarization	CPU time (s) per iteration	No. of iterations	CPU time (s)
SCG	10	2000	TE	14.5	416	6064.8
SDG	30	6000	TE	75.0	383	28757.0
PBTG-BMIA/CAG	10 and 30	6000	TE	11.6	662	7708.1
SCG	10	2000	TM	14.51	108	1595.0
SDG	30	6000	TM	77.0	108	8333.2
PBTG-BMIA/CAG	10 and 30	6000	TM	12.2	135	1655.0

Table 5.2.2 Comparison of CPU based on PBTG-BMIA/CAG and single grid method (1 realization and $L = 100$ wavelength).

b) Comparison Between a PBTG Combined with BMIA/CAG and a Single Dense Grid

In Figs. 5.2.4a and 5.2.4b, we compare the results of the bistatic scattering coefficients respectively obtained from a single realization of rough surface and averaged over 20 realizations of rough surfaces using SDG and PBTG-BMIA/CAG for a TE wave. For PBTG-BMIA/CAG, the two grids are used with $n_{cg} = 10$ and $n_{dg} = 30$. The results obtained by PBTG-BMIA/CAG are almost identical to the SDG results. In Figs. 5.2.5a and 5.2.5b, the comparisons are made for a TM wave which also show that PBTG-BMIA/CAG can give almost the same results as SDG. The comparisons of the surface fields between SDG and PBTG-BMIA/CAG for TE and TM cases are shown in Figs. 5.2.6a and 5.2.6b, respectively. The agreements are good. The emissivities calculated by SDG and PBTG-BMIA/CAG are compared in Table 5.2.1. The emissivities calculated by PBTG-BMIA/CAG are very close to those of SDG for TE and TM waves. The difference of emissivities averaged over 20 realizations between SDG and PBTG-BMIA/CAG is -0.007889 for a TE wave and -0.002933 for a TM wave. This leads to maximum differences of 2.3667 K and 0.8799 K in brightness temperatures, respectively. We also compare the CPU between PBTG-BMIA/CAG, SDG, and SCG. In Table 5.2.2, we give the comparisons of the total CPU and CPU per iteration based on PBTG-BMIA/CAG and single grid methods for one realization. The total CPU of PBTG-BMIA/CAG is slightly larger that of SCG. But the CPU of PBTG-BMIA/CAG is still several times smaller than that of SDG.

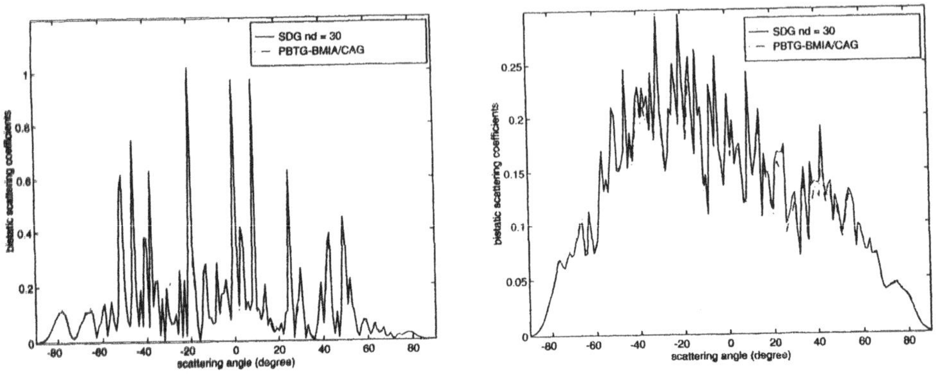

(a) (b)

Figure 5.2.4 Comparison of the bistatic scattering coefficients between the single dense grid of 30 points per wavelength and the PBTG-BMIA/CAG with $r_f = 1\lambda$. TE wave, rms $h = 0.5\lambda$, correlation length of $l = 0.6\lambda$, dielectric constant of $\epsilon_r = 25 + i$, surface length of $L = 100\lambda$, and tapering parameter of $g = L/4$ at incidence angle of $\theta_i = 30°$. (a) One realization, (b) 20 realizations.

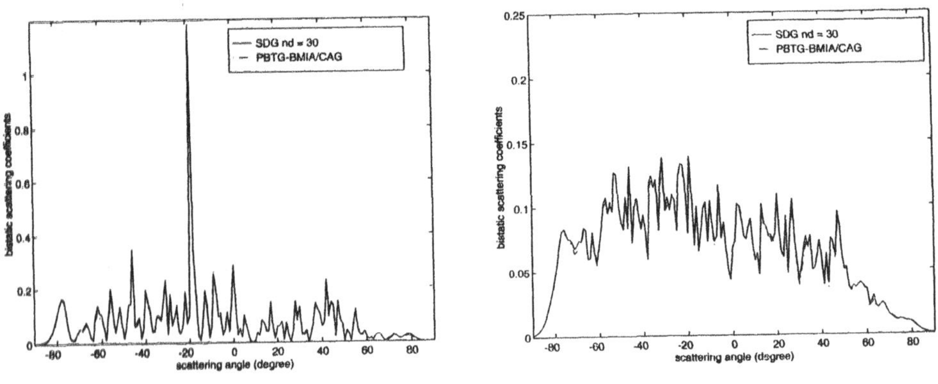

(a) (b)

Figure 5.2.5 Comparison of the bistatic scattering coefficients between the single dense grid of 30 points per wavelength and the PBTG-BMIA/CAG with $r_f = 1\lambda$. TM wave, rms $h = 0.5\lambda$, correlation length of $l = 0.6\lambda$, dielectric constant of $\epsilon_r = 25 + i$, surface length of $L = 100\lambda$, and tapering parameter of $g = L/4$ at incidence angle of $\theta_i = 30°$. (a) One realization, (b) 20 realizations.

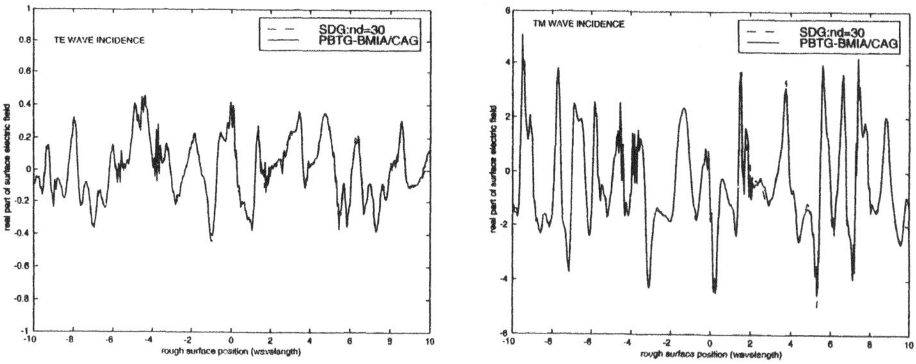

(a) (b)

Figure 5.2.6 Comparison of the surface fields between the single dense grid of 30 points per wavelength and the PBTG-BMIA/CAG with $r_f = 1\lambda$. rms $h = 0.5\lambda$, correlation length of $l = 0.6\lambda$, dielectric constant of $\epsilon_r = 25 + i$, surface length of $L = 100\lambda$, and tapering parameter of $g = L/4$ at incidence angle of $\theta_i = 30°$. (a) TE wave, (b) TM wave.

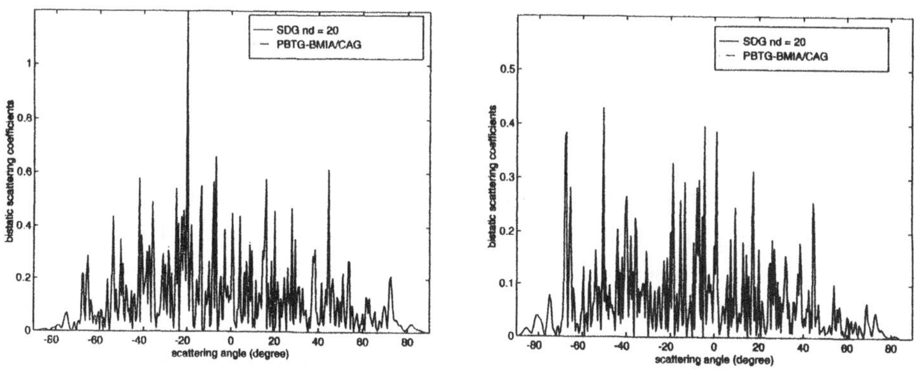

(a) (b)

Figure 5.2.7 Comparison of the bistatic scattering coefficients between the single dense grid of 20 points per wavelength and the PBTG-BMIA/CAG with $r_f = 1\lambda$ for one realization. rms $h = 0.3\lambda$, correlation length of $l = 0.5\lambda$, dielectric constant of $\epsilon_r = 17 + i$, surface length of $L = 500\lambda$, and tapering parameter of $g = L/8$ at incidence angle of $\theta_i = 30°$. (a) TE wave, (b) TM wave.

Method	Number of surface unknowns	Incident angle	Polari-zation	Emissivity	CPU time (s) per iterations	Number of itera-tions	CPU time (mins)
SDG	20,000	30	TE	0.639483	42.87	792	566
PBTG-BMIA/CAG	20,000	30	TE	0.645528	9.84	1579	259
SDG	20,000	30	TM	0.803430	34.52	252	145
PBTG-BMIA/CAG	20,000	30	TM	0.804980	9.57	326	52
SDG	20,000	85	TE	0.140942	40.72	358	243
PBTG-BMIA/CAG	20,000	85	TE	0.073925	9.23	604	93
SDG	20,000	85	TM	0.508661	35.58	145	86
PBTG-BMIA/CAG	20,000	85	TM	0.475900	9.83	183	30
PBTG-BMIA/CAG	30,000	85	TE		13.10	1323	290
PBTG-BMIA/CAG	30,000	85	TM		13.60	292	68

Table 5.2.3 Comparison of emissivity and CPU based on PBTG-BMIA/CAG and SDG method (1 realization and $L = 500$ wavelength).

c) Comparison Between a PBTG-BMIA/CAG and a SDG for the Large Surface Length Case

In Figs. 5.2.7a and 5.2.7b, the bistatic scattering coefficients obtained by PBTG-BMIA/CAG and SDG, respectively, are compared for the case of a large surface length of 500λ, an rms height of 0.3λ, a correlation length of 0.5λ, and a dielectric constant of $17 + i$ at incidence angle of 30° for one realization for both TE and TM waves. In this case, SDG has 20 points per wavelength. PBTG-BMIA/CAG is with $n_{cg} = 10$ and $n_{dg} = 20$. The agreements are good. The comparisons of emissivities and CPU are shown in Table 5.2.3. The uses of PBTG-BMIA/CAG with $n_{cg} = 10$ and $n_{dg} = 40$ are shown shown.

d) Backscattering Coefficients from a Rough Surface with a Large Dielectric Constant at Near-Grazing Incidence Angle

We also compare the bistatic scattering coefficients between a PBTG-BMIA/CAG and an SDG at an incidence angle of 85° in Figs. 5.2.8a and 5.2.8b. In this case, other parameters used are the same as the Figs. 5.2.7a and 5.2.7b. The agreements are good except for a small difference in the forward scattering directions. It is important to note that PBTG-BMIA/CAG gives accurate results in backscattering direction.

In Fig. 5.2.9a, we show the bistatic scattering coefficients of TE wave at 85° incidence angle averaged over various number of realizations. In

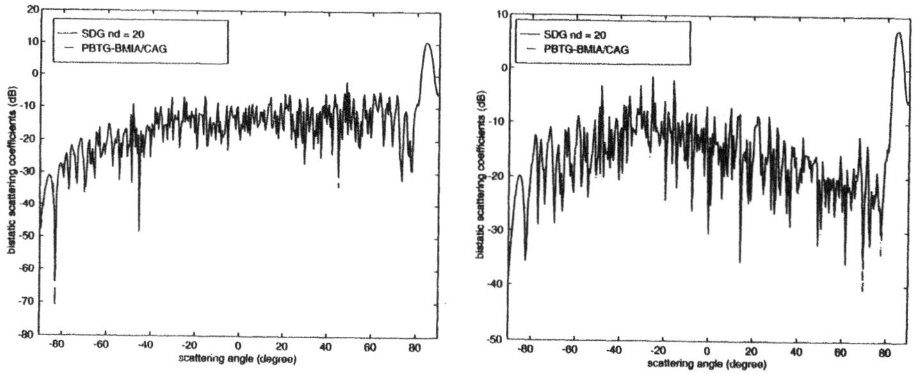

(a) (b)

Figure 5.2.8 Comparison of the bistatic scattering coefficients between the single dense grid of 20 points per wavelength and the PBTG-BMIA/CAG with $r_f = 1\lambda$ for one realization. rms $h = 0.3\lambda$, correlation length of $l = 0.5\lambda$, dielectric constant of $\epsilon_r = 17 + i$, surface length of $L = 500\lambda$, and tapering parameter of $g = L/8$ at incidence angle of $\theta_i = 85°$. (a) TE wave, (b) TM wave.

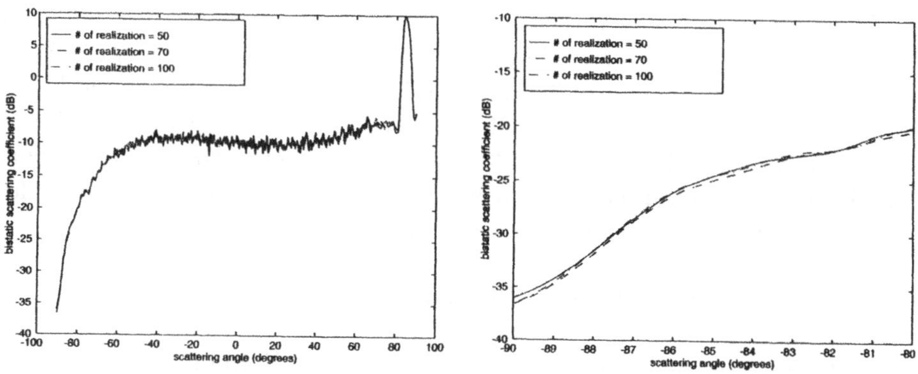

(a) (b)

Figure 5.2.9 Comparison of the bistatic scattering coefficients averaged over various number of realizations calculated by PBTG-BMIA/CAG with $r_f = 1\lambda$ and using the dense grid of 30 points per wavelength and coarse grid of 10 points per wavelength. TE wave, rms $h = 0.5\lambda$, correlation length of $l = 0.6\lambda$, dielectric constant of $\epsilon_r = 25 + i$, surface length of $L = 500\lambda$, and tapering parameter of $g = L/8$ at incidence angle of $\theta_i = 85°$. (a) Entire range of scattering angles, (b) vicinity of backscattering direction.

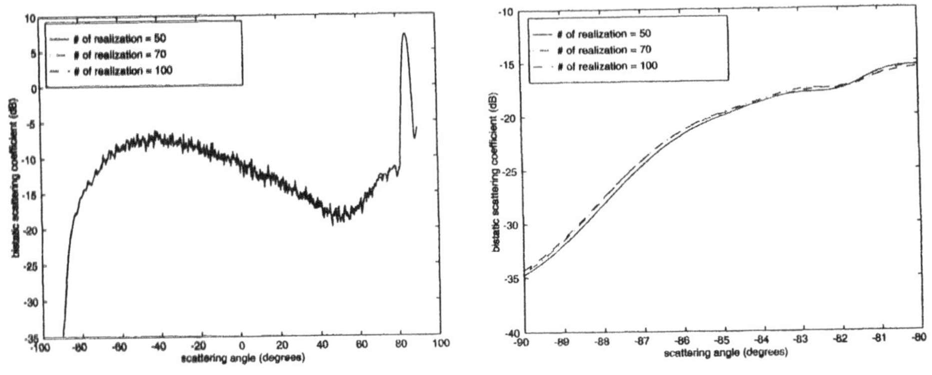

(a) (b)

Figure 5.2.10 Comparison of the bistatic scattering coefficients averaged over various number of realizations calculated by PBTG-BMIA/CAG with $r_f = 1\lambda$ and using the dense grid of 30 points per wavelength and coarse grid of 10 points per wavelength. TM wave, rms $h = 0.5\lambda$, correlation length of $l = 0.6\lambda$, dielectric constant of $\epsilon_r = 25 + i$, surface length of $L = 500\lambda$, and tapering parameter of $g = L/8$ at incidence angle of $\theta_i = 85°$. (a) Entire range of scattering angles, (b) vicinity of backscattering direction.

Fig. 5.2.9b, we zoom in and show the bistatic scattering coefficients in the vicinity of backscattering direction. In Figs. 5.2.10a and 5.2.10b, we show the corresponding results for a TM wave. We take the surface length of 500 wavelengths, an rms height of 0.5λ, a correlation length of 0.6λ, and a dielectric constant of $25 + i$. In PBTG-BMIA/CAG, we use two grids of $n_{cg} = 10$ points per wavelength and $n_{dg} = 30$ points per wavelength. We found that 50 realizations are required for convergence of backscattering coefficients for TE waves and 70 realizations are required for TM waves.

3 Steepest Descent Fast Multipole Method

The fast multipole method (FMM) was invented by Rohklin [Rohklin, 1990; Coifman et al. 1993]. Michielssen and Chew [1996] used the steepest descent method to express the product of impedance matrix and column vector. Jandhyala et al. [1998a,b] applied the steepest decent fast multipole method (SDFMM) to 3-D rough surface scattering problem. Recently, FMM has been applied extensively to large scale electromagnetic boundary value problems. SDFMM has also been combined with the PBTG method [Li, 2000]. In the following, we apply the method for one-dimensional rough surface for the Dirichlet problem. In Sections 3.1–3.4, the theoretical analysis is presented. In Section 3.5, the computational algorithm and the computational complexity of the approach are given.

3.1 Steepest Descent Path for Green's Function

The Green's function is

$$g(\bar{r}, \bar{r}') = \frac{i}{4} H_0^{(1)}(k|\bar{r} - \bar{r}'|) \tag{5.3.1}$$

with plane wave representation

$$g(x, z) = \frac{i}{4\pi} \int_{-\infty}^{\infty} dk_z e^{ik_x|x|} e^{ik_z z} \frac{1}{k_x} \tag{5.3.2}$$

Later on $x \to x - x'$, $z \to z - z'$.

The integration contour is on the real k_z axis. Next we make transformation to complex angle.

$$k_z = k\cos\alpha = k\cos(\alpha' + i\alpha'') = k_z' + ik_z'' \tag{5.3.3}$$

$$k_x = k\sin\alpha \tag{5.3.4}$$

Balancing real and imaginary parts

$$k_z' = k\cos\alpha' \cosh\alpha'' \tag{5.3.5}$$

$$k_z'' = -k\sin\alpha' \sinh\alpha'' \tag{5.3.6}$$

Similarly for k_x

$$k_x' + ik_x'' = k\sin\alpha' \cosh\alpha'' + ik\cos\alpha' \sinh\alpha'' \tag{5.3.7}$$

We also have $dk_z = -k\sin\alpha\, d\alpha$.

The original contour integral is now in the complex α plane and is denoted the Sommerfeld integration path (SIP)

$$g(x, z) = \frac{i}{4\pi} \int_{\text{SIP}} d\alpha\, e^{ik|x|\sin\alpha + ikz\cos\alpha} \tag{5.3.8}$$

The Sommerfeld integration path extends from $\alpha = 0 + i\infty$ to $\alpha = 0$ along the imaginary axis. It then goes from $\alpha = 0$ to $\alpha = \pi$ along the real α axis. Finally it goes from $\alpha = \pi$ to $\alpha = \pi - i\infty$ (Fig. 5.3.1). Since $|x| \gg |z|$, the dominant exponential term is

$$\begin{aligned} e^{ik|x|\sin\alpha} &= e^{i(k\sin\alpha' \cosh\alpha'' + ik|x|\cos\alpha' \sinh\alpha'')} \\ &= e^{ik|x| - k|x|\cos\alpha' \sinh\alpha''} \end{aligned} \tag{5.3.9}$$

The steepest descent path Γ is defined by

$$\text{Re}(\sin\alpha) = 1 \tag{5.3.10}$$

so that

$$\sin\alpha' \cosh(\alpha'') = 1 \tag{5.3.11}$$

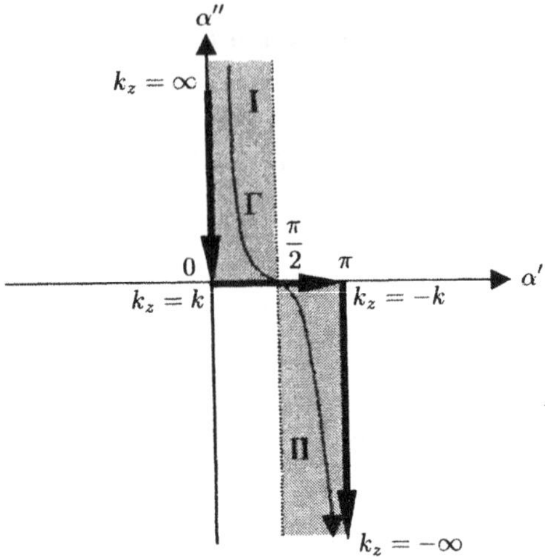

Figure 5.3.1 Steepest descent path and Sommerfeld integration path (SIP).

The saddle point is at $\alpha = \pi/2$. The steepest descent path Γ is a contour of constant phase for the dominant exponential term, the amplitude of which decreases rapidly away from the saddle point.

In the vicinity of $(\alpha', \alpha'') = (\frac{\pi}{2}, 0)$, by letting $\alpha' = \frac{\pi}{2} + \beta$, we have $\sin(\frac{\pi}{2} + \beta)\cosh\alpha'' = 1$ or $\cos\beta\cosh\alpha'' = 1$. For small β and α'', the equation becomes $\left(1 - \frac{\beta^2}{2}\right)\left(1 + \frac{(\alpha'')^2}{2}\right) = 1$, giving $\alpha'' = -\beta$. Thus the steepest descent path Γ has slope $= -1$ in the vicinity of $(\alpha', \alpha'') = (\frac{\pi}{2}, 0)$ as shown in Fig. 5.3.1.

Since the major contribution comes from the vicinity of $\alpha' = \frac{\pi}{2}$, the integration can be discretized as follows:

$$\begin{aligned} g(x, z) &= \frac{i}{4\pi} \int_\Gamma d\alpha \; e^{ik|x|\sin\alpha + ikz\cos\alpha} \\ &= \frac{i}{4\pi} \sum_{q=1}^{Q} e^{ik|x|\sin\alpha_q + ikz\cos\alpha_q} \Delta\alpha_q \end{aligned} \tag{5.3.12}$$

If we use the transformation $\alpha_{new} = \pi - \alpha$ and then let $\alpha_{new} \to \alpha$, we have

$$\begin{aligned} g(x, z) &= \frac{i}{4\pi} \int_\Gamma d\alpha e^{ik|x|\sin\alpha - ikz\cos\alpha} \\ &= \frac{i}{4\pi} \sum_{q=1}^{Q} e^{ik|x|\sin\alpha_q - ikz\cos\alpha_q} \Delta\alpha_q \end{aligned} \tag{5.3.13}$$

where Q is the number of angles. This means that the sign of the cosine term in the exponent in (5.3.12) can be switched. Let BW be the bandwidth on either side of $\alpha' = \frac{\pi}{2}$. We sample evenly on the α' axis with interval $\Delta\alpha'$. Let the number of angles Q be an odd integer.

$$\Delta\alpha' = \frac{2(BW)}{Q-1} \quad \text{(sample evenly on real } \alpha'\text{-axis)} \tag{5.3.14}$$

$$\alpha_q = \alpha'_q + i\alpha''_q \tag{5.3.15}$$

$$\alpha'_q = \frac{\pi}{2} - (BW) + (q-1)\Delta\alpha' \tag{5.3.16}$$

$$\cosh\alpha''_q = \frac{1}{\sin\alpha'_q} = \frac{1}{\cos\beta} \tag{5.3.17}$$

Note that α''_q is an odd function of β, $\alpha''_q > 0$ for $\alpha'_q < \frac{\pi}{2}$ and $\alpha''_q < 0$ for $\alpha'_q > \frac{\pi}{2}$. The selection of BW must be large enough to ensure that integrand becomes small enough. Note that BW depends on $|x|$. When $|x|$ is small the contribution is from larger range of α. The worst case and the largest BW corresponds to $|x| = x_{\text{min}}$, where x_{min} is the minimum separation in x. It is required that

$$kx_{\text{min}}(BW)^2 \gg 1 \tag{5.3.18}$$

Let

$$BW = \frac{C_l}{\sqrt{kx_{\text{min}}}} \quad (C_l = \text{large constant}) \tag{5.3.19}$$

On the other hand, the selection of sampling $\Delta\alpha'$ must be small enough to ensure enough sampling of the integrand. The worst case and the smallest $\Delta\alpha'$ corresponds to $|x| = x_{\text{max}}$, where x_{max} is the maximum separation in x. It is required that

$$kx_{\text{max}}(\Delta\alpha')^2 \ll 1 \tag{5.3.20}$$

Let

$$kx_{\text{max}}(\Delta\alpha')^2 = C_s \quad (C_s = \text{small constant}) \tag{5.3.21}$$

$$\Delta\alpha' = \frac{C_s}{\sqrt{k\,|x_{\text{max}}|}} \tag{5.3.22}$$

Thus the number of sampling points is on the order of

$$Q = \mathcal{O}\left(\frac{BW}{\Delta\alpha'}\right) = \mathcal{O}\left(\frac{C_l}{C_s}\sqrt{\frac{x_{\text{max}}}{x_{\text{min}}}}\right) \tag{5.3.23}$$

Note that Q can be a fixed constant if

$$\frac{x_{\text{max}}}{x_{\text{min}}} = \text{constant} \tag{5.3.24}$$

In the multilevel grouping, the division of groups is such that $x_{\max}/x_{\min}$ is the same for each level.

3.2 Multi-Level Impedance Matrix Decomposition and Grouping

In the fast multipole method, the impedance matrix elements are decomposed into various levels based on multilevel grouping. To illustrate, let the number of elements in the 1st level group be M and the number of groups of the 1st level be L. Then the total number of elements is $N = LM$ (e.g., $M = 20$, $L = 64$, then $N = 1280$ — We use this case as an illustration in this section. Generalization can readily be made to other values of M and L). Let $L = 2^{p+1}$, where p corresponds to the level of decomposition of the impedance matrix. The first level groups have M elements each. Beyond the first level, the number of elements increase by a factor of 2 for each higher level. Level 1 group has M elements, level 2 group has $2M$ elements, level 3 group has 2^2M elements, and so on. For $L = 64$, $p = 5$

level of group	# of elements	# of groups
1	M	64
2	$2M$	32
3	$4M$	16
4	$8M$	8
5	$16M$	4
6	$32M$	2

(5.3.25)

The grouping as shown in Fig. 5.3.2. Group m at level n is denoted as m_n.

The impedance matrix is decomposed as ($p = 5$ for this case)

$$\begin{aligned}\overline{\overline{Z}} = \overline{\overline{Z}}^{(0)} + \overline{\overline{Z}}^{(1)U} + \overline{\overline{Z}}^{(2)U} + \overline{\overline{Z}}^{(3)U} + \overline{\overline{Z}}^{(4)U} + \overline{\overline{Z}}^{(5)U} \\ + \overline{\overline{Z}}^{(1)L} + \overline{\overline{Z}}^{(2)L} + \overline{\overline{Z}}^{(3)L} + \overline{\overline{Z}}^{(4)L} + \overline{\overline{Z}}^{(5)L}\end{aligned} \quad (5.3.26)$$

where the number in the parenthesis stands for the level of the group, superscript U and L denote the upper matrix and lower matrix respectively. The upper matrix has column index larger than row index for nonzero elements. It is the reverse for the lower matrix.

Let $\overline{\overline{Z}}^{(i)}_{m_i n_i}$ denote the interaction of elements between group m_i and group n_i in level i. They are all full matrices.

$$\overline{\overline{Z}}^{(1)}_{m_i n_i} = \text{dimension } M \times M \quad (5.3.27)$$

$$\overline{\overline{Z}}^{(2)}_{m_i n_i} = \text{dimension } 2M \times 2M \quad (5.3.28)$$

Note that $\overline{\overline{Z}}^{(i)}_{m_i n_i}$ are defined differently from $\overline{\overline{Z}}^{(i)}$. For example, $\overline{\overline{Z}}^{(2)}$ is of di-

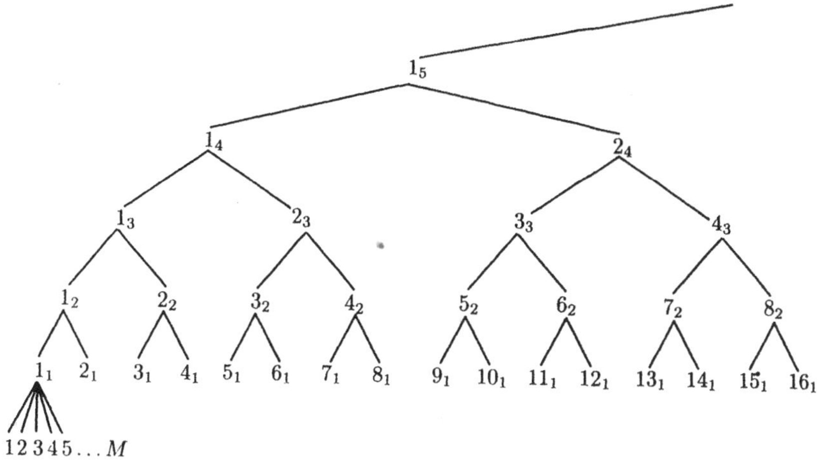

Figure 5.3.2 Multilevel fast multipoles structure.

mension $N \times N$ while $\overline{\overline{Z}}^{(2)}_{m_i n_i}$ is of dimension $2M \times 2M$. The various definitions of impedance matrix elements should be distinguished.

$$\overline{\overline{0}}^{(1)} = \text{zero matrix of dimension } M \times M \tag{5.3.29}$$

$$\overline{\overline{0}}^{(2)} = \text{zero matrix of dimension } 2M \times 2M \tag{5.3.30}$$

$$\overline{\overline{0}}^{(i)} = \text{zero matrix of dimension } (iM) \times (iM) \tag{5.3.31}$$

Suppose we use $M = 20$. Then examples are

$$\overline{\overline{Z}}^{(1)}_{3_1 9_1} = \begin{bmatrix} Z_{41,161} & \cdots & Z_{41,180} \\ \vdots & \ddots & \vdots \\ Z_{60,161} & \cdots & Z_{60,180} \end{bmatrix} = 20 \times 20 \tag{5.3.32}$$

$$\overline{\overline{Z}}^{(2)}_{3_2 9_2} = \begin{bmatrix} \overline{\overline{Z}}^{(1)}_{5_1 17_1} & \overline{\overline{Z}}^{(1)}_{5_1 18_1} \\ \overline{\overline{Z}}^{(1)}_{6_1 17_1} & \overline{\overline{Z}}^{(1)}_{6_1 18_1} \end{bmatrix} = 40 \times 40 \tag{5.3.33}$$

Thus $\overline{\overline{Z}}^{(1)}_{m_1 n_1}$, $\overline{\overline{Z}}^{(2)}_{m_2 n_2}$, $\overline{\overline{Z}}^{(3)}_{m_3 n_3}$, etc., keep track of all the individual impedance matrix elements.

The 0th level impedance matrix $\overline{\overline{Z}}^{(0)}$ represent interaction at level 1 between itself and its neighbors on the two sides.

$$\overline{\overline{Z}}^{(0)} = \begin{bmatrix} \overline{\overline{Z}}^{(1)}_{1_1 1_1} & \overline{\overline{Z}}^{(1)}_{1_1 2_1} & \overline{\overline{0}}^{(1)} & \overline{\overline{0}}^{(1)} & \cdots \\ \overline{\overline{Z}}^{(1)}_{2_1 1_1} & \overline{\overline{Z}}^{(1)}_{2_1 2_1} & \overline{\overline{Z}}^{(1)}_{2_1 3_1} & \overline{\overline{0}}^{(1)} & \cdots \\ \overline{\overline{0}}^{(1)} & \overline{\overline{Z}}^{(1)}_{3_1 2_1} & \overline{\overline{Z}}^{(1)}_{3_1 3_1} & \overline{\overline{Z}}^{(1)}_{3_1 4_1} & \cdots \\ \overline{\overline{0}}^{(1)} & \overline{\overline{0}}^{(1)} & \cdots & \cdots & \cdots \\ \cdots & \cdots & \cdots & \cdots & \cdots \end{bmatrix} = N \times N \tag{5.3.34}$$

For $\overline{\overline{Z}}^{(0)}$, the impedance matrix elements can be calculated accurately by using near field integrations.

In the multilevel fast multipole method, the impedance matrix of each level is generated by following rules:

(i) In each level, the impedance matrices only interact with itself or its nearest neighbors.

(ii) Each impedance matrix element in $\overline{\overline{Z}}$ can occur only once in the matrix decomposition. If a pair had interacted previously in lower level groups, it cannot interact again in the current level. That entry has to be set to zero at the current level.

To generate the next level, i.e., 1st level impedance matrix $\overline{\overline{Z}}^{(1)}$, we first apply rule (i) to get the $N \times N$ matrix

$$\begin{bmatrix} \overline{\overline{Z}}^{(2)}_{1_2 1_2} & \overline{\overline{Z}}^{(2)}_{1_2 2_2} & \overline{\overline{0}}^{(2)} & \overline{\overline{0}}^{(2)} & \cdots \\ \overline{\overline{Z}}^{(2)}_{2_2 1_2} & \overline{\overline{Z}}^{(2)}_{2_2 2_2} & \overline{\overline{Z}}^{(2)}_{2_2 3_2} & \overline{\overline{0}}^{(2)} & \cdots \\ \overline{\overline{0}}^{(2)} & \overline{\overline{Z}}^{(2)}_{3_2 2_2} & \overline{\overline{Z}}^{(2)}_{3_2 3_2} & \overline{\overline{Z}}^{(2)}_{3_2 4_2} & \cdots \\ \overline{\overline{0}}^{(2)} & \overline{\overline{0}}^{(2)} & \cdots & \cdots & \cdots \\ \cdots & \cdots & \cdots & \cdots & \cdots \end{bmatrix} \tag{5.3.35}$$

However, according to rule (ii), the diagonal matrix elements should be set to zero because they have already interacted in $\overline{\overline{Z}}^{(0)}$. Setting them to zero then gives

$$\begin{bmatrix} \overline{\overline{0}}^{(2)} & \overline{\overline{Z}}^{(2)}_{1_2 2_2} & \overline{\overline{0}}^{(2)} & \overline{\overline{0}}^{(2)} & \cdots \\ \overline{\overline{Z}}^{(2)}_{2_2 1_2} & \overline{\overline{0}}^{(2)} & \overline{\overline{Z}}^{(2)}_{2_2 3_2} & \overline{\overline{0}}^{(2)} & \cdots \\ \overline{\overline{0}}^{(2)} & \overline{\overline{Z}}^{(2)}_{3_2 2_2} & \overline{\overline{0}}^{(2)} & \overline{\overline{Z}}^{(2)}_{3_2 4_2} & \cdots \\ \overline{\overline{0}}^{(2)} & \overline{\overline{0}}^{(2)} & \cdots & \cdots & \cdots \\ \cdots & \cdots & \cdots & \cdots & \cdots \end{bmatrix} \tag{5.3.36}$$

However, in

$$\overline{\overline{Z}}^{(2)}_{1_2 2_2} = \begin{bmatrix} \overline{\overline{Z}}^{(1)}_{1_1 3_1} & \overline{\overline{Z}}^{(1)}_{1_1 4_1} \\ \overline{\overline{Z}}^{(1)}_{2_1 3_1} & \overline{\overline{Z}}^{(1)}_{2_1 4_1} \end{bmatrix} \tag{5.3.37}$$

$\overline{\overline{Z}}^{(1)}_{2_1 3_1}$ has already been included in $\overline{\overline{Z}}^{(0)}$. Thus, we set that entry to zero also and define, with a prime superscript,

$$\overline{\overline{Z}}^{(2)\prime}_{1_2 2_2} = \begin{bmatrix} \overline{\overline{Z}}^{(1)}_{1_1 3_1} & \overline{\overline{Z}}^{(1)}_{1_1 4_1} \\ \overline{\overline{0}}^{(1)} & \overline{\overline{Z}}^{(1)}_{2_1 4_1} \end{bmatrix} = \text{dimension } 2M \times 2M \tag{5.3.38}$$

Similarly, we define

$$\overline{\overline{Z}}^{(2)\prime}_{2_2 1_2} = \begin{bmatrix} \overline{\overline{Z}}^{(1)}_{3_1 1_1} & \overline{\overline{0}}^{(1)} \\ \overline{\overline{Z}}^{(1)}_{4_1 1_1} & \overline{\overline{Z}}^{(1)}_{4_1 2_1} \end{bmatrix} \tag{5.3.39}$$

Thus, the first level impedance matrix assumes the final form:

$$\overline{\overline{Z}}^{(1)} = \overline{\overline{Z}}^{(1)U} + \overline{\overline{Z}}^{(1)L} \tag{5.3.40}$$

$$\overline{\overline{Z}}^{(1)U} = \begin{bmatrix} \overline{\overline{0}}^{(2)} & \overline{\overline{Z}}^{(2)\prime}_{1_2 2_2} & \overline{\overline{0}}^{(2)} & \overline{\overline{0}}^{(2)} & \cdots \\ \overline{\overline{0}}^{(2)} & \overline{\overline{0}}^{(2)} & \overline{\overline{Z}}^{(2)\prime}_{2_2 3_2} & \overline{\overline{0}}^{(2)} & \cdots \\ \overline{\overline{0}}^{(2)} & \overline{\overline{0}}^{(2)} & \overline{\overline{0}}^{(2)} & \overline{\overline{Z}}^{(2)\prime}_{3_2 4_2} & \cdots \\ \overline{\overline{0}}^{(2)} & \overline{\overline{0}}^{(2)} & \cdots & \cdots & \cdots \\ \cdots & \cdots & \cdots & \cdots & \cdots \end{bmatrix}$$

$$= \begin{bmatrix} \overline{\overline{0}}^{(1)} & \overline{\overline{0}}^{(1)} & \overline{\overline{Z}}^{(1)}_{1_1 3_1} & \overline{\overline{Z}}^{(1)}_{1_1 4_1} & \overline{\overline{0}}^{(1)} & \cdots & \cdots & \cdots & \cdots \\ \overline{\overline{0}}^{(1)} & \overline{\overline{0}}^{(1)} & \overline{\overline{0}}^{(1)} & \overline{\overline{Z}}^{(1)}_{2_1 4_1} & \overline{\overline{0}}^{(1)} & \overline{\overline{0}}^{(1)} & \cdots & \cdots & \cdots \\ \cdots & \cdots & \overline{\overline{0}}^{(1)} & \overline{\overline{0}}^{(1)} & \overline{\overline{Z}}^{(1)}_{3_1 5_1} & \overline{\overline{Z}}^{(1)}_{3_1 6_1} & \overline{\overline{0}}^{(1)} & \cdots & \cdots \\ \cdots & \cdots & \cdots & \overline{\overline{0}}^{(1)} & \overline{\overline{0}}^{(1)} & \overline{\overline{Z}}^{(1)}_{4_1 6_1} & \overline{\overline{0}}^{(1)} & \overline{\overline{0}}^{(1)} & \cdots \\ \cdots & \cdots & \cdots & \cdots & \overline{\overline{0}}^{(1)} & \overline{\overline{0}}^{(1)} & \overline{\overline{Z}}^{(1)}_{5_1 7_1} & \overline{\overline{Z}}^{(1)}_{5_1 8_1} & \cdots \\ \cdots & \cdots & \cdots & \cdots & \cdots & \overline{\overline{0}}^{(1)} & \overline{\overline{0}}^{(1)} & \overline{\overline{Z}}^{(1)}_{6_1 8_1} & \cdots \\ \cdots & \cdots & \cdots & \cdots & \cdots & \cdots & \cdots & \cdots & \cdots \end{bmatrix} \tag{5.3.41}$$

$$\overline{\overline{Z}}^{(1)L} = \begin{bmatrix} \overline{\overline{0}}^{(2)} & \overline{\overline{0}}^{(2)} & \overline{\overline{0}}^{(2)} & \overline{\overline{0}}^{(2)} & \cdots \\ \overline{\overline{Z}}^{(2)\prime}_{2_2 1_2} & \overline{\overline{0}}^{(2)} & \overline{\overline{0}}^{(2)} & \overline{\overline{0}}^{(2)} & \cdots \\ \overline{\overline{0}}^{(2)} & \overline{\overline{Z}}^{(2)\prime}_{3_2 2_2} & \overline{\overline{0}}^{(2)} & \overline{\overline{0}}^{(2)} & \cdots \\ \overline{\overline{0}}^{(2)} & \overline{\overline{0}}^{(2)} & \cdots & \cdots & \cdots \\ \cdots & \cdots & \cdots & \cdots & \cdots \end{bmatrix}$$

$$= \begin{bmatrix} \overline{\overline{0}}^{(1)} & \cdots & \cdots & \cdots & \cdots & \cdots & \cdots \\ \overline{\overline{0}}^{(1)} & \overline{\overline{0}}^{(1)} & \cdots & \cdots & \cdots & \cdots & \cdots \\ \overline{\overline{Z}}^{(1)}_{3_1 1_1} & \overline{\overline{0}}^{(1)} & \overline{\overline{0}}^{(1)} & \cdots & \cdots & \cdots & \cdots \\ \overline{\overline{Z}}^{(1)}_{4_1 2_1} & \overline{\overline{Z}}^{(1)}_{4_1 2_1} & \overline{\overline{0}}^{(1)} & \overline{\overline{0}}^{(1)} & \cdots & \cdots & \cdots \\ \overline{\overline{0}}^{(1)} & \overline{\overline{0}}^{(1)} & \overline{\overline{Z}}^{(1)}_{5_1 3_1} & \overline{\overline{0}}^{(1)} & \overline{\overline{0}}^{(1)} & \cdots & \cdots \\ \cdots & \overline{\overline{0}}^{(1)} & \overline{\overline{Z}}^{(1)}_{6_1 3_1} & \overline{\overline{Z}}^{(1)}_{6_1 4_1} & \overline{\overline{0}}^{(1)} & \cdots & \cdots \\ \cdots & \cdots & \cdots & \cdots & \cdots & \cdots & \cdots \end{bmatrix} \tag{5.3.42}$$

For the 2nd level impedance matrix, we have

$$\overline{\overline{Z}}^{(2)} = \overline{\overline{Z}}^{(2)U} + \overline{\overline{Z}}^{(2)L} \tag{5.3.43}$$

Note that

$$\overline{\overline{Z}}^{(3)}_{1_3 2_3} = \begin{bmatrix} \overline{\overline{Z}}^{(2)}_{1_2 3_2} & \overline{\overline{Z}}^{(2)}_{1_2 4_2} \\ \overline{\overline{Z}}^{(2)}_{2_2 3_2} & \overline{\overline{Z}}^{(2)}_{2_2 4_2} \end{bmatrix} \tag{5.3.44}$$

But $\overline{\overline{Z}}^{(2)}_{2_2 3_2}$ has already been included in previous level impedance matrices $\overline{\overline{Z}}^{(0)}$ or $\overline{\overline{Z}}^{(1)U} + \overline{\overline{Z}}^{(1)L}$. Thus, we set that part to zero and define, with a prime superscript,

$$\overline{\overline{Z}}^{(3)\prime}_{1_3 2_3} = \begin{bmatrix} \overline{\overline{Z}}^{(2)}_{1_2 3_2} & \overline{\overline{Z}}^{(2)}_{1_2 4_2} \\ \overline{\overline{0}}^{(2)} & \overline{\overline{Z}}^{(2)}_{2_2 4_2} \end{bmatrix} = \text{dimension } 2^2M \times 2^2M$$

$$= \begin{bmatrix} \overline{\overline{Z}}^{(1)}_{1_1 5_1} & \overline{\overline{Z}}^{(1)}_{1_1 6_1} & \overline{\overline{Z}}^{(1)}_{1_1 7_1} & \overline{\overline{Z}}^{(1)}_{1_1 8_1} \\ \overline{\overline{Z}}^{(1)}_{2_1 5_1} & \overline{\overline{Z}}^{(1)}_{2_1 6_1} & \overline{\overline{Z}}^{(1)}_{2_1 7_1} & \overline{\overline{Z}}^{(1)}_{2_1 8_1} \\ \overline{\overline{0}} & \overline{\overline{0}} & \overline{\overline{Z}}^{(1)}_{3_1 7_1} & \overline{\overline{Z}}^{(1)}_{3_1 8_1} \\ \overline{\overline{0}} & \overline{\overline{0}} & \overline{\overline{Z}}^{(1)}_{4_1 7_1} & \overline{\overline{Z}}^{(1)}_{4_1 8_1} \end{bmatrix} \tag{5.3.45}$$

Similarly, let

$$\overline{\overline{Z}}^{(3)\prime}_{2_3 1_3} = \begin{bmatrix} \overline{\overline{Z}}^{(2)}_{3_2 1_2} & \overline{\overline{0}}^{(2)} \\ \overline{\overline{Z}}^{(2)}_{4_2 1_2} & \overline{\overline{Z}}^{(2)}_{4_2 2_2} \end{bmatrix} \tag{5.3.46}$$

Thus the 2nd level upper matrix $\overline{\overline{\overline{Z}}}^{(2)U}$ is

$$
\begin{aligned}
\overline{\overline{\overline{Z}}}^{(2)U} &= \begin{bmatrix}
\overline{\overline{0}}^{(3)} & \overline{\overline{\overline{Z}}}^{(3)\prime}_{1_3 2_3} & \overline{\overline{0}}^{(3)} & \overline{\overline{0}}^{(3)} & \cdots \\
\overline{\overline{0}}^{(3)} & \overline{\overline{0}}^{(3)} & \overline{\overline{\overline{Z}}}^{(3)\prime}_{2_3 3_3} & \overline{\overline{0}}^{(3)} & \cdots \\
\overline{\overline{0}}^{(3)} & \overline{\overline{0}}^{(3)} & \overline{\overline{0}}^{(3)} & \overline{\overline{\overline{Z}}}^{(3)\prime}_{3_3 4_3} & \cdots \\
\overline{\overline{0}}^{(3)} & \overline{\overline{0}}^{(3)} & \cdots & \cdots & \cdots \\
\cdots & \cdots & \cdots & \cdots & \cdots
\end{bmatrix} \\
&= \begin{bmatrix}
\overline{\overline{0}}^{(2)} & \overline{\overline{0}}^{(2)} & \overline{\overline{\overline{Z}}}^{(2)}_{1_2 3_2} & \overline{\overline{\overline{Z}}}^{(2)}_{1_2 4_2} & \overline{\overline{0}}^{(2)} & \cdots & \cdots & \cdots & \cdots \\
\overline{\overline{0}}^{(2)} & \overline{\overline{0}}^{(2)} & \overline{\overline{0}}^{(2)} & \overline{\overline{\overline{Z}}}^{(2)}_{2_2 4_2} & \overline{\overline{0}}^{(2)} & \overline{\overline{0}}^{(2)} & \cdots & \cdots & \cdots \\
\cdots & \cdots & \overline{\overline{0}}^{(2)} & \overline{\overline{0}}^{(2)} & \overline{\overline{\overline{Z}}}^{(2)}_{3_2 5_2} & \overline{\overline{\overline{Z}}}^{(2)}_{3_2 6_2} & \overline{\overline{0}}^{(2)} & \cdots & \cdots \\
\cdots & \cdots & \cdots & \overline{\overline{0}}^{(2)} & \overline{\overline{0}}^{(2)} & \overline{\overline{\overline{Z}}}^{(2)}_{4_2 6_2} & \overline{\overline{0}}^{(2)} & \overline{\overline{0}}^{(2)} & \cdots \\
\cdots & \cdots & \cdots & \cdots & \overline{\overline{0}}^{(2)} & \overline{\overline{0}}^{(2)} & \overline{\overline{\overline{Z}}}^{(2)}_{5_2 7_2} & \overline{\overline{\overline{Z}}}^{(2)}_{5_2 8_2} & \cdots \\
\cdots & \cdots & \cdots & \cdots & \cdots & \overline{\overline{0}}^{(2)} & \overline{\overline{0}}^{(2)} & \overline{\overline{\overline{Z}}}^{(2)}_{6_2 8_2} & \cdots \\
\cdots & \cdots & \cdots & \cdots & \cdots & \cdots & \cdots & \cdots & \cdots
\end{bmatrix} \\
&= \begin{bmatrix}
\overline{\overline{0}}^{(1)} & \overline{\overline{0}}^{(1)} & \overline{\overline{0}}^{(1)} & \overline{\overline{0}}^{(1)} & \overline{\overline{\overline{Z}}}^{(1)}_{1_1 5_1} & \overline{\overline{\overline{Z}}}^{(1)}_{1_1 6_1} & \overline{\overline{\overline{Z}}}^{(1)}_{1_1 7_1} & \overline{\overline{\overline{Z}}}^{(1)}_{1_1 8_1} & \overline{\overline{0}}^{(1)} & \cdots \\
\overline{\overline{0}}^{(1)} & \overline{\overline{0}}^{(1)} & \overline{\overline{0}}^{(1)} & \overline{\overline{0}}^{(1)} & \overline{\overline{\overline{Z}}}^{(1)}_{2_1 5_1} & \overline{\overline{\overline{Z}}}^{(1)}_{2_1 6_1} & \overline{\overline{\overline{Z}}}^{(1)}_{2_1 7_1} & \overline{\overline{\overline{Z}}}^{(1)}_{2_1 8_1} & \overline{\overline{0}}^{(1)} & \cdots \\
\cdots & \cdots & \overline{\overline{0}}^{(1)} & \overline{\overline{0}}^{(1)} & \overline{\overline{0}}^{(1)} & \overline{\overline{0}}^{(1)} & \overline{\overline{\overline{Z}}}^{(1)}_{3_1 7_1} & \overline{\overline{\overline{Z}}}^{(1)}_{3_1 8_1} & \overline{\overline{0}}^{(1)} & \cdots \\
\cdots & \cdots & \cdots & \overline{\overline{0}}^{(1)} & \overline{\overline{0}}^{(1)} & \overline{\overline{0}}^{(1)} & \overline{\overline{\overline{Z}}}^{(1)}_{4_1 7_1} & \overline{\overline{\overline{Z}}}^{(1)}_{4_1 8_1} & \overline{\overline{0}}^{(1)} & \cdots \\
\cdots & \cdots & \cdots & \cdots & \overline{\overline{0}}^{(1)} & \overline{\overline{0}}^{(1)} & \overline{\overline{0}}^{(1)} & \overline{\overline{0}}^{(1)} & \overline{\overline{\overline{Z}}}^{(1)}_{5_1 9_1} & \cdots \\
\cdots & \cdots & \cdots & \cdots & \cdots & \overline{\overline{0}}^{(1)} & \overline{\overline{0}}^{(1)} & \overline{\overline{0}}^{(1)} & \overline{\overline{\overline{Z}}}^{(1)}_{6_1 9_1} & \cdots \\
\cdots & \cdots & \cdots & \cdots & \cdots & \cdots & \cdots & \overline{\overline{0}}^{(1)} & \overline{\overline{0}}^{(1)} & \cdots \\
\cdots & \cdots & \cdots & \cdots & \cdots & \cdots & \cdots & \cdots & \cdots & \cdots
\end{bmatrix}
\end{aligned}
\tag{5.3.47}
$$

A similar expression can be found for $\overline{\overline{\overline{Z}}}^{(2)L}$.

Similar derivation can be applied to higher levels. For level 3, $\overline{\overline{\overline{Z}}}^{(3)} = \overline{\overline{\overline{Z}}}^{(3)U} + \overline{\overline{\overline{Z}}}^{(3)L}$, with

$$
\overline{\overline{Z}}^{(3)U} = \begin{bmatrix}
\overline{\overline{0}}^{(4)} & \overline{\overline{Z}}^{(4)\prime}_{1_4 2_4} & \overline{\overline{0}}^{(4)} & \overline{\overline{0}}^{(4)} & \cdots \\
\overline{\overline{0}}^{(4)} & \overline{\overline{0}}^{(4)} & \overline{\overline{Z}}^{(4)\prime}_{2_4 3_4} & \overline{\overline{0}}^{(4)} & \cdots \\
\overline{\overline{0}}^{(4)} & \overline{\overline{0}}^{(4)} & \overline{\overline{0}}^{(4)} & \overline{\overline{Z}}^{(4)\prime}_{3_4 4_4} & \cdots \\
\overline{\overline{0}}^{(4)} & \overline{\overline{0}}^{(4)} & \cdots & \cdots & \cdots \\
\cdots & \cdots & \cdots & \cdots & \cdots
\end{bmatrix}
$$

$$
= \begin{bmatrix}
\overline{\overline{0}}^{(3)} & \overline{\overline{0}}^{(3)} & \overline{\overline{Z}}^{(3)}_{1_3 3_3} & \overline{\overline{Z}}^{(3)}_{1_3 4_3} & \overline{\overline{0}}^{(3)} & \cdots & \cdots & \cdots & \cdots \\
\overline{\overline{0}}^{(3)} & \overline{\overline{0}}^{(3)} & \overline{\overline{0}}^{(3)} & \overline{\overline{Z}}^{(3)}_{2_3 4_3} & \overline{\overline{0}}^{(3)} & \overline{\overline{0}}^{(3)} & \cdots & \cdots & \cdots \\
\cdots & \cdots & \overline{\overline{0}}^{(3)} & \overline{\overline{0}}^{(3)} & \overline{\overline{Z}}^{(3)}_{3_3 5_3} & \overline{\overline{Z}}^{(3)}_{3_3 6_3} & \overline{\overline{0}}^{(3)} & \cdots & \cdots \\
\cdots & \cdots & \cdots & \overline{\overline{0}}^{(3)} & \overline{\overline{0}}^{(3)} & \overline{\overline{Z}}^{(3)}_{4_3 6_3} & \overline{\overline{0}}^{(3)} & \overline{\overline{0}}^{(3)} & \cdots \\
\cdots & \cdots & \cdots & \cdots & \overline{\overline{0}}^{(3)} & \overline{\overline{0}}^{(3)} & \overline{\overline{Z}}^{(3)}_{5_3 7_3} & \overline{\overline{Z}}^{(3)}_{5_3 8_3} & \cdots \\
\cdots & \cdots & \cdots & \cdots & \cdots & \overline{\overline{0}}^{(3)} & \overline{\overline{0}}^{(3)} & \overline{\overline{Z}}^{(3)}_{6_3 8_3} & \cdots \\
\cdots & \cdots & \cdots & \cdots & \cdots & \cdots & \cdots & \cdots & \cdots
\end{bmatrix}
\tag{5.3.48}
$$

A similar expression can be found for $\overline{\overline{Z}}^{(3)L}$.

Continuing in this manner, the original impedance matrix is decomposed up to $\overline{\overline{Z}}^{(5)} = \overline{\overline{Z}}^{(5)U} + \overline{\overline{Z}}^{(5)L}$.

3.3 Multi-Level Discretization of Angles and Interpolation

Because of the manner the impedance matrix elements are grouped at various levels, a constant $x_{\max}/x_{\min}$ ratio in different levels of $\overline{\overline{Z}}^{(i)}$ is maintained.

Figure 5.3.3 illustrates the ratio $x^{(1)}_{\max}/x^{(1)}_{\min}$ for level 1 group in $\overline{\overline{Z}}^{(1)}$. Note that $x^{(1)}_{\max}$ and $x^{(1)}_{\min}$ are as shown because we have deleted those elements that belong to $\overline{\overline{Z}}^{(0)}$.

In $\overline{\overline{Z}}^{(1)}$

$$x^{(1)}_{\max} \simeq 4M\Delta x \tag{5.3.49}$$

$$x^{(1)}_{\min} \simeq M\Delta x \tag{5.3.50}$$

$$\frac{x^{(1)}_{\max}}{x^{(1)}_{\min}} = 4 \tag{5.3.51}$$

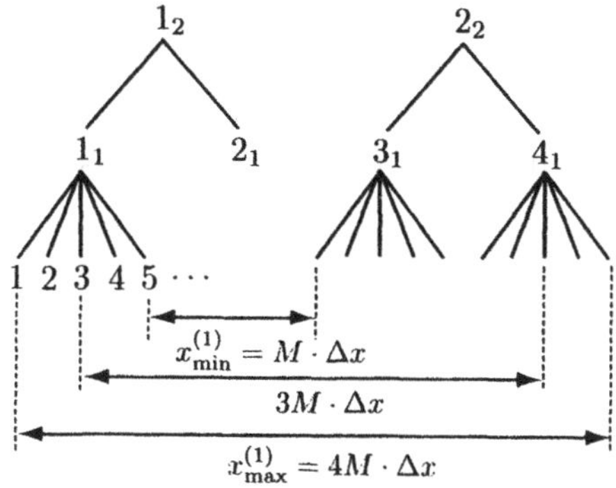

Figure 5.3.3 The x_{max}/x_{min} ratio for level 1 group. $x_{max}/x_{min} = 4$.

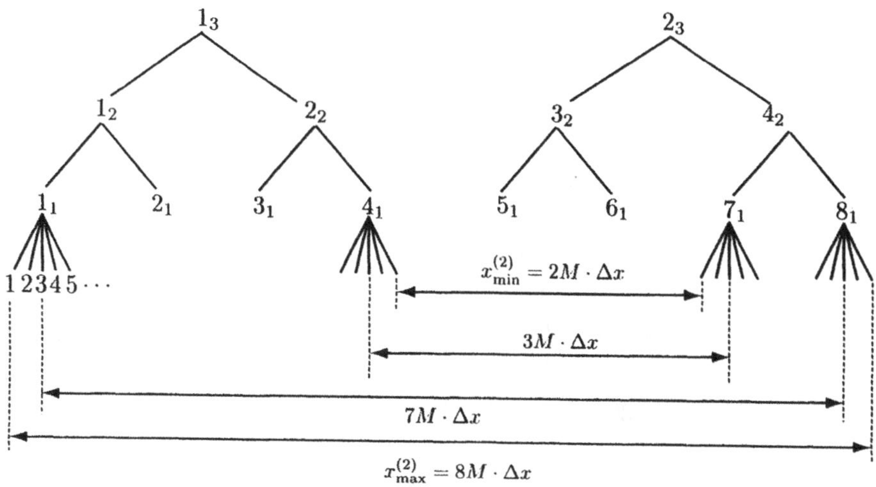

Figure 5.3.4 The x_{max}/x_{min} ratio for level 2 group.

Figure 5.3.4 illustrates the ratio $x_{max}^{(2)}/x_{min}^{(2)}$ for level 2 group in $\overline{\overline{Z}}^{(2)}$. Note that $x_{max}^{(2)}$ and $x_{min}^{(2)}$ are as shown because we have deleted those elements that belong to $\overline{\overline{Z}}^{(0)}$ and $\overline{\overline{Z}}^{(1)}$. In $\overline{\overline{Z}}^{(2)}$

$$x_{max}^{(2)} \simeq 4(2M)\Delta x \tag{5.3.52}$$

$$x_{min}^{(2)} \simeq 2M\Delta x \tag{5.3.53}$$

$$\frac{x_{max}^{(2)}}{x_{min}^{(2)}} = 4 \tag{5.3.54}$$

Similarly, $x_{\max}^{(i)}/x_{\min}^{(i)} = 4$ for general level i. Because of the ratio of $x_{\max}^{(i)}/x_{\min}^{(i)}$ is equal to 4 for all levels, the number of angles Q in each level is kept constant by (5.3.23).

Thus from (5.3.22), the sampling interval of angles in level i is,

$$\Delta\alpha'^{(i)} = \frac{C_s}{\sqrt{k\left|x_{\max}^{(i)}\right|}} \tag{5.3.55}$$

$$Q \simeq \frac{2C_l}{C_s}\sqrt{\frac{\left|x_{\max}^{(i)}\right|}{\left|x_{\min}^{(i)}\right|}} \simeq \frac{2C_l}{C_s}\cdot 2 = \frac{4C_l}{C_s} \tag{5.3.56}$$

Thus if we choose $C_l \simeq 3$, $C_s = \frac{1}{3}$ then Q can be rounded to 37. Let

$$Q = 37 \tag{5.3.57}$$

$$x_{\max}^{(i)} = 2^{i-1}(4M\Delta x) \tag{5.3.58}$$

From (5.3.22),

$$\Delta\alpha'^{(i)} = \frac{\frac{1}{3}}{\sqrt{k\left|x_{\max}^{(i)}\right|}} \tag{5.3.59}$$

$$BW^{(i)} = \left(\frac{Q-1}{2}\right)\Delta\alpha'^{(i)} = 18\Delta\alpha'^{(i)} \tag{5.3.60}$$

For example, let $M = 20$, $\Delta x = \frac{\lambda}{10}$, then for $i = 1$

$$x_{\max}^{(1)} = 4(20)\frac{\lambda}{10} = 8\lambda \tag{5.3.61}$$

$$\Delta\alpha'^{(i)} = \frac{\frac{1}{3}}{\sqrt{\frac{2\pi}{\lambda}8\lambda}} = \frac{1}{3(7.08)} = 0.047 \text{ rad} \tag{5.3.62}$$

$$BW^{(1)} = 0.846 \text{ rad} \tag{5.3.63}$$

For $i = 2$

$$x_{\max}^{(2)} = 16\lambda \tag{5.3.64}$$

$$\Delta\alpha'^{(2)} = 0.033 \text{ rad} \tag{5.3.65}$$

$$BW^{(2)} = 0.6 \text{ rad} \tag{5.3.66}$$

We tabulate $\alpha_q^{(1)}$, $\alpha_q^{(2)}$, and $\alpha_q^{(3)}$ in Table 5.3.1. Note that the set of angles

q	$\alpha_q^{(1)}$	$\alpha_q^{(2)}$	$\alpha_q^{(3)}$
1	$0.7245 + 0.9703i$	$0.9724 + 0.6377i$	$1.1477 + 0.4364i$
2	$0.7715 + 0.9011i$	$1.0056 + 0.5979i$	$1.1712 + 0.4107i$
3	$0.8185 + 0.8353i$	$1.0389 + 0.5589i$	$1.1947 + 0.3853i$
4	$0.8656 + 0.7722i$	$1.0721 + 0.5207i$	$1.2182 + 0.3602i$
5	$0.9126 + 0.7117i$	$1.1054 + 0.4832i$	$1.2417 + 0.3352i$
6	$0.9596 + 0.6533i$	$1.1386 + 0.4463i$	$1.2652 + 0.3105i$
7	$1.0066 + 0.5967i$	$1.1719 + 0.4100i$	$1.2887 + 0.2859i$
8	$1.0536 + 0.5419i$	$1.2051 + 0.3741i$	$1.3122 + 0.2615i$
9	$1.1006 + 0.4885i$	$1.2383 + 0.3388i$	$1.3357 + 0.2373i$
10	$1.1477 + 0.4364i$	$1.2716 + 0.3038i$	$1.3592 + 0.2132i$
11	$1.1947 + 0.3853i$	$1.3048 + 0.2692i$	$1.3827 + 0.1892i$
12	$1.2417 + 0.3352i$	$1.3381 + 0.2348i$	$1.4062 + 0.1653i$
13	$1.2887 + 0.2859i$	$1.3713 + 0.2008i$	$1.4297 + 0.1415i$
14	$1.3357 + 0.2373i$	$1.4046 + 0.1670i$	$1.4533 + 0.1178i$
15	$1.3827 + 0.1892i$	$1.4378 + 0.1334i$	$1.4768 + 0.0942i$
16	$1.4297 + 0.1415i$	$1.4711 + 0.0999i$	$1.5003 + 0.0706i$
17	$1.4768 + 0.0942i$	$1.5043 + 0.0665i$	$1.5238 + 0.0470i$
18	$1.5238 + 0.0470i$	$1.5376 + 0.0333i$	$1.5473 + 0.0235i$
19	1.5708	1.5708	1.5708

Table 5.3.1 The α_q angles for the first three levels. Parameters used: $M = 20$, $\Delta x = \frac{\lambda}{10}$, $Q = 37$.

for $q > \frac{Q+1}{2} = 19$ is symmetrically placed on the lower half of the complex α plane.

Interpolation rules have to be established for interpolation functions of one level of angles to another level of angles. The set of angles for lower level is of larger range than the set of higher level angles. Clearly $BW^{(i)} > BW^{(i+1)}$. Thus a function of $f(\alpha_q^{(i+1)})$ can be obtained from interpolation of the set $f(\alpha_q^{(i)})$.

Let $In_u^{(i)}(\alpha_q^{(i+1)}, \alpha_{q'}^{(i)})$ be the interpolation functions. Then

$$f(\alpha_q^{(i+1)}) = \sum_{q'=1}^{Q} In_u^{(i)}(\alpha_q^{(i+1)}, \alpha_{q'}^{(i)}) f(\alpha_{q'}^{(i)}) \tag{5.3.67}$$

A simple way is to choose a two point linear interpolation. For example, from Table 5.3.1, the $\alpha_6^{(2)}$ falls between $\alpha_9^{(1)}$ and $\alpha_{10}^{(1)}$ in the steepest descent path.

Thus the linear interpolation is

$$b(\alpha_6^{(2)}) = \frac{f(\alpha_{10}^{(1)})(\alpha_6^{(2)} - \alpha_9^{(1)})}{\alpha_{10}^{(1)} - \alpha_9^{(1)}} + \frac{f(\alpha_9^{(1)})(\alpha_6^{(2)} - \alpha_{10}^{(1)})}{\alpha_9^{(1)} - \alpha_{10}^{(1)}} \tag{5.3.68}$$

The linear interpolation function is

$$In_u^{(1)}(\alpha_6^{(2)}, \alpha_{q'}^{(1)}) = \begin{cases} \dfrac{\alpha_6^{(2)} - \alpha_{10}^{(1)}}{\alpha_9^{(1)} - \alpha_{10}^{(1)}} & q' = 9 \\ \dfrac{\alpha_6^{(2)} - \alpha_9^{(1)}}{\alpha_{10}^{(1)} - \alpha_9^{(1)}} & q' = 10 \\ 0 & \text{all other } q' \text{ values} \end{cases} \tag{5.3.69}$$

Other $In_u^{(1)}(\alpha_q^{(2)}, \alpha_{q'}^{(1)})$ can be obtained similarly.

3.4 Steepest Descent Expression of Multi-Level Impedance Matrix Elements

The location of group centers of each level can be defined. For example in the first level

$$X_{1_1} = \frac{x_1 + x_2 + x_3 + \cdots + x_M}{M} \tag{5.3.70}$$

$$X_{2_1} = \frac{x_{M+1} + x_{M+2} + x_{M+3} + \cdots + x_{2M}}{M} \tag{5.3.71}$$

For higher levels, we can define, for example

$$X_{1_2} = \frac{X_{1_1} + X_{2_1}}{2} \tag{5.3.72}$$

$$X_{2_2} = \frac{X_{3_1} + X_{4_1}}{2} \tag{5.3.73}$$

$$X_{2_3} = \frac{X_{3_2} + X_{4_2}}{2} \tag{5.3.74}$$

In general, for a level higher than the first level,

$$X_{m_n} = \frac{X_{(2m-1)_{n-1}} + X_{(2m)_{n-1}}}{2} \tag{5.3.75}$$

Note that the centers do not have z coordinates. Next, steepest descent path representation of matrix elements are derived. For example, $Z_{1,41}$ is in $\overline{\overline{Z}}^{(1)U}$ because it is in $\overline{\overline{Z}}^{(1)}_{1_1 3_1}$.

$$Z_{1,41} = \frac{i}{4}\Delta x H_0^{(1)}(k\sqrt{(x_{41} - x_1)^2 + (z_{41} - z_1)^2}) \tag{5.3.76}$$

Elements not in $\overline{\overline{Z}}^{(0)}$ are not in the near field, thus we do not need near field integration for matrix elements.

Using steepest descent and level 1 angles and noting that for $\overline{\overline{Z}}^{(U)}$ matrix elements, $x_j > x_i$,

$$\begin{aligned}
Z_{1,41} &= \Delta x \frac{i}{4\pi} \sum_{q=1}^{Q} \Delta\alpha_q^{(1)} e^{ik \sin\alpha_q^{(1)}(x_{41}-x_1)+ik\cos\alpha_q^{(1)}(z_{41}-z_1)} \\
&= \Delta x \frac{i}{4\pi} \sum_{q=1}^{Q} \Delta\alpha_q^{(1)} e^{ik \sin\alpha_q^{(1)}(X_{1_1}-x_1)-ik\cos\alpha_q^{(1)}(z_1)} \\
&\quad \cdot e^{ik\sin\alpha_q^{(1)}(X_{3_1}-X_{1_1})} \cdot e^{ik\sin\alpha_q^{(1)}(x_{41}-X_{3_1})+ik\cos\alpha_q^{(1)}(z_{41})} \\
&= \sum_{q=1}^{Q} V_{1,1_1}^{(1)U}(\alpha_q^{(1)}) T_{1_1 3_1}^{(1)U}(\alpha_q^{(1)}) W_{3_1,41}^{(1)U}(\alpha_q^{(1)})
\end{aligned} \tag{5.3.77}$$

where

$V_{iI_1}^{(1)U}(\alpha_q) = e^{ik\sin\alpha_q(X_{I_1}-x_i)-ik\cos\alpha_q(z_i)}$

(the ith element relative to the center of the Ith group of level 1) (5.3.78)

$T_{I_1J_1}^{(1)U}(\alpha_q^{(1)}) = \Delta\alpha_q \Delta x \frac{i}{4\pi} e^{ik\sin\alpha_q^{(1)}(X_{J_1}-X_{I_1})}$

(interaction between the Ith group and the Jth group of level 1) (5.3.79)

$W_{J_1j}^{(1)U}(\alpha_q) = e^{-ik\sin\alpha_q(X_{J_1}-x_j)-ik\cos\alpha_q(-z_j)}$

(the jth element relative to the center of the Jth group of level 1) (5.3.80)

Note that the above definitions of $V^{(1)U}(\alpha_q)$ and $W^{(1)U}(\alpha_q)$ are for general angle directions α_q of any level.

Then the 1st level upper matrix $\overline{\overline{Z}}^{(1)U}$ can be written as

$$\overline{\overline{Z}}^{(1)U} = \sum_{q=1}^{Q} \overline{\overline{V}}^{(1)U}(\alpha_q^{(1)}) \overline{\overline{T}}^{(1)U}(\alpha_q^{(1)}) \overline{\overline{W}}^{(1)U}(\alpha_q^{(1)}) \tag{5.3.81}$$

where

$$
\overline{\overline{V}}^{(1)U}(\alpha_q^{(1)}) = \begin{bmatrix}
V_{1,1_1}^{(1)U} & 0 & & 0 \\
V_{2,1_1}^{(1)U} & 0 & & \vdots \\
\vdots & \vdots & & \vdots \\
V_{M,1_1}^{(1)U} & 0 & & \vdots \\
0 & V_{M+1,2_1}^{(1)U} & & \vdots \\
\vdots & \vdots & & \vdots \\
\vdots & V_{2M,2_1}^{(1)U} & & \vdots \\
\vdots & 0 & \ddots & 0 \\
\vdots & \vdots & & V_{M(L-1)+1,L_1}^{(1)U} \\
\vdots & \vdots & & \vdots \\
0 & 0 & & V_{ML,L_1}^{(1)U}
\end{bmatrix} = N \times L \quad (5.3.82)
$$

$$
\overline{\overline{T}}^{(1)U}(\alpha_q^{(1)}) = \begin{bmatrix}
0 & 0 & T_{1_1,3_1}^{(1)U} & T_{1_1,4_1}^{(1)U} & 0 & 0 & \vdots \\
0 & 0 & 0 & T_{2_1,4_1}^{(1)U} & 0 & 0 & \vdots \\
0 & \vdots & \vdots & 0 & T_{3_1,5_1}^{(1)U} & T_{3_1,6_1}^{(1)U} & \vdots \\
0 & \vdots & \vdots & \vdots & 0 & T_{4_1,6_1}^{(1)U} & \vdots \\
\vdots & \vdots & \vdots & \vdots & \vdots & \vdots & \ddots
\end{bmatrix}
$$

$$
= L \times L \quad (5.3.83)
$$

$$
\overline{\overline{W}}^{(1)U}(\alpha_q^{(1)}) = \left[\begin{array}{ccccc}
W_{1_1,1}^{(1)U} & W_{1_1,2}^{(1)U} & \cdots & W_{1_1,M}^{(1)U} & 0 \\
0 & 0 & \cdots & 0 & W_{2_1,M+1}^{(1)U} \\
\cdots & \cdots & \cdots & \cdots & \cdots \\
0 & 0 & \cdots & \cdots & \cdots
\end{array}\right.
$$

$$
\left.\begin{array}{cccccc}
\cdots & \cdots & \cdots & \cdots & \cdots & 0 \\
\cdots & W_{2_1,2M}^{(1)U} & 0 & \cdots & \cdots & \cdots \\
\cdots & & \ddots & \cdots & \cdots & 0 \\
\cdots & \cdots & 0 & W_{L_1,M(L-1)+1}^{(1)U} & \cdots & W_{L_1,ML}^{(1)U}
\end{array}\right]
$$

$$
= L \times N \quad (5.3.84)
$$

In (5.3.84), the second half is connected to the first half to make it a $L \times N$ matrix. The awkward arrangement is a result of the lack of space.

Note that although the matrix notation is used for "matrix products" in (5.3.81), it is actually an element by an element product. Also note that $\overline{\overline{W}}^{(1)U}$ resembles the matrix transpose of $\overline{\overline{V}}^{(1)U}$.

Similarly the element $Z_{1,117}$ is contained in $\overline{\overline{Z}}^{(2)U}$ because it is in $\overline{\overline{Z}}^{(1)U}_{1_1 6_1}$ and in $\overline{\overline{Z}}^{(2)U}_{1_2 3_2}$. Thus, we have the following example of level 2 element.

$$Z_{1,117} = \Delta x \frac{i}{4\pi} \sum_{q=1}^{Q} \Delta\alpha_q^{(2)} e^{ik\sin\alpha_q^{(2)}(x_{117}-x_1)+ik\cos\alpha_q^{(2)}(z_{117}-z_1)}$$

$$= \Delta x \frac{i}{4\pi} \sum_{q=1}^{Q} \Delta\alpha_q^{(2)} e^{ik\sin\alpha_q^{(2)}(X_{1_1}-x_1)-ik\cos\alpha_q^{(2)}(z_1)+ik\sin\alpha_q^{(2)}(X_{1_2}-X_{1_1})}$$

$$\cdot\, e^{ik\sin\alpha_q^{(2)}(X_{3_2}-X_{1_2})} \cdot e^{ik\sin\alpha_q^{(2)}(X_{6_1}-X_{3_2})+ik\sin\alpha_q^{(2)}(x_{117}-X_{6_1})+ik\cos\alpha_q^{(2)}(z_{117})}$$

$$= \sum_{q=1}^{Q} V_{1,1_1}^{(1)U}(\alpha_q^{(2)}) V_{1_1,1_2}^{(2)U}(\alpha_q^{(2)}) T_{1_2 3_2}^{(2)U}(\alpha_q^{(2)}) W_{3_2,6_1}^{(2)U}(\alpha_q^{(2)}) W_{6_1,117}^{(1)U}(\alpha_q^{(2)}) \quad (5.3.85)$$

Note that $V^{(1)U}$ and $W^{(1)U}$ in (5.3.85) are as defined earlier in (5.3.78) and (5.3.80) for general α_q's. In (5.3.85),

$$V_{i_1 I_2}^{(2)U}(\alpha_q) = e^{ik\sin\alpha_q(X_{I_2}-X_{i_1})}$$

(the ith element relative to the center of the Ith group of level 2) (5.3.86)

$$T_{I_2 J_2}^{(2)U}(\alpha_q^{(2)}) = \Delta\alpha_q^{(2)} \Delta x \frac{i}{4\pi} e^{ik\sin\alpha_q^{(2)}(X_{J_2}-X_{I_2})}$$

(interaction between the Ith group and the Jth group of level 2) (5.3.87)

$$W_{J_2 j_1}^{(2)U}(\alpha_q) = e^{-ik\sin\alpha_q(X_{J_2}-X_{j_1})}$$

(the jth element relative to the center of the Jth group of level 2) (5.3.88)

Note also that $V^{(2)U}$ and $W^{(2)U}$, like $V^{(1)U}$ and $W^{(1)U}$, are defined for general α_q.

Hence, the 2nd level upper matrix $\overline{\overline{Z}}^{(2)U}$ can be written as

$$\overline{\overline{Z}}^{(2)U} = \sum_{q=1}^{Q} \overline{\overline{V}}^{(1)U}(\alpha_q^{(2)})\, \overline{\overline{V}}^{(2)U}(\alpha_q^{(2)})\, \overline{\overline{T}}^{(2)U}(\alpha_q^{(2)})\, \overline{\overline{W}}^{(2)U}(\alpha_q^{(2)})\, \overline{\overline{W}}^{(1)U}(\alpha_q^{(2)})$$

(5.3.89)

where

$$\overline{\overline{V}}^{(2)U}(\alpha_q^{(2)}) = \begin{bmatrix} V_{1_1 1_2}^{(2)U} & 0 & 0 & 0 \\ V_{2_1 1_2}^{(2)U} & 0 & \vdots & \vdots \\ 0 & V_{3_1 2_2}^{(2)U} & \vdots & \vdots \\ 0 & V_{4_1 2_2}^{(2)U} & 0 & \vdots \\ \vdots & 0 & V_{5_1 3_2}^{(2)U} & \vdots \\ \vdots & \vdots & V_{6_1 3_2}^{(2)U} & \vdots \\ \vdots & \vdots & \vdots & \ddots \end{bmatrix} \tag{5.3.90}$$

$$\overline{\overline{T}}^{(2)U}(\alpha_q^{(2)}) = \begin{bmatrix} 0 & 0 & T_{1_2,3_2}^{(2)U} & T_{1_2,4_2}^{(2)U} & 0 & 0 & \vdots \\ 0 & 0 & 0 & T_{2_2,4_2}^{(2)U} & 0 & 0 & \vdots \\ 0 & \vdots & \vdots & 0 & T_{3_2,5_2}^{(2)U} & T_{3_2,6_2}^{(2)U} & \vdots \\ 0 & \vdots & \vdots & \vdots & 0 & T_{4_2,6_2}^{(2)U} & \vdots \\ \vdots & \vdots & \vdots & \vdots & \vdots & \vdots & \ddots \end{bmatrix} \tag{5.3.91}$$

$$\overline{\overline{W}}^{(2)U}(\alpha_q^{(2)}) = \begin{bmatrix} W_{1_2,1_1}^{(2)U} & W_{1_2,2_1}^{(2)U} & 0 & 0 & \cdots & \cdots & \cdots \\ 0 & 0 & W_{2_2,3_1}^{(2)U} & W_{2_2,4_1}^{(2)U} & 0 & \cdots & \cdots \\ 0 & \cdots & \cdots & 0 & W_{3_2,5_1}^{(2)U} & W_{3_2,6_1}^{(2)U} & \cdots \\ \vdots & \cdots & \cdots & \cdots & \cdots & \cdots & \ddots \end{bmatrix} \tag{5.3.92}$$

The matrices have the same pattern for higher levels. To generalize, the mth level upper matrices ($I < J$) can be written as follows

$$\begin{aligned} \overline{\overline{Z}}^{(m)U} &= \sum_{q=1}^{Q} \overline{\overline{V}}^{(1)U}(\alpha_q^{(m)})\, \overline{\overline{V}}^{(2)U}(\alpha_q^{(m)}) \cdots \overline{\overline{V}}^{(m)U}(\alpha_q^{(m)})\, \overline{\overline{T}}^{(m)U}(\alpha_q^{(m)}) \\ &\quad \overline{\overline{W}}^{(m)U}(\alpha_q^{(m)}) \cdots \overline{\overline{W}}^{(2)U}(\alpha_q^{(m)})\, \overline{\overline{W}}^{(1)U}(\alpha_q^{(m)}) \end{aligned}$$

where

$$\overline{\overline{V}}^{(m)U}(\alpha_q^{(m)}) = \begin{bmatrix} V_{1_{m-1}1_m}^{(m)U} & 0 & 0 & 0 \\ V_{2_{m-1}1_m}^{(m)U} & 0 & \vdots & \vdots \\ 0 & V_{3_{m-1}2_m}^{(m)U} & \vdots & \vdots \\ 0 & V_{4_{m-1}2_m}^{(m)U} & 0 & \vdots \\ \vdots & 0 & V_{5_{m-1}3_m}^{(m)U} & \vdots \\ \vdots & \vdots & V_{6_{m-1}3_m}^{(m)U} & \vdots \\ \vdots & \vdots & \vdots & \ddots \end{bmatrix}$$

$$\text{for } m = 2, 3, 4, 5, \ldots \tag{5.3.93}$$

$$\overline{\overline{T}}^{(m)U}(\alpha_q^{(m)}) = \begin{bmatrix} 0 & 0 & T_{1_m,3_m}^{(m)U} & T_{1_m,4_m}^{(m)U} & 0 & 0 & \vdots \\ 0 & 0 & 0 & T_{2_m,4_m}^{(m)U} & 0 & 0 & \vdots \\ 0 & \vdots & \vdots & 0 & T_{3_m,5_m}^{(m)U} & T_{3_m,6_m}^{(m)U} & \vdots \\ 0 & \vdots & \vdots & \vdots & 0 & T_{4_m,6_m}^{(m)U} & \vdots \\ \vdots & \vdots & \vdots & \vdots & \vdots & \vdots & \ddots \end{bmatrix}$$

$$\text{for } m = 2, 3, 4, 5, \ldots \tag{5.3.94}$$

$$\overline{\overline{W}}^{(m)U}(\alpha_q^{(m)}) =$$

$$\begin{bmatrix} W_{1_m,1_{m-1}}^{(m)U} & W_{1_m,2_{m-1}}^{(m)U} & 0 & 0 & \cdots & \cdots & \cdots \\ 0 & 0 & W_{2_m,3_{m-1}}^{(m)U} & W_{2_m,4_{m-1}}^{(m)U} & 0 & \cdots & \cdots \\ 0 & \cdots & \cdots & 0 & W_{3_m,5_{m-1}}^{(m)U} & W_{3_m,6_{m-1}}^{(m)U} & \cdots \\ \vdots & \cdots & \cdots & \cdots & \cdots & \cdots & \ddots \end{bmatrix}$$

$$\text{for } m = 2, 3, 4, 5, \ldots \tag{5.3.95}$$

The matrix elements are as follows

$$V_{i_{m-1}I_m}^{(m)U}(\alpha_q) = e^{ik\sin\alpha_q(X_{I_m} - X_{i_{m-1}})} \quad \text{for } m = 2, 3, 4, \ldots \tag{5.3.96}$$

$$T_{I_m J_m}^{(m)U}(\alpha_q^{(m)}) = \Delta\alpha_q^{(m)} \Delta x \frac{i}{4\pi} e^{ik\sin\alpha_q^{(m)}(X_{J_m} - X_{I_m})} \tag{5.3.97}$$

$$W_{J_m j_{m-1}}^{(m)U}(\alpha_q) = e^{-ik\sin\alpha_q(X_{J_m} - X_{j_{m-1}})} \quad \text{for } m = 2, 3, 4, \ldots \tag{5.3.98}$$

The reason for separating the lower L and upper U matrices is because in the steepest descent representation of Green's function, the definition in $e^{ik\sin\alpha|x|}$ have absolute value of x.

For lower matrix, note that

$$g(x,z) = \frac{i}{4\pi}\int_\Gamma d\alpha e^{ik\sin(\alpha)|x|+ik\cos(\alpha)z} = \frac{i}{4\pi}\int_\Gamma d\alpha e^{ik\sin(\alpha)|x|-ik\cos(\alpha)z} \tag{5.3.99}$$

That is, the sign of $\cos(\alpha)z$ factor can be changed. Note that for lower impedance matrix $\overline{\overline{Z}}^L$, $x_i > x_j$. For example, analogous to (5.3.77),

$$\begin{aligned} Z_{41,1} &= \Delta x \frac{i}{4\pi}\sum_{q=1}^{Q}\Delta\alpha_q^{(1)} e^{ik\sin\alpha_q^{(1)}(x_{41}-x_1)+ik\cos\alpha_q^{(1)}(z_{41}-z_1)} \\ &= \Delta x \frac{i}{4\pi}\sum_{q=1}^{Q}\Delta\alpha_q^{(1)} e^{ik\sin\alpha_q^{(1)}(x_{41}-X_{3_1})+ik\cos\alpha_q^{(1)}z_{41}} \\ &\quad \cdot e^{ik\sin\alpha_q^{(1)}(X_{3_1}-X_{1_1})} \cdot e^{ik\sin\alpha_q^{(1)}(X_{1_1}-x_1)-ik\cos\alpha_q^{(1)}z_1} \\ &= \sum_{q=1}^{Q} V_{41,3_1}^{(1)L}(\alpha_q^{(1)}) T_{3_1 1_1}^{(1)L}(\alpha_q^{(1)}) W_{1_1,1}^{(1)L}(\alpha_q^{(1)}) \end{aligned} \tag{5.3.100}$$

The mth level lower matrices can be written $(I > J)$

$$\begin{aligned} \overline{\overline{Z}}^{(m)L} &= \sum_{q=1}^{Q} \overline{\overline{V}}^{(1)L}(\alpha_q^{(m)})\,\overline{\overline{V}}^{(2)L}(\alpha_q^{(m)}) \cdots \overline{\overline{V}}^{(m)L}(\alpha_q^{(m)})\,\overline{\overline{T}}^{(m)L}(\alpha_q^{(m)}) \\ &\quad \overline{\overline{W}}^{(m)L}(\alpha_q^{(m)}) \cdots \overline{\overline{W}}^{(2)L}(\alpha_q^{(m)})\,\overline{\overline{W}}^{(1)L}(\alpha_q^{(m)}) \end{aligned} \tag{5.3.101}$$

where the $\overline{\overline{V}}$, $\overline{\overline{T}}$, and $\overline{\overline{W}}$ matrices are given as follows

$$\overline{\overline{V}}^{(1)L}(\alpha_q) = \begin{bmatrix} V_{1,1_1}^{(1)L} & 0 & & 0 \\ V_{2,1_1}^{(1)L} & 0 & & \vdots \\ \vdots & \vdots & & \vdots \\ V_{M,1_1}^{(1)L} & 0 & & \vdots \\ 0 & V_{M+1,2_1}^{(1)L} & & \vdots \\ \vdots & \vdots & & \vdots \\ \vdots & V_{2M,2_1}^{(1)L} & & \vdots \\ \vdots & 0 & \ddots & 0 \\ \vdots & \vdots & & V_{M(L-1)+1,L_1}^{(1)L} \\ \vdots & \vdots & & \vdots \\ 0 & 0 & & V_{ML,L_1}^{(1)L} \end{bmatrix} = N \times L \tag{5.3.102}$$

$$\overline{\overline{T}}^{(1)L}(\alpha_q^{(1)}) = \begin{bmatrix} 0 & \vdots & \vdots & \vdots & \vdots & 0 \\ 0 & 0 & \vdots & \vdots & \vdots & \vdots \\ T^{(1)L}_{3_1,1_1} & 0 & \vdots & \vdots & \vdots & \vdots \\ T^{(1)L}_{4_1,1_1} & T^{(1)L}_{4_1,2_1} & 0 & \vdots & \vdots & \vdots \\ 0 & 0 & T^{(1)L}_{5_1,3_1} & 0 & \vdots & \vdots \\ \vdots & \vdots & T^{(1)L}_{6_1,3_1} & T^{(1)L}_{6_1,4_1} & 0 & \vdots \\ 0 & \vdots & \vdots & \vdots & \vdots & 0 \end{bmatrix} = L \times L \tag{5.3.103}$$

$$\overline{\overline{W}}^{(1)L}(\alpha_q) = \left[\begin{matrix} W^{(1)L}_{1_1,1} & W^{(1)L}_{1_1,2} & \cdots & W^{(1)L}_{1_1,M} & 0 \\ 0 & 0 & \cdots & 0 & W^{(1)L}_{2_1,M+1} \\ \cdots & \cdots & \cdots & \cdots & \cdots \\ 0 & 0 & \cdots & \cdots & \cdots \end{matrix} \right.$$

$$\left. \begin{matrix} \cdots & \cdots & \cdots & \cdots & \cdots & 0 \\ \cdots & W^{(1)L}_{2_1,2M} & 0 & \cdots & \cdots & \cdots \\ \cdots & & \ddots & \cdots & \cdots & 0 \\ \cdots & \cdots & 0 & W^{(1)L}_{L_1,M(L-1)+1} & \cdots & W^{(1)L}_{L_1,ML} \end{matrix} \right]$$

$$= L \times N \tag{5.3.104}$$

For $m = 2, 3, 4, \ldots$, we have

$$\overline{\overline{V}}^{(m)L}(\alpha_q) = \begin{bmatrix} V^{(m)L}_{1_{m-1}1_m} & 0 & 0 & 0 \\ V^{(m)L}_{2_{m-1}1_m} & 0 & \vdots & \vdots \\ 0 & V^{(m)L}_{3_{m-1}2_m} & \vdots & \vdots \\ 0 & V^{(m)L}_{4_{m-1}2_m} & 0 & \vdots \\ \vdots & 0 & V^{(m)L}_{5_{m-1}3_m} & \vdots \\ \vdots & \vdots & V^{(m)L}_{6_{m-1}3_m} & \vdots \\ \vdots & \vdots & \vdots & \ddots \end{bmatrix} \tag{5.3.105}$$

$$\overline{\overline{T}}^{(m)L}(\alpha_q^{(m)}) = \begin{bmatrix} 0 & \vdots & \vdots & \vdots & \vdots & \vdots & 0 \\ 0 & 0 & \vdots & \vdots & \vdots & \vdots & \vdots \\ T^{(m)L}_{3_m,1_m} & 0 & \vdots & \vdots & \vdots & \vdots & \vdots \\ T^{(m)L}_{4_m,1_m} & T^{(m)L}_{4_m,2_m} & 0 & \vdots & \vdots & \vdots & \vdots \\ 0 & 0 & T^{(m)L}_{5_m,3_m} & 0 & \vdots & \vdots & \vdots \\ \vdots & 0 & T^{(m)L}_{6_m,3_m} & T^{(m)L}_{6_m,4_m} & 0 & \vdots & \vdots \\ \vdots & \vdots & \vdots & \vdots & \vdots & \vdots & \vdots \end{bmatrix} \tag{5.3.106}$$

$$\overline{\overline{W}}^{(m)L}(\alpha_q) =$$
$$\begin{bmatrix} W^{(m)L}_{1_m,1_{m-1}} & W^{(m)L}_{1_m,2_{m-1}} & 0 & 0 & \cdots & \cdots & \cdots \\ 0 & 0 & W^{(m)L}_{2_m,3_{m-1}} & W^{(m)L}_{2_m,4_{m-1}} & 0 & \cdots & \cdots \\ 0 & \cdots & \cdots & 0 & W^{(m)L}_{3_m,5_{m-1}} & W^{(m)L}_{3_m,6_{m-1}} & \cdots \\ \vdots & \cdots & \cdots & \cdots & \cdots & \cdots & \ddots \end{bmatrix} \tag{5.3.107}$$

The matrix elements are

$$\begin{aligned} V^{(1)L}_{iI_1}(\alpha_q) &= e^{(ik\sin\alpha_q(x_i - X_{I_1}) + ik\cos\alpha_q(z_i))} \\ &= W^{(1)U}_{I_1,i}(\alpha_q) \end{aligned} \tag{5.3.108}$$

$$\begin{aligned} V^{(m)L}_{i_{m-1}I_m}(\alpha_q) &= e^{ik\sin\alpha_q(X_{i_{m-1}} - X_{I_m})} \\ &= W^{(m)U}_{I_m,i_{m-1}}(\alpha_q) \qquad \text{for } m = 2, 3, 4, \ldots \end{aligned} \tag{5.3.109}$$

$$\begin{aligned} T^{(m)L}_{I_mJ_m}(\alpha_q) &= \Delta\alpha_q^{(m)}\Delta x \frac{i}{4\pi} e^{ik\sin\alpha_q^{(m)}(X_{I_m} - X_{J_m})} \\ &\qquad \text{for } m = 1, 2, 3, 4, \ldots \end{aligned} \tag{5.3.110}$$

$$\begin{aligned} W^{(1)L}_{J_1j}(\alpha_q) &= e^{(-ik\sin\alpha_q(x_j - X_{J_1}) - ik\cos\alpha_q(z_j))} \\ &= V^{(1)U}_{j,J_1}(\alpha_q) \qquad \text{for } m = 1 \end{aligned} \tag{5.3.111}$$

$$\begin{aligned} W^{(m)L}_{J_mj_{m-1}}(\alpha_q) &= e^{-ik\sin\alpha_q(X_{j_{m-1}} - X_{J_m})} \\ &= V^{(m)U}_{j_{m-1},J_m}(\alpha_q) \qquad \text{for } m = 2, 3, 4, \ldots \end{aligned} \tag{5.3.112}$$

3.5 SDFMM Algorithm

In the following we describe the SDFMM algorithm and the computational steps. Each computational step ends with //. In an iterative solution to matrix equation, the important step is to compute the product of impedance matrix $\overline{\overline{Z}}$ and a column vector $\overline{b}$ from surface field. Continuing the same example where $M = 20$, $L = 64$ and $N = 1280$, we have

$$\begin{aligned}\overline{\overline{Z}}\,\overline{b} &= \left[\overline{\overline{Z}}^{(0)} + \overline{\overline{Z}}^{(1)} + \overline{\overline{Z}}^{(2)} + \overline{\overline{Z}}^{(3)} + \overline{\overline{Z}}^{(4)} + \overline{\overline{Z}}^{(5)}\right]\overline{b} \\ &= \left[\overline{\overline{Z}}^{(0)} + \overline{\overline{Z}}^{(1)U} + \overline{\overline{Z}}^{(1)L} + \overline{\overline{Z}}^{(2)U} + \overline{\overline{Z}}^{(2)L}\right. \\ &\quad \left. + \overline{\overline{Z}}^{(3)U} + \overline{\overline{Z}}^{(3)L} + \overline{\overline{Z}}^{(4)U} + \overline{\overline{Z}}^{(4)L} + \overline{\overline{Z}}^{(5)U} + \overline{\overline{Z}}^{(5)L}\right]\overline{b} \qquad (5.3.113)\end{aligned}$$

(1) *Computational step*: Compute $\overline{\overline{Z}}^{(0)}\overline{b}$ by direct multiplication //.
The number of computational steps for near field interactions N_{near} is

$$\begin{aligned}N_{\text{near}} &= 3M^2L - 2M^2 = 3MN - 2M^2 \\ &= M(3N - 2M) \qquad (5.3.114)\end{aligned}$$

For non-near field interactions, the two processes are aggregation and dis-aggregation. Define the upper matrix

$$\begin{aligned}\overline{p}^U &= \overline{\overline{Z}}^{(1)U}\overline{b} + \overline{\overline{Z}}^{(2)U}\overline{b} + \overline{\overline{Z}}^{(3)U}\overline{b} + \overline{\overline{Z}}^{(4)U}\overline{b} + \overline{\overline{Z}}^{(5)U}\overline{b} \\ &= \sum_{q=1}^{Q} \overline{\overline{V}}^{(1)U}(\alpha_q^{(1)})\,\overline{\overline{T}}^{(1)U}(\alpha_q^{(1)})\,\overline{\overline{W}}^{(1)U}(\alpha_q^{(1)})\,\overline{b} \\ &+ \sum_{q=1}^{Q} \overline{\overline{V}}^{(1)U}(\alpha_q^{(2)})\,\overline{\overline{V}}^{(2)U}(\alpha_q^{(2)})\,\overline{\overline{T}}^{(2)U}(\alpha_q^{(2)})\,\overline{\overline{W}}^{(2)U}(\alpha_q^{(2)})\,\overline{\overline{W}}^{(1)U}(\alpha_q^{(2)})\,\overline{b} + \cdots \\ &+ \sum_{q=1}^{Q} \overline{\overline{V}}^{(1)U}(\alpha_q^{(5)}) \cdots \overline{\overline{V}}^{(5)U}(\alpha_q^{(5)})\,\overline{\overline{T}}^{(5)U}(\alpha_q^{(5)})\,\overline{\overline{W}}^{(5)U}(\alpha_q^{(5)}) \cdots \overline{\overline{W}}^{(1)U}(\alpha_q^{(5)})\,\overline{b}\end{aligned}$$

(5.3.115)

For general α_q, not restricted to angular direction of a particular level, let

$$\overline{b}^{(1)U}(\alpha_q) = \overline{\overline{W}}^{(1)U}(\alpha_q)\,\overline{b} = L \times 1 \qquad (5.3.116)$$

$$\bar{b}^{(2)U}(\alpha_q) = \overline{\overline{W}}^{(2)U}(\alpha_q)\,\bar{b}^{(1)}(\alpha_q) = \overline{\overline{W}}^{(2)U}(\alpha_q)\,\overline{\overline{W}}^{(1)U}(\alpha_q)\,\bar{b}$$
$$= \frac{L}{2} \times 1 \tag{5.3.117}$$

$$\bar{b}^{(3)U}(\alpha_q) = \overline{\overline{W}}^{(3)U}(\alpha_q)\,\bar{b}^{(2)}(\alpha_q) = \overline{\overline{W}}^{(3)U}(\alpha_q)\,\overline{\overline{W}}^{(2)U}(\alpha_q)\,\overline{\overline{W}}^{(1)U}(\alpha_q)\,\bar{b}$$
$$= \frac{L}{2^2} \times 1 \tag{5.3.118}$$

$$\vdots$$

$$\bar{b}^{(5)U}(\alpha_q) = \overline{\overline{W}}^{(5)U}(\alpha_q)\,\bar{b}^{(4)}(\alpha_q) = \overline{\overline{W}}^{(5)U}(\alpha_q)\cdot\ldots\cdot\overline{\overline{W}}^{(1)U}(\alpha_q)\,\bar{b}$$
$$= \frac{L}{2^4} \times 1 = 4 \times 1 \tag{5.3.119}$$

(2) *Computational step*: Compute and store $\overline{\overline{W}}^{(1)U}(\alpha_q^{(1)})$, $\overline{\overline{W}}^{(2)U}(\alpha_q^{(2)})$, $\overline{\overline{W}}^{(3)U}(\alpha_q^{(3)})$, $\ldots$, $\overline{\overline{W}}^{(5)U}(\alpha_q^{(5)})$. Note that there is no need to compute $\overline{\overline{W}}^{(1)U}(\alpha_q^{(2)})$, $\overline{\overline{W}}^{(1)U}(\alpha_q^{(3)})$, $\overline{\overline{W}}^{(1)U}(\alpha_q^{(4)})$, etc. Also compute and store the interpolation functions $In_u^{(i)}(\alpha_q^{(i+1)}, \alpha_{q'}^{(i)})$ //
Then we have

$$\bar{p}^U = \sum_{q=1}^{Q} \overline{\overline{V}}^{(1)U}(\alpha_q^{(1)})\,\overline{\overline{T}}^{(1)U}(\alpha_q^{(1)})\,\bar{b}^{(1)}(\alpha_q^{(1)})$$
$$+ \sum_{q=1}^{Q} \overline{\overline{V}}^{(1)U}(\alpha_q^{(2)})\,\overline{\overline{V}}^{(2)U}(\alpha_q^{(2)})\,\overline{\overline{T}}^{(2)U}(\alpha_q^{(2)})\,\bar{b}^{(2)}(\alpha_q^{(2)})$$
$$+ \sum_{q=1}^{Q} \overline{\overline{V}}^{(1)U}(\alpha_q^{(3)})\,\overline{\overline{V}}^{(2)U}(\alpha_q^{(3)})\,\overline{\overline{V}}^{(3)U}(\alpha_q^{(3)})\,\overline{\overline{T}}^{(3)U}(\alpha_q^{(3)})\,\bar{b}^{(3)}(\alpha_q^{(3)})$$
$$+ \cdots + \sum_{q=1}^{Q} \overline{\overline{V}}^{(1)U}(\alpha_q^{(5)})\cdots\overline{\overline{V}}^{(5)U}(\alpha_q^{(5)})\,\overline{\overline{T}}^{(5)U}(\alpha_q^{(5)})\,\bar{b}^{(5)}(\alpha_q^{(5)}) \tag{5.3.120}$$

(3) *Computational step*: Compute $\bar{b}^{(1)}(\alpha_q^{(1)}) = \overline{\overline{W}}^{(1)U}(\alpha_q^{(1)})\,\bar{b}$ //
The number of computational steps for calculating $\bar{b}^{(1)}(\alpha_q^{(1)})$ is

$$QN$$

(4) *Computational step*: Compute $\bar{b}^{(1)}(\alpha_q^{(2)})$ from $\bar{b}^{(1)}(\alpha_q^{(1)})$ by interpolation

$$\bar{b}^{(1)}(\alpha_q^{(2)}) = \sum_{q'=1}^{Q} In_u^{(1)}(\alpha_q^{(2)}, \alpha_{q'}^{(1)})\bar{b}^{(1)}(\alpha_{q'}^{(1)}) \ // \tag{5.3.121}$$

The number of computational steps for interpolation is

$$Q^2 \frac{L}{2^{(2-2)}}$$

(5) *Computational step*: Compute $\bar{b}^{(2)}(\alpha_q^{(2)})$ by

$$\bar{b}^{(2)}(\alpha_q^{(2)}) = \overline{\overline{W}}^{(2)U}(\alpha_q^{(2)})\,\bar{b}^{(1)}(\alpha_q^{(2)}) \tag{5.3.122}$$

The number of computational steps for the above operation is

$$Q \frac{L}{2^{(2-2)}}$$

Compute also $\bar{b}^{(2)}(\alpha_q^{(3)})$ from $\bar{b}^{(2)}(\alpha_q^{(2)})$ by interpolation

$$\bar{b}^{(2)}(\alpha_q^{(3)}) = \sum_{q'=1}^{Q} In_u^{(2)}(\alpha_q^{(3)}, \alpha_{q'}^{(2)})\,\bar{b}^{(2)}(\alpha_{q'}^{(2)}) \tag{5.3.123}$$

The number of computational steps for the above interpolation is

$$Q^2 \frac{L}{2^{(3-2)}}$$

Then compute

$$\bar{b}^{(3)}(\alpha_q^{(3)}) = \overline{\overline{W}}^{(3)U}(\alpha_q^{(3)})\,\bar{b}^{(2)}(\alpha_q^{(3)}) \; // \tag{5.3.124}$$

The number of computational steps for the above operation is

$$Q \frac{L}{2^{(3-2)}}$$

In this manner, we move up the levels. For each level there is one product and one interpolation.

(6) *Computational step*: The computational steps for the mth level:

1. Interpolation

$$\bar{b}^{(m-1)}(\alpha_q^{(m)}) = \sum_{q'=1}^{Q} In_u^{(m-1)}(\alpha_q^{(m)}, \alpha_{q'}^{(m-1)})\,\bar{b}^{(m-1)}(\alpha_{q'}^{(m-1)}) \tag{5.3.125}$$

The number of computational steps for the above operation is

$$Q^2 \frac{L}{2^{(m-2)}}$$

2. Multiplication

$$\bar{b}^{(m)}(\alpha_q^{(m)}) = \overline{\overline{W}}^{(m)U}(\alpha_q^{(m)})\,\bar{b}^{(m-1)}(\alpha_q^{(m)}) \tag{5.3.126}$$

The number of computational steps for the above operation is

$$Q\frac{L}{2^{(m-2)}}$$

Steps repeat up to $\overline{b}^{(5)}(\alpha_q^{(5)})$ //

When $\overline{b}^{(p)}(\alpha_q^{(p)})$ is completed, the process of aggregation is computed. The total number of steps in aggregation is

$$\begin{aligned} N_A &= QN + \sum_{m=2}^{N}(Q^2+Q)\frac{L}{2^{(m-2)}} \\ &\approx QN + 2(Q^2+Q)\frac{N}{M} \end{aligned} \tag{5.3.127}$$

(7) *Computational step*: multiplication of $\overline{\overline{T}}$ matrices

$$\overline{c}^{(1)U}(\alpha_q^{(1)}) = \overline{\overline{T}}^{(1)U}(\alpha_q^{(1)})\,\overline{b}^{(1)}(\alpha_q^{(1)}) \tag{5.3.128}$$

$$\overline{c}^{(2)U}(\alpha_q^{(2)}) = \overline{\overline{T}}^{(2)U}(\alpha_q^{(2)})\,\overline{b}^{(2)}(\alpha_q^{(2)}) \tag{5.3.129}$$

$$\vdots$$

$$\overline{c}^{(5)U}(\alpha_q^{(5)}) = \overline{\overline{T}}^{(5)U}(\alpha_q^{(5)})\,\overline{b}^{(5)}(\alpha_q^{(5)}) \quad // \tag{5.3.130}$$

The number of computational steps for the product of $\overline{\overline{T}}$ matrices with column vectors is N_T:

$$\begin{aligned} N_T &= 3LQ + 3\frac{L}{2}Q + \cdots + 3\frac{L}{2^{(p-1)}}Q = \sum_{m=1}^{p} 3\frac{L}{2^{(m-1)}}Q \\ &\approx 6LQ = 6Q\frac{N}{M} \end{aligned} \tag{5.3.131}$$

Next, we start the process of dis-aggregation.

$$\begin{aligned} \overline{p}^U &= \sum_{q=1}^{Q}\overline{\overline{V}}^{(1)U}(\alpha_q^{(1)})\,\overline{c}^{(1)}(\alpha_q^{(1)}) + \sum_{q=1}^{Q}\overline{\overline{V}}^{(1)U}(\alpha_q^{(2)})\,\overline{\overline{V}}^{(2)U}(\alpha_q^{(2)})\,\overline{c}^{(2)}(\alpha_q^{(2)}) \\ &+ \sum_{q=1}^{Q}\overline{\overline{V}}^{(1)U}(\alpha_q^{(3)})\,\overline{\overline{V}}^{(2)U}(\alpha_q^{(3)})\,\overline{\overline{V}}^{(3)U}(\alpha_q^{(3)})\,\overline{c}^{(3)}(\alpha_q^{(3)}) \\ &+ \sum_{q=1}^{Q}\overline{\overline{V}}^{(1)U}(\alpha_q^{(4)})\,\overline{\overline{V}}^{(2)U}(\alpha_q^{(4)})\,\overline{\overline{V}}^{(3)U}(\alpha_q^{(4)})\,\overline{\overline{V}}^{(4)U}(\alpha_q^{(4)})\,\overline{c}^{(4)}(\alpha_q^{(4)}) \end{aligned}$$

$$+\sum_{q=1}^{Q}\overline{\overline{V}}^{(1)U}(\alpha_q^{(5)})\,\overline{\overline{V}}^{(2)U}(\alpha_q^{(5)})\,\overline{\overline{V}}^{(3)U}(\alpha_q^{(5)})\,\overline{\overline{V}}^{(4)U}(\alpha_q^{(5)})\,\overline{\overline{V}}^{(5)U}(\alpha_q^{(5)})\,\overline{c}^{(5)}(\alpha_q^{(5)})$$

$$(5.3.132)$$

For general α_q, not restricted to angle direction of any particular level, let

$$\overline{\overline{J}}^{(4)U}(\alpha_q)=\overline{\overline{V}}^{(1)U}(\alpha_q)\,\overline{\overline{V}}^{(2)U}(\alpha_q)\,\overline{\overline{V}}^{(3)U}(\alpha_q)\,\overline{\overline{V}}^{(4)U}(\alpha_q) \qquad (5.3.133)$$

Then the sum of last two terms in $\overline{p}^U$ in (5.3.132) is

$$I_{45}=\sum_{q=1}^{Q}\overline{\overline{V}}^{(1)U}(\alpha_q^{(4)})\,\overline{\overline{V}}^{(2)U}(\alpha_q^{(4)})\,\overline{\overline{V}}^{(3)U}(\alpha_q^{(4)})\,\overline{\overline{V}}^{(4)U}(\alpha_q^{(4)})\,\overline{c}^{(4)}(\alpha_q^{(4)})$$

$$+\sum_{q=1}^{Q}\overline{\overline{V}}^{(1)U}(\alpha_q^{(5)})\,\overline{\overline{V}}^{(2)U}(\alpha_q^{(5)})\,\overline{\overline{V}}^{(3)U}(\alpha_q^{(5)})\,\overline{\overline{V}}^{(4)U}(\alpha_q^{(5)})\,\overline{\overline{V}}^{(5)U}(\alpha_q^{(5)})\,\overline{c}^{(5)}(\alpha_q^{(5)})$$

$$=\sum_{q=1}^{Q}\overline{\overline{J}}^{(4)U}(\alpha_q^{(4)})\,\overline{c}^{(4)}(\alpha_q^{(4)})+\sum_{q=1}^{Q}\overline{\overline{J}}^{(4)U}(\alpha_q^{(5)})\,\overline{\overline{V}}^{(5)U}(\alpha_q^{(5)})\,\overline{c}^{(5)}(\alpha_q^{(5)})$$

$$(5.3.134)$$

Note that $\overline{\overline{J}}^{(4)U}(\alpha_q^{(5)})$ is related to $\overline{\overline{J}}^{(4)U}(\alpha_q^{(4)})$ by interpolation

$$\overline{\overline{J}}^{(4)U}(\alpha_q^{(5)})=\sum_{q'=1}^{Q}In_u^{(4)}(\alpha_q^{(5)},\alpha_{q'}^{(4)})\,\overline{\overline{J}}^{(4)U}(\alpha_{q'}^{(4)}) \qquad (5.3.135)$$

Substitute (5.3.135) in the expression for I_{45} in (5.3.134)

$$I_{45}=\sum_{q=1}^{Q}\overline{\overline{J}}^{(4)U}(\alpha_q^{(4)})\,\overline{c}^{(4)}(\alpha_q^{(4)})$$

$$+\sum_{q=1}^{Q}\sum_{q'=1}^{Q}In_u^{(4)}(\alpha_q^{(5)},\alpha_{q'}^{(4)})\,\overline{\overline{J}}^{(4)U}(\alpha_{q'}^{(4)})\,\overline{\overline{V}}^{(5)U}(\alpha_q^{(5)})\,\overline{c}^{(5)}(\alpha_q^{(5)}) \qquad (5.3.136)$$

In (5.3.136), interchanging dummy variable $q'=q$ in the last term and then combining with the first term, we have

$$I_{45}=\sum_{q=1}^{Q}\overline{\overline{J}}^{(4)U}(\alpha_q^{(4)})\Big[\overline{c}^{(4)}(\alpha_q^{(4)})+$$

$$\sum_{q'=1}^{Q}In_u^{(4)}(\alpha_{q'}^{(5)},\alpha_q^{(4)})\,\overline{\overline{V}}^{(5)U}(\alpha_{q'}^{(5)})\,\overline{c}^{(5)}(\alpha_{q'}^{(5)})\Big] \qquad (5.3.137)$$

(8) *Computational step*:

$$\overline{\gamma}^{(4)}(\alpha_q^{(5)}) = \overline{\overline{V}}^{(5)U}(\alpha_q^{(5)})\,\overline{c}^{(5)}(\alpha_q^{(5)}) \quad // \tag{5.3.138}$$

The number of computational steps for the above operation is

$$Q\frac{L}{2^{(p-2)}}$$

(9) *Computational step*:

$$\overline{\beta}^{(4)}(\alpha_q^{(4)}) = \sum_{q'=1}^{Q} In_u^{(4)}(\alpha_{q'}^{(5)}, \alpha_q^{(4)})\,\overline{\gamma}^{(4)}(\alpha_{q'}^{(5)}) \tag{5.3.139}$$

This step is known as anterpolation because summation is on the first argument of the interpolation function instead of the second argument in usual interpolation. Use the stored interpolation functions to perform this step //
The number of computational steps for the above operation is

$$Q^2\frac{L}{2^{(p-2)}}$$

$$I_{45} = \sum_{q=1}^{Q} \overline{\overline{J}}^{(4)U}(\alpha_q^{(4)})\left[\overline{c}^{(4)}(\alpha_q^{(4)}) + \overline{\beta}^{(4)}(\alpha_q^{(4)})\right] \tag{5.3.140}$$

Thus for each level, we need to multiply, anterpolate, and add. Also note that we do not calculate I_{45} in (5.3.140) but continue on to consider I_{345}, I_{2345}, I_{12345}, etc., where I_{345} means the summation of the third, fourth, and fifth terms. Now

$$\begin{aligned}\overline{p}^U &= \sum_{q=1}^{Q} \overline{\overline{V}}^{(1)U}(\alpha_q^{(1)})\,\overline{c}^{(1)}(\alpha_q^{(1)}) + \sum_{q=1}^{Q} \overline{\overline{V}}^{(1)U}(\alpha_q^{(2)})\,\overline{\overline{V}}^{(2)U}(\alpha_q^{(2)})\,\overline{c}^{(2)}(\alpha_q^{(2)}) \\ &+ \sum_{q=1}^{Q} \overline{\overline{V}}^{(1)U}(\alpha_q^{(3)})\,\overline{\overline{V}}^{(2)U}(\alpha_q^{(3)})\,\overline{\overline{V}}^{(3)U}(\alpha_q^{(3)})\,\overline{c}^{(3)}(\alpha_q^{(3)}) \\ &+ \sum_{q=1}^{Q} \overline{\overline{V}}^{(1)U}(\alpha_q^{(4)})\cdots\overline{\overline{V}}^{(4)U}(\alpha_q^{(4)})\left[\overline{c}^{(4)}(\alpha_q^{(4)}) + \overline{\beta}^{(4)}(\alpha_q^{(4)})\right]\end{aligned} \tag{5.3.141}$$

(10) *Computational step*: Addition and multiplication

$$\overline{\gamma}^{(3)}(\alpha_q^{(4)}) = \overline{\overline{V}}^{(4)U}(\alpha_q^{(4)})\left[\overline{c}^{(4)}(\alpha_q^{(4)}) + \overline{\beta}^{(4)}(\alpha_q^{(4)})\right] \tag{5.3.142}$$

The number of computational steps for the above operation is

$$Q\frac{L}{2^{(p-3)}}$$

Anterpolation

$$\overline{\beta}^{(3)}(\alpha_{q'}^{(3)}) = \sum_{q'=1}^{Q} In_u^{(3)}(\alpha_{q'}^{(4)}, \alpha_q^{(3)})\,\overline{\gamma}^{(3)}(\alpha_{q'}^{(4)}) \;// \tag{5.3.143}$$

The number of computational steps for the above operation is

$$Q^2\frac{L}{2^{(p-3)}}$$

Then

$$\overline{p}^U = \sum_{q=1}^{Q} \overline{\overline{V}}^{(1)U}(\alpha_q^{(1)})\,\overline{c}^{(1)}(\alpha_q^{(1)}) + \sum_{q=1}^{Q} \overline{\overline{V}}^{(1)U}(\alpha_q^{(2)})\,\overline{\overline{V}}^{(2)U}(\alpha_q^{(2)})\,\overline{c}^{(2)}(\alpha_q^{(2)})$$
$$+ \sum_{q=1}^{Q} \overline{\overline{V}}^{(1)U}(\alpha_q^{(3)})\,\overline{\overline{V}}^{(2)U}(\alpha_q^{(3)})\,\overline{\overline{V}}^{(3)U}(\alpha_q^{(3)})\left[\overline{c}^{(3)}(\alpha_q^{(3)}) + \overline{\beta}^{(3)}(\alpha_q^{(3)})\right] \tag{5.3.144}$$

(11) *Computational step*: we move down the level in this manner. Addition, multiplication and anterpolation.

$$\overline{\gamma}^{(m-1)}(\alpha_q^{(m)}) = \overline{\overline{V}}^{(m)U}(\alpha_q^{(m)})\left[\overline{c}^{(m)}(\alpha_q^{(m)}) + \overline{\beta}^{(m)}(\alpha_q^{(m)})\right] \tag{5.3.145}$$

$$\overline{\beta}^{(m-1)}(\alpha_{q'}^{(m-1)}) = \sum_{q=1}^{Q} In_u^{(m-1)}(\alpha_q^{(m)}, \alpha_{q'}^{(m-1)})\,\overline{\gamma}^{(m-1)}(\alpha_q^{(m)}) \;// \tag{5.3.146}$$

The number of computational steps for calculating $\overline{\gamma}^{(m-1)}(\alpha_q^{(m)})$ and $\overline{\beta}^{(m-1)}(\alpha_{q'}^{(m-1)})$ is respectively,

$$Q\frac{L}{2^{(m-2)}} \quad \text{and} \quad Q^2\frac{L}{2^{(m-2)}}$$

The same procedure goes on until we reach the first level, which does not need anterpolation. The first level needs QN steps. Thus the total number of steps for dis-aggregation is

$$N_D = QN + \sum_{m=2}^{p}(Q^2+Q)\frac{L}{2^{(m-2)}}$$
$$\approx QN + 2(Q^2+Q)\frac{N}{M} \tag{5.3.147}$$

(12) *Computational step*: Repeat in the same manner for $\overline{p}^L$.//
The total number of steps is, including $\overline{p}^U$ and $\overline{p}^L$,

$$\begin{aligned} N_{\text{total}} &= N_{\text{near}} + 2(N_A + N_T + N_D) \\ &= M(3N - 2M) + 2\left[QN + 2(Q^2 + Q)\frac{N}{M} + 6Q\frac{N}{M}\right. \\ &\qquad \left. +QN + 2(Q^2 + Q)\frac{N}{M}\right] \\ &= M(3N - 2M) + 4N\left[Q + \frac{Q^2 + 4Q}{M}\right] \end{aligned}$$

Since Q and M are fixed, the CPU requirement is $\mathcal{O}(N)$.

3.6 Numerical Results

In Fig. 5.3.5, we illustrate the numerical reslts of the SDFMM method. We consider a single realization of random rough surface with $h = 0.3\lambda$ and $l = \lambda$, $\theta_i = 50°$. The frequency is 19 GHz so that $\lambda = 0.0158$ m. We use $\Delta x = \lambda/10$, $g = L/4$, $L = N\Delta x$. In SDFMM, we use $N = 320$ and $M = 10$. In 5.3.5, we compare the results of direct multiplication and multilevel SDFMM. The two results are indistinguishable.

4 Method of Ordered Multiple Interactions (MOMI)

In previous sections, we have used the banded matrix iterative approach and the conjugate gradient approach for treating matrix equations. The method was also enhanced by using canonical grid together with FFT (BMIA/CAG). In this section, a different iterative approach known as method of ordered multiple interactions (MOMI) is discussed. The method was first proposed in connection with using the parabolic equation method for rough surface scattering [Spivak, 1990; Holliday et al. 1996]. and was later extended to treat higher order scattering in random rough surface [Kapp and Brown, 1996].

4.1 Matrix Equations Based on MFIE for TE and TM Waves for PEC

The method used MFIE for both TE and TM waves for perfect electric conductor. As discussed in Chapter 4, using MFIE for TM waves has been standard treatment. For TE waves, usually, EFIE is utilized. In Chapter 4, Section 3.2, we also derived the MFIE for TE case. We use pulse basis

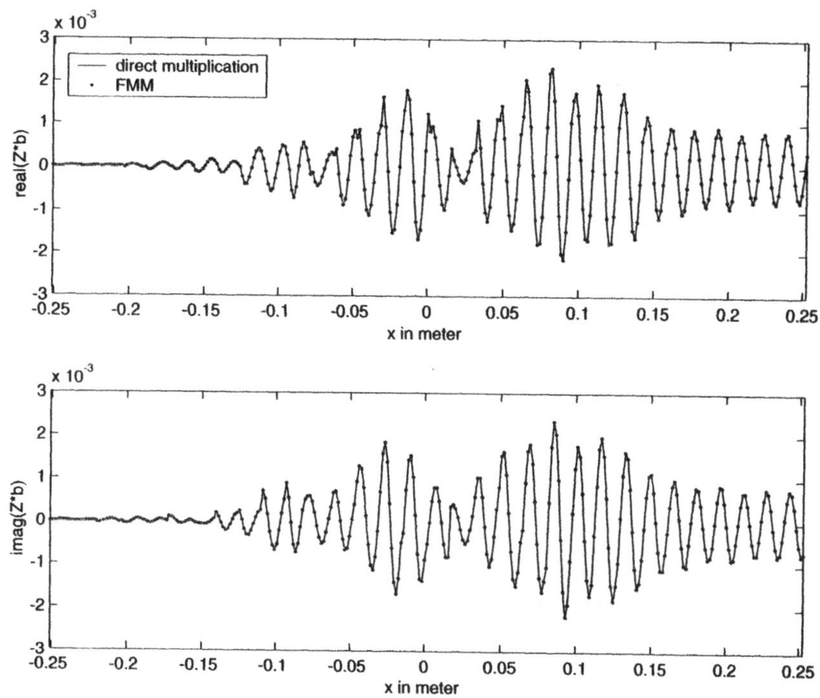

Figure 5.3.5 Comparison of SDFMM and direct multiplication.

functions and point matching.

(A) For TE case, the integral equation is given by Eq. (4.3.32) with ψ representing the electric field. In matrix form,

$$\sum_{n=1}^{N} Z'_{mn} v_n = b'_m, \qquad m = 1, 2, \ldots, N \tag{5.4.1}$$

where

$$v_n = [\hat{n} \cdot \nabla \psi(\bar{r})]_{x=x_n, z=f(x_n)} \tag{5.4.2}$$

$$b'_m = 2\,[\hat{n} \cdot \nabla \psi_{inc}(\bar{r}))]_{x=x_m, z=f(x_m)} \tag{5.4.3}$$

$$Z'_{mn} = 2 \int_{x_n - \frac{\Delta x}{2}}^{x_n + \frac{\Delta x}{2}} dx K_D^{(M)}(x_m, x) \qquad (m \neq n) \tag{5.4.4a}$$

$$Z'_{mm} = 1 + 2 ⨍_{x_m - \frac{\Delta x}{2}}^{x_m + \frac{\Delta x}{2}} dx K_D^{(M)}(x_m, x) \tag{5.4.4b}$$

where $⨍$ is the principal value integral. Note that we have multiplied

Eq. (4.3.32) by 2. Thus the right hand side has a factor of 2 and the impedance matrix elements also have a factor of 2. The quantity $K_D^{(M)}(x', x)$ is given in (4.3.33). We have seen that for energy conservation, it is necessary to perform numerical integration of near-field impedance matrix elements including the self-patch. As a result of the numerical integration, the diagonal element of the impedance matrix element is not equal to unity, as shown in (5.4.4b). To apply MOMI, we divide (5.4.1) by Z'_{mm} so that

$$\sum_{n=1}^{N} Z_{mn} v_n = b_m \tag{5.4.5}$$

where

$$b_m = \frac{b'_m}{Z'_{mm}} \tag{5.4.6}$$

$$Z_{mn} = \frac{Z'_{mn}}{Z'_{mm}} \tag{5.4.7}$$

Thus the diagonal elements Z_{mm} are now equal to unity.

After the matrix equation is solved, the scattered wave can be computed by

$$\psi_s^{(N)}(\theta_s) = \sum_{n=1}^{N} \Delta x \sqrt{1 + (f'(x_n))^2} v_n e^{-ik(\sin\theta_s x_n + f(x_n)\cos\theta_s)} \tag{5.4.8}$$

(B) For TM case, the integral equation is given by Eq. (4.1.151) with ψ representing the magnetic field. In matrix form,

$$\sum_{n=1}^{N} Z'_{mn} \psi_n = b'_m \tag{5.4.9}$$

$$\psi_n = [\psi(\bar{r})]_{x=x_n, z=f(x_n)} \tag{5.4.10}$$

$$b'_m = 2\,[\psi_{inc}(\bar{r})]_{x=x_m, z=f(x_m)} \tag{5.4.11}$$

$$Z'_{mn} = -2 \int_{x_n - \frac{\Delta x}{2}}^{x_n + \frac{\Delta x}{2}} dx K_N(x_m, x) \qquad (m \neq n) \tag{5.4.12a}$$

$$Z'_{mm} = 1 - 2 \,⨍_{x_m - \frac{\Delta x}{2}}^{x_m + \frac{\Delta x}{2}} dx K_N(x_m, x) \tag{5.4.12b}$$

where $K_N(x', x)$ is given in (4.1.158). To make diagonal elements equal to unity, we again divide (5.4.9) by Z'_{mm} to get

$$\sum_{n=1}^{N} Z'_{mn} \psi_n = b'_m \tag{5.4.13}$$

where

$$b_m = \frac{b'_m}{Z'_{mm}} \tag{5.4.14}$$

$$Z_{mn} = \frac{Z'_{mn}}{Z'_{mm}} \tag{5.4.15}$$

After the matrix equation is solved, the far field scattered field is computed from

$$\psi_s^{(N)}(\theta_s) = \sum_{n=1}^{N} \Delta x i k [f'(x_n) \sin\theta_s - \cos\theta_s] \psi_n e^{-ik(\sin\theta_s x_n + f(x_n)\cos\theta_s)} \tag{5.4.16}$$

4.2 Iterative Approach

To apply the method of ordered multiple interactions (MOMI), the following iterative approach is used. The impedance matrix is decomposed into

$$\overline{\overline{Z}} = \overline{\overline{I}} - \overline{\overline{L}} - \overline{\overline{U}} \tag{5.4.17}$$

where

$$L_{mn} = \begin{cases} -Z_{mn} & \text{for } m > n \\ 0 & \text{otherwise} \end{cases} \tag{5.4.18}$$

$$U_{mn} = \begin{cases} -Z_{mn} & \text{for } m < n \\ 0 & \text{otherwise} \end{cases} \tag{5.4.19}$$

Note that $\overline{\overline{L}}$ and $\overline{\overline{U}}$ are lower and upper triangular matrix, respectively. The diagonal elements are also zero. Note that for both TE and TM waves, the diagonal elements of the impedance matrix $Z_{mm} = 1$, with $m = 1, 2, \ldots, N$. We treat the case of TE waves (the TM waves can be treated in a similar manner). The matrix equation is

$$\overline{\overline{Z}}\,\overline{v} = \overline{b} \tag{5.4.20}$$

It is rewritten as

$$\overline{\overline{A}}\,\overline{\overline{B}}\,\overline{v} = \overline{b} + \overline{\overline{L}}\,\overline{\overline{U}}\,\overline{v} \tag{5.4.21}$$

where

$$\overline{\overline{A}} = \overline{\overline{I}} - \overline{\overline{L}} \tag{5.4.22}$$

$$\overline{\overline{B}} = \overline{\overline{I}} - \overline{\overline{U}} \tag{5.4.23}$$

Note that $\overline{\overline{A}}$ and $\overline{\overline{B}}$ are triangular matrices so that the product of the inverse and a column vector can be solved by back-substitution and not by inversion. The back-substitutions require $\mathcal{O}(N^2)$ operations. For example, $\overline{\overline{B}}^{-1}\,\overline{v}$

is solved by back-substitution. Let

$$\overline{\overline{B}}\,\overline{v} = \overline{q} \tag{5.4.24}$$

Then

$$\overline{\overline{A}}\,\overline{q} = \overline{b} + \overline{\overline{L}}\,\overline{\overline{U}}\,\overline{\overline{B}}^{-1}\overline{q} \tag{5.4.25}$$

Thus we use (5.4.25) to solve for $\overline{q}$, and then we use (5.4.24) to solve for $\overline{v}$.

The zeroth order solution is

$$\overline{\overline{A}}\,\overline{q}^{(0)} = \overline{b} \tag{5.4.26}$$

$$\overline{\overline{B}}\,\overline{v}^{(0)} = \overline{q}^{(0)} \tag{5.4.27}$$

The zeroth order solution, $\overline{v}^{(0)}$, is solved by using (5.4.26) and (5.4.27). The zeroth order solution is also known as the "new" Born term.

For higher order solutions, they obey the equation,

$$\overline{\overline{A}}\,\overline{q}^{(n)} = \overline{b} + \overline{\overline{L}}\,\overline{\overline{U}}\,\overline{\overline{B}}^{-1}\overline{q}^{(n-1)} \tag{5.4.28}$$

$$\overline{q}^{(n)} = \overline{q}^{(0)} + \overline{\overline{A}}^{-1}\overline{\overline{L}}\,\overline{\overline{U}}\,\overline{\overline{B}}^{-1}\overline{q}^{(n-1)} \tag{5.4.29}$$

$$\overline{\overline{B}}\,\overline{v}^{(n)} = \overline{q}^{(n)} \tag{5.4.30}$$

Then, using the recurrence relation,

$$\overline{q}^{(n)} = \overline{q}^{(0)} + \overline{\overline{A}}^{-1}\overline{\overline{L}}\,\overline{\overline{U}}\,\overline{\overline{B}}^{-1}\overline{q}^{(n-1)} \tag{5.4.31}$$

$$\overline{v}^{(n)} = \overline{\overline{B}}^{-1}\overline{q}^{(n)} \tag{5.4.32}$$

This means

$$\overline{v}^{(1)} = \overline{v}^{(0)} + \overline{\overline{B}}^{-1}\overline{\overline{A}}^{-1}\overline{\overline{L}}\,\overline{\overline{U}}\,\overline{v}^{(0)} \tag{5.4.33}$$

$$\begin{aligned}\overline{v}^{(2)} &= \overline{v}^{(0)} + \overline{\overline{B}}^{-1}\overline{\overline{A}}^{-1}\overline{\overline{L}}\,\overline{\overline{U}}\,\overline{v}^{(1)} \\ &= \overline{v}^{(0)} + \overline{\overline{B}}^{-1}\overline{\overline{A}}^{-1}\overline{\overline{L}}\,\overline{\overline{U}}\,\overline{v}^{(0)} + \overline{\overline{B}}^{-1}\overline{\overline{A}}^{-1}\overline{\overline{LUB}}^{-1}\overline{\overline{A}}^{-1}\overline{\overline{L}}\,\overline{\overline{U}}\,\overline{v}^{(0)}\end{aligned} \tag{5.4.34}$$

In general

$$\overline{v}^{(n)} = \overline{v}^{(0)} + \sum_{j=1}^{n}\left[\overline{\overline{B}}^{-1}\overline{\overline{A}}^{-1}\overline{\overline{L}}\,\overline{\overline{U}}\right]^{j}\overline{v}^{(0)} \tag{5.4.35}$$

Note that

$$\overline{\overline{U}}\,\overline{\overline{B}}^{-1} = \left(\overline{\overline{I}} - \overline{\overline{B}}\right)\overline{\overline{B}}^{-1} = \overline{\overline{B}}^{-1} - \overline{\overline{I}} \tag{5.4.36}$$

$$\overline{\overline{A}}^{-1}\overline{\overline{L}} = \overline{\overline{A}}^{-1} - \overline{\overline{I}} \tag{5.4.37}$$

In the numerical algorithm [Kapp and Brown, 1996], the procedure is as follows. From (5.4.36)–(5.4.37) and (5.4.29), the updating algorithm is,

$$\overline{q}^{(n)} = \overline{q}^{(0)} + \left(\overline{\overline{A}}^{-1} - \overline{\overline{I}}\right)\left(\overline{\overline{B}}^{-1} - \overline{\overline{I}}\right)\overline{q}^{(n-1)} \tag{5.4.38}$$

When all the orders are completed, we get the final $\overline{q}^{(final)}$. Thus, the final solution is obtained by solving

$$\overline{v} = \overline{\overline{B}}^{-1}\overline{q}^{(final)} \tag{5.4.39}$$

Note that we calculate $\overline{c}^{(n-1)} = \left(\overline{\overline{B}}^{-1} - \overline{\overline{I}}\right)\overline{q}^{(n-1)}$ by

$$\overline{c}^{(n-1)} = \overline{\overline{B}}^{-1}\overline{q}^{(n-1)} - \overline{q}^{(n-1)} \tag{5.4.40}$$

This is done by back-substitution and a subtraction of a column vector. Similarly, compute $\overline{d}^{(n-1)} = \left(\overline{\overline{A}}^{-1} - \overline{\overline{I}}\right)\overline{c}^{(n-1)}$ by

$$\overline{d}^{(n-1)} = \overline{\overline{A}}^{-1}\overline{c}^{(n-1)} - \overline{c}^{(n-1)} \tag{5.4.41}$$

4.3 Numerical Results

In Figs. 5.4.1 and 5.4.2, we plot the surface currents for a single realization of random rough surface. The real part of the surface current v_n for the case of TE wave incidence with $\theta_i = 30°$, $h = 0.4\lambda$, $l = 0.2\lambda$ using 80 points per wavelength. Comparisons are made between matrix inversion, zeroth-order MOMI and tenth-order MOMI. The tenth order solution is in good agreement with matrix inversion. In Table 5.4.1, we test energy conservation using

$$P = \int_{-\frac{\pi}{2}}^{\frac{\pi}{2}} d\theta_s \sigma(\theta_s)$$

which should be equal to unity. As noted in Section 3.2 of Chapter 4, MFIE of TE case has inadequate energy conservation test.

No. of points/λ	Matrix inversion	0th-order MOMI	10th-order MOMI
40	0.9158	0.8262	0.9126
80	0.9530	0.8272	0.9526

Table 5.4.1 Comparison of the energy conservation for TE case using MFIE at $\theta_i = 30°$ with $h = 0.4\,\lambda$, $l = 0.2\,\lambda$, and $L = 25.6\lambda$.

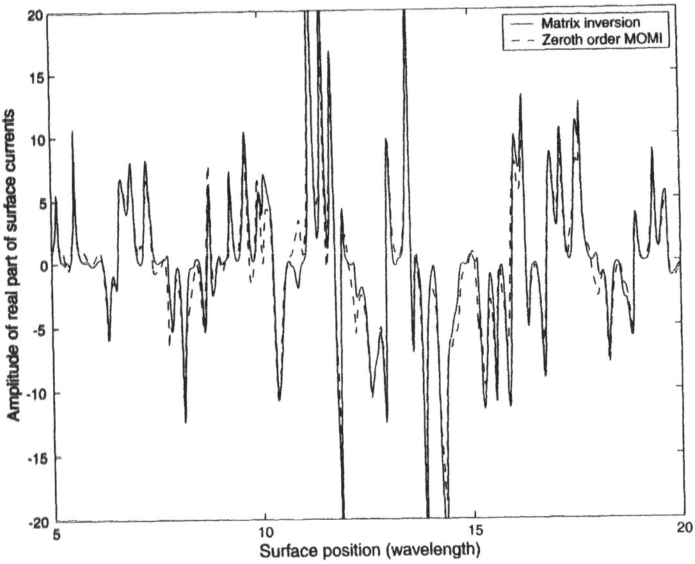

Figure 5.4.1 Comparison of the surface currents by matrix inversion and zeroth-order MOMI for TE case at $\theta_i = 30°$ with $h = 0.4\lambda$ and $l = 0.2\lambda$, 80 points/λ discretization and near-field integration are used in the simulation. $L = 25.6\lambda$.

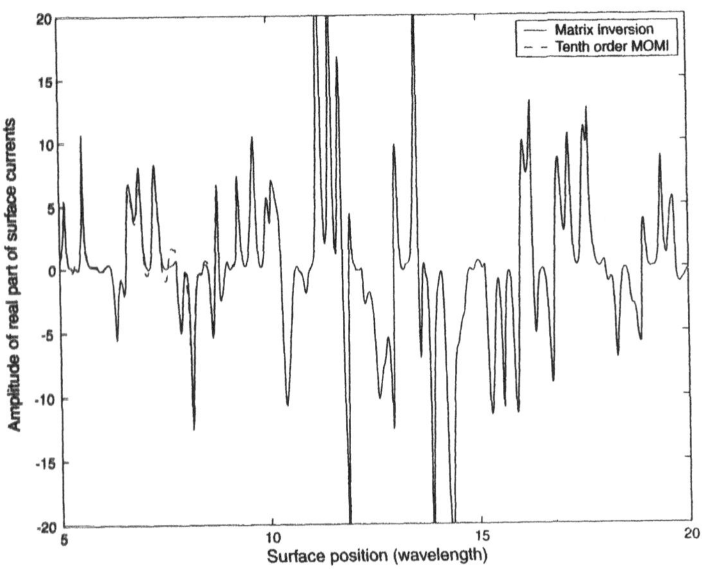

Figure 5.4.2 Comparison of the surface currents by matrix inversion and tenth-order MOMI for TE case at $\theta_i = 30°$ with $h = 0.4\lambda$ and $l = 0.2\lambda$, 80 points/λ discretization and near-field integration are used in the simulation. $L = 25.6\lambda$.

5 Physics-Based Two-Grid Method Combined with the Multilevel Fast Multipole Method

For scattering by lossy dielectric rough surfaces with large permittivities, we have introduced the PBTG method in Section 2. The PBTG method was further combined with BMIA/CAG in Section 2. In this section, the PBTG is used in conjunction with the steepest descent multilevel fast multipole method (SDFMM) to solve wave scattering from one-dimensional random lossy dielectric rough surfaces. The proposed algorithm has the computational complexities of $\mathcal{O}(N_{dg})$ for near-field interactions and $\mathcal{O}(N_{cg})$ for non-near field interactions, where N_{dg} and N_{cg} are the number of sampling points on the dense and coarse grid, respectively. Using the proposed algorithm, wave scattering from Gaussian and non-Gaussian rough surfaces is investigated and illustrated. Special emphasis is put on checking the accuracy of the algorithm and energy conservation.

We illustrate the single grid method and the PBTG in Section 5.1. We discuss the computational complexity of the combined algorithm of the PBTG with the SDFMM in Section 5.2. In Section 5.3, we apply the algorithm to Gaussian rough surfaces and discuss the accuracy of the algorithm. In Section 5.4, we apply the method to study scattering by the modified power-law spectrum.

5.1 Single Grid and PBTG

Consider a tapered plane wave $\psi_{inc}(\bar{r})$ incident on a random dielectric rough surface defined by $z = f(x)$. The surface fields satisfy the dual surface integral equations. let $G(\bar{r}, \bar{r}')$ and $G_1(\bar{r}, \bar{r}')$ be the 2-dimensional Green's functions of free space and the medium, respectively. Let ρ be equal to μ_1/μ and ϵ_1/ϵ for TE and TM polarization, respectively. Using the method of moments (MoM), the integral equations are cast into the matrix equations:

$$\sum_{j=1}^{N} a_{ij} u(x_j) + \sum_{j=1}^{N} b_{ij} \psi(x_j) = \psi_{inc}(x_i) \tag{5.5.1}$$

$$\sum_{j=1}^{N} a_{ij}^{(1)} \rho u(x_j) + \sum_{j=1}^{N} b_{ij}^{(1)} \psi(x_j) = 0 \tag{5.5.2}$$

where $u(x) = \sqrt{1 + [f'(x)]^2} \dfrac{\partial \psi}{\partial n}$. Expressions a_{ij}, b_{ij}, $a_{ij}^{(1)}$, and $b_{ij}^{(1)}$ are as in (5.2.6)–(5.2.9).

The quantity N is the number of sampling points on the surfaces. The matrix elements a_{ij}, b_{ij}, $a_{ij}^{(1)}$, and $b_{ij}^{(1)}$ are determined by the Green's functions.

We let Roman numeral subscripts i, j denote indexing with the dense grid and $\tilde{i}$, $\tilde{j}$ with the coarse grid. Assume that the upper medium is free space and the lower medium is lossy with a relative complex permittivity ϵ_1. We can define a distance limit r_λ as determined by the complex permittivity of the lower medium. Outside this limit the field interaction between the ith and the jth point is vanishingly small, and the lower medium Green's function can be set equal to zero. Therefore we can approximate

$$a_{ij}^{(1)} \approx \tilde{a}_{ij}^{(1)} = \begin{cases} a_{ij}^{(1)} & r_{ij} \leq r_l \\ 0 & r_{ij} \geq r_l \end{cases} \tag{5.5.3}$$

$$b_{ij}^{(1)} \approx \tilde{b}_{ij}^{(1)} = \begin{cases} b_{ij}^{(1)} & r_{ij} \leq r_l \\ 0 & r_{ij} \geq r_l \end{cases} \tag{5.5.4}$$

where r_{ij} is the distance between the ith point and the jth point on the dense grid. Thus $\tilde{a}_{ij}^{(1)}$ and $\tilde{b}_{ij}^{(1)}$ are banded matrices and Equation (5.5.2) becomes

$$\sum_{j=1}^{N} \tilde{a}_{ij}^{(1)} \rho u(x_j) + \sum_{j=1}^{N} \tilde{b}_{ij}^{(1)} \psi(x_j) = 0 \tag{5.5.5}$$

Based on the observation that the upper medium Green's function is slowly varying on the dense grid, we decompose the upper medium Green's function into near field and non-near field interactions

$$\sum_{j=1}^{N} a_{ij} u(x_j) = \sum_{j=1}^{N} a_{ij}^{s} u_j + \sum_{j=1}^{N} a_{ij}^{ns} u_j \tag{5.5.6}$$

$$\sum_{j=1}^{N} b_{ij} \psi(x_j) = \sum_{j=1}^{N} b_{ij}^{s} \psi_j + \sum_{j=1}^{N} b_{ij}^{ns} \psi_j \tag{5.5.7}$$

where a_{ij}^{s}, b_{ij}^{s}, a_{ij}^{ns}, and b_{ij}^{ns} are determined by

$$a_{ij}^{s} = \begin{cases} a_{ij} & r_{ij} \leq r_f \\ 0 & r_{ij} > r_f \end{cases} \tag{5.5.8}$$

$$b_{ij}^{s} = \begin{cases} b_{ij} & r_{ij} \leq r_f \\ 0 & r_{ij} > r_f \end{cases} \tag{5.5.9}$$

$$a_{ij}^{ns} = \begin{cases} 0 & r_{ij} \leq r_f \\ a_{ij} & r_{ij} > r_f \end{cases} \tag{5.5.10}$$

$$b_{ij}^{ns} = \begin{cases} o & r_{ij} \leq r_f \\ b_{ij} & r_{ij} > r_f \end{cases} \tag{5.5.11}$$

Thus r_f is the distance separating near field and non-near field. For non-near-field interactions, the Green's function of free space is slowly varying on the dense grid. We can use the coarse grid to sample it. Assume the number of sampling points on the coarse grid is smaller than that on the dense grid by a factor of n_1, where $n_1 = \text{integer}(\text{Re}(\sqrt{\epsilon_1}))$. Thus the $\tilde{i}$th point on the coarse grid corresponds to n_1 points $\tilde{i}_1, \tilde{i}_2, \ldots, \tilde{i}_{n_1}$ on the dense grid. The $\tilde{i}$ with no subscript, it refers to the same coarse point $\tilde{i}$. For $\tilde{i}_p$, $p = 1, 2, \ldots, n_1$, it refers to the n_1 dense grid points associated with the coarse grid point $\tilde{i}$. The $\tilde{i}_p$th ($p = 1, 2, \ldots, n_1$) point is the ith dense grid point where i is given by $i = (\tilde{i} - 1)n_1 + p$. When we calculate the convolution of a_{ij}^{ns} and surface fields of u_j on the dense grid, the following approximation can be made.

$$\sum_{j=1}^{N} a_{ij}^{ns} u(x_j) = \sum_{\tilde{j}=1}^{N/n_1} \sum_{q=1}^{n_1} a_{\tilde{i}_p \tilde{j}_q}^{ns} u(x_{\tilde{j}_q}) \approx \sum_{\tilde{j}=1}^{N/n_1} a_{\tilde{i}_p \tilde{j}}^{ns} \sum_{q=1}^{n_1} u(x_{\tilde{j}_q}) \tag{5.5.12}$$

In getting (5.5.12), we need the property that the Green's function of free space is essentially constant over an interval of n_1 points. Furthermore, the elements $a_{\tilde{i}_p \tilde{j}}^{ns}$ of $p = 1, 2, \ldots, n_1$ can be found by interpolating from the coarse grid to the dense grid,

$$a_{\tilde{i}_p \tilde{j}}^{ns} = \sum_{\tau=-T}^{T} In(\tilde{i}_p, \tilde{i} + \tau) a_{(\tilde{i}+\tau)\tilde{j}}^{ns} \tag{5.5.13}$$

where $In(\tilde{i}_p, \tilde{i}+\tau)$ is the interpolation operator and T is the number of points of the coarse grid we use to interpolate. Then

$$\sum_{j=1}^{N} a_{ij}^{ns} u(x_j) = \sum_{\tilde{j}=1}^{N/n_1} \sum_{\tau=-T}^{T} In(\tilde{i}_p, \tilde{i}+\tau) a_{(\tilde{i}+\tau)\tilde{j}}^{ns} \sum_{q=1}^{n_1} u(x_{\tilde{j}_q}) = \sum_{\tau=-T}^{T} In(\tilde{i}_p, \tilde{i}+\tau) g_{\tilde{i}+\tau} \tag{5.5.14}$$

where

$$g_{\tilde{i}+\tau} = \sum_{\tilde{j}=1}^{N/n_1} a_{(\tilde{i}+\tau)\tilde{j}}^{ns} \sum_{q=1}^{n_1} u(x_{\tilde{j}_q}) \tag{5.5.15}$$

What is done is that the surface fields on the dense grid are first averaged before being convolved with the free space Green's function on the coarse grid. Then we use interpolation to find N values on the dense grid. Similarly, we can obtain:

$$\sum_{j=1}^{N} b_{ij}^{ns} \psi(x_j) = \sum_{\tau=-T}^{T} In(\tilde{i}_p, \tilde{i} + \tau) h_{\tilde{i}+\tau} \tag{5.5.16}$$

where

$$h_{\tilde{i}+\tau} = \sum_{\tilde{j}=1}^{N/n_1} b^{ns}_{(\tilde{i}+\tau)\tilde{j}} \sum_{q=1}^{n_1} \psi(x_{\tilde{j}_q}) \tag{5.5.17}$$

Thus, Eq. (5.5.1) can be rewritten as

$$\left[\sum_{j=1}^{N} a^{s}_{\tilde{i}j} u(x_j) + \sum_{j=1}^{N} b^{s}_{\tilde{i}j}\psi(x_j)\right] + \sum_{\tau=-T}^{T} In(\tilde{i}_p, \tilde{i}+\tau) g_{\tilde{i}+\tau} + \sum_{\tau=-T}^{T} In(\tilde{i}_p, \tilde{i}+\tau) h_{\tilde{i}+\tau} = \psi_{inc}(x_i) \tag{5.5.18}$$

We can write Eqs. (5.5.5) and (5.5.18) as the following matrix equations.

$$\overline{\overline{Z}}^{(1)}_{A,sdg} \cdot \overline{u}_{sdg} + \overline{\overline{Z}}^{(1)}_{B,sdg} \cdot \overline{\psi}_{sdg} = \overline{0}_{sdg} \tag{5.5.19}$$

$$\overline{\overline{Z}}_{A,sdg} \cdot \overline{u}_{sdg} + \overline{\overline{Z}}_{B,sdg} \cdot \overline{\psi}_{sdg} + \tilde{\overline{\overline{Z}}}_{A,scg} \cdot \tilde{\overline{u}}_{scg}\Big|_{\text{intp}} + \tilde{\overline{\overline{Z}}}_{B,scg} \cdot \tilde{\overline{\psi}}_{scg}\Big|_{\text{intp}} = \overline{\psi}_{inc,sdg} \tag{5.5.20}$$

Note that the elements of $\tilde{a}^{(1)}_{ij}$, $\tilde{b}^{(1)}_{ij}$, $\tilde{a}^{s}_{\tilde{i}j}$, and $\tilde{b}^{s}_{\tilde{i}j}$ consist only of banded matrices. The main CPU requirements are to calculate the values of $g_{\tilde{i}}$ and $h_{\tilde{i}}$ with $\tilde{i}$ from 1 to N/n_1. For direct matrix and column vector multiplication, it takes approximately $(N/n_1)^2$ operations. Note that N/n_1 is the number of unknowns on the coarse grid. The computational steps can be further reduced using the BMIA or the FMM. In Section 2, we have illustrated the PBTG-BMIA. In the following section, we will illustrate how to combine the PBTG with the multilevel steepest descent path FMM.

5.2 Computational Complexity of the Combined Algorithm of the PBTG with the MLFMM

For the multilevel FMM, multi-sized groups are formed. At the lowest level, the N elements are decomposed into L groups. Each group includes M elements where $N = ML$. Then each two subgroups at the level form an upper level group (large group) until the highest level. The interactions of groups at each level are calculated only for the non-near groups at this level inside the neighboring groups of an upper level. Thus, the impedance matrix can be written as the sum of the following matrices.

$$\overline{\overline{Z}} = \overline{\overline{Z}}^{(0)} + \overline{\overline{Z}}^{(1)} + \cdots + \overline{\overline{Z}}^{(n)} + \cdots + \overline{\overline{Z}}^{(\tilde{p})} \tag{5.5.21}$$

where the matrix $\overline{\overline{Z}}^{(n)}$ includes only the elements that would be computed at the nth level.

As illustrated in Section 3.2, the procedure of the multilevel SDFMM is composed of three steps. First, the surface fields at each element are translated to the group centers at each level. When transferring the field from the lower level group center to the upper level group center, an interpolation is required to find the values of the fields from the coarser angles to the finer angles. Second, the interactions of group centers at each level are calculated. Only those of the non-near groups inside the neighboring groups of the upper level are calculated at each level. Lastly, the receiving fields at each group center are distributed to its subgroup centers/elements. The last step is performed from the highest level to the lowest level with anterpolation.

The number of computational steps for the first step is

$$N_1 = 2\left[QN + \sum_{n=2}^{p} Q\frac{L}{2^{(n-2)}}\right] + 2\sum_{n=2}^{p} QQ_1\frac{L}{2^{(n-2)}} \approx 2QN + 4Q(Q_1+1)\frac{N}{M} \tag{5.5.22}$$

where the first term comes from aggregating the surface fields to group centers, and the second one comes from interpolating the fields from the coarse angles to the dense angles for all the levels. The integer Q_1 is the number of operations used to interpolate for a single angle.

The number of computational steps for the second step is

$$N_2 = \sum_{n=1}^{p} 3\frac{L}{2^{(n-1)}}Q \approx 6\frac{Q}{M}N, \tag{5.5.23}$$

and the number of computational steps for the last one is

$$N_3 = 2\left[QN + \sum_{n=2}^{p} Q\frac{L}{2^{(n-2)}}\right] + 2\sum_{n=2}^{p} QQ_1\frac{L}{2^{(n-2)}} \approx 2QN + 4Q(Q_1+1)\frac{N}{M}. \tag{5.5.24}$$

The first term is for disaggregating and the second is for anterpolating.

Thus, the total number of computational steps for the matrix-vector multiplication is

$$N_{\text{total}} = M(3N - 2M) + 2Q\left[2 + \frac{4Q_1+7}{M}\right]N, \tag{5.5.25}$$

where the first term is from near field interactions and last one is from non-near field interactions. As long as Q is constant, the computation steps will increase only with linear N.

To combine the PBTG with the multilevel SDFMM, the near field interactions have to be calculated directly on the dense grid, and the non-near field interactions are calculated on the coarse grid through the MLFMM.

In Eq. (5.5.19), the number of computational steps for $\overline{\overline{Z}}^{(1)}_{A,sdg} \cdot \overline{u}_{sdg}$ and $\overline{\overline{Z}}^{(1)}_{B,sdg} \cdot \overline{\psi}_{sdg}$ is $2M_1 N_{sdg}$, where M_1 is determined by the permittivity of the lossy medium and the number of sampling points per wavelength on the dense grid and is a constant for a given case because no non-near field interactions are calculated through it. In Eq. (5.5.20), the MLSDFMM is used. We treat the matrices $\overline{\overline{Z}}_{A,sdg}$ and $\overline{\overline{Z}}_{B,sdg}$ as $\overline{\overline{Z}}^{(0)}$ in Eq. (5.5.21) and $\tilde{\overline{\overline{Z}}}_{A,scg}$ and $\tilde{\overline{\overline{Z}}}_{B,scg}$ as $\overline{\overline{Z}}^{(1)} + \Lambda + \overline{\overline{Z}}^{(n)} + \Lambda + \overline{\overline{Z}}^{(\tilde{p})}$. Thus the number of computational steps for this part is $2M_{sdg}(3N_{sdg} - 2M_{sdg}) + 4Q[2 + \frac{4Q_1+7}{M_{scg}}]N_{scg}$ plus the number of operations used to interpolate from the coarse grid to the dense grid, which is proportional to N_{sdg}. The total number of operations for the PBTG-MLSDFMM is $2M_1 N_{sdg} + 2M_{sdg}(3N_{sdg} - 2M_{sdg}) + 4Q[2 + \frac{4Q_1+7}{M_{scg}}]N_{scg} + Q_2 N_{sdg}$. Thus the computational complexity of the PBTG-MLSDFMM is $\mathcal{O}(N_{sdg} + N_{scg})$.

Numerical simulation results will be presented in terms of the normalized bistatic scattering coefficients and emissivities or brightness temperatures. The bistatic scattering coefficients are defined as, using a tapered plane wave in the spectral domains,

$$\sigma(\theta_s, \theta_i) = \frac{\left| \int_{-\infty}^{\infty} dx \left[ik\psi(x) \left(\frac{df}{dx} \sin\theta_s - \cos\theta_s \right) - u(x) \right] \exp(-ik\alpha(\theta_s, x)) \right|^2}{4\pi g^2 \int_{-k}^{k} dk_x k_z \exp\left[-(k_x - k\sin\theta_i)^2 g^2/2 \right]} \tag{5.5.26}$$

In (5.5.26), $\alpha(\theta_s, x) = \sin\theta_s x + \cos\theta_s f(x)$, and $k_z = \sqrt{k^2 - k_x^2}$.

5.3 Gaussian Rough Surfaces and CPU Comparison

In this section, we illustrate the accuracy of the algorithm for the energy conservation, and compare the results and CPU requirements between the PBTG-MLSDFMM and the BMIA with the single dense grid. We use 10 points per wavelength as the coarse grid and 30 points per wavelength as the dense grid. We also plot the results from the PBTG-BMIA. All the results are run on a Pentium II with 450 MHz and 256 Mbytes.

In Fig. 5.5.1, the bistatic scattering coefficients computed from the BMIA, PBTG-FMM, and PBTG-BMIA are shown for one realization. The

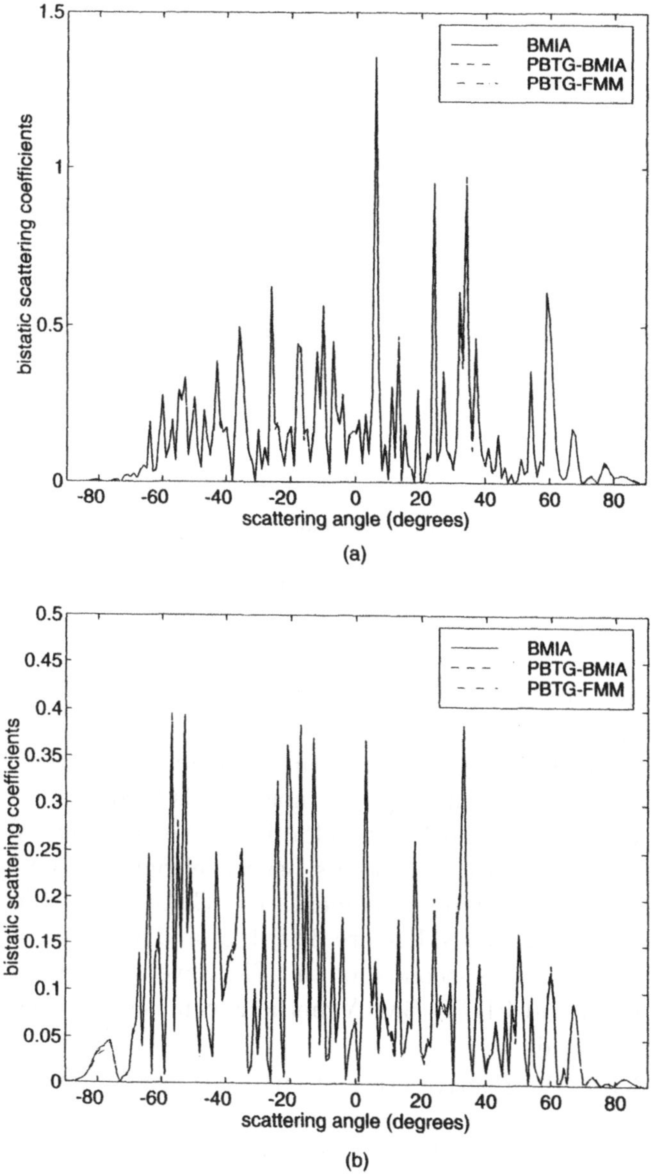

Figure 5.5.1 Comparison of the bistatic scattering coefficients computed by the PBTG-FMM, PBTG-BMIA, and BMIA. The rms height is 0.3 wavelength, correlation length is 0.5 wavelength, angle of incidence is 30 degrees, permittivity is $25 + 2i$, and surface length is 128 wavelengths. (a) TE wave incidence; (b) TM wave incidence.

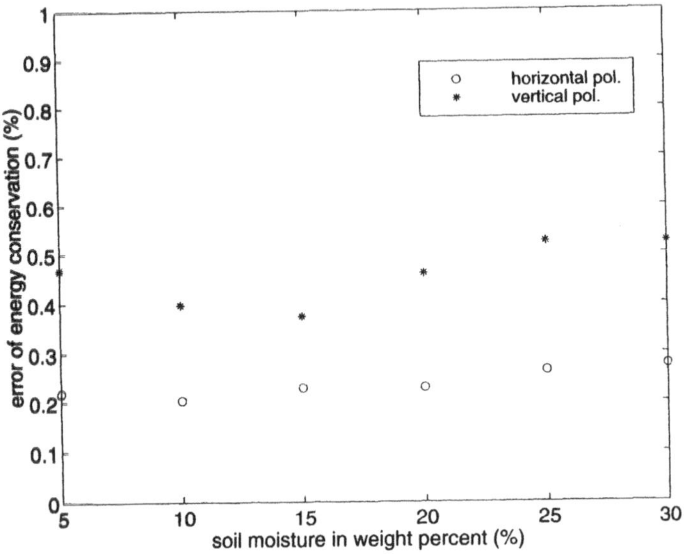

Figure 5.5.2 Error of the energy conservation check as a function of soil moisture.

relative permittivity of the lower medium is $25 + 2i$, and the rms height and correlation length are 0.3 and 0.5 wavelengths. We use 30 points per wavelength as the dense grid and the surface length is 128 wavelengths with 7,680 surface unknowns. The angle of incidence is 30 degrees. Figure 5.5.1a is for TE wave incidence and 5.5.1b is for TM wave incidence. We regard the results from the BMIA with 30 points per wavelength as the accurate one. The three results are in good agreement. The CPU required for the BMIA, PBTG-FMM, and PBTG-BMIA are 372, 163, and 157 seconds for the TE wave and 123, 49, and 48 seconds for the TM wave, respectively. Thus both the PBTG-FMM and PBTG-BMIA methods can save CPU compared with the single dense grid.

In Fig. 5.5.2, the relative error of energy conservation check is plotted as a function of the relative permittivity of the lower medium. The results are from the PBTG-MLSDFMM. The relative permittivities of the lower medium are changed according to the soil moistures from 5 to 30 percent in weight. The correlation length is 1.0 wavelength in this case and the other parameters are same as in Fig. 5.5.1. The energy conservation is less than 0.6% for all the cases.

In Fig. 5.5.3, the CPU per iteration in the conjugate gradient method is plotted against the number of surface unknowns for the three methods.

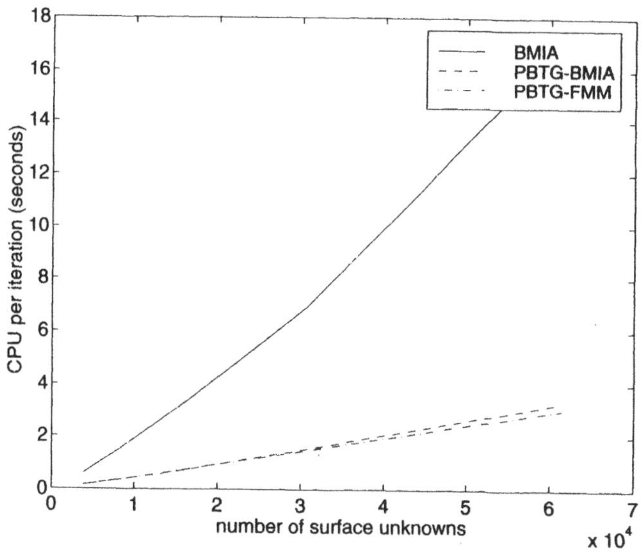

Figure 5.5.3 Comparisons of CPU time per iteration in the conjugate gradient method required by the PBTG-FMM, PBTG-BMIA, and BMIA. N is the number of surface unknowns.

The dense grid is fixed at 30 points per wavelength and we change the surface length. It is shown that the PBTG-BMIA and PBTG-MLSDFMM have similar performance for the cases we compute. The first algorithm is an $\mathcal{O}(N \log N)$ algorithm and the latter is a linear algorithm with N. Both of them take less CPU than the BMIA with the dense grid.

5.4 Non-Gaussian Surfaces

There are two types of correlation functions often used [Chen and Ishimaru, 1990], Gaussian and exponential correlation functions. The spectral densities of the Gaussian and exponential are given, respectively, by

$$W(k) = \frac{h^2 l}{\sqrt{4\pi}} \exp\left(-\frac{k^2 l^2}{4}\right) \tag{5.5.27}$$

$$W(k) = \frac{h^2 l}{\pi} \frac{1}{1 + k^2 l^2} \tag{5.5.28}$$

where h is rms height, l is correlation length, and k is surface wavenumber. It has been found that the surfaces with Gaussian spectral density are far away from real natural rough surfaces such as soil and ocean whereas the surfaces with exponential correlation function are without rms slope, which is required for numerical simulations of wave scattering from random rough surfaces. The third type of surface roughness spectrum, power-law spectral density, is proposed as the following [Chen and Ishimaru, 1990; Kuga et al. 1993].

$$W(k) = \frac{h^2 l}{\sqrt{4\pi}} \left\{ 1 + \pi \left[\frac{(2n-3)!!}{(2n-2)!!} \right]^2 \frac{k^2 l^2}{4} \right\}^{-n} \tag{5.5.29}$$

where $(2n-2)!! = 2 \times 4 \times \cdots \times (2n-2)$, $(2n-3)!! = 1 \times 3 \times \cdots \times (2n-3)$, and $(-1)!! = 1$. The above spectrum becomes a Gaussian spectrum when the power index of n goes to infinity and is very similar to the spectrum with an exponential correlation function when n is one. The parameters h and l are supposed to be the rms height and correlation length in the above spectrum, respectively. But if we compare the power-law spectrum with a power index of one with the spectrum of the exponential correlation function, we find that the real correlation length of power-law spectrum is actually $\sqrt{\pi} l/2$. This can be seen by rewriting the power-law spectrum with the power of 1 as:

$$W(k) = \frac{h^2 \left(\sqrt{\pi} l/2\right)}{\pi} \frac{1}{1 + k^2 \left(\sqrt{\pi} l/2\right)^2} \tag{5.5.30}$$

Thus, a coefficient varying with the power index is needed and introduced to overcome this problem. The modified power-law spectrum is the following:

$$W(k) = \frac{h^2 l}{\sqrt{4\pi}} \frac{1}{b_p} \left[1 + \left(\frac{a_p}{b_p} \right)^2 \frac{k^2 l^2}{4} \right]^{-p} \tag{5.5.31}$$

where $a_p = \Gamma(p-0.5)/\Gamma(p)$ and Γ is the Gamma function and $b_1 = \sqrt{\pi}/2$, $b_2 = 0.95$, $b_3 = 0.97, b_4 = 0.98, \ldots$, and $b_\infty = 1.0$ are determined numerically, and h is the rms height and l is the correlation length. The modified power-law spectrum becomes a Gaussian spectrum when the power index n goes to infinity, and is the spectrum of an exponential correlation function when n is one. The important feature of the proposed spectrum is that it gives various spectra but with fixed rms height and correlation length, which are physical parameters usually used to describe the rough surfaces.

We next show some numerical results of the bistatic scattering coefficients and the brightness temperatures from wet soil with the power law spectrum. The rms height and correlation length are fixed at 0.3 and 1.0 wavelength, respectively. The surface length is 64 wavelengths and the dense grid is 30 points per wavelength. The simulation was performed by the PBTG-MLSDFMM.

In Fig. 5.5.4, the comparisons of the bistatic scattering coefficients between surfaces with a power law spectrum with different power indices are

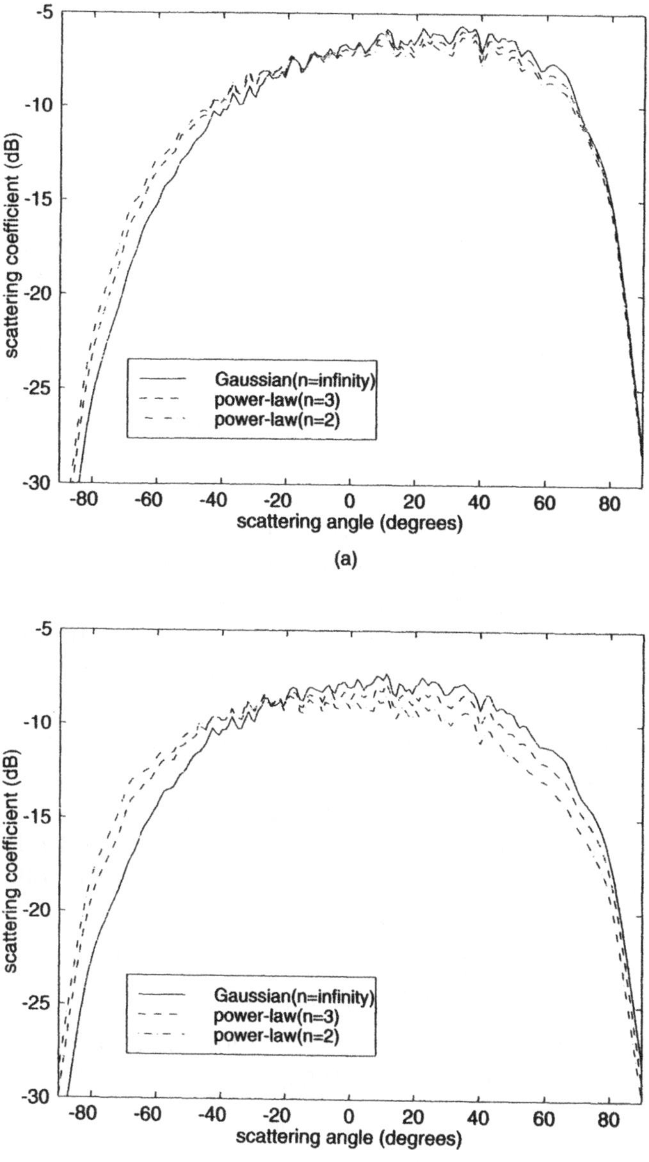

Figure 5.5.4 Comparisons of the bistatic scattering coefficients from various spectra but with fixed rms height of 0.3 wavelength and correlation length of 1.0 wavelength at a angle of incidence of 30 degrees. The relative permittivity is $17.7 + i2.26$. (a) TE wave; (b) TM wave.

shown. The relative permittivity of the lower medium is $17.7 + i2.26$ and the angle of incidence is 30 degrees. Figure 5.5.4(a) is for TE wave incidence and 5.5.4(b) is for TM wave incidence. For scattering angles smaller than 0 degrees, the surface with the power index of 2 has the strongest scattering and the Gaussian surface has the weakest. For scattering angle larger than 0 degree, the situation is the opposite. The reason is that the surface with smaller power index has a larger rms slope, which is an important factor for increasing scattering in the back directions. Unlike the results presented by Maradudin et al. [Maradudin et al. 1990; Maradudin and Mendez, 1996], there is no obvious backscattering enhancement that is shown in these cases. A possible reason is that rough surface used here is smoother than theirs.

In Fig. 5.5.5, we plot the brightness temperature as a function of the observation angle. The physical temperature is 300K and the other parameters are the same as in Fig. 5.5.4. The emissivity is calculated as one minus the reflectivity. Figure 5.5.5(a) is for horizontal polarization and 5.5.5(b) is for vertical polarization. For the same roughness parameters of rms height and correlation length, the differences in brightness temperature for the various spectra can be as large as 15 K. For the horizontal polarization, this difference is almost the same for observation angles between 10 and 50 degrees. For the vertical polarization, the difference has the largest value at the observation angle of 10 degrees and the smallest value at the observation angle of 50 degrees. In general, the surfaces with the smaller power index have the larger brightness temperature.

In Fig. 5.5.6, the brightness temperature is presented as a function of the soil moistures. The relationship between soil moisture and relative permittivity is from Tsang and Newton [1982] which assumes the following mixing formula at a wavelength of $\lambda = 21$ cm:

$$\frac{\epsilon'}{\epsilon_o} = \begin{cases} 2.56 + 0.30 s_m & \text{for } s_m < 11.5 \\ -9.90 + 1.38 s_m & \text{for } s_m \geq 11.5 \end{cases} \tag{5.5.32a}$$

$$\frac{\epsilon''}{\epsilon_o} = \begin{cases} 0.06 s_m & \text{for } s_m < 11.5 \\ -1.44 + 0.185 s_m & \text{for } s_m \geq 11.5 \end{cases} \tag{5.5.32b}$$

where ϵ' and ϵ'' are respectively the real and imaginary parts of the soil permittivity and s_m is the percent of soil moisture by weight.

The observation angle is fixed at 30 degrees and the other parameters are the same as in the preceding figure. Again, for the horizontal polarization, the differences of the brightness temperatures among the different spectra are essentially the same for soil moistures between 5 and 30 percent. For the vertical polarization, this difference increases with increasing soil moisture.

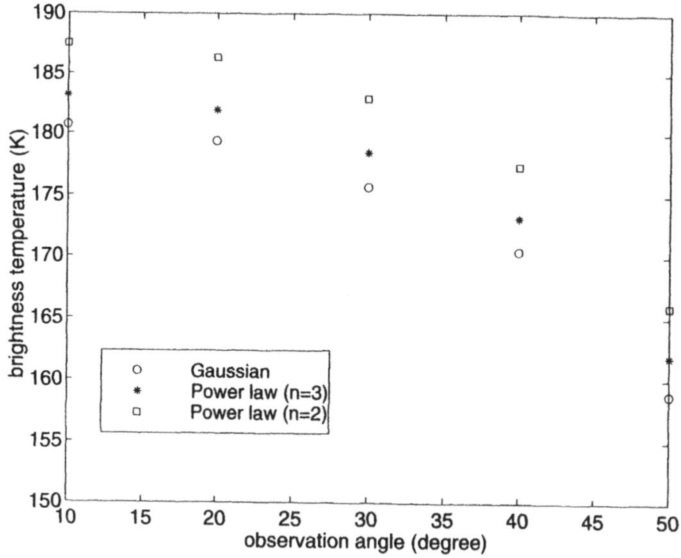

(a)

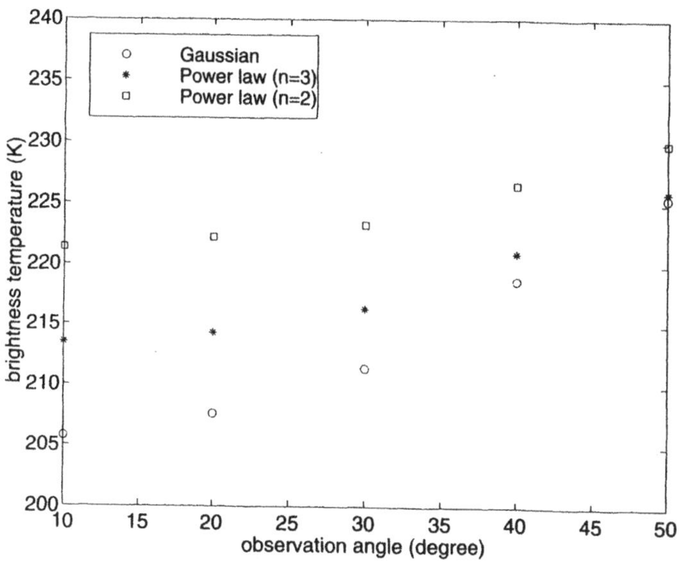

(b)

Figure 5.5.5 Brightness temperature of a rough surface as a function of observation angles and comparisons between various spectra. The rms height is 0.3 wavelength and the correlation length is 1.0 wavelength. The relative permittivity is $17.7 + i2.26$.

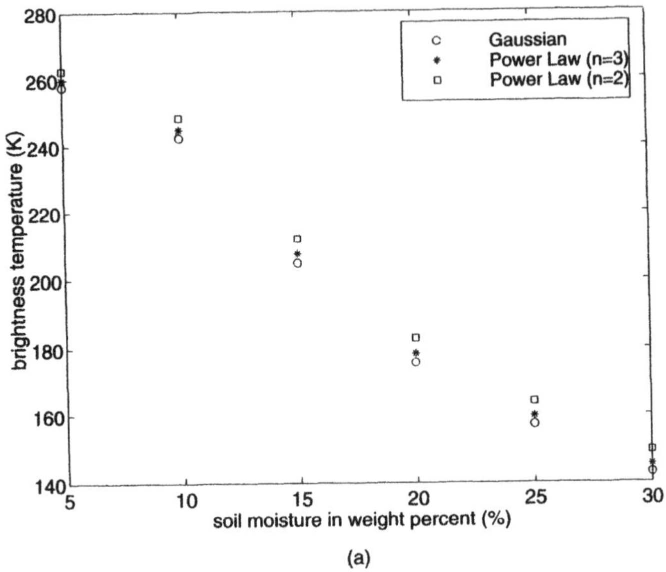

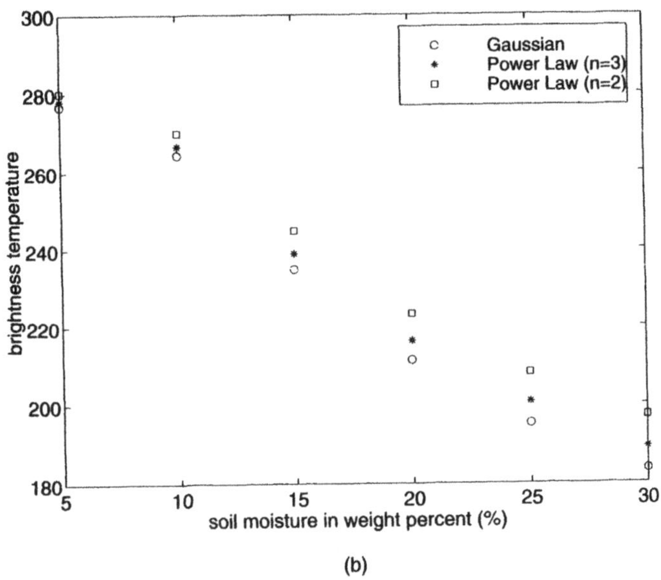

Figure 5.5.6 Brightness temperature of a rough surface as a function of soil moisture and comparisons between various spectra. The rms height is 0.3 wavelength and correlation length is 1.0 wavelength. The observation angle is 30 degrees.

REFERENCES AND ADDITIONAL READINGS

Anastassiu, H. T., M. Smelyanskiy, S. Bindiganavale, and J. L. Volakis (1998), Scattering from relatively flat surfaces using the adaptive integral method, *Radio Sci.*, **33(1)**, 7–16.

Axline, R. M. and A. K. Fung (1978), Numerical computation of scattering from a perfectly conducting random surface, *IEEE Trans. Antennas Propagat.*, **26(3)**, 482–488.

Belszynski, E., M. Belszynski, and T. Jaroszewicz (1994), A fast integral-equation solver for electromagnetic scattering problems, *IEEE Ant. and Propagat. Soc. Int. Sym.*, **1**, 416–419.

Briggs, W. L. (1987), *A Multigrid Tutorial*, SIAM, Philadelphia.

Chan, C. H., S. H. Lou, L. Tsang, and J. A. Kong (1991), Electromagnetic scattering of waves by random rough surface: A finite-difference time-domain approach, *Microwave Opt. Technol. Lett.*, **4(9)**, 355–359.

Chan, C. H., L. Tsang, and Q. Li (1998), Monte-Carlo simulations of large-scale one-dimensional random rough surface scattering at near grazing incidence: Penetrable case, *IEEE Trans. Antennas Propagation*, **46(1)**, 142–149.

Chen, J. S. and A. Ishimaru (1990), Numerical simulation of the second order Kirchhoff approximation from very rough surfaces and study of backscattering enhancement, *J. Acous. Soc. Am.*, **88**, 1846–1850.

Chou, H. T. (2000), Extension of the forward-backward method using spectral acceleration for the fast analysis of large array problems, *IEE Proc. – Microw., Antennas, and Propag.*, **147(3)**, 167–172.

Chou, H. T. and J. T. Johnson (1998), A novel acceleration algorithm for the computation of scattering from rough surfaces with the forward-backward method, *Radio Sci.*, **33(5)**, 1277–1287.

Coifman, R., V. Rohklin, and S. Wandzura (1993), The fast multipole method for the wave equation: A pedestrian prescription, *IEEE Antennas Propag. Mag.*, **35(3)**, 7–12.

Devayya, R. and D. J. Wingham (1992), The numerical calculation of rough surface scattering by the conjugate gradient method, *IEEE Trans. Geosci. Remote Sens.*, **30(3)**, 645–648.

Donohue, D. J., H. C. Ku, and D. R. Thompson (1998), Application of iterative moment method solutions to ocean surface radar scattering, *IEEE Trans. Antennas Propagat.*, **46**, 121–132.

Fung, A. K., Z. Li, and K. S. Chen (1992), Backscattering from a randomly rough dielectric surface, *IEEE Trans. Geosci. Remote Sens.*, **30(2)**, 356–369.

Holliday, D., L. L. DeRaad, Jr., and G. J. St-Cyr (1996), Forward-backward: A new method for computing low-grazing angle scattering, *IEEE Trans. Antennas Propagat.*, **44**, 722–729.

Jandhyala, V., E. Michielssen, S. Balasubramaniam, and W. C. Chew (1998a), A combined steepest descent-fast multipole algorithm for the fast analysis of three-dimensional scattering by rough surfaces, *IEEE Trans. Geosci. Remote Sens.*, **36(3)**, 738–748.

Jandhyala, V., B. Shanker, E. Michielssen, and W. C. Chew (1998b), A fast algorithm for the analysis of scattering by dielectric rough surface, *J. Opt. Soc. Am. A*, 1877–1885.

Johnson, J. T. (1996), Applications of numerical models for rough surface scattering, Ph.D. thesis, Massachusetts Institute of Technology.

Kapp, D. A. and G. S. Brown (1996), A new numerical method for rough surface scattering calculations, *IEEE Trans. Antennas Propagat.*, **44**, 711–721.

Kuga, Y., J. S. Colburn, and P. Phu (1993), Millimeter-wave scattering from one-dimensional surfaces of different surface correlation functions, *Waves in Random Media*, **3**, 101–110.

Li, Q. (2000), Numerical simulation of interactions of electromagnetic waves with lossy dielectric surfaces using fast computational methods, Ph.D. thesis, University of Washington, Seattle.

Li, Q., C. H. Chan, and L. Tsang (1999), Monte-Carlo simulations of wave scattering from lossy dielectric random rough surfaces using the physics-based two-grid method and canonical grid method, *IEEE Trans. Antennas Propagat.*, **47(4)**, 752–763.

Li, Q. and L. Tsang (2001), Wave scattering from lossy dielectric random rough surfaces using the physics-based two grid method in conjunction with the mutilevel fast multipole method, *Radio Sci.*, in press.

Li, Q., L. Tsang, K. S. Pak, and C. H. Chan (2000), Bistatic scattering and emissivities of random rough dielectric lossy surfaces with the physics-based two-grid method in conjunction with the sparse-matrix canonical grid method, *IEEE Trans. Antennas Propagat.*, **48(1)**, 1–11.

Li, S., C. H. Chan, L. Tsang, Q. Li, and L. Zhou (2000), Parallel implementation of the sparse-matrix/canonical grid method for the analysis of two-dimensional random rough surfaces (Three-dimensional scattering problem) on a Beowulf System, *IEEE Trans. Geosci. Remote Sens.*, **38**, 1600–1608.

Lin, C. M., C. H. Chan, and L. Tsang (1999), Conical diffraction of electromagnetic waves from one-dimensional lossy dielectric rough surfaces by combined wavelet transform and banded-matrix iterative approach/canonical grid methods, *IEEE Trans. Geosci. Remote Sens.*, **37(5)**, 2295–2304.

Liu, C. C. and W.C. Chew (1994), A multilevel algorithm for solving a boundary integral equation of wave scattering, *Microwave Opt. Technol. Lett.*, **7**, 466–470.

Lou, S. H., L. Tsang, and C. H. Chan (1991), Application of finite element method to Monte Carlo simulations of scattering of waves by random rough surfaces: penetrable case, *Waves in Random Media*, **1(4)**, 287–307.

Maradudin, A. A. and E. R. Mendez (1996), The utility of an impedance boundary condition in the scattering of light from one-dimensional randomly rough dielectric surfaces, *Optics and Spectroscopy,* **80**, 409–420.

Maradudin, A. A., T. Michel, A. R. McGurn, and E. R. Mendez (1990), Enhanced backscattering of light from a random grating, *Ann. Phys.*, **203(2)**, 255–307.

Michelssen, E., A. Boag, and W. C. Chew (1996), Scattering from elongated objects, *IEE Proceedings Microwave Ant. and Propag.*, **143**, 277–283.

Michielssen, E. and W. C. Chew (1996), The fast steepest descent path algorithm for analyzing scattering from two-dimensional objects, *Radio Sci.*, **31(5)**, 1215–1224.

Pak, K. (1996), Studies of large-scale random rough surface scattering problems based on Monte Carlo simulations with efficient computation integral equations methods, Ph.D. thesis, University of Washington, Seattle.

Pak, K., L. Tsang, C. H. Chan, and J. Johnson (1995), Backscattering enhancement of vector electromagnetic waves from two-dimensional perfectly conducting random rough surfaces based on Monte Carlo simulations, *J. Opt. Soc. Am. A*, **12(11)**, 2491–2499.

Pak, K., L. Tsang, and J. Johnson (1997), Numerical simulations and backscattering enhancement of electromagnetic waves from two-dimensional dielectric random rough surfaces with the sparse matrix canonical grid method, *J. Opt. Soc. Am. A*, **14(7)**, 1515–1529.

Phillips, J. R. and J. K. White (1997), A precorrected-FFT method for electrostatic analysis of complicated 3-D structures, *IEEE Transactions on Computer-Aided Design of Integrated Circuits and Systems*, **16(10)**, 1059–1072.

Rohklin, V. (1990), Rapid solution of integral equations of scattering theory in two dimensions, *J. of Comp. Phys.*, **36**, 414–439.

Spivak, M. (1990), A numerical approach to rough surface scattering by the parabolic method, *J. Acous. Soc. Am.*, **87(5)**, 1999–2004.

Thorsos, E. I. (1988), The validity of the Kirchhoff approximation for rough surface scattering using a Gaussian roughness spectrum, *J. Acous. Soc. Am.*, **83(1)**, 78–92.

Tsang, L., C. H. Chan, K. Pak, and H. Sangani (1994), A BMIA/FFT algorithm for the Monte Carlo simulations of large scale random rough surface scattering, *IEEE Ant. and Propagat. Soc. Int. Sym.*, **3**, 2028–2031.

Tsang, L., C. H. Chan, K. Pak, and H. Sangani (1995), Monte Carlo simulations of large-scale problems of random rough surface scattering and applications to grazing incidence with the BMIA/canonical grid method, *IEEE Trans. Antennas Propagat.*, **43(8)**, 851–859.

Tsang, L., C. H. Chan, and H. Sangani (1993a), A banded matrix iterative approach to Monte Carlo simulations of scattering of waves by large-scale random rough surface problems: TM case, *Electronics Lett.*, **29(2)**, 166–168.

Tsang, L., C. H. Chan, and H. Sangani (1993b), Application of a banded matrix iterative approach to Monte Carlo simulations of scattering of waves by a random rough surface: TM Case, *Microwave Opt. Technol. Lett.*, **6(2)**, 148–151.

Tsang, L., C. H. Chan, H. Sangani, A. Ishimaru, and P. Phu (1993c), A banded matrix iterative approach to Monte-Carlo simulations of large-scale random rough surface scattering: TE case, *J. Electromag. Waves and Appl.*, **7(9)**, 1185–1200.

Tsang, L. and Q. Li (1997), Numerical solution of scattering of waves by lossy dielectric surfaces using a physics-based two-grid method, *Microwave Opt. Technol. Lett.*, **16(6)**, 356–364.

Tsang, L. and R. W. Newton (1982), Microwave emissions from soils with rough surfaces, *J. Geophys. Res.*, **87(11)**, 9017–9024.

West, J. C. and J. M. Sturm (1999), On iterative approaches for electromagnetic rough-surface scattering problems, *IEEE Trans. Antennas Propagat.*, **47(8)**, 1281–1288.

Chapter 6

THREE-DIMENSIONAL WAVE SCATTERING FROM TWO-DIMENSIONAL ROUGH SURFACES

In Chapter 5, we have applied fast computation methods to two-dimensional scattering with one-dimensional random rough surface. In this chapter, we extend the algorithms to three-dimensional problems with two-dimensional random rough surfaces. Study of three-dimensional scattering problem is an active research area. For applications, three-dimensional simulations represent the physical reality and can be used for comparison with real-life data. With the advent of modern computers, new results of large-scale simulations are reported continually. We emphasize that the numerical results presented in this chapter are limited to cases when they were first reported in the literature using computer resources available at that time. In Section 1, we apply the sparse-matrix flat-surface iterative approach (SMFSIA) to solve the problem of scalar wave scattering from a 2-D random rough surface and extend the method to study electromagnetic wave scattering by a two-dimensional perfect electric conductor. Numerical results are illustrated for incident angles of 10° and 20° and with areas between $256\lambda^2$ to $1024\lambda^2$ and up to 1000 surface realizations. The cases of rms heights of 0.5λ and 1λ are considered. Backscattering enhancement is exhibited for both co-polarized and cross-polarized components. Comparisons are made with controlled laboratory millimeter wave experimental data at 20° incident angle. The advantages of millimeter wave scattering experiments are that the calibration allows the comparison of scattered power to incident power [Kuga and Phu, 1996]. Thus, the absolute value of the bistatic scattering coefficient as normalized by the incident power is measured. Thus, we are able to compare the absolute values of the bistatic scattering coefficient between Monte Carlo simulations and experiments. The comparison is without adjustable parameters. It is shown that the co-polarized bistatic coefficient is in good agreement and the cross-polarized bistatic coefficient is in excellent agreement. Agreement is in terms of both absolute magnitude and angular dependence. In Section 2, we discuss integral equations for dielectric surfaces. In Section 3, electromagnetic wave scattering by 2-D dielectric surfaces are studied. In Section 4, we consider bistatic electromagnetic wave scattering from a two-dimensional lossy dielectric random rough surfaces (3-D scattering problem) with large permittivity. For media with large permittivities, the fields can have large spatial variations on the surface. Thus a dense discretization of the surface is required to implement the method of moments (MoM) for the surface integral equations. Such a dense discretization is also required to ensure that the emissivity can be calculated accurately for passive remote sensing applications. We used the physics-based two-grid (PBTG) method that can give the accurate results of the surface fields on the dense grid and also the emissivities. The numerical results are calculated by using the PBTG in conjunction with the sparse-matrix canonical grid method (SMCG). The numerical results are compared with that of the second order small perturbation method (SPM) which is described in Chapter 1 of Volume III. The comparisons show that a large difference in brightness temperature exists between the SPM and numerical simulation results for cases with moderate rms slope.

We also model real-life applications in microwave emission from soils. The results are calculated at two frequencies, viz., L- and C-bands, using the same physical roughness parameters for a variety of soil moisture conditions. This has important application because in practice, the same soil can be measured at different frequencies, different incident angles, and using both active and passive sensors. The physical roughness parameters of the medium are independent of these sensor characteristics and should be used to characterize soils. In the last section, we describe the implementation of the SMCG/PBTG method in a low cost parallel Beowulf system that was completed at the City University of Hong Kong [Li et al. 1999, 2000b].

1 Scattering by Non-Penetrable Media

1.1 Scalar Wave Scattering

In this section, we give results of scattering of scalar waves by 2-D non-penetrable surfaces (3-D scattering problem). The computation is based on a numerical method called the sparse-matrix flat-surface iterative approach. The approach decomposes the matrix of the integral equation as a sum of a sparse matrix, a flat-surface block Toeplitz matrix, and a weak remainder that is followed by an iterative solution until convergence is achieved.

1.1.1 Formulation and Numerical Method

Consider a tapered scalar plane wave $\psi_{inc}(x, y, z)$ impinging upon a 2-D random rough surface with Dirichlet boundary condition and with a random height profile $z = f(x, y)$. The incident direction is $\hat{k}_i = \sin\theta_0 \cos\phi_0 \hat{x} + \sin\theta_0 \sin\phi_0 \hat{y} - \cos\theta_0 \hat{z}$. It is tapered so that the illuminated rough surface can be confined to the surface area $L \times L$. The incident field is

$$\psi_{inc}(x, y, z) = \exp[-ik(\cos\theta_0 z - x\sin\theta_0\cos\phi_0 - y\sin\theta_0\sin\phi_0)(1+w)]\exp(-t) \tag{6.1.1}$$

where $t = t_x + t_y$, and

$$t_x = \frac{(\cos\theta_0 \cos\phi_0 x + \cos\theta_0 \sin\phi_0 y + \sin\theta_0 z)^2}{g^2 \cos^2\theta_0} \tag{6.1.2}$$

$$t_y = \frac{(-\sin\phi_0 x + \cos\phi_0 y)^2}{g^2} \tag{6.1.3}$$

$$w = \frac{1}{k^2}\left[\frac{(2t_x - 1)}{g^2\cos^2\theta_0} + \frac{(2t_y - 1)}{g^2}\right] \tag{6.1.4}$$

and g is the parameter that controls the tapering of the incident wave. The Dirichlet boundary condition is that the wave function ψ is equal to zero on the random surface. A Fredholm integral equation of the first kind can be formed. Let $\bar{r}' = \hat{x}x' + \hat{y}y' + \hat{z}f(x', y')$ denote a field point and let $\bar{r} = \hat{x}x + \hat{y}y + \hat{z}z$ denote a source point on the rough surface. Then the integral equation is

$$0 = \psi_{inc}(\bar{r}) - \iint dxdy G_o(x, y, f(x,y); x', y', f(x', y'))U(x,y) \tag{6.1.5}$$

where $G_o(\bar{r}, \bar{r}') = \dfrac{\exp(ik|\bar{r} - \bar{r}'|)}{4\pi|\bar{r} - \bar{r}'|}$ is the free-space Green's function and the unknown surface field $U(x, y)$ is

$$U(x,y) = \frac{\partial\psi(\bar{r})}{\partial n}\left[1 + \left(\frac{\partial f}{\partial x}\right)^2 + \left(\frac{\partial f}{\partial y}\right)^2\right]^{\frac{1}{2}} \tag{6.1.6}$$

To apply the numerical method, we decompose the integral of (6.1.5) into three terms. We first identify the strong interaction neighborhood of r_d. Let $\rho = [(x - x')^2 + (y - y')^2]^{\frac{1}{2}}$ be the separation between the field point and the source point on the horizontal plane. Let the flat-surface Green's function G_{FS} be

$$G_{FS}(x - x', y - y') = G_o(x, y, z = 0; x', y', z' = 0) = \frac{\exp(ik\rho)}{4\pi\rho} \tag{6.1.7}$$

Then the integral equation (6.1.5) becomes

$$\iint_{\rho < r_d} dxdy G_o(\bar{r}, \bar{r}')U(x,y) + \iint_{\rho \geq r_d} dxdy G_{FS}(x - x', y - y')U(x,y)$$

$$= \psi_{inc}(\bar{r}') - \iint_{\rho \geq r_d} dxdy \left[G_o(\bar{r}, \bar{r}') - G_{FS}(x - x', y - y')\right] U(x,y) \tag{6.1.8}$$

The second term is a 2-D flat-surface convolution integral, whereas the last integral is a weak interaction because G_o and G_{FS} are approximately equal when r_d is large. In terms of matrix notation, we write (6.1.5) and (6.1.8), respectively, as

$$\overline{\overline{Z}}\,\overline{X} = \overline{b} \tag{6.1.9}$$

$$\overline{\overline{Z}} = \overline{\overline{Z}}^{(s)} + \overline{\overline{Z}}^{(FS)} + \overline{\overline{Z}}^{(w)} \tag{6.1.10}$$

where $\overline{\overline{Z}}$ is the matrix of the original integral equation (6.1.5). In (6.1.10), $\overline{\overline{Z}}^{(s)}$ is the strong interaction matrix corresponding to the first integral of (6.1.8). It is a sparse matrix consisting of interactions within a neighborhood of r_d. The flat-surface matrix of G_{FS} is $\overline{\overline{Z}}^{(FS)}$. It is a block Toeplitz matrix so that the product of $\overline{\overline{Z}}^{(FS)}$ with a column vector can be computed by a 2-D fast-Fourier-transform algorithm. $\overline{\overline{Z}}^{(w)}$ is the weak matrix corresponding to the last integral on the right-hand side of (6.1.8). The calculation procedure is, for the first-order and higher-order solutions,

$$\left[\overline{\overline{Z}}^{(s)} + \overline{\overline{Z}}^{(FS)}\right]\overline{X}^{(1)} = \overline{b} \tag{6.1.11}$$

$$\left[\overline{\overline{Z}}^{(s)} + \overline{\overline{Z}}^{(FS)}\right]\overline{X}^{(n+1)} = \overline{b}^{(n+1)} \tag{6.1.12}$$

where

$$\overline{b}^{(n+1)} = \overline{b} - \overline{\overline{Z}}^{(w)}\overline{X}^{(n)} \tag{6.1.13}$$

for $n \geq 1$. For each order the generic equation to be solved is

$$\left[\overline{\overline{Z}}^{(s)} + \overline{\overline{Z}}^{(FS)}\right]\overline{X}^{(u)} = \overline{b}^{(u)} \tag{6.1.14}$$

where the superscript (u) stands for updated. Note that the flat surface $\overline{\overline{Z}}^{(FS)}$ as well as the sparse matrix $\overline{\overline{Z}}^{(s)}$ is on the left-hand side of the general equation.

The matrix equation (6.1.14) is solved by the conjugate gradient method (CGM). For each order, we use CGM to solve (6.1.12). The error of the original exact matrix equation (6.1.9) is defined by the error norm

$$E^{(n)} = \left[\frac{\|\overline{\overline{Z}}\,\overline{X}^{(n)} - \overline{b}\|}{\|\overline{b}\|}\right]^{\frac{1}{2}} \times 100\% \tag{6.1.15}$$

In the iteration procedure the iteration stops when a certain error criterion is reached. We call the method SMFSIA (sparse matrix flat surface iterative approach). In Section 1.1.3 we examine the convergence of this iterative

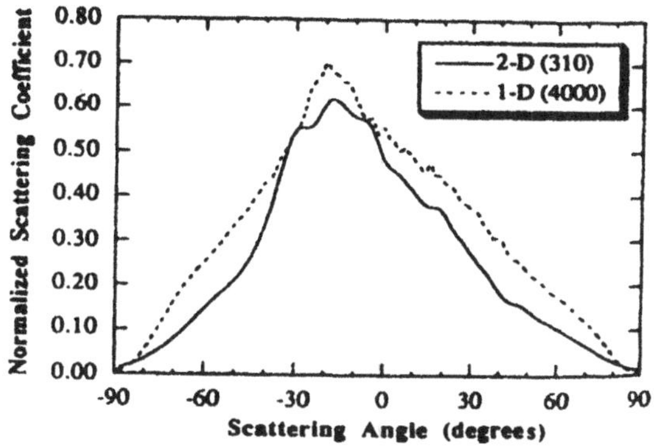

Figure 6.1.1 Comparison of normalized bistatic scattering coefficients of 2-D and 1-D random rough surfaces (310 and 4000 realizations, respectively) for an rms height of 0.5λ and correlation length of 0.707λ with incident angle $\theta_0 = 20°$.

approach. After the unknown $U(x, y)$ is computed, the normalized bistatic scattering coefficient $\sigma(\hat{k}_s)$ in the direction $\hat{k}_s$ is calculated by a weighted integration of the surface field. We have

$$\sigma(\hat{k}_s) = \frac{|F(\hat{k}_s)|^2}{8\pi^3 g^2 \cos\theta_0 \left[1 - \dfrac{(1 + \cos^2\theta_0 + 2\tan^2\theta_0)}{2k^2 g^2 \cos^2\theta_0}\right]} \tag{6.1.16}$$

where

$$F(\hat{k}_s) = -\iint dx dy U(x, y) \exp[-ikx \sin\theta_s \cos\phi_s - iky \sin\theta_s \sin\phi_s - ikf(x, y) \cos\theta_s] \tag{6.1.17}$$

The normalized bistatic scattering coefficient is such that the integration of it over the 2π solid angle of the upper half-space is equal to unity for conservation of energy.

1.1.2 Results and Discussion

We use the method to compute the solution of a random rough surface with a surface area of $80\lambda^2$ and 4096 surface unknowns. The area is L^2, with $L = 8.94\lambda$. This represents a sampling of 51 points per λ^2, and a total of 310 realizations are used. The rough surface has an rms height of 0.5λ and a correlation length of 0.707λ so that the rms slope is 1. The neighborhood distance r_d is chosen to be 3λ.

Figure 6.1.1 shows a comparison of the normalized bistatic scattering coefficient for the 2-D random rough surface and that of the 1-D random

Angle (deg)	Bistatic coefficient	Angle (deg)	Bistatic coefficient
−88.0	1.5238238×10^{-2}	0.0	0.4824193
−86.0	1.8758761×10^{-2}	2.0	0.4650497
−84.0	2.3060517×10^{-2}	4.0	0.4557824
−82.0	2.8241135×10^{-2}	6.0	0.4430476
−80.0	3.4370523×10^{-2}	8.0	0.4276016
−78.0	4.1479971×10^{-2}	10.0	0.4114577
−76.0	4.9557414×10^{-2}	12.0	0.3945898
−74.0	5.8554430×10^{-2}	14.0	0.3824328
−72.0	6.8408959×10^{-2}	16.0	0.3791174
−70.0	7.9079978×10^{-2}	18.0	0.3775388
−68.0	9.0578713×10^{-2}	20.0	0.3663619
−66.0	0.1029617	22.0	0.3450787
−64.0	0.1162529	24.0	0.3236764
−62.0	0.1303005	26.0	0.3070970
−60.0	0.1446681	28.0	0.2907895
−58.0	0.1587106	30.0	0.2721548
−56.0	0.1719020	32.0	0.2546927
−54.0	0.1842503	34.0	0.2400115
−52.0	0.1964813	36.0	0.2244638
−50.0	0.2098318	38.0	0.2052310
−48.0	0.2256480	40.0	0.1853912
−46.0	0.2450749	42.0	0.1709639
−44.0	0.2689258	44.0	0.1640919
−42.0	0.2978484	46.0	0.1608696
−40.0	0.3328274	48.0	0.1558608
−38.0	0.3750854	50.0	0.1472065
−36.0	0.4240414	52.0	0.1369274
−34.0	0.4746332	54.0	0.1276718
−32.0	0.5177182	56.0	0.1202371
−30.0	0.5446179	58.0	0.1137029
−28.0	0.5533673	60.0	0.1068778
−26.0	0.5529661	62.0	9.9265866×10^{-2}
−24.0	0.5589816	64.0	9.1046810×10^{-2}
−22.0	0.5788418	66.0	8.2607575×10^{-2}
−20.0	0.6030874	68.0	7.4216217×10^{-2}
−18.0	0.6167714	70.0	6.5988846×10^{-2}
−16.0	0.6151025	72.0	5.7991304×10^{-2}
−14.0	0.6033526	74.0	5.0314512×10^{-2}
−12.0	0.5891802	76.0	4.3079440×10^{-2}
−10.0	0.5787917	78.0	3.6408279×10^{-2}
−8.0	0.5735879	80.0	3.0396651×10^{-2}
−6.0	0.5672148	82.0	2.5101034×10^{-2}
−4.0	0.5485466	84.0	2.0536819×10^{-2}
−2.0	0.5150658	86.0	1.6683213×10^{-2}
		88.0	1.3490823×10^{-2}

Table 6.1.1 Bistatic coefficients of the 2-D rough surface of Fig. 6.1.1

rough surface with the angle of incidence of $\theta_0 = 20°$ and $\phi_0 = 0°$. The results of bistatic scattering for 2-D random rough surface results are also tabulated in Table 6.1.1.

For 3-D scattering, the bistatic scattering coefficients are shown in the plane of incidence with $\phi_s = 0°$ or $\phi_s = 180°$. This convention is also used in Table 6.1.1. The scattering angle is labeled as positive for $\phi_s = 0°$ and negative for $\phi_s = 180°$. The tapering parameter g is equal to $L/3$. For the 1-D case the surface length is set at 40λ and the results are averaged over 4000 realizations. Backscattering enhancements are observed for both 2-D and 1-D surfaces, with a peak at $\theta_s = -20°$. Note that because the surface area is only $(8.94\lambda)^2$, the incident wave has a half beamwidth of 6°. Thus there is associated broadening that is due to the beam pattern. From the results it is also seen that there is no forward peak that is due to the forward scattering. It should be noted that the integration of the normalized bistatic scattering coefficient curve of the 1-D surface over 1-D angle θ_s is equal to unity, whereas the normalized bistatic scattering coefficient of the 2-D surface is only plotted for observation on the x-z plane. The two curves follow a similar trend, with the 2-D curve always below the 1-D curve. The 2-D bistatic scattering coefficient is lower because scattering can occur for directions off the plane of incidence.

Figure 6.1.2 shows the convergence of the bistatic scattering coefficients for the 2-D surface versus the number of realizations. The coefficients converge in approximately 310 realizations, which is evident from the fact that the curves for 310 and 275 realizations overlie each other. The coefficient also shows that the backscattering peak requires more realizations to converge than other scattering directions. In Table 6.1.2, we list the error norm and the number of iterations for each order. For each order of solution in (6.1.12), a required number of iterations is used for the CGM. For each order of the SMFSIA the solution is solved by the CGM. The convergence criteria for both the SMFSIA and the CGM are set at 1%. The number of iterations required for each order of the SMFSIA decreases with the increase of the solution order, and at the sixth order the solution error is less than 0.1%. In Table 6.1.1 we list the bistatic scattering coefficients for the results averaged over 310 realizations to facilitate comparisons with other methods. Figure 6.1.3 shows a comparison of the second-order Kirchhoff method with the exact solution. The second order Kirchhoff method uses the Neumann iteration method and iterate up to second order. The results clearly show that the second-order Kirchhoff method fails for this case.

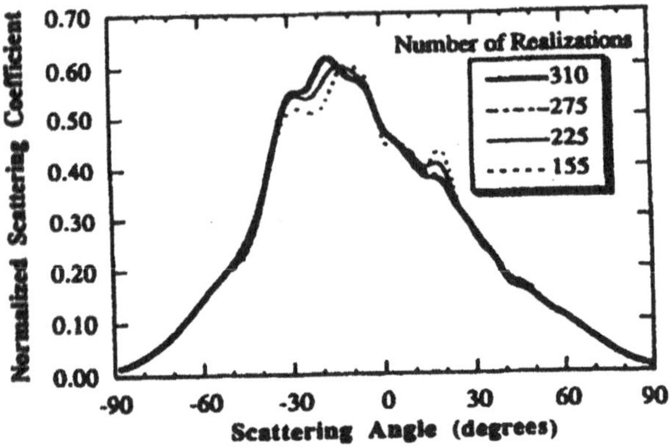

Figure 6.1.2 Convergence of bistatic scattering coefficient with the number of realizations for a 2-D rough surface. Four cases are shown: 155, 225, 275, and 310 realizations. The cases of 275 and 310 realizations overlap each other.

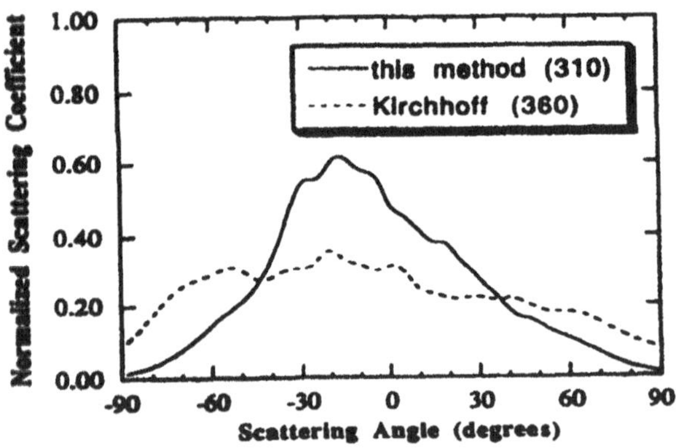

Figure 6.1.3 Comparison between the SMFSIA and the second-order Kirchhoff method. The second-order Kirchhoff method result is based on 360 realizations.

Order Number	Number of CGM Iterations	Error Norm %
1	27	29.31
2	20	6.74
3	13	2.56
4	7	1.04
5	3	0.12
6	1	0.01

Table 6.1.2 Convergence of the SMFSIA of 2-D surface of a single realization.

For each order of solution the error norm $E^{(n)}$ can be easily computed as follows. From (6.1.12) and (6.1.15) we obtain

$$\|\overline{\overline{Z}}\,\overline{X}^{(n)} - \overline{b}\| = \|(\overline{\overline{Z}}^{(s)} + \overline{\overline{Z}}^{(FS)})\overline{X}^{(n)} - (\overline{b} - \overline{\overline{Z}}^{(w)}\overline{X}^{(n)})\| = \|\overline{b}^{(n)} - \overline{b}^{(n+1)}\| \tag{6.1.18}$$

Table 6.1.2 shows the convergence of the SMFSIA for a single realization.

1.1.3 Convergence of SMFSIA

In actual implementation of the SMFSIA the iteration stops when the error norm of the original matrix equation has reached the established smallness criterion. In this section we examine the convergence of the SMFSIA. We note that the right-hand side $\overline{b}$ corresponds to the incident wave and is of the order of $\mathcal{O}(1)$, where $\mathcal{O}$ stands for the order. The column vector $\overline{\overline{Z}}^{(w)}\overline{X}$ corresponds to the last term of (6.1.8). It is the original impedance matrix with the removal of the near field sparse matrix and the flat surface impedance matrix.

$$\mathcal{O}(\overline{\overline{Z}}^{(w)}\overline{X}) = \mathcal{O}\left\{\iint_{\rho>r_d} dxdy\left[\frac{\exp[ik(x_d^2+y_d^2+z_d^2)^{\frac{1}{2}}]}{4\pi(x_d^2+y_d^2+z_d^2)^{\frac{1}{2}}} - \frac{\exp(ik\rho)}{4\pi\rho}\right]U(x,y)\right\} \tag{6.1.19}$$

where $x_d = x - x'$, $y_d = y - y'$, $z_d = f(x,y) - f(x',y')$ and $\rho = (x_d^2 + y_d^2)^{\frac{1}{2}}$. We note that $U(x,y)$ is of the order $k\mathcal{O}(1)$ and has a phase that is randomly fluctuating, whereas z_d is of the order of rms height h. Thus in the limit of large r_d, with $r_d \gg h$, the first term in the square brackets can be Taylor-expanded. Therefore we have

$$\begin{aligned}
\mathcal{O}\left[\overline{\overline{Z}}^{(w)}\overline{X}\right] &= \mathcal{O}\left[\iint_{\rho>r_d} dxdy\frac{\exp(ik\rho)}{4\pi\rho}\left(\frac{ikz_d^2}{2\rho}\right)U(x,y)\right] \\
&= \mathcal{O}\left[\int_{r_d}^{\infty} d\rho\rho\int_0^{2\pi} d\phi\frac{\exp(ik\rho)}{4\pi\rho}\left(\frac{ikz_d^2}{2\rho}\right)U(x,y)\right] \\
&= \mathcal{O}\left[\int_{r_d}^{\infty} d\rho\frac{\exp(ik\rho)}{4\rho}ikh^2U(x,y)\right]
\end{aligned} \tag{6.1.20}$$

By performing integration by parts of (6.1.20), we get

$$\mathcal{O}\left[\overline{\overline{Z}}^{(w)}\overline{X}\right] = \mathcal{O}\left(\frac{kh^2}{r_d}\right) \tag{6.1.21}$$

which is much smaller than $\mathcal{O}(1)$ in the limit of large r_d.

1.2 Electromagnetic Wave Scattering by Perfectly Conducting Surfaces

1.2.1 Surface Integral Equation

Consider an electromagnetic wave impinging upon a rough surface that is perfectly conducting. Then the integral equation for $\overline{r}$ above the rough surface is

$$\overline{H}(\overline{r}) = \int_{S'} dS' \left[\nabla \times \overline{\overline{G}}(\overline{r}, \overline{r}') \cdot \hat{n}' \times \overline{H}(\overline{r}')\right] + \overline{H}_{inc}(\overline{r}) \qquad (6.1.22)$$

Letting $\overline{r} \to \overline{r}_+$ which is a point infinitesimally above the rough surface, we have

$$\hat{n}\times\overline{H}(\overline{r}) = \hat{n}\times\overline{H}_{inc}(\overline{r}) + \lim_{\overline{r}\to\overline{r}_+} \hat{n}\times \int_{S'} dS' \left[\nabla \times \overline{\overline{G}}(\overline{r}, \overline{r}') \cdot \hat{n}' \times \overline{H}(\overline{r}')\right] \qquad (6.1.23)$$

For a constant vector $\overline{a}$

$$\begin{aligned} \nabla \times \overline{\overline{G}}(\overline{r}, \overline{r}') \cdot \overline{a} &= \nabla \times \left(\overline{\overline{I}} + \frac{1}{k^2}\nabla\nabla\right) g(\overline{r}, \overline{r}') \cdot \overline{a} \\ &= \nabla g(\overline{r}, \overline{r}') \times \overline{a} \end{aligned} \qquad (6.1.24)$$

Using (6.1.24) in (6.1.23), we have

$$\hat{n}\times\overline{H}(\overline{r}) = \hat{n}\times\overline{H}_{inc}(\overline{r}) + \lim_{\overline{r}\to\overline{r}_+} \hat{n}\times \int_{S'} dS' \left[\nabla g(\overline{r}, \overline{r}') \times (\hat{n}' \times \overline{H}(\overline{r}'))\right] \qquad (6.1.25)$$

The integral in (6.1.25) is singular at $\overline{r} = \overline{r}'$. To handle this problem, we perform similarly to the Neumann case of Chapter 4, Section 1.5. Let

$$\overline{r} = \overline{\rho} + \delta\hat{n} \qquad (6.1.26)$$

where $\overline{\rho}$ is a point on the rough surface. If $\delta > 0, \overline{r} = \overline{r}_+$, and if $\delta < 0, \overline{r} = \overline{r}_-$. Thus, $\overline{r}_+$ is infinitesimally above the surface while $\overline{r}_-$ is infinitesimally below the surface. In (6.1.25) and (6.1.26) $\hat{n}$ is the normal to the surface at point $\overline{\rho}$. Then

$$\hat{n} \times \overline{H}(\overline{\rho}) = \hat{n} \times \overline{H}_{inc}(\overline{\rho}) + \lim_{\delta\to 0+} \hat{n} \times \int_{S'} dS' \left[\left(\nabla g(\overline{r}, \overline{r}')\right)_{\overline{r}=\overline{r}_+} \times \hat{n}' \times \overline{H}(\overline{r}')\right] \qquad (6.1.27)$$

The $\int_{S'}$ integration is divided into a circular disk S_a of small radius a about $\overline{\rho}$ and the rest which is known as the principal value integral

$$\int_S dS' = \int_{S_a} dS' + \int_{S-S_a} dS' = \int_{S_a} dS' + ⨍ dS' \qquad (6.1.28)$$

where $⨍$ represents principal value integral, which is the integration over S with an infinitesimal circular disk of radius a, S_a, removed from S.

We next examine

$$I = \lim_{\delta\to 0} \hat{n} \times \int_{S_a} dS' \left(\nabla g(\overline{r}, \overline{r}')\right)_{\overline{r}=\overline{r}_\pm} \times \hat{n}' \times \overline{H}(\overline{r}')$$

$$= \lim_{\delta\to 0} \hat{n} \times \left[\left(\int_{S_a} dS' \nabla g(\overline{r}, \overline{r}') \right)_{\overline{r}=\overline{r}_\pm} \times \left(\hat{n} \times \overline{H}(\overline{r})\right) \right] \tag{6.1.29}$$

where $\delta > 0$ for $\overline{r} = \overline{r}_+$ and $\delta < 0$ for $\overline{r} = \overline{r}_-$. Let (ρ', ϕ') be the polar coordinates of $\overline{r}'$ of circular disk of radius a centered about $\overline{\rho}$

$$\lim_{\overline{r}\to\overline{r}_\pm} \int_{S_a} dS' \nabla g(\overline{r}, \overline{r}') = \int_0^a d\rho' \rho' \int_0^{2\pi} d\phi' \nabla \frac{1}{4\pi|\overline{r} - \overline{r}'|}$$

$$= \int_0^a d\rho' \rho' \int_0^{2\pi} d\phi' \left[-\frac{1}{4\pi|\overline{r} - \overline{r}'|^3} (\overline{r} - \overline{r}') \right]_{\overline{r}=\overline{r}_\pm}$$

$$= \int_0^a d\rho' \rho' \int_0^{2\pi} d\phi' \left[-\frac{(\hat{n}\delta - (\rho' \cos\phi' \hat{x} + \rho' \sin\phi' \hat{y})}{4\pi(\rho'^2 + \delta^2)^{3/2}} \right]$$

The $\hat{x}$ and $\hat{y}$ components integrate to zero because of the $\sin\phi'$ and $\cos\phi'$ dependence. Thus

$$\int_{S_a} dS' \nabla g(\overline{r}, \overline{r}') = -\hat{n}\frac{\delta}{2} \int_0^a d\rho' \rho' \frac{1}{(\rho'^2 + \delta^2)^{3/2}}$$

$$= -\hat{n}\frac{1}{2} \left[-\frac{\delta}{(\rho'^2 + \delta^2)^{1/2}} \right]_0^a = -\frac{\hat{n}}{2} \left[-\frac{\delta}{(a^2 + \delta^2)^{1/2}} \pm 1 \right]$$

where the + sign is for $\delta > 0$ and − sign is for $\delta < 0$. Hence

$$\int_{S_a} dS' \nabla g(\overline{r}, \overline{r}') = \begin{cases} -\dfrac{\hat{n}}{2} & \text{for } \overline{r}_+ \\ \dfrac{\hat{n}}{2} & \text{for } \overline{r}_- \end{cases} \tag{6.1.30}$$

Thus

$$I = \hat{n} \times \left[\left(-\frac{\hat{n}}{2} \right) \times \left(\hat{n} \times \overline{H}(\overline{r})\right) \right] = \pm \frac{\hat{n} \times \overline{H}(\overline{r})}{2}$$

with the + sign for $\overline{r} = \overline{r}_+$ and − sign for $\overline{r} = \overline{r}_-$. Equation (6.1.30) is a useful integral identity. We have the integral equation known as MFIE:

$$\hat{n} \times \overline{H}(\overline{r}) = 2\hat{n} \times \overline{H}_{inc}(\overline{r}) + 2\hat{n} \times \mathop{\int\!\!\!\!\!\!-}_S dS' \nabla g(\overline{r}, \overline{r}') \times \hat{n}' \times \overline{H}(\overline{r}') \tag{6.1.31}$$

1.2.2 Surface Integral Equation for Rough Surface Scattering

The 2-D surface spectrum is given by

$$W(K_x, K_y) = \frac{l_x l_y h^2}{4\pi} \exp\left(-\frac{K_x^2 l_x^2}{4} - \frac{K_y^2 l_y^2}{4}\right) \tag{6.1.32}$$

where l_x, l_y, are the correlation lengths in x and y directions respectively, and K_x and K_y are the spatial frequency in x and y directions respectively. The incident direction is $\hat{k}_i = \sin\theta_i \cos\phi_i \hat{x} + \sin\theta_i \sin\phi_i \hat{y} - \cos\theta_i \hat{z}$ and has a perpendicular polarization (TE incidence). The incident wave is tapered so that the illuminated rough surface can be confined to the surface area $L_x \times L_y$. The incident magnetic field is then

$$\overline{H}_i(x, y, z) = -\frac{1}{\eta} \int_{-\infty}^{+\infty} dk_x \int_{-\infty}^{+\infty} dk_y \exp(ik_x x + ik_y y - ik_z z) E_{TE}(k_x, k_y) \hat{h}(-k_z) \tag{6.1.33}$$

where

$$\hat{h}(-k_z) = \frac{k_z}{kk_\rho}(\hat{x}k_x + \hat{y}k_y) + \frac{k_\rho}{k}\hat{z} \tag{6.1.34}$$

where $k_z = \sqrt{k^2 - k_x^2 - k_y^2}$ and $k_\rho = \sqrt{k_x^2 + k_y^2}$. In (6.1.33), $E_{TE}(k_x, k_y)$ is the spectrum of the incident wave and we use the following spectrum.

$$E_{TE}(k_x, k_y) = \frac{1}{4\pi^2} \int_{-\infty}^{+\infty} dx \int_{-\infty}^{+\infty} dy \exp(-ik_x x - ik_y y) \cdot \exp(i(k_{ix} x + k_{iy} y)(1 + w)) \exp(-t) \tag{6.1.35}$$

where $t = t_x + t_y = (x^2 + y^2)/g^2$ and

$$t_x = \frac{(\cos\theta_i \cos\phi_i x + \cos\theta_i \sin\phi_i y)^2}{g^2 \cos^2\theta_i} \tag{6.1.36}$$

$$t_y = \frac{(-\sin\phi_i x + \cos\phi_i y)^2}{g^2} \tag{6.1.37}$$

$$w = \frac{1}{k^2}\left[\frac{(2t_x - 1)}{g^2 \cos^2\theta_i} + \frac{(2t_y - 1)}{g^2}\right] \tag{6.1.38}$$

The w and t terms are introduced to approximate the tapered wave solution that was previously used for the scalar wave case. However, in (6.1.33), we have a spectrum of electromagnetic plane waves so that the incident waves obey the Maxwell equations exactly. Let $\overline{r}' = \hat{x}x' + \hat{y}y' + \hat{z}f(x', y')$ denote a source point and $\overline{r} = \hat{x}x + \hat{y}y + \hat{z}f(x, y)$ denote a field point on the rough surface. We have the magnetic field integral equation (MFIE) on the perfectly conducting rough surface is, as derived in Eq. (6.1.31). Note that

$$\nabla g = (\overline{r} - \overline{r}')G(R) \tag{6.1.39}$$

$$G(R) = \frac{(ikR-1)\exp(ikR)}{4\pi R^3} \tag{6.1.40}$$

and $R = \sqrt{(x-x')^2 + (y-y')^2 + (f(x,y) - f(x',y'))^2}$. The MFIE can be reduced to two coupled scalar integral equations

$$\begin{aligned}
&\frac{F_x(\overline{r})}{2} + \frac{\partial f(x,y)}{\partial y}\int dx'dy'G(R)[(x-x')F_y(\overline{r}') - (y-y')F_x(\overline{r}')] \\
&\quad + \int dx'dy'G(R)\left\{\left[-(x-x')\frac{\partial f(x',y')}{\partial x'} + (f(x,y) - f(x',y'))\right]F_x(\overline{r}')\right. \\
&\quad \left. -(x-x')\frac{f(x',y')}{\partial y'}F_y(\overline{r}')\right\} = -\frac{\partial f(x,y)}{\partial y}H_{iz}(\overline{r}) - H_{iy}(\overline{r})
\end{aligned} \tag{6.1.41}$$

and

$$\begin{aligned}
&\frac{F_y(\overline{r})}{2} - \frac{\partial f(x,y)}{\partial x}\int dx'dy'G(R)[(x-x')F_y(\overline{r}') - (y-y')F_x(\overline{r}')] \\
&\quad - \int dx'dy'G(R)\left\{(y-y')\frac{\partial f(x',y')}{\partial x'}F_x(\overline{r}')\right. \\
&\quad \left. + \left[(y-y')\frac{\partial f(x',y')}{\partial y'} - (f(x,y) - f(x',y'))\right]F_y(\overline{r}')\right\} \\
&\quad = \frac{\partial f(x,y)}{\partial x}H_{iz}(\overline{r}) + H_{ix}(\overline{r})
\end{aligned} \tag{6.1.42}$$

where

$$F_x(\overline{r}) = \sqrt{1 + \left(\frac{\partial f(x,y)}{\partial x}\right)^2 + \left(\frac{\partial f(x,y)}{\partial y}\right)^2}\ \hat{n} \times \overline{H}(\overline{r}) \cdot \hat{x} \tag{6.1.43}$$

and

$$F_y(\overline{r}) = \sqrt{1 + \left(\frac{\partial f(x,y)}{\partial x}\right)^2 + \left(\frac{\partial f(x,y)}{\partial y}\right)^2}\ \hat{n} \times \overline{H}(\overline{r}) \cdot \hat{y} \tag{6.1.44}$$

are proportional respectively to the x and y components of $\hat{n} \times \overline{H}$ on the surface.

1.2.3 Computation Methods

(A) SMFSIA

In SMFSIA, we choose r_d, the neighborhood distance. Let

$$\rho_R = \sqrt{(x-x')^2 + (y-y')^2} \tag{6.1.45}$$

be the horizontal separation between two points $(x, y, f(x, y))$ and $(x', y', f(x', y'))$. If the neighborhood distance r_d is chosen to be much greater than the rms height, then the Green's function $G(R)$ of (6.1.40) is asymptotically equal to

$$G_{FS}(\rho_R) = \frac{(ik\rho_R - 1)\exp(ik\rho_R)}{4\pi\rho_R^3} \tag{6.1.46}$$

Note that $G_{FS}(\rho_R)$ is translationally invariant in the horizontal plane. Thus, we write $G(R)$ in the integrands of the integral equations of (6.1.41) and (6.1.42) for $\rho_R > r_d$ by

$$G(R) = G_{FS}(\rho_R) + (G(R) - G_{FS}(\rho_R)) \tag{6.1.47}$$

Then the integral equation (6.1.41)–(6.1.42) becomes

$$\begin{aligned}
&\frac{F_x(\overline{r})}{2} + \frac{\partial f(x,y)}{\partial y}\int_{\rho_R<r_d} dx'dy' G(R)[(x-x')F_y(\overline{r}') - (y-y')F_x(\overline{r}')] \\
&+ \int_{\rho_R<r_d} dx'dy' G(R)\left\{\left[-(x-x')\frac{\partial f(x',y')}{\partial x'} + (f(x,y) - f(x',y'))\right] F_x(\overline{r}')\right. \\
&\left. - (x-x')\frac{f(x',y')}{\partial y'} F_y(\overline{r}')\right\} \\
&+ \frac{\partial f(x,y)}{\partial y}\int_{\rho_R\geq r_d} dx'dy' G_{FS}(\rho_R)[(x-x')F_y(\overline{r}') - (y-y')F_x(\overline{r}')] \\
&+ \int_{\rho_R\geq r_d} dx'dy' G_{FS}(\rho_R)\left\{\left[-(x-x')\frac{\partial f(x',y')}{\partial x'}\right.\right. \\
&\left.\left. + (f(x,y) - f(x',y'))\right] F_x(\overline{r}') - (x-x')\frac{f(x',y')}{\partial y'} F_y(\overline{r}')\right\} \\
&= -\frac{\partial f(x,y)}{\partial y} H_{iz}(\overline{r}) - H_{iy}(\overline{r}) \\
&\quad - \frac{\partial f(x,y)}{\partial y}\int_{\rho_R\geq r_d} dx'dy'(G(R) - G_{FS}(\rho_R))[(x-x')F_y(\overline{r}') - (y-y')F_x(\overline{r}')] \\
&\quad - \int_{\rho_R\geq r_d} dx'dy'(G(R) - G_{FS}(\rho_R))\left\{\left[-(x-x')\frac{\partial f(x',y')}{\partial x'}\right.\right. \\
&\quad \left.\left. + (f(x,y) - f(x',y'))\right] F_x(\overline{r}') - (x-x')\frac{f(x',y')}{\partial y'} F_y(\overline{r}')\right\}
\end{aligned} \tag{6.1.48}$$

and

$$\frac{F_y(\overline{r})}{2} - \frac{\partial f(x,y)}{\partial x}\int_{\rho_R<r_d} dx'dy'G(R)[(x-x')F_y(\overline{r}') - (y-y')F_x(\overline{r}')]$$
$$-\int_{\rho_R<r_d} dx'dy'G(R)\left\{(y-y')\frac{\partial f(x',y')}{\partial x'}F_x(\overline{r}')\right.$$
$$\left.+\left[(y-y')\frac{\partial f(x',y')}{\partial y'} - (f(x,y)-f(x',y'))\right]F_y(\overline{r}')\right\}$$
$$-\frac{\partial f(x,y)}{\partial x}\int_{\rho_R\geq r_d} dx'dy'G_{FS}(\rho_R)[(x-x')F_y(\overline{r}') - (y-y')F_x(\overline{r}')]$$
$$-\int_{\rho_R\geq r_d} dx'dy'G_{FS}(\rho_R)\left\{(y-y')\frac{\partial f(x',y')}{\partial x'}F_x(\overline{r}')\right.$$
$$\left.+\left[(y-y')\frac{\partial f(x',y')}{\partial y'} - (f(x,y)-f(x',y'))\right]F_y(\overline{r}')\right\}$$
$$=\frac{\partial f(x,y)}{\partial x}H_{iz}(\overline{r}) + H_{ix}(\overline{r})$$
$$+\frac{\partial f(x,y)}{\partial x}\int_{\rho_R\geq r_d} dx'dy'(G(R)-G_{FS}(\rho_R))[(x-x')F_y(\overline{r}') - (y-y')F_x(\overline{r}')]$$
$$+\int_{\rho_R\geq r_d} dx'dy'(G(R)-G_{FS}(\rho_R))\left\{(y-y')\frac{\partial f(x',y')}{\partial x'}F_x(\overline{r}')\right.$$
$$\left.+\left[(y-y')\frac{\partial f(x',y')}{\partial y'} - (f(x,y)-f(x',y'))\right]F_y(\overline{r}')\right\} \tag{6.1.49}$$

The last terms on the right-hand side of the integral equations of (6.1.48)–(6.1.49) are small since $r_d > h$. In terms of matrix notation the SMFSIA procedure is as follows. The surface integral equations (6.1.41) and (6.1.42) are cast into a matrix equation by the method of moments. This gives

$$\overline{\overline{Z}}\,\overline{x} = \overline{b} \tag{6.1.50}$$

Then, the original matrix is decomposed into the sum of a strong matrix, a block Toeplitz flat-surface part, and a weak remainder as

$$(\overline{\overline{Z}}^{(s)} + \overline{\overline{Z}}^{(FS)} + \overline{\overline{Z}}^{(w)})\overline{x} = \overline{b} \tag{6.1.51}$$

In (6.1.51), $\overline{\overline{Z}}^{(s)}$ is a matrix corresponding to the integrals of (6.1.48) and (6.1.49) with $\rho_R < r_d$. The strong matrix is a sparse matrix. The flat surface matrix corresponds to the second term with $G_{FS}(\rho_R)$ in (6.1.48) and (6.1.49) with $\rho_R \geq r_d$. The flat surface matrix is a block Toeplitz matrix. The weak remainder matrix elements consist of the differences of the Green's function $G(R) - G_{FS}(\rho_R)$ connecting the two points whose horizontal distance is greater than r_d and corresponds to the last terms on the right hand sides of (6.1.48) and (6.1.49). The weak and the flat surface matrix elements are nonzero only for those points whose horizontal interaction distances are greater than r_d. Next, the matrix equation is rearranged to take an iterative form. The calculation procedure is, for the first-order and higher order solutions

$$(\overline{\overline{Z}}^{(s)} + \overline{\overline{Z}}^{(FS)})\, \overline{x}^{(1)} = \overline{b} \tag{6.1.52}$$

$$(\overline{\overline{Z}}^{(s)} + \overline{\overline{Z}}^{(FS)})\, \overline{x}^{(n+1)} = \overline{b}^{(n+1)} \tag{6.1.53}$$

$$\overline{b}^{(n+1)} = \overline{b} - \overline{\overline{Z}}^{(w)} \overline{x}^{(n)} \tag{6.1.54}$$

The flat surface impedance matrix $\overline{\overline{Z}}^{(FS)}$ must be on the left-hand side of (6.1.53) for the SMFSIA to work for 2-D surfaces. For each order of solution $\overline{x}^{(n)}$, the matrix equations (6.1.52) or (6.1.53) is solved by the conjugate gradient method. Note that the product of $\overline{\overline{Z}}^{(FS)}$ with $\overline{x}$ can be computed by a 2-D fast Fourier transform (FFT) algorithm which makes conjugate gradient iteration more efficient. The iteration of $\overline{\overline{Z}}^{(w)} \overline{X}$ through (6.1.53) and (6.1.54) is carried out until the error norm

$$\sqrt{\frac{\|\overline{\overline{Z}}\, \overline{X}^{(n)} - \overline{b}\|}{\|\overline{b}\|}} \tag{6.1.55}$$

falls below a threshold. In this section, an error norm of 1% is used for all numerical simulations.

(B) SMFSIA/CAG

We can further improve the SMFSIA by using the flat surface as a canonical grid (CAG). This method is called SMFSIA/CAG. For the weak remainder matrix elements, Green's function is approximately equal to the Green's function of the horizontal distance between the two points. Green's function can be expanded in a Taylor's series about the horizontal distance between the two points.

$$G(R) - G_{FS}(\rho_R) = \frac{(ikR - 1)\exp(ikR)}{4\pi R^3} - \frac{(ik\rho_R - 1)\exp(ik\rho_R)}{4\pi \rho_R^3}$$

$$= \sum_{m=1}^{M} a_m(\rho_R) \left(\frac{z_d^2}{\rho_R^2} \right)^m \tag{6.1.56}$$

where $z_d = f(x, y) - f(x', y')$. The larger the r_d the less number of terms we need in (6.1.56). In this section, we keep up to the sixth term in the Taylor series. In the following, the first 3 coefficients are listed for reference.

$$a_1(\rho_R) = -k^2\frac{\exp(ik\rho_R)}{4\pi\rho_R} - 3ik\frac{\exp(ik\rho_R)}{4\pi\rho_R^2} + 3\frac{\exp(ik\rho_R)}{4\pi\rho_R^3} \tag{6.1.57}$$

$$a_2(\rho_R) = -ik^3\frac{\exp(ik\rho_R)}{32\pi} + 6k^2\frac{\exp(ik\rho_R)}{32\pi\rho_R} + 15ik\frac{\exp(ik\rho_R)}{32\pi\rho_R^2} - 15\frac{\exp(ik\rho_R)}{32\pi\rho_R^3} \tag{6.1.58}$$

$$a_3(\rho_R) = k_4\rho_R\frac{\exp(ik\rho_R)}{192\pi} + 10ik^3\frac{\exp(ik\rho_R)}{192\pi} - 42k^2\frac{\exp(ik\rho_R)}{192\pi\rho_R} - 96ik\frac{\exp(ik\rho_R)}{196\pi\rho_R^2} + 96\frac{\exp(ik\rho_R)}{196\pi\rho_R^3} \tag{6.1.59}$$

The important property of above coefficients is that they are translationally invariant. In terms of the matrix equation, the iterative procedure is then

$$(\overline{\overline{Z}}^{(s)} + \overline{\overline{Z}}^{(FS)})\,\overline{x}^{(n+1)} = \overline{b}^{(n+1)} \tag{6.1.60}$$

$$\overline{b}^{(n+1)} = \overline{b} - \sum_{m=1}^{6} \overline{\overline{Z}}_m^{(w)}\,\overline{x}^{(n)} \tag{6.1.61}$$

where $\overline{\overline{Z}}_m^{(w)}$ is the expanded form of the weak matrix. The updated right-hand side is calculated by the FFT.

Like the SMFSIA, SMFSIA/CAG has an adjustable parameter r_d (the neighborhood distance). Furthermore, in SMFSIA/CAG there is a second adjustable parameter, which is the number of Taylor series coefficients of (6.1.56). These two adjustable parameters of r_d and the number of Taylor series terms are interdependent. They are chosen to optimize the CPU.

The numerical simulation results are presented in terms of the normalized bistatic scattering coefficient as normalized by the incident power

$$\gamma_{\alpha h}(\theta_s, \phi_s) = \frac{|\mathcal{E}_\alpha^s|^2}{2\eta P_h^{inc}} \tag{6.1.62}$$

where $\alpha = h$ (horizontal polarization) and $\alpha = v$ (vertical polarization). The incident power is

$$P_h^{inc} = \frac{2\pi^2}{\eta}\int_{k_\rho<k} dk_x dk_y |E_{TE}(k_x, k_y)|^2 \frac{k_z}{k} \tag{6.1.63}$$

and the co-polarized and cross-polarized components of $\mathcal{E}_\alpha^s$ are respectively

$$\mathcal{E}_h^s = \frac{\eta i k}{4\pi}\int dx'dy' \exp(-ik\beta')\{F_x(x', y')\sin\phi_s - F_y(x', y')\cos\phi_s\} \tag{6.1.64}$$

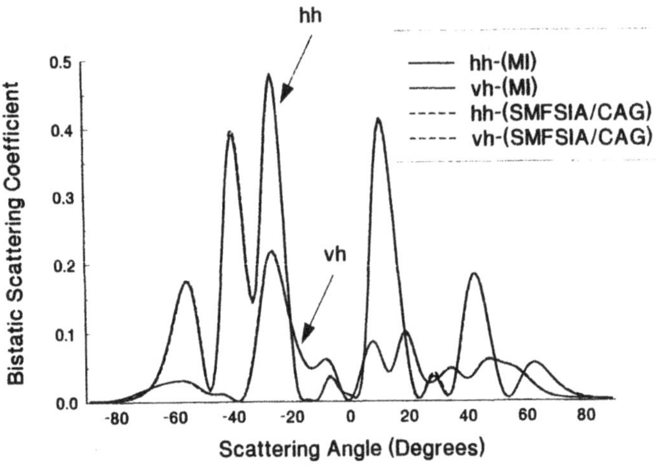

Figure 6.1.4 Comparison between the SMFSIA/CAG and the full matrix inversion (MI) result.

and

$$\mathcal{E}_v^s = \frac{\eta ik}{4\pi} \int dx'dy' \exp(-ik\beta') \Big\{ F_x(x',y') \left[\frac{\partial f(x',y')}{\partial x'} \sin\theta_s - \cos\theta_s \cos\phi_s \right]$$

$$+ F_y(x',y') \left[\frac{\partial f(x',y')}{\partial y'} sin\theta_s - \cos\theta_s \sin\phi_s \right] \Big\} \tag{6.1.65}$$

where $\beta' = x' \sin\theta_s \cos\phi_s + y' \sin\theta_s \sin\phi_s + f(x',y')\cos\theta_s$. The normalization is such that energy conservation requires the integration of γ_{hh} and γ_{vh} over solid angles to give unity.

1.2.4 Numerical Simulation Results

We first describe the various numerical accuracy tests of the methods. In the MoM implementation, we use the pulse basis function and point matching technique.

To confirm the accuracy of the SMFSIA/CAG, comparison is made in Fig. 6.1.4 between the normalized bistatic scattering coefficient obtained by the exact matrix inversion (MI) with the solution obtained by the SMF-SIA/CAG. The results are for a single surface realization. Surface lengths in the x and y directions are $L_x = L_y = 8.0\lambda$. The rms height is $h = 0.5\lambda$, with correlation lengths of $l_x = l_y = 1.0\lambda$. The surface is sampled at 16 points per λ^2 to give 2048 surface unknowns. The neighborhood distance r_d is 3.5λ. It can be observed from Fig. 6.1.4 that both co-polarized and cross-polarized components of MI and SMFSIA/CAG curves completely overlap each other.

Method	CPU
SMFSIA	42 hours
SMFSIA/CAG	8 hours

Table 6.1.3 CPU comparison (based on a SUN SPARC10).

Figure	$r_d(\lambda)$	g	$L_x = L_y$ (λ)	pts/λ^2	N	realization	$h(\lambda)$	$l(\lambda)$
6.1.4	3.5	$L_x/3$	8.0	16	2048	1	0.5	1.0
6.1.5a	1.0	$L_x/3$	16.0	64	32768	1	0.2	0.6
6.1.5b	1.0	$L_x/3$	16.0	64	32768	280	0.2	0.6
6.1.6	2.6	$L_x/2$	32.0	16	32768	272–1000	0.5	1.0
6.1.7	2.6	$L_x/2$	16.0	16	8192	650	0.5	1.0
6.1.8	2.6	$L_x/2$	16.0	64	32768	769	0.5	1.0
6.1.9	2.6	∞	16.0	64	32768	495	0.5	1.0
6.1.10	2.6	$L_x/2$	32.0	16	32768	600	1.0	2.0

Table 6.1.4 List of parameters for the simulations.

The SMFSIA/CAG represents an improvement over the SMFSIA in terms of the matrix solving time. In Table 6.1.3, the matrix solving time is given for the case of 32768 surface unknowns for a single surface realization. The run times are for a SUN SPARC10 workstation. Surface lengths are $L_x = L_y = 16\lambda$ with an area of $L^2 = 256\lambda^2$. This represents 64 sample points per λ^2. The rms height is $h = 0.2\lambda$, with correlation lengths of $l_x = l_y = 0.6\lambda$. For these surface parameters, r_d is 1.0λ. From Table 6.1.3, it is observed that SMFSIA/CAG is 5 times faster than the SMFSIA. The results of Figs. 6.1.5a show that both methods give the same bistatic scattering coefficient. It should be noted that when compared to the SMFSIA, the efficiency of the SMFSIA/CAG will be improved even more for those cases involving larger surface areas than the $256\lambda^2$ given in this example. This is because for given rough surface statistics, as the surface area increases CPU for the weak matrix multiplication step becomes more dominant. SMFSIA/CAG speeds up the computation of the product of the weak matrix and the column vector. In Fig. 6.1.5b, a Monte Carlo simulation result of Fig. 6.1.5a is presented for an ensemble average of 280 realizations.

We will show convergence of bistatic scattering with respect to the number of surface realizations, surface size, surface sampling rate and tapering parameter. In Table 6.1.4, all the numerical parameters used for the results illustrated in various figures are tabulated.

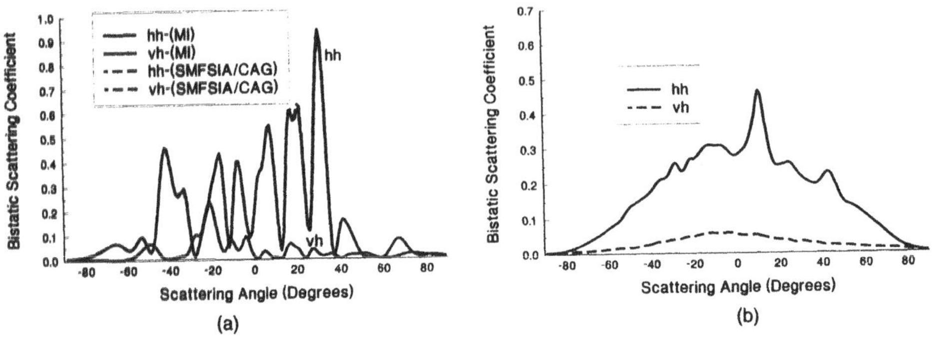

Figure 6.1.5 **(a)** Comparison between the original SMFSIA and SMFSIA/CAG for 1 realization. Rough surface parameters $L_x = L_y = 16\lambda$, $h = 0.2\lambda$, $l_x = l_y = 0.6\lambda$ and incident angle $\theta_i = 10°$ and $\phi_i = 0°$. **(b)** A Monte Carlo simulation result with SMFSIA/CAG and 280 realizations.

The convergence of the solution of the matrix equation is defined with respect to the error norm of the matrix equation (6.1.55). The error norm criteria of $< 1\%$ is satisfied for each surface realization.

In Fig. 6.1.6, the averaged normalized bistatic scattering coefficients are plotted for various numbers of realizations (272, 432, and 1000). Surface lengths in the x and y directions are $L_x = L_y = 32\lambda\,(L_x \times L_y = 1024\lambda^2)$ with a surface rms height $h_x = h_y = 0.5\lambda$ and correlation lengths of $l_x = l_y = 1.0\lambda$. The surface is sampled at 16 points per λ^2 to give 32768 surface unknowns. The neighborhood distance r_d is 2.6λ. The incident angle is $\theta_i = 10°$ and $\phi_i = 0°$. The tapering parameter g is equal to $L_x/2$. The computations were carried out using the IBM SP/2 at the Maui High Performance Computing Center. On the average, for the case of 32768 surface unknowns, it takes approximately 2.4 CPU minutes per realization. Figure 6.1.6 shows that after 272 realizations, for scattering angles away from the backscattering angle of $-10°$, the result shows convergence. Note that the backscattering enhancement peak converges last. It takes many more realizations, usually between 400 to 1000 realizations as shown in Fig. 6.1.6, for the enhancement peak to be well defined.

Figure 6.1.7 shows a comparison between a smaller surface area of $256\lambda^2$ (650 realizations) and $1024\lambda^2$ case (1000 realizations). The result shows good agreement including the backscattering direction. A convergence of scattering coefficient with respect to the surface area is a necessary condition in simulations. Other important convergence tests are with respect to the number of sample points representing the rough surface (Fig. 6.1.8) and with respect to the tapering parameter g (Fig. 6.1.9). In Fig. 6.1.8, the cases of $256\lambda^2$ area

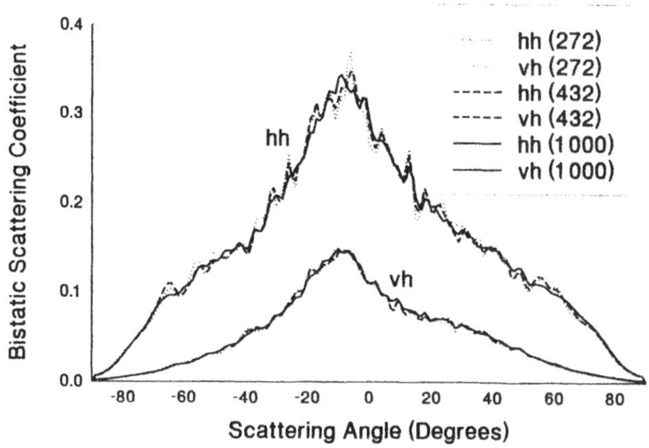

Figure 6.1.6 Backscattering enhancement of cases with $h = 0.5\lambda$ and $l_x = l_y = 1.0\lambda$. Surface parameters are $L_x = L_y = 32\lambda$ ($L_x \times L_y = 1024\lambda^2$) and the surface is sampled at 16 points/λ^2 ($r_d = 2.6\lambda$, $\theta_i = 10°$, $\phi_i = 0°$ and $g = L_x/2$). Convergence with respect to the number of realizations is shown.

sampled at 64 points per λ^2 and $256\lambda^2$ area sampled at 16 points per λ^2 are given. Good agreement between the two cases establishes the sampling rate needed in Monte Carlo simulations with the moderate incident angle and moderate rms slope of this example. In Fig. 6.1.9, a tapered plane wave with a tapering parameter $g = L_x/2$ and a plane wave ($g = \infty$) incident on a rough surface with area of $256\lambda^2$ are compared. A plane wave incident case has diffraction contributions from the boundary edges of the rough surface. However, the result agrees with a tapered plane wave case in Fig. 6.1.9. This illustrates that for these combinations of rms height and correlation length, the bistatic scattering is large for the scattering angles of interest and the edge diffraction levels are negligible. Furthermore, it shows that the spot size of the incident tapered wave is large enough for the two cases to agree.

For $h = 0.5\lambda$ and $l = 1.0\lambda$ cases of Table 6.1.4, we compare in Table 6.1.5 the average error defined by the equation

$$\text{average error}(\%) = \frac{1}{N_{\theta_s}} \sum_{\theta_s} \left| \frac{\gamma(\theta_s) - \gamma(\theta_s)_{1000}}{\gamma(\theta_s)} \right| \times 100 \tag{6.1.66}$$

for Figs. 6.1.7, 6.1.8, and 6.1.9. In the above equation, N_r is the number of realizations, $\gamma(\theta_s)_{1000}$ is the bistatic scattering coefficient of the 1000 realization case of Fig. 6.1.6, $\gamma(\theta_s)$ is the bistatic scattering coefficient of Figs. 6.1.7, 6.1.8 or 6.1.9, and N_θ is the number of scattering angles.

In Fig. 6.1.10, the normalized bistatic scattering coefficient in the plane of incidence for the 1-D random rough surface, 2-D surface with scalar inci-

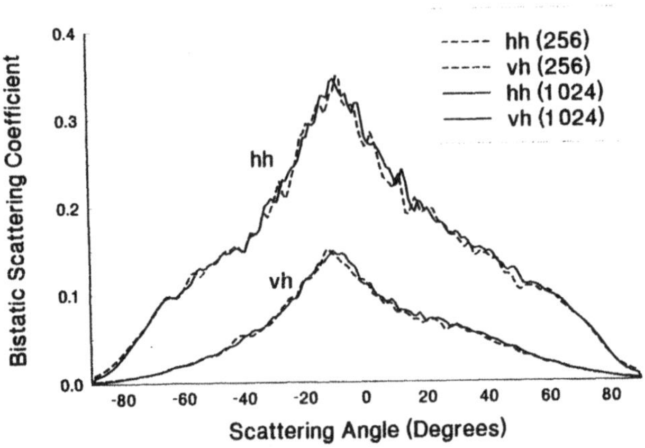

Figure 6.1.7 Convergence with respect to area ($L_{1x} = L_{1y} = 16\lambda$, $L_{1x} \times L_{1y} = 256\lambda^2$ and $L_{2x} = L_{2y} = 32\lambda$, $L_{2x} \times L_{2y} = 1024\lambda^2$). Other parameters are those of Fig. 6.1.6. The surfaces are sampled at 16 points/λ^2.

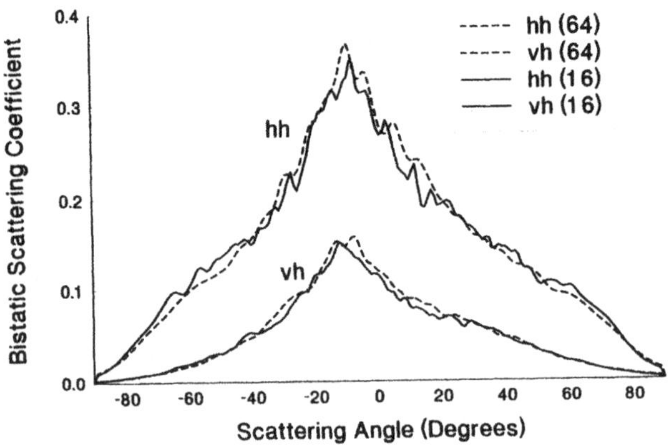

Figure 6.1.8 Convergence with respect to the surface sampling rate of 64 points/λ^2 ($L_{1x} = L_{1y} = 16\lambda$, $L_{1x} \times L_{1y} = 256\lambda^2$) and 16 points/$\lambda^2$ ($L_{2x} = L_{2y} = 16\lambda$, $L_{2x} \times L_{2y} = 256\lambda^2$). Other parameters are those of Fig. 6.1.6.

Figure	number of realizations	% error co-pol.	% error cross-pol.
6.1.7	650	5.8	6.2
6.1.8	769	8.8	7.0
6.1.9	495	9.3	14.2

Table 6.1.5 Average error respect to 1000 realization case of Fig. 6.1.6.

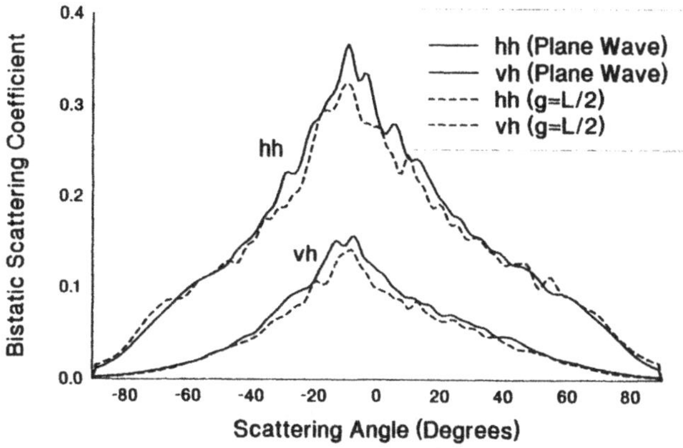

Figure 6.1.9 Convergence with respect to the tapering parameter g for 64 points per square wavelength sampling. For $g = L_x/2$, the $L_x = L_y = 16\lambda$ result of Fig. 6.1.8 is used. The $g = \infty$ case is for 495 realizations with $L_x = L_y = 16\lambda$. Other parameters are those of Fig. 6.1.6.

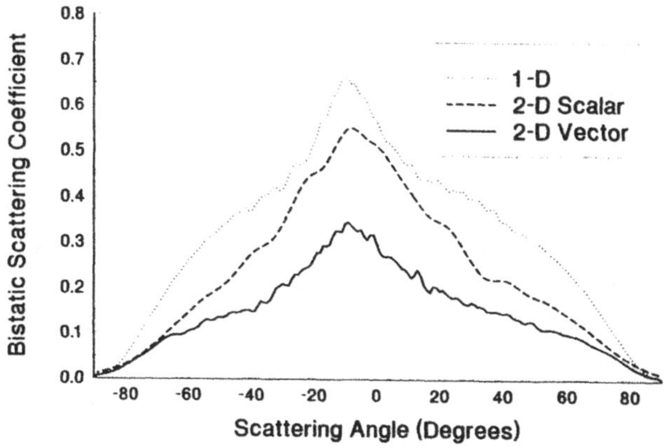

Figure 6.1.10 Backscattering enhancement of three cases: 1-D, 2-D scalar wave incidence, and 2-D electromagnetic wave incidence. Comparison of normalized bistatic scattering coefficients for rms height of 0.5λ and correlation length of 1.0λ with incident angle $\theta_i = 10°$ and $\phi_i = 0°$.

dent wave, and the 2-D surface with electromagnetic incident wave are compared. The incidence angle is $\theta_i = 10°$ and $\phi_i = 0°$. The tapering parameter $g = L/3$ for the 2-D surface with scalar wave incidence, $g = L/2$ for the 2-D surface with a electromagnetic wave incidence, and $L/4$ for the 1-D surface. In the 1-D surface simulations, the surface length is set at 40λ for an ensemble average of 4000 realizations. For the scalar wave scattering from a 2-D

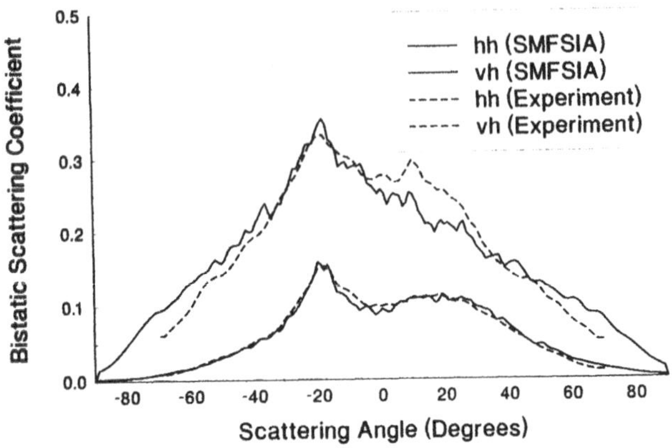

Figure 6.1.11 Monte Carlo simulation comparison of SMFSIA and experimental data of backscattering enhancement. For SMFSIA an area of 1024 square wavelength with tapering $g = L_x/2$. The surface has an rms height of 1λ and correlation length of 2λ.

surface with area of $81\lambda^2$, an ensemble average of 674 realizations is taken. For the electromagnetic wave scattering case, the co-polarized component of Fig. 6.1.6 is given as a comparison. Three curves follow a similar trend for all scattering angles. Backscattering enhancement at $-10°$ is observed for all three cases. For the 2-D surface cases, the scattering coefficient is lower at all angles than the 1-D case since the scattering can occur for directions off the plane of incidence. The co-polarized scattering coefficient of the electromagnetic case is less than that of the scalar wave due to a significant amount of energy converted into the cross-polarization component.

We next compare with a controlled millimeter wave laboratory experiment [Kuga et. al. 1993; Kuga and Phu, 1996]. The rough surface is a Gaussian random rough surface with a Gaussian correlation function. It has a rms height of 1λ and correlation of 2λ. The incident angle is $20°$ from the normal. In Fig. 6.1.11, the Monte Carlo simulation results of both experimental and numerical results are given. The numerical simulation is carried out with tapering $g = L_x/2$, $L_x = L_y = 32\lambda\,(L_x \times L_y = 1024\lambda^2)$, and 16 points per λ^2 sampling (32768 surface unknowns). In the experiment, the calibration is done by using a electric conductor so that the scattered power can be compared with total power. Thus, the absolute order of the bistatic scattering coefficient as defined by (6.1.62) is measured in the experiment. Therefore, with this set-up, it is possible to compare the absolute order of bistatic scattering coefficient between numerical simulation and experiment. Both Monte Carlo simulation and experiment show backscattering enhancement for both co-polarized and cross-polarized components. The co-polarized scattering is in good agreement. The cross-polarized scattering is in excellent agreement. The comparison is excellent in view of the fact that the absolute values are compared and there are no adjustable parameters. Backscattering enhancement is a result of the contributions of the cyclical scattering processes (see Chapter 8 of Volume III) which begin at the second order. For co-polarization, the second-order terms can be obscured by the presence of first-order scattering. However, for cross-polarization, the first-order scattering is zero. Thus, the second-order backscattering enhancement is more clearly exhibited as seen in the simulations and the experimental data.

2 Integral Equations for Dielectric Surfaces

2.1 Electromagnetic Fields with Electric and Magnetic Sources

In formulating integral equations for electromagnetic wave scattering, a convenient method is through the use of equivalent electric and magnetic currents [Harrington, 1961; Wang, 1991].

For the Maxwell equations with electric sources of current density $\overline{J}$, volume charge density ρ_v, and surface charge density ρ_s

$$\nabla \times \overline{E} = i\omega\mu\overline{H} \tag{6.2.1}$$

$$\nabla \times \overline{H} = -i\omega\epsilon\overline{E} + \overline{J} \tag{6.2.2}$$

$$\nabla \cdot \epsilon\overline{E} = \rho_v \tag{6.2.3}$$

$$\nabla \cdot \mu\overline{H} = 0 \tag{6.2.4}$$

$$\nabla \cdot \overline{J} - i\omega\rho_v = 0 \tag{6.2.5}$$

$$\nabla_s \cdot \overline{J}_s - i\omega\rho_s = 0 \tag{6.2.6}$$

where $\nabla_s \cdot \overline{J}_s$ is the surface divergence of $\overline{J}_s$. Vector potential $\overline{A}$ and scalar potential Φ can be employed.

$$\overline{H} = \frac{1}{\mu}\nabla \times \overline{A} \tag{6.2.7}$$

$$\overline{E} = i\omega\overline{A} - \nabla\Phi \tag{6.2.8}$$

Using the gauge condition

$$\frac{1}{\mu}\nabla \cdot \overline{A} = i\omega\epsilon\Phi \tag{6.2.9}$$

Then

$$\left(\nabla^2 + k^2\right)\overline{A} = -\mu\overline{J} \tag{6.2.10}$$

$$\left(\nabla^2 + k^2\right)\Phi = -\frac{\rho_v}{\epsilon} \tag{6.2.11}$$

Using the Green's function

$$g(\overline{r},\overline{r}') = \frac{e^{ik|\overline{r}-\overline{r}'|}}{4\pi|\overline{r}-\overline{r}'|} \tag{6.2.12}$$

we have

$$\overline{A}(\overline{r}) = \mu \int d\overline{r}' g(\overline{r},\overline{r}')\, \overline{J}(\overline{r}') \tag{6.2.13}$$

$$\Phi(\overline{r}) = \frac{1}{\epsilon} \int d\overline{r}' g(\overline{r},\overline{r}')\, \rho_v(\overline{r}') \tag{6.2.14}$$

The magnetic field is

$$\overline{H} = \nabla \times \int d\overline{r}' g(\overline{r},\overline{r}')\overline{J}(\overline{r}') = -\int d\overline{r}' \nabla g(\overline{r},\overline{r}') \times \overline{J}(\overline{r}') \tag{6.2.15}$$

The electric field is

$$\begin{aligned}\overline{E}(\overline{r}) &= i\omega\mu \int d\overline{r}' g(\overline{r},\overline{r}')\, \overline{J}(\overline{r}') - \nabla \frac{1}{\epsilon} \int d\overline{r}' g(\overline{r},\overline{r}')\, \rho_v(\overline{r}') \\ &= i\omega\mu \left[\int d\overline{r}' g(\overline{r},\overline{r}')\, \overline{J}(\overline{r}') + \frac{1}{k^2} \nabla \int d\overline{r}' g(\overline{r},\overline{r}')\, \nabla' \cdot \overline{J}(\overline{r}') \right]\end{aligned} \tag{6.2.16}$$

The fields are expressed in terms of electric current or in terms of electric current and electric charge.

For the case of equivalent magnetic sources of magnetic current density $\overline{M}$, volume charge density m, and surface charge density m_s, the Maxwell equations are

$$\nabla \times \overline{E} = i\omega\mu\overline{H} - \overline{M} \tag{6.2.17}$$

$$\nabla \times \overline{H} = -i\omega\epsilon\overline{E} \tag{6.2.18}$$

$$\nabla \cdot \epsilon\overline{E} = 0 \tag{6.2.19}$$

$$\nabla \cdot \mu\overline{H} = m \tag{6.2.20}$$

$$\nabla \cdot \overline{M} - i\omega m = 0 \tag{6.2.21}$$

$$\nabla_s \cdot \overline{M}_s - i\omega m_s = 0 \tag{6.2.22}$$

Vector potential $\overline{F}$ and scalar potential ψ can be employed. Derivation is analogous to that of the electric source case

$$\overline{E} = -\frac{1}{\epsilon} \nabla \times \overline{F} \tag{6.2.23}$$

$$\overline{H} = -i\omega\overline{F} - \nabla\psi \tag{6.2.24}$$

The gauge condition is

$$\frac{1}{\epsilon} \nabla \cdot \overline{F} = i\omega\mu\psi \tag{6.2.25}$$

Then

$$\left(\nabla^2 + k^2\right)\overline{F} = -\epsilon\overline{M} \tag{6.2.26}$$

$$\left(\nabla^2 + k^2\right)\psi = -\frac{m}{\mu} \tag{6.2.27}$$

$$\overline{F}(\overline{r}) = \epsilon \int d\overline{r}' g\left(\overline{r},\overline{r}'\right)\overline{M}\left(\overline{r}'\right) \tag{6.2.28}$$

$$\psi(\overline{r}) = \frac{1}{\mu}\int d\overline{r}' g\left(\overline{r},\overline{r}'\right) m\left(\overline{r}'\right) \tag{6.2.29}$$

The electric field is

$$\overline{E}(\overline{r}) = -\nabla \times \int d\overline{r}' g\left(\overline{r},\overline{r}'\right)\overline{M}\left(\overline{r}'\right) = -\int d\overline{r}' \nabla g\left(\overline{r},\overline{r}'\right) \times \overline{M}\left(\overline{r}'\right) \tag{6.2.30}$$

and the magnetic field is

$$\begin{aligned}\overline{H}(\overline{r}) &= i\omega\epsilon \int d\overline{r}' g\left(\overline{r},\overline{r}'\right)\overline{M}\left(\overline{r}'\right) - \nabla \int d\overline{r}' g\left(\overline{r},\overline{r}'\right)\frac{m\left(\overline{r}'\right)}{\mu} \\ &= i\omega\epsilon\left[\int d\overline{r}' g\left(\overline{r},\overline{r}'\right)\overline{M}\left(\overline{r}'\right) + \frac{1}{k^2}\nabla \int d\overline{r}' g\left(\overline{r},\overline{r}'\right)\nabla' \cdot \overline{M}\left(\overline{r}'\right)\right]\end{aligned} \tag{6.2.31}$$

The fields are expressed entirely in terms of magnetic current or in terms of magnetic current and magnetic charge.

When both electric and magnetic sources are present, Eqs. (6.2.16)–(6.2.15) and (6.2.30)–(6.2.31) are combined to give

$$\begin{aligned}\overline{E}(\overline{r}) = &-\int d\overline{r}' \nabla g\left(\overline{r},\overline{r}'\right) \times \overline{M}\left(\overline{r}'\right) \\ &+ i\omega\mu\left[\int d\overline{r}' g\left(\overline{r},\overline{r}'\right)\overline{J}\left(\overline{r}'\right) + \frac{1}{k^2}\nabla \int d\overline{r}' g\left(\overline{r},\overline{r}'\right)\nabla' \cdot \overline{J}\left(\overline{r}'\right)\right]\end{aligned} \tag{6.2.32}$$

$$\begin{aligned}\overline{H}(\overline{r}) = &\, i\omega\epsilon\left[\int d\overline{r}' g\left(\overline{r},\overline{r}'\right)\overline{M}\left(\overline{r}'\right) + \frac{1}{k^2}\nabla \int d\overline{r}' g\left(\overline{r},\overline{r}'\right)\nabla' \cdot \overline{M}\left(\overline{r}'\right)\right] \\ &+ \int d\overline{r}' \nabla g\left(\overline{r},\overline{r}'\right) \times \overline{J}\left(\overline{r}'\right)\end{aligned} \tag{6.2.33}$$

The boundary conditions separating two media of μ_1, ϵ_1 and μ_2, ϵ_2 with surface electric and magnetic sources are

$$\hat{n} \times \overline{E}_1 - \hat{n} \times \overline{E}_2 = -\overline{M}_s \tag{6.2.34}$$

$$\hat{n} \times \overline{H}_1 - \hat{n} \times \overline{H}_2 = \overline{J}_s \tag{6.2.35}$$

$$\hat{n} \cdot \epsilon_1\overline{E}_1 - \hat{n} \cdot \epsilon_2\overline{E}_2 = \rho_s \tag{6.2.36}$$

$$\hat{n} \cdot \mu_1\overline{H}_1 - \hat{n} \cdot \mu_2\overline{H}_2 = m_s \tag{6.2.37}$$

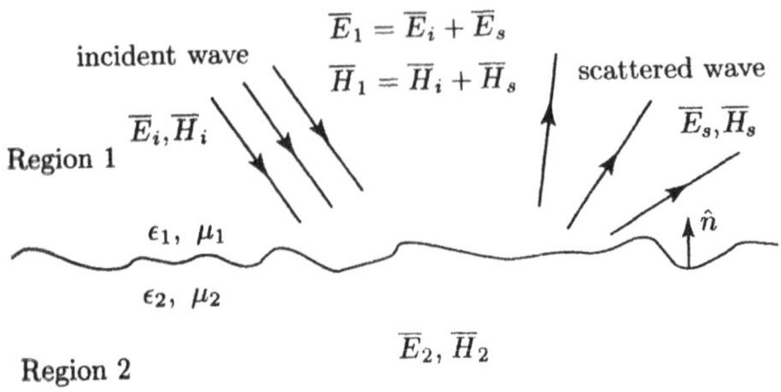

Figure 6.2.1 Scattering of waves by dielectric surface: Physical problem, no surface sources.

2.2 Physical Problem and Equivalent Exterior and Interior Problems

Consider an incident wave with electric field $\overline{E}_i$ and magnetic field $\overline{H}_i$ (Fig. 6.2.1) on an dielectric surface. Let ϵ_1, μ_1 be the permittivity and permeability respectively above the surface and ϵ_2, μ_2 be the permittivity and permeability respectively below the surface. This is the physical problem.

In region 1, the electromagnetic fields are

$$\overline{E}_1 = \overline{E}_i + \overline{E}_s \tag{6.2.38}$$

$$\overline{H}_1 = \overline{H}_i + \overline{H}_s \tag{6.2.39}$$

In region 2, the electromagnetic fields are $\overline{E}_2$ and $\overline{H}_2$. The boundary conditions are

$$\hat{n} \times \overline{E}_1 - \hat{n} \times \overline{E}_2 = 0 \tag{6.2.40}$$

$$\hat{n} \times \overline{H}_1 - \hat{n} \times \overline{H}_2 = 0 \tag{6.2.41}$$

$$\hat{n} \cdot \epsilon_1 \overline{E}_1 - \hat{n} \cdot \epsilon_2 \overline{E}_2 = 0 \tag{6.2.42}$$

$$\hat{n} \cdot \mu_1 \overline{H}_1 - \hat{n} \cdot \mu_2 \overline{H}_2 = 0 \tag{6.2.43}$$

Because this is a dielectric surface, there are neither surface currents nor surface charges at the boundary separating regions 1 and 2. Hence we have the boundary condition of (6.2.40)–(6.2.43). The physical problem can be represented by the exterior problem and the interior problem using equivalent surface currents and charges.

2.2.1 Equivalent Exterior Problem, Equivalent Currents and Integral Equations

Consider the equivalent exterior problem A with the same incident electric field $\overline{E}_i$ and magnetic field $\overline{H}_i$ and same electric field $\overline{E}_1$ and magnetic field $\overline{H}_1$ in region 1 as in the physical problem. In the equivalent problem A, let

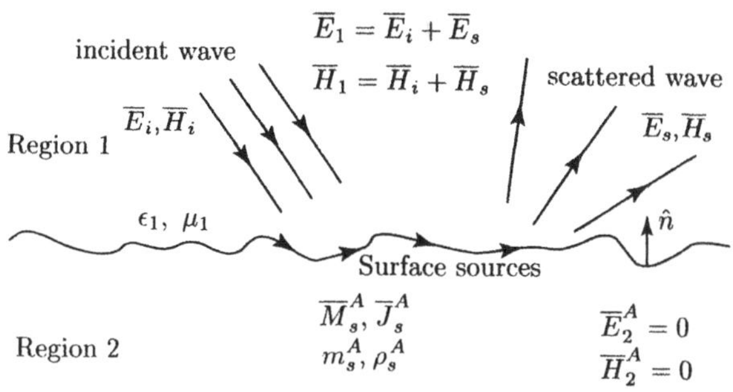

Figure 6.2.2 Equivalent problem A, exterior problem with $\overline{M}_s^A$, $\overline{J}_s^A$, ρ_s^A, m_s^A.

there be equivalent sources on the boundary (Fig. 6.2.2)

$$-\overline{M}_s^A = \hat{n} \times \overline{E}_1 \tag{6.2.44}$$

$$\overline{J}_s^A = \hat{n} \times \overline{H}_1 \tag{6.2.45}$$

$$\rho_s^A = \hat{n} \cdot \epsilon_1 \overline{E}_1 \tag{6.2.46}$$

$$m_s^A = \hat{n} \cdot \mu_1 \overline{H}_1 \tag{6.2.47}$$

Because of these equivalent sources, we have

$$\hat{n} \times \overline{E}_1 - \hat{n} \times \overline{E}_2^A = -\overline{M}_s^A \tag{6.2.48}$$

$$\hat{n} \times \overline{H}_1 - \hat{n} \times \overline{H}_2^A = \overline{J}_s^A \tag{6.2.49}$$

$$\hat{n} \cdot \epsilon_1 \overline{E}_1 - \hat{n} \cdot \epsilon_2 \overline{E}_2^A = \rho_s^A \tag{6.2.50}$$

$$\hat{n} \cdot \mu_1 \overline{H}_1 - \hat{n} \cdot \mu_2 \overline{H}_2^A = m_s^A \tag{6.2.51}$$

Using (6.2.44)–(6.2.47) in (6.2.48)–(6.2.51), we have $\overline{E}_2^A = 0$, and $\overline{H}_2^A = 0$ at the boundary. By Huygen's principle

$$\overline{E}_2^A = 0 \tag{6.2.52}$$

$$\overline{H}_2^A = 0 \tag{6.2.53}$$

everywhere in region 2.

The scattered field generated by these equivalent sources are

$$\overline{E}_s(\overline{r}) = -\int dS' \nabla g_1\left(\overline{r}, \overline{r}'\right) \times \overline{M}_s^A\left(\overline{r}'\right) + i\omega\mu_1\left[\int dS' g_1\left(\overline{r}, \overline{r}'\right) \overline{J}_s^A\left(\overline{r}'\right) + \frac{1}{k_1^2} \nabla \int dS' g_1\left(\overline{r}, \overline{r}'\right) \nabla'_s \cdot \overline{J}_s^A\left(\overline{r}'\right)\right] \tag{6.2.54}$$

$$\overline{H}_s(\overline{r}) = i\omega\epsilon_1 \left[\int dS' g_1(\overline{r},\overline{r}')\,\overline{M}_s^A(\overline{r}') + \frac{1}{k_1^2}\nabla \int dS' g_1(\overline{r},\overline{r}')\,\nabla'_s\cdot\overline{M}_s^A(\overline{r}')\right] + \int dS' \nabla g_1(\overline{r},\overline{r}') \times \overline{J}_s^A(\overline{r}') \tag{6.2.55}$$

where

$$g_1(\overline{r},\overline{r}') = \frac{e^{ik_1|\overline{r}-\overline{r}'|}}{4\pi|\overline{r}-\overline{r}'|} \tag{6.2.56}$$

The equivalent sources radiative into all of space. As these equivalent sources radiate into region 2, these fields will cancel the incident fields to give $\overline{E}_2^A = 0$ and $\overline{H}_2^A = 0$. Note that region 2 also has ϵ_1 and μ_1 as region 1.

For $\overline{r}$ in region 2, we thus have the E equation from the extinction theorem, $\overline{E}_i + \overline{E}_s = 0$:

$$-\int dS' \nabla g(\overline{r},\overline{r}') \times \overline{M}_s^A(\overline{r}') + i\omega\mu_1 \left[\int dS' g_1(\overline{r},\overline{r}')\,\overline{J}_s^A(\overline{r}') + \frac{1}{k_1^2}\nabla \int dS' g_1(\overline{r},\overline{r}')\,\nabla'_s\cdot\overline{J}_s^A(\overline{r}')\right] = -\overline{E}_i(\overline{r}) \tag{6.2.57}$$

and for $\overline{r}$ in region 2 the H equation from the extinction theorem, $\overline{H}_i + \overline{H}_s = 0$:

$$i\omega\epsilon_1 \left[\int dS' g_1(\overline{r},\overline{r}')\,\overline{M}_s^A(\overline{r}') + \frac{1}{k_1^2}\nabla \int dS' g_1(\overline{r},\overline{r}')\,\nabla'_s\cdot\overline{M}_s^A(\overline{r}')\right] + \int dS' \nabla g_1(\overline{r},\overline{r}') \times \overline{J}_s^A(\overline{r}') = -\overline{H}_i(\overline{r}) \tag{6.2.58}$$

We can verify that these equations are identical to those based on vector Green's theorem and dyadic Green's functions in previous chapters. Note that the Green's function g_1 is used in Eqs. (6.2.57) and (6.2.58) for $\overline{r}$ in region 2. There are 6 scalar integral equations above. However, they are not independent and just by themselves cannot be used to calculate the unknown surfaces fields. This is obvious because μ_2, ϵ_2, and g_2 are not involved in the equations. Thus we also need the interior equivalent problem.

2.2.2 Equivalent Interior Problem, Equivalent Currents and Integral Equations

Consider the equivalent problem B with the same electric field $\overline{E}_2$ and magnetic field $\overline{H}_2$ in region 2 as in the physical problem. At the boundary we introduce equivalent sources (Fig. 6.2.3). With $\hat{n}_i = -\hat{n}$

$$-\overline{M}_s^B = \hat{n}_i \times \overline{E}_2 \tag{6.2.59}$$

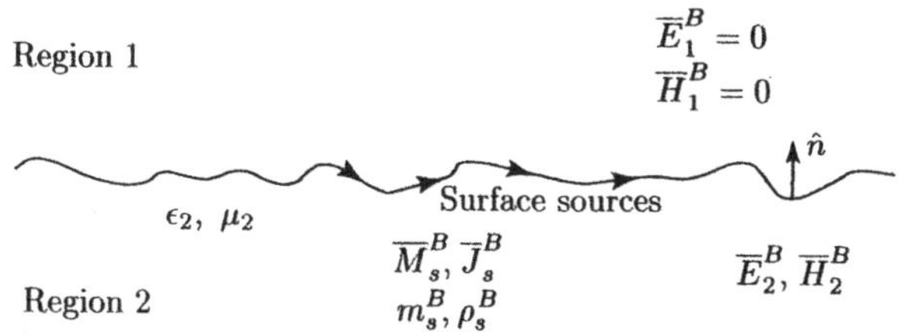

Figure 6.2.3 Equivalent problem B, interior problem with $\overline{M}_s^B$, $\overline{J}_s^B$, ρ_s^B, m_s^B.

$$\overline{J}_s^B = \hat{n}_i \times \overline{H}_2 \tag{6.2.60}$$

$$\rho_s^B = \hat{n}_i \cdot \epsilon_2 \overline{E}_2 \tag{6.2.61}$$

$$m_s^B = \hat{n}_i \cdot \mu_2 \overline{H}_2 \tag{6.2.62}$$

Because of these equivalent sources, we have

$$\overline{E}_1^B = 0 \tag{6.2.63}$$

$$\overline{H}_1^B = 0 \tag{6.2.64}$$

in region 1.

The field generated by these equivalent sources are

$$\begin{aligned}\overline{E}^B(\overline{r}) = &-\int dS' \nabla g_2(\overline{r}, \overline{r}') \times \overline{M}_s^B(\overline{r}') + i\omega\mu_2 \left[\int dS' g_2(\overline{r}, \overline{r}') \overline{J}_s^B(\overline{r}') \right. \\ &\left. + \frac{1}{k_2^2} \nabla \int dS' g_2(\overline{r}, \overline{r}') \nabla'_s \cdot \overline{J}_s^B(\overline{r}') \right]\end{aligned} \tag{6.2.65}$$

$$\begin{aligned}\overline{H}^B(\overline{r}) = &\, i\omega\epsilon_2 \left[\int dS' g_1(\overline{r}, \overline{r}') \overline{M}_s^B(\overline{r}') + \frac{1}{k_2^2} \nabla \int dS' g_1(\overline{r}, \overline{r}') \nabla'_s \cdot \overline{M}_s^B(\overline{r}') \right] \\ &+ \int dS' \nabla g_2(\overline{r}, \overline{r}') \times \overline{J}_s^B(\overline{r}')\end{aligned} \tag{6.2.66}$$

where

$$g_2(\overline{r}, \overline{r}') = \frac{e^{ik_2|\overline{r}-\overline{r}'|}}{4\pi|\overline{r}-\overline{r}'|} \tag{6.2.67}$$

These equivalent sources radiative with the Green's function g_2 into all of space. As these equivalent sources radiate into region 1, these fields will give $\overline{E}_1^B = 0$ and $\overline{H}_1^B = 0$. Note that we also have ϵ_2 and μ_2 for region 1 as well.

Thus, for $\overline{r}$ in region 1, we have the E equation from $\overline{E}_1^B = 0$

$$\begin{aligned}&-\int dS' \nabla g_2(\overline{r}, \overline{r}') \times \overline{M}_s^B(\overline{r}') \\ &+ i\omega\mu_2 \left[\int dS' g_2(\overline{r}, \overline{r}') \overline{J}_s^B(\overline{r}') + \frac{1}{k_2^2} \nabla \int dS' g_2(\overline{r}, \overline{r}') \nabla'_s \cdot \overline{J}_s^B(\overline{r}') \right] = 0\end{aligned} \tag{6.2.68}$$

and for $\overline{r}$ in region 1 the H equation from $\overline{H}_1^B = 0$

$$i\omega\epsilon_2 \left[\int dS' g_2(\overline{r}, \overline{r}') \overline{M}_s^B (\overline{r}') + \frac{1}{k_2^2} \nabla \int dS' g_2(\overline{r}, \overline{r}') \nabla'_s \cdot \overline{M}_s^B (\overline{r}') \right]$$

$$+ \int dS' \nabla g_2(\overline{r}, \overline{r}') \times \overline{J}_s^B (\overline{r}') = 0 \qquad (6.2.69)$$

There are six scalar integral equations above. Note that in (6.2.68) and (6.2.69), the Green's function g_2 is used for $\overline{r}$ in region 1. However, they are not independent and just by themselves cannot be used to calculate the unknown surfaces fields. This is obvious because μ_1, ϵ_1, and g_1 are not involved in the equations. Neither is the incident field involved in (6.2.68) and (6.2.69). Thus we also need the exterior equivalent problem.

Because of the continuity of tangential E and H, and continuity of normal D and B for the physical problem, and $\hat{n}_i = -\hat{n}$, we have the following relations between equivalent sources A and B.

$$\overline{M}_s^A = -\overline{M}_s^B \qquad (6.2.70)$$

$$\overline{J}_s^A = -\overline{J}_s^B \qquad (6.2.71)$$

$$\rho_s^A = -\rho_s^B \qquad (6.2.72)$$

$$m_s^A = -m_s^B \qquad (6.2.73)$$

for the equivalent sources between problems A and B.

2.3 Surface Integral Equations for Equivalent Surface Currents, Tangential and Normal Components of Fields

Surface integral equations are obtained by taking various combinations of the interior and exterior integral equations. As we discuss earlier, both integral equations from the exterior problem and from the interior problem need to be included. Let

$\overline{r}_+$ = a point on the surface but in region 1 (infinitesimally close) (6.2.74)

$\overline{r}_-$ = a point on the surface but in region 2 (infinitesimally close) (6.2.75)

Define

$$\overline{M}_s = \overline{M}_s^A = -\overline{M}_s^B = \overline{E}_1 \times \hat{n} = \overline{E}_2 \times \hat{n} \qquad (6.2.76)$$

$$\overline{J}_s = \overline{J}_s^A = -\overline{J}_s^B = \hat{n} \times \overline{H}_1 = \hat{n} \times \overline{H}_2 \qquad (6.2.77)$$

$$\rho_s = \rho_s^A = -\rho_s^B = \hat{n} \cdot \epsilon_1 \overline{E}_1 = \hat{n} \cdot \epsilon_2 \overline{E}_2 \qquad (6.2.78)$$

$$m_s = m_s^A = -m_s^B = \hat{n} \cdot \mu_1 \overline{H}_1 = \hat{n} \cdot \mu_2 \overline{H}_2 \qquad (6.2.79)$$

Thus $\overline{M}_s$ and $\overline{J}_s$ represent tangential electric field and magnetic field for the physical problem, where ρ_s and m_s represent normal components of $\overline{D}$

and $\overline{B}$ for the physical problem. Suppose we take the tangential E equation for exterior problem equation (6.2.57) and the tangential H equation for the interior problem equation (6.2.69),

$$\left[-\int dS'\nabla g_1(\overline{r}_-,\overline{r}')\times\overline{M}_s(\overline{r}')\right]_{\tan}+i\omega\mu_1\left[\int dS' g_1(\overline{r}_-,\overline{r}')\overline{J}_s(\overline{r}')\right.$$
$$\left.+\frac{1}{k_1^2}\nabla\int dS' g_1(\overline{r}_-,\overline{r}')\nabla'_s\cdot\overline{J}_s(\overline{r}')\right]_{\tan}=\left[-\overline{E}_i(\overline{r}_-)\right]_{\tan} \tag{6.2.80}$$

$$-i\omega\epsilon_2\left[\int dS' g_2(\overline{r}_+,\overline{r}')\overline{M}_s(\overline{r}')+\frac{1}{k_2^2}\nabla\int dS' g_2(\overline{r}_+,\overline{r}')\nabla'_s\cdot\overline{M}_s(\overline{r}')\right]_{\tan}$$
$$-\left[\int dS'\nabla g_2(\overline{r}_+,\overline{r}')\times\overline{J}_s(\overline{r}')\right]_{\tan}=0 \tag{6.2.81}$$

where subscript "tan" means the tangential vector components. From (6.1.30), we know that for a small circular disk S_a around the point $\overline{r}$ on the surface, we have

$$\int_{S_a} dS'\nabla g(\overline{r}_+,\overline{r}')=-\frac{\hat{n}}{2}=-\int_{S_a} dS'\nabla g(\overline{r}_-,\overline{r}') \tag{6.2.82}$$

Then

$$\left[-\frac{\hat{n}}{2}\times\overline{M}_s(\overline{r})\right]+\left[-\int_P dS'\nabla g(\overline{r},\overline{r}')\times\overline{M}_s(\overline{r}')\right]_{\tan}$$
$$+i\omega\mu_1\left[\int dS' g_1(\overline{r},\overline{r}')\overline{J}_s(\overline{r}')+\frac{1}{k_1^2}\nabla\int dS' g_1(\overline{r},\overline{r}')\nabla'_s\cdot\overline{J}_s(\overline{r}')\right]_{\tan}$$
$$=\left[-\overline{E}_i(\overline{r})\right]_{\tan} \tag{6.2.83}$$

$$-i\omega\epsilon_2\left[\int dS' g_2(\overline{r},\overline{r}')\overline{M}_s(\overline{r}')+\frac{1}{k_2^2}\nabla\int dS' g_2(\overline{r},\overline{r}')\nabla'_s\cdot\overline{M}_s(\overline{r}')\right]_{\tan}$$
$$-\left[\int_P dS'\nabla g_2(\overline{r},\overline{r}')\times\overline{J}_s(\overline{r}')\right]_{\tan}+\frac{\hat{n}}{2}\times\overline{J}_s(\overline{r})=0 \tag{6.2.84}$$

where $\int_P$ = principal value integral. Also

$$\hat{n}\times\overline{M}_s=\overline{E}-\hat{n}\left(\hat{n}\cdot\overline{E}\right)=\overline{E}_{\tan} \tag{6.2.85}$$
$$\hat{n}\times\overline{J}_s=-\overline{H}+\hat{n}\left(\hat{n}\cdot\overline{H}\right)=-\overline{H}_{\tan} \tag{6.2.86}$$

Thus the above contains 4 scalar equations for the four scalar unknowns of the surface electric and surface magnetic currents. Standard MoM techniques can then be applied. For example, using Galerkin's method and the Rao-Wilton-Glisson (RWG) basis functions [Rao et al. 1982], matrix equations can be readily obtained.

Surface integral equations can also be obtained for the 6 electromagnetic field components which is also known as the Stratton-Chu formulation. We take the the cross product of $\hat{n}$ and the H exterior equation (6.2.58), and the cross product of $\hat{n}$ and the E interior equation (6.2.68)

$$i\omega\epsilon_1\hat{n}\times\left[\int dS' g_1(\overline{r}_-,\overline{r}')\overline{M}_s(\overline{r}')+\frac{1}{k_1^2}\nabla\int dS' g_1(\overline{r}_-,\overline{r}')\nabla'_s\cdot\overline{M}_s(\overline{r}')\right]$$
$$+\hat{n}\times\int dS'\nabla g_1(\overline{r}_-,\overline{r}')\times\overline{J}_s(\overline{r}') = -\hat{n}\times\overline{H}_i(\overline{r}) \tag{6.2.87}$$

$$\hat{n}\times\int dS'\nabla g_2(\overline{r}_+,\overline{r}')\times\overline{M}_s(\overline{r}')$$
$$-i\omega\mu_2\hat{n}\times\left[\int dS' g_2(\overline{r}_+,\overline{r}')\overline{J}_s(\overline{r}')+\frac{1}{k_2^2}\nabla\int dS' g_2(\overline{r}_+,\overline{r}')\nabla'_s\cdot\overline{J}_s(\overline{r}')\right]=0 \tag{6.2.88}$$

Note that

$$\hat{n}\times\int_{S_a} dS'\nabla g(\overline{r}_+,\overline{r}')\times\overline{Q}(\overline{r}')$$
$$= -\hat{n}\times\left(\frac{\hat{n}}{2}\times\overline{Q}(\overline{r})\right) = -\hat{n}\times\int_{S_a} dS'\nabla g(\overline{r}_-,\overline{r}')\times\overline{Q}(\overline{r}') \tag{6.2.89}$$

$$\hat{n}\times\int_{S_a} dS'\nabla g(\overline{r}_+,\overline{r}')P(\overline{r}')$$
$$= -\hat{n}\times\int_{S_a} dS'\nabla g(\overline{r}_-,\overline{r}')P(\overline{r}') = \hat{n}\times\left(-\frac{\hat{n}}{2}P(\overline{r})\right) = 0 \tag{6.2.90}$$

Thus

$$\hat{n}\times\int_{S_a} dS'\nabla g_1(\overline{r}_-,\overline{r}')\times\overline{J}_s(\overline{r}')$$
$$= \hat{n}\times\left(\frac{\hat{n}}{2}\times\overline{J}_s\right) = \frac{1}{2}\hat{n}(\hat{n}\cdot\overline{J}_s)-\frac{\overline{J}_s}{2} = -\frac{\hat{n}\times\overline{H}}{2} \tag{6.2.91}$$

$$\hat{n}\times\int_{S_a} dS'\nabla g_2(\overline{r}_+,\overline{r}')\times\overline{M}_s(\overline{r}') = -\hat{n}\times\left(\frac{\hat{n}}{2}\times\overline{M}_s\right) = -\frac{\hat{n}\times\overline{E}}{2} \tag{6.2.92}$$

Using the surface integral equations from (6.2.89)–(6.2.92) in (6.2.87)–(6.2.88),

$$i\omega\epsilon_1\hat{n}\times\left[\int dS' g_1(\overline{r}_-,\overline{r}')\overline{M}_s(\overline{r}')+\frac{1}{k_1^2}\int_P dS'\nabla g_1(\overline{r}_-,\overline{r}')\nabla'_s\cdot\overline{M}_s(\overline{r}')\right]$$
$$-\frac{\hat{n}\times\overline{H}(\overline{r})}{2}+\hat{n}\times\int_P dS'\left[\nabla g_1(\overline{r}_-,\overline{r}')\times\overline{J}_s(\overline{r}')\right] = -\hat{n}\times\overline{H}_i(\overline{r}) \tag{6.2.93}$$

$$
-\frac{\hat{n}\times\overline{E}(\overline{r})}{2}+\hat{n}\times\int_P dS'\left[\nabla g_2(\overline{r}_+,\overline{r}')\times\overline{M}_s(\overline{r}')\right]
$$
$$
-i\omega\mu_2\hat{n}\times\left[\int dS' g_2(\overline{r}_+,\overline{r}')\overline{J}_s(\overline{r}')+\frac{1}{k_2^2}\int_P dS'\nabla g_2(\overline{r}_+,\overline{r}')\nabla'_s\cdot\overline{J}_s(\overline{r}')\right]=0 \tag{6.2.94}
$$

We next introduce normal components of surface fields that correspond to surface charges

$$
\nabla'_s\cdot\overline{M}_s(\overline{r}')=i\omega m_s(\overline{r}')=i\omega\hat{n}'\cdot\mu_1\overline{H}_1(\overline{r}')=i\omega\hat{n}'\cdot\mu_2\overline{H}_2(\overline{r}') \tag{6.2.95}
$$
$$
\nabla'_s\cdot\overline{J}_s(\overline{r}')=i\omega\rho_s(\overline{r}')=i\omega\hat{n}'\cdot\epsilon_1\overline{E}_1(\overline{r}')=i\omega\hat{n}'\cdot\epsilon_2\overline{E}_2(\overline{r}') \tag{6.2.96}
$$

Remember that tangential electric fields are continuous and normal components are discontinuous. Rearranging terms and also using $\nabla g=-\nabla' g$ we get from (6.2.93) and (6.2.94),

$$
\hat{n}\times\left[\int dS'\,i\omega\epsilon_1 g_1(\overline{r},\overline{r}')\left[\hat{n}'\times\overline{E}(\overline{r}')\right]-\int_P dS'\nabla' g_1(\overline{r},\overline{r}')\hat{n}'\cdot\overline{H}_1(\overline{r}')\right]
$$
$$
-\frac{\hat{n}\times\overline{H}(\overline{r})}{2}+\hat{n}\times\int_P dS'\left[\nabla' g_1(\overline{r},\overline{r}')\times\left(\hat{n}'\times\overline{H}(\overline{r}')\right)\right]=\hat{n}\times\overline{H}_i(\overline{r}) \tag{6.2.97}
$$

$$
-\frac{\hat{n}\times\overline{E}(\overline{r})}{2}+\hat{n}\times\int_P dS'\left[\nabla' g_2(\overline{r},\overline{r}')\times\left[\hat{n}'\times\overline{E}(\overline{r}')\right]\right]
$$
$$
-\hat{n}\times\left[\int dS' i\omega\mu_2 g_2(\overline{r},\overline{r}')\left(\hat{n}'\times\overline{H}(\overline{r}')\right)+\frac{1}{\epsilon_2}\int_P dS'\nabla' g_2(\overline{r},\overline{r}')\hat{n}'\cdot\epsilon_1\overline{E}_1(\overline{r}')\right]
$$
$$
=0 \tag{6.2.98}
$$

To complete the picture, we need normal E and H equations. To do that, we take normal dot product with the E exterior equation (6.2.57), and the normal dot product with the H interior equation (6.2.69)

$$
-\hat{n}\cdot\int dS'\nabla g(\overline{r}_-,\overline{r}')\times\overline{M}_s^A(\overline{r}')
$$
$$
+i\omega\mu_1\hat{n}\cdot\left[\int dS' g_1(\overline{r},\overline{r}')\overline{J}_s^A(\overline{r}')+\frac{1}{k_1^2}\nabla\int dS' g_1(\overline{r}_-,\overline{r}')\nabla'_s\cdot\overline{J}_s^A(\overline{r}')\right]
$$
$$
=-\hat{n}\cdot\overline{E}_i(\overline{r}) \tag{6.2.99}
$$

$$
i\omega\epsilon_2\hat{n}\cdot\left[\int dS' g_2(\overline{r},\overline{r}')\overline{M}_s^B(\overline{r}')+\frac{1}{k_2^2}\nabla\int dS' g_2(\overline{r}_+,\overline{r}')\nabla'_s\cdot\overline{M}_s^B(\overline{r}')\right]
$$
$$
+\hat{n}\cdot\int dS'\nabla g_2(\overline{r}_+,\overline{r}')\times\overline{J}_s^B(\overline{r}')=0 \tag{6.2.100}
$$

Using similar techniques to the tangential components case, and separating out the principal values, (6.2.99) and (6.2.100) become

$$\hat{n}\cdot\int_P dS'\nabla' g_1(\overline{r},\overline{r}')\times\left[\hat{n}'\times\overline{E}\left(\overline{r}'\right)\right]+\frac{\left[\hat{n}\cdot\overline{E}_1\left(\overline{r}\right)\right]}{2}$$
$$-\hat{n}\cdot\left[\int dS' i\omega\mu_1 g_1(\overline{r},\overline{r}')\left[\hat{n}'\times\overline{H}\left(\overline{r}'\right)\right]+\int_P dS'\nabla' g_1(\overline{r},\overline{r}')\left[\hat{n}'\cdot\overline{E}_1\left(\overline{r}'\right)\right]\right]$$
$$=\hat{n}\cdot\overline{E}_i\left(\overline{r}\right) \qquad (6.2.101)$$

$$-\hat{n}\cdot\left[\int dS' i\omega\epsilon_2 g_2(\overline{r},\overline{r}')\left[-\hat{n}'\times\overline{E}\left(\overline{r}'\right)\right]+\int_P dS'\nabla' g_2(\overline{r},\overline{r}')\left[\hat{n}'\cdot\frac{\mu_1}{\mu_2}\overline{H}_1\left(\overline{r}'\right)\right]\right]$$
$$+\hat{n}\cdot\int_P dS'\nabla' g_2(\overline{r},\overline{r}')\times\left[\hat{n}'\times\overline{H}\left(\overline{r}'\right)\right]-\frac{1}{2}\left[\hat{n}\cdot\frac{\mu_1}{\mu_2}\overline{H}_1\left(\overline{r}\right)\right]=0 \qquad (6.2.102)$$

Thus we have six scalar equations in (6.2.97)–(6.2.98) and (6.2.101)–(6.2.102) for the six field components of $\hat{n}\times\overline{E}_1$, $\hat{n}\times\overline{H}_1$, $\hat{n}\cdot\overline{E}_1$, and $\hat{n}\cdot\overline{H}_1$.

3 Two-Dimensional Rough Dielectric Surfaces with Sparse Matrix Canonical Grid Method

3.1 Integral Equation and SMCG Method

Consider an electromagnetic wave with fields $\overline{E}_i(x,y,z)$ and $\overline{H}_i(x,y,z)$ impinging on a 2-D rough surface with a random height profile $z = f(x,y)$. Above the rough surface is a free space (region 1) while the subsurface is characterized by permittivity ϵ_2 and permeability μ_2 (region 2). The height function $z = f(x,y)$ is a random process with zero mean. The incident direction is $\hat{k}_i = \sin\theta_i\cos\phi_i\hat{x} + \sin\theta_i\sin\phi_i\hat{y} - \cos\theta_i\hat{z}$. The incident field is given as

$$\overline{E}_i(x,y,z)=\int_{-\infty}^{+\infty} dk_x\int_{-\infty}^{+\infty} dk_y \exp(ik_x x+ik_y y-ik_z z)E_{TE}(k_x,k_y)\hat{e}(-k_z) \qquad (6.3.1)$$

where

$$\hat{e}(-k_z)=\frac{1}{k_\rho}(\hat{x}k_y-\hat{y}k_x) \qquad (6.3.2)$$

and incident magnetic field is given by (6.1.33). We use the same spectrum $E_{TE}(k_x,k_y)$ as given by Eqs. (6.1.35)–(6.1.38). Let $\overline{r}' = \hat{x}x' + \hat{y}y' + \hat{z}f(x',y')$ denote a source point and $\overline{r} = \hat{x}x + \hat{y}y + \hat{z}f(x,y)$ denote a field point on the rough surface. Then the integral equations on the dielectric rough surface using the 6 scalar components are, from Section 2 and repeated below for convenience,

$$0 = -\frac{\hat{n} \times \overline{E}(\overline{r})}{2} - \hat{n} \times \left[\int i\omega \hat{n}' \times \overline{H}(\overline{r}')\mu_2 g_2 dS' \right.$$
$$\left. + \int \left\{ (\hat{n}' \times \overline{E}(\overline{r}')) \times \nabla' g_2 + (\hat{n}' \cdot \overline{E}(\overline{r}') \frac{\epsilon_1}{\epsilon_2} \nabla' g_2) \right\} dS' \right] \quad (6.3.3)$$

$$\hat{n} \times \overline{H}^{inc}(\overline{r}) = \frac{\hat{n} \times \overline{H}(\overline{r})}{2} - \hat{n} \times \left[\int -i\omega \hat{n}' \times \overline{E}(\overline{r}')\epsilon_1 g_1 dS' \right.$$
$$\left. + \int \{ (\hat{n}' \times \overline{H}(\overline{r}')) \times \nabla' g_1 + \hat{n}' \cdot \overline{H}(\overline{r}') \nabla' g_1 \} dS' \right] \quad (6.3.4)$$

$$\hat{n} \cdot \overline{E}^{inc}(\overline{r}) = \frac{\hat{n} \cdot \overline{E}(\overline{r})}{2} - \hat{n} \cdot \left[\int \hat{n}' \times \overline{H}(\overline{r}') i\omega \mu_1 g_1 dS' \right.$$
$$\left. + \int \{ (\hat{n}' \times \overline{E}(\overline{r}')) \times \nabla' g_1 + \nabla' g_1 \hat{n}' \cdot \overline{E}(\overline{r}') \} dS' \right] \quad (6.3.5)$$

$$0 = -\frac{\hat{n} \cdot \overline{H}(\overline{r})}{2} - \hat{n} \cdot \left[\int -\hat{n}' \times \overline{E}(\overline{r}') i\omega \epsilon_2 g_2 dS' \right.$$
$$\left. + \int \left\{ (\hat{n}' \times \overline{H}(\overline{r}')) \times \nabla' g_2 + \nabla' g_2 \hat{n}' \cdot \frac{\mu_1}{\mu_2} \overline{H}(\overline{r}') \right\} dS' \right] \quad (6.3.6)$$

where $\overline{E}(\overline{r})$, $\overline{H}(\overline{r})$ represent surface fields when approaching the surface from region 1. In (6.3.3)–(6.3.6), g_1 and g_2 are the scalar Green's function in region 1 (air) and region 2 (lossy dielectric medium), respectively

$$g_{1,2} = \frac{\exp(ik_{1,2}R)}{4\pi R} \quad (6.3.7)$$

and

$$\nabla' g_{1,2} = (\overline{r} - \overline{r}')G_{1,2}(R) \quad (6.3.8)$$

where

$$G_{1,2}(R) = \frac{(1 - ik_{1,2}R)\exp(ik_{1,2}R)}{4\pi R^3} \quad (6.3.9)$$

and $R = \sqrt{(x-x')^2 + (y-y')^2 + (f(x,y) - f(x',y'))^2}$. The unit normal vectors $\hat{n}$ and $\hat{n}'$ refer respectively to unprimed and primed coordinates and point away from the second medium. The vector integral equations are put in the form of 6 coupled scalar integral equations. Let F_x and F_y represent the x and y components of $\hat{n} \times \overline{H}$ and F_n represent the normal component of $\overline{H}$. To be specific, let

$$F_x(\overline{r}) = \sqrt{1 + \left(\frac{\partial f(x,y)}{\partial x}\right)^2 + \left(\frac{\partial f(x,y)}{\partial y}\right)^2}\ \hat{n} \times \overline{H}(\overline{r}) \cdot \hat{x}$$
$$= \left(-\frac{\partial f}{\partial y}\hat{z} - \hat{y}\right) \cdot \overline{H} \tag{6.3.10}$$

$$F_y(\overline{r}) = \sqrt{1 + \left(\frac{\partial f(x,y)}{\partial x}\right)^2 + \left(\frac{\partial f(x,y)}{\partial y}\right)^2}\ \hat{n} \times \overline{H}(\overline{r}) \cdot \hat{y}$$
$$= \left(\frac{\partial f}{\partial x}\hat{z} + \hat{x}\right) \cdot \overline{H} \tag{6.3.11}$$

$$F_n(\overline{r}) = \sqrt{1 + \left(\frac{\partial f(x,y)}{\partial x}\right)^2 + \left(\frac{\partial f(x,y)}{\partial y}\right)^2}\ \hat{n} \cdot \overline{H}(\overline{r})$$
$$= \left(-\frac{\partial f}{\partial x}\hat{x} - \frac{\partial f}{\partial y}\hat{y} + \hat{z}\right) \cdot \overline{H} \tag{6.3.12}$$

Similarly, for the x, y components of the tangential electric field and the normal component of the electric field, they are represented by I_x, I_y, and I_z as follows:

$$I_x(\overline{r}) = \sqrt{1 + \left(\frac{\partial f(x,y)}{\partial x}\right)^2 + \left(\frac{\partial f(x,y)}{\partial y}\right)^2}\ \hat{n} \times \overline{E}(\overline{r}) \cdot \hat{x}$$
$$= \left(-\frac{\partial f}{\partial y}\hat{z} - \hat{y}\right) \cdot \overline{E} \tag{6.3.13}$$

$$I_y(\overline{r}) = \sqrt{1 + \left(\frac{\partial f(x,y)}{\partial x}\right)^2 + \left(\frac{\partial f(x,y)}{\partial y}\right)^2}\ \hat{n} \times \overline{E}(\overline{r}) \cdot \hat{y}$$
$$= \left(\frac{\partial f}{\partial x}\hat{z} + \hat{x}\right) \cdot \overline{E} \tag{6.3.14}$$

$$I_n(\overline{r}) = \sqrt{1 + \left(\frac{\partial f(x,y)}{\partial x}\right)^2 + \left(\frac{\partial f(x,y)}{\partial y}\right)^2}\ \hat{n} \cdot \overline{E}(\overline{r})$$
$$= \left(-\frac{\partial f}{\partial x}\hat{x} - \frac{\partial f}{\partial y}\hat{y} + \hat{z}\right) \cdot \overline{E} \tag{6.3.15}$$

Then the six scalar coupled integral equations are

$$0 = -\frac{I_x(\overline{r})}{2} + \fint dx'dy' \Bigg\{ I_x(\overline{r}')G_2(R)$$
$$\cdot \left[\frac{\partial f(x,y)}{\partial y}(y - y') + \frac{\partial f(x',y')}{\partial x'}(x - x') - (z - z')\right]$$

$$
\begin{aligned}
&+ I_y(\overline{r}')G_2(R)\left[-\frac{\partial f(x,y)}{\partial y}(x-x') + \frac{\partial f(x',y')}{\partial y'}(x-x')\right] \\
&+ I_n(\overline{r}')\frac{\epsilon_1}{\epsilon_2}G_2(R)\left[\frac{\partial f(x,y)}{\partial y}(z-z') + (y-y')\right]\Bigg\} \\
&+ \int dx'dy'\Bigg\{ik_1\eta_1 g_2 F_x(\overline{r}')\frac{\partial f(x,y)}{\partial y}\frac{\partial f(x',y')}{\partial x'} \\
&+ ik_1\eta_1 g_2 F_y(\overline{r}')\left[\frac{\partial f(x,y)}{\partial y}\frac{\partial f(x',y')}{\partial y'} + 1\right]\Bigg\}
\end{aligned}
\tag{6.3.16}
$$

$$
\begin{aligned}
0 = &-\frac{I_y(\overline{r})}{2} + \fint dx'dy'\Bigg\{I_x(\overline{r}')G_2(R) \\
&\cdot\left[\frac{\partial f(x',y')}{\partial x'}(y-y') - \frac{\partial f(x,y)}{\partial x}(y-y')\right] \\
&+ I_y(\overline{r}')G_2(R)\left[-(z-z') + \frac{\partial f(x',y')}{\partial y'}(y-y') + \frac{\partial f(x,y)}{\partial x}(x-x')\right] \\
&+ I_n(\overline{r}')\frac{\epsilon_1}{\epsilon_2}G_2(R)\left[-\frac{\partial f(x,y)}{\partial x}(z-z') - (x-x')\right]\Bigg\} \\
&+ \int dx'dy'\Bigg\{ik_1\eta_1 g_2 F_x(\overline{r}')\left[-1 - \frac{\partial f(x,y)}{\partial x}\frac{\partial f(x',y')}{\partial x'}\right] \\
&+ iF_y(\overline{r}')k_1\eta_1 g_2\left[-\frac{\partial f(x,y)}{\partial x}\frac{\partial f(x',y')}{\partial y'}\right]\Bigg\}
\end{aligned}
\tag{6.3.17}
$$

$$
\begin{aligned}
I_n^{inc}(\overline{r}) = &\frac{I_n(\overline{r})}{2} + \fint dx'dy'\Bigg\{G_1(R)I_x(\overline{r}')\Bigg[-\frac{\partial f(x,y)}{\partial x}\frac{\partial f(x',y')}{\partial x'}(y-y') \\
&+ \frac{\partial f(x,y)}{\partial y}\left[\frac{\partial f(x',y')}{\partial x'}(x-x') - (z-z')\right] - (y-y')\Bigg] \\
&+ G_1(R)I_y(\overline{r}')\Bigg[\frac{\partial f(x,y)}{\partial x}\left[(z-z') - \frac{\partial f(x',y')}{\partial y'}(y-y')\right] \\
&+ \frac{\partial f(x,y)}{\partial y}\frac{\partial f(x',y')}{\partial y'}(x-x') + (x-x')\Bigg] \\
&+ G_1(R)I_n(\overline{r}')\left[\frac{\partial f(x,y)}{\partial x}(x-x') + \frac{\partial f(x,y)}{\partial y}(y-y') - (z-z')\right]\Bigg\} \\
&+ \int dx'dy'\Bigg\{ik_1\eta_1 g_1 F_x(\overline{r}')\left[\frac{\partial f(x,y)}{\partial x} - \frac{\partial f(x',y')}{\partial x'}\right] \\
&+ ik_1\eta_1 g_1 F_y(\overline{r}')\left[\frac{\partial f(x,y)}{\partial y} - \frac{\partial f(x',y')}{\partial y'}\right]\Bigg\}
\end{aligned}
\tag{6.3.18}
$$

$$
\begin{aligned}
F_x^{inc}(\overline{r}) = \frac{F_x(\overline{r})}{2} &+ \int dx'dy' \left\{ -i\frac{k_1}{\eta_1} g_1 I_x(\overline{r}') \left[\frac{\partial f(x,y)}{\partial y} \frac{\partial f(x',y')}{\partial x'} \right] \right. \\
&\left. - i\frac{k_1}{\eta_1} g_1 I_y(\overline{r}') \left(\frac{\partial f(x,y)}{\partial y} \frac{\partial f(x',y')}{\partial y'} + 1 \right) \right\} + \fint dx'dy' \left\{ G_1(R) F_x(\overline{r}') \right. \\
&\cdot \left[\frac{\partial f(x,y)}{\partial y}(y-y') + \frac{\partial f(x',y')}{\partial x'}(x-x') - (z-z') \right] \\
&+ G_1(R) F_y(\overline{r}') \left[-\frac{\partial f(x,y)}{\partial y}(x-x') + \frac{\partial f(x',y')}{\partial y'}(x-x') \right] \\
&\left. + G_1(R) F_n(\overline{r}') \left[\frac{\partial f(x,y)}{\partial y}(z-z') + (y-y') \right] \right\}
\end{aligned} \tag{6.3.19}
$$

$$
\begin{aligned}
F_y^{inc}(\overline{r}) = \frac{F_y(\overline{r})}{2} &+ \int dx'dy' \left\{ -i\frac{k_1}{\eta_1} g_1 I_x(\overline{r}') \left[-1 - \frac{\partial f(x',y')}{\partial x'} \frac{\partial f(x,y)}{\partial x} \right] \right. \\
&\left. + i\frac{k_1}{\eta_1} g_1 I_y(\overline{r}') \frac{\partial f(x',y')}{\partial y'} \frac{\partial f(x,y)}{\partial x} \right\} \\
&+ \fint dx'dy' \left\{ G_1(R) F_x(\overline{r}') \left[\frac{\partial f(x',y')}{\partial x'}(y-y') - \frac{\partial f(x,y)}{\partial x}(y-y') \right] \right. \\
&+ G_1(R) F_y(\overline{r}') \left[-(z-z') + \frac{\partial f(x',y')}{\partial y'}(y-y') + \frac{\partial f(x,y)}{\partial x}(x-x') \right] \\
&\left. + G_1(R) F_n(\overline{r}') \left[-\frac{\partial f(x,y)}{\partial x}(z-z') - (x-x') \right] \right\}
\end{aligned} \tag{6.3.20}
$$

$$
\begin{aligned}
0 = -\frac{F_n(\overline{r})}{2} &+ \int dx'dy' \left\{ -ik_1\eta_1 \frac{\epsilon_2}{\epsilon_1} g_2 I_x(\overline{r}') \left[\frac{\partial f(x,y)}{\partial x} - \frac{\partial f(x',y')}{\partial x'} \right] \right. \\
&\left. - ik_1\eta_1 \frac{\epsilon_2}{\epsilon_1} I_y(\overline{r}') g_2 \left[\frac{\partial f(x,y)}{\partial y} - \frac{\partial f(x',y')}{\partial y'} \right] \right\} \\
&+ \fint dx'dy' \left\{ G_2(R) F_x(\overline{r}') \left[-\frac{\partial f(x,y)}{\partial x} \frac{\partial f(x',y')}{\partial x'}(y-y') \right. \right. \\
&\left. + \frac{\partial f(x,y)}{\partial y} \left[\frac{\partial f(x',y')}{\partial x'}(x-x') - (z-z') \right] - (y-y') \right] \\
&+ G_2(R) F_y(\overline{r}') \left[\frac{\partial f(x,y)}{\partial x}(z-z') + \frac{\partial f(x',y')}{\partial y'} \frac{\partial f(x,y)}{\partial y}(x-x') \right. \\
&\left. - \frac{\partial f(x',y')}{\partial y'} \frac{\partial f(x,y)}{\partial x}(y-y') + (x-x') \right] + G_2(R) F_n(\overline{r}') \\
&\left. \cdot \left[\frac{\partial f(x,y)}{\partial x}(x-x') + \frac{\partial f(x,y)}{\partial y}(y-y') - (z-z') \right] \right\}
\end{aligned} \tag{6.3.21}
$$

In terms of a matrix notation, the SMCG procedure is as follows. First, the above 6 scalar surface integral equations are discretized into a matrix equation by the moment method. Then, we choose the neighborhood distance r_d as the distance which defines the boundary between the weak and strong element of the impedance matrix $\overline{\overline{Z}}$ (for example, $r_d = 2\lambda$). Let

$$\rho_R = \sqrt{(x-x')^2 + (y-y')^2} \tag{6.3.22}$$

represent the horizontal separation between two points on the rough surface $(x, y, f(x, y))$ and $(x', y', f(x', y'))$. The strong matrix is a sparse matrix. For the weak matrix elements, we expand the Green's function in a Taylor's series about the flat surface, $f(x, y) = 0$.

$$G_{1,2}(R) = \frac{(1 - ik_{1,2}R)\exp(ik_{1,2}R)}{4\pi R^3} = \sum_{m=0}^{M} a_m^{(1,2)}(\rho_R)\left(\frac{z_d^2}{\rho_R^2}\right)^m \tag{6.3.23}$$

$$g_{1,2} = \frac{\exp(ik_{1,2}R)}{4\pi R} = \sum_{m=0}^{M} b_m^{(1,2)}(\rho_R)\left(\frac{z_d^2}{\rho_R^2}\right)^m \tag{6.3.24}$$

where $z_d = f(x, y) - f(x', y')$. The above coefficients $a_m^{(1,2)}(\rho_R)$ and $b_m^{(1,2)}(\rho_R)$ are translationally invariant in the horizontal directions. In the numerical results of this section, we keep the expansion terms at 6 ($M = 5$) in (6.3.23) and (6.3.24). In the following the first 4 coefficients are listed for reference

$$a_0^{(1,2)}(\rho_R) = (1 - ik_{1,2}\rho_R)\frac{\exp(ik_{1,2}\rho_R)}{4\pi\rho_R^3} \tag{6.3.25}$$

$$a_1^{(1,2)}(\rho_R) = \frac{\exp(ik_{1,2}\rho_R)}{4\pi}\left\{\frac{k_{1,2}^2}{2\rho_R} + \frac{3ik_{1,2}}{2\rho_R^2} - \frac{3}{2\rho_R^3}\right\} \tag{6.3.26}$$

$$a_2^{(1,2)}(\rho_R) = \frac{\exp(ik_{1,2}\rho_R)}{4\pi}\left\{\frac{ik_{1,2}^3}{8} - \frac{6k_{1,2}^2}{8\rho_R} - \frac{15ik_{1,2}}{8\rho_R^2} + \frac{15}{8\rho_R^3}\right\} \tag{6.3.27}$$

$$a_3^{(1,2)}(\rho_R) = \frac{\exp(ik_{1,2}\rho_R)}{4\pi}\left\{-\frac{k_{1,2}^4\rho_R}{48} - \frac{10ik_{1,2}^3}{48} + \frac{105k_{1,2}^2}{112\rho_R} + \frac{35ik_{1,2}}{16\rho_R^2} - \frac{35}{16\rho_R^3}\right\} \tag{6.3.28}$$

$$b_0^{(1,2)}(\rho_R) = \frac{\exp(ik_{1,2}\rho_R)}{4\pi\rho_R} \tag{6.3.29}$$

$$b_1^{(1,2)}(\rho_R) = \exp(ik_{1,2}\rho_R)\left\{\frac{ik_{1,2}}{8\pi} - \frac{1}{8\pi\rho_R}\right\} \tag{6.3.30}$$

$$b_2^{(1,2)}(\rho_R) = \exp(ik_{1,2}\rho_R)\left\{-\frac{k_{1,2}^2\rho_R}{32\pi} - \frac{3ik_{1,2}}{32\pi} + \frac{3}{32\pi\rho_R}\right\} \tag{6.3.31}$$

$$b_3^{(1,2)}(\rho_R) = \exp(ik_{1,2}\rho_R)\left\{-i\frac{k_{1,2}^3\rho_R^2}{192\pi} + \frac{k_{1,2}^2\rho_R}{32\pi} + \frac{13ik_{1,2}}{192\pi} - \frac{13}{192\pi\rho_R}\right\} \tag{6.3.32}$$

The impedance matrix is decomposed into the sum of a strong and a weak matrix.

$$\overline{\overline{Z}} = \overline{\overline{Z}}^{(s)} + \overline{\overline{Z}}^{(w)} \tag{6.3.33}$$

where $\overline{\overline{Z}}^{(s)}$ represents near field strong interaction and $\overline{\overline{Z}}^{(w)}$ represents non-near field weak interaction. Next, the weak matrix elements are expanded in a Taylor's series about the horizontal distance between the two points

$$\overline{\overline{Z}}^{(w)} = \sum_{m=0}^{M} \overline{\overline{Z}}_m^{(w)} \tag{6.3.34}$$

The zeroth term in (6.3.34) is called the flat surface contribution

$$\overline{\overline{Z}}^{(FS)} = \overline{\overline{Z}}_0^{(w)} \tag{6.3.35}$$

The iterative matrix-solving procedure is, for the first-order and higher order solutions

$$(\overline{\overline{Z}}^{(s)} + \overline{\overline{Z}}^{(FS)})\overline{x}^{(1)} = \overline{b} \tag{6.3.36}$$

$$(\overline{\overline{Z}}^{(s)} + \overline{\overline{Z}}^{(FS)})\overline{x}^{(n+1)} = \overline{b}^{(n+1)} \tag{6.3.37}$$

$$\overline{b}^{(n+1)} = \overline{b} - \sum_{m=1}^{M} \overline{\overline{Z}}_m^{(w)}\overline{x}^{(n)} \tag{6.3.38}$$

Equations (6.3.36) and (6.3.37) are solved using the conjugate gradient method (CGM). The flat surface matrix $\overline{\overline{Z}}^{(FS)}$ which represents the lowest order Taylor expansion term is on the left-hand side of the matrix equation. Without the flat-surface matrix on the left-hand side, we have observed that the iteration does not converge for rough surfaces with moderate rms heights. Thus, the terms strong and weak refer to the magnitude of the matrix elements, instead of their total contributions to the iterative matrix equation.

The product of $\overline{\overline{Z}}^{(FS)}$ with $\overline{x}$ can be computed using a 2-D FFT algorithm. Updating the right-hand side is also calculated using the FFT. An additional advantage of the SMCG is that only the Taylor expanded coefficients need to be stored.

With the number of Taylor series coefficient fixed at $M = 5$, for a given rough surface the computational complexity will depend on the number of CGM iterations (6.3.37), SMCG iterations (6.3.38) and the neighborhood distance r_d. The total number of operations (multiplications) is approximately

$$N_{\text{CGM}}\left[256{r_d}^2 nN + 2N\log(N)m_{\text{FFT}}\right] + N_{SMCG}\left[72N\log(N)m_{\text{FFT}}\right] \tag{6.3.39}$$

where N_{CGM} and N_{SMCG} are the number of iterations in CGM matrix solver and the number of right-hand side updates, respectively, n is the number of sample points per λ^2, and m_{FFT} is the total number of FFT's and inverse FFT's. Note that in the above equation, the total number of iterations required for convergence depends on r_d. As an example, consider a simulation with the following number of iterations $N_{\text{CGM}} = 110$ and $N_{\text{SMCG}} = 4$, with sampling of $n = 64$, $r_d = 2.5\lambda$, $N = 16384$, and $m_{\text{FFT}} = 120$. The term containing the r_d dominates with 1.55×10^{11} computational steps. Therefore, SMCG is effective for a moderate rms height so that the number of Taylor series expansion terms can be manageable and the neighborhood distance r_d can be much smaller than the surface length. In this section, an error norm of 0.01 is used for all numerical simulations.

Integral Form of Matrix Equations

For completeness, we list the detailed integral form of the six matrix equations in the following

$$\begin{aligned}
&-\sum_{m=1}^{5}\int_{\rho_R>r_d} dx'dy'\left\{ I_x^{(n)}(\overline{r}')a_m^{(2)}(\rho_R)\left[\frac{z_d^2}{\rho_R}\right]^m \left[\frac{\partial f(x,y)}{\partial y}(y-y')\right.\right.\\
&\left.+\frac{\partial f(x',y')}{\partial x'}(x-x')-(z-z')\right]\\
&+I_y^{(n)}(\overline{r}')a_m^{(2)}(\rho_R)\left[\frac{z_d^2}{\rho_R}\right]^m\left[-\frac{\partial f(x,y)}{\partial y}(x-x')+\frac{\partial f(x',y')}{\partial y'}(x-x')\right]\\
&\left.+I_n^{(n)}(\overline{r}')\frac{\epsilon_1}{\epsilon_2}a_m^{(2)}\left(\frac{z_d^2}{\rho_R}\right)^m\left[\frac{\partial f(x,y)}{\partial y}(z-z')+(y-y')\right]\right\}\\
&-\sum_{m=1}^{5}\int_{\rho_R>r_d} dx'dy'\left\{ik_1\eta_1 b_m^{(2)}(\rho_R)\left[\frac{z_d^2}{\rho_R}\right]^m F_x^{(n)}(\overline{r}')\frac{\partial f(x,y)}{\partial y}\frac{\partial f(x',y')}{\partial x'}\right.
\end{aligned}$$

$$+ ik_1\eta_1 b_m^{(2)}(\rho_R)\left[\frac{z_d^2}{\rho_R}\right]^m F_y^{(n)}(\overline{r}')\left[\frac{\partial f(x,y)}{\partial y}\frac{\partial f(x',y')}{\partial y'}+1\right]\Bigg\}$$

$$= -\frac{I_x^{(n+1)}(\overline{r})}{2} + \fint_{\rho_R\le r_d} dx'dy'\Bigg\{I_x^{(n+1)}(\overline{r}')G_2(R)$$

$$\cdot\left[\frac{\partial f(x,y)}{\partial y}(y-y') + \frac{\partial f(x',y')}{\partial x'}(x-x') - (z-z')\right]$$

$$+ I_y^{(n+1)}(\overline{r}')G_2(R)\left[-\frac{\partial f(x,y)}{\partial y}(x-x') + \frac{\partial f(x',y')}{\partial y'}(x-x')\right]$$

$$+ I_n^{(n+1)}(\overline{r}')\frac{\epsilon_1}{\epsilon_2}G_2(R)\left[\frac{\partial f(x,y)}{\partial y}(z-z') + (y-y')\right]\Bigg\}$$

$$+ \int_{\rho_R\le r_d} dx'dy'\Bigg\{ik_1\eta_1 g_2 F_x^{(n+1)}(\overline{r}')\frac{\partial f(x,y)}{\partial y}\frac{\partial f(x',y')}{\partial x'}$$

$$+ ik_1\eta_1 g_2 F_y^{(n+1)}(\overline{r}')\left[\frac{\partial f(x,y)}{\partial y}\frac{\partial f(x',y')}{\partial y'}+1\right]\Bigg\}$$

$$+ \int_{\rho_R> r_d} dx'dy'\Bigg\{I_x^{(n+1)}(\overline{r}')a_0^{(2)}(\rho_R)$$

$$\cdot\left[\frac{\partial f(x,y)}{\partial y}(y-y') + \frac{\partial f(x',y')}{\partial x'}(x-x') - (z-z')\right]$$

$$+ I_y^{(n+1)}(\overline{r}')a_0^{(2)}(\rho_R)\left(-\frac{\partial f(x,y)}{\partial y}(x-x') + \frac{\partial f(x',y')}{\partial y'}(x-x')\right)$$

$$+ I_n^{(n+1)}(\overline{r}')\frac{\epsilon_1}{\epsilon_2}a_0^{(2)}(\rho_R)\left[\frac{\partial f(x,y)}{\partial y}(z-z') + (y-y')\right]\Bigg\}$$

$$+ \int dx'dy'\Bigg\{ik_1\eta_1 b_0^{(2)}(\rho_R)F_x^{(n+1)}(\overline{r}')\frac{\partial f(x,y)}{\partial y}\frac{\partial f(x',y')}{\partial x'}$$

$$+ ik_1\eta_1 b_0^{(2)}(\rho_R)F_y^{(n+1)}(\overline{r}')\left[\frac{\partial f(x,y)}{\partial y}\frac{\partial f(x',y')}{\partial y'}+1\right]\Bigg\} \qquad (6.3.40)$$

$$-\sum_{m=1}^{5}\int_{\rho_R>r_d} dx'dy'\Bigg\{I_x^{(n)}(\overline{r}')a_m^{(2)}(\rho_R)\left[\frac{z_d}{\rho_R}\right]^m$$

$$\cdot\left[\frac{\partial f(x',y')}{\partial x'}(y-y') - \frac{\partial f(x,y)}{\partial x}(y-y')\right] + I_y^{(n)}(\overline{r}')a_m^{(2)}(\rho_R)\left[\frac{z_d}{\rho_R}\right]^m$$

$$\cdot\left[-(z-z') + \frac{\partial f(x',y')}{\partial y'}(y-y') + \frac{\partial f(x,y)}{\partial x}(x-x')\right]$$

$$+ I_n^{(n)}(\overline{r}')\frac{\epsilon_1}{\epsilon_2}a_m^{(2)}(\rho_R)\left[\frac{z_d}{\rho_R}\right]^m\left[-\frac{\partial f(x,y)}{\partial x}(z-z')+(x-x')\right]\Bigg\}$$

$$-\sum_{m=1}^{5}\int_{\rho_R>r_d}dx'dy'\Bigg\{ik_1\eta_1 b_m^{(2)}(\rho_R)\left[\frac{z_d}{\rho_R}\right]^m F_x^{(n)}(\overline{r}')\left[-1-\frac{\partial f(x,y)}{\partial x}\frac{\partial f(x',y')}{\partial x'}\right]$$

$$\cdot +iF_y^{(n)}(\overline{r}')k_1\eta_1 b_m^{(2)}(\rho_R)\left[\frac{z_d}{\rho_R}\right]^m\left[-\frac{\partial f(x,y)}{\partial x}\frac{\partial f(x',y')}{\partial y'}\right]\Bigg\}$$

$$= -\frac{I_y^{(n+1)}(\overline{r})}{2}+⨍_{\rho_R<r_d}dx'dy'\Bigg\{I_x^{(n+1)}(\overline{r}')G_2(R)$$

$$\cdot\left[\frac{\partial f(x',y')}{\partial x'}(y-y')-\frac{\partial f(x,y)}{\partial x}(y-y')\right]$$

$$+ I_y^{(n+1)}(\overline{r}')G_2(R)\left[-(z-z')+\frac{\partial f(x',y')}{\partial y'}(y-y')+\frac{\partial f(x,y)}{\partial x}(x-x')\right]$$

$$+ I_n^{(n+1)}(\overline{r}')\frac{\epsilon_1}{\epsilon_2}G_2(R)\left[-\frac{\partial f(x,y)}{\partial x}(z-z')+(x-x')\right]\Bigg\}$$

$$+\int_{\rho_R<r_d}dx'dy'\Bigg\{ik_1\eta_1 g_2 F_x^{(n+1)}(\overline{r}')\left[-1-\frac{\partial f(x,y)}{\partial x}\frac{\partial f(x',y')}{\partial x'}\right]$$

$$+ iF_y^{(n+1)}(\overline{r}')k_1\eta_1 g_2\left[-\frac{\partial f(x,y)}{\partial x}\frac{\partial f(x',y')}{\partial y'}\right]\Bigg\}$$

$$+\int_{\rho_R>r_d}dx'dy'\Bigg\{I_x^{(n+1)}(\overline{r}')a_0^{(2)}(\rho_R)\left[\frac{\partial f(x',y')}{\partial x'}(y-y')-\frac{\partial f(x,y)}{\partial x}(y-y')\right]$$

$$+ I_y^{(n+1)}(\overline{r}')a_0^{(2)}(\rho_R)\left[-(z-z')+\frac{\partial f(x',y')}{\partial y'}(y-y')+\frac{\partial f(x,y)}{\partial x}(x-x')\right]$$

$$+ I_n^{(n+1)}(\overline{r}')\frac{\epsilon_1}{\epsilon_2}a_0^{(2)}(\rho_R)\left[-\frac{\partial f(x,y)}{\partial x}(z-z')+(x-x')\right]\Bigg\}$$

$$+\int_{\rho_R>r_d}dx'dy'\Bigg\{ik_1\eta_1 b_0^{(2)}(\rho_R)F_x^{(n+1)}(\overline{r}')\left[-1-\frac{\partial f(x,y)}{\partial x}\frac{\partial f(x',y')}{\partial x'}\right]$$

$$+ iF_y^{(n+1)}(\overline{r}')k_1\eta_1 b_0^{(2)}(\rho_R)\left[-\frac{\partial f(x,y)}{\partial x}\frac{\partial f(x',y')}{\partial y'}\right]\Bigg\} \qquad (6.3.41)$$

$$-\sum_{m=1}^{5}\int_{\rho_R>r_d}dx'dy'\Bigg\{a_m^{(1)}(\rho_R)\left[\frac{z_d}{\rho_R}\right]^m I_x(\overline{r}')\Bigg[-\frac{\partial f(x,y)}{\partial x}\frac{\partial f(x',y')}{\partial x'}(y-y')$$

$$+\frac{\partial f(x,y)}{\partial y}\left[\frac{\partial f(x',y')}{\partial x'}(x-x')-(z-z')\right]-(y-y')\Bigg]$$

$$+ a_m^{(1)}(\rho_R)\left[\frac{z_d}{\rho_R}\right]^m I_y(\bar{r}')\left[\frac{\partial f(x,y)}{\partial x}\left[(z-z') - \frac{\partial f(x',y')}{\partial y'}(y-y')\right]\right.$$

$$\left.+ \frac{\partial f(x,y)}{\partial y}\frac{\partial f(x',y')}{\partial y'}(x-x') + (x-x')\right]$$

$$\left.+ a_m^{(1)}(\rho_R)\left[\frac{z_d}{\rho_R}\right]^m I_n(\bar{r}')\left[\frac{\partial f(x,y)}{\partial x}(x-x') + \frac{\partial f(x,y)}{\partial y}(y-y') - (z-z')\right]\right\}$$

$$- \sum_{m=1}^{5}\int_{\rho_R > r_d} dx'dy'\left\{ik_1\eta_1 b_m^{(1)}(\rho_R)\left[\frac{z_d}{\rho_R}\right]^m F_x(\bar{r}')\left[\frac{\partial f(x,y)}{\partial x} - \frac{\partial f(x',y')}{\partial x'}\right]\right.$$

$$\left.+ ik_1\eta_1 b_m^{(1)}(\rho_R)\left[\frac{z_d}{\rho_R}\right]^m F_y(\bar{r}')\left[\frac{\partial f(x,y)}{\partial y} - \frac{\partial f(x',y')}{\partial y'}\right]\right\} + I_n^{inc}(\bar{r})$$

$$= \frac{I_n(\bar{r})}{2} + ⨍ dx'dy'\left\{G_1(R)I_x(\bar{r}')\left[-\frac{\partial f(x,y)}{\partial x}\frac{\partial f(x',y')}{\partial x'}(y-y')\right.\right.$$

$$\left.+ \frac{\partial f(x,y)}{\partial y}\left[\frac{\partial f(x',y')}{\partial x'}(x-x') - (z-z')\right] - (y-y')\right]$$

$$+ G_1(R)I_y(\bar{r}')\left[\frac{\partial f(x,y)}{\partial x}\left[(z-z') - \frac{\partial f(x',y')}{\partial y'}(y-y')\right]\right.$$

$$\left.+ \frac{\partial f(x,y)}{\partial y}\frac{\partial f(x',y')}{\partial y'}(x-x') + (x-x')\right]$$

$$\left.+ G_1(R)I_n(\bar{r}')\left[\frac{\partial f(x,y)}{\partial x}(x-x') + \frac{\partial f(x,y)}{\partial y}(y-y') - (z-z')\right]\right\}$$

$$+ \int dx'dy'\left\{ik_1\eta_1 g_1 F_x(\bar{r}')\left[\frac{\partial f(x,y)}{\partial x} - \frac{\partial f(x',y')}{\partial x'}\right]\right.$$

$$\left.+ ik_1\eta_1 g_1 F_y(\bar{r}')\left[\frac{\partial f(x,y)}{\partial y} - \frac{\partial f(x',y')}{\partial y'}\right]\right\}$$

$$+ \int_{\rho_R > r_d} dx'dy'\left\{a_0^{(1)}(\rho_R)I_x(\bar{r}')\left[-\frac{\partial f(x,y)}{\partial x}\frac{\partial f(x',y')}{\partial x'}(y-y')\right.\right.$$

$$\left.+ \frac{\partial f(x,y)}{\partial y}\left[\frac{\partial f(x',y')}{\partial x'}(x-x') - (z-z')\right] - (y-y')\right]$$

$$+ a_0^{(1)}(\rho_R)I_y(\bar{r}')\left[\frac{\partial f(x,y)}{\partial x}\left[(z-z') - \frac{\partial f(x',y')}{\partial y'}(y-y')\right]\right.$$

$$\left.+ \frac{\partial f(x,y)}{\partial y}\frac{\partial f(x',y')}{\partial y'}(x-x') + (x-x')\right]$$

$$\left.+ a_0^{(1)}(\rho_R)I_n(\bar{r}')\left[\frac{\partial f(x,y)}{\partial x}(x-x') + \frac{\partial f(x,y)}{\partial y}(y-y') - (z-z')\right]\right\}$$

$$+\int_{\rho_R>r_d} dx'dy'\left\{ik_1\eta_1 b_0^{(1)}(\rho_R)F_x(\bar{r}')\left[\frac{\partial f(x,y)}{\partial x}-\frac{\partial f(x',y')}{\partial x'}\right]\right.$$
$$\left.+ik_1\eta_1 b_0^{(1)}(\rho_R)F_y(\bar{r}')\left[\frac{\partial f(x,y)}{\partial y}-\frac{\partial f(x',y')}{\partial y'}\right]\right\} \qquad (6.3.42)$$

$$F_x^{inc}(\bar{r})-\sum_{m=1}^{5}\int_{\rho_R>r_d} dx'dy'\left\{-i\frac{k_1}{\eta_1}b_m^{(1)}(\rho_R)\left[\frac{z_d^2}{\rho_R^2}\right]^m I_x^{(n)}(\bar{r}')\left[\frac{\partial f(x,y)}{\partial y}\frac{\partial f(x',y')}{\partial x'}\right]\right.$$
$$\left.-i\frac{k_1}{\eta_1}b_m^{(1)}(\rho_R)\left[\frac{z_d^2}{\rho_R^2}\right]^m I_y^{(n)}(\bar{r}')\left(\frac{\partial f(x,y)}{\partial y}\frac{\partial f(x',y')}{\partial y'}+1\right)\right\}$$
$$-\sum_{m=1}^{5}\int_{\rho_R>r_d} dx'dy'\left\{a_m^{(1)}(\rho_R)\left[\frac{z_d^2}{\rho_R^2}\right]^m F_x^{(n)}(\bar{r}')\left[\frac{\partial f(x,y)}{\partial y}(y-y')\right.\right.$$
$$\left.+\frac{\partial f(x',y')}{\partial x'}(x-x')-(z-z')\right]$$
$$+a_m^{(1)}(\rho_R)\left[\frac{z_d^2}{\rho_R^2}\right]^m F_y^{(n)}(\bar{r}')\left[-\frac{\partial f(x,y)}{\partial y}(x-x')+\frac{\partial f(x',y')}{\partial y'}(x-x')\right]$$
$$\left.+a_m^{(1)}(\rho_R)\left[\frac{z_d^2}{\rho_R^2}\right]^m F_n^{(n)}(\bar{r}')\left[\frac{\partial f(x,y)}{\partial y}(z-z')+(y-y')\right]\right\}$$
$$=\frac{F_x^{(n+1)}(\bar{r})}{2}+\int_{\rho\le r_d} dx'dy'\left\{-i\frac{k_1}{\eta_1}g_1 I_x^{(n+1)}(\bar{r}')\left[\frac{\partial f(x,y)}{\partial y}\frac{\partial f(x',y')}{\partial x'}\right]\right.$$
$$\left.-i\frac{k_1}{\eta_1}g_1 I_y^{(n+1)}(\bar{r}')\left(\frac{\partial f(x,y)}{\partial y}\frac{\partial f(x',y')}{\partial y'}+1\right)\right\}$$
$$+\fint_{\rho_R\le r_d} dx'dy'\left\{G_1(R)F_x^{(n+1)}(\bar{r}')\right.$$
$$\cdot\left[\frac{\partial f(x,y)}{\partial y}(y-y')+\frac{\partial f(x',y')}{\partial x'}(x-x')-(z-z')\right]$$
$$+G_1(R)F_y^{(n+1)}(\bar{r}')\left[-\frac{\partial f(x,y)}{\partial y}(x-x')+\frac{\partial f(x',y')}{\partial y'}(x-x')\right]$$
$$\left.+G_1(R)F_n^{(n+1)}(\bar{r}')\left[\frac{\partial f(x,y)}{\partial y}(z-z')+(y-y')\right]\right\}$$
$$+\int_{\rho_R>r_d} dx'dy'\left\{-i\frac{k_1}{\eta_1}b_0^{(1)}(\rho_R)I_x^{(n+1)}(\bar{r}')\left[\frac{\partial f(x,y)}{\partial y}\frac{\partial f(x',y')}{\partial x'}\right]\right.$$
$$\left.-i\frac{k_1}{\eta_1}b_0^{(1)}(\rho_R)I_y^{(n+1)}(\bar{r}')\left(\frac{\partial f(x,y)}{\partial y}\frac{\partial f(x',y')}{\partial y'}+1\right)\right\}$$

$$+\int_{\rho_R>r_d} dx'dy'\left\{a_0^{(1)}(\rho_R)F_x^{(n+1)}(\overline{r}')\right.$$
$$\cdot\left[\frac{\partial f(x,y)}{\partial y}(y-y')+\frac{\partial f(x',y')}{\partial x'}(x-x')-(z-z')\right]$$
$$+a_0^{(1)}(\rho_R)F_y^{(n+1)}(\overline{r}')\left[-\frac{\partial f(x,y)}{\partial y}(x-x')+\frac{\partial f(x',y')}{\partial y'}(x-x')\right]$$
$$\left.+a_0^{(1)}(\rho_R)F_n^{(n+1)}(\overline{r}')\left[\frac{\partial f(x,y)}{\partial y}(z-z')+(y-y')\right]\right\} \qquad (6.3.43)$$

$$F_y^{inc}(\overline{r})-\sum_{m=1}^{5}\int_{\rho_R>r_d} dx'dy'\left\{-i\frac{k_1}{\eta_1}b_m^{(1)}(\rho_R)\right.$$
$$\cdot\left[\frac{z_d^2}{\rho_R^2}\right]^m I_x^{(n)}(\overline{r}')\left[-1-\frac{\partial f(x',y')}{\partial x'}-\frac{\partial f(x,y)}{\partial x}\right]$$
$$\left.+i\frac{k_1}{\eta_1}b_m^{(1)}(\rho_R)\left[\frac{z_d^2}{\rho_R^2}\right]^m I_y^{(n)}(\overline{r}')\frac{\partial f(x',y')}{\partial y'}\frac{\partial f(x,y)}{\partial x}\right\}$$
$$-\sum_{m=1}^{5}\int_{\rho_R>r_d} dx'dy'\left\{a_m^{(1)}(\rho_R)\left[\frac{z_d^2}{\rho_R^2}\right]^m F_x^{(n)}(\overline{r}')\right.$$
$$\cdot\left[\frac{\partial f(x',y')}{\partial x'}(y-y')-\frac{\partial f(x,y)}{\partial x}(y-y')\right]+a_m^{(1)}(\rho_R)\left[\frac{z_d^2}{\rho_R^2}\right]^m F_y^{(n)}(\overline{r}')$$
$$\cdot\left[-(z-z')+\frac{\partial f(x',y')}{\partial y'}(y-y')+\frac{\partial f(x,y)}{\partial x}(x-x')\right]$$
$$\left.+a_m^{(1)}(\rho_R)\left[\frac{z_d^2}{\rho_R^2}\right]^m F_n^{(n)}(\overline{r}')\left[-\frac{\partial f(x,y)}{\partial x}(z-z')-(x-x')\right]\right\}$$
$$=\frac{F_y^{(n+1)}(\overline{r})}{2}+\int_{\rho_R\le r_d} dx'dy'\left\{-i\frac{k_1}{\eta_1}g_1 I_x^{(n+1)}(\overline{r}')\left[-1-\frac{\partial f(x',y')}{\partial x'}-\frac{\partial f(x,y)}{\partial x}\right]\right.$$
$$\left.+i\frac{k_1}{\eta_1}g_1 I_y^{(n+1)}(\overline{r}')\frac{\partial f(x',y')}{\partial y'}\frac{\partial f(x,y)}{\partial x}\right\}$$
$$+\fint_{\rho_R\le r_d} dx'dy'\left\{G_1(R)F_x^{(n+1)}(\overline{r}')\left[\frac{\partial f(x',y')}{\partial x'}(y-y')-\frac{\partial f(x,y)}{\partial x}(y-y')\right]\right.$$
$$+G_1(R)F_y^{(n+1)}(\overline{r}')\left[-(z-z')+\frac{\partial f(x',y')}{\partial y'}(y-y')+\frac{\partial f(x,y)}{\partial x}(x-x')\right]$$
$$\left.+G_1(R)F_n^{(n+1)}(\overline{r}')\left[-\frac{\partial f(x,y)}{\partial x}(z-z')-(x-x')\right]\right\}$$

$$+\int_{\rho_R>r_d} dx'dy'\left\{-i\frac{k_1}{\eta_1}b_0^{(1)}(\rho_R)I_x^{(n+1)}(\bar{r}')\left[-1-\frac{\partial f(x',y')}{\partial x'}-\frac{\partial f(x,y)}{\partial x}\right]\right.$$
$$\left.+i\frac{k_1}{\eta_1}b_0^{(1)}(\rho_R)I_y^{(n+1)}(\bar{r}')\frac{\partial f(x',y')}{\partial y'}\frac{\partial f(x,y)}{\partial x}\right\}$$
$$+\int_{\rho_R>r_d} dx'dy'\left\{a_0^{(1)}(\rho_R)F_x^{(n+1)}(\bar{r}')\left[\frac{\partial f(x',y')}{\partial x'}(y-y')-\frac{\partial f(x,y)}{\partial x}(y-y')\right]\right.$$
$$+a_0^{(1)}(\rho_R)F_y^{(n+1)}(\bar{r}')\left[-(z-z')+\frac{\partial f(x',y')}{\partial y'}(y-y')+\frac{\partial f(x,y)}{\partial x}(x-x')\right]$$
$$\left.+a_0^{(1)}(\rho_R)F_n^{(n+1)}(\bar{r}')\left[-\frac{\partial f(x,y)}{\partial x}(z-z')-(x-x')\right]\right\} \qquad (6.3.44)$$

$$-\sum_{m=1}^{5}\int_{\rho_R>r_d} dx'dy'\left\{-ik_1\eta_1\frac{\epsilon_2}{\epsilon_1}b_m^{(2)}(\rho_R)\left[\frac{z_d^2}{\rho_R^2}\right]^m I_x^{(n)}(\bar{r}')\left[\frac{\partial f(x,y)}{\partial x}-\frac{\partial f(x',y')}{\partial x'}\right]\right.$$
$$\left.-ik_1\eta_1\frac{\epsilon_2}{\epsilon_1}I_y^{(n)}(\bar{r}')b_m^{(2)}(\rho_R)\left[\frac{z_d^2}{\rho_R^2}\right]^m\left[\frac{\partial f(x,y)}{\partial y}-\frac{\partial f(x',y')}{\partial y'}\right]\right\}$$
$$-\sum_{m=1}^{5}\int_{\rho_R>r_d} dx'dy'\left\{a_m^{(2)}(\rho_R)\left[\frac{z_d^2}{\rho_R^2}\right]^m F_x^{(n)}(\bar{r}')\left[-\frac{\partial f(x,y)}{\partial x}\frac{\partial f(x',y')}{\partial x'}(y-y')\right.\right.$$
$$\left.+\frac{\partial f(x,y)}{\partial y}\left[\frac{\partial f(x',y')}{\partial x'}(x-x')-(z-z')\right]-(y-y')\right]$$
$$+a_m^{(2)}(\rho_R)\left[\frac{z_d^2}{\rho_R^2}\right]^m F_y^{(n)}(\bar{r}')\left[\frac{\partial f(x,y)}{\partial x}(z-z')+\frac{\partial f(x',y')}{\partial y'}\frac{\partial f(x,y)}{\partial y}(x-x')\right.$$
$$\left.-\frac{\partial f(x',y')}{\partial y'}\frac{\partial f(x,y)}{\partial x}(y-y')+(x-x')\right]+a_m^{(2)}(\rho_R)\left[\frac{z_d^2}{\rho_R^2}\right]^m F_n^{(n)}(\bar{r}')$$
$$\left.\cdot\left[\frac{\partial f(x,y)}{\partial x}(x-x')+\frac{\partial f(x,y)}{\partial y}(y-y')-(z-z')\right]\right\}$$
$$=-\frac{F_n(\bar{r})}{2}+\int_{\rho_R\le r_d} dx'dy'\left\{-ik_1\eta_1\frac{\epsilon_2}{\epsilon_1}g_2I_x^{(n+1)}(\bar{r}')\left[\frac{\partial f(x,y)}{\partial x}-\frac{\partial f(x',y')}{\partial x'}\right]\right.$$
$$\left.-ik_1\eta_1\frac{\epsilon_2}{\epsilon_1}I_y^{(n+1)}(\bar{r}')g_2\left[\frac{\partial f(x,y)}{\partial y}-\frac{\partial f(x',y')}{\partial y'}\right]\right\}$$
$$+⨍_{\rho_R\le r_d} dx'dy'\left\{G_2F_x^{(n+1)}(\bar{r}')\left[-\frac{\partial f(x,y)}{\partial x}\frac{\partial f(x',y')}{\partial x'}(y-y')\right.\right.$$
$$\left.+\frac{\partial f(x,y)}{\partial y}\left[\frac{\partial f(x',y')}{\partial x'}(x-x')-(z-z')\right]-(y-y')\right]$$

$$+ G_2(R)F_y^{(n+1)}(\overline{r}')\left[\frac{\partial f(x,y)}{\partial x}(z-z') + \frac{\partial f(x',y')}{\partial y'}\frac{\partial f(x,y)}{\partial y}(x-x')\right.$$
$$\left. - \frac{\partial f(x',y')}{\partial y'}\frac{\partial f(x,y)}{\partial x}(y-y') + (x-x')\right]$$
$$\left.+ G_2(R)F_n^{(n+1)}(\overline{r}')\left[\frac{\partial f(x,y)}{\partial x}(x-x') + \frac{\partial f(x,y)}{\partial y}(y-y') - (z-z')\right]\right\}$$
$$+ \int_{\rho_R > r_d} dx'dy'\left\{-ik_1\eta_1\frac{\epsilon_2}{\epsilon_1}b_0^{(2)}(\rho_R)I_x^{(n+1)}(\overline{r}')\left[\frac{\partial f(x,y)}{\partial x} - \frac{\partial f(x',y')}{\partial x'}\right]\right.$$
$$\left. - ik_1\eta_1\frac{\epsilon_2}{\epsilon_1}I_y^{(n+1)}(\overline{r}')b_0^{(2)}(\rho_R)\left[\frac{\partial f(x,y)}{\partial y} - \frac{\partial f(x',y')}{\partial y'}\right]\right\}$$
$$+ \int_{\rho_R > r_d} dx'dy'\left\{a_0^{(2)}(\rho_R)F_x^{(n+1)}(\overline{r}')\left[-\frac{\partial f(x,y)}{\partial x}\frac{\partial f(x',y')}{\partial x'}(y-y')\right.\right.$$
$$\left. + \frac{\partial f(x,y)}{\partial y}\left[\frac{\partial f(x',y')}{\partial x'}(x-x') - (z-z')\right] - (y-y')\right]$$
$$+ a_0^{(2)}(\rho_R)F_y^{(n+1)}(\overline{r}')\left[\frac{\partial f(x,y)}{\partial x}(z-z') + \frac{\partial f(x',y')}{\partial y'}\frac{\partial f(x,y)}{\partial y}(x-x')\right.$$
$$\left. - \frac{\partial f(x',y')}{\partial y'}\frac{\partial f(x,y)}{\partial x}(y-y') + (x-x')\right] + a_0^{(2)}(\rho_R)F_n^{(n+1)}(\overline{r}')$$
$$\left.\cdot\left[\frac{\partial f(x,y)}{\partial x}(x-x') + \frac{\partial f(x,y)}{\partial y}(y-y') - (z-z')\right]\right\} \tag{6.3.45}$$

3.2 Numerical Results of Bistatic Scattering Coefficient

The numerical simulation results are presented in terms of the bistatic scattering coefficient as normalized by the incident power. For an incident wave with a horizontal polarization (TE) we have $\gamma_{\alpha h}(\theta_s, \phi_s)$ and P_h^{inc} is given by (6.1.62) and (6.1.63) respectively. In medium 1, the co-polarized and cross-polarized scattered components of $\mathcal{E}_\alpha^s$ for $\alpha = v$ (vertical polarization) and h (horizontal polarization) expressed in terms of the surface field components, F_x, F_y, I_x, I_y, are respectively

$$\mathcal{E}_h^s = \frac{ik}{4\pi}\int dx'dy' \exp(-ik\beta')\left[\left\{I_x(x',y')\cos\theta_s\cos\phi_s + I_y(x',y')\cos\theta_s\sin\phi_s\right.\right.$$
$$\left. - I_x(x',y')\frac{\partial f(x',y')}{\partial x'}\sin\theta_s - I_y(x',y')\frac{\partial f(x',y')}{\partial y'}\sin\theta_s\right\}$$
$$\left. - \eta\{F_x(x',y')\sin\phi_s - F_y(x',y')\cos\phi_s\}\right] \tag{6.3.46}$$

and

$$\mathcal{E}_v^s = \frac{ik}{4\pi}\int dx'dy' \exp(-ik\beta')\Bigg[\{I_x(x',y')\sin\phi_s - I_y(x',y')\cos\phi_s\}$$
$$+\eta\Bigg\{F_x(x',y')\cos\theta_s\cos\phi_s + F_y\cos\theta_s\sin\phi_s - F_x(x',y')\frac{\partial f(x',y')}{\partial x'}\sin\theta_s$$
$$-F_y(x',y')\frac{\partial f(x',y')}{\partial y'}\sin\theta_s\Bigg\}\Bigg] \tag{6.3.47}$$

where $\beta' = x'\sin\theta_s\cos\phi_s + y'\sin\theta_s\sin\phi_s + f(x',y')\cos\theta_s$. The expressions for the transmitted waves are similar. The normalization is that the integration of reflected and transmitted power over solid angles gives unity for a lossless second medium.

The rough surface has a 2-D Gaussian power spectrum given by

$$W(K_x, K_y) = \frac{l_x l_y h^2}{4\pi}\exp\left(-\frac{K_x^2 l_x^2}{4} - \frac{K_y^2 l_y^2}{4}\right) \tag{6.3.48}$$

We describe various accuracy tests using a 2-D dielectric rough surface. Then, we compute the solution of an electromagnetic wave scattering problem with up to 98,304 surface unknowns and up to 300 realizations. We first compare against matrix inversion (MI) for a small problem. In Figs. 6.3.1 and 6.3.2 the normalized bistatic scattering coefficient obtained by the MI is compared to the solutions obtained by the SMCG for co- and cross-polarized components, respectively, using a single surface realization. Surface lengths in the x and y directions are $L_x = L_y = 4.0\lambda$ which gives an area of $L^2 = 16\lambda^2$. The rms height is $h = 0.2\lambda$, with a correlation length of $l_x = l_y = 1.0\lambda$ and a relative permittivity of $(3.0 + i0.2)$. The surface is sampled at 16 points per λ^2 to give 1536 surface unknowns. The neighborhood distance $r_d = 2.0\lambda$. The incident angles for all illustrations in this paper are $\theta_i = 10°$ and $\phi_i = 0°$. It can be observed that both polarized components of MI and SMCG curves almost completely overlap each other in Fig. 6.3.1.

(A) Accuracy Tests: Choice of r_d and Taylor Series Terms

The convergence of the solution is defined with respect to the error norm of the matrix equation. The error norm criteria is set at $< 1\%$ for all simulations in this section. This is satisfied for all realizations.

Next, the minimum neighborhood distance is shown to be a function of rough surface statistics. In Figs. 6.3.3 and 6.3.4, a convergence test is performed with respect to the size of r_d. Surface lengths are $L_x = L_y = 16\lambda$, which gives an area of $L^2 = 256\lambda^2$. The rms height is $h = 0.2\lambda$, correlation

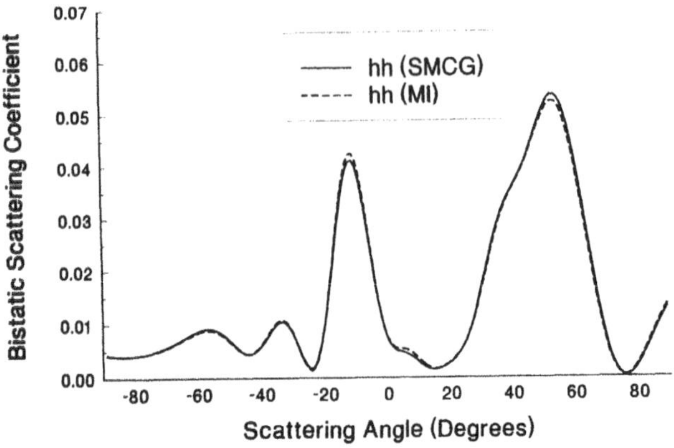

Figure 6.3.1 Comparison between SMCG and full matrix inversion (MI): Co-polarized result. The simulation parameters are $L_x = L_y = 4.0\lambda$, $h = 0.2\lambda$, $l_x = l_y = 1.0\lambda$, $\epsilon_r = (3.0 + i0.2)$, and $r_d = 2.0\lambda$. The incident angles are $\theta_i = 10^\circ$ and $\phi_i = 0^\circ$.

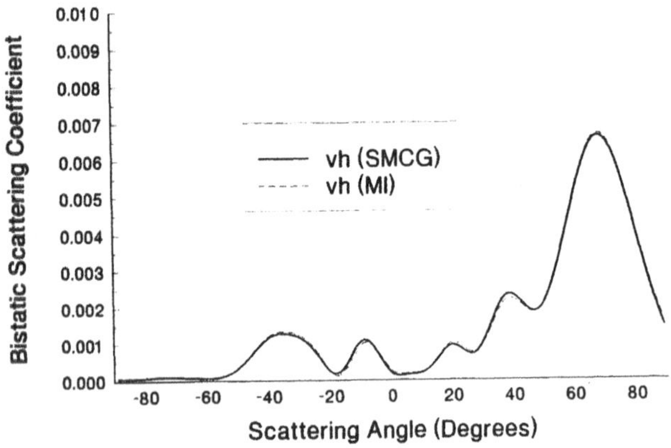

Figure 6.3.2 Comparison between SMCG and full matrix inversion (MI): Cross-polarized result. The simulation parameters are those of Fig. 6.3.1.

lengths are $l_x = l_y = 0.6\lambda$, and the surface is sampled at 64 points per λ^2 to give 98,304 surface unknowns. The neighborhood distance r_d varies from 2.0 to 5.0 λ. It can be seen that for an accurate simulation of the co-polarized scattering, $r_d = 2.0\lambda$ can be used for this surface. Note that $r_d > 2.0\lambda$ appears sufficient for the cross-polarized scattering coefficient. This demonstrates that the choice of neighborhood distance is a function of rough surface statistics, and many different r_d work for a given surface. Therefore, the neighborhood distance r_d can be chosen to optimize CPU.

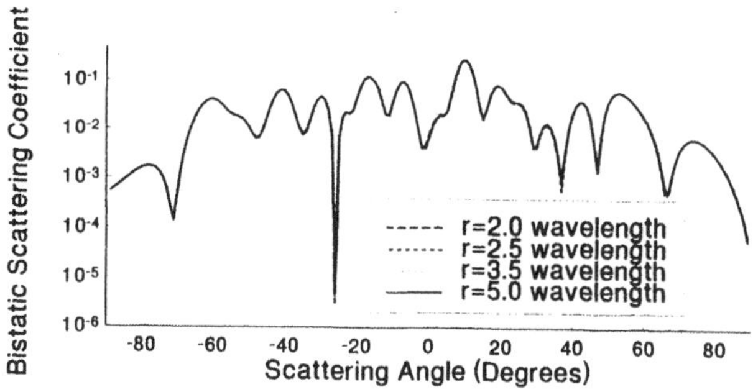

Figure 6.3.3 Convergence with respect to the neighborhood distance r_d for co-polarized component. Simulation parameters are $L_x = L_y = 16\lambda$, $h = 0.2\lambda$, $l_x = l_y = 0.6\lambda$, $\epsilon_r = (6.5 + i1.0)$, and $\theta_i = 10°$ and $\phi_i = 0°$.

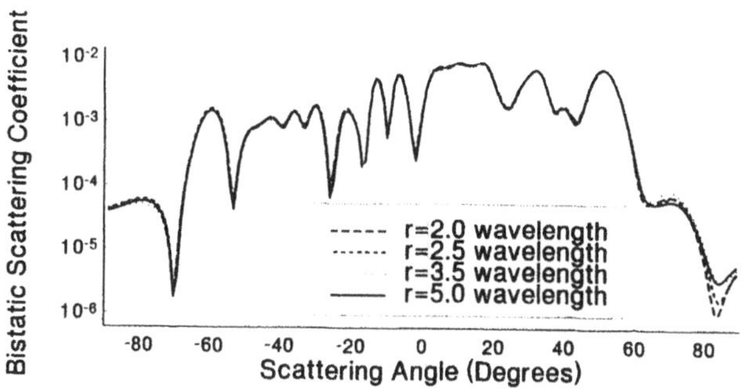

Figure 6.3.4 Convergence with respect to the neighborhood distance r_d for cross-polarized component. The simulation parameters are those of Fig. 6.3.3.

(B) CPU Dependence on r_d and ϵ_2

Next, the matrix solving time dependence on the neighborhood distance (r_d) and dielectric constant of the second medium are illustrated. In Fig. 6.3.5, total CPU time (in hours) is plotted as a function of neighborhood distance. The CPU times are based on a DEC Alpha workstation. The simulation parameters are those of Fig. 6.3.3. It can be seen that the CPU time has an approximately $r_d^{1.2}$ dependence. In Fig. 6.3.6, the CPU dependence with respect to the relative permittivity is illustrated. For this figure, the simulation parameters are as follows: Surface length in the $\hat{x}$ and $\hat{y}$ directions are $L_x = L_y = 8\lambda$ ($L_x \times L_y = 64\lambda^2$) with a surface rms height $h = 0.5\lambda$ and correlation lengths of $l_x = l_y = 1.0\lambda$. The surface is sampled at 64 points per λ^2 to give 24,576 surface unknowns. The neighborhood distance $r_d = 3.5\lambda$. The incident angle is $\theta_i = 10°$ and $\phi_i = 0°$. The tapering parameter g is equal to $L_x/3$. The CPU time increases approximately as $\epsilon_r^{1.5}$. Since in this

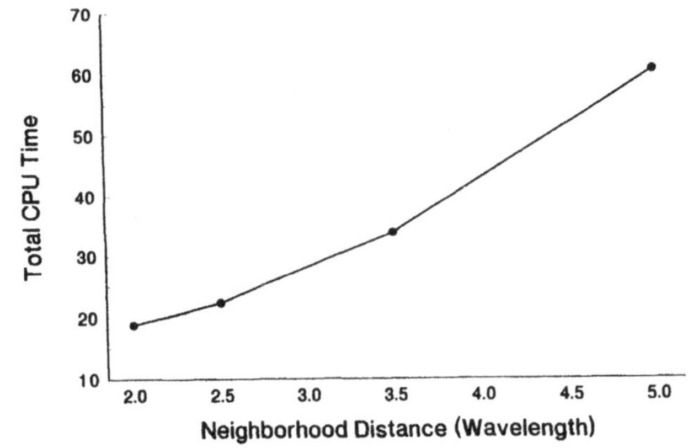

Figure 6.3.5 Dependence of CPU time on neighborhood distance r_d. The simulation parameters are those of Fig. 6.3.3.

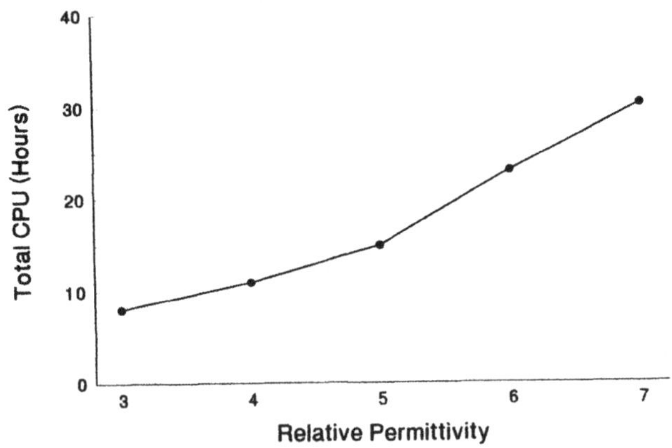

Figure 6.3.6 Dependence of CPU time on ϵ_r. $L_x = L_y = 8.0\lambda$, $h = 0.5\lambda$, $l_x = l_y = 1.0\lambda$, $r_d = 3.5\lambda$, $\theta_i = 10°$ and $\phi_i = 0°$.

example the second medium relative permittivity varies, it is important to check power conservation (Fig. 6.3.7) with respect to permittivity. A general decrease in power conservation from 99.8 to 98.9 is seen as relative permittivity increases from $\epsilon_r = 3$ to $\epsilon_r = 7$. One reason for the increase of CPU with ϵ_r is that as ϵ_r increases, the sampling density of 64 points per square wavelength is not enough, and it takes more iterations to converge. Thus for large permittivity, we need to sample more points per square wavelength on the surface.

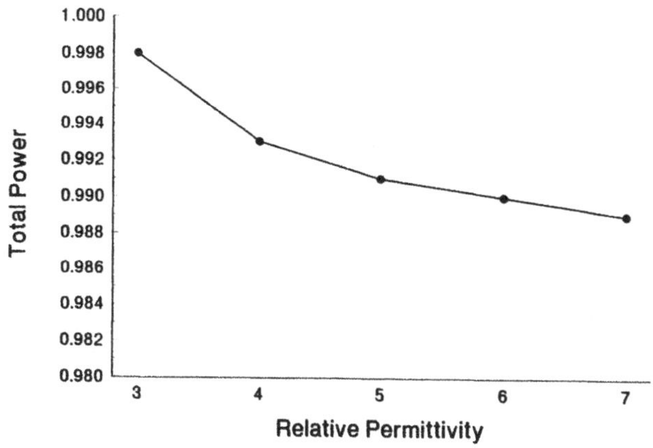

Figure 6.3.7 Dependence of power conservation error on ϵ_r. The simulation parameters are those of Fig. 6.3.6.

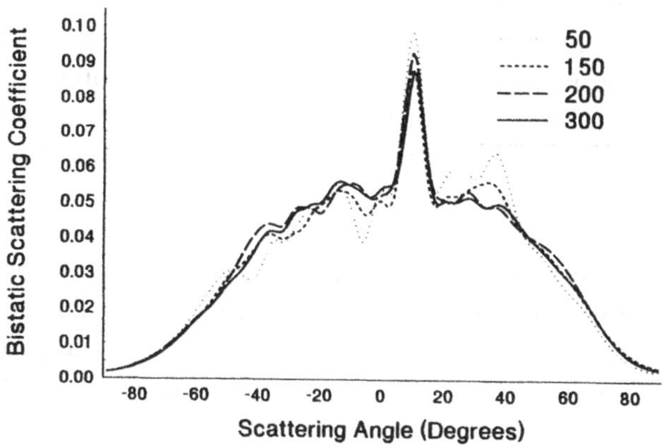

Figure 6.3.8 Bistatic co-polarized scattering coefficients for $h = 0.2\lambda$, $l_x = l_y = 0.6\lambda$, $L_x = L_y = 16\lambda$, $\epsilon_r = (6.5 + i1.0)$, $r_d = 3.0\lambda$, $\theta_i = 10°$ and $\phi_i = 0°$. Result illustrates convergence with respect to the number of realizations.

(C) Monte Carlo Simulation Results

In Fig. 6.3.8, the average normalized bistatic scattering coefficients are plotted for up to 300 realizations. Surface length in the x and y directions are $L_x = L_y = 16\lambda$ ($L_x \times L_y = 256\lambda^2$) with a surface rms height $h = 0.2\lambda$ and correlation lengths of $l_x = l_y = 0.6\lambda$. The surface is sampled at 64 points per λ^2 to give 98,304 surface unknowns. The neighborhood distance $r_d = 3.0\lambda$. The incident angle is $\theta_i = 10°$ and $\phi_i = 0°$. The tapering parameter g is equal to $L_x/3$. Note that the sharp forward specular peak at

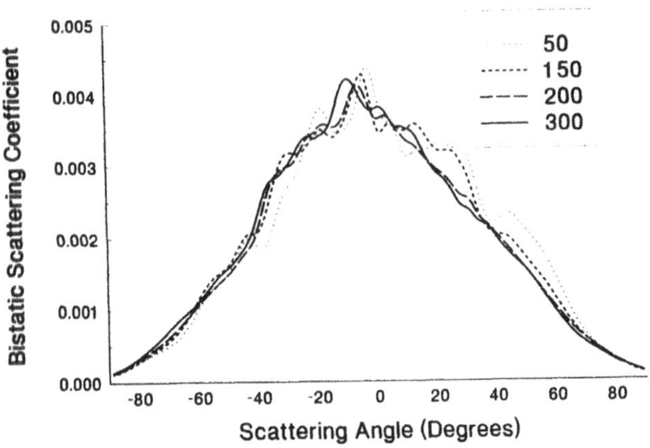

Figure 6.3.9 Bistatic cross-polarized scattering coefficients with simulation parameters of Fig. 6.3.8. Result illustrates convergence with respect to the number of realizations.

$+10°$ is present in the co-polarized scattering component due to the coherent wave. In Fig. 6.3.9, the cross-polarized components of Fig. 6.3.8 are plotted for up to 300 realizations. Unlike the co-polarized component, a peak in the backscattering direction can be seen. As in the perfect electric conductor case, the second-order backscattering enhancement is more clearly exhibited for the cross-polarized case.

The calculations needed for the Monte Carlo simulations were carried out on a supercomputer located in Hawaii. The CPU time required on the Maui supercomputer depended on the number of nodes. The supercomputer is a parallel computer with 400 nodes. Compared to a serial processing, each realization requires much less clock time when using parallel processing. Thus, the required CPU time depends on the availability of nodes. However, it is informative to note that on the average approximately 25 CPU hours are required for one realization on DEC Alpha workstation. The average number of iterations required for Fig. 6.3.8 are $N_{CGM} = 105$ and $N_{SMCG} = 4$. With $N = 16384$ and $m_{FFT} = 120$, Eq. (6.3.39) gives approximately 2.5×10^{11} computational steps.

A Monte Carlo simulation of the backscattering enhancement is illustrated for both co- and cross-polarization in Fig. 6.3.10 for the case of larger rms height. Surface lengths in the x and y directions are $L_x = L_y = 8\lambda$ ($L_x \times L_y = 64\lambda^2$) with a surface rms height $h = 0.5\lambda$ and correlation lengths of $l_x = l_y = 1.0\lambda$. The surface is sampled at 64 points per λ^2 to give 24,576 surface unknowns. The neighborhood distance $r_d = 3.5\lambda$. The incident angle is $\theta_i = 10°$ and $\phi_i = 0°$. The tapering parameter g is equal to $L_x/2$. The surface parameters of $h = 0.5\lambda$ and $l = 1.0\lambda$ are moderately rough

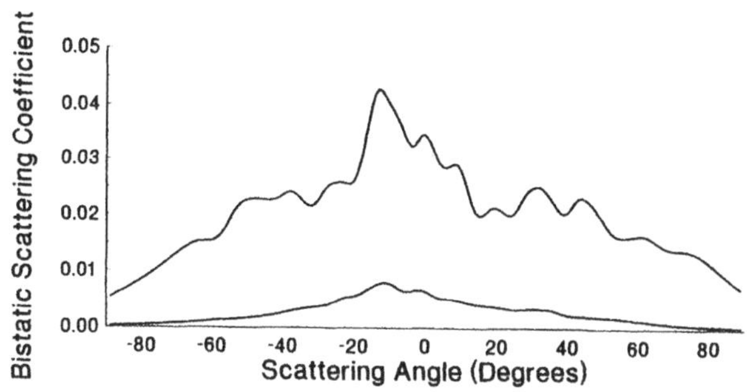

Figure 6.3.10 Bistatic scattering coefficients for co-polarization and cross polarization for $h = 0.5\lambda$, $l_x = l_y = 1.0\lambda$, $L_x = L_y = 16\lambda$, $\epsilon_r = (6.5 + i1.0)$, $r_d = 3.5\lambda$, $\theta_i = 10°$, and $\phi_i = 0°$.

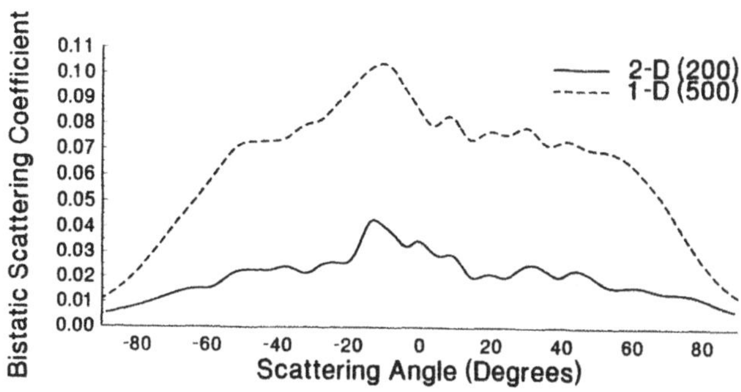

Figure 6.3.11 Backscattering enhancement of two cases: 1-D and 2-D electromagnetic wave incidence on dielectric surface ($\epsilon_r = 6.5 + i1.0$). Comparison of normalized bistatic scattering coefficients for $h = 0.5\lambda$, $l = 1.0\lambda$, $\theta_i = 10°$, and $\phi_i = 0°$.

with an incident angle of 10°. The peaks near $-10°$ are clearly visible. In the simulation, the co-polarized component requires many realizations for the backscattering peak to converge. This is because in co-polarization, the first-order scattering obscures the second-order scattering contribution, which is mostly responsible for the backscattering enhancement peak. The cross-polarized component of the backscattering enhancement peak converges after 50 realization averages. This clearly demonstrates the dominance of the second-order scattering contribution to the cross-polarized result. The average number of iterations for this example are $N_{CGM} = 220$ and $N_{SMCG} = 6$. With $N = 4096$ and $m_{FFT} = 120$, Eq. (6.3.39) gives approximately 1.8×10^{11} computational steps.

In Fig. 6.3.11, the normalized bistatic scattering coefficient in the plane of incidence for the 1-D random rough surface and the 2-D dielectric surface with electromagnetic incident wave are compared. The incidence angle is $\theta_i = 10°$ and $\phi_i = 0°$. In the 1-D surface simulations, the tapering parameter is $g = L/4$, surface length is 30λ, and an ensemble average of 500 realizations are used. For the 2-D dielectric rough surface simulation, the co-polarized result of Fig. 6.3.10 is used. Two curves follow a similar trend for all scattering angles. The backscattering enhancement at near $-10°$ is observed in both co-polarization and cross-polarization.

4 Scattering by Lossy Dielectric Surfaces with PBTG Method

4.1 Introduction

In this section, we study the scattering of electromagnetic waves by lossy dielectric surfaces with large permittivity. This has broad applications in natural media. For wet soil, the relative permittivity can be as high as $17 + i2.0$ at 1.4 GHz. For ocean surfaces, the permittivity can be as high as $39.7 + i40.2$ at 14 GHz. For lossy dielectric rough surfaces with high permittivity, there can be rapid spatial variations of the dielectric medium Green's function and surface fields. Also the wave can propagate from a point on the rough surface to another point on the rough surface through the lossy medium. Based on physical reasons, the surface fields can have large spatial variations within short distances. These large spatial variations can also be attributed to fine scale geometric irregularities on the surface or due to large permittivity of the lower medium. From the point of view of spatial frequency, the high spatial frequency components can interact with the low spatial frequency components. Thus a dense grid is needed to discretize the surface fields.

In Chapter 5, we have developed the physics-based two-grid (PBTG) method. In PBTG, two grids are used: a dense grid and a sparse grid. The sparse grid is that of the usual 8 to 10 points per wavelength. The dense grid ranges from 16 or higher number of points per wavelength depending on the relative permittivity of the lossy dielectric medium. The surface fields are calculated on the dense grid. In the formulation of the surface integral equations, two Green's functions are used. The free space Green's function and the Green's function of the lossy dielectric medium. Although the surface fields have large spatial variations, what need to be calculated are the convolution of the surface fields with the two Green's functions. The PBTG is based on two observations: (1) the Green's function of the lossy dielectric is attenuative (spatial limited), and (2) the Green's function of free-space is slowly varying on the dense grid (spatial frequency limited). The first observation results in a sparse matrix for the Green's function of the lossy dielectric. When this Green's function convolves with the surface fields on the dense grid, it will be just the product of a sparse matrix and a column vector. The second observation allows us, when using the free-space Green's function to convolve with the surface fields of dense grid, to first average the values of surface unknowns on the dense grid and then place them on the coarse grid.

In Chapter 5, the PBTG method was implemented for 1-D surface (2-D scattering problem). In this section, we (i) extend the PBTG to 2-D rough surface (3-D scattering problem), (ii) combine the PBTG method with the sparse matrix canonical grid method (SMCG) for improving CPU and memory requirements, and (iii) study bistatic scattering coefficients and emissivity for wave scattering from 2-D dielectric rough surface with high permittivity. The wave interaction in the rough surface is divided into: (1) very near field, (2) near field, and (3) non-near field. (1) Very near field is of distance of separation less than half a wavelength. (2) Near field separation is between half a wavelength and r_d wavelengths. (3) Non-near field is beyond r_d wavelengths. For very near field interactions, we use the usual product of sparse matrix and column vector. For near-field and non-near field interactions, the free space Green's function is slowly varying on the dense grid. We first average the fields on the dense grid to get fields on the coarse grid. For the non-near field interactions, we further expand free space Green's function on a canonical grid of a horizontal surface so that the fast Fourier Transform (FFT) can be applied. In the lower medium, the non-near field interactions are neglected because of lossy properties of the lower medium. The approach is denoted as PBTG/SMCG. The computational complexity and the memory requirements for the algorithm are $\mathcal{O}(N_{scg} \log(N_{scg}))$ and $\mathcal{O}(N_{scg})$, respectively, where N_{scg} is the number of grid points on the coarse grid. The second-order small perturbation method (SPM) will be studied in Volume III. Also, SPM agrees with the small slope approximation in emissivity calculation for half-space case [Irisov, 1997]. Monte Carlo simulations of emissivities are compared with those of the second-order SPM.

In Section 4.2, the formulation of the problem of wave impinging upon a 2-D dielectric surface (3-D scattering problem) is described, and the surface integral equations are converted into a matrix equation using a single grid discretization. In Section 4.3, we describe the physics-based two-grid algorithm and combine it with the sparse matrix canonical grid method. The mathematical expressions of the bistatic scattering coefficients and the emissivity are given. In Sections 4.4 and 4.5, numerical results are illustrated.

4.2 Formulation and Single Grid Implementation

Consider an electromagnetic wave, $\overline{E}_i(\overline{r})$ and $\overline{H}_i(\overline{r})$, impinging upon a 2-D dielectric rough surface with a random height profile $z = f(x, y)$. It is tapered so that the illuminated rough surface can be confined to the surface area $L_x \times L_y$. The direction of incident wave is $\hat{k}_i = \sin\theta_i \cos\phi_i \hat{x} + \sin\theta_i \sin\phi_i \hat{y} - \cos\theta_i \hat{z}$.

The incident fields for TE wave incidence are given as

$$\overline{E}_i(\overline{r}) = \int_{-\infty}^{+\infty} dk_x \int_{-\infty}^{+\infty} dk_y \exp(ik_x x + ik_y y - ik_z z) E(k_x, k_y) \hat{e}(-k_z) \tag{6.4.1}$$

$$\overline{H}_i(\overline{r}) = -\frac{1}{\eta_1} \int_{-\infty}^{+\infty} dk_x \int_{-\infty}^{+\infty} dk_y \exp(ik_x x + ik_y y - ik_z z) E(k_x, k_y) \hat{h}(-k_z) \tag{6.4.2}$$

where

$$\hat{e}(-k_z) = \frac{1}{k_\rho} (\hat{x}k_y - \hat{y}k_x) \tag{6.4.3}$$

$$\hat{h}(-k_z) = \frac{k_z}{kk_\rho} (\hat{x}k_x + \hat{y}k_y) + \frac{k_\rho}{k}\hat{z} \tag{6.4.4}$$

and for TM wave incidence

$$\overline{E}_i(\overline{r}) = \int_{-\infty}^{+\infty} dk_x \int_{-\infty}^{+\infty} dk_y \exp(ik_x x + ik_y y - ik_z z) E(k_x, k_y) \hat{h}(-k_z) \tag{6.4.5}$$

$$\overline{H}_i(\overline{r}) = \frac{1}{\eta_1} \int_{-\infty}^{+\infty} dk_x \int_{-\infty}^{+\infty} dk_y \exp(ik_x x + ik_y y - ik_z z) E(k_x, k_y) \hat{e}(-k_z) \tag{6.4.6}$$

The spectrum of the incident wave, $E(k_x, k_y)$, is given as

$$E(k_x, k_y) = \frac{1}{4\pi^2} \int_{-\infty}^{\infty} dx \int_{-\infty}^{\infty} dy \exp(-ik_x x - ik_y y) \cdot \exp\left[i\left(k_{ix}x + k_{iy}y\right)(1 + w)\right] \exp(-t) \tag{6.4.7}$$

where $t = t_x + t_y = (x^2 + y^2)/g^2$ and

$$t_x = \frac{(\cos\theta_i \cos\phi_i x + \cos\theta_i \sin\phi_i y)^2}{g^2 \cos^2\theta_i} \tag{6.4.8}$$

$$t_y = \frac{(-\sin\phi_i x + \cos\phi_i y)^2}{g^2} \tag{6.4.9}$$

$$w = \frac{1}{k_1^2}\left(\frac{2t_x - 1}{g^2\cos^2\theta_i} + \frac{2t_y - 1}{g^2}\right) \tag{6.4.10}$$

The six component surface integral equations are given by (6.3.3)–(6.3.6). As in Section 3.1, we also take the x and y components of the tangential field equations and the normal component of the field equation. The method of moments (MoM) is used to discretize the integral equation. The resulting matrix equations are

$$\sum_{n=1}^{N}\left[Z_{mn}^{p1}I_n^{(1)} + Z_{mn}^{p2}I_n^{(2)} + Z_{mn}^{p3}I_n^{(3)} + Z_{mn}^{p4}I_n^{(4)} + Z_{mn}^{p5}I_n^{(5)} + Z_{mn}^{p6}I_n^{(6)}\right] = I_m^{(p)inc} \tag{6.4.11}$$

for $p = 1, 2, 3$ which correspond the surface integral equation when approaching the surface from free space and for $p = 4, 5, 6$ when approaching the surface from the lower medium. The quantities of $I_m^{(p)inc}$ are zero for $p = 4, 5, 6$. Also

$$I_n^{(1)} = F_x(\overline{r}) = S_{xy}(\overline{r}_n)\left[\hat{n}\times\overline{H}(\overline{r}_n)\right]\cdot\hat{x} \tag{6.4.12}$$

$$I_n^{(2)} = F_y(\overline{r}) = S_{xy}(\overline{r}_n)\left[\hat{n}\times\overline{H}(\overline{r}_n)\right]\cdot\hat{y} \tag{6.4.13}$$

$$I_n^{(3)} = I_n(\overline{r}) = S_{xy}(\overline{r}_n)\hat{n}\cdot\overline{E}(\overline{r}_n) \tag{6.4.14}$$

$$I_n^{(4)} = I_x(\overline{r}) = S_{xy}(\overline{r}_n)\left[\hat{n}\times\overline{E}(\overline{r}_n)\right]\cdot\hat{x} \tag{6.4.15}$$

$$I_n^{(5)} = I_y(\overline{r}) = S_{xy}(\overline{r}_n)\left[\hat{n}\times\overline{E}(\overline{r}_n)\right]\cdot\hat{y} \tag{6.4.16}$$

$$I_n^{(6)} = F_n(\overline{r}) = S_{xy}(r_n)\hat{n}\cdot\overline{H}(\overline{r}_n) \tag{6.4.17}$$

are surface unknowns and $S_{xy} = \left\{1 + \left[\frac{\partial f(x,y)}{\partial x}\right]^2 + \left[\frac{\partial f(x,y)}{\partial y}\right]^2\right\}^{1/2}$. The Z_{mn}^{pq} are the impedance matrix elements and are determined by the free space Green's function and the dielectric medium Green's function. The parameter N is the number of points we use to discretize the rough surface.

4.3 Physics-Based Two-Grid Method

In this section, we describe the physics-based two-grid method. We assume that the upper medium is the free space and the lower medium is lossy with large permittivity.

$$\epsilon_2 = \epsilon_2'(1 + i\tan\delta) \tag{6.4.18}$$

where $\tan\delta$ stands for loss tangent. Let λ_1 and λ_2 represent the wavelength of the wave in the free space and the lower medium, respectively, and

$$n_2 = \text{integer}\left(\sqrt{\epsilon_2'}\right) \tag{6.4.19}$$

Then, the relationship between λ_1 and λ_2 can be expressed approximately by

$$\lambda_2 \approx \frac{\lambda_1}{n_2} \tag{6.4.20}$$

The number of sampling points needed in the lower medium should be n_2 times that of the free space.

In the physics-based two-grid method, we use two grids with samplings per wavelength of n_{scg} (coarse grid) and n_{sdg} (dense grid), respectively. Let N_{sdg} and N_{scg} be respectively the total number of points on the dense grid and the coarse grid.

$$N_{sdg} = \left(n_{sdg}\frac{L_x}{\lambda_1}\right)\left(n_{sdg}\frac{L_y}{\lambda_1}\right) \tag{6.4.21}$$

$$N_{scg} = \left(n_{scg}\frac{L_x}{\lambda_1}\right)\left(n_{scg}\frac{L_y}{\lambda_1}\right) \tag{6.4.22}$$

For example $n_{scg} = 8$ and $n_{sdg} = 8n_2$. We first re-write equation (20) using the dense grid.

$$\sum_{n=1}^{N_{sdg}}\left[Z_{mn}^{p1}I_n^{(1)} + Z_{mn}^{p2}I_n^{(2)} + Z_{mn}^{p3}I_n^{(3)} + Z_{mn}^{p4}I_n^{(4)} + Z_{mn}^{p5}I_n^{(5)} + Z_{mn}^{p6}I_n^{(6)}\right] = I_m^{(p)inc} \tag{6.4.23}$$

The Roman numeral subscripts m, n denote indexing with the dense grid. In Fig. 6.4.1, we plot the real part of the products of distance and the two Green's functions. We make the following three observations:

(1) The Green's function in the lower region is heavily attenuative. Let k_2'' be the imaginary part of k_2. If $k_2''r > C$, where C is a constant, then the field interaction between the mth and the nth point is vanishingly small. We can define a distance limit as dictated by dissipative loss:

$$r_l = \frac{C}{k_2''} \tag{6.4.24}$$

outside of which the lower medium Green's function can be set equal to zero. Based on comparisons with the results from SMCG, C is fixed at 1.5.

Based on this observation, we calculate the left-hand sides of (6.4.23) for as follows by approximating

$$Z_{mn}^{pq} \approx \tilde{Z}_{mn}^{pq} = \begin{cases} Z_{mn}^{pq} & r_{mn} \le r_l \\ 0 & r_{mn} \ge r_l \end{cases} \tag{6.4.25}$$

where r_{mn} is the distance between the mth point and the nth point on the dense grid. Thus $\tilde{Z}_{mn}^{pq}$ $(p = 4, 5, 6)$ are sparse matrices and Eq. (6.4.23) for

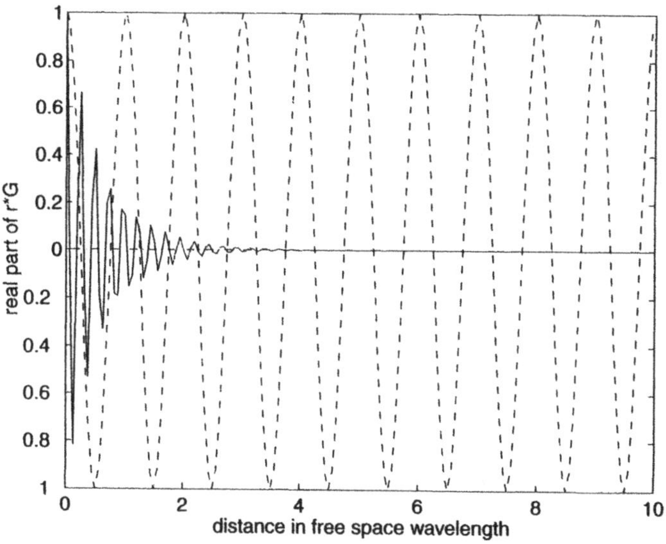

Figure 6.4.1 The variation of rG of free space (dash line) and lossy medium (solid line) with relative permittivity of $17 + i2.0$ as a function of distance.

$p = 4, 5, 6$ becomes

$$\sum_{n=1}^{N_{sdg}} \left[\widetilde{Z}^{p1}_{mn} I^{(1)}_n + \widetilde{Z}^{p2}_{mn} I^{(2)}_n + \widetilde{Z}^{p3}_{mn} I^{(3)}_n + \widetilde{Z}^{p4}_{mn} I^{(4)}_n + \widetilde{Z}^{p5}_{mn} I^{(5)}_n + \widetilde{Z}^{p6}_{mn} I^{(6)}_n \right] = I^{(p)inc}_m \tag{6.4.26}$$

(2) For non-near field interaction, Green's function for the upper medium is slowly varying on the dense grid. Thus when performing matrix and column vector multiplication on the dense grid as indicated in (6.4.23), the Green's function of the upper medium is essentially constant over an area of $n_2 \times n_2$ points on the dense grid. Thus we can write

$$\sum_{l=1}^{n_2^2} Z^{pq}_{(m+l')(n+l)} I^{(q)}_{n+l} \approx Z^{pq}_{m_{mp} n_{mp}} \sum_{l=1}^{n_2^2} I^{(q)}_{n+l} = n_2^2 Z^{pq}_{m_{mp} n_{mp}} \left(\frac{1}{n_2^2} \sum_{l=1}^{n_2^2} I^{(q)}_{n+l} \right) \tag{6.4.27}$$

where $l' = 1, 2, \ldots, n_2^2$ and the points with indexes m_{mp} and n_{mp} are the central point of the n_2^2 dense grid points of $m+1,\ m+2,\ \ldots,\ m+n_2^2$ and $n+1,\ n+2,\ \ldots,\ n+n_2^2$, respectively. What is performed in (6.4.27) is that the surface fields on the dense grid are first averaged before multiplied by the upper medium Green's function.

(3) The slowly varying nature of Green's function of the upper medium

only applies to non-near field interaction. For near field interaction, Green's functions G_1 and G_2 have similar rate of variations. Thus we separate out a distance, say 1λ, outside of which G_2 is much more rapidly varying spatially than G_1.

Based on the observations above, we decompose the upper medium Green's function into near field and non-near field interactions

$$\sum_{n=1}^{N_{sdg}} Z_{mn}^{pq} I_n^{(q)} = \sum_{n=1}^{N_{sdg}} Z_{mn}^{pq(s)} I_n^{(q)} + \sum_{n=1}^{N_{sdg}} Z_{mn}^{pq(ns)} I_n^{(q)} \tag{6.4.28}$$

where $Z_{mn}^{pq(s)}$ and $Z_{mn}^{pq(ns)}$ are determined by

$$Z_{mn}^{pq(s)} = \begin{cases} Z_{mn}^{pq} & r_{mn} \leq r_f \\ 0 & r_{mn} \geq r_f \end{cases} \tag{6.4.29}$$

$$Z_{mn}^{pq(ns)} = \begin{cases} 0 & r_{mn} \leq r_f \\ Z_{mn}^{pq} & r_{mn} \geq r_f \end{cases} \tag{6.4.30}$$

Thus r_f is the distance outside which the Green's function of the lower medium is fast varying compared with that of free space Green's function.

Let $\tilde{m}$ and $\tilde{n}$ denote the coarse grid indices. The coarse grid has surface unknowns $\widetilde{I}_{\tilde{n}}^{(q)}$, which are averages of the dense grid surface unknowns. Thus if $\bar{r}_{\tilde{n}}$ is centered in the group of the n_2^2 dense grid points of $n+1$, $n+2$, ..., $n+n^2$, we have

$$\widetilde{I}_{\tilde{n}}^{(q)} = \frac{I_{n+1}^{(q)} + I_{n+2}^{(q)} + \cdots + I_{n+n_2^2}^{(q)}}{n_2^2} \tag{6.4.31}$$

The Green's function of the upper medium on the coarse grid is represented by $\widetilde{Z}_{\tilde{m}\tilde{n}}^{pq}$. Then Eq. (6.4.23) for $p = 1, 2, 3$ becomes

$$\sum_{q=1}^{6}\sum_{n=1}^{N_{sdg}} Z_{mn}^{pq(s)} I_n^{(q)} + \sum_{q=1}^{6}\left[\sum_{\tilde{n}=1}^{N_{scg}} \widetilde{Z}_{\tilde{m}\tilde{n}}^{pq(ns)} \widetilde{I}_{\tilde{n}}^{(q)}\right]_{intp} = I_m^{(p)inc} \tag{6.4.32}$$

Note in Eq. (6.4.32) that $\sum_{n=1}^{N_{sdg}} Z_{mn}^{pg(s)} I_n^{(q)}$ includes N_{sdg} values of $m = 1, 2, \ldots, N_{sdg}$ on the dense grid, while $\sum_{\tilde{n}=1}^{N_{scg}} \widetilde{Z}_{\tilde{m}\tilde{n}}^{pq(ns)} \widetilde{I}_{\tilde{n}}^{(q)}$ only has values of $\tilde{m} = 1, 2, \ldots, N_{scg}$ on the coarse grid. Thus we first compute $\sum_{\tilde{n}=1}^{N_{scg}} \widetilde{Z}_{\tilde{m}\tilde{n}}^{pq(ns)}$ $\widetilde{I}_{\tilde{n}}^{(q)}$. Then we use linear interpolation of $\sum_{\tilde{n}=1}^{N_{scg}} \widetilde{Z}_{\tilde{m}\tilde{n}}^{pq(ns)} \widetilde{I}_{\tilde{n}}^{(q)}$ on the coarse grid to find N_{sdg} values on the dense grid. In Eq. (6.4.32), we use subscript *intp* to represent that interpolation. Thus the computational steps for matrix-vector multiplication are associated with the number of surface unknowns on the coarse grid. The algorithm is described pictorially in Fig. 6.4.2. The PBTG

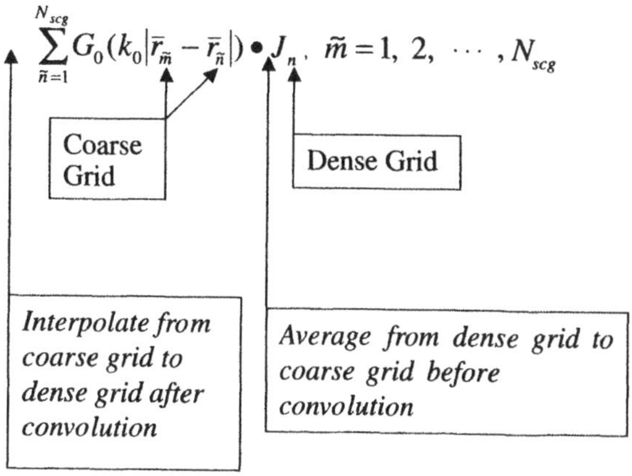

Figure 6.4.2 Illustration of convolution of upper medium Green's function with surface fields on dense grid.

is also used in conjunction with the SMCG. The computational complexity of the combined algorithm of PBTG/SMCG is $\mathcal{O}(N_{scg}\log(N_{scg}))$.

The numerical simulation results are presented in terms of the bistatic scattering coefficients normalized by the incident power. For an incident wave with a polarization β, we have

$$\gamma_{\alpha\beta}(\theta_s,\phi_s;\,\theta_i,\varphi_i)=\frac{|E_\alpha^s|^2}{2\eta_1 P_\beta^{inc}} \tag{6.4.33}$$

The incident power is

$$P_\beta^{inc}=\frac{2\pi^2}{\eta}\int_{k_\rho<k} dk_x dk_y\,|E(k_x,k_y)|^2\,\frac{k_z}{k} \tag{6.4.34}$$

where $k_\rho=\sqrt{k_x^2+k_y^2}$.

The horizontal and vertical polarized scattered components of E_α^s are, respectively, given by (6.3.46) and (6.3.47). The emissivity of the rough surface at incident angle (θ_i,ϕ_i) is

$$e_\beta(\theta_i,\phi_i)=1-\frac{1}{4\pi}\iint\left[\gamma_{h\beta}(\theta_s,\phi_s;\,\theta_i,\varphi_i)+\gamma_{v\beta}(\theta_s,\phi_s;\,\theta_i,\varphi_i)\right]\sin\theta_s d\theta_s d\phi_s \tag{6.4.35}$$

4.4 Numerical Results and Comparison with Second Order Perturbation Method

In this section, we illustrate the numerical simulation results of wave scattering from 2-D lossy dielectric rough surface. Simulations are based on Gaussian random rough surfaces with Gaussian correlation functions. All the cases are computed with the relative dielectric constants of $17 + i2$ and $4.06 + i0.3$, surface area of 8 free space wavelengths by 8 free space wavelengths, rms height from 0.05 to 0.3 wavelengths, and correlation length of 1.0 wavelength.

The results of three methods will be shown: (i) single coarse grid of 8 points per free space wavelength with SMCG (SCG/SMCG), (ii) single dense gird of 16 points per free space wavelength with SMCG (SDG/SMCG), and (iii) PBTG/SMCG. Based on the experience of 1-D dielectric rough surfaces, the sampling density of 16 points per wavelength for the permittivity of $17 + i2.0$ gives convergent results. Thus we regard (ii) as accurate results. We will show that (iii) is as accurate as (ii) and takes much less CPU time. First, we compare the bistatic scattering coefficients for one realization based on different methods. Next, the emissivities and CPU are compared. The results show that the dense grid is required for the case with large dielectric constant. Then the variation of brightness temperatures with observation angles is illustrated and compared with the second order small perturbation method. Finally, the brightness temperature as a function of rms height is shown for different observation angles. We use $T = 300\,\mathrm{K}$ as the physical temperature. All the numerical results were computed on a DEC Alpha workstation.

(A) Comparisons of Bistatic Scattering Coefficients Computed by Various Methods

In Figs. 6.4.3a and 6.4.3b, the comparisons of bistatic scattering coefficients in the plane of incidence $\phi_i = 0°$, $\phi_s = 0°$ and $180°$ of a single realization of rough surface obtained by the SCG/SMCG, SDG/SMCG, and PBTG/SMCG are shown. Figure 6.4.3a is for co-polarization and 6.4.3b for cross-polarization. The incidence wave is TE wave with the incidence angle of 10 degrees, and the rms height is 0.3 free space wavelength. The relative permittivity of lower medium is $17+i2$. There are some small differences among three results. That means it is not strictly necessary to use dense grid for the calculation of the bistatic scattering coefficients for this case of TE incidence. But the PBTG/SMCG can give better results than the SCG/SMCG. In Figs. 6.4.4a and 6.4.4b, the results are shown for TM wave incidence. It

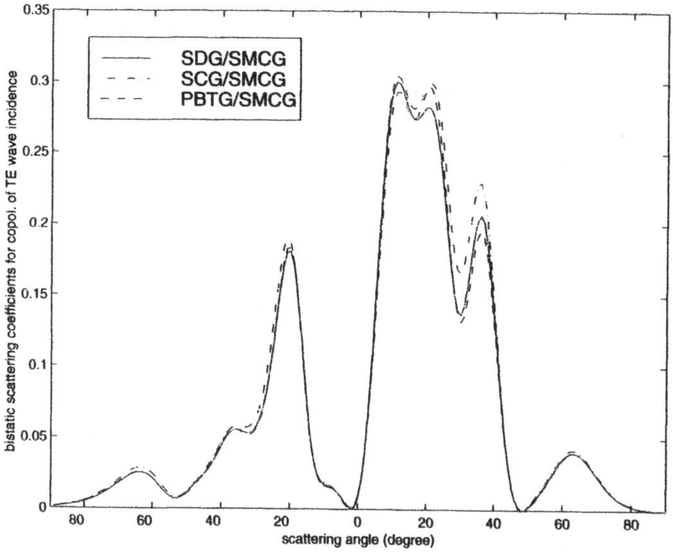

(a)

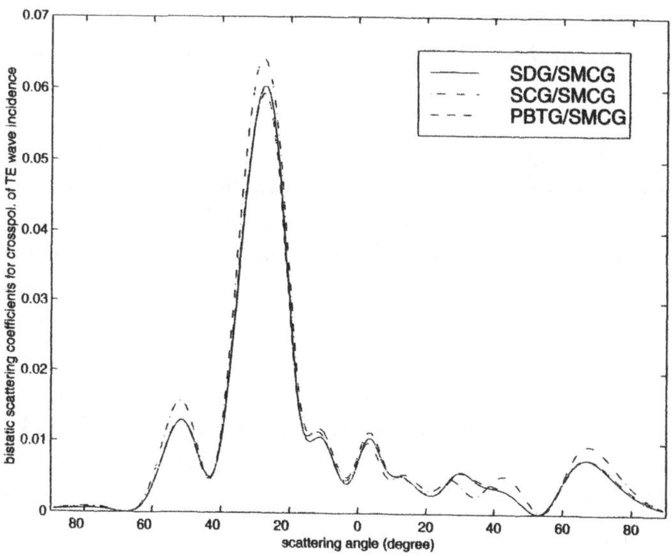

(b)

Figure 6.4.3 Comparison of the bistatic scattering coefficients between the SDG/SMCG, SCG/SMCG, and PBTG/SMCG for the TE wave incidence. The case is with rms height of 0.3 wavelengths, correlation lengths of 1 wavelength, surface lengths of 8 by 8 wavelengths, and relative permittivity of $17 + 2i$ at incidence angle of 10 degrees. a) co-polarization b) cross-polarization.

Method	Incidence Wave	CPU time (hours)	Emissivity	Difference of emissivity*	Difference of Temperature
SDG/SMCG	TE	45.05	0.6612		
SCG/SMCG	TE	9.11	0.6165	−0.044	13.4
PBTG/SMCG	TE	8.48	0.6645	0.0033	0.99
SDG/SMCG	TM	48.83	0.6690		
SCG/SMCG	TM	10.16	0.6224	−0.046	13.97
PBTG/SMCG	TM	8.68	0.6729	0.0039	1.17

* The difference of emissivity means the emissivity minus the emissivity of corresponding SDG/MCG.

Table 6.4.1 Comparison of emissivities and CPU based on different methods.

is seen that the coarse grid leads to a larger error for TM wave than for TE wave. Comparing the results of Figs. 6.4.3 and 6.4.4 indicates that the PBTG/SMCG can give better results than the SCG/SMCG particularly for cross-polarization component.

(B) Comparisons of Emissivities and CPU Requirements from Various Methods

In Table 6.4.1, the emissivities obtained by various methods are shown for one realization. The input parameters of rough surface are the same as before. The difference of emissivities between the SDG/SMCG and the SCG/SMCG for TE wave incidence is 0.0447. It will lead to a difference of 13.41 K in brightness temperature and is unacceptable in passive remote sensing applications. The emissivities obtained by the PBTG/SMCG are also shown. The difference between the SDG/SMCG and the PBTG/SMCG is only 0.003316. That will give a small difference of 0.99 K in brightness temperature. The emissivities for TM wave incidence are also shown in Table 6.4.1. It is seen that the PBTG/SMCG can give almost the same results as the SDG/SMCG while the SCG/SMCG cannot. The CPU requirements for various methods are also shown in Table 6.4.1. It is clear that the SDG/SMCG requires the most CPU. On the other hand the PBTG/SMCG is five times faster than the SDG/SMCG and takes even less CPU than the SCG/SMCG. The fact that the PBTG/SMCG requires less CPU than SCG/SMCG is because the former requires less number of conjugate gradient iterations. Thus PBTG/SMCG can obtain the accurate results and require much less CPU than that of the single dense grid.

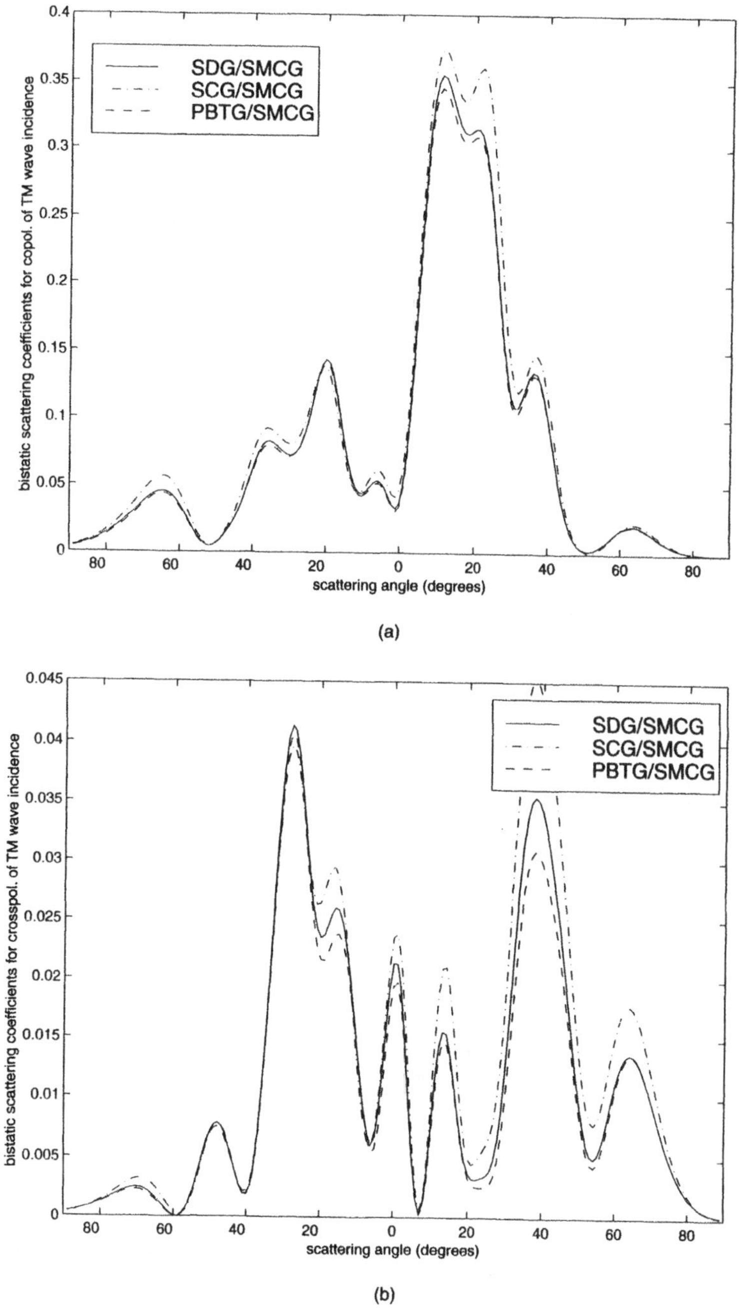

Figure 6.4.4 Same as Fig. 6.4.3 except for TM wave incidence.

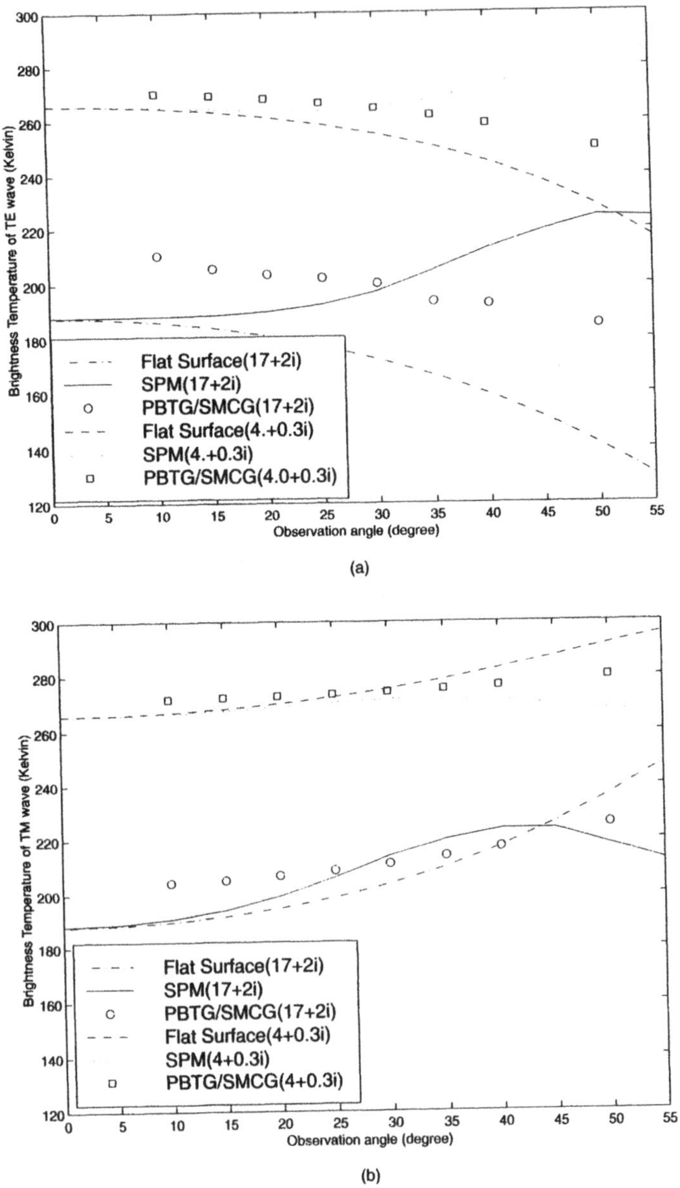

Figure 6.4.5 The brightness temperature of Monte Carlo simulation averaged over 5 realizations as a function of observation angles and comparisons with that from the second order small perturbation method and flat surface. The case is with rms height of 0.3 wavelengths, correlation lengths of 1.0 wavelength, relative permittivities of $17 + i2.0$ and $4.06 + i0.3$, and physical temperature of 300 K. (a) TE wave (b) TM wave.

(C) Variation of Brightness Temperature with Observation Angles and Comparison with Results from the Second Order Small Perturbation Method

The brightness temperature results based on averaging over 5 realizations are shown in Figs. 6.4.5a and 6.4.5b for horizontal and vertical polarization, respectively. The rough surface is with rms height of 0.3 wavelengths and correlation length of 1.0 wavelength. The observation angles are varied from 10 degrees to 50 degrees. The relative permittivities is $17.0 + i2.0$. We note that in the simulation of emissivity in passive remote sensing, only a small number of realizations are required. This is because in passive remote sensing an integration of scattered angles is used and that has built in smoothing. For the case of permittivity of $17 + i2$ at the observation angle of 10 degrees, the horizontal emissivity averaged over 10 realizations is 0.699 and is 0.701 for averaging over 5 realizations. The difference between them is 0.002. That means that averaging over 5 realizations can give accurate results. Figure 6.4.5a is for TE wave and 6.4.5b for TM wave. The brightness temperature shown in solid line is from the second order small perturbation method (SPM), in dash-dot line is from flat surface, and in circle is from numerical simulation results. It is shown that surface roughness increases the brightness temperature over flat surface for horizontal polarization. It can increase or decrease the brightness temperature for vertical polarization depending on observation angles. It is well-known that the SPM cannot give the correct results of emissivities for moderate to large rms slope. For TE wave, the numerical results show that the brightness temperature decreases with observation angles. On the other hand the SPM results show that brightness temperature increases with observation angles. The brightness temperature for the permittivity of $4.06 + i0.3$ is also plotted in the figures. Similar features are exhibited in this case. It can be seen that surface roughness has a larger influence in brightness temperature for large permittivity than for the small permittivity. For horizontal polarization, more energy is reflected with the increase of incident angles. Thus the brightness temperature decreases with the observation angles. For vertical polarization, more energy is transmitted into lower medium with the increase of incident angles if angle is less than the Brewster angle. Thus the brightness temperature increases with the observation angles.

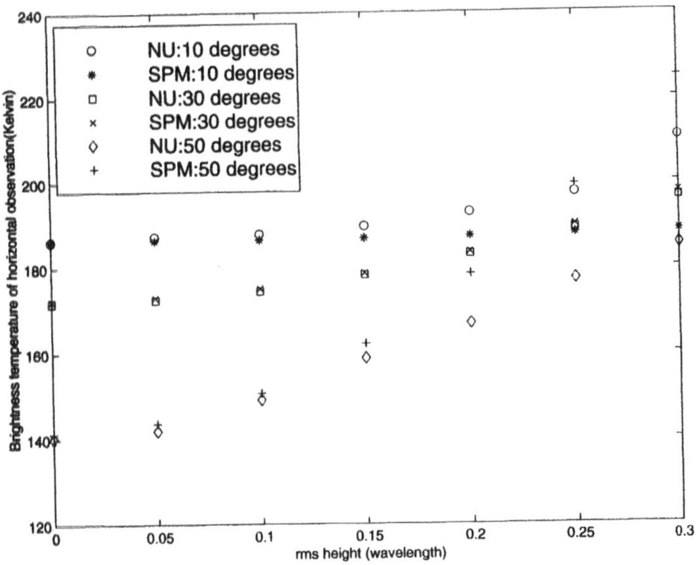

(a)

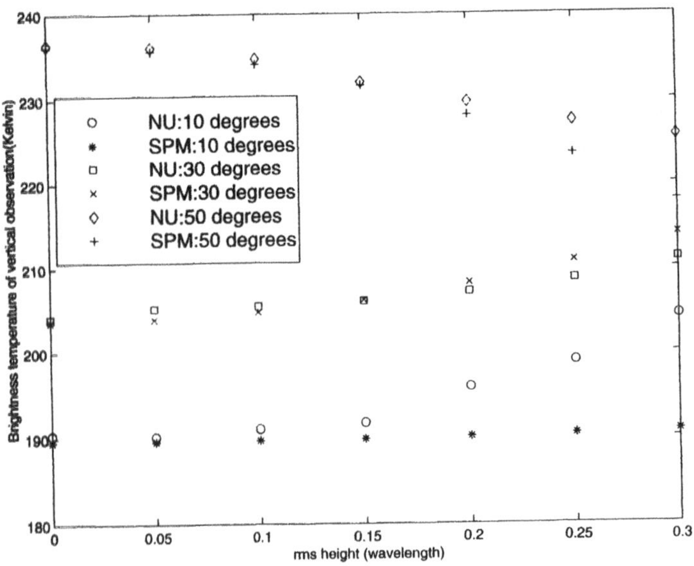

(b)

Figure 6.4.6 The brightness temperature of Monte Carlo simulation averaged over 5 realizations as a function of rms heights and comparisons with that from the second order small perturbation method. The rms height of zero means flat surface. The case is with correlation lengths of 1.0 wavelength, relative permittivity of $17 + 2i$, and physical temperature of 300 K at observation angles of 10, 30, and 50 degrees. a) TE wave b) TM wave.

(D) Variation of Brightness Temperatures with rms Height and Comparison with that from the Second Order Small Perturbation Method

The brightness temperatures as functions of rms height are plotted in Fig. 6.4.6a for horizontal polarization and in Fig. 6.4.6b for vertical polarization for the observation angles of 10, 30, and 50 degrees, respectively. The correlation length is 1 wavelength and the permittivity is $17 + 2i$. We also show the results from the second order SPM. The numerical results are averaged over 5 realizations. For the small rms height, the results of simulations and SPM are in good agreement. It also illustrates that the numerical algorithm can give the correct emissivity calculations. For the case of flat surface, one needs to use many angles in integrating near the specular direction to give correct emissivity. With the increase of rms height, the differences between them get large, especially for the observation angles of 10 and 50 degrees. Because the numerical results and SPM results cross each other around the observation angle of 30 degrees as shown in Fig. 6.4.5, the differences between them is small. For the rms height of 0.3 wavelength, the difference between SPM and numerical simulation in horizontal polarization can be as large as 30 K in the brightness temperatures at the observation angle of 50 degrees. Another feature shown in the figures is that the surface roughness increases the brightness temperature for all the cases except for the vertical polarization at observation angle of 50 degrees. The reason is that the observation angle of 50 degrees is close to the Brewster angle. At the Brewster angle, the emissivity of flat surface is the maximum for vertical polarization. Thus surface roughness will lead to a decrease in the emissivity near the Brewster angle for vertical polarization.

(E) Comparisons with Empirical Formula

In passive remote sensing with soil, an empirical formula, which has been used for many years, is as follows [Wang et al. 1983; Njoku and Li, 1999; Jackson, 2001]

$$T_{B\nu}(\theta_i, \varphi_i) = \left\{1 - \left[r_{\nu 0}(\theta_i, \varphi_i)(1 - Q) + r_{h0}(\theta_i, \varphi_i)Q\right] e^{-H\cos^2\theta_i}\right\} T \quad (6.4.36)$$

$$T_{Bh}(\theta_i, \varphi_i) = \left\{1 - \left[r_{h0}(\theta_i, \varphi_i)(1 - Q) + r_{\nu 0}(\theta_i, \varphi_i)Q\right] e^{-H\cos^2\theta_i}\right\} T \quad (6.4.37)$$

where $T_{B\nu}$ and T_{Bh} are brightness temperatures for vertical and horizontal polarizations, respectively, T is physical temperature of dielectric medium, and Q and H are empirical constants that are used to fit the data. The parameters $r_{\nu 0}$ and r_{h0} are the flat surface Fresnel reflectivity of vertical

Observation angle(degrees)		Flat Surface	SPM	PBTG/SMCG	Q^*	H^*
10	TE	186.2	188.4	210.4	1.5997	0.1971
	TM	189.6	190.7	204.4		
15	TE	184.1	188.9	205.7	0.5362	0.1825
	TM	191.7	194.3	205.2		
20	TE	181.1	190.1	203.7	0.3592	0.1907
	TM	194.7	199.3	206.9		
25	TE	177.2	192.8	202.3	0.3198	0.2081
	TM	198.7	205.8	208.8		
30	TE	172.3	197.7	200.2	0.2943	0.2284
	TM	203.7	213.9	211.1		
35	TE	166.3	204.5	193.6	0.2287	0.2242
	TM	209.8	221.1	213.9		
40	TE	159.2	212.3	192.7	0.2519	0.2759
	TM	217.3	224.4	217.2		
50	TE	141.1	225.3	185.1	0.2497	0.3927
	TM	236.3	218.2	225.7		

* Parameters Q and H are computed from the brightness temperatures from the numerical simulation results.

Table 6.4.2 The brightness temperatures of Fig. 6.4.5.

and horizontal polarized waves, respectively. The empirical parameter Q represents coupling between polarizations. On the other hand, the factor $\exp(-H\cos^2\theta_i)$ represents depletion of reflectivity. Both Q and H are assumed independent of observation angles and soil moisture.

In Table 6.4.2, we tabulate the brightness temperatures of Fig. 6.4.5 based on numerical simulations for the permittivity of $17 + i2.0$ and also use the simulated results to compute Q and H parameters from (6.4.36) and (6.4.37). The table shows that Q and H are actually functions of observation angles. In the original proposal [Wang et al. 1983], H is supposed to be $4k^2h^2$ where h is the rms height. For this case, $4k^2h^2 = 14.19$. From Table 6.4.2, H is much smaller than 14.19. The table shows that Q and H are empirical parameters that do not have physical meaning.

4.5 Numerical Simulations of Emissivity of Soils with Rough Surfaces at Microwave Frequencies

In this section, we illustrate the numerical simulation results of emissivities of wet soils with 2-D rough surface (3-D problem). A key result is that emissivities are calculated for soils at two frequencies using the same values of physical roughness parameters of rms height and correlation length in centimeters. The simulations have important applications in microwave remote sensing of soils. In practice, active and passive sensors are used at different frequencies and incident angles. The same physical roughness parameters are independent of sensor characteristics and should be used to characterize the soil surface. Simulations in this section are based on Gaussian random rough surfaces with Gaussian correlation functions. In the numerical results, the surface area used is 64 square wavelengths. We use 64 points per square wavelength as coarse grid and 256 points per square wavelength as dense grid, which leads to a total of 98,304 surface unknowns. For the rough surfaces with rms height of 0.3 wavelength, correlation length of 1.0 wavelength, and relative permittivity of $17 + 2i$, the total CPUs for the single dense grid with SMCG are about 45 hours for TE wave and 49 hours for TM wave, respectively, on DEC 3000/700 workstation. By using the PBTG with SMCG, the CPUs are about 8.5 hours for TE and 9.0 hours for TM, respectively, for the same case. The permittivities of wet soil by weight of the soil moisture at L-band (1.4 GHz) are listed in Table 6.4.3. The permittivities of soil by the soil moisture in volumetric water contents (cm^3/cm^3) for L- and C-bands are given in Table 6.4.4. C-band is at 5 GHz. Permittivities are based on the results of [Wang and Schmugge, 1980] and are also given in Eq. (5.4.1) of Volume I. All the cases are computed at the observation angle of 50 degrees, unless otherwise indicated. In Tables 6.4.5 to 6.4.9, emissivities are tabulated for ease of comparisons by interested readers. We also illustrate the brightness temperatures in Figs. 6.4.7 to 6.4.11. The physical temperature is taken as 300 K for all cases.

(A) Variation of Brightness Temperatures at L-Band of Wet Soil with the Soil Moisture and Correlation length at Fixed rms Height

In Figs. 6.4.7 and 6.4.8, we plot the brightness temperatures as a function of soil moisture by weight at L-band for horizontal and vertical polarizations, respectively. The corresponding permittivities are taken from Table 6.4.3. The rms height is fixed at 0.1 wavelength and the correlation lengths are varied from 0.33 to 1.0 wavelengths. It is shown that brightness tem-

Moisture (%)	5	7.5	10.0	12.5	15.0	17.5	20.0	22.5	15.0
Permittivity	4.06+ i0.300	4.81+ i0.45	5.56+ i0.600	7.35+ i0.872	10.8+ i1.335	14.25+ i1.797	17.7+ i2.260	21.15+ i2.722	24.6+ i3.185

Table 6.4.3 Relative permittivities of soil moistures in weight at L-band.

Moisture	0.05	0.10	0.15	0.20	0.25	0.30	0.35	0.40
Permittivity at L-band	3.66+ i0.15	4.66+ i0.29	6.26+ i0.52	8.45+ i0.85	11.3+ i1.27	15.2+ i1.82	19.2+ i2.41	23.1+ i3.04
Permittivity at C-band	3.62+ i0.19	4.52+ i0.44	5.94+ i0.85	7.90+ i1.42	10.4+ i2.17	13.9+ i3.22	17.4+ i4.30	20.8+ i5.30

Table 6.4.4 Relative permittivities of soil moistures in volumetric water content (cm^3/cm^3) at L- and C-bands.

Moisture	Correlation Length (in wavelengths) horizontal polarization							
(%)	0.33	0.4	0.5	0.6	0.7	0.8	0.9	1.0
5.00	0.8060	0.8011	0.7918	0.7841	0.7777	0.7720	0.7671	0.7636
7.50	0.7758	0.7693	0.7591	0.7514	0.7442	0.7377	0.7330	0.7286
10.0	0.7464	0.7393	0.7290	0.7205	0.7132	0.7072	0.7017	0.6973
12.5	0.7092	0.6964	0.6855	0.6743	0.6676	0.6614	0.6569	0.6520
15.0	0.6385	0.6236	0.6088	0.5990	0.5919	0.5862	0.5758	0.5766
17.5	0.5819	0.5668	0.5504	0.5393	0.5365	0.5248	0.5193	0.5149
20.0	0.5453	0.5309	0.5149	0.5042	0.4961	0.4896	0.4840	0.4792
22.5	0.5220	0.5077	0.4925	0.4827	0.4755	0.4695	0.4642	0.4596
25.0	0.4986	0.4826	0.4678	0.4577	0.4508	0.4454	0.4409	0.4366

Table 6.4.5 Computed soil emissivities at L-band. Horizontal polarization; rms height = 0.1 wavelength, observation angle = 50 degrees.

Moisture	Correlation Length (in wavelengths) vertical polarization							
(%)	0.33	0.4	0.5	0.6	0.7	0.8	0.9	1.0
5.00	0.9565	0.9545	0.9541	0.9550	0.9563	0.9578	0.9592	0.9603
7.50	0.9437	0.9410	0.9400	0.9408	0.9423	0.9439	0.9454	0.9468
10.0	0.9308	0.9276	0.9261	0.9268	0.9284	0.9302	0.9318	0.9333
12.5	0.9081	0.9024	0.8995	0.8995	0.9006	0.9018	0.9031	0.9043
15.0	0.8568	0.8508	0.8469	0.8468	0.8480	0.8493	0.8510	0.8524
17.5	0.8202	0.8104	0.8036	0.8041	0.8054	0.8073	0.8091	0.8107
20.0	0.7841	0.7744	0.7684	0.7655	0.7696	0.7716	0.7734	0.7714
22.5	0.7535	0.7412	0.7361	0.7363	0.7382	0.7402	0.7423	0.7440
25.0	0.7272	0.7123	0.7048	0.7073	0.7092	0.7113	0.7134	0.7130

Table 6.4.6 Computed soil emissivities at L-band. Vertical polarization; rms height = 0.1 wavelength, observation angle = 50 degrees.

Emissivity	Soil Moisture in Weight (%)								
	5.00	7.50	10.0	12.5	15.0	17.5	20.0	22.5	25.0
h (case 1)	0.8060	0.7758	0.7464	0.7092	0.6385	0.5819	0.5453	0.5220	0.4986
h (case 2)	0.8392	0.8053	0.7946	0.7519	0.6868	0.6376	0.6054	0.5804	0.5535
h (case 3)	0.8107	0.7820	0.7686	0.7252	0.6593	0.6110	0.5802	0.5653	0.5333
v (case 1)	0.9565	0.9437	0.9308	0.9081	0.8568	0.8202	0.7841	0.7535	0.7272
v (case 2)	0.9289	0.9123	0.8990	0.8667	0.8143	0.7698	0.7345	0.7064	0.6806
v (case 3)	0.9302	0.9154	0.9052	0.8758	0.8266	0.7840	0.7503	0.7227	0.6961

Case 1: rms height = 0.1 wavelength, correlation length = 0.333 wavelength, observation angle = 50 degrees
Case 2: rms height = 0.3 wavelength, correlation length = 1.000 wavelength, observation angle = 50 degrees
Case 3: rms height = 0.3 wavelength, correlation length = 1.000 wavelength, observation angle = 55 degrees

Table 6.4.7 Computed soil emissivities for different soil moistures at L-band.

Emissivity	Volumetric Water Content (cm^3/cm^3)							
	0.05	0.10	0.15	0.20	0.25	0.30	0.35	0.40
h (L-band)	0.8434	0.8008	0.7490	0.6996	0.6446	0.5868	0.5553	0.5278
h (C-band)	0.8626	0.8285	0.7900	0.7613	0.7160	0.6658	0.6313	0.6032
v (L-band)	0.9615	0.9444	0.9181	0.8817	0.8424	0.7995	0.7593	0.7256
v (C-band)	0.9393	0.9178	0.8867	0.8441	0.8045	0.7630	0.7284	0.6989

rms height = 2.45 cm, correlation length = 8.0 cm, observation angle = 50 degrees;
L-band: f = 1.4 GHz; C-band: f = 5.0 GHz.

Table 6.4.8 Computed soil emissivities for different volumetric water contents.

Emissivity	Volumetric Water Content (cm^3/cm^3)							
	0.05	0.10	0.15	0.20	0.25	0.30	0.35	0.40
h (L-band)	0.7807	0.7258	0.6684	0.6198	0.5597	0.4915	0.4560	0.4354
h (C-band)	0.8202	0.7793	0.7290	0.6778	0.6234	0.5656	0.5272	0.4990
v (L-band)	0.9854	0.9747	0.9503	0.9132	0.8766	0.8426	0.8042	0.7633
v (C-band)	0.9604	0.9430	0.9169	0.8820	0.8427	0.7968	0.7581	0.7254

rms height = 0.73 cm, correlation length = 3.5 cm, observation angle = 50 degrees;
L-band: f = 1.4 GHz; C-band: f = 5.0 GHz.

Table 6.4.9 Computed soil emissivities for different volumetric water contents.

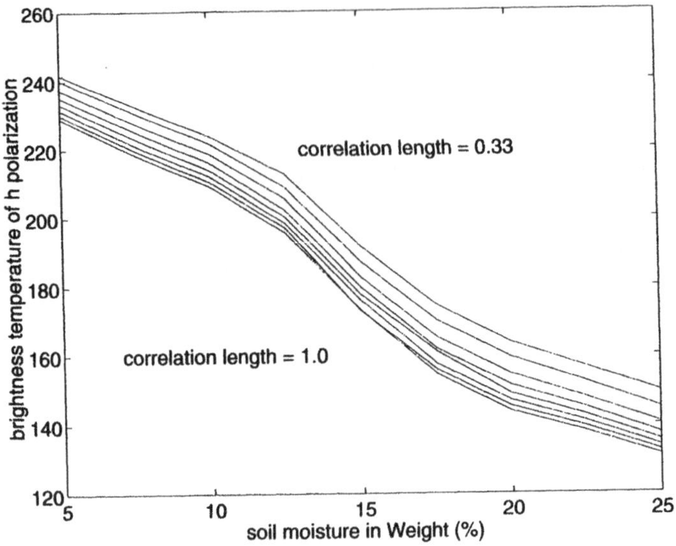

Figure 6.4.7 The variation of brightness temperatures at L-band with the soil moistures in weight. The rough surface is with rms height of 0.1 wavelength and correlation lengths from 0.33 to 1.0 wavelength. The observation angle is at 50 degrees. Horizontal polarization.

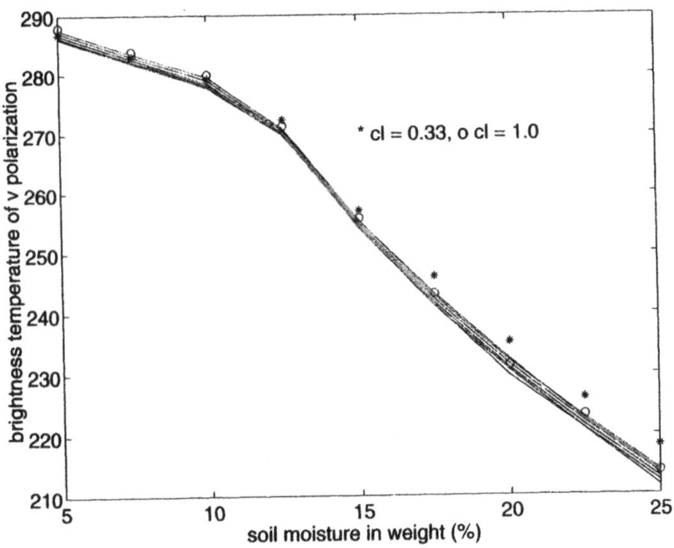

Figure 6.4.8 The variation of brightness temperatures at L-band with the soil moistures in weight. The rough surface is with rms height of 0.1 wavelength and correlation lengths of from 0.33 to 1.0 wavelength. The observation angle is at 50 degrees. Vertical polarization.

perature decreases with the increase of soil moistures for both horizontal and vertical polarizations. For horizontal polarization, soils with largest correlation length of 1.0 wavelength have the smallest brightness temperature and vice versa. For vertical polarization, this feature is not exhibited. The surface roughness has a larger influence in brightness temperature for horizontal polarization than the vertical. The brightness temperature in vertical polarization is mainly determined by the rms height for this case. With the increase of soil moisture, the differences in brightness temperatures between different correlation lengths increase. For horizontal polarization, the biggest difference in brightness temperature increases from 12.7 K degrees for 5% of soil moisture to 18.6 K degrees for 25% of soil moisture. It increases from 1.86 K degrees to 6.72 K degrees for the corresponding vertical polarization. This shows that the effects of surface roughness become larger for large soil moisture than small one.

(B) Variation of Brightness Temperature at L-Band with the Soil Moistures for the Rough Surfaces with the Fixed rms Slope and Different rms Height

In Fig. 6.4.9, the brightness temperatures are plotted as a function of soil moisture in both horizontal and vertical polarizations. Two sets of simulation results are used. One is with rms height of 0.1 wavelength and correlation length of 0.33 wavelength. The other is with rms height of 0.3 wavelength and correlation length of 1.0 wavelength. The rms slopes for these two cases are the same. It can be seen that the increase of rms height decreases the difference in brightness temperature between horizontal and vertical polarizations. As scattering by rough surface increases, there is "mixing" of polarizations. The roughness effect is more pronounced for horizontal polarization.

(C) Variation of Brightness Temperatures with the Soil Moistures at Both L- and C-Bands Using the Same Physical Roughness Parameters

The brightness temperatures in passive microwave remote sensing are measured at L- and C-bands. So it is useful to compare the brightness temperatures of soils at L- and C-bands using the same values in centimeters of physical roughness parameters of rms height and correlation length. We also use the same physical values of moisture for the two frequencies. In Fig. 6.4.10, we plot the brightness temperatures as a function of soil moisture in volumetric water content at both L- and C-bands. The corresponding permittivities are taken from Table 6.4.4. The physical parameters are that

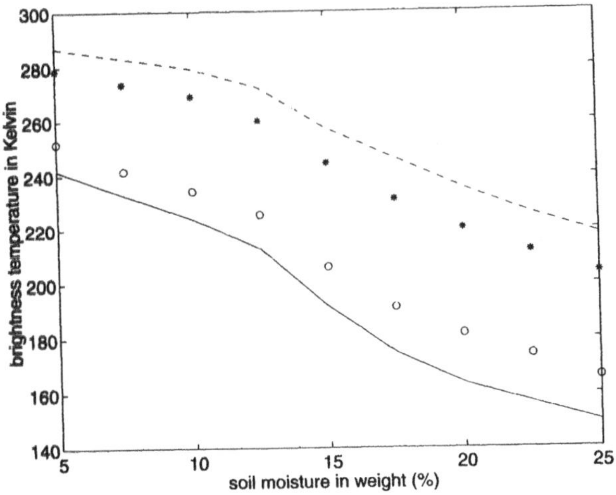

Figure 6.4.9 The variation of brightness temperatures with the soil moistures in weight. The rough surfaces are with rms height of 0.1 and 0.3 wavelength and correlation lengths of 0.33 and 1.0 wavelength, respectively. The two surfaces have the same rms slope. The observation angle is at 50 degrees. The Solid line and dash line is h- and v-polarization with rms height of 0.1 wavelength, respectively. The circle and star is h- and v-polarization with rms height of 0.3 wavelength, respectively.

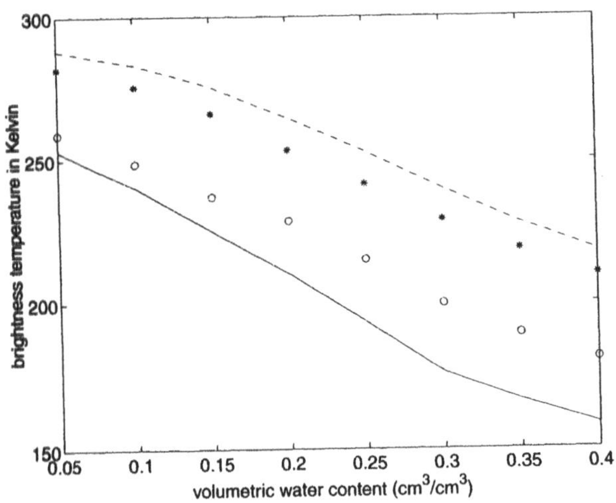

Figure 6.4.10 The variation of brightness temperatures with the soil moistures in volumetric water content at L- and C-bands using the permittivities in Table 6.4.4. The physical rough surface parameters are rms height of 2.45 cm and correlation length of 8.0 cm. The observation angle is at 50 degrees. The Solid line and dash line is h- and v-polarization at L-band, respectively. The circle and star is h- and v-polarization at C-band, respectively.

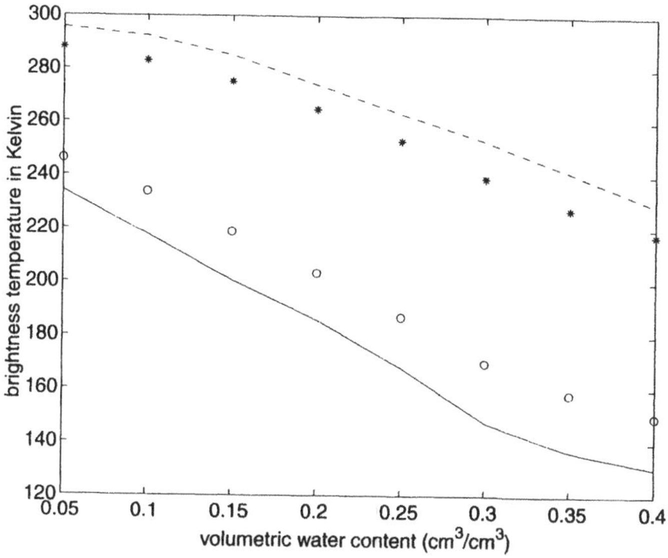

Figure 6.4.11 The variation of brightness temperatures with the soil moistures in volumetric water content at L- and C-bands using the permittivities in Table 6.4.4. The physical rough surface parameters are rms height of 0.73 cm and correlation length of 3.5 cm. The observation angle is at 50 degrees. The Solid line and dash line is h- and v-polarization at L-band, respectively. The circle and star is h- and v-polarization at C-band, respectively.

the rms height is 2.45 cm and correlation length is 8.0 cm. We can see that the brightness temperatures are higher at C-band than at L-band for horizontal polarization and lower at C-band than at L-band for vertical polarization. Since the permittivities between L- and C-bands are comparable, the results indicate that roughness has a larger effect on C-band than at L-band. In Fig. 6.4.11, the results are shown with different rough surface parameters. The rms height is 0.73 cm and correlation length is 3.5 cm in this figure. Comparing the results of Figs. 6.4.10 and 6.4.11 indicates that as roughness decreases, the brightness temperatures of L- and C-bands are closer to each other.

(D) Comparisons with Empirical Formula

In passive remote sensing with soil, a popular model for calculating the brightness temperatures from soil is the Q and H empirical model given in Eqs. (6.4.36) and (6.4.37). In Table 6.4.2 we have shown that Q and H are dependent on observation angles.

In Tables 6.4.10 and 6.4.11, we tabulate the values of Q and H com-

Q and H	Volumetric Water Content (cm^3/cm^3)							
	0.05	0.10	0.15	0.20	0.25	0.30	0.35	0.40
Q (L-band)	0.1304	0.1242	0.1190	0.1292	0.1258	0.1144	0.1305	0.1407
H (L-band)	0.4661	0.4450	0.4337	0.4189	0.3718	0.3273	0.3289	0.3148
Q (C-band)	0.2644	0.2675	0.2793	0.3248	0.3255	0.3157	0.3173	0.3200
H (C-band)	0.4013	0.3956	0.4246	0.4812	0.4520	0.4284	0.4233	0.4084

rms height = 2.45 cm, correlation length = 8.0 cm, observation angle = 50 degrees;
L-band: $f = 1.4$ GHz; C-band: $f = 5.0$ GHz.

Table 6.4.10 Parameters Q and H determined from the numerical simulation results of Fig. 6.4.10.

Q and H	Volumetric Water Content (cm^3/cm^3)							
	0.05	0.10	0.15	0.20	0.25	0.30	0.35	0.40
Q (L-band)	-0.0344	-0.0541	-0.0545	-0.0356	-0.0457	-0.0866	-0.0844	-0.0549
H (L-band)	0.0271	0.0538	0.1051	0.1547	0.1437	0.1297	0.1441	0.1436
Q (C-band)	0.1111	0.1106	0.1085	0.1127	0.1115	0.1074	0.1107	0.1169
H (C-band)	0.1541	0.1769	0.2044	0.2165	0.1919	0.1622	0.1573	0.1529

rms height = 0.73 cm, correlation length = 3.5 cm, observation angle = 50 degrees;
L-band: $f = 1.4$ GHz; C-band: $f = 5.0$ GHz.

Table 6.4.11 Parameters Q and H determined from the numerical simulation results of Fig. 6.4.11.

puted from (6.4.36) and (6.4.37) by using the brightness temperatures of Figs. 6.4.10 and 6.4.11 based on numerical simulations. The tables show that parameters of Q and H are dependent on soil moistures, frequencies, and observation angles. The tables show that Q and H are empirical parameters that do not have physical meaning. On the other hand, the physical roughness parameters are rms height and correlation length in absolute value of centimeters. They are physical characteristics of the roughness and are independent of angles, frequencies and soil moistures.

5 Four Stokes Parameters Based on Tangential Surface Fields

The four Stokes parameters of passive remote sensing are described in Section 5.4 of Chapter 3 in Volume I. In previous sections, we calculate the four Stokes parameters by using one minus the integration of the bistatic scatter-

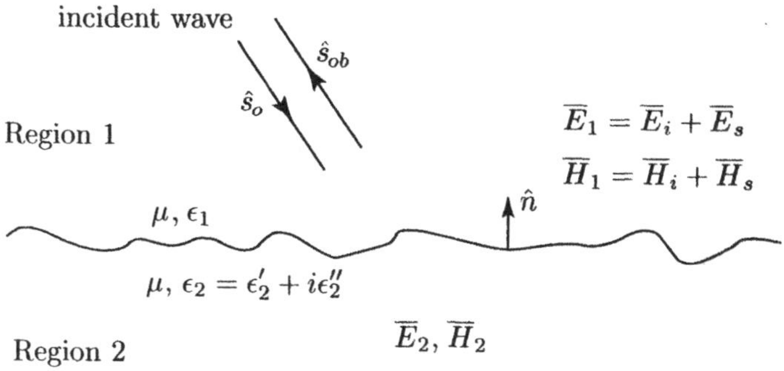

Figure 6.5.1 Incident wave in direction $\hat{s}_{ob}$ and brightness temperatures in direction $\hat{s}_o = -\hat{s}_{ob}$.

ing coefficients. On Monte Carlo simulations, the emissivity and absorptivity can also be calculated directly from the surface fields.

Consider a dielectric medium with a surface S, normal $\hat{n}$, with a physical temperature T. Consider an incident wave in direction $\hat{s}_{ob}$, with incident power P_i (Fig. 6.5.1). Let $a_\beta(\hat{s}_{ob})$ be the absorptivity when the incident wave has polarization β. The absorptivity is the fraction of the incident power that is absorbed by the medium. Then from (3.5.75a–d) of Volume I, the brightness temperatures of the four Stokes parameters in the observation direction of $\hat{s}_o = -\hat{s}_{ob}$ are

$$T_{Bv}(\hat{s}_o) = Ta_v(\hat{s}_{ob}) \tag{6.5.1}$$

$$T_{Bh}(\hat{s}_o) = Ta_h(\hat{s}_{ob}) \tag{6.5.2}$$

$$U_B(\hat{s}_o) = T_{Bv}(\hat{s}_o) + T_{Bh}(\hat{s}_o) - 2Ta_P(\hat{s}_{ob}) \tag{6.5.3}$$

$$V_B(\hat{s}_o) = T_{Bv}(\hat{s}_o) + T_{Bh}(\hat{s}_o) - 2Ta_R(\hat{s}_{ob}) \tag{6.5.4}$$

where subscript v means vertical polarized, h means horizontal polarized, subscript P refers to linear polarization with incident polarization vector

$$\widehat{P} = \frac{1}{\sqrt{2}}\left[\hat{v}(\hat{s}_{ob}) + \hat{h}(\hat{s}_{ob})\right] \tag{6.5.5}$$

and subscript R is right-hand circular polarized with incident polarization vector

$$\widehat{R} = \frac{1}{\sqrt{2}}\left[\hat{v}(\hat{s}_{ob}) + i\hat{h}(\hat{s}_{ob})\right] \tag{6.5.6}$$

Let the incident wave of β polarization with $\beta = v,\ h,\ P,\ R$ has electric field $\overline{E}_i$ and magnetic field $\overline{H}_i$ and the scattered wave has fields $\overline{E}_s$ and $\overline{H}_s$. In region 1,

$$\overline{E}_1 = \overline{E}_i + \overline{E}_s \tag{6.5.7}$$

$$\overline{H}_1 = \overline{H}_i + \overline{H}_s \tag{6.5.8}$$

and in region 2, the fields are $\overline{E}_2$ and $\overline{H}_2$. The boundary conditions are

$$\hat{n} \times \overline{E}_1 = \hat{n} \times \overline{E}_2 \tag{6.5.9}$$

$$\hat{n} \times \overline{H}_1 = \hat{n} \times \overline{H}_2 \tag{6.5.10}$$

Let P_i, P_s, and P_a be respectively the incident power, scattered power, and the absorption power. Then

$$a_\beta = \frac{P_a}{P_i} \tag{6.5.11}$$

By conservation of energy

$$P_i = P_a + P_s \tag{6.5.12}$$

so that

$$a_\beta = 1 - \frac{P_s}{P_i} \tag{6.5.13}$$

The incident Poynting's vector is

$$\overline{S}_i = \frac{1}{2}\text{Re}\left(\overline{E}_i \times \overline{H}_i^*\right) \tag{6.5.14}$$

and the scattered and transmitted wave Poynting's vector are respectively

$$\overline{S}_s = \frac{1}{2}\text{Re}\left(\overline{E}_s \times \overline{H}_s^*\right) \tag{6.5.15}$$

$$\overline{S}_2 = \frac{1}{2}\text{Re}\left(\overline{E}_2 \times \overline{H}_2^*\right) \tag{6.5.16}$$

Both $\overline{S}_i$ and $\overline{S}_s$ are divergence free. In view of the divergence theorem, the surface integral over Poynting's vector to get power for $\overline{S}_i$ and $\overline{S}_s$ can be taken over surfaces in region 1. In particular, we can use the the rough surface (Fig. 6.5.1). Thus

$$P_i = -\int_S dS\overline{S}_i \cdot \hat{n} = -\int_S dS\frac{1}{2}\text{Re}\left(\overline{E}_i \times \overline{H}_i^*\right) \cdot \hat{n} \tag{6.5.17}$$

$$P_s = \int_S dS\overline{S}_s \cdot \hat{n} = \int_S dS\frac{1}{2}\text{Re}\left(\overline{E}_s \times \overline{H}_s^*\right) \cdot \hat{n} \tag{6.5.18}$$

Thus the absorptivity is

$$a_\beta = 1 - \frac{1}{P_i}\int_S dS\frac{1}{2}\text{Re}\left(\overline{E}_s \times \overline{H}_s^*\right) \cdot \hat{n} \tag{6.5.19}$$

$$= 1 - \frac{1}{P_i}\int_S dS\frac{1}{2}\text{Re}\left[\left(\overline{E}_1 - \overline{E}_i\right) \times \left(\overline{H}_1^* - \overline{H}_i^*\right)\right] \cdot \hat{n} \tag{6.5.20}$$

In Monte Carlo simulations, we calculate the tangential surface fields $\hat{n} \times \overline{E}_1$ and $\hat{n} \times \overline{H}_1$. Absorptivity can be calculated by using (6.5.20). Another formula can be derived as follows. The power absorbed is obtained from the integration of dissipation over region 2

$$P_a = \int_{V_2} d\overline{r} \frac{1}{2} \omega \epsilon_2'' \overline{E}_2 \cdot \overline{E}_2^* \tag{6.5.21}$$

Using Maxwell's equations, we have

$$\nabla \cdot \left[\frac{1}{2} \text{Re} \left(\overline{E}_2 \times \overline{H}_2^* \right) \right] = \nabla \cdot \overline{S}_2 = \frac{1}{2} \omega \epsilon_2'' \overline{E}_2 \cdot \overline{E}_2^* \tag{6.5.22}$$

Thus

$$P_a = \int_{V_2} d\overline{r} \nabla \cdot \overline{S}_2 = \int_S dS \overline{S}_2 \cdot \hat{n} \tag{6.5.23}$$

$$= \int_S dS \hat{n} \cdot \left[\frac{1}{2} \text{Re} \left(\overline{E}_2 \times \overline{H}_2^* \right) \right] \tag{6.5.24}$$

However, since tangential electric and magnetic fields are continuous, we have

$$P_a = \int_S dS \hat{n} \cdot \left[\frac{1}{2} \text{Re} \left(\overline{E}_1 \times \overline{H}_1^* \right) \right] \tag{6.5.25}$$

The absorptivity is

$$a_\beta = \frac{1}{P_i} \int_S dS \hat{n} \cdot \left[\frac{1}{2} \text{Re} \left(\overline{E}_1 \times \overline{H}_1^* \right) \right] \tag{6.5.26}$$

Using Monte Carlo simulations, the surface fields are calculated. Thus we can use the two formulas derived in this section to calculate the absorptivity from the tangential surface fields. Once the absorptivity for the four different polarization incidences, v, h, P, and R, are determined, the brightness temperatures of the four Stokes parameters can be calculated. We can also verify conservation of energy. Reflectivity r_β is integration of scattered power. Thus we can use the surface fields to get r_β.

$$r_\beta = \frac{1}{P_i} \int_S dS \frac{1}{2} \text{Re} \left(\overline{E}_s \times \overline{H}_s^* \right) \cdot \hat{n}$$

$$= \frac{1}{P_i} \int_S dS \frac{1}{2} \text{Re} \left[\left(\overline{E}_1 - \overline{E}_i \right) \times \left(\overline{H}_1^* - \overline{H}_i^* \right) \right] \cdot \hat{n} \tag{6.5.27}$$

Energy conservation can be verified by checking that

$$r_\beta \text{ of } (6.5.27) + a_\beta \text{ of } (6.5.26) = 1 \tag{6.5.28}$$

6 Parallel Implementation of SMCG on Low Cost Beowulf System

6.1 Introduction

The need for solving MoM for large-scale rough surface problems is apparent. (1) For example, in low grazing problems, the surface area has to be very large in order to include the multiple scattering and shadowing effects. (2) For scattering and emission by soil surfaces or ice surfaces, it is important to solve larger scale problem of several thousand square wavelengths to better study the target response. (3) For the case of ocean surfaces, the roughness is of many scales from several wavelengths to sub-wavelengths. In order to simulate the overall response, a larger surface is needed. (4) Fine discretization is needed for surfaces with lossy dielectric. All these problems require millions or tens of millions of surface unknowns in the MoM implementation of surface integral equations. With the advent of fast computation methods, new frontiers are opened for large-scale simulations. Furthermore, the use of parallel computing will enhance the capabilities. In the past, parallel computing is hampered by the cost of massively parallel processors (MPP). High performance computing has been static because of the high price/performance ratio. However, recently the parallel computing is made more possible with low cost Beowulf system of PC clusters [Sterling et al. 1999]. For a Beowulf cluster, a node is a personal computer from the commercial mass market with LINUX as the operating system. Also critical to the recent success of Beowulf was the development of low cost moderate bandwidth local area network (LAN). The Beowulf project was started at the Center of Excellence in Space data and Information Science (CESDIS), NASA, Washington DC in 1994 to provide affordable parallel computing capability. It received worldwide acceptance among academic and research institutions.

In the previous sections, we have used the SMCG method for solving large-scale problem. A salient feature of SMCG is that it is FFT based. Since MPI (message-passing interface) of FFT is available [Frigo and Johnson, 1997], this facilitates the implementation of SMCG on low cost Beowulf system. This has been accomplished at the City University of Hong Kong [Li et al. 1999, 2000b].

In this section, we present a cost-effective solution by implementing the SMCG method on a Beowulf system that consists of PC's (processors) connected by 100 Base TX Ethernet switch. The workloads of computing the sparse-matrix-vector multiplication corresponding to the near interactions and the fast Fourier transform (FFT) operations corresponding to the far interactions in the SMCG method can be easily distributed among all the processors. Both perfectly conducting and lossy dielectric surfaces of Gaussian spectrum are analyzed thereafter. When possible, speedup factors against a single processor are given. It is shown that the SMCG method for a single realization of rough surface scattering can be efficiently adapted for parallel implementation. The largest number of surface unknowns solved in this section is over 1.5 million. On the other hand, a problem of 131,072 surface unknowns for a PEC random rough surface of 1024 square wavelengths only requires a CPU time of less than 20 minutes.

6.2 Low-Cost Beowulf Cluster

We describe the Beowulf system that was used to perform the computations presented in this section.

A. The Hardware

The Beowulf system at City University of Hong Kong that is used to compute results in this section consists of 17 computing processors (nodes) which are connected to two 3Com Superstack LinkSwitch 300 Fast Ethernet switch with twelve 100 Base TX ports. A 24-port switch would provide faster networking speed. A 9-node system was first installed to try out the concept using a 12-port switch. This system was later expanded to 17 nodes with an addition of 12-port switch. Each computing node has an Intel Pentium II 450 MHz Processor, 256 MByte SDRAM and a 3Com 3C905 Network Interface for connection to the communication switch. One of the nodes is designated as the server and it has 12 GBytes of hard disk. All the other nodes either has 1 or 2 GBytes of hard disk so that they can be used individually. The whole set up was under US $30,000 in 1998. Although the 1000 Base TX Ethernet card and switch are now available, the price is still too expensive. Optical fiber will be required for the connections between the PC's and the switch. It is estimated that an additional cost of approximately US $50,000 is needed in 1999 to upgrade the Beowulf to 1000 Base TX connection using the existing 17 PC's. It should be noted that the speed of the computer code highly depends on the hardware. When PC's were upgraded from 233 MHz to 450 MHz while all other hardware remains unchanged, a speed up factor of 1.6 was obtained. The communication time among processors for rough surface simulations using integral equation approach is small compared to that of the matrix-vector multiplication. It is more cost effective to upgrade the system using faster processors than to upgrade the Ethernet cards and communication switch.

B. The Software

The system software requires Redhat Linux Version 6.0 and the Message Passing Interface (MPI) Version 1.1.2. Detailed system software set up can be found in the CESDIS web page. The aim of MPI is to develop a widely used standard for writing message-passing programs [Gropp et al. 1994]. The message-passing model consists of a set of processes that have only local memory but are able to communicate with other processes by sending and receiving messages from one node to the other and vice versa. Data transfer from the local memory of one process to the local memory of another requires operations to be performed by both processes.

A key to the numerical efficiency and computer memory reduction of the SMCG and PBTG methods is the use of FFT's for performing the convolution of the weak part of the matrix-vector multiplication. The MPI version of the fast Fourier transform code was adopted — the Fastest Fourier Transform in the West (FFTW) developed by the Laboratory of Computer Science, MIT [Frigo and Johnson, 1997] was used. This subroutine is written in C. A Fortran wrapper was written so that this subroutine was incorporated in the parallel Fortran codes. In addition, a real-time monitoring system was also developed so that all the MPI commands can be monitored on the fly.

6.3 Parallel Implementation of the SMCG Method and the PBTG Method

In applying the SMCG method, we rewrite the MoM matrix equation $[Z]\{x\} = \{b\}$ as

$$[Z^s]\{x\} + [Z^w]\{x\} = \{b\} \tag{6.6.1}$$

where $[Z^s]$ and $[Z^w]$ correspond to the strong near-field interaction and the weak far-field interaction, respectively. The distance chosen to distinguish the near and far interactions is defined so that the Green's functions can converge rapidly within a few orders of Taylor series expansion about $z = 0$ when the separation between the source and field point is beyond r_d. In this section, we use 49 terms in the expansion. The convolution of $[Z^s]\{x\}$ is computed by direct multiplication. In contrast, for the far interactions $[Z^w]$, because of the expansion of the Green's function with the Taylor series, $[Z^w]\{x\}$ can be computed using FFT's as

$$[Z^w]\{x\} = \sum_{i=0} [T_{i2}]\mathcal{F}^{-1} \cdot [\tilde{G}_i]\mathcal{F} \cdot [T_{i1}]\{x\} \tag{6.6.2}$$

where $[T_{i1,i2}]$ is a block diagonal matrix, $[\tilde{G}_i]$ is a block Toeplitz matrix, and quantity with a tilde means that it is in the Fourier domain. The number of terms in the summation depends on the order of the Taylor series expansion. Substituting (6.6.2) into (6.6.1), the unknown vector $\{x\}$ is solved iteratively as

$$[Z^s]\{x^0\} + [T_{02}]\mathcal{F}^{-1} \cdot [\tilde{G}_0]\mathcal{F} \cdot [T_{01}]\{x^0\} = \{b\} \tag{6.6.3}$$

$$\begin{aligned}[Z^s]\{x^{n+1}\} &+ [T_{02}]\mathcal{F}^{-1} \cdot [\tilde{G}_0]\mathcal{F} \cdot [T_{01}]\{x^{n+1}\} \\ &= \{b\} - \sum_{i=1}[T_{i2}]\mathcal{F}^{-1} \cdot [\tilde{G}_i] \cdot [T_{i1}]\{x^n\} = \{b^n\}\end{aligned} \tag{6.6.4}$$

In all the examples shown, the maximum value of n is 4 while the typical value is 2. The number of CG iteration steps reduces significantly when n increases. Initially, the error norm of the CG method may be reduced rapidly. This reduction may eventually be flattened out. Therefore, the maximum number of CG iterative steps was set at 100 for each value of n. After modifying the right-hand side, the CG will converges rapidly again. All terms in both (6.6.3) and (6.6.4) can be parallelized easily as will be discussed.

For a lossy dielectric rough surface with high permittivity, we use the PBTG method. The Green's function in the lower region is heavily attenuative. Thus, matrix elements of the Green's function for the lower medium correspond to source and field point separation beyond a radial distance r_c can simply be set to zero.

(1) The matrix-vector multiplication corresponds to the Green's function in the lossy medium is a sparse-matrix-vector multiplication operation on the dense grid.
(2) For the near interactions of the free-space Green's function, the near-interaction range r_d defined in the SMCG method is further divided into two regions. Within a radial distance r_c ($< r_d$) of the source point, we have sparse-matrix-vector multiplication on the dense grid.
(3) For the interactions between r_c and r_d, we have averaging from the dense grid to the coarse grid, sparse-matrix-vector multiplication on the coarse grid and interpolation from the coarse grid to the dense grid.
(4) Beyond r_d, the convolution between the far interactions $[Z^w]$ corresponds to the free-space Green's function and the vector $\{x\}$ will be computed by averaging from the dense grid to the coarse grid, using FFT's for matrix vector multiplication on the coarse grid and interpolation from the coarse grid to the dense grid.

In the lossy surface simulations, we select $r_d = 3\lambda_o$ and $r_c = 0.5\lambda_o$, which make the Taylor expansion converges quickly and the Green's function in the lower medium sufficiently attenuated beyond . The procedure for averaging and interpolation in the convolution of the free space Green's function is described by the equation

$$y_m = \sum_{n=1}^{N} G_o(k_o|\bar{r}_m - \bar{r}_n|) \cdot x_{\tilde{n}}, \quad m = 1, 2, \ldots, N \tag{6.6.5}$$

where m and n are indices in the coarse grid and $\tilde{n}$ in the dense grid. A coarse grid can be used for the non-near field convolution of the free space Green's function because the free space Green's function is spatial frequency limited on the dense grid. The value of x at $\tilde{n}$ is obtained from the average of the values of x at the dense grid. After the convolution using the coarse grid, the values of y at the dense grid will be interpolated from the coarse grid. It is not difficult to see that the steps of averaging and interpolation in the PBTG method would not cause significant additional computation to the SMCG method when analyzing lossy dielectric rough surfaces.

Parallel implementation of the SMCG method has been reported [Johnson et al. 1996, 1999; Pak et al. 1997; Johnson and Chou, 1998] on the multimillion-dollar IBM SP/2-based parallel computer at the Maui High Performance Computing Center. However, the parallelization that is implemented in this section is for a single realization of rough surface and for one incident angle with a large number of surface unknowns. Previous efforts are for different realizations and/or different incident angles.

In this section, parallel versions of both the SMCG method and the PBTG method are implemented on the low-cost Beowulf system. The sparse-matrix-vector multiplications are implemented using parallelization so that both the CPU time and memory storage are distributed among the processors. The use of the MPI version of FFTW, however, requires less computer memory than that of distributing the multiple FFT calls among the processors. For an $N \times N$ FFT, the implementation only requires to store N/N_P rows of the $N \times N$ array where N_P is the number of processors used. There is no significant difference in the CPU time required in these implementations of the weak part of the matrix-vector multiplication. However, the reduced memory requirement may be important in implementing the multi-level SMCG method [Chan et al. 1998; Li et al. 2001]. In this multi-level SM/CM scheme, the strong interaction range and hence the memory storage of the sparse matrix can be reduced significantly. On the other hand, 3-D FFT's will be required and the memory reduction in the FFT's become important. Apparently, the use of the MPI version of FFTW can distribute the computation evenly among the processors and can better synchronize the parallel code as opposed to distributing the multiple FFT calls among the processors.

In the following, we briefly describe the parallelization of the key steps in the SMCG and PBTG methods.

A. Sparse-Matrix-Vector Multiplication

For a large rough surface, the number of the non-zero elements of the strong matrix $[Z^s]$ may still be too large to store, so we recalculate them when necessary. When the number of surface unknown is manageable, we only compute the strong matrix once and store them in an array instead. The strong matrix $[Z^s]$ is computed in parallel and distributively stored among the processors. The unknown vector $\{x\}$ is stored in all processors for the convenience of performing the conjugate gradient procedure. Note that storing an array of a couple of million elements does not require much memory storage but it can reduce some communication time among the processors. Each processor performs a different part of the sparse-matrix-vector multiplication, then the results in each processor are combined and broadcasted back to all processors.

B. Diagonal Matrix Pre-Multiplication, Fast Fourier Transform, and Diagonal Matrix Post-Multiplication

The convolution of the weak matrix with the vector is performed by FFT's. In order to perform the convolution, the array should be zero padded before applying the FFT's, and the original array is only one quarter of the zero-padded one. The zero-padded array is uniformly distributed among all 16 processors, while only half of the processors, in which one quarter of the zero-padded array is stored, perform the pre-multiplication. Upon applying the forward FFT, only 1/16 of the array $\mathcal{F}\cdot[T_{01}]\{x\}$ are stored in each processor. Similarly, the corresponding 1/16 of the array $[\tilde{G}_0]$ are stored in the same processor. The products between the elements of these two arrays are carried out in parallel. Upon taking the inverse FFT, only one quarter of the full array residing in half of the processors require the post-multiplication.

In performing the FFT's, we use the MPI_FFT transposed command which is much faster than the normal FFT operation. We perform the transposed FFT command twice, one for the forward FFT and one for the inverse FFT as shown in (6.6.3) and (6.6.4). The result will be stored in the original sequence as depicted in Fig. 6.6.1.

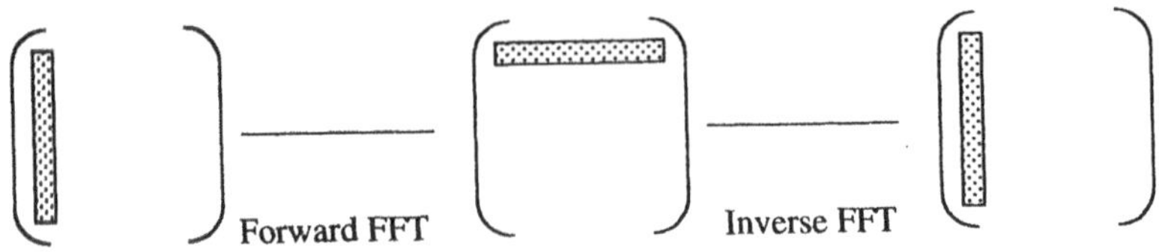

Figure 6.6.1 Storage of the partitioned array in each step of the FFT's.

6.4 Numerical Results

In this section, we present results for PEC and lossy dielectric surfaces with Gaussian spectrum. For the PEC and lossy dielectric surfaces, we compare the bistatic scattering coefficients obtained from a single processor and 16 processors to verify the parallel version codes. We only use 16 instead of 17 processors in our present parallel codes because of the radix-2 FFT requirement although we can slightly improve the computation by using 17 nodes for all the operations other than the FFT's. We also give the comparison between the results obtained with near-field integration and those obtained by the collocation method to demonstrate the accuracy improvement after using the near-field integration. The brightness temperature is calculated by multiplying the emissivity with the surface physical temperature, which is chosen to be 283K in this paper. The emissivity is obtained by integrating the scattering coefficients above the half space. For the simulations in this section, we choose the radius of the incident tapered wave spot size to be 1/3 of the edge length to provide sufficient field attenuation at surface edges. For PEC cases, we use the SMCG method, while the SMCG/PBTG method is used for lossy dielectric surfaces. In the SMCG/PBTG method, we choose $r_c = 0.5\lambda_0$, $r_d = 3\lambda_0$ and 64 points per square wavelength for the coarse grid and 256 points per square wavelength for the dense grid. For lossy dielectric surfaces, the permittivity $\epsilon_r = 45 + i30$ is used. The CG iterations are stopped when the error is at 1%.

For Gaussian surfaces of rms height, h, up to $0.3\lambda_o$ and size up to 32 by $32\lambda_o^2$ with a correlation length of $1\lambda_o$, we present the bistatic scattering coefficients and/or brightness temperatures. In Figs. 6.6.2 and 6.6.3, we give the bistatic scattering coefficients of the PEC and the lossy dielectric surfaces. The results using a single processor and those using 16 processors overlie each other. It is important to compare the CPU time between a single processor and 16 processors. In Tables 6.6.1 and 6.6.2, we show the comparison of CPU time for calculating PEC and lossy dielectric surfaces, respectively. Speedup factor is also tabulated. The tables indicate that the speedup factor increases

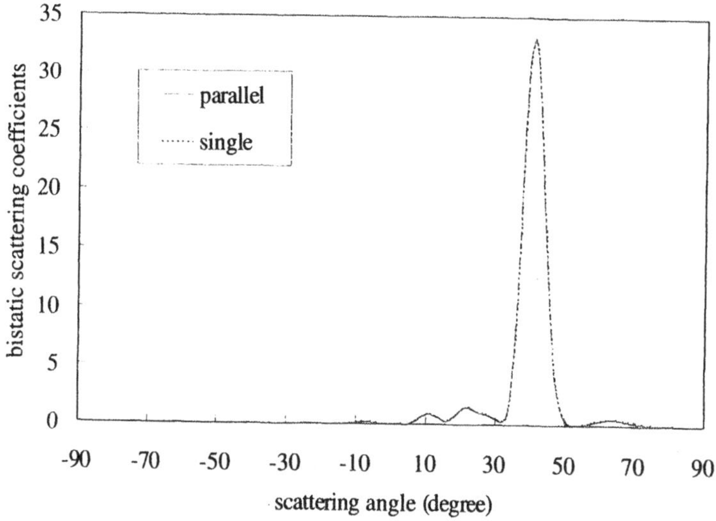

Figure 6.6.2 Comparison of the bistatic scattering coefficients for PEC rough surface obtained by the single code and parallel code. $8 \times 8\lambda_o^2$, $h = 0.1\lambda_o$ incident angle $= 40°$.

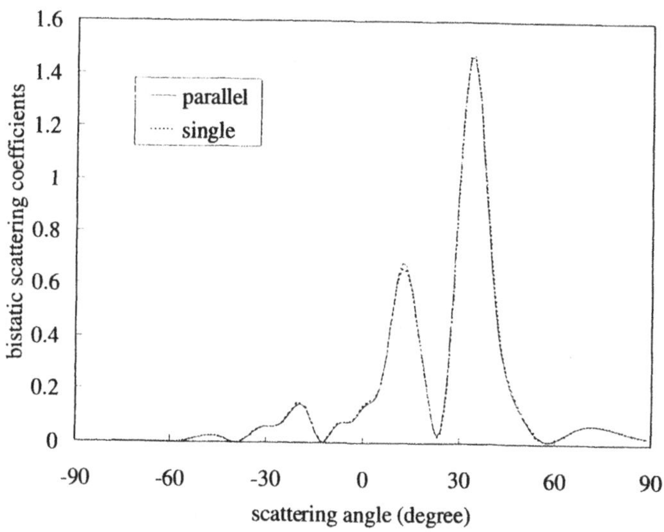

Figure 6.6.3 Comparison of the bistatic scattering coefficients for lossy dielectric rough surface obtained by the single code and parallel code. $8 \times 8\lambda_o^2$, $h = 0.2\lambda_o$ incident angle $=$ $30°$, $\epsilon_r = 45 + i30$.

Surface size (λ_o^2)	Number of unknowns	CPU time (Sec.)		
		Sequential code	Speedup factor	Parallel code
8 x 8	8,192	518	7.85	66
16 x 16	32,768	3020	13.73	220
32 x 32	131,072	------	------	1,120
64 x 64	524,288	------	------	6,045
32 x 32	(one CG term)	382.48	41.57	9.2
64 x 64	(one CG term)	4381.74	115.61	37.9

Table 6.6.1 Comparison of CPU time for PEC rough surface $h = 0.1\lambda_o$.

Surface size (λ_0^2)	Number of unknowns	rms height (λ_0)	CPU time (Sec.)		
			Sequential code	Speedup factor	Parallel code
8 x 8	98,340	0.05	49,368	9.77	5,051
		0.10	77,643	11.57	6,712
		0.15	114,385	11.55	9,906
		0.20	161,313	11.79	13,682
		0.25	207,095	11.92	17,371
		0.30	275,485	12.23	22,517
16 x 16	393,216	0.05	270,204	13.20	20,477
		0.10	469,831	13.43	34,981
32 x 32	1,572,864	0.05	-------	-------	133,798
		0.10	-------	-------	249,460

Table 6.6.2 Comparison of CPU time for lossy dielectric rough surface $\epsilon_r = 45 + i30$.

when the number of unknowns increases. For the surfaces with the same size, the CPU time increases with h because of the slow convergence of the CGM for rougher surfaces. For example, it only needs 2 iterations ($n = 1$) with 40 CG terms in the first iteration and 1 CG term in the second iteration for the $8 \times 8\lambda_o^2$ surface with $h = 0.05\lambda_o$. In contrast, it needs 3 iterations ($n = 2$) with 100 CG terms in the first iteration, 41 CG terms in the second iteration and 4 CG terms in the last iteration when h increases to $0.3\lambda_o$. The speedup factors become large when the CPU time increases. It is shown that the best speedup factor is around 13.7 for the PEC surface with 32,768 surface unknowns and $h = 0.1\lambda_o$, and 13.4 for the lossy dielectric surface with 0.4 million surface unknowns for the same h. For the PEC surfaces, CPU times of the parallel code are obtained when the strong matrix is computed once

Total time	4584.8 s
Communication time	120.8 s
Near interaction calculation time	2918.6 s
Far interaction calculation time	289.4 s
Emissivity calculation time	1225.7 s

Table 6.6.3 CPU time distribution for the case of $8 \times 8\lambda_o^2$ surface with $h = 0.05\lambda_o$.

and stored. When the number of unknowns exceeds 131,000, we recalculate the strong matrix in the sequential code when necessary. In the last two rows of Table 6.6.1, we give the comparison of CPU time for one CG term when the surface sizes are 32×32 and $64 \times 64\lambda_o^2$, respectively. The strong matrix is recalculated in the sequential code. One can see that the speedup factor can be up to 115.6 for 1 CG term. To complete the conjugate gradient solution of sequential code will be prohibitive long. For dielectric surfaces, when the number of unknowns exceeds 1.5 millions, the code on a single processor cannot be run and therefore parallel computation is the only alternative. In the 1.5 million unknown example, the computer code requires about 78% of the 4.0 GBytes of RAM. We should emphasize that the CPU time comparisons shown in Table 6.6.1 and 6.6.2 are based on the collocation method. For the lossy dielectric surface, it is found that near-field integration is necessary. Although this near-field integration requires additional CPU time, the resulting matrix is better conditioned so that the number of CG terms may be reduced. For examples, with near-field integration, the CPU time for the 8×8 and $16 \times 16\lambda_o^2$ surface with $h = 0.05\lambda_o$ reduces from 5,051 to 4,585 and 20,474 to 19,085 seconds, respectively. It is expected that similar speed-up factors are obtained when replacing the collocation method with the near-field integration. Table 6.6.3 shows the CPU time distribution for a $8 \times 8\lambda_o^2$ surface with near-field integration, where $h = 0.05\lambda_o$. The communication time is only a small part of the total CPU time. Most of the CPU time is spent on calculating the sparse matrix and sparse-matrix-vector multiplication. Communication time only attributes to 2.6% of the total CPU time while that of the far interaction calculation based on FFT is 6.3%.

Table 6.6.4 shows the comparison of the computed brightness temperature versus surface size and rms height, and the comparison between the results obtained with and without the near-field integration. The incident angle for the $8 \times 8\lambda_o^2$ surfaces is 30° while that of the remaining ones is 50°. The result of an infinite flat surface with the same permittivity is also given for reference. It should be pointed out that the brightness temperatures of the rough surfaces are obtained from one realization. Different realizations

Surface size (λ_0^2)	rms height (λ_0)	Temperature (K) (without integration)	Temperature (K) (with integration)
8 x 8 ($\theta_i = 30°$)	0.05	108.45	106.59
	0.10	109.92	108.26
	0.15	112.64	111.49
	0.20	116.44	116.30
	0.25	122.15	122.58
	0.30	126.66	129.61
16 x 16 ($\theta_i = 50°$)	0.05	81.86	81.08
	0.10	86.51	86.09
	0.15	94.51	94.69
	0.20	102.79	103.38
	0.25	108.08	110.43
32 x 32 ($\theta_i = 50°$)	0.05	82.06	81.23
	0.10	86.53	86.19
	0.15	94.09	94.54
Infinite flat plane	$\theta_i = 30°$, Temperature = 103.07 K; $\theta_i =$ 0°, Temperature = 80.85 K		

Table 6.6.4 Brightness temperatures of lossy dielectric rough surfaces. Incident angle is 30° for $8 \times 8\lambda_o^2$, 50° for 16×16 and $32 \times 32\lambda_o^2$ surface, $\epsilon_r = 45 + i30$.

may give different results varying within about 4 K which is illustrated in Table 6.6.5. From Table 6.6.4, it is observed that when h is small, the brightness temperature resembles that of the flat surface provided that the surface size is large enough, e.g., $16 \times 16\lambda_o^2$. The brightness temperature increases with increasing h. The difference between the temperatures of the $16 \times 16\lambda_o^2$ and those of $32 \times 32\lambda_o^2$ surfaces is less than 1 K for all the cases with different h considered. It implies that a surface with $16 \times 16\lambda_o^2$ is sufficiently large in numerical simulations at least in the cases considered. We also plot the results in Fig. 6.6.4 to give a better illustration. From the comparison between results obtained with and without the near-field integration, one can see that the accuracy is improved after using the near-field integration. The maximum improvement may be up to about 3 K. Table 6.6.5 shows the ensemble average of the brightness temperature versus the number of realizations for surfaces of $16 \times 16\lambda_o^2$. The difference among different realizations is small, and the result converges within five realizations.

Realization	Temperature of one realization (K)	Average temperature (K)
1	126.33	-------
2	125.12	125.725
3	123.49	124.980
4	124.86	124.950
5	123.04	124.568

Table 6.6.5 Ensemble average of brightness temperatures versus number of realizations. Incident angle = 50°. The surface size is $16 \times 16\lambda_o^2$, $h = 0.3\lambda_o$, $\epsilon_r = 45 + i30$.

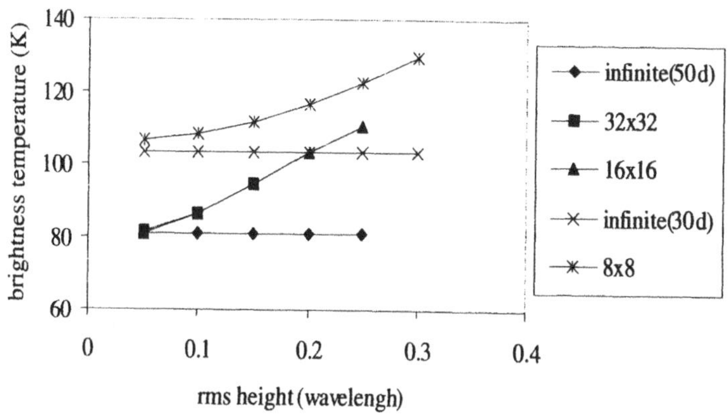

Figure 6.6.4 Brightness temperatures of lossy dielectric rough surfaces with different rms heights and sizes. Incident angle is 30° for $8 \times 8\lambda_o^2$ surface, 50° for 16×16 and $32 \times 32\lambda_o^2$ surface, $\epsilon_r = 45 + i30$.

REFERENCES AND ADDITIONAL READINGS

Adams, R. J. and G. S. Brown (1998), Use of fast multipole method with method of ordered multipole interactions, *Electronics Lett.*, **34(23)**, 2219–2220.

Arias-González, J. R., M. Nieto-Vesperinas, and A. Madrazo (1999), Morphology-dependent resonances in the scattering of electromagnetic waves from an object buried beneath a plane or a random rough surface, *J. Opt. Soc. Am. A*, **16(12)**, 2928–2934.

Belszynski, E., M. Belszynski, and T. Jaroszewicz (1994), A fast integral-equation solver for electromagnetic scattering problems, *IEEE Ant. and Propagat. Soc. Int. Sym.*, **1**, 416–419.

Brown, G. S., Ed. (1998), Special issue on low grazing-angle backscattering from rough surfaces, *IEEE Trans. Antennas Propagat.*, **46**.

Chan, C. H., L. Tsang, and Q. Li (1998), Monte-Carlo simulations of large-scale one-dimensional random rough surface scattering at near grazing incidence: Penetrable case, *IEEE Trans. Antennas Propagat.*, **46(1)**, 142–149.

Chew, W. C. and Q. H. Liu, Eds. (2000), Special Issue on computational wave issues in remote sensing, imaging and target identification, propagation and inverse scattering, *IEEE Trans. Geosci. Remote Sens.*, **38**.

Chou, H. T. (2000), Extension of the forward-backward method using spectral acceleration for the fast analysis of large array problems, *IEE Proc. – Microw., Antennas, and Propag.*, **147(3)**, 167–172.

Chou, H. T. and J. T. Johnson (1998), A novel acceleration algorithm for the computation of scattering from rough surfaces with the forward-backward method, *Radio Sci.*, **33(5)**, 1277–1287.

Coifman, R., V. Rokhlin, and S. Wandzura (1993), The fast multipole method for the wave equation: A pedestrian prescription, *IEEE Ant. Propagat. Mag.*, **35(3)**, 7–12.

Frigo, M. and S. G. Johnson (1997), *The Fastest Fourier Transform in the West*, Cambridge, MA. Available online at `http://www.fftw.org`.

Greenbaum, A. (1997), *Iterative Methods for Solving Linear Systems*, SIAM Frontiers in Applied Mathematics, vol. 17, SIAM, Philadephia.

Gropp, W., E. Lusk, and A. Skiellum (1994), *Using MPI: Portable Parallel Programming with the Message-Passing Interface*, MIT Press, Cambridge, MA.

Harrington, R. F. (1961), *Time-Harmonic Electromagnetic Fields*, McGraw-Hill, New York.

Harrington, R. F. (1968), *Field Computation by Moment Method*, Macmillan, New York.

Holliday, D., L. L. DeRaad, Jr., and G. J. St-Cyr (1996), Forward-backward: A new method for computing low-grazing angle scattering, *IEEE Trans. Antennas Propagat.*, **44**, 722–729.

Holliday, D., L. L. DeRaad, Jr., and G. J. St-Cyr (1998), Forward-backward method for scattering from imperfect conductors, *IEEE Trans. Antennas Propagat.*, **46**, 101–107.

Irisov, V. G. (1997), Small-slope expansion for thermal and reflected radiation from a rough surface, *Waves in Random Media*, **7(1)**, 1–10.

Jackson, T. J. (2001), Multiple resolution analysis of I-band brightness temperature for soil moisture, *IEEE Trans. Geosci. Remote Sens.*, **39(1)**, 151–164.

Janaswamy, R. (1994), A fast finite difference method for propagation predictions over irregular, inhomogeneous terrain, *IEEE Trans. Antennas Propagat.*, **AP-42**, 1257–1267.

Jandhyala, V., E. Michielssen, S. Balasubramaniam, and W. C. Chew (1998a), A combined steepest descent-fast multipole algorithm for the fast analysis of three-dimensional scattering by rough surfaces, *IEEE Trans. Geosci. Remote Sens.*, **36(3)**, 738–748.

Jandhyala, V., B. Shanker, E. Michielssen, and W. C. Chew (1998b), A fast algorithm for the analysis of scattering by dielectric rough surface, *J. Opt. Soc. Am. A*, 1877–1885.

Johnson, J. T. (1996), Applications of numerical models for rough surface scattering, Ph.D. thesis, Massachusetts Institute of Technology.

Johnson, J. T. and H. T. Chou (1998), Numerical studies of low grazing angle backscatter from 1-D and 2-D impedance surfaces, *Proc. IGRASS 1998*, Seattle WA, 2295–2297.

Johnson, J. T., R. T. Shin, J. C. Eidson, L. Tsang, and J. A. Kong (1997), A method of moments model for VHF propagation, *IEEE Trans. Antennas Propagat.*, **45**, 115–125.

Johnson, J. T., R. T. Shin, J. A. Kong, and L. Tsang (1999), A numerical study of ocean polarimetric thermal emission, *IEEE Trans. Geosci. Remote Sens.*, **37(1)**, 8–20.

Johnson, J. T., L. Tsang, R. Shin, K. Pak, C. H. Chan, A. Ishimaru, and Y. Kuga (1996), Backscattering enhancement of electromagnetic waves from two-dimensional perfectly conducting random rough surfaces: A comparison of Monte Carlo simulations with experimental data, *IEEE Trans. Antennas Propagat.*, **44**, 748–756.

Kapp, D. A. and G. S. Brown (1996), A new numerical method for rough surface scattering calculations, *IEEE Trans. Antennas Propagat.*, **44**, 711–721.

Krause, K. C., S. H. Lou, L. Tsang, and C. H. Chan (1991), Application of the finite element method to Monte Carlo simulations of random rough surface scattering with Neumann boundary conditions, *Microwave Opt. Technol. Lett.*, **4(7)**, 255–258.

Kuga, Y., J. S. Colburn, and P. Phu (1993), Millimeter-wave scattering from one-dimensional surface of different surface correlation functions, *Waves in Random Media*, **3**, 101–110.

Kuga, Y. and P. Phu (1996), Experimental techniques in random media and rough surface, *Progress in Electromag. Res.*, **PIER 14**, Chapter 2, Elsevier Science Publishers, Cambridge.

Li, Q. (2000), Numerical simulation of interactions of electromagnetic waves with lossy dielectric surfaces using fast computational methods, Ph.D. thesis, University of Washington, Seattle.

Li, Q., L. Tsang, K. S. Pak, and C. H. Chan (2000a), Bistatic scattering and emissivities of random rough dielectric lossy surfaces with the physics-based two-grid method in conjunction with the sparse-matrix canonical grid method, *IEEE Trans. Antennas Propagat.*, **48(1)**, 1–11.

Li, S. Q., C. H. Chan, L. Tsang, and Q. Li (1999), Parallel implementation of the sparse-matrix canonical grid method for two-dimensional lossy dielectric random rough surfaces (3D scattering problems) on a Beowulf system, *IEEE Ant. and Propagat. Soc. Int. Sym.*, **1**, 522–525.

Li, S. Q., C. H. Chan, L. Tsang, and Q. Li (2000b), Parallel implementation of the sparse-matrix/canonical grid method for the analysis of two-dimensional random rough surfaces (Three-dimensional scattering problem) on a Beowulf System, *IEEE Trans. Geosci. Remote Sens.*, **38**, 1600–1608.

Li, S. Q., C. H. Chan, and L. Tsang (2001), Multilevel expansion of the sparse-matrix canonical grid method for two-dimensional random rough surfaces, *IEEE Trans. Antennas Propagat.*, in press.

Lin, C.-M. and C. H. Chan (1998), Monte Carlo simulations for the electromagnetic scattering of rough surfaces by the combined wavelet transform and banded-matrix iterative approach/canonical grid method, *Microwave Opt. Technol. Lett.*, **19(4)**, 274–279.

Liou, Y.-A., K.-S. Chen, K.-S., and T.-D. Wu (2001), Reanalysis of L-band brightness predicted by the LSP/R model for prairie grassland: incorporation of rough surface scattering, *IEEE Trans. Geosci. Remote Sens.*, **39(1)**, 129–135.

Lou, S. H. (1991), Application of numerical methods to Monte Carlo simulations of scattering of waves by random rough surfaces, Ph.D. thesis, University of Washington, Seattle.

Lou, S. H., L. Tsang, C. H. Chan, and A. Ishimaru (1990), Monte Carlo simulations of scattering waves by a random rough surfaces with the finite element method and the finite difference method, *Microwave Opt. Technol. Lett.*, **3(5)**, 150–154.

Lou, S. H., L. Tsang, C. H. Chan, and A. Ishimaru (1991), Application of the finite element method to Monte Carlo simulations of scattering of waves by random rough surface with the periodic boundary conditions, *J. Electromag. Waves and Appl.*, **5**, 835–855.

Macaskill, C. and B. J. Kachoyan (1988), Numerical evaluation of the statistics of acoustic scattering from a rough surface, *J. Acous. Soc. Am. A*, **84**, 1825–1835.

Michielssen, E., A. Boag, and W. C. Chew (1996), Scattering from elongated objects: Direct solution in $\mathcal{O}(N \log 2N)$ Operations, *IEE Proc. Microwave Antennas Propagation*, **143(4)**, 277–283.

Michielssen, E. and W. C. Chew (1996), Fast steepest descent path algorithm for analyzing scattering from two-dimensional objects, *Radio Sci.*, **31(5)**, 1215–1224.

Njoku, E. G. and L. Li (1999), Retrieval of land surface parameters using passive microwave measurements at 6–18 GHz, *IEEE Trans. Geosci. Remote Sens.*, **37(1)**, 79–93.

Ngo, H. D. and C. Rino (1994), Applications of beam simulation to scattering at low grazing angles: Part 1. Methodology and validation, *Radio Sci.*, **29(6)**, 1365–1379.

Pak, K. (1996), Studies of large-scale random rough surface scattering problems based on Monte Carlo simulations with efficient computation integral equations methods, Ph.D. thesis, University of Washington, Seattle.

Pak, K., L. Tsang, and J. Johnson (1997), Numerical simulations and backscattering enhancement of electromagnetic waves from two-dimensional dielectric random rough surfaces with the sparse matrix canonical grid method, *J. Opt. Soc. Am. A*, **14(7)**, 1515–1529.

Peterson, A. F., S. L. Ray, and R. Mittra (1998), *Computational Methods for Electromagnetics*, Oxford University Press, New York.

Rao, S. M., D. R. Wilton, and A. W. Glisson (1982), Electromagnetic scattering by surfaces of arbitrary shape, *IEEE Trans. Antennas Propagat.*, **30(3)**, 409–418.

Rino, C. L. and H. D. Ngo (1994), Application of beam simulation to scattering at low grazing angles. 2. Oceanlike surfaces, *Radio Sci.*, **29(6)**, 1381–1391.

Rino, C. L. and H. D. Ngo (1998), Numerical simulation of low-grazing angle ocean microwave backscatter and its relation to sea spikes, *IEEE Trans. Antennas Propagat.*, **46(1)**, 133–141.

Rokhlin, V. (1983), Rapid solution of integral equations of classic potential theory, *J. of Comp. Phys.*, **60**, 187–207.

Rokhlin, V. (1990), Rapid solution of integral equations of scattering theory in two dimensions, *J. of Comp. Phys.*, **86**, 414–439.

Sarabandi, K. and T. Chiu (1997), Electromagnetic scattering from slightly rough surfaces with inhomogeneous dielectric profiles, *IEEE Trans. Antennas Propagat.*, **45(9)**, 1419–1430.

Schneider, J. B. and S. L. Broschat (1995), The measured equation of invariance method applied to randomly rough surfaces, *Applied Computational Electromagnetics Society Journal*, **10(1)**, 19–30.

Sterling, T. L., J. Salmon, D. J. Becker, and D. F. Savarese (1999), *How to Build a Beowulf, A Guide to the Implementation and Application of PC Clusters*, The MIT Press, Cambridge, MA.

Torrungrueng, D., H.-T. Chou, J. T. Johnson (2000), A novel acceleration algorithm for the computation of scattering from two-dimensional large-scale perfectly conducting random rough surfaces with the forward-backward method, *IEEE Trans. Geosci. Remote Sens.*, **38(4)**, 1656–1668.

Tran, P. (1997), Calculation of the scattering of electromagnetic waves from a two-dimensional perfectly conducting surface using the method of ordered multiple interaction, *Waves in Random Media*, **7(3)**, 295–302.

Tran, P. and J. M. Elson (1998), Banded method of ordered multiple interaction for the scattering of electromagnetic waves from a rough surface, *J. Opt. Soc. Am. A*, **15(6)**, 1643–1646.

Tran, P. and A. A. Maradudin (1993), Scattering of a scalar beam from a two-dimensional randomly rough hard wall: Dirichlet and Neumann boundary conditions, *Appl. Optics*, **32(15)**, 2848–2851.

Voronovich, A. G. (1994), *Wave Scattering from Rough Surfaces*, Springer-Verlag, Berlin.

Wagner, R. L., J. S. Song, and W. C. Chew (1997), Monte Carlo simulation of electromagnetic scattering from two-dimensional random rough surfaces, *IEEE Trans. Antennas Propagat.*, **45(2)**, 235–245.

Wang, J. J. H. (1991), *Generalized Moment Methods in Electromagnetics : Formulation and Computer Solution of Integral Equations*, Wiley-Interscience, New York, NY.

Wang, J. R., P. E. O'Neill, T. J. Jackson, and E. T. Engman (1983), Multi-frequency measurements of the effects of soil moisture, soil texture and surface roughness, *IEEE Trans. Geosci. Remote Sens.*, **21(1)**, 44–51.

Wang, J. R. and T. J. Schmugge (1980), An empirical model for the complex dielectric permittivity of soils as a function of water content, *IEEE Trans. Geosci. Remote Sens.*, **18(4)**, 288–295.

Chapter 7

VOLUME SCATTERING SIMULATIONS

In Chapters 4 to 6 we have studied extensively the simulation of rough surface scattering. In those problems, we have simulated the final results of bistatic scattering coefficients, backscattering coefficients, and emissivities. In this chapter we will perform volume scattering simulations. However, for volume scattering simulations, there is one more dimension than rough surface scattering, and it becomes difficult to simulate the final results of bistatic and backscattering coefficients. Nevertheless, it is still possible to simulate important features of volume scattering. We decompose the problem into two steps. In step 1 we perform volume scattering simulation on a test volume of many scatterers to calculate the extinction coefficients and the phase matrix. In this step 1, the coherent multiple scattering effects among the many scatterers that are within the test volume are included. In step 2 the simulated values of extinction coefficients and phase matrix elements are used in radiative transfer equation to calculate the bistatic scattering coefficients, backscattering coefficients, and emissivities.

Classical vector radiative transfer theory which is studied in Volume I has been used extensively in studying the electromagnetic scattering effects. However, classical radiative transfer theory assumes that the particles scatter independently. This assumption is based on the random phases of scattering by different particles and is valid if the particle positions are independent and the randomness of relative positions is comparable to or larger than a wavelength. Such an assumption can be invalid at microwave frequencies. For example, in dense media with particles closely packed together as in grains in snow and ice, and for scatterers clustered together as in branches and leaves in a vegetation canopy, the randomness in the relative positions of scatterers can be less than a wavelength. In such a case the scatterers scatter collectively, and the effects of scattering have to be taken into account.

In this chapter, we study the collective scattering effects of a conglomeration of particles. The radiative transfer theory is modified by defining the phase matrix and extinction coefficient as, respectively, the bistatic cross section per unit volume of space and the extinction cross section per unit volume of space. The limit of the volume of space is taken to be larger or much larger than a wavelength so that the collective and coherent scattering effects of N particles within the test volume are taken into account. We test the concept of collective scattering by studying the case of point scatterers. The advantage of point scatterers is that the exact solutions of Monte Carlo simulations can be readily computed based on the Foldy-Lax self-consistent multiple scattering equations. We compare the case when the point scatterers are uniformly distributed and the case when the point scatterers are

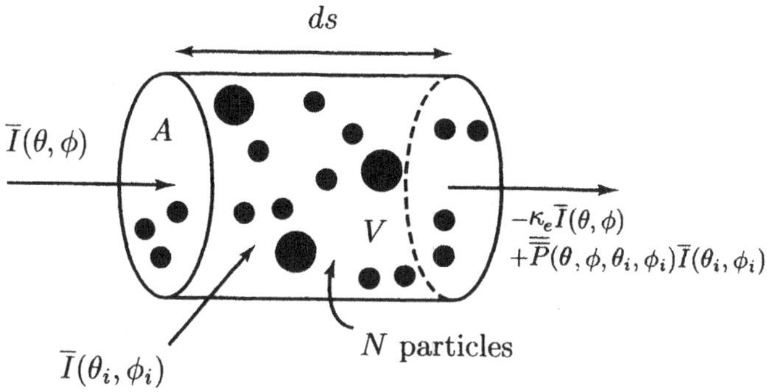

Figure 7.1.1 An elemental volume element V with arbitrary cross section area A and length ds for a random medium containing discrete scatterers. The elemental volume element contains many particles.

in clusters. We also describe the iterative approach for solving the multiple scattering equations. Forward scattering amplitude in conjunction with Foldy's approximation has been used to calculate the extinction rate. It is shown that this is only applicable if the N-particle forward scattering amplitude is calculated to the second order. It is also shown that the coherent wave approaches a delta function in the forward direction in the limit of a large volume and large number of scatterers (large N). Thus, the coherent field intensity has to be subtracted from the total scattered intensity to give (a) the phase matrix which describes the scattering of energy away from the forward direction and (b) the extinction coefficient which describes the wave decay in the forward direction. We perform Monte Carlo simulations by solving the wave equations and averaging over many realizations.

1 Combining Simulations of Collective Volume Scattering Effects with Radiative Transfer Theory

In this section we redefine the scattering and extinction parameters of radiative transfer theory to include the collective scattering behavior of particles. This distinguishes the definition from the conventional definition of averaged single-particle scattering behavior.

In radiative transfer theory we consider a specific intensity incident upon an elemental volume with cross-sectional area A and length ds (Fig. 7.1.1). The cross-sectional shape is arbitrary. It is an elemental test volume in the development of the radiative transfer equation. It is not small in the sense of the wave equation. The elemental volume $V = Ads$ needs to obey the following criteria:

(i) $V \gg \lambda^3$ or $V \gtrsim \lambda^3$. In other words, $V = N_V \lambda^3$, where N_V is a number of moderate size.

(ii) There is a large number of particles, N, contained in V. The particles are randomly positioned.

Criteria (i) and (ii) are necessary to ensure that the coherent multiple scattering in volume V will include random phase fluctuations so that the energy transfer in and out of volume V can be averaged. Furthermore, if the particles are of different sizes, shapes, and permittivities in volume V, averaging of particle scattering behavior will also be included in the radiative energy transfer within elemental test volume V. Criteria (i) and (ii) establish a lower limit of volume V.

Let E_i be the incident field in the direction $\hat{k}_i = (\theta_i, \phi_i)$, and let E_s be the scattered field from volume V of N particles.

$$E_i = \exp(i\overline{k}_i \cdot \overline{r}) \tag{7.1.1}$$

The scattered field can be decomposed into coherent field $\langle E_s \rangle$ and incoherent field $\mathcal{E}_s$.

$$E_s = \langle E_s \rangle + \mathcal{E}_s \tag{7.1.2}$$

The scattered field E_s is a statistical field that contains the coherent multiple scattering within volume V.

From one elemental volume V to most other elemental volumes, the far field of E_s can be taken. We have

$$E_s = F(\hat{k}_s, \hat{k}_i)\frac{\exp(ikr)}{r} \tag{7.1.3}$$

where $\hat{k}_s = (\theta_s, \phi_s)$ is the scattered direction. In (7.1.3), $F(\hat{k}_s, \hat{k}_i)$ represents the N-particle collective scattering amplitude. It can be decomposed into coherent component $\langle F(\hat{k}_s, \hat{k}_i)\rangle$ and incoherent component $\mathcal{F}(\hat{k}_s, \hat{k}_i)$.

$$F(\hat{k}_s, \hat{k}_i) = \langle F(\hat{k}_s, \hat{k}_i)\rangle + \mathcal{F}(\hat{k}_s, \hat{k}_i) \tag{7.1.4}$$

The optical theorem can be applied to N scatterers, which can be viewed as a single object called N-scatterer. Thus, for the case of nonabsorptive scatterers we have

$$\frac{4\pi}{k}\mathrm{Im}\left\{F(\hat{k}_i, \hat{k}_i)\right\} = \int_{4\pi} d\Omega_s |F(\hat{k}_s, \hat{k}_i)|^2 \tag{7.1.5}$$

The phase function $P(\hat{k}_s, \hat{k}_i)$ and the scattering coefficient κ_s are next defined to take into account the collective scattering behavior of N particles.

$$P(\hat{k}_s, \hat{k}_i) = \lim_{V \to \text{large}} \frac{\langle |\mathcal{F}(\hat{k}_s, \hat{k}_i)|^2 \rangle}{V} \tag{7.1.6}$$

$$\kappa_s = \int d\Omega_s P(\hat{k}_s, \hat{k}_i) \tag{7.1.7}$$

The limit in (7.1.6) is taken for fixed n_o, where $n_o = N/V$ is the number of particles per unit volume. Thus the definition of (7.1.6) and (7.1.7) contains collective scattering effects. The results of (7.1.6) and (7.1.7) should also be independent of the shape of elemental test volume V. On the other hand, the coherent N-particle scattering amplitude $\langle F \rangle$ as given by the coherent wave is dependent on the shape of the elemental volume V. Thus the phase function in (7.1.6) only includes the incoherent wave. The phase function and the scattering coefficient defined in (7.1.6) and (7.1.7) are per unit volume quantities in the limit of large volume. Note that the limit with large N and large V is taken for convergence. The coherent field is excluded from the equations defining the extinction coefficient and the phase function. The absorption coefficient is

$$\kappa_a = \frac{P_a}{V \times (\text{incident flux})} \tag{7.1.8}$$

where P_a is the power absorbed by the N particles in volume V that includes coherent wave interaction among the particles. Thus, the absorption coefficient is the absorption cross section per unit volume in the limit of large V. The extinction coefficient is

$$\kappa_e = \kappa_a + \kappa_s \tag{7.1.9}$$

Once the phase matrix and the extinction coefficient are calculated, the radiative transfer equation assumes the following standard form

$$\frac{d}{ds} I(\bar{r}, \hat{s}) = -\kappa_e I(\bar{r}, \hat{s}) + \int d\Omega' P(\hat{s}, \hat{s}') I(\bar{r}, \hat{s}') \tag{7.1.10}$$

To establish the upper limit of elemental volume V for the definition of the phase function and the scattering coefficient, we note that if V is too large, then the definition of (7.1.6) will include the radiative transfer theory of (7.1.10) and is actually the bistatic cross section of the overall medium (rather than the elemental test volume V as used) with multiple incoherent scattering included. Thus, the upper limit of elemental volume V is that V needs to be small enough so that multiple scattering of incoherent waves can be excluded. Multiple scattering of incoherent waves will take place in the distance scale of the mean free path l_{mfp} where

$$l_{mfp} = \frac{1}{\kappa_e} \tag{7.1.11}$$

Thus the third criterion of V is an upper limit and is

(iii) $V \ll \left(\frac{1}{\kappa_e}\right)^3$

The volume V can simultaneously satisfy criteria (i) and (iii) if the mean free path is much larger than the wavelength or the wavenumber is much larger than the extinction coefficient.

$$l_{mfp} \gg \lambda \tag{7.1.12}$$

$$k \gg \kappa_e \tag{7.1.13}$$

The conditions as exhibited in (7.1.12) and (7.1.13) are actually the conditions for transport-type equations to be valid. When the condition is violated so that $l_{mfp} \simeq \lambda$ (known as the Ioffe-Regel criterion) [Sheng, 1990], strong photon localization may take place and radiative transfer-type equations are inappropriate. For real-life applications, conditions (7.1.12) and (7.1.13) are usually obeyed. However, even though (7.1.12) and (7.1.13) are obeyed, there are still cases when backscattering enhancement (weak photon localization) will take place. Backscattering enhancement is a result of cyclical scattering diagrams and is not included in the radiative transfer equations which only include ladder diagrams. We discuss backscattering enhancement in Volume III. For the case of independent scattering of the N scatterers, $\langle|\mathcal{F}(\hat{k}_s, \hat{k}_i)|^2\rangle = N\langle|f|^2\rangle$ so that $P(\hat{k}_s, \hat{k}_i) = n_o\langle|f|^2\rangle$, where $n_o = N/V$ is the number of particles per unit volume and $\langle|f|^2\rangle$ is the averaged single-particle scattering cross section. Thus, (7.1.6) and (7.1.7) reduce to the result of conventional radiative transfer theory. In the following sections we illustrate the applications of these new definitions by using examples of two different types of random distributions of point scatterers.

2 Foldy-Lax Self-Consistent Multiple Scattering Equations

2.1 Final Exciting Field and Multiple Scattering Equation

Consider a distribution of N particles in a volume V (Fig. 7.2.1). Let ψ_{inc} be the incident wave. The wave generally will be scattered by all the particles. Each particle will have incidence upon it, the incident wave and the multiply scattered wave from all other particles. How do we write an expression for the field ψ that is the final answer after multiple scattering? A self-consistent result can be constructed in a manner due to Foldy [1945] and Lax [1952].

Each particle j has a single-particle scattering transition operator T_j. This is the response of the particle when it exists alone. In Volume I we have calculated single-particle scattering properties of scatterers with simple shapes. The operator T_j is an exact transition operator for a single particle

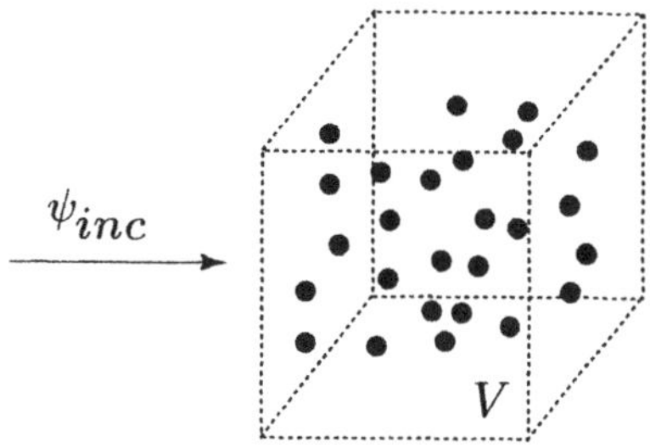

Figure 7.2.1 An incident field $\psi_{inc}(\bar{r})$ impinges on a collection of N particles in volume V.

and includes near-field and far-field effects.

Thus given a single particle j and a field ψ_j^E impinging upon a particle j, the scattered field from particle j is

$$\psi_j^s = G_0 T_j \psi_j^E \tag{7.2.1}$$

where the superscript E denotes exciting field and G_0 is the Green's function (propagator).

Next consider two particles j and l with the incident field ψ_{inc} upon them. The total field includes

$$\psi = \psi_{inc} + G_0 T_l \psi_{inc} + G_0 T_j \psi_{inc} + \cdots \tag{7.2.2}$$

where $G_0 T_l \psi_{inc}$ and $G_0 T_j \psi_{inc}$ are the scattered fields from particle l and particle j, respectively, that "directly" scatters ψ_{inc}.

However, there can be second-order scattering which are $G_0 T_l G_0 T_j \psi_{inc}$ and $G_0 T_j G_0 T_l \psi_{inc}$. The second-order field $G_0 T_l G_0 T_j \psi_{inc}$ scatters the incident field from j to l which further scatters the field. Similarly, $G_0 T_j G_0 T_l \psi_{inc}$ consists of scattering first from particle l and then from particle j. Thus we have

$$\psi = \psi_{inc} + G_o T_l \psi_{inc} + G_0 T_j \psi_{inc} + G_0 T_l G_0 T_j \psi_{inc} + G_0 T_j G_0 T_l \psi_{inc} + \cdots \tag{7.2.3}$$

We can keep repeating for the third order, the fourth order and so on, up to an infinite number of terms

$$\begin{aligned}\psi = {} & \psi_{inc} + G_0 T_l \psi_{inc} + G_0 T_j \psi_{inc} + G_0 T_l G_0 T_j \psi_{inc} + G_0 T_j G_0 T_l \psi_{inc} \\ & + G_0 T_l G_0 T_j G_0 T_l \psi_{inc} + G_0 T_j G_0 T_l G_0 T_j \psi_{inc} \\ & + G_0 T_l G_0 T_j G_0 T_l G_0 T_j \psi_{inc} + G_0 T_j G_0 T_l G_0 T_j G_0 T_l \psi_{inc} + \cdots \end{aligned} \tag{7.2.4}$$

The infinite series in (7.2.4) can be rearranged as follows

$$\psi = \psi_{inc} + G_0T_l\Big(\psi_{inc} + G_0T_j\psi_{inc} + G_0T_jG_0T_l\psi_{inc} + G_0T_jG_0T_lG_0T_j\psi_{inc} + \ldots\Big) + G_0T_j\Big(\psi_{inc} + G_0T_l\psi_{inc} + G_0T_lG_0T_j\psi_{inc} + G_0T_lG_0T_jG_0T_l\psi_{inc} + \cdots\Big) \tag{7.2.5}$$

The first sum in (7.2.5) went through particle l as the last particle, and the second sum in (7.2.5) went through particle j as the last particle. We define the exciting field ψ_l^E to be the sum in the first parentheses and define the exciting field ψ_j^E to be the sum in the second parenthesis. Thus

$$\psi_j^E = \psi_{inc} + G_0T_l\psi_{inc} + G_0T_lG_0T_j\psi_{inc} + G_0T_lG_0T_jG_oT_l\psi_{inc} + G_0T_lG_0T_jG_0T_lG_0T_j\psi_{inc} + \cdots \tag{7.2.6}$$

$$\psi_l^E = \psi_{inc} + G_0T_j\psi_{inc} + G_0T_jG_0T_l\psi_{inc} + G_0T_jG_0T_lG_0T_j\psi_{inc} + G_0T_jG_0T_lG_0T_jG_0T_l\psi_{inc} + \cdots \tag{7.2.7}$$

We also let

$$\psi_l^s = G_0T_l\psi_l^E \tag{7.2.8a}$$

$$\psi_j^s = G_0T_j\psi_j^E \tag{7.2.8b}$$

be the scattered fields from particles l and j, respectively. Then,

$$\psi = \psi_{inc} + \psi_l^s + \psi_j^s \tag{7.2.9}$$

We also have

$$\psi = \psi_{inc} + \psi_s \tag{7.2.10}$$

$$\psi_s = \psi_l^s + \psi_j^s \tag{7.2.11}$$

Thus ψ_j^E of (7.2.6) represents the "final" exciting field that excites particle j. It expresses the idea that after going through multiple scattering between the two particles, this is the field that is finally exciting the particle j. Similarly ψ_l^s is the "final" scattered field from particle l.

Next, the infinite series in (7.2.6) can be manipulated as follows:

$$\psi_j^E = \psi_{inc} + G_0T_l(\psi_{inc} + G_0T_j\psi_{inc} + G_0T_jG_0T_l\psi_{inc} + G_0T_jG_0T_lG_0T_j\psi_{inc} + \cdots) \tag{7.2.12}$$

Comparing with (7.2.7), it is clear that the term in the parentheses in (7.2.12) is ψ_l^E. Thus

$$\psi_j^E = \psi_{inc} + G_0T_l\psi_l^E \tag{7.2.13}$$

Similarly from (7.2.7) we obtain

$$\psi_l^E = \psi_{inc} + G_0T_j\psi_j^E \tag{7.2.14}$$

As derived above, the concepts of final exciting field and final scattered field are rigorous.

Note that $G_0 T_j \psi_j^E$ does not appear on the right-hand side of (7.2.13) because the final exciting field of particle j does not excite itself. Equations (7.2.13) and (7.2.14) are the self-consistent multiple scattering equations for the exciting fields of two particles. After the exciting fields are solved, (7.2.8) and (7.2.11) are used to calculate the total scattered field. Note that the Foldy-Lax multiple scattering equations for the two particles as given by (7.2.13) and (7.2.14) are exact. It is also important to recognize that T_l and T_j are only the single-particle scattering transition operator.

The equations of two-particle scattering can be readily generalized to N particles. For N particles

$$\psi_j^E = \psi_{inc} + \sum_{\substack{l=1 \\ l \neq j}}^{N} G_0 T_l \psi_l^E \tag{7.2.15}$$

where $j = 1, \ldots, N$. The total scattered field is

$$\psi_s = \sum_{j=1}^{N} \psi_j^s \tag{7.2.16}$$

$$\psi_j^s = G_0 T_j \psi_j^E \tag{7.2.17}$$

The total field is

$$\psi = \psi_{inc} + \psi_s \tag{7.2.18}$$

Equations (7.2.15)–(7.2.18) are the Foldy-Lax self-consistent multiple scattering equations. They are derivable from Maxwell's equations and are exact relations without approximations [Peterson and Strom, 1973; Tsang et al. 1985]. Generally, (7.2.15) consists of N equations with N unknowns ψ_j^E, $j = 1, \ldots, N$. In principle they can be solved numerically to yield the exact result of the multiple scattering equations.

2.2 Foldy-Lax Equations for Point Scatterers

For the cases of point scatterers, the transition operator T_j is simple so that the Foldy-Lax multiple scattering equations assume a simple form. The point scatterer has a simple single-particle scattering relation of

$$f \frac{e^{ikr}}{r} \tag{7.2.19}$$

where f is an isotropic quantity. To obey the optical theorem, f must be complex. The scattering cross section is

$$\sigma_s = \int d\Omega_s |f|^2 = 4\pi |f|^2 \tag{7.2.20}$$

Hence, applying the forward scattering theorem, we have

$$\sigma_a + 4\pi |f|^2 = \frac{4\pi}{k} \text{Im} f \tag{7.2.21}$$

where σ_a is the absorption cross section of the particle. Thus we assume that

$$f = f' + i f'' \tag{7.2.22}$$

and $|f''| \ll |f'|$ so that

$$f'' = k f'^2 + \frac{k\sigma_a}{4\pi} \tag{7.2.23}$$

If $\sigma_a = 0$, then $f'' = k f'^2$. For a point scatterer,

$$T_j = -4\pi f \delta(\overline{r} - \overline{r}_j) \tag{7.2.24}$$

so that

$$G_o T_j \psi_j^E = f \frac{e^{ik|\overline{r} - \overline{r}_j|}}{|\overline{r} - \overline{r}_j|} \psi_j^E(\overline{r}_j) \tag{7.2.25}$$

For N point particles centered at $\overline{r}_1, \ldots, \overline{r}_N$, the Foldy-Lax multiple scattering equations assume the following form

$$\psi_j^E(\overline{r}) = \psi_{inc}(\overline{r}) + \sum_{\substack{l=1 \\ l \neq j}}^{N} f \frac{e^{ik|\overline{r} - \overline{r}_l|}}{|\overline{r} - \overline{r}_l|} \psi_l^E(\overline{r}_l) \tag{7.2.26}$$

$$\psi(\overline{r}) = \psi_{inc}(\overline{r}) + \sum_{l=1}^{N} f \frac{e^{ik|\overline{r} - \overline{r}_l|}}{|\overline{r} - \overline{r}_l|} \psi_l^E(\overline{r}_l) \tag{7.2.27}$$

$$\psi_j^s(\overline{r}) = f \frac{e^{ik|\overline{r} - \overline{r}_j|}}{|\overline{r} - \overline{r}_j|} \psi_j^E(\overline{r}_j) \tag{7.2.28}$$

The total scattered field and the total field are, respectively,

$$\psi_s(\overline{r}) = \sum_{l=1}^{N} f \frac{e^{ik|\overline{r} - \overline{r}_l|}}{|\overline{r} - \overline{r}_l|} \psi_l^E(\overline{r}_l) \tag{7.2.29}$$

$$\psi(\overline{r}) = \psi_{inc}(\overline{r}) + \psi_s(\overline{r}) \tag{7.2.30}$$

If an iterative approach is used to solve (7.2.26), then the first-order solution is

$$\psi_l^{E(1)}(\overline{r}_l) = \psi_{inc}(\overline{r}_l) \tag{7.2.31}$$

for the exciting field. For the first-order scattered field we have

$$\psi_s^{(1)}(\overline{r}) = \sum_{l=1}^{N} f \frac{e^{ik|\overline{r}-\overline{r}_l|}}{|\overline{r}-\overline{r}_l|} \psi_{inc}(\overline{r}_l) \tag{7.2.32a}$$

$$\psi_s^{(1)}(\overline{r}) = \sum_{l=1}^{N} \psi_l^{s(1)}(\overline{r}) \tag{7.2.32b}$$

where

$$\psi_l^{s(1)}(\overline{r}) = f \frac{e^{ik|\overline{r}-\overline{r}_l|}}{|\overline{r}-\overline{r}_l|} \psi_{inc}(\overline{r}_l) \tag{7.2.32c}$$

The higher-order solutions are, for $n \geq 1$,

$$\psi_j^{E(n+1)}(\overline{r}) = \psi_{inc}(\overline{r}) + \sum_{\substack{l=1 \\ l\neq j}}^{N} f \frac{e^{ik|\overline{r}-\overline{r}_l|}}{|\overline{r}-\overline{r}_l|} \psi_l^{E(n)}(\overline{r}_l) \tag{7.2.33}$$

$$\psi_s^{(n+1)}(\overline{r}) = \sum_{l=1}^{N} f \frac{e^{ik|\overline{r}-\overline{r}_l|}}{|\overline{r}-\overline{r}_l|} \psi_l^{E(n+1)}(\overline{r}_l) \tag{7.2.34}$$

The Foldy-Lax multiple scattering equation (7.2.26) can be cast in a matrix form. Let

$$\psi_j = \psi_j^E(\overline{r}_j) \tag{7.2.35}$$

$$A_{ij} = f \frac{e^{ik|\overline{r}_i-\overline{r}_j|}}{|\overline{r}_i-\overline{r}_j|} \tag{7.2.36}$$

$$b_j = \psi_{inc}(\overline{r}_j) \tag{7.2.37}$$

Then (7.2.26) becomes

$$\psi_j = b_j + \sum_{\substack{l=1 \\ l\neq j}}^{N} A_{jl}\psi_l \tag{7.2.38}$$

If we further let

$$Z_{ij} = \begin{cases} 1 & \text{if } i = j \\ -A_{ij} & \text{if } i \neq j \end{cases} \tag{7.2.39}$$

then we have the following matrix equation:

$$\sum_{l=1}^{N} Z_{jl}\psi_l = b_j \tag{7.2.40}$$

In matrix notation

$$\overline{\overline{Z}} \cdot \overline{\psi} = \overline{b} \tag{7.2.41}$$

where $\overline{\overline{Z}}$ is the impedance matrix, $\overline{\psi}$ is the unknown column vector of the exciting fields of the particles, and $\overline{b}$ is the right-hand side. The iterative approach that is used in (7.2.31)–(7.2.34) is called the direct iteration, and the Neumann series or the Born series. The first-order term is known as the Born approximation.

2.3 The N-Particle Scattering Amplitude

The matrix equation (7.2.40) can be solved exactly and the exciting fields can be calculated. The final scattered field is, in the far field, from (7.2.29),

$$\psi_s = \sum_{l=1}^{N} f \frac{e^{ik|\overline{r}-\overline{r}_l|}}{|\overline{r}-\overline{r}_l|} \psi_l^E(\overline{r}_l) = \frac{e^{ikr}}{r} \sum_{l=1}^{N} f e^{-i\overline{k}_s \cdot \overline{r}_l} \psi_l^E(\overline{r}_l) \tag{7.2.42}$$

where $\overline{k}_s = k\hat{k}_s$ and $\hat{k}_s$ is the scattered direction. Thus if we regard the N particles as a single scattering object, the N-particle, then

$$\psi_s = F(\hat{k}_s, \hat{k}_i) \frac{e^{ikr}}{r} \tag{7.2.43}$$

where $F(\hat{k}_s, \hat{k}_i)$ is the N-particle bistatic scattering amplitude including collective scattering effects. Comparing (7.2.43) and (7.2.42) gives

$$F(\hat{k}_s, \hat{k}_i) = \sum_{l=1}^{N} f e^{-i\overline{k}_s \cdot \overline{r}_l} \psi_l^E(\overline{r}_l) \tag{7.2.44}$$

3 Analytical Solutions of Point Scatterers

We shall consider two cases of point scatterers. In the first case, the point scatterers are uniformly randomly distributed. In the second case, the point scatterers form clusters. We study scattering by a collection of clusters. We shall show that even though the two cases have the same average number of scatterers per unit volume, their scattering properties are quite different.

3.1 Phase Function and Extinction Coefficient for Uniformly Distributed Point Scatterers

In this section we illustrate the phase function and the extinction coefficient for nonabsorptive point scatterers using the definition of (7.1.6) and (7.1.7). The Foldy-Lax self-consistent multiple scattering equations will be solved to second order.

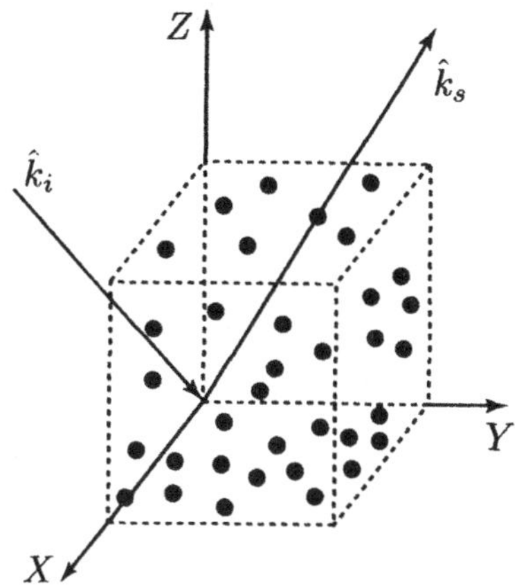

Figure 7.3.1 Monte Carlo simulation geometry. Many point particles are randomly and uniformly distributed in a cubic box with volume V.

Consider an incident plane wave E_{inc} in the direction $\hat{k}_i$ impinging upon a volume V, the size of which obeys the three criteria of Section 1. The volume V contains N number of nonabsorptive point scatterers located at $\overline{r}_1, \overline{r}_2, \ldots, \overline{r}_N$ (Fig. 7.3.1). The point scatterers are uniformly and randomly distributed in volume V.

The multiple scattering equations for the "final" exciting field E^j_{ex} are, from (7.2.26),

$$E^j_{ex}(\overline{r}_j) = E_{inc}(\overline{r}_j) + \sum_{\substack{l=1 \\ l\neq j}}^{N} f \frac{\exp(ik|\overline{r}_j - \overline{r}_l|)}{|\overline{r}_j - \overline{r}_l|} E^l_{ex}(\overline{r}_l) \tag{7.3.1}$$

where $j = 1, 2, 3, \ldots, N$. After the exciting fields $E^j_{ex}, j = 1, 2, \ldots, N$, are solved, then the total field is given by

$$E(\overline{r}) = E_{inc}(\overline{r}) + \sum_{l=1}^{N} f \frac{\exp(ik|\overline{r} - \overline{r}_l|)}{|\overline{r} - \overline{r}_l|} E^l_{ex}(\overline{r}_l) \tag{7.3.2}$$

From (7.2.44), the scattering amplitude for N particles is

$$F(\hat{k}_s, \hat{k}_i) = \sum_{l=1}^{N} f \ \exp(-i\overline{k}_s \cdot \overline{r}_l) E^l_{ex}(\overline{r}_l) \tag{7.3.3}$$

The conventional radiative transfer theory gives the phase function and scattering coefficient as

$$P(\hat{k}_s, \hat{k}_i) = n_o |f|^2 \tag{7.3.4}$$

$$\kappa_s = 4\pi n_o |f|^2 \tag{7.3.5}$$

The particle positions are discussed in Chapter 4, Section 5 of Volume I. From (4.5.18) of Volume I

$$\langle n^{(1)}(\overline{r})\rangle = Np(\overline{r}) = n_o \tag{7.3.6}$$

so that

$$p(\overline{r}) = \frac{1}{V} \tag{7.3.7}$$

is the single-particle probability density function. From (4.5.21) and (4.5.22) of Volume I

$$\langle n^{(2)}(\overline{r},\overline{r}')\rangle = N(N-1)p_2(\overline{r},\overline{r}') \tag{7.3.8a}$$

where $p_2(\overline{r},\overline{r}')$ is the joint probability density function. Furthermore,

$$\langle n^{(2)}(\overline{r},\overline{r}')\rangle = \langle n^{(1)}(\overline{r})\rangle\langle n^{(1)}(\overline{r}')\rangle g(\overline{r},\overline{r}') \tag{7.3.8b}$$

where g is the pair distribution function. Hence

$$p_2(\overline{r},\overline{r}') = \frac{n_o^2}{N(N-1)}g(\overline{r},\overline{r}') = \frac{N}{(N-1)V^2}g(\overline{r},\overline{r}') \tag{7.3.8c}$$

As the particle separation becomes large, their joint probability density function should be independent. Thus

$$\lim_{r\to\infty} g(\overline{r}) = 1 \tag{7.3.9}$$

If we view the N particles as a single object, the N-particle, the forward scattering theorem should also be applicable to the N-particle scattering amplitude. Thus

$$\frac{4\pi}{k}\mathrm{Im}F(\hat{k}_s,\hat{k}_i) = \int d\Omega_s|F(\hat{k}_s,\hat{k}_i)|^2 \tag{7.3.9a}$$

First-Order Solution

For the first-order solution we have

$$E_{ex}^j(\overline{r}_j) = \exp(i\overline{k}_i\cdot\overline{r}_j) \tag{7.3.10}$$

Then, the first-order N-particle scattering amplitude is

$$F^{(1)}(\hat{k}_s,\hat{k}_i) = f\sum_{j=1}^{N}\exp(i\overline{k}_d\cdot\overline{r}_j) \tag{7.3.11}$$

where $\overline{k}_d = \overline{k}_i - \overline{k}_s$. From (7.3.11), we see that the random positions of $\overline{r}_j$ will give random phase fluctuations except in the forward direction of $\overline{k}_d = 0$.

Taking the configurational average of (7.3.11) gives

$$\langle F^{(1)}(\hat{k}_s,\hat{k}_i)\rangle = f\sum_{j=1}^{N}\int_V d\overline{r}_j \exp(i\overline{k}_d\cdot\overline{r}_j)p(\overline{r}_j) = \frac{f}{V}\sum_{j=1}^{N}\int_V d\overline{r}_j \exp(i\overline{k}_d\cdot\overline{r}_j) \tag{7.3.12}$$

Note that before averaging, each $\overline{r}_j$ is distinct as given by (7.3.11). However, $\overline{r}_j$ becomes a dummy integration variable on averaging. The summation over j can now be replaced by N. We thus have

$$\langle F^{(1)}(\hat{k}_s,\hat{k}_i)\rangle = n_o f\int_V d\overline{r}_j \exp(i\overline{k}_d\cdot\overline{r}_j) \tag{7.3.13}$$

Since the volume V is much larger than a wavelength, the coherent field as given by (7.3.13) is sharply peaked in the forward direction of $\hat{k}_s = \hat{k}_i$. This concept of strong forward scattering of a coherent wave is evident in wave theory.

$$\lim_{V\to\infty}\langle F^{(1)}(\hat{k}_s,\hat{k}_i)\rangle = n_o f(2\pi)^3\delta(\overline{k}_d) \tag{7.3.14}$$

The absolute value of the amplitude squared is

$$|F^{(1)}(\hat{k}_s,\hat{k}_i)|^2 = \sum_{j=1}^{N}\sum_{l=1}^{N}|f|^2\exp(i\overline{k}_d\cdot(\overline{r}_j-\overline{r}_l)) \tag{7.3.15}$$

The double summation is next separated into $j = l$ and $j \neq l$, that is scattering from the same particle and scattering from two different particles.

$$|F^{(1)}(\hat{k}_s,\hat{k}_i)|^2 = N|f|^2 + \sum_{j=1}^{N}\sum_{\substack{l=1\\ l\neq j}}^{N}|f|^2\exp(i\overline{k}_d\cdot(\overline{r}_j-\overline{r}_l)) \tag{7.3.16}$$

Next the average of (7.3.16) is taken with the second term averaged over the two-particle joint probability density function of (7.3.8c)

$$\langle|F^{(1)}(\hat{k}_s,\hat{k}_i)|^2\rangle = N|f|^2 + \sum_{j=1}^{N}\sum_{\substack{l=1\\ l\neq j}}^{N}\int d\overline{r}_j\int d\overline{r}_l|f|^2 e^{i\overline{k}_d\cdot(\overline{r}_j-\overline{r}_l)}p_2(\overline{r}_j,\overline{r}_l)$$

$$= N|f|^2 + \sum_{j=1}^{N}\sum_{\substack{l=1\\ l\neq j}}^{N}\int d\overline{r}_j\int d\overline{r}_l|f|^2 e^{i\overline{k}_d\cdot(\overline{r}_j-\overline{r}_l)}\frac{N}{(N-1)V^2}g(\overline{r}_j-\overline{r}_l) \tag{7.3.17}$$

Note that on integration both $\overline{r}_j$ and $\overline{r}_l$ become dummy variables. Then

$$\langle|F^{(1)}(\hat{k}_s,\hat{k}_i)|^2\rangle = N|f|^2 + \frac{N^2}{V^2}|f|^2\int d\overline{r}_j\int d\overline{r}_l e^{i\overline{k}_d\cdot(\overline{r}_j-\overline{r}_l)}g(\overline{r}_j-\overline{r}_l)$$

$$= N|f|^2 + \frac{N^2}{V^2}|f|^2 \int d\overline{r}_j \int d\overline{r}_l e^{i\overline{k}_d\cdot(\overline{r}_j-\overline{r}_l)} [g(\overline{r}_j - \overline{r}_l) - 1]$$

$$+ \frac{N^2}{V^2}|f|^2 \int d\overline{r}_j \int d\overline{r}_l e^{i\overline{k}_d\cdot(\overline{r}_j-\overline{r}_l)} \tag{7.3.18}$$

The first term in (7.3.18) represents that of conventional radiative transfer theory and the second term represents correlation effects. The last term in (7.3.18) corresponds to the coherent intensity that is in the forward direction. However, from (7.3.11) we have

$$\frac{1}{k}\text{Im}\{F^{(1)}(\hat{k}_i, \hat{k}_i)\} = \frac{N}{k}\text{Im}f = N|f|^2 \tag{7.3.19}$$

Thus, (7.3.19) only contains the first term of (7.3.18) and does not contain the pair distribution function as in (7.3.18), nor does it contain the sharply peaked forward scattering. Thus, the first-order solution alone does not obey the optical theorem of (7.1.5). For the optical scattering theorem to be obeyed, it shall be shown that the ImF has to be carried out to the second order in scattering. Coherent forward scattering does not contribute to the phase function nor to the scattering coefficient. Next we calculate the incoherent bistatic scattering intensity. Define incoherent scattering amplitude by

$$\mathcal{F}(\hat{k}_s, \hat{k}_i) = F(\hat{k}_s, \hat{k}_i) - \langle F(\hat{k}_s, \hat{k}_i)\rangle \tag{7.3.20a}$$

and

$$\langle|\mathcal{F}(\hat{k}_s, \hat{k}_i)|^2\rangle = \langle|F(\hat{k}_s, \hat{k}_i)|^2\rangle - |\langle F(\hat{k}_s, \hat{k}_i)\rangle|^2 \tag{7.3.20b}$$

Note that from (7.3.13) we ave

$$|\langle F^{(1)}(\hat{k}_s, \hat{k}_i)\rangle|^2 = \frac{N^2}{V^2}|f|^2 \int_V d\overline{r}_j \int_V d\overline{r}_l \exp(i\overline{k}_d \cdot (\overline{r}_j - \overline{r}_l)) \tag{7.3.20c}$$

Note that (7.3.20c) is identical to the last term of (7.3.18). To first order we have

$$\mathcal{F}^{(1)}(\hat{k}_s, \hat{k}_i) = F^{(1)}(\hat{k}_s, \hat{k}_i) - \langle F^{(1)}(\hat{k}_s, \hat{k}_i)\rangle \tag{7.3.20d}$$

Thus, if we take the average of the absolute value squared of $\mathcal{F}^{(1)}$ by subtracting (7.3.20c) from (7.3.18), we obtain

$$\langle|\mathcal{F}^{(1)}(\hat{k}_s, \hat{k}_i)|^2\rangle = \langle|F^{(1)}(\hat{k}_s, \hat{k}_i)|^2\rangle - |\langle F^{(1)}(\hat{k}_s, \hat{k}_i)\rangle|^2$$

$$= N|f|^2 + \frac{N^2}{V^2}|f|^2 \int d\overline{r}_j \int d\overline{r}_l \exp(i\overline{k}_d \cdot (\overline{r}_j - \overline{r}_l))[g(\overline{r}_j - \overline{r}_l) - 1]$$

$$= N|f|^2 + \frac{N^2}{V}|f|^2 \int d\overline{r} \exp(i\overline{k}_d \cdot \overline{r})[g(\overline{r}) - 1] \tag{7.3.21}$$

We further divide (7.3.21) by V to calculate the phase function:

$$P(\hat{k}_s, \hat{k}_i) = \frac{\langle|\mathcal{F}^{(1)}(\hat{k}_s, \hat{k}_i)|^2\rangle}{V}$$
$$= n_o|f|^2 + n_o^2|f|^2 \int d\bar{r} \exp(i\bar{k}_d \cdot \bar{r})[g(\bar{r}) - 1] \qquad (7.3.22)$$

In the conventional radiative transfer theory, $P(\hat{k}_s, \hat{k}_i) = n_o|f|^2$. Thus, the phase function of collective scattering is not the same as conventional theory except when the particle positions are independent, that is, $g = 1$.

We next carry out an angular integration over the scattered directions in (7.3.22). We make use of the integral identity

$$\int_{4\pi} d\Omega_s \exp(-i\bar{k}_s \cdot \bar{r}) = 4\pi \frac{\sin(kr)}{kr} \qquad (7.3.23)$$

Integrating over the scattered angles, we obtain

$$\int P(\hat{k}_s, \hat{k}_i) d\Omega_s = 4\pi n_o|f|^2 + n_o^2 \int d\bar{r} |f|^2 \Big(g(\bar{r}) - 1\Big) \frac{4\pi \sin(kr)}{kr} \exp(i\bar{k}_i \cdot \bar{r}) \qquad (7.3.24)$$

We next examine the consequence if the coherent intensity is included in the phase matrix. Let the phase matrix of coherent wave be P_c :

$$P_c(\hat{k}_s, \hat{k}_i) = \frac{1}{V}|\langle F(\hat{k}_s, \hat{k}_i)\rangle|^2 = \frac{n_o^2|f|^2}{V} \int_V d\bar{r}_j \exp(i\bar{k}_d \cdot \bar{r}_j) \int_V d\bar{r}_l \exp(-i\bar{k}_d \cdot \bar{r}_l)$$
$$= n_o^2|f|^2 \int d\bar{r} \exp(i\bar{k}_d \cdot \bar{r}) \qquad (7.3.25)$$

Integration of (7.3.25) over scattered directions gives

$$\int d\Omega_s P_c(\hat{k}_s, \hat{k}_i) = n_o^2|f|^2 \int_V d\bar{r} \exp(i\bar{k}_i \cdot \bar{r}) 4\pi \frac{\sin(kr)}{kr}$$
$$= n_o^2|f|^2 \int_V d\bar{r} \cos(\bar{k}_i \cdot \bar{r}) 4\pi \frac{\sin(kr)}{kr} \qquad (7.3.26)$$

The integration of P_c over directions gives the power contained in the coherent wave. Summation of the integration of P and P_c then gives, from (7.3.26) and (7.3.24)

$$\int d\Omega_s \Big(P(\hat{k}_s, \hat{k}_i) + P_c(\hat{k}_s, \hat{k}_i)\Big)$$
$$= 4\pi n_o|f|^2 + n_o^2 \int_V d\bar{r} |f|^2 \cos(\bar{k}_i \cdot \bar{r}) \Big(g(\bar{r}) - 1\Big) 4\pi \frac{\sin(kr)}{kr}$$
$$+ n_o^2 \int_V d\bar{r} |f|^2 \cos(\bar{k}_i \cdot \bar{r}) 4\pi \frac{\sin(kr)}{kr} \qquad (7.3.27)$$

As clear from (7.3.25), the phase matrix of the coherent wave is a Dirac delta function in the forward direction in the limit of large V. Its power as given by (7.3.26) is nonconvergent with large V and also depends on the shape of V. Purely forward scattering does not affect radiative transfer, which describes the redistribution of radiative energy in different directions. This further justifies the exclusion of the coherent wave in the phase matrix.

Second-Order Solution

Next, we show that energy conservation is obeyed if we include second-order scattering amplitude in the forward direction. In the second-order solution, the exciting field is

$$E_j^{ex} = \exp(i\overline{k}_i \cdot \overline{r}_j) + \sum_{\substack{l=1 \\ l \neq j}}^{N} f \frac{\exp(ik|\overline{r}_j - \overline{r}_l|)}{|\overline{r}_j - \overline{r}_l|} \exp(i\overline{k}_i \cdot \overline{r}_l) \tag{7.3.28}$$

Putting (7.3.28) into (7.3.3) gives the N-particle scattering amplitude as

$$\begin{aligned} F(\hat{k}_s, \hat{k}_i) &= \sum_{l=1}^{N} f \ \exp(i\overline{k}_i \cdot \overline{r}_l) \exp(-i\overline{k}_s \cdot \overline{r}_l) \\ &+ \sum_{l=1}^{N} f \ \exp(-i\overline{k}_s \cdot \overline{r}_l) \sum_{\substack{j=1 \\ j \neq l}}^{N} f \frac{\exp(ik|\overline{r}_j - \overline{r}_l|)}{|\overline{r}_j - \overline{r}_l|} \exp(i\overline{k}_i \cdot \overline{r}_j) \end{aligned} \tag{7.3.29}$$

The forward scattering amplitude to second order is

$$F^{(2)}(\hat{k}_i, \hat{k}_i) = Nf + \sum_{l=1}^{N} \sum_{\substack{j \neq l \\ j=1}}^{N} f^2 \frac{\exp(ik|\overline{r}_j - \overline{r}_l|)}{|\overline{r}_j - \overline{r}_l|} \exp(-i\overline{k}_i \cdot (\overline{r}_l - \overline{r}_j)) \tag{7.3.30}$$

where $F^{(2)}$ is the sum of the first-order term and the second-order term. Taking the average using $p_2(\overline{r}_j, \overline{r}_l)$ as given by (7.3.8c)

$$\begin{aligned} \langle F^{(2)}(\hat{k}_i, \hat{k}_i) \rangle &= Nf + \sum_{l=1}^{N} \sum_{\substack{j \neq l \\ j=1}}^{N} f^2 \int d\overline{r}_j \int d\overline{r}_l \frac{\exp(ik|\overline{r}_j - \overline{r}_l|)}{|\overline{r}_j - \overline{r}_l|} \\ &\quad \cdot \exp(i\overline{k}_i \cdot (\overline{r}_l - \overline{r}_j)) \frac{N}{(N-1)V^2} g(\overline{r}_l, \overline{r}_j) \\ &= Nf + \frac{N^2}{V^2} f^2 \int d\overline{r}_j \int d\overline{r}_l \frac{\exp(ik|\overline{r}_j - \overline{r}_l|)}{|\overline{r}_j - \overline{r}_l|} \\ &\quad \cdot \exp(i\overline{k}_i \cdot (\overline{r}_l - \overline{r}_j)) g(\overline{r}_l - \overline{r}_j) \end{aligned} \tag{7.3.31}$$

Using the property that $g(\overline{r}) = g(-\overline{r})$ gives

$$\begin{aligned}
&\langle F^{(2)}(\hat{k}_i, \hat{k}_i)\rangle \\
&= Nf + \frac{N^2}{V^2} f^2 \int d\overline{r}_j \int d\overline{r}_l \frac{\exp(ik|\overline{r}_j - \overline{r}_l|)}{|\overline{r}_j - \overline{r}_l|} \cos(\overline{k}_i \cdot (\overline{r}_l - \overline{r}_j)) g(\overline{r}_l - \overline{r}_j) \\
&= Nf + \frac{N^2}{V} f^2 \int d\overline{r} \frac{\exp(ikr)}{r} \cos(\overline{k}_i \cdot \overline{r}) g(\overline{r})
\end{aligned} \tag{7.3.32}$$

To verify optical theorem, we take the imaginary part of (7.3.32)

$$\begin{aligned}
&\frac{4\pi}{k} \frac{\text{Im}\{\langle F^{(2)}(\hat{k}_i, \hat{k}_i)\rangle\}}{V} \\
&= \frac{4\pi}{k} n_o \text{Im}\{f\} + n_o^2 f^2 \int_V d\overline{r} \frac{4\pi \sin(kr)}{kr} \cos(\overline{k}_i \cdot \overline{r})[g(\overline{r}) - 1] \\
&\quad + n_o^2 f^2 \int_V d\overline{r} \frac{4\pi \sin(kr)}{kr} \cos(\overline{k}_i \cdot \overline{r})
\end{aligned} \tag{7.3.33}$$

Comparing the right-hand sides of (7.3.33) and (7.3.27) shows that they agree with each other. Thus

$$\begin{aligned}
&\int d\Omega_s \left(P(\hat{k}_s, \hat{k}_i) + P_c(\hat{k}_s, \hat{k}_i)\right) \\
&= \frac{4\pi}{k} \frac{\text{Im}\{\langle F^{(2)}(\hat{k}_i, \hat{k}_i)\rangle\}}{V} \\
&= \int d\Omega_s \left[\frac{|\mathcal{F}^{(1)}(\hat{k}_s, \hat{k}_i)|^2}{V} + \frac{|\langle F^{(1)}(\hat{k}_s, \hat{k}_i)\rangle|^2}{V}\right]
\end{aligned} \tag{7.3.34}$$

Thus, to apply the optical theorem, the N-particle forward scattering amplitude has to be calculated to the second order so that

$$\begin{aligned}
\kappa_e &= \lim_{V \to large} \int d\Omega_s \frac{|\mathcal{F}^{(1)}(\hat{k}_s, \hat{k}_i)|^2}{V} \\
&= \lim_{V \to large} \left[\frac{4\pi \text{Im}\{\langle F^{(2)}(\hat{k}_i, \hat{k}_i)\rangle\}}{kV} - \frac{1}{V} \int d\Omega_s |\langle F^{(1)}(\hat{k}_s, \hat{k}_i)\rangle|^2\right]
\end{aligned} \tag{7.3.35}$$

3.2 Scattering by Collection of Clusters

We next consider scattering by collection of clusters. Let each cluster be labeled as a primary scatterer, and let the point scatterers within each cluster be labeled as secondary scatterers. Then the phase function and extinction coefficients depend on g_s, which is the pair distribution function among secondary scatterers within a primary scatterer as well as the pair function between clusters g_p.

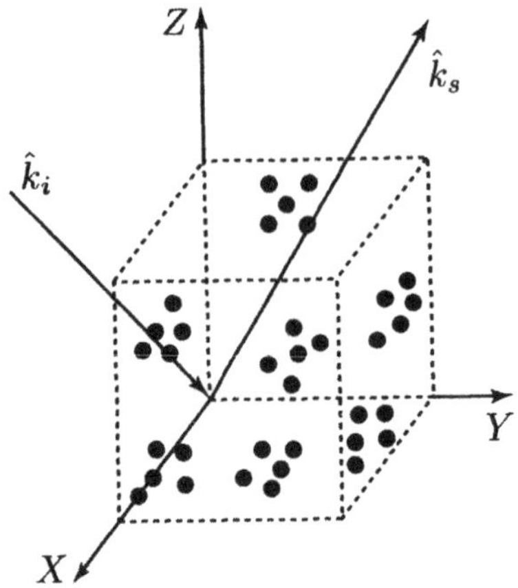

Figure 7.3.2 Clustered point particles are randomly distributed in a cubic box of size L with volume V. The clusters are randomly distributed, and within each cluster of size l_c the particles are randomly distributed.

Consider a volume element V as defined in Section 1. The volume contains N_p primary scatterers (clusters), each of which consists of N_s secondary point scatterers (Fig. 7.3.2). The N_p clusters are centered at $\overline{r}_\alpha$, $\alpha = 1, 2, \ldots, N_p$, and within each cluster α the secondary scatterers are centered at $\overline{r}_{\alpha j}$ with respect to the center of the αth cluster, $j = 1, 2, \ldots, N_s$. Thus

$$N = N_s N_p \tag{7.3.36}$$

is the total number of particles in volume V. Then

$$n_o = \frac{N}{V} = N_s n_p \tag{7.3.37}$$

is the number of particles per unit volume and $n_p = N_p/V$ is the number of clusters per unit volume. Note that N and N_p are large numbers in V while N_s does not have to be large. Then, from (7.3.11), the first-order collective scattering amplitude is

$$F^{(1)}(\hat{k}_s, \hat{k}_i) = f \sum_{j=1}^{N} \exp(i\overline{k}_d \cdot \overline{r}_j) = f \sum_{\alpha=1}^{N_p} \sum_{j=1}^{N_s} \exp(i\overline{k}_d \cdot (\overline{r}_\alpha + \overline{r}_{\alpha j})) \tag{7.3.38}$$

where $\overline{r}_j = \overline{r}_\alpha + \overline{r}_{\alpha j}$. This can be expressed in terms of the scattering amplitude of the primary scatterers with

$$f_\alpha^P(\hat{k}_s, \hat{k}_i) = f \sum_{j=1}^{N_s} \exp(i\overline{k}_d \cdot \overline{r}_{\alpha j}) \tag{7.3.39a}$$

so that

$$F^{(1)}(\hat{k}_s, \hat{k}_i) = \sum_{\alpha=1}^{N_p} \exp(i\overline{k}_d \cdot \overline{r}_\alpha) f_\alpha^P(\hat{k}_s, \hat{k}_i) \tag{7.3.39b}$$

The phase function is

$$P(\hat{k}_s, \hat{k}_i) = \frac{\langle |F^{(1)}(\hat{k}_s, \hat{k}_i)|^2\rangle - |\langle F^{(1)}(\hat{k}_s, \hat{k}_i)\rangle|^2}{V} \tag{7.3.40a}$$

From (7.3.39b) we obtain

$$\frac{\langle |F^{(1)}(\hat{k}_s, \hat{k}_i)|^2\rangle}{V} = \frac{1}{V}\Bigg\{\sum_{\alpha=1}^{N_p} \langle |f_\alpha^P(\hat{k}_s, \hat{k}_i)|^2\rangle$$

$$+ \sum_{\alpha=1}^{N_p}\sum_{\substack{\beta=1\\ \beta\neq\alpha}}^{N_p} \exp(i\overline{k}_d\cdot(\overline{r}_\alpha - \overline{r}_\beta))\langle f_\alpha^P(\hat{k}_s, \hat{k}_i) f_\beta^{P*}(\hat{k}_s, \hat{k}_i)\rangle\Bigg\}$$

$$= \frac{1}{V}\Bigg\{N_p\langle |f_\alpha^P(\hat{k}_s, \hat{k}_i)|^2\rangle$$

$$+ \frac{N_p^2}{V}\int d\overline{r} \exp(i\overline{k}_d\cdot\overline{r})\langle f_\alpha^P(\hat{k}_s, \hat{k}_i) f_\beta^{P*}(\hat{k}_s, \hat{k}_i)\rangle g_p(\overline{r})\Bigg\} \tag{7.3.40b}$$

On the other hand,

$$\langle F^{(1)}(\hat{k}_s, \hat{k}_i)\rangle = n_p \int d\overline{r} \exp(i\overline{k}_d\cdot\overline{r})\langle f_\alpha^P(\hat{k}_s, \hat{k}_i)\rangle \tag{7.3.40c}$$

In (7.3.40b) we have made use of the joint probability density function of primary scatterers, that is analogous to (7.3.8c),

$$p_p(\overline{r}_\alpha, \overline{r}_\beta) = \left(\frac{g_p(\overline{r}_\alpha, \overline{r}_\beta)}{V^2}\right)\frac{N_p}{N_p - 1} \tag{7.3.41}$$

and g_p is the pair distribution function of primary scatterers. Note that from this point of view, the primary scatterer is a "unit" scatterer. Putting (7.3.40b) and (7.3.40c) into (7.3.40a) gives

$$P(\hat{k}_s, \hat{k}_i) = n_p\langle |f_\alpha^P(\hat{k}_s, \hat{k}_i)|^2\rangle$$

$$+ n_p^2\langle f_\alpha^P(\hat{k}_s, \hat{k}_i)\rangle\langle f_\beta^P(\hat{k}_s, \hat{k}_i)\rangle \int d\overline{r} \exp(i\overline{k}_d\cdot\overline{r})\,(g_p(\overline{r}) - 1) \tag{7.3.42}$$

Let the primary scatterer size be confined to v_p. Then the first moment of primary scatterer scattering amplitude is from (7.3.39a)

$$\langle f_\alpha^P(\hat{k}_s, \hat{k}_i)\rangle = \frac{N_s f}{v_p}\int_{v_p} d\overline{r} \exp(i\overline{k}_d\cdot\overline{r}) \tag{7.3.43}$$

From (7.3.39a) we obtain

$$\begin{aligned}&\langle|f_\alpha^P(\hat{k}_s,\hat{k}_i)|^2\rangle\\&=|f|^2N_s+|f|^2\sum_{j=1}^{N_s}\sum_{\substack{l=1\\l\neq j}}^{N_s}\langle\exp(i\overline{k}_d\cdot(\overline{r}_{\alpha j}-\overline{r}_{\alpha l}))\rangle \qquad (7.3.44a)\\&=N_s|f|^2+\frac{N_s^2}{v_p^2}\int_{v_p}d\overline{r}_{\alpha j}\int_{v_p}d\overline{r}_{\alpha l}|f|^2\exp(i\overline{k}_d\cdot(\overline{r}_{\alpha j}-\overline{r}_{\alpha l}))g_s(\overline{r}_{\alpha j},\overline{r}_{\alpha l}) \qquad (7.3.44b)\end{aligned}$$

where the joint probability density function of secondary scatterers is, analogous to (7.3.8c),

$$p_s(\overline{r}_{\alpha j},\overline{r}_{\alpha l})=\left(\frac{g_s(\overline{r}_{\alpha j},\overline{r}_{\alpha l})}{v_p^2}\right)\frac{N_s}{N_s-1} \qquad (7.3.45)$$

with g_s being the pair distribution function of secondary scatterers. To illustrate, we consider a special case with $g_p=1, g_s=1$, and let v_p be a cubic volume of l_c^3. Then the integrals in (7.3.43) and (7.3.44) can be carried out readily. Putting the results into (7.3.42) gives

$$P(\hat{k}_s,\hat{k}_i)=n_o|f|^2+n_oN_s|f|^2\left\{\frac{\sin\left(k_{dx}\frac{l_c}{2}\right)}{k_{dx}\frac{l_c}{2}}\right\}^2\left\{\frac{\sin\left(k_{dy}\frac{l_c}{2}\right)}{k_{dy}\frac{l_c}{2}}\right\}^2\left\{\frac{\sin\left(k_{dz}\frac{l_c}{2}\right)}{k_{dz}\frac{l_c}{2}}\right\}^2 \qquad (7.3.46)$$

where k_{dz}, k_{dy}, and k_{dz} are respectively the x-, y-, and z-components of $\overline{k}_d$. Note that the result in (7.3.46) is substantially different from that of (7.3.22), where the particles are not in clusters. In some cases, the result of (7.3.46) can be much larger than that of (7.3.22). This shows that when small point scatterers cluster to form "larger" particles, the scattering can be much larger even when the total number of small point scatterers remain the same.

4 Monte Carlo Simulation Results of Point Scatterers

In this section we illustrate Monte Carlo simulations of scattering by point scatterers. We calculate the extinction coefficients and phase functions for the cases of uniformly random distribution and the case of clustered random distributions.

Consider an incident wave impinging upon N scatterers

$$E_{inc}(\overline{r})=e^{i\overline{k}_i\cdot\overline{r}} \qquad (7.4.1)$$

where

$$\overline{k}_i = k(\sin\theta_i \cos\phi_i \hat{x} + \sin\theta_i \sin\phi_i \hat{y} + \cos\theta_i \hat{z}) \tag{7.4.2}$$

We start by putting N particles in a cubic box of size $V = L^3$. The coordinates $\overline{r}_j = (x_j, y_j, z_j), j = 1, 2, \ldots, N$, are determined by choosing three random numbers between 0 and 1 and then multiplying them by L. Once the positions of the N particles are given, we can solve the Foldy-Lax multiple scattering equations

$$E_{ex}^j = E_{inc}(\overline{r}_j) + \sum_{\substack{l=1 \\ l\neq j}}^{N} f \frac{\exp(ik|\overline{r}_j - \overline{r}_l|)}{|\overline{r}_j - \overline{r}_l|} E_{ex}^l(\overline{r}_l) \tag{7.4.3}$$

$j = 1, 2, \ldots, N$. After the exciting field $E_{ex}^j(\overline{r}_j)$ are calculated, the "final" scattered field can be calculated in the far-field region. Let

$$\overline{k}_s = k(\sin\theta_s \cos\phi_s \hat{x} + \sin\theta_s \sin\phi_s \hat{y} + \cos\theta_s \hat{z}) \tag{7.4.4}$$

be the observation direction of the scattered field. Hence,

$$E_s(\overline{r}) = \frac{e^{ikr}}{r} F_N(\hat{k}_s, \hat{k}_i) \tag{7.4.5}$$

where

$$F_N(\hat{k}_s, \hat{k}_i) = \sum_{l=1}^{N} f \ \exp(-i\overline{k}_s \cdot \overline{r}_l) E_{ex}^l(\overline{r}_l) \tag{7.4.6}$$

is the N-particle scattering amplitude. These can be calculated for many realizations. We then calculate the realization averages. Let angular bracket $\langle\ \rangle$ denote realization average. The coherent scattering amplitude is

$$\langle F_N(\hat{k}_s, \hat{k}_i)\rangle = \frac{1}{N_r} \sum_{r=1}^{N_r} F_N^r(\hat{k}_s, \hat{k}_i) \tag{7.4.7}$$

where r is the realization index and N_r is the number of realizations. The incoherent scattering amplitude for each realization is

$$\mathcal{F}_N^r(\hat{k}_s, \hat{k}_i) = F_N^r(\hat{k}_s, \hat{k}_i) - \langle F_N(\hat{k}_s, \hat{k}_i)\rangle \tag{7.4.8}$$

The phase function is calculated by

$$P(\hat{k}_s, \hat{k}_i) = \frac{\langle|\mathcal{F}_N(\hat{k}_s, \hat{k}_i)|^2\rangle}{V} = \frac{1}{N_r V} \sum_{r=1}^{N_r} |\mathcal{F}_N^r(\hat{k}_s, \hat{k}_i)|^2 \tag{7.4.9}$$

The extinction coefficient is calculated by

$$\kappa_e = \int_0^{\pi} d\theta_s \sin\theta_s \int_0^{2\pi} d\phi_s P(\hat{k}_s, \hat{k}_i) \tag{7.4.10}$$

We present the Monte Carlo simulation results of two types of configurations as shown in Figs. 7.3.1 and 7.3.2. In Fig. 7.3.1 the nonabsorptive particles are uniformly and randomly distributed in a cubic box of size L^3. In Fig. 7.3.2 the same number of particles are distributed in clusters in the cubic box of size L^3. The clusters contain N_s particles each. In each cluster the particles are randomly distributed in a cubic volume of size l_c^3. All distance dimensions are in wavelengths (λ). In the Monte Carlo simulations, (7.4.3) is solved exactly to find the incoherent phase function as defined by (7.4.8). Then the scattering coefficient defined by (7.4.10) is found by integrating over all scattered angles. To make comparisons, we also find the scattering coefficient and the phase function by approximating the multiple scattering equation to the second order

$$F_N^{(2)}(\hat{k}_s, \hat{k}_i) = \sum_{j=1}^{N} f e^{i(\bar{k}_i - \bar{k}_s)\cdot\bar{r}_j} + \sum_{j=1}^{N} \sum_{\substack{l=1 \\ l\neq j}}^{N} f^2 e^{-i\bar{k}_s\cdot\bar{r}_j} \frac{e^{ik|\bar{r}_j - \bar{r}_l|}}{|\bar{r}_j - \bar{r}_l|} e^{i\bar{k}_i\cdot\bar{r}_l} \tag{7.4.11}$$

The second-order iterative approach is then compared with the exact solution.

The cubic box is of volume V and the length of each side L is 50λ. A scalar plane wave is incident at $\theta_i = 10°$ and $\phi_i = 10°$. The incident and scattering angles are defined as in Fig. 7.3.1 such that the backscattering direction is $\theta_s = \theta_i$ and $\phi_s = \pi + \phi_i$. For illustrations, let scattering amplitude $f = (0.008905 + i0.0005)\lambda$. The value of f is typical of a small particle. The real and imaginary parts are chosen to obey the optical theorem of (7.2.23) of a single non-absorptive particle. In different realizations, we only change the positions of the particles. The positions of 500 scatterers are generated randomly. For the uniform random case, $N = 500$ center positions are generated randomly within the cubic box (Fig. 7.3.1) of side length L. For the case of clustered random distribution, $N = 500$ particle positions are generated in two stages (Fig. 7.3.2). The 500 particles are of 50 clusters with each cluster containing 10 particles. The cluster volume is a cube with length l_c. Initially, 50 center positions are generated within the cube of length L and they are randomly chosen with the constraint that there is no interpenetration of the different cluster cubes. Then in the second stage, we randomly generate 10 particles inside each cluster of volume l_c^3. We vary l_c and investigate its effects. The results are calculated for 100 realizations.

In Monte Carlo simulations, it is important to demonstrate convergence with respect to the number of particles N in the cubic box and the number of realizations N_r for a fixed particle number density n_o. We first present the convergence of extinction coefficients as we increase the number of realiza-

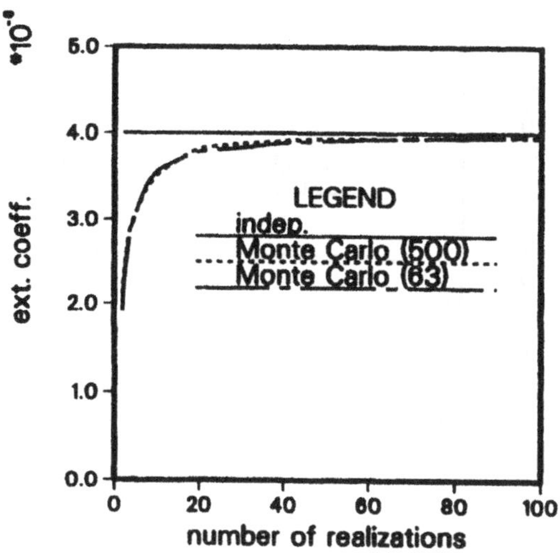

Figure 7.4.1 Convergence test for the Monte Carlo simulations with respect to realizations and the number of particles for $N = 63$ ($L = 25\lambda$) and $N = 500$ ($L = 50\lambda$) point particles uniformly and randomly distributed as in Fig. 7.3.1. They are compared with the independent scattering result.

tions and the number of particles in Fig. 7.4.1. The results in Fig. 7.4.1 are for the uniform random case. Initially, we put $N = 63$ particles in a cubic box of size $L = 25\lambda$, so that the particle density is $n_o = 0.004\lambda^{-3}$. This is to be compared with the independent scattering theory, $\kappa_{ei} = 4\pi n_o|f|^2$, where the subscript i stands for independent scattering. The independent scattering as defined by (7.3.5) predicts $\kappa_{ei} = 4.0 \times 10^{-6}\lambda^{-1}$, while in the Monte Carlo simulations we get $\kappa_e = 3.958 \times 10^{-6}\lambda^{-1}$ after 100 realizations. To demonstrate convergence with respect to the number of particles, we increase the number of particles to $N = 500$ while n_o remains at $0.004\lambda^{-3}$. This is done by increasing the number of particles to 500 and increasing the volume to $V = L^3 = (50\lambda)^3$. This result is also given in Fig. 7.4.1, where $\kappa_e = 3.944 \times 10^{-6}\lambda^{-1}$ after 100 realizations. Figure 7.4.1 indicates convergence of the results with respect to the number of realizations and the number of particles. All the Monte Carlo results to be presented from this point on are given for $N = 500$ particles with $L = 50\lambda$, $\theta_i = 10°$, and $\phi_i = 10°$.

Next, we consider the case of clustered random distribution with 10 particles in each cluster. Figure 7.4.2 shows the convergence of the extinction coefficients versus realizations. The relative extinction coefficient (κ_{re}) is the collective extinction coefficient divided by that of independent scattering

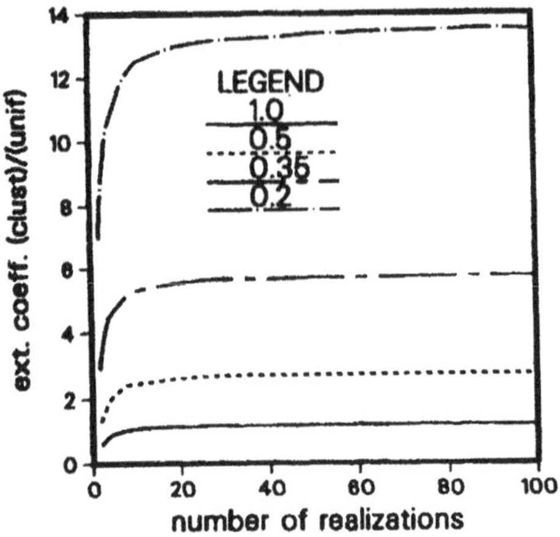

Figure 7.4.2 Monte Carlo simulations for the clustered scatterers of Fig. 7.3.2. l_c is decreased from 1.0λ to 0.2λ. There are 50 clusters containing 10 scatterers each ($N = 500, L = 50\lambda$).

($\kappa_{re} = \frac{\kappa_e}{\kappa_{ei}}$ with $\kappa_{ei} = 4.0 \times 10^{-6}\lambda^{-1}$). There are 50 clusters, each containing 10 particles in all plots given. The cluster cube size l_c is varied from 1.0λ to 0.2λ. For the cubic box size of 1.0λ there is very little difference between the uniform and clustered distribution as can be seen from the solid curve in Fig. 7.4.2. However, as the cluster volume is made smaller than a wavelength, κ_{re} can become large. We see from Fig. 7.4.2 that the result of κ_{re} for the case of $l_c = 0.2\lambda$ is much larger than κ_{re} for the case of $l_c = 1\lambda$. This shows that when considering random media problems, the significance of the clustered geometry must be considered. In Figs. 7.4.3 and 7.4.4, the phase functions of the Monte Carlo results of Fig. 7.4.1 (uniform random $N = 500$) and the $l_c = 0.2\lambda$ of Fig. 7.4.2 are presented, respectively. Both the incoherent phase function P_{inc}(dotted line) and the combined phase function $P_{inc} + P_{coherent}$ (solid line) are given. The results are averaged over 100 realizations. In the polar plots, the top half-plane gives $\phi_s = 10°$ and θ_s ranges from $0 \to 180°$ going clockwise. In the bottom half-plane $\phi_s = 190°$ and θ_s ranges from $0 \to 180°$ going counterclockwise. Comparing Figs. 7.4.3 and 7.4.4, we see that the clustered scattering case gives a larger incoherent contribution than does the uniform case, while the opposite is true for the coherent contribution. In both figures, the coherent part has a sharp peak of large amplitude (many decibels larger) in the forward direction ($\theta_s = 170°, \phi_s = 10°$). Note that the polar plot is in the decibel scale, so that the coherent forward peak has much larger amplitude than that of the other directions. As V approaches infinity,

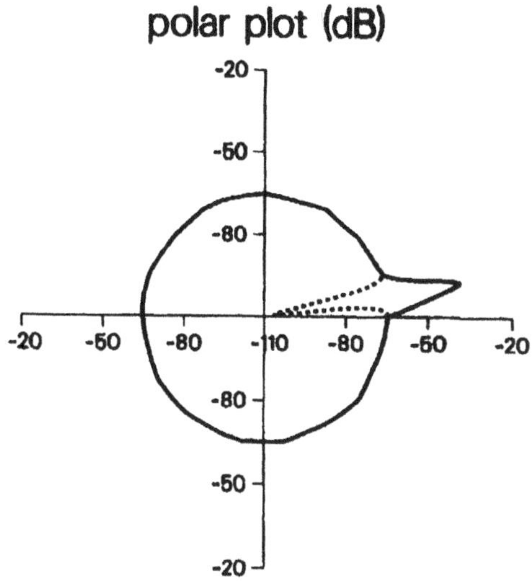

Figure 7.4.3 Polar plot (dB) of the phase function of Fig. 7.4.1 of the uniform random case for $N = 500$ and $L = 50\lambda$ with the geometry given as in Fig. 7.3.1. Upper half-plane shows $\phi_s = 10°$ and θ_s is $0° \to 180°$ going clockwise. Lower half-plane shows $\phi_s = 190°$ and θ_s is $0° \to 180°$ going counterclockwise. Incident angles are $\theta_i = 10°$ and $\phi_i = 10°$.

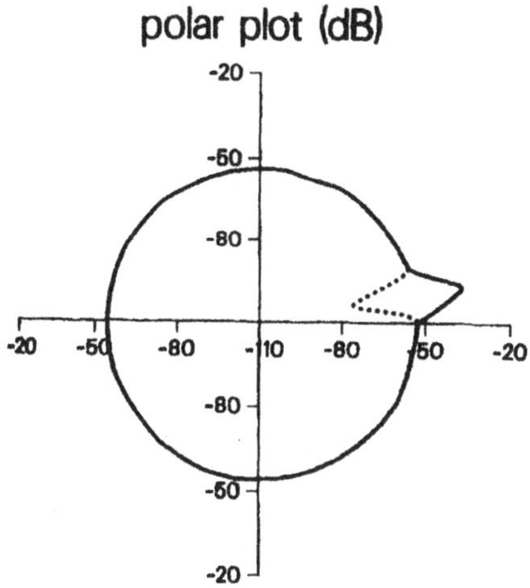

Figure 7.4.4 Polar plot (dB) of the phase function of Fig. 7.4.2 of the clustered random case for $N = 500$, $l_c = 0.2\lambda$, and $L = 50\lambda$ with incident angles at $\theta_i = 10°$ and $\phi_i = 10°$. Upper half-plane shows $\phi_s = 10°$ and θ_s is $0° \to 180°$ going clockwise. Lower half-plane shows $\phi_s = 190°$ and θ_s is $0° \to 180°$ going counterclockwise.

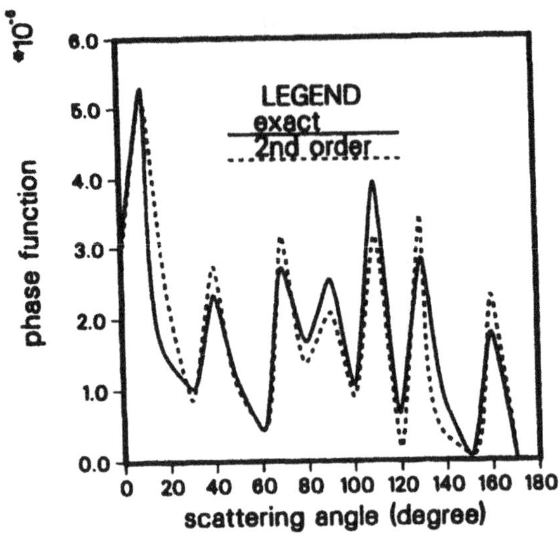

Figure 7.4.5 Comparison test between the second-order approach and the exact solution. Phase function of Fig. 7.4.2 with $l_c = 0.35\lambda$, $N = 500$ and $L = 50\lambda$ for 1 realization ($\phi_i = 10°$, $\theta_i = 10°$, $\phi_s = 20°$ and θ_s ranges from 0° to 180°).

the coherent wave will approach a Dirac delta function. Thus, the coherent peak has to be excluded from the phase matrix and extinction coefficient calculations.

All the Monte Carlo results given so far are obtained by solving the multiple scattering equation (7.4.3) exactly. To compare the iterative method solution and the exact solution, we compare the incoherent phase functions for a single realization using the second-order approximation and the exact method. The comparison is shown in Fig. 7.4.5. In the figure, $l_c = 0.35\lambda$, $N = 500$, $\phi_s = 20°$, and θ_s is varied from 0° to 180°. For one realization, the second-order solution shows significant differences from the exact solution for most scattering angles. However, when 100 realizations are averaged, the second-order solution agrees with the exact method (Fig. 7.4.6). This shows that for a single realization, higher-order scattering effects are important. However, these higher-order scattering effects tend to cancel out on averaging over realizations, and the second-order theory gives a quite accurate average. The iteration approach can be very attractive for more complicated problems of vector electromagnetic scattering by a large number of dielectric scatterers of arbitrary size. In Fig. 7.4.7 the comparison between the analytic collective scattering theory of (7.3.46), the independent scattering of (7.3.4), and the Monte Carlo simulation of Fig. 7.4.6 with

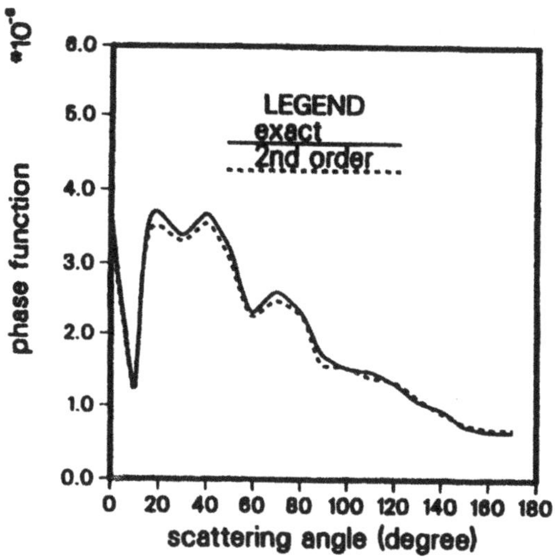

Figure 7.4.6 Comparison test between the second-order approach and the exact solution of Fig. 7.4.5. Results are averaged over 100 realizations ($\phi_i = 10°$, $\theta_i = 10°$, $\phi_s = 20°$, and θ_s ranges from 0° to 180°).

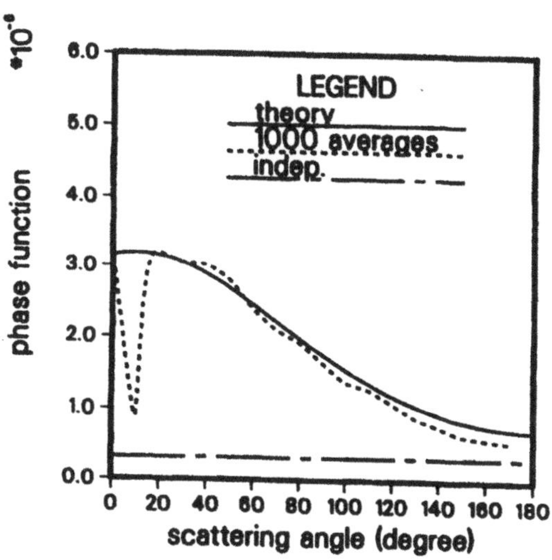

Figure 7.4.7 Comparison test between the analytical result from (7.3.46), the independent scattering, and the Monte Carlo simulation of Fig.7.4.6 (1000 realizations). The parameters are: $l_c = 0.35\lambda$, $N_s = 10$, $N_p = 50$, $N = 500$, $n_o = 0.004\lambda^{-3}$ and $L = 50\lambda$ ($\phi_i = 10°$, $\theta_i = 10°$, $\phi_s = 20°$, and θ_s ranges from 0° and 180°).

1000 realizations is given. While the Monte Carlo result and the collective scattering theory agrees, the independent scattering predicts the wrong result. The dip in the Monte Carlo result is due to the subtraction of the coherent part from the total phase function.

REFERENCES AND ADDITIONAL READINGS

Au, W. C., J. A. Kong, and L. Tsang (1994), Absorption enhancement of scattering of electromagnetic waves by dielectric cylinder clusters, *Microwave Opt. Technol. Lett.*, **7(10)**, 454–457.

Au, W. C., L. Tsang, R. T. Shin, and J. A. Kong (1996), Collective scattering and absorption in microwave interaction with vegetation canopies, *Progress in Electromag. Res.*, **14**, 182–231, EMW Publishers, Cambridge, Massachusetts.

Chuah, H. T. and H. S. Tan (1992), A microwave emission model for vegetation medium using the Monte Carlo method, Part I and Part II, *J. Electromag. Waves and Appl.*, **6**, 799–852.

de Vries, P., D. V. van Coevorden, and A. Lagendijk (1998), Point scatterers for classical waves, *Rev. Modern Phys.*, **70**, 447–466.

Fikioris, J. G. and P. C. Waterman (1964), Multiple scattering of waves, II. Hole corrections in the scalar case, *J. Math. Phys.*, **5**, 1413–1420.

Foldy, L. L. (1945), The multiple scattering of waves, *Phys. Rev.*, **67**, 107–119.

Lax, M. (1951), Multiple scattering of waves, *Rev. Modern Phys.*, **23**, 287–310.

Lax, M. (1952), Multiple scattering of waves II. The effective field in dense systems, *Phys. Rev.*, **85**, 261–269.

Manchuck, G. I., G. A. Mikhailou, M. A. Nagaralieu, R. A. Darbinjn, B. A. Kargin, and B. S. Elepou (1980), *The Monte Carlo Methods in Atmosphere Optics*, Springer-Verlag, New York.

Peterson, B. and S. Strom (1973), T matrix for electromagnetic scattering from an arbitrary number of scatterers and representation of E(3), *Phys. Rev. D*, **8**, 3661–3678.

Sheng, P., Ed. (1990), *Scattering and Localization of Classical Waves in Random Media*, World Scientific, Singapore.

Tsang, L., J. A. Kong, Z. Chen, K. Pak, and C. Hsu (1995), Theory of microwave scattering from vegetation based on the collective scattering effects of discrete scatterers, *Passive Microwave Remote Sensing of Land-Atmosphere Interaction*, B. J. Choudhury, Y. H. Kerr, E. G. Njoku, and P. Pampaloni, Eds., 117–154.

Tsang, L., J. A. Kong, and R. T. Shin (1985), *Theory of Microwave Remote Sensing*, Wiley-Interscience, New York.

Tsang, L., C. Mandt, and K. H. Ding (1992), Monte Carlo simulations of the extinction rate of dense media with randomly distributed dielectric spheres based on solution of Maxwell's equations, *Optics Lett.*, **17(5)**, 314–316.

Yueh, S. H., J. A. Kong, J. K. Jao, R. T. Shin, and T. Le Toan (1992), Branching model for vegetation, *IEEE Trans. Geosci. Remote Sens.*, **30**, 390–402.

Chapter 8

PARTICLE POSITIONS FOR DENSE MEDIA CHARACTERIZATIONS AND SIMULATIONS

The scattering of electromagnetic waves by random media depends on the positions of the particles. A dense medium denotes a medium where the particles are densely packed and occupy appreciable fractional volume. In the random medium, the probability density function (pdf) of particle positions $p(\overline{r}_i)$ is uniformly random. However, because of the finiteness of particle size, the joint probability density function of two particle positions cannot be independent. If $p(\overline{r}_i, \overline{r}_j)$ is the joint probability density function of two particles centered at $\overline{r}_i$ and $\overline{r}_j$, then $p(\overline{r}_i, \overline{r}_j) = 0$ for $|\overline{r}_i - \overline{r}_j|$ smaller than the minimum separation which is the diameter if the two particles are spheres of the same radius. The pair distribution function is proportional to the joint probability density functions of two particles. The finiteness of the particle sizes creates nontrivial pair distribution functions. The Fourier transform of the pair distribution function is the structure factor. We have discussed pair distribution function in Chapter 4, Section 5.2 of Volume I. In this chapter, we consider two methods of studying random particle positions. One method is based on analytic theory to derive analytic pair distribution functions. The other method is to use Monte Carlo simulations to generate particle positions. The Monte Carlo procedure of generating particle positions is also important as numerical solutions of the Maxwell equations can be computed based on the generated realizations. On the other hand, analytic pair distribution function is useful for analytic scattering theory of dense media. Analytic scattering theory of dense media will be treated in Volume III.

In Section 1, we describe the pair distribution function under the Percus–Yevick approximation. In Section 2, we consider the case where the particles in the same medium can have different sizes. In Section 3, we describe Monte Carlo simulations of spherical particles. In Section 4, we consider the case of sticky particles. This represents the case when particles have adhesive force. The adhesive force means that the particles can form aggregates. We consider the case of collections of aggregates. In Section 5, the Monte Carlo technique is extended to particles of spheroidal shapes.

1 Pair Distribution Functions and Structure Factors

1.1 Introduction

Let N be the number of particles. They are centered at $\overline{r}_1, \overline{r}_2, \ldots, \overline{r}_N$. Let the particles be put in a volume V. Then

$$p(\overline{r}_i) = \frac{1}{V} \tag{8.1.1}$$

is the single-particle pdf, and from Volume I, Eqs. (4.5.21)–(4.5.23),

$$p(\overline{r}_i, \overline{r}_j) = \frac{g(\overline{r}_i, \overline{r}_j)}{V^2} \frac{N}{N-1} \tag{8.1.2}$$

is the joint pdf of two particles, and g is the pair distribution function. In the limit of large N, $p(\overline{r}_i, \overline{r}_j) \approx g(\overline{r}_i, \overline{r}_j)/V^2$.

Analytic theory of volume scattering is discussed in Volume III, where it is shown that in applying the quasi-crystalline approximation and the quasi-crystalline approximation with coherent potential, the pair distribution function of particle positions must be specified. In the special case of independent particle position, $g(\overline{r}) = 1$. Another approximation to the pair distribution function is the hole-correction (HC) approximation, given by $g(r) = 0$ for $r < b$ and $g(r) = 1$ for $r \geq b$, where b is the diameter of the circumscribing sphere of the particle. For the case of spherical particles with radius a, b is equal to $2a$. The hole-correction approximation takes into account the fact that the particles cannot interpenetrate each other. Neither the independent position approximation nor the hole-correction approximation is correct when the fractional volume of scatterers, f, is appreciable. It is easier to visualize this for the case of one-dimensional scatterers. The hole-correction approximation is illustrated in Fig. 8.1.1A. Next we imagine that f is equal to unity so that the entire volume V is occupied by scatterers. In such a case, the centers of these one-dimensional particles will be separated by integral multiples of b from each other. The pair distribution function $g(r)$ will be zero for $r \neq mb$ where m is any nonzero integer. It consists of delta functions at the position of r equal to an integral multiple of b (Fig. 8.1.1B). Thus, the hole-correction approximation is poor in such a limit. When f is not equal to 1 but appreciable, the pair distribution function will be of a form between A and B in Fig. 8.1.1. We also note that as the two-particle separation r approaches infinity, the positions of the particles should be independent of each other. Hence $\lim_{r\to\infty} g(r) = 1$ for f not equal to maximum concentration.

The study of pair distribution functions is a subject of interest in statistical mechanics [McQuarrie, 1976]. Based on the form of the pair distribution function, substances can be classified into three different types: (1) gas; (2) liquid and amorphous solid; and (3) crystalline solid. The three forms of the pair distribution function are illustrated in Fig. 8.1.2. The case of gas with particles sparsely distributed is considered to be a system of extreme disorder, so that the hole-correction approximation or the independent position approximation is a good description of the pair distribution function. In the opposite extreme, the case of crystalline solid with relative positions

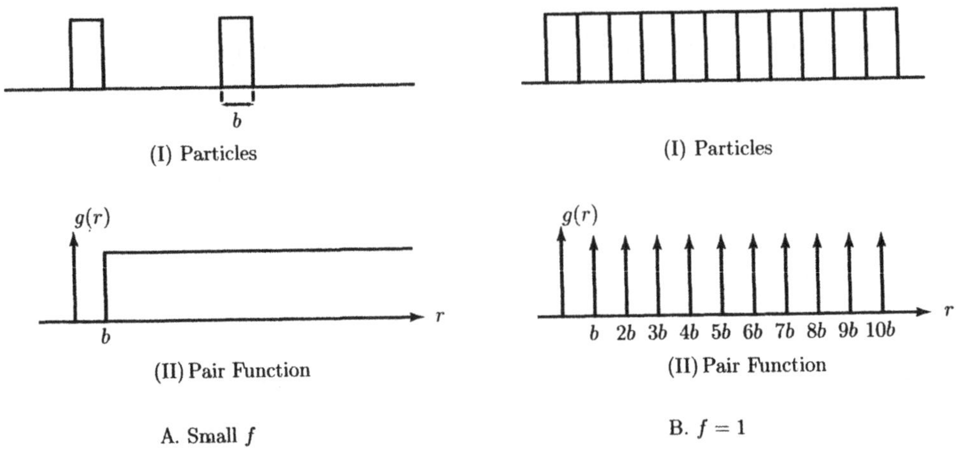

Figure 8.1.1 Pair distribution function for one-dimensional particles: (A) Small f; particles and pair function. (B) $f = 1$; particles and pair function.

of the particles fixed is a case of extreme order. The pair distribution function exhibits sharp peaks (Fig. 8.1.2c). The case of a liquid and amorphous solid is a system of partial order and is an interpolation between the two extreme cases of gas and crystalline solid (Fig. 8.1.2d). Extensive experimental and theoretical investigations have been carried out for the pair distribution function of liquid and amorphous solid [Waseda, 1980]. To the first-order approximation, the bistatic scattering intensity is proportional to the structure factor which is related to the Fourier transform of the pair distribution functions. The study of pair distribution functions is an important subject in molecular theory of fluids as well as in random media [Wertheim, 1963, 1964; McQuarrie, 1976; Ziman, 1979; Perram et al. 1984; Perla et al. 1986; Penders and Vrij, 1990; Shi et al. 1993; Zurk et al. 1997].

1.2 Percus–Yevick Equation and Pair Distribution Function for Hard Spheres

A total influence h of a particle 1 on another particle 2 can be defined as

$$h(\overline{r}_{12}) = g(\overline{r}_{12}) - 1 \tag{8.1.3}$$

The total influence is decomposed into a sum of direct and indirect correlation functions. The direct correlation function or influence is denoted by $c(\overline{r})$:

$$h(\overline{r}_{12}) = c(\overline{r}_{12}) + \text{indirect} \tag{8.1.4}$$

The direct correlation function is defined such that it satisfies the following

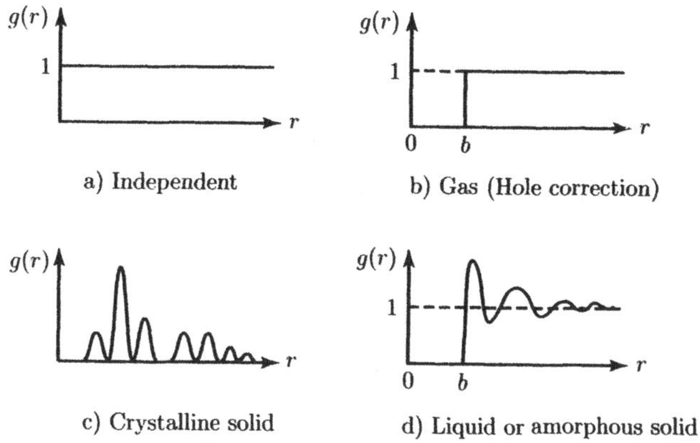

Figure 8.1.2 Pair functions for (a) independent particle position, (b) gas, (c) crystalline solid, and (d) liquid and amorphous solid.

integral equation:

$$h(\bar{r}_{12}) = c(\bar{r}_{12}) + n_o \int d\bar{r}_3 c(\bar{r}_{13}) h(\bar{r}_{32}) \tag{8.1.5a}$$

From (8.1.3),

$$h(\bar{r}_{12}) = c(\bar{r}_{12}) + n_o \int d\bar{r}_3 c(\bar{r}_{13}) [g(\bar{r}_{32}) - 1] \tag{8.1.5b}$$

which is known as the Ornstein-Zernike equation. The physical interpretation of the second term of (8.1.5a), which is the indirect correlation function, is that the indirect influence of particle 1 on particle 2 is a result of particle 1 acting directly on a particle at $\bar{r}_3$, which in turn exerts total influence on particle 2. The indirect influence is averaged over particle positions $\bar{r}_3$ and weighted by the number of particles per unit volume n_o as indicated in (8.1.5a) and (8.1.5b).

The Ornstein-Zernike equation consists of two unknowns $c(\bar{r})$ and $h(\bar{r})$ in one equation. An approximation is to be made on the relation between $c(\bar{r})$ and $h(\bar{r})$, reducing (8.1.5) to one equation and one unknown. The Percus–Yevick approximation [Percus and Yevick, 1958] can be introduced in the following heuristic manner.

The potential energy between two particles is governed by $u(\bar{r})$ where r is their separation. For the case of hard sphere potential, we have

$$u(\bar{r}) = \begin{cases} \infty & \text{for } r < b \\ 0 & \text{for } r \geq b \end{cases} \tag{8.1.6}$$

Equation (8.1.6) says that in the absence of other particles, the potential energy between the two particles is infinite when they overlap each other (thus disallowing interpenetration) and is zero otherwise. For this case, we define $y(\overline{r})$ so that

$$g(\overline{r}) = \begin{cases} 0 & \text{for } r < b \\ y(\overline{r}) & \text{for } r \geq b \end{cases} \tag{8.1.7}$$

The function $y(\overline{r})$ is defined for both $r < b$ and $r \geq b$. Equation (8.1.7) defines it for $r \geq b$. Later we will define it for $r < b$.

In (8.1.7), we let $g(\overline{r}) = y(\overline{r})$ for $r \geq b$. Then, for hard-sphere potential we have

$$h(\overline{r}) = g(\overline{r}) - 1 = \begin{cases} -1 & \text{for } r < b \\ y(\overline{r}) - 1 & \text{for } r \geq b \end{cases} \tag{8.1.8}$$

When $y = 1$, there is no indirect influence. Thus $y - 1$ is a measure of indirect influence. Also $h - c$ is equal to to indirect influence. The Percus–Yevick approximation consists of equating $h - c$ to $y - 1$ for all $\overline{r}$.

$$h(\overline{r}) - c(\overline{r}) = y(\overline{r}) - 1 \tag{8.1.9}$$

This equation then extends the definition of $y(\overline{r})$ to $r < b$.

From (8.1.8) to (8.1.9) and (8.1.3) we obtain

$$c(\overline{r}) = h(\overline{r}) + 1 - y(\overline{r}) = g(\overline{r}) - y(\overline{r}) \tag{8.1.10a}$$

From (8.1.10a) and (8.1.7) we have

$$y(\overline{r}) = \begin{cases} -c(\overline{r}) & \text{for } r < b \\ g(\overline{r}) & \text{for } r \geq b \end{cases} \tag{8.1.10b}$$

Also from (8.1.10a) and (8.1.10b) we have

$$c(\overline{r}) = \begin{cases} -y(\overline{r}) & \text{for } r < b \\ 0 & \text{for } r \geq b \end{cases} \tag{8.1.10c}$$

Let $\overline{r}_{12} = \overline{r}_1 - \overline{r}_2 = \overline{r}$, $\overline{r}_{13} = \overline{r}_1 - \overline{r}_3 = \overline{r}'$ so that $\overline{r}_{32} = \overline{r}_3 - \overline{r}_2 = \overline{r} - \overline{r}'$ in (8.1.5b). Note that from (8.1.8) and (8.1.10c), we have both h and c expressed in terms of y. Thus expressing (8.1.5b) in terms of y, we have

$$y(\overline{r}) = 1 - n_o \int_{r'<b} d\overline{r}' y(\overline{r}') \left[g(\overline{r} - \overline{r}') - 1\right] \tag{8.1.11}$$

We further decompose the integral of (8.1.11) into two parts of $|\overline{r} - \overline{r}'| < b$ and $|\overline{r} - \overline{r}'| \geq b$. The integral equation becomes:

$$y(\overline{r}) = 1 + n_o \int_{\substack{r'<b \\ |\overline{r}-\overline{r}'|<b}} d\overline{r}' y(\overline{r}') - n_o \int_{\substack{r'<b \\ |\overline{r}-\overline{r}'|>b}} d\overline{r}' y(\overline{r}')[y(\overline{r} - \overline{r}') - 1] \tag{8.1.12}$$

A closed-form solution can be obtained for $c(\overline{r})$ via the integral equation (8.1.12) [Wertheim, 1963, 1964]. The solution is given below. Let the distance be normalized by b:

$$x = \frac{r}{b} \tag{8.1.13}$$

The solution for $c(\overline{r})$ is a cubic polynomial

$$-c(x) = \alpha + \beta x + \delta x^3 \tag{8.1.14}$$

for $x \leq 1$ and $c(x) = 0$ for $x > 1$. In (8.1.14) we have

$$\alpha = \frac{(1+2f)^2}{(1-f)^4} \tag{8.1.15}$$

$$\beta = -6f\,\frac{(1+f/2)^2}{(1-f)^4} \tag{8.1.16}$$

$$\delta = \frac{f(1+2f)^2}{2(1-f)^4} \tag{8.1.17}$$

$$f = \frac{n_o \pi b^3}{6} \tag{8.1.18}$$

For spheres, $b = 2a$ and f given in (8.1.18) is the fractional volume of spherical scatterers.

1.3 Calculation of Structure Factor and Pair Distribution Function

The Born approximation and the distorted Born approximation give a bistatic scattering cross section that is proportional to the structure factor. The structure factor is related to the Fourier transform of the total correlation function.

Define the Fourier transform of the total correlation function as

$$H(\overline{p}) = \frac{1}{(2\pi)^3}\int_{-\infty}^{\infty} d\overline{r}\, e^{-i\overline{p}\cdot\overline{r}}\, h(\overline{r}) \tag{8.1.19}$$

and let $C(\overline{p})$ denote the Fourier transform of the direct correlation function $c(\overline{r})$.

$$C(\overline{p}) = \frac{1}{(2\pi)^3}\int_{-\infty}^{\infty} d\overline{r}\, e^{-i\overline{p}\cdot\overline{r}}\, c(\overline{r}) \tag{8.1.20}$$

The integral in the Ornstein-Zernike equation of (8.1.5) is a convolution integral. Thus, in the Fourier transform domain, the integral in the Ornstein-Zernike equation is proportional to the product of $C(\overline{p})$ and $H(\overline{p})$. Solving the equation gives

$$H(\overline{p}) = \frac{C(\overline{p})}{1 - n_o(2\pi)^3 C(\overline{p})} \tag{8.1.21}$$

The structure factor $S(\overline{p})$ is defined as

$$S(\overline{p}) = 1 + n_o(2\pi)^3 H(\overline{p}) \tag{8.1.22}$$

The Quantity $H(\overline{p})$ for the Percus–Yevick Hard-Sphere Pair Function

The closed-form solution for the direct correlation function of Percus–Yevick hard spheres is given in (8.1.14) through (8.1.18). The Fourier transform $C(\overline{p})$ can be calculated readily and is only a function of pb, b, and f :

$$C(\overline{p}) = C_{PY}(pb, b, f) \tag{8.1.23}$$

with

$$\begin{aligned}(2\pi)^3 n_o C_{PY}(pb, b, f) \\ = 24f \Bigg\{ \frac{(\alpha+\beta+\delta)}{u^2}\cos u - \frac{(\alpha+2\beta+4\delta)}{u^3}\sin u \\ - \frac{2(\beta+6\delta)}{u^4}\cos u + \frac{2\beta}{u^4} + \frac{24\delta}{u^5}\sin u + \frac{24\delta}{u^6}(\cos u - 1)\Bigg\}\end{aligned} \tag{8.1.24}$$

where $u = pb$ and α, β, δ are given in (8.1.15) through (8.1.17). The closed-form solution of $H(\overline{p} = 0)$ is obtained by letting $p = 0$ in (8.1.21), (8.1.23), and (8.1.24). We get

$$n_o(2\pi)^3 H(\overline{p} = 0) = n_o \int_{-\infty}^{\infty} d\overline{r}\,[g(\overline{r}) - 1] = -1 + \frac{(1-f)^4}{(1+2f)^2} \tag{8.1.25}$$

To numerically compute the pair distribution function, the following steps are taken.

Step 1: Use (8.1.24) to compute $C(\overline{p})$. Because of spherical symmetry, it is a function of $p = |\overline{p}|$ only.

Step 2: Use (8.1.21) to compute $H(\overline{p}) = H(p)$.

Step 3: Take inverse Fourier transform of $H(\overline{p})$ to calculate $h(\overline{r})$ in accordance with (8.1.19). Because of spherical symmetry, we only need to carry out a one-dimensional integral:

$$\begin{aligned}h(\overline{r}) = h(r) = \int_{-\infty}^{\infty} d\overline{p} e^{i\overline{p}\cdot\overline{r}} H(\overline{p}) = 2\pi \int_0^{\pi} d\theta_p \sin\theta_p \int_0^{\infty} dp p^2 e^{ipr\cos\theta_p} H(p) \\ = 4\pi \int_0^{\infty} dp p^2 \left(\frac{\sin pr}{pr}\right) H(p)\end{aligned} \tag{8.1.26}$$

Step 4: The pair function is $g(r) = h(r) + 1$.

In Fig. 8.1.3 we plot the $g(\overline{r})$ for hard spheres for $f = 0.2$ and $f = 0.4$. We note that the pair distribution functions are equal to zero for r less

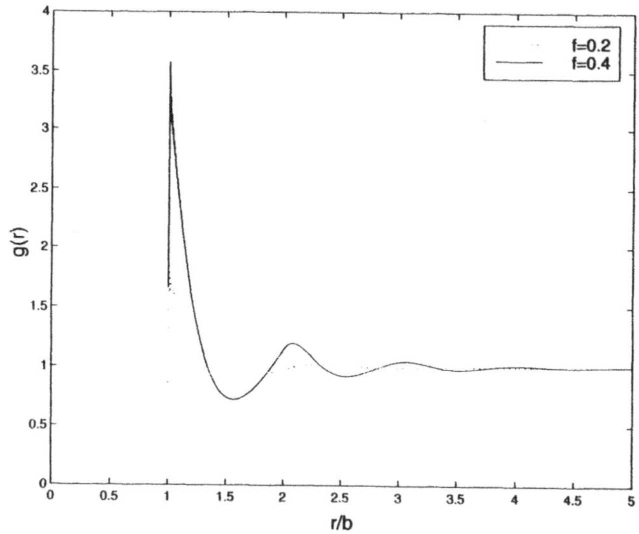

Figure 8.1.3 The Percus–Yevick pair distribution function $g(r)$ for hard spheres with distance normalized by the diameter of the spheres.

than a diameter. They assume maximum values at $r = b$. As r increases, they fluctuate and asymptotically approach unity. The case with larger f fluctuates more. If we treat $g = 1$ as the independent position result, we can see that positions are less independent as f increases.

2 Percus–Yevick Pair Distribution Functions for Multiple Sizes

Many media consist of particles with multiple sizes. For a system of spherical particles of L different sizes with radii $a_1, a_2, \ldots, a_L$ respectively, the pair distribution function $g_{ij}(r)$ for a pair of particles of sizes a_i and a_j is proportional to the conditional probability of finding a particle of size a_j at a distance r from the origin given that there is a particle of size a_i at the origin. The pair distribution functions depend on the sizes $a_1, a_2, \ldots, a_L$ and on the number densities $n_1, n_2, \ldots, n_L$ of the L species of spheres. The number density n_l is the number of particles per unit volume of radius a_l. The conditional probability is in the presence of other particles. Similar to the case of single size, the total correlation function $h_{ij}(r)$ between a pair of particles is defined as [Baxter, 1970]

$$h_{ij}(r) = g_{ij}(r) - 1 \tag{8.2.1}$$

The direct correlation function $c_{ij}(r)$ is related to $h_{ij}(r)$ by means of the generalized Ornstein-Zernike relation:

$$h_{ij}(r) = c_{ij}(r) + \sum_{l=1}^{L} n_l \int d\overline{r}'\, c_{il}(r') h_{lj}(|\overline{r} - \overline{r}'|) \tag{8.2.2}$$

which is an extension of (8.1.5). Under the Percus–Yevick approximation, and applying to the special case of zero interparticle force except for non-interpenetration, we have, as an extension of (8.1.10c) and (8.1.8),

$$c_{ij}(\overline{r}) = \begin{cases} -y_{ij}(\overline{r}) & \text{for } r < a_i + a_j \\ 0 & \text{for } r \geq a_i + a_j \end{cases} \tag{8.2.3}$$

$$h_{ij}(\overline{r}) = \begin{cases} -1 & \text{for } r < a_i + a_j \\ y_{ij}(\overline{r}) - 1 & \text{for } r \geq a_i + a_j \end{cases} \tag{8.2.4}$$

When (8.2.3) and (8.2.4) are substituted into (8.2.2), (8.2.2) becomes a set of nonlinear integral equations for $y_{ij}(\overline{r}), i, j = 1, 2, \ldots, L$. The total correlation function $h_{ij}(\overline{r})$ describes the total influence of a particle of radius a_i on another particle of radius a_j with respect to their spatial arrangements. From (8.2.2) the total correlation is separated into two contributions. The first contribution is the direct effect between two particles which is short ranged, as indicated in (8.2.3), and is described by $c_{ij}(\overline{r})$. The second contribution is an indirect effect, in which a particle of size a_i influences some other particles of size a_l at $\overline{r}'$, which in turn affects the particle of size a_j. The solution for impenetrable spherical particles has been obtained by Baxter [1970] using a factorization technique.

Let $H_{ij}(\overline{p})$ and $C_{ij}(\overline{p})$ be proportional to the respective 3-D Fourier transform of $h_{ij}(\overline{r})$ and $c_{ij}(\overline{r})$:

$$H_{ij}(\overline{p}) = (n_i n_j)^{1/2} \int d\overline{r}\, e^{i\overline{p}\cdot\overline{r}}\, h_{ij}(\overline{r}) \tag{8.2.5}$$

$$C_{ij}(\overline{p}) = (n_i n_j)^{1/2} \int d\overline{r}\, e^{i\overline{p}\cdot\overline{r}}\, c_{ij}(\overline{r}) \tag{8.2.6}$$

In matrix form, let H_{ij} and C_{ij} denote the (i, j) element of matrices $\overline{\overline{H}}$ and $\overline{\overline{C}}$, respectively. The Fourier transform of Ornstein-Zernike relation becomes

$$\overline{\overline{H}}(p) = \overline{\overline{C}}(p) + \overline{\overline{C}}(p)\overline{\overline{H}}(p) \tag{8.2.7}$$

Because of the spherical particle assumption, the transform only depends on $p = |\overline{p}|$. Based on the generalized Wiener–Hopf technique, matrix $\overline{\overline{C}}(p)$ can be factorized into

$$\overline{\overline{C}}(p) = \overline{\overline{I}} - \overline{\overline{Q}}^T(-p)\overline{\overline{Q}}(p) \tag{8.2.8}$$

where $\overline{\overline{I}}$ is a unit matrix, and the superscript T denotes the transpose of matrix. The matrix $\overline{\overline{Q}}$ which has components Q_{ij} as

$$Q_{ij}(p) = \delta_{ij} - \int_{S_{ij}}^{R_{ij}} dr\, e^{ipr} Q_{ij}(r) \tag{8.2.9}$$

where δ_{ij} is the Kronecker delta, $S_{ij} = a_i - a_j$, and $R_{ij} = a_i + a_j$, with a_i and a_j being the radii of particle species i and j, respectively. The function $Q_{ij}(r)$ has been solved and is given by

$$Q_{ij}(r) = 2\pi(n_i n_j)^{1/2} q_{ij}(r) \tag{8.2.10}$$

The solution of $q_{ij}(r)$ is, for $S_{ij} \leq r \leq R_{ij}$,

$$q_{ij}(r) = A_i \frac{r^2}{2} + B_i r + D_{ij} \tag{8.2.11}$$

with

$$A_i = \frac{1 - \xi_3 + 6a_i\xi_2}{(1-\xi_3)^2} \tag{8.2.12}$$

$$B_i = -\frac{6a_i^2\xi_2}{(1-\xi_3)^2} \tag{8.2.13}$$

$$D_{ij} = -A_i \frac{R_{ij}^2}{2} - B_i R_{ij} \tag{8.2.14}$$

$$\xi_\alpha = \frac{\pi}{6} \sum_{j=1}^{L} n_j (2a_j)^\alpha \tag{8.2.15}$$

and $\alpha = 0, 1, 2, 3$.

From (8.2.8)–(8.2.15), the matrix elements $C_{ij}(p)$ can be obtained as

$$\begin{aligned} C_{ij}(p) &= -\frac{\pi}{6}\frac{\sqrt{n_i n_j}}{1-\xi_3} \Bigg\{ M_j \left[\cos X_i + X_i \sin X_i + \frac{3\xi_2 R_i \cos X_i}{1-\xi_3} + \frac{3\xi_1 N_i}{1-\xi_3} + \frac{9\xi_2^2 N_i}{(1-\xi_3)^2} \right] \\ &+ M_i \left[\cos X_j + X_j \sin X_j + \frac{3\xi_2 R_j \cos X_j}{1-\xi_3} + \frac{3\xi_1 N_j}{1-\xi_3} + \frac{9\xi_2^2 N_j}{(1-\xi_3)^2} \right] \\ &+ M_i M_j \left[\frac{\xi_o}{1-\xi_3} + \frac{p^2\xi_2}{4(1-\xi_3)} + \frac{6\xi_1\xi_2}{(1-\xi_3)^2} + \frac{9\xi_2^3}{(1-\xi_3)^3} \right] \\ &+ 3N_i R_j \cos X_j + 3N_j R_i \cos X_i + \frac{9\xi_2 N_i N_j}{1-\xi_3} \Bigg\} \end{aligned} \tag{8.2.16}$$

where

$$M_i = 3R_i^3 \left[\frac{\sin X_i}{X_i^3} - \frac{\cos X_i}{X_i^2} \right] \tag{8.2.17}$$

$$N_i = R_i^2 \frac{\sin X_i}{X_i} \tag{8.2.18}$$

$$X_i = \frac{pR_i}{2} \tag{8.2.19}$$

and $R_i = 2a_i$ is the diameter of a particle of size a_i. Thus, the procedure for calculating the pair distribution function is as follows. First $C_{ij}(p)$ is calculated from (8.2.16)–(8.2.19). Next, $\overline{\overline{H}}(p)$ is computed by solving the linear matrix equation (8.2.7). Then $h_{ij}(r)$ is obtained by taking an inverse Fourier transform of $\overline{\overline{H}}(p)$ according to (8.2.5). Finally, the pair distribution function $g_{ij}(r)$ can be calculated using (8.2.1). Numerical results of the Percus–Yevick pair distribution function of multiple sizes will be shown in Section 3.3.

3 Monte Carlo Simulations of Particle Positions

In this section we describe Monte Carlo simulations of the pair distribution function of dense discrete random media with multiple sizes of particles. Two Monte Carlo methods, namely, the Metropolis technique and the sequential addition of particles, are commonly used. Particle positions are generated randomly in a cubic box. The number of particles used is between 400 and 4000. In the Metropolis Monte Carlo technique the particles are next shuffled to create different realizations. On the other hand, the sequential addition method adds particles serially and randomly into the box. Thus, in the sequential addition method the particles are not shuffled. In positioning the particles, we assume that there are no interparticle forces except that interpenetration is not allowed. Many realizations are used. We have used up to 3000 realizations. The pair distribution functions are calculated by using the definition of joint probability density functions and by counting the occurrence of pair separation of particles as a function of separation distance. The counting is averaged over these realizations. The edge effects of the cubic box are taken into account by adopting the periodic boundary conditions. The Monte Carlo results for the pair distribution function are illustrated and compared with the results of the Percus–Yevick approximation for the single-size case and for the case of multiple sizes. They are found to be in good agreement. The Monte Carlo simulations show that the Percus–Yevick approximations of single size and multiple sizes can be used for macroscopic objects such as particles in geophysical terrain and composite materials.

The computer experiments are performed on a system with N number of particles. These N spheres with different sizes are placed inside a cubic cell of side length ℓ. Let N_i be the number of particles of radius a_i. Then number density n_i and fractional volume f_i for particles of radius a_i are given by

$$n_i = \frac{N_i}{\ell^3} \tag{8.3.1}$$

$$f_i = n_i \frac{4\pi}{3} a_i^3 \tag{8.3.2}$$

$$N = \sum_{i=1}^{L} N_i \tag{8.3.3}$$

$$f = \sum_{i=1}^{L} f_i \tag{8.3.4}$$

where f is the total fractional volume occupied by the particles.

We note that the pair distribution function $g_{ij}(\overline{r})$ approaches unity as $\overline{r}$ becomes large, meaning that the particle positions are independent if they are far apart. For a fractional volume of less than 40%, the pair distribution functions are practically equal to unity for pair separations larger than five diameters. Thus taking N between 400 and 4000 is generally sufficient to simulate the pair distribution function. In the simulations, periodic boundary conditions are employed. The N particles of interest are placed inside the central cell, and this cental cell is taken to be surrounded by the periodic images of itself. Each image cell contains N particles with exactly the same geometric arrangement as that in the central primary cell.

3.1 Metropolis Monte Carlo Technique

We use the technique of Metropolis et al. [1953] for both the 3-D case and 2-D case. The 2-D case is described in Chapter 9, Section 2. For the 3-D case, we consider a cubic box containing N particles. Initially, the N spheres are placed randomly inside the primary cell with no overlap. To generate new realizations, the particles are shuffled as follows. In each cycle every particle is subject to be randomly displaced once. The acceptance of its new position is according to whether it overlaps another sphere or not. The displacement is random and is not governed by any interparticle force except that it cannot penetrate other particles. We should note that the displacement of particles introduced in the simulation is for the purpose of creating new configurations and realizations. It does not mean that the particles move physically. The steps are as follows.

Step 1: Set an initial configuration of the system. The initial configuration may be produced by assigning the particle coordinates within the cell. All the coordinates of spheres lie in the range $[0, \ell]$. The initial configuration may be obtained by placing all the particles in a periodic spatial function in the box.

Step 2: Change the system configuration by random displacement of particles. The particles are displaced sequentially in the simulation. The displacement of a particle to a new trial position is determined according to the following prescription:

$$x \to x + \Delta\eta_1 \qquad y \to y + \Delta\eta_2 \qquad z \to z + \Delta\eta_3 \tag{8.3.5}$$

where x, y, and z are the particle coordinates and Δ is the maximum displacement allowed in each movement. The η_i, $i = 1, 2, 3$, are independent random numbers uniformly distributed in the range $[-\ell, \ell]$. Thus after a displacement, the particle is equally likely to be anywhere in a little cube of side $2\ell\Delta$ centered about its original position. The prescription for displacing a particle given by (8.3.5) must satisfy the periodic boundary conditions; that is, if a particle is displaced past the cell on one side during the simulation, it actually reenters the cell from the opposite side. For example, if the center of a sphere changes to a new position (x', y', z') in this step, and if $x' > \ell$, or $x' < 0$, then its new x coordinate will be $x' - \ell$, or $x' + \ell$, respectively. The same applies to the $y-$ and $z-$ coordinates. If Δ is too small, displacements are usually accepted. If Δ is too large, most displacements will be rejected. The rule of thumb is to choose Δ such that the acceptance rate A_c is from 30% to 70%.

Step 3: Check whether the displacements are accepted. It is possible that the new position of the displaced particle may overlap one of the other particles in the system. If not, accept the displacement and update the coordinates of the displaced particle. Otherwise, reject the displacement and return the particle to its original position. The value of maximum displacement Δ influences the acceptance percentage of new positions. Even though checking the overlap of particles is straightforward, care must be taken in cases where a pair of particles do not overlap in the central cell but where the image of one of them in the next cell is within the forbidden region of the other.

Step 4: Update the number of configurations generated. Add one to the number of configurations counted. It should be emphasized that even if a displacement of the particle in a particular stage of *Step 2* is not allowed, we still consider ourselves to be in a new configuration. A new configuration means that every particle has to be subject to a single attempted displacement even if some of the displacements are not allowed. We shall distinguish between configurations and realizations. Every N_c configuration is called a realization. Thus between two realizations, every particle has been displaced an average of $N_c A_c$ times where A_c is the acceptance rate.

Step 5: Count the frequency of occurrence of different pair separations. For each configuration so reached, we will count, $C_{ij}^{It}(r)$, the number of particles of size a_j having their centers located in a spherical shell with inner radius r and outer radius $r + dr$, where r is measured from a tagged particle I of size a_i during the simulation cycle of realization t. The subscripts i and j are indices of particle sizes, the superscript I is for labeling the particle, and the superscript t is the index of realization. These tabulations are carried out for a number of divisions of r between R_{ij} and $5R_{ij}$, where $R_{ij} = a_i + a_j$. The thickness of the spherical shell dr will then be the smallest length over which the pair distribution functions are resolved.

The above process is repeated for a large number of times to generate many different realizations and to record the frequency of occurrence of different pair separations. Thus, the average number of particles of size a_j contained within a spherical shell of thickness dr at a distance r from a particle of size a_i, averaged over all particles of size a_i and total T realizations, can be expressed as

$$\langle C_{ij}(r)\rangle = \frac{1}{T}\frac{1}{N_i}\sum_{t=1}^{T}\sum_{I=1}^{N_i} C_{ij}^{It}(r) \tag{8.3.6}$$

On the other hand, if the particles were totally uncorrelated, the average number of particles of size a_j in the spherical shell surrounding a particle of size a_i at a distance r away would be

$$\langle C_{ij}(r)\rangle_{unc} = \frac{n_j\,(N_j - \delta_{ij})}{N_j} 4\pi r^2 dr \tag{8.3.7}$$

where <u>unc</u> stands for uncorrelated, $4\pi r^2 dr$ is the volume of a spherical shell of thickness dr, and δ_{ij} is the Kronecker delta function which corrects the inability of the tagged particle to be in the spherical shell. The pair distribution functions $g_{ij}(r)$ are then calculated from (8.3.6) and (8.3.7) by means of the relation

$$g_{ij}(r) = \frac{\langle C_{ij}(r)\rangle}{\langle C_{ij}(r)\rangle_{unc}} = \frac{N_j}{TN_i n_j\,(N_j - \delta_{ij})\,(4\pi r^2 dr)}\sum_{t=1}^{T}\sum_{I=1}^{N_i} C_{ij}^{It} \tag{8.3.8}$$

Generally, a pair of particles is considered to be uncorrelated when their separation is greater than $5R_{ij}$, and the values of pair distribution functions $g_{ij}(r)$ are taken to be unity for $r > 5R_{ij}$. The pair distribution functions vanish, $g_{ij}(r) = 0$ for $r < R_{ij}$, due to the non-interpenetrable characteristic of macroscopic spheres.

3.2 Sequential Addition Method

The simulation of random distribution of particles using sequential addition method is different from the Metropolis technique in that the realizations of discrete random media are constructed by depositing spheres one by one into the box subject to the same condition of no overlap with all other particles. Thus, in the sequential addition method the particles do not move, and there is no shuffling of particles. The procedures adopted are as follows.

Step 1: Generate a random test site to accommodate a new particle. The x, y, and z coordinates of this trial position within the cube are determined by three independent random number generators as

$$x = \zeta_1 \qquad y = \zeta_2 \qquad z = \zeta_3 \tag{8.3.9}$$

where ζ_i, $i = 1, 2, 3$ are random numbers uniformly distributed in the range $[0, \ell]$ and $0 \leq x, y, z \leq \ell$.

Step 2: Check the separation of trial site from those particles already established in the primary cell. A new particle will be left on the test position if it does not overlap with any other particles; otherwise it is rejected and the procedure goes back to *Step 1* to generate another test site.

Step 3: Introduce the next particle and return to *Step 1*. The additional sphere is joined one at a time to the group of particles until all N particles are filled in the primary cell.

Step 4: Count the frequency of occurrence of different pair separations. This step is the same as *Step 5* of the Metropolis technique.

The above process is repeated to generate many different realizations. The procedures for obtaining the pair distribution functions are as in equations (8.3.6)–(8.3.8) for the Metropolis method.

3.3 Numerical Results

The Monte Carlo techniques are used to generate the random distribution of particles of multiple sizes and to obtain the pair distribution functions $g_{ij}(r)$ from the computer generated samples. The Monte Carlo simulation results of pair distribution functions are compared with those under Percus–Yevick approximation. For the computer experimental results shown below, a unit cell, $\ell = 1$, containing spheres with different sizes was chosen as our model system. The numerical results of pair distribution functions using the Metropolis technique are illustrated in Figs. 8.3.1–8.3.6, and the simulation results using the sequential addition method are shown in Fig. 8.3.7. In all the figures except Figs. 8.3.2 and 8.3.7, we have used $T = 3000$ configurations

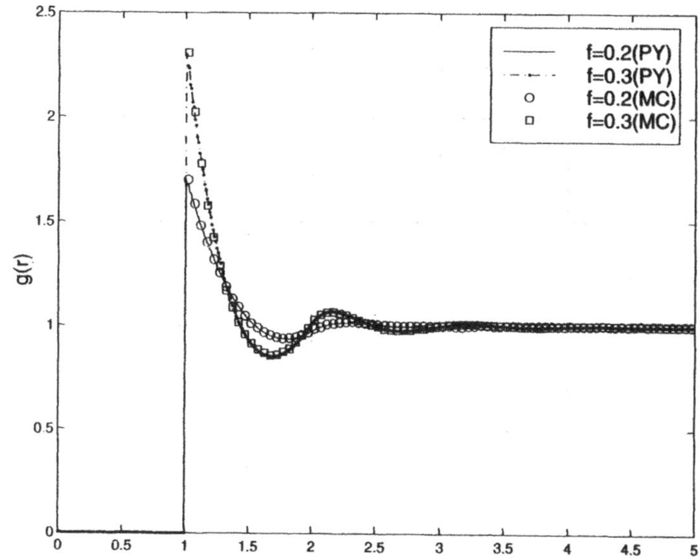

Figure 8.3.1 Pair distribution functions $g(r)$ for media with particles of single size as a function of r normalized by $2a$. The simulation parameters are (i) $N = 400$, $f = 0.2$, $\Delta = 0.06105$ and (ii) $N = 600$, $f = 0.3$, $\Delta = 0.02757$. The data points are Metropolis Monte Carlo simulation results, and the curves are obtained from Percus–Yevick calculations.

and, assuming $N_c = 1$, also 3000 realizations. The choice of displacement Δ is dictated by the following considerations. The choice of maximum displacement Δ in the Metropolis technique for each movement will affect the percentage of accepted displacements. If Δ is too big, almost all the displacements will be rejected, so the ensemble of system realizations will be composed of nearly the same configurations repeated many times and provides very little new information at each step. On the other hand, if Δ is too small, the displacement of a particle is always accepted, but the configuration changes very slowly and gives inefficient sampling in the configuration space. The size of Δ has been adjusted by trials to give about 30% to 70% acceptance of the new position.

In Fig. 8.3.1 the pair distribution functions are shown for systems with particles of identical size for two cases: (i) $f = 0.2$, $N = 400$ and (ii) $f = 0.3$, $N = 600$. The pair distribution function $g(r)$ is at its maximum at the separation equal to one diameter of particle, which shows a strong likelihood that the particles will clump together in a dense medium. The pair distribution function behaves more oscillatory and has higher peaks for the case of larger concentration of particles. The result also shows that the pair

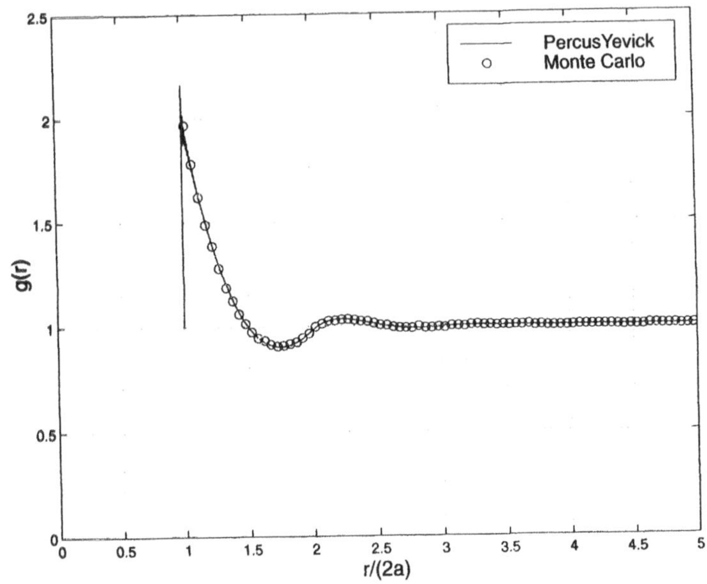

Figure 8.3.2 Pair distribution function $g(r)$ for media with particles of single size as a function of r normalized by $2a$. The simulation parameters are $f = 0.25$, $N = 4000$, $N_c = 10$ configurations per realization, and $T = 30$ realizations. The simulated results are compared with Percus–Yevick calculations. The acceptance rate A_c is 42.5%.

distribution function asymptotically approaches unity as the interparticle distance increases. The PY results are in good agreement with Monte Carlo simulations.

In Fig. 8.3.2 we consider a volume with $f = 0.25$ and $N = 4000$. We take $N_c = 10$ configurations before we obtain a realization, but we take only $T = 30$ realizations. The simulated pair distribution functions, as shown in the figure, still compare very well with the Percus–Yevick approximation. This example shows that the number of required realizations can be small if the number of particles is large.

In Figs. 8.3.3 and 8.3.4 we illustrate the results of pair distribution functions for a medium with particles of two different sizes. For Fig. 8.3.3 we placed $N = 558$ spheres inside a unit cell in which $N_1 = 486$ of them are smaller particles and $N_2 = 72$ are larger in size with $a_2 = 1.5a_1$. The volume fractions for each size are $f_1 = 0.16$ and $f_2 = 0.08$, respectively. In Fig. 8.3.4 we consider a system having a higher concentration of particles with a larger size ratio, but with a smaller number of larger particles. The model parameters are $N = 816$, $N_1 = 768$, $N_2 = 48$, $a_2 = 2a_1$, $f_1 = 0.2$, and $f_2 = 0.1$. In Figs. 8.3.5 and 8.3.6 the pair distribution functions are for

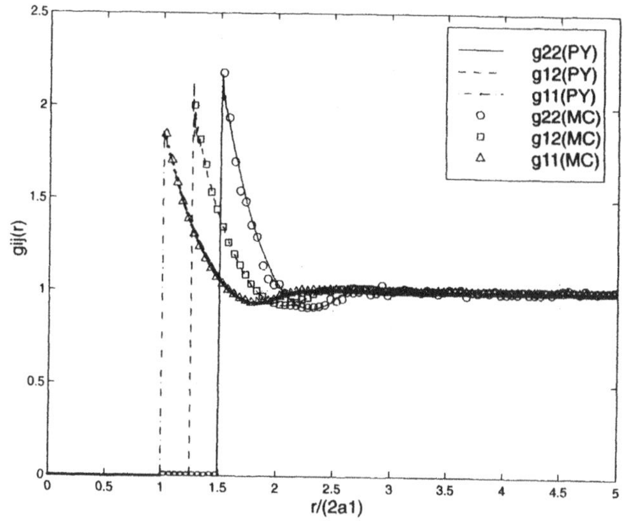

Figure 8.3.3 Pair distribution functions $g_{ij}(r)$ for media with particles of two different sizes as a function of r normalized by $2a_1$. The simulation parameters are $N = 558$, $N_1 = 486$, $N_2 = 72$, $a_2 = 1.5a_1$, $f_1 = 0.16$, $f_2 = 0.08$, and $\Delta = 0.04498$. The data points are Metropolis Monte Carlo simulation results, and the curves are obtained from Percus–Yevick calculations.

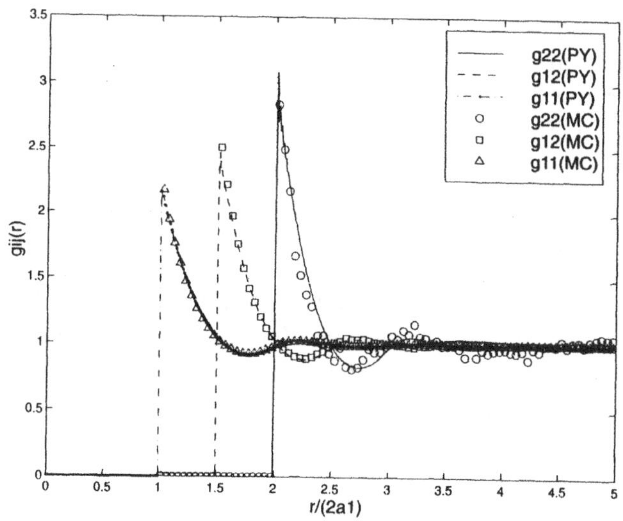

Figure 8.3.4 Pair distribution functions $g_{ij}(r)$ for media with particles of two different sizes as a function of r normalized by $2a_1$. The simulation parameters are $N = 816$, $N_1 = 768$, $N_2 = 48$, $a_2 = 2a_1$, $f_1 = 0.2$, $f_2 = 0.1$, and $\Delta = 0.03169$. The data points are Metropolis Monte Carlo simulation results, and the curves are obtained from Percus–Yevick calculations.

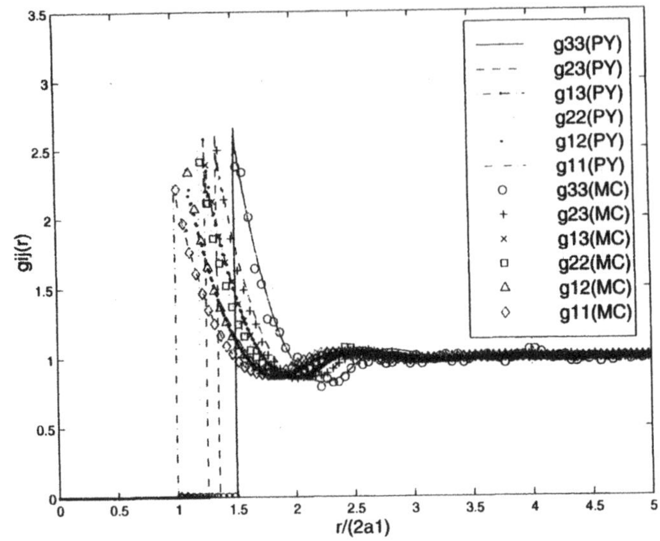

Figure 8.3.5 Pair distribution functions $g_{ij}(r)$ for media with particles of three different sizes as a function of r normalized by $2a_1$. The simulation parameters are $N = 481$, $N_1 = 324$, $N_2 = 125$, $N_3 = 32$, $a_3 = 1.5a_1$, $a_2 = 1.2a_1$, $f_1 = 0.15$, $f_2 = 0.1$, $f_3 = 0.05$, and $\Delta = 0.03167$. The data points are Metropolis Monte Carlo simulation results, and the curves are obtained from Percus–Yevick calculations.

a mixture of three particle sizes with $a_3 > a_2 > a_1$. The particle sizes and volume fractions are (i) $a_3 = 1.5a_1$, $a_2 = 1.2a_1$, $f_1 = 0.15$, $f_2 = 0.1$, and $f_3 = 0.05$ in Fig. 8.3.5 and (ii) $a_3 = 2a_1$, $a_2 = 1.2a_1$, $f_1 = 0.12$, $f_2 = 0.09$, and $f_3 = 0.05$ in Fig. 8.3.6. We observe that it usually requires more particles in the simulation processes to obtain better results when the ratios of particle size are increased. The maxima of pair distribution functions for multiple size cases are at the separation equal to the sum of the radii of two particles. Thus $g_{ij}(r)$ is maximum at $r = a_i + a_j$. We note that $g_{ij}(r)$ exhibits the most oscillation when both i and j are of the largest size, while there is less oscillation in $g_{ij}(r)$ for smaller sizes. This shows that in a dense medium with particles of different sizes, the larger particles show stronger correlation and less freedom than smaller particles in positioning themselves in the presence of all other particles. Generally, the features of the function of g_{ij} are between g_{ii} and g_{jj}. The results in Figs. 8.3.3–8.3.6 are in good agreement with the Percus–Yevick approximation.

In Fig. 8.3.7, we use the sequential addition method to generate random distributions of particles and re-examine the cases of the particles of same size for (i) $f = 0.2$, $N = 400$, and (ii) $f = 0.3$, $N = 600$. The simulated pair distribution functions obtained are averaged over $T = 30$ realizations. The

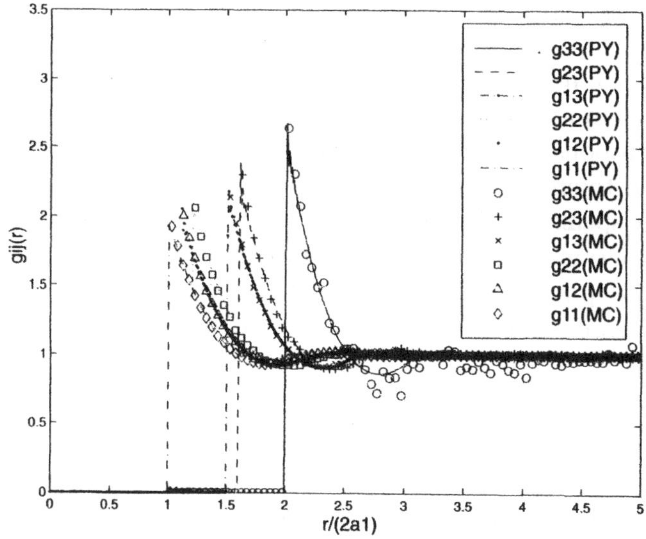

Figure 8.3.6 Pair distribution functions $g_{ij}(r)$ for media with particles of three different sizes as a function of r normalized by $2a_1$. The simulation parameters are $N = 1284$, $N_1 = 864$, $N_2 = 375$, $N_3 = 45$, $a_3 = 2a_1$, $a_2 = 1.2a_1$, $f_1 = 0.12$, $f_2 = 0.09$, $f_3 = 0.05$, and $\Delta = 0.03213$. The data points are Metropolis Monte Carlo simulation results, and the curves are obtained from Percus–Yevick calculations.

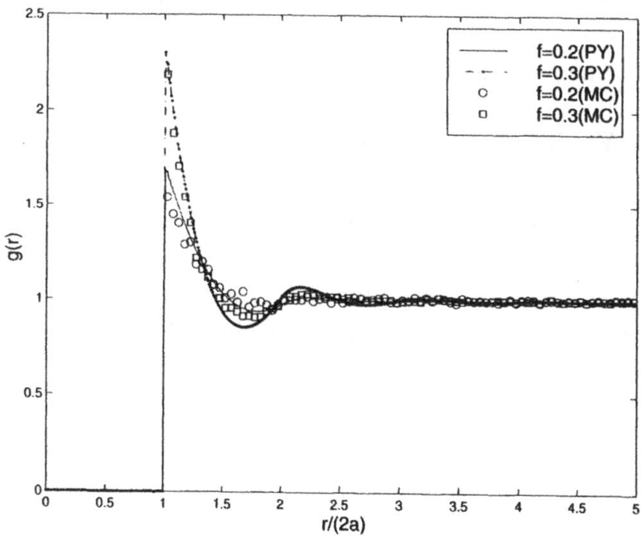

Figure 8.3.7 Pair distribution functions $g(r)$ for media with particles of single size as a function of r normalized by $2a$. The simulation parameters are (i) $N = 400$, $f = 0.2$ and (ii) $N = 600$, $f = 0.3$, and $T = 30$ realizations. The data points are Monte Carlo simulation results using sequential addition method, and the curves are obtained from Percus–Yevick calculations.

simulated pair distribution functions compare well with the Percus–Yevick approximation. This example shows that both the numbers of required realizations and the number of particles may be small if using the method of sequential random addition.

Both Metropolis shuffling approach and sequential deposition method have been used to create ensembles of randomly distributed spherical particles. Generally speaking, Metropolis method is a good candidate for creating high volume fraction configurations. Because of the rising rate of overlapping, the sequential deposition method will become difficult by depositing particles one by one into the test volume when the density of particles becomes higher (> 30%).

4 Sticky Particles

In this section we consider particle with surface adhesion. This means that when the particles are randomly packed together, they have a tendency to form clusters and bonds with each other. This is an appropriate model for certain types of geophysical media and composite materials. For example, terrain snow consists of ice grains that form bridges as metamorphism occurs. The clustering potential is modeled in the Percus–Yevick pair distribution formulation with spheres displaying surface adhesion. The adhesive force is parameterized using a variable τ which governs the degree of clustering. The structure factor has a closed form analytical solution.

4.1 Percus–Yevick Pair Distribution Function for Sticky Spheres

We shall consider a system consisting of non-penetrable, spherical particles of diameter d with a non-zero surface adhesive force, i.e. sticky hard spheres (SHS). In this model, the interaction between two particles is of very short range, with the nature of surface adhesion, and it is strong enough to bind the two particles when they contact each other. The SHS model is characterized by an interparticle potential, $u(r)$, given by:

$$u(r) = \begin{cases} \infty & \text{for } 0 < r < s \\ \ln \dfrac{12\tau(d-s)}{d} & \text{for } s < r < d \\ 0 & \text{for } r > d \end{cases} \tag{8.4.1}$$

A limit is taken in which the range of interaction becomes infinitesimal and, simultaneously, its well depth infinite, in such a way that

$$\lim_{s \to d} (d-s) e^{-u(r)} = \frac{d}{12\tau} < \infty \tag{8.4.2}$$

with d and τ being held fixed. The parameter τ in (8.4.1) is dimensionless, and its inverse is a measure of the attraction or stickiness between particles. The case of $\tau^{-1} = \infty$ corresponds to infinite stickiness and $\tau^{-1} = 0$ corresponds to non-sticky particles.

The total correlation function $h(r)$ between a pair of particle is as in (8.1.3). The direct correlation function $c(r)$ between a pair of particles, which is short-ranged, is related to $h(r)$ by means of the Ornstein-Zernike relation as in (8.1.5a).

Under the Percus–Yevick (PY) approximation, when the potential $u(r)$ vanishes, so does $c(r)$, thus:

$$c(r) = 0 \qquad \text{for} \quad r > d \tag{8.4.3}$$

Also, since the particles are non-penetrable, the pair distribution function $g(r)$ is zero when $r < s$. Hence, $h(r) = -1$ as $r < s$. The behavior of function $h(r)$ is more complicated in the region $s < r < d$. It can be seen from (8.4.1) that $h(r)$ will have a delta-function singularity when $s \to d$. In this limit, we have

$$h(r) = -1 + \frac{td}{12\tau}\delta(r-d) \qquad \text{for } 0 < r \le d \tag{8.4.4}$$

$$g(r) = h(r) + 1 = \frac{td}{12\tau}\delta(r-d) \qquad \text{for } 0 < r \le d \tag{8.4.5}$$

where t is a dimensionless parameter to be determined later. The parameter t tends to zero in the limit $\tau^{-1} = 0$. The PY approximation of the pair distribution function for the sticky hard spherical particles can be solved analytically using the factorization method of Baxter [1968a,b].

The Ornstein-Zernike relationship can be Fourier transformed to obtain a convenient algebraic equation:

$$1 - n_o\widetilde{C}(p) = \{1 + n_o\widetilde{H}(p)\}^{-1} \tag{8.4.6}$$

where n_o is the number of particles per unit volume. $\widetilde{H}(\overline{p})$ and $\widetilde{C}(\overline{p})$ are respectively the 3-D Fourier transforms of $h(\overline{r})$ and $c(\overline{r})$, i.e.,

$$\widetilde{H}(\overline{p}) = \int d\overline{r} e^{i\overline{p}\cdot\overline{r}} h(\overline{r}) = (2\pi)^3 H(p) \tag{8.4.7}$$

$$\widetilde{C}(\overline{p}) = \int d\overline{r} e^{i\overline{p}\cdot\overline{r}} c(\overline{r}) = (2\pi)^3 C(p) \tag{8.4.8}$$

According to the Wiener-Hopf technique due to Baxter, the left-hand side of (8.4.6) can be factorized into the form

$$1 - n_o\widetilde{C}(p) = \widetilde{Q}(p)\widetilde{Q}(-p) \tag{8.4.9}$$

where $\widetilde{Q}(p)$ is defined by

$$\widetilde{Q}(p) = 1 - 2\pi n_o \int_0^d dr e^{ipr} Q(r) \tag{8.4.10}$$

Here $Q(r)$ is a real function, and $Q(r) = 0$ for $r > d$. Explicit relations between $Q(r)$ and $c(r)$ and $h(r)$ can be obtained by substituting (8.4.7)–(8.4.8) and (8.4.10) into (8.4.6) and (8.4.9), and taking the inverse Fourier transforms. It is found that [Baxter, 1971]

$$rc(r) = -Q'(r) + 2\pi n_o \int_r^d dx Q'(x) Q(x - r) \tag{8.4.11}$$

for $0 < r < d$, and

$$rh(r) = -Q'(r) + 2\pi n_o \int_0^d dx (r - x) h(|r - x|) Q(x) \tag{8.4.12}$$

for $r > 0$, where $Q'(r)$ is the derivative of $Q(r)$.

For sticky particles, using (8.4.4), (8.4.5), and (8.4.12), in the range $0 < r < d$, gives rise to the closed-form expression for $Q(r)$.

$$Q(r) = A\frac{r^2}{2} + Br + D \tag{8.4.13}$$

where

$$A = \frac{1 + 2f - \mu}{(1 - f)^2} \tag{8.4.14}$$

$$B = \frac{(-3f + \mu)d}{2(1 - f)^2} \tag{8.4.15}$$

$$D = -A\frac{d^2}{2} - Bd + \frac{td^2}{12} \tag{8.4.16}$$

$$f = \frac{\pi}{6} n_o d^3 \tag{8.4.17}$$

$$\mu = tf(1 - f) \tag{8.4.18}$$

The direct correlation function $c(r)$ can be evaluated from (8.4.11) by using (8.4.13). The explicit expression of $1 - n_o\widetilde{C}(p)$ in (8.4.9) is

$$1 - n_o\widetilde{C}(p) = \left\{\frac{f}{1 - f}\left[\left(1 - tf + \frac{3f}{1 - f}\right)\Phi(X) + [3 - t(1 - f)]\,\Psi(X)\right] + \cos X\right\}^2 + \left\{\frac{f}{1 - f}[X\Phi(X)] + \sin X\right\}^2 \tag{8.4.19}$$

where $X = pd/2$ and

$$\Psi(X) = \frac{\sin X}{X} \tag{8.4.20}$$

$$\Phi(X) = 3\left[\frac{\sin X}{X^3} - \frac{\cos X}{X^2}\right] \tag{8.4.21}$$

For a given volume fraction f and stickiness parameter τ, the parameter t is determined by solving the quadratic equation

$$\frac{f}{12}t^2 - \left(\tau + \frac{f}{1-f}\right)t + \frac{1+f/2}{(1-f)^2} = 0 \tag{8.4.22}$$

When τ is greater than a value

$$\tau_c = \frac{2-\sqrt{2}}{6} \tag{8.4.23}$$

there are two real solutions for t throughout the permissible fractional volume range $0 < f < 1$. Below this value there exists an intermediate range of volume fractions within which there are no real solutions for t. Moreover, a further condition to determine the solution of t is that $\widetilde{Q}(0)$ must be positive, or

$$\mu < 1 + 2f \tag{8.4.24}$$

The procedure for calculating the pair function is as follows. Given the particle diameter d, particle concentration f, and particle stickiness τ, the parameter t is first determined from (8.4.22)–(8.4.24). By using the parameter t and Eqs. (8.4.19)–(8.4.21), $\widetilde{C}(p)$ can be computed. Next, $\widetilde{H}(p)$ is computed by solving (8.4.6). Then $h(r)$ is obtained by taking an inverse Fourier transform of $\widetilde{H}(p)$ according to (8.4.7). Finally, the pair distribution function $g(r)$ can be calculated from $h(r) + 1$.

In Fig. 8.4.1, the pair distribution functions are shown for systems with sticky particles of identical size, $\tau = 0.2$ and $\tau = 0.5$, and compared with the non-sticky case for $f = 0.3$. In the figure, we do not include the Dirac delta functions of (8.4.5). The major features for the pair distribution functions of sticky particles are the occurrence of discontinuities when the particle separation equals one and two diameters. The discontinuity at $r = d$ arises because the particles cannot penetrate each other. The height of the peak at $r = d$ grows rapidly with the increase of particle stickiness while the width of this peak is reduced. This shows a stronger connectedness between particles as particles become more sticky, and other particles are more likely to be excluded from the region $d < r < 2d$. Therefore, in a dense medium with sticky particles, the sticky particles tend to aggregate together. The

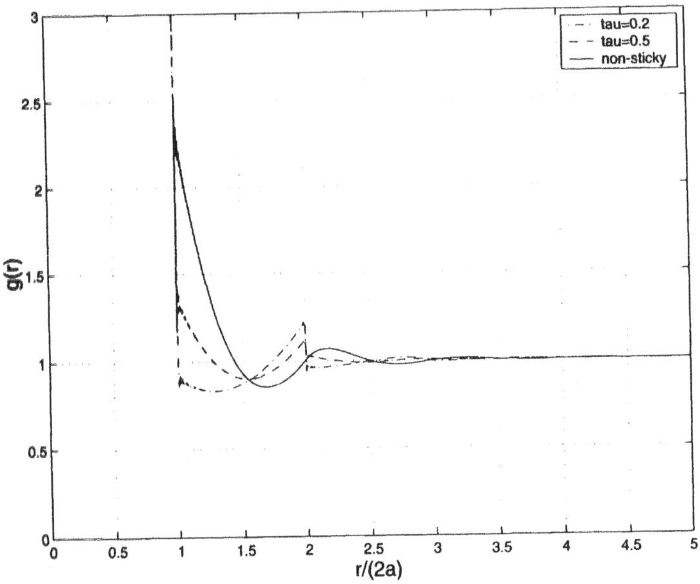

Figure 8.4.1 Percus–Yevick Pair distribution function for sticky spheres with $\tau = 0.2$ and $\tau = 0.5$, and for non-sticky spheres. Fractional volume is $f = 0.3$.

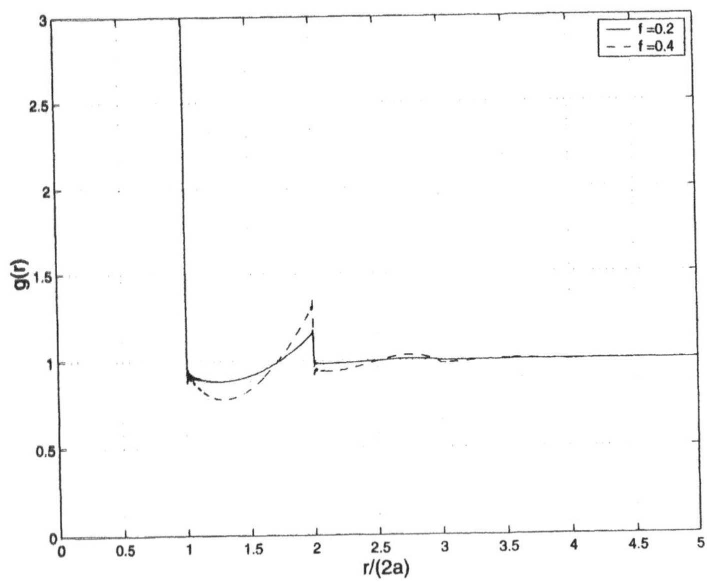

Figure 8.4.2 Percus–Yevick Pair distribution function for sticky spheres with $f = 0.2$ and $f = 0.4$ and stickiness $\tau = 0.2$.

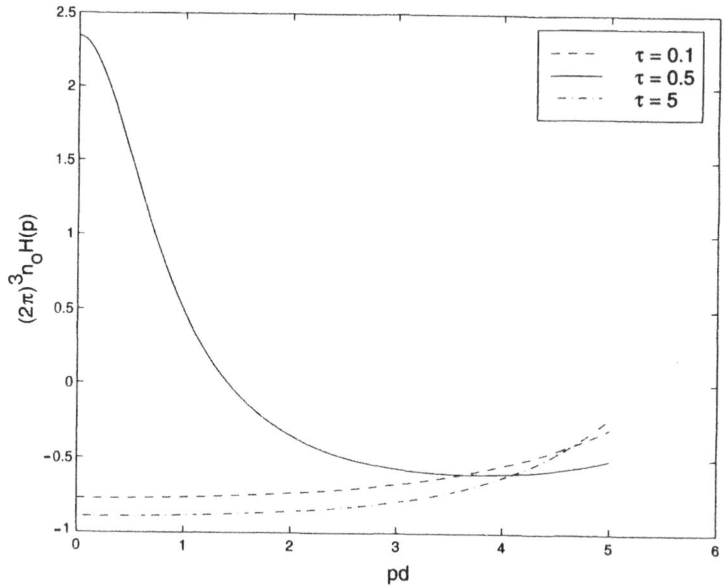

Figure 8.4.3 $(2\pi)^3 n_o H(p)$ as a function of pd for $\tau = 0.1,\ 0.5$ and 5.

discontinuity at $r = 2d$ arises from the fact that for $r > 2d$, the integrand in the integral of (8.4.12) does not include the contribution of the delta function of (8.4.4) and (8.4.5), while for $r < 2d$, the delta function is included. Physically, when the separation between the two particles is larger than twice the diameter, the probability of these two particles being bound or connected to a third particle drops to zero. In Fig. 8.4.2, the pair distribution functions are plotted for $f = 0.2$ and $f = 0.4$, with stickiness $\tau = 0.2$. For higher concentration of sticky particles, the pair distribution function displays more fluctuations just as the case of non-sticky particles. In Fig. 8.4.3, we plot $n_o(2\pi)^3 H(p)$ as a function of pd for $f = 0.3$ and for $\tau = 0.1,\ 0.5$ and 5. We note that for large pd, the values of the three curves are comparable. For small pd, they are very different, with $n_o(2\pi)^3 H(p)$ positive for small τ and negative for large τ.

4.2 Pair Distribution Functions of Adhesive Sphere Mixture

In this section, the model of spherical particles with identical size and interaction is extended to be polydisperse with respect to both sizes and interactions. Consider an M-component mixture system of spherical particles of diameter d_i, $i = 1, 2, \ldots, M$, with the interparticle adhesive potential $u_{ij}(r)$

given by:

$$u_{ij}(r) = \lim_{\sigma_{ij}\to d_{ij}} \begin{cases} \infty & \text{for } 0 < r < \sigma_{ij} \\ \ln \dfrac{12\tau_{ij}(d_{ij}-\sigma_{ij})}{d_{ij}} & \text{for } \sigma_{ij} < r < d_{ij} \\ 0 & \text{for } r > d_{ij} \end{cases} \tag{8.4.25}$$

where the limit is taken with d_{ij} and τ_{ij} held fixed, and $d_{ij} = (d_i+d_j)/2$ is the cross-term hard sphere diameter. The parameter τ_{ij} in (8.4.25) is dimensionless, and its inverse is a measure of the attraction or stickiness between particles of species i and j. The case of $\tau_{ij}^{-1} = \infty$ corresponds to infinite stickiness and $\tau_{ij}^{-1} = 0$ corresponds to non-sticky particles.

Under the Percus–Yevick (PY) approximation, the direct correlation function $c_{ij}(r)$ and the total correlation function $h_{ij}(r)$ are respectively approximated by

$$c_{ij}(r) = 0 \qquad \text{for } r > d_{ij} \tag{8.4.26}$$

$$h_{ij}(r) = -1 + \frac{t_{ij}\, d_{ij}}{12\tau_{ij}}\,\delta(r-d_{ij}) \qquad \text{for } 0 < r < d_{ij} \tag{8.4.27}$$

The parameters t_{ij} in (8.4.27) can be obtained by solving the following system of $M(M+1)/2$ coupled quadratic equations

$$t_{ij}\tau_{ij} = A_i + \frac{B_i}{d_{ij}} + \frac{\pi}{6}\sum_{\alpha=1}^{M} n_\alpha \frac{t_{j\alpha}\, d_{j\alpha}^2}{d_{ij}}\, q_{i\alpha}(s_{i\alpha}) \tag{8.4.28}$$

where n_α is the number density of α species particles, $s_{ij} = (d_i - d_j)/2$, and

$$q_{ij}(r) = \frac{A_i}{2}(r^2 - d_{ij}^2) + b_i(r - d_{ij}) + \frac{t_{ij}d_{ij}^2}{12} \tag{8.4.29a}$$

$$A_i = \frac{1-\xi_3+3d_i\xi_2}{(1-\xi_3)^2} - \frac{X_i}{(1-\xi_3)} \tag{8.4.29b}$$

$$B_i = -\frac{3d_i^2\xi_2}{2(1-\xi_3)^2} + \frac{d_iX_i}{2(1-\xi_3)} \tag{8.4.29c}$$

$$X_i = \frac{\pi}{6}\sum_{\alpha=1}^{M} n_\alpha t_{i\alpha} d_\alpha d_{i\alpha}^2 \tag{8.4.29d}$$

$$\xi_\mu = \frac{\pi}{6}\sum_{\alpha=1}^{M} n_\alpha (d_\alpha)^\mu \tag{8.4.29e}$$

with $\mu = 0, 1, 2, 3$. Insertion of (8.4.29a)–(8.4.29e) into (8.4.28) yields

$$t_{ij}\tau_{ij} = \frac{1-\xi_3}{(1-\xi_3)^2} + \frac{3d_id_j\xi_2}{(1-\xi_3)^2}$$
$$-\frac{\pi d_i(1-\xi_3)}{24(1-\xi_3)^2(d_i+d_j)}\sum_{\alpha=1}^{M} n_\alpha t_{j\alpha}\,(d_j+d_\alpha)^2 d_\alpha$$
$$-\frac{\pi d_j}{24(1-\xi_3)(d_i+d_j)}\sum_{\alpha=1}^{M} n_\alpha t_{i\alpha}\,(d_j+d_\alpha)^2 d_\alpha$$
$$+\frac{\pi}{576(d_i+d_j)}\sum_{\alpha=1}^{M} n_\alpha t_{i\alpha}\,t_{j\alpha}\,(d_j+d_\alpha)^2(d_i+d_\alpha)^2 \tag{8.4.30}$$

which is useful in solving the parameters t_{ij}.

The PY approximation of the pair distribution function $g_{ij}(r) = h_{ij}(r) + 1$ for sticky spherical particles can be solved analytically using the factorization method of Baxter [1968a,b]. The multi-species Ornstein-Zernike (OZ) relation, as in (8.2.2), can be Fourier transformed to obtain a convenient algebraic equation $\overline{\overline{H}}(\overline{p}) = \overline{\overline{C}}(\overline{p}) + \overline{\overline{C}}(\overline{p}) \cdot \overline{\overline{H}}(\overline{p})$. The matrix elements $H_{ij}(\overline{p})$ and $C_{ij}(\overline{p})$ are

$$H_{ij}(\overline{p}) = (n_in_j)^{1/2}\int d\overline{r}\; e^{i\overline{p}\cdot\overline{r}} h_{ij}(\overline{r}) \tag{8.4.31}$$

$$C_{ij}(\overline{p}) = (n_in_j)^{1/2}\int d\overline{r}\; e^{i\overline{p}\cdot\overline{r}} c_{ij}(\overline{r}) \tag{8.4.32}$$

Because of the spherical particle assumption, the transform only depends on $p = |\overline{p}|$. The Wiener-Hopf technique can be used to factorize the transformed OZ equation. The matrix elements $C_{ij}(p)$ are

$$-C_{ij}(p) = \frac{\pi}{6}\sqrt{n_in_j}\,\cos x_j\left[-t_{ji}d_{ji}^2 d_i\, j_0(x_i) + \frac{3d_i^3\, j_1(x_i)}{(1-\xi_3)x_i}\right.$$
$$\left. + \frac{3d_jd_i^2\, j_0(x_i)}{(1-\xi_3)} - \frac{6B_jd_i^3\, j_1(x_i)}{d_jx_i}\right]$$
$$+\frac{\pi}{6}\sqrt{n_in_j}\,\cos x_i\left[-t_{ij}d_{ij}^2 d_j\, j_0(x_j) + \frac{3d_j^3\, j_1(x_j)}{(1-\xi_3)x_j}\right.$$
$$\left. + \frac{3d_id_j^2\, j_0(x_j)}{(1-\xi_3)} - \frac{6B_id_j^3\, j_1(x_j)}{d_ix_j}\right]$$
$$+\frac{\pi}{6}\sqrt{n_in_j}\left[\sin x_j\frac{3d_jd_i^2\, j_1(x_i)}{(1-\xi_3)} + \sin x_i\frac{3d_id_j^2\, j_1(x_j)}{(1-\xi_3)}\right]$$

$$
\begin{aligned}
&+\left(\frac{6}{\pi}\right)^2 \sum_{l=1}^{M} n_l \sqrt{n_i n_j} \left\{ \frac{9d_l^2 d_i^2 d_j^2 \, j_1(x_i) j_1(x_j)}{(1-\xi_3)} \right. \\
&+\left[-t_{li} d_{li}^2 d_i \, j_0(x_i) + \frac{3d_i^3 \, j_1(x_i)}{(1-\xi_3)x_i} \right. \\
&\qquad \left. + \frac{3d_l d_i^2 \, j_0(x_i)}{(1-\xi_3)} - \frac{6B_l d_i^3 \, j_1(x_i)}{d_l x_i} \right] \\
&+\left[-t_{lj} d_{lj}^2 d_j \, j_0(x_j) + \frac{3d_j^3 \, j_1(x_j)}{(1-\xi_3)x_j} \right. \\
&\qquad \left. \left. + \frac{3d_l d_j^2 \, j_0(x_j)}{(1-\xi_3)} - \frac{6B_l d_j^3 \, j_1(x_j)}{d_l x_j} \right] \right\}
\end{aligned}
\tag{8.4.33}
$$

where

$$x_i = \frac{pd_i}{2} \tag{8.4.34}$$

$$j_0(x_i) = \frac{\sin x_i}{x_i} \tag{8.4.35}$$

$$j_1(x_i) = \frac{\sin x_i}{x_i^2} - \frac{\cos x_i}{x_i} \tag{8.4.36}$$

The procedure of calculating the pair function is as described in Section 8.2. First $C_{ij}(p)$ is calculated from (8.4.33). Next, $\overline{\overline{H}}(p)$ is computed by solving the linear matrix equation (8.2.7). Then $h_{ij}(r)$ is obtained by taking an inverse Fourier transform of $\overline{\overline{H}}(p)$ according to (8.4.31). Finally, the pair distribution function $g_{ij}(r)$ can be calculated by using (8.2.1).

In Fig. 8.4.4, the pair distribution functions for a binary mixture of adhesive spheres are illustrated with $a_2 = 0.6a_1$, $f_1 = 0.1$, $f_2 = 0.15$, and $\tau_{11} = \tau_{12} = \tau_{22} = 0.2$. We note that, besides the discontinuities at the particle separation equals to the sum of two particles' radii, the discontinuities of the pair distribution functions also occur at other particle separations which depend on the various co-linear arrangements of three particles. For example, the arrangement of a larger sphere between two small spheres gives the discontinuity at $r = 1.6d_1$, while the arrangement of a larger particle with two consecutive small spheres gives the discontinuity at $r = 1.4d_1$. Because of the symmetry relation $g_{ij} = g_{ji}$, only three pair distribution functions are illustrated. The pair distribution functions for another case of binary mixture of adhesive spheres are shown in Fig. 8.4.5 with $a_2 = 0.4a_1$, $f_1 = 0.2$, $f_2 = 0.04$, and $\tau_{11} = 1$, $\tau_{12} = 0.2$, and $\tau_{22} = \infty$. For this case, since the particle species 2 is non-adhesive, there is no discontinuity at $r = 0.8d_1$.

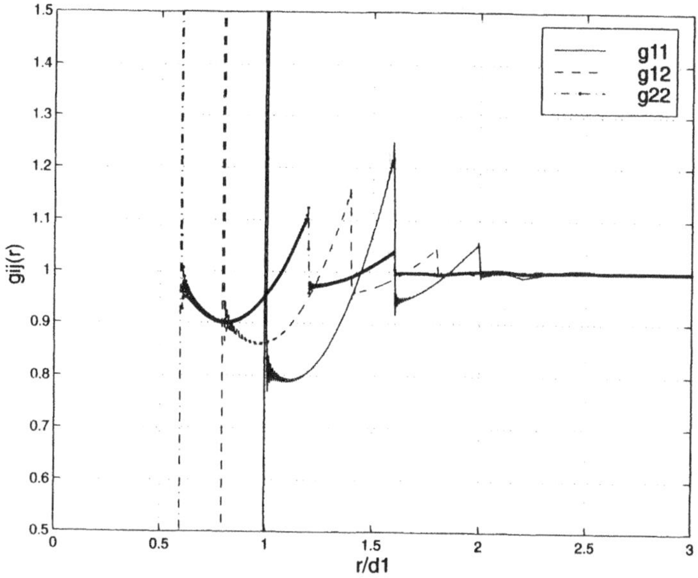

Figure 8.4.4 Percus–Yevick pair distribution function for a binary mixture of sticky spheres with $a_2 = 0.6a_1$, $f_1 = 0.1$, $f_2 = 0.15$, and $\tau_{11} = \tau_{12} = \tau_{22} = 0.2$.

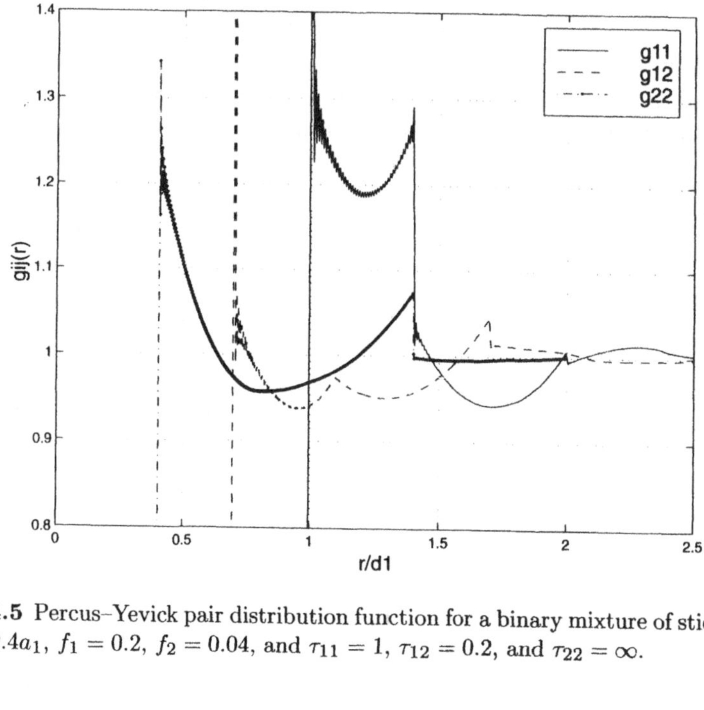

Figure 8.4.5 Percus–Yevick pair distribution function for a binary mixture of sticky spheres with $a_2 = 0.4a_1$, $f_1 = 0.2$, $f_2 = 0.04$, and $\tau_{11} = 1$, $\tau_{12} = 0.2$, and $\tau_{22} = \infty$.

4.3 Monte Carlo Simulation of Adhesive Spheres

The Metropolis shuffling technique as described in Section 3.1 is usually employed for particles without surface adhesion. This conventional method generates new configurations by giving random displacements to particles and then accepting or rejecting the configuration based on particle overlap. However, this method cannot be directly applied to particles interacting via the adhesive potential as given by (8.4.1). By taking the limit, $s \to d$ in the exponential of (8.4.1), we have

$$e^{-u(r)} = \begin{cases} \frac{d}{12\tau}\delta(r-d) & \text{for } r \le d \\ 1 & \text{for } r > d \end{cases} \tag{8.4.37}$$

here d is the particle diameter and τ is the stickiness parameter related to the adhesion strength. Thus, the adhesive potential (8.4.1) implies that there exists a finite probability of contact, proportional to $d/(12\tau)$, for adhesive particle pairs. As τ decreases, the adhesion strength increases, and favors more the formation of clusters. This feature cannot be captured by using the shuffling technique since it is very unlikely to sample any surfaces of other particles while each particle is displaced in a three-dimensional space.

In order to explore the available contact states of adhesive particles, Seaton and Glandt [1986, 1987] and Kranendonk and Frenkel [1988] have developed alternative sampling procedures which allow a particle to break or form bonds with other particles during its Monte Carlo displacement. They have considered only transitions between four binding states $\alpha = 0, 1, 2, 3$, corresponding to the unbounded, single-bond, double-bond, and triple-bond configurations, respectively. The α-bond state indicates that a particle is in contact with α number of other particles. The maximum number of bonds is 12. The binding states $\alpha = 0, 1, 2, 3$ of a particle, represented by a filled circle, are illustrated in Fig. 8.4.6. We will describe the Kranendonk-Frenkel algorithm in this section for the three-dimensional simulations of adhesive particles. The description of Seaton-Glandt algorithm will be given in Chapter 9 for the two-dimensional simulations.

In the Kranendonk-Frenkel algorithm, a total effective volume $V_{\text{eff}}^{(\alpha)}$ is assigned to the α-bond state of the test particle k as given by

$$V_{\text{eff}}^{(\alpha)} = \sum_i \left(\frac{d}{12\tau}\right)^{\alpha} \int dq_i \tag{8.4.38}$$

where dq_i represents all degrees of freedom of the test particle k, remaining after making the α bonds. The summation in (8.4.38) is over all possible combinations of the test particle with all the other particles that can realize the type of α-bond. Possible excluded volume effects are ignored in com-

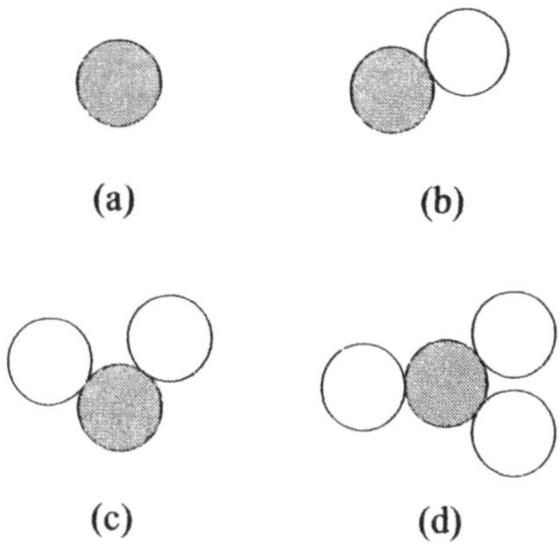

Figure 8.4.6 The particle binding states: (a) unbounded, (b) single-bond, (c) double-bond, and (d) triple-bond.

puting the integral in (8.4.38). The effective volume $V_{\text{eff}}^{(\alpha)}$ represents a direct measure of the probability of the α-bond configuration, with a contributing factor $(\frac{d}{12\tau})^{\alpha}$.

The total effective volume associated with the unbounded state of a test particle k is

$$V_{\text{eff}}^{(0)} = V \tag{8.4.39}$$

which is simply the total volume available to the test particle, including overlapping regions. Next we examine the single bond (Fig. 8.4.7a). Let r be the distance between particle i and test particle k. Then $\int_0^{2\pi}\int_0^{\pi} r^2 \sin\theta d\theta d\phi$ is the surface area that the center $\bar{r}_k$ of particle k can be located such that particle k is attached to particle i. The total effective volume of the single-bond state of a test particle k is

$$\begin{aligned} V_{\text{eff}}^{(1)} &= \sum_{i \neq k}^{N} \left(\frac{d}{12\tau}\right) \int_0^{2\pi}\int_0^{\pi} r^2 \sin\theta \; d\theta d\phi \\ &= \sum_{i \neq k}^{N} v_{\text{eff}}^{(1)i} = (N-1)\left(\frac{d}{12\tau}\right)(4\pi d^2) \end{aligned} \tag{8.4.40}$$

As shown in Fig. 8.4.7a, the test particle k can be placed anywhere on the surface of another particle i. Thus, as indicated in (8.4.40), $v_{\text{eff}}^{(1)i}$ is the effective volume associated with the single bond between particles k and i.

$$v_{\text{eff}}^{(1)i} = \left(\frac{d}{12\tau}\right) 4\pi d^2 \tag{8.4.41}$$

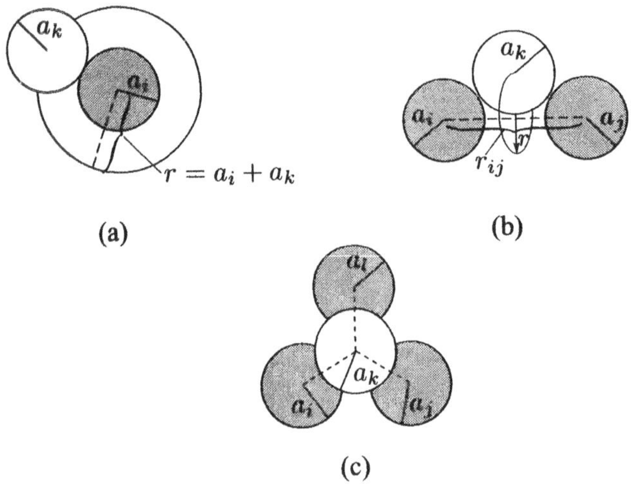

Figure 8.4.7 The effective volumes of the binding states of a test particle k: (a) single-bond, (b) double-bond, and (c) triple-bond.

The total effective volume of one bond is then equal to the surface area of a sphere with radius d times $\frac{d}{12\tau}$ and $(N-1)$, excluding the displaced particle itself.

Next, we consider $V_{\text{eff}}^{(2)}$ for the double-bond state of a particle. From Fig. 8.4.7b, r is the radius of the circle on which the center of the test particle k can be placed. This circle is formed by all points that have a distance d form both particles i and j. Thus the total effective volume of the double-bond state associated with the test particle k is given by

$$\begin{aligned} V_{\text{eff}}^{(2)} &= \sum_{\substack{i<j \\ i,j\neq k}}^{N} \left(\frac{d}{12\tau}\right)^2 \int_0^{2\pi} r \, d\phi \\ &= \sum_{\substack{i<j \\ i,j\neq k}}^{N} \left(\frac{d}{12\tau}\right)^2 \int_0^{2\pi} \sqrt{d^2 - \left(\frac{r_{ij}}{2}\right)^2} \, d\phi \\ &= \sum_{\substack{i<j \\ i,j\neq k}}^{N} v_{\text{eff}}^{(2)ij} = \left(\frac{d}{12\tau}\right)^2 \sum_{\substack{i<j \\ i,j\neq k}}^{N} 2\pi\sqrt{d^2 - \left(\frac{r_{ij}}{2}\right)^2} \end{aligned} \tag{8.4.42}$$

here $v_{\text{eff}}^{(2)ij}$ is the effective volume for the double-bond configuration of the particle k with particles i and i.

$$v_{\text{eff}}^{(2)ij} = \left(\frac{d}{12\tau}\right)^2 2\pi\sqrt{d^2 - \left(\frac{r_{ij}}{2}\right)^2} \tag{8.4.43}$$

Note that $v_{\text{eff}}^{(2)ij} = 0$ when $r_{ij} > 2d$. This is because particle k can not form a double-bond with particles i and j simultaneously when their distance is greater than two diameters. The effective volume for a triple-bond configuration of particle k with particles i, j, and l is

$$v_{\text{eff}}^{(3)ijl} = \left(\frac{d}{12\tau}\right)^3 \frac{1}{|\hat{e}_{ik} \cdot (\hat{e}_{jk} \times \hat{e}_{lk})|} \tag{8.4.44}$$

and the total effective volume of the triple-bond state of the test particle k is given by

$$V_{\text{eff}}^{(3)} = \sum_{\substack{i<j<l \\ i,j,l \neq k}}^{N} \left(\frac{d}{12\tau}\right)^3 \frac{1}{|\hat{e}_{ik} \cdot (\hat{e}_{jk} \times \hat{e}_{lk})|} \tag{8.4.45}$$

where $\hat{e}_{ik}$, $\hat{e}_{jk}$, and $\hat{e}_{lk}$ are unit vectors along the directions joining the center of particle k to the centers of particles i, j, and l, that form a triple-bond with particle k. A necessary condition is that the diameter of the inscribed circle within the triangle, whose vertices are the centers of particles i, j, and l, must be be less than the diameter of the test particle k, as shown in Fig. 8.4.7c.

The calculations of the total effective volumes $V_{\text{eff}}^{(\alpha)}$, $\alpha = 0, 1, 2, 3$ of the test particle k do not depend on its current binding state. In the trial move of sticky particles, the final move consists of changing the binding state. The binding state may change from its present β-bond configuration to a new β'-bond configuration. Thus, each move will involve a transition among zero bond, single bond, double bond, and triple bond states. We calculate $V_{\text{eff}}^{(2)}$ and $V_{\text{eff}}^{(3)}$ by doing the following. We first make a list of all pairs of particles with which the test particle k can form a double-bond. This double-bond list is then used to compute the total effective volume $V_{\text{eff}}^{(2)}$ according to (8.4.42). From the list of pairs we further select those triplets that can form a triple-bond with the test particle, and these possible triplets are used to calculate the total effective volume $V_{\text{eff}}^{(3)}$ using (8.4.45). In this algorithm, we let the test particle free to move over the entire volume, every particle, except itself, is possible to form a single-bond with the test particle. Thus the total effective volumes $V_{\text{eff}}^{(0)}$ and $V_{\text{eff}}^{(1)}$ are constants during the whole simulation process. The new binding state β' for the test particle is selected with the probability of

$$P(\beta') = \frac{V_{\text{eff}}^{(\beta')}}{\sum_{\alpha=0}^{3} V_{\text{eff}}^{(\alpha)}} \tag{8.4.46}$$

It is noted that the magnitude of $P(\beta')$ depends on the binding structure of the system prior to the particle k being moved.

The three-dimensional Monte Carlo simulation of adhesive particles begins with N unbounded spheres randomly or regularly placed inside a cubic box with no overlap. In the first stage, the Monte Carlo steps are similar to the ones described in Section 8.3.1 for non-sticky particles. The spheres are shuffled to create an initial realization. In the second stage, however, instead of moving the test particle randomly within a small cube of side Δ to determine its trial location, the test adhesive sphere is displaced according to the breaking of its current binding state and the formation of new binding state with other particles. Thus the second stage consists of transitions among binding states. If the test particle is found to bind with four or more particles, it will not be moved. If not, the attempt of changing the test particle's number of bonds is described in the following.

Before changing the test particle k's binding state, we have to establish a catalog of the current system's binding structure, that is, to build the lists of pairs and triplets as described above, and to compute the probabilities $P(\beta')$ for $\beta'=0,1,2,3$. A random number generator, between 0 and 1, is then used to generate a random number and compared with the obtained $P(\beta')$ of the current system to decide the new binding state of the test particle. Obviously, the transition involves all four states. If $\beta' = 0$ is favored, the new location of the test particle is determined from three random numbers, for its x, y, and z coordinates. If $\beta'=1$, an integer random number i with equal probabilities, $i=1,2,\ldots,N$, $i\neq k$, is generated to select the potential single-bond candidate particle. Random polar and azimuthal angles are then generated to determine the location of particle k's center on a spherical surface of radius d centered at particle i (Fig. 8.4.7a). If $\beta'=2$, another random number, between 0 and 1, is generated to select a candidate pair of particles (i,j) from the double-bond list according to the following probability to form a double-bond with the test particle k.

$$p^{(2)}(i,j) = \frac{v_{\text{eff}}^{(2)ij}}{V_{\text{eff}}^{(2)}} \tag{8.4.47}$$

Another random angle between 0 and 2π is then generated to select the location of particle k's center on a circle centered at the midpoint of the vector $\bar{r}_{ij}$ joining the centers of particles i and j (Fig. 8.4.7b). Similarly, when $\beta' = 3$, a random number, between 0 and 1, is generated to select a candidate triple of particles (i, j, l) from the triple-bond list according to the following probability to form a triple-bond with the test particle k.

$$p^{(3)}(i,j,l) = \frac{v_{\text{eff}}^{(3)ijl}}{V_{\text{eff}}^{(3)}} \tag{8.4.48}$$

For the triple-bond, there are only two choices for the positions of particle k, either above or under the surface containing the triangle formed by the line segments connecting the centers of particles i, j, and l. It is possible that the new position of the test particle may overlap with one of the other particles in the system. If not, we will accept the displacement and update the coordinates of the displaced particle. Otherwise, reject the displacement and return the test particle to its original position. The calculation of total effective volumes requires the maintenance of an updated catalog of the system's binding configuration, so that we need to rebuild the pair-list and the triplet-list whenever a trial move of the test particle is successful.

The algorithm is applied to a system of 300 sticky spheres of identical size and stickiness. The volume fraction is $f = 0.3$. The Monte Carlo results are obtained by averaging over 10,000 passes. Fig. 8.4.8 illustrates the Monte Carlo results of pair distribution functions for particles with stickiness $\tau =$ 0.5. A Comparison with Percus–Yevick (PY) approximation calculations is also made in Fig. 8.4.8, which shows good agreements between these two approaches. In Fig. 8.4.9, the particle sticky strength is increased to $\tau = 0.2$. Compared with PY approximation, a slight dephasing appears in the interval $d < r < 1.5d$ between Monte Carlo data and PY results. Moreover, another discontinuity of the pair distribution function near $r = \sqrt{3}d$ is predicted in this case as discussed in the work of Seaton and Glandt [1987].

The effective volumes described in the Kranendonk-Frenkel algorithm can be easily generalized to the case of multi-species of adhesive particles. Consider a system that has M-species of spherical particles of diameter d_i, $i = 1, 2, \ldots, M$, with the interparticle adhesive potential $u_{ij}(r)$ as given in (8.4.25). Similarly, the effective volume $v_{\text{eff}}^{(1)i}$ associated with the single-bond between particle k and particle i becomes

$$\begin{aligned} v_{\text{eff}}^{(1)i} &= \left(\frac{d_{ik}}{12\tau_{ik}}\right) \int_0^{2\pi} \int_0^{\pi} r^2 \sin\theta \; d\theta d\phi \\ &= \left(\frac{d_{ik}}{12\tau_{ik}}\right) 4\pi \; d_{ik}^2 \end{aligned} \tag{8.4.49}$$

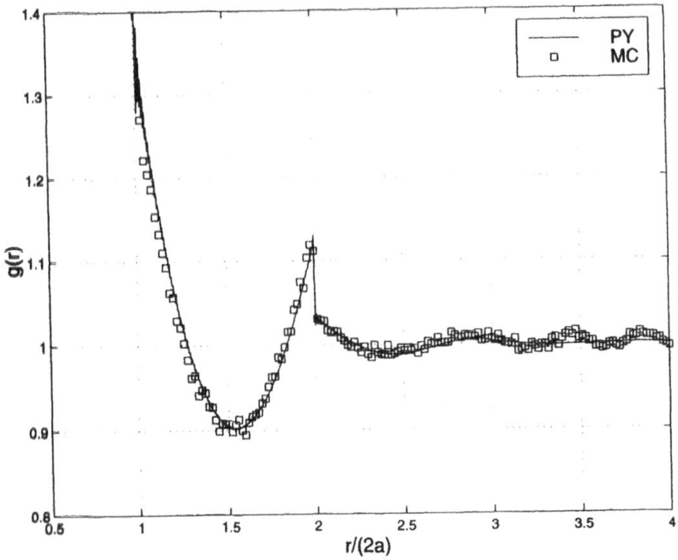

Figure 8.4.8 Pair distribution functions obtained from Monte Carlo simulations and Percus–Yevick approximation for a system of single size sticky spheres with $\tau = 0.5$ and $f_v = 0.3$. The symbols are Monte Carlo simulation results and the curve is for Percus–Yevick calculations.

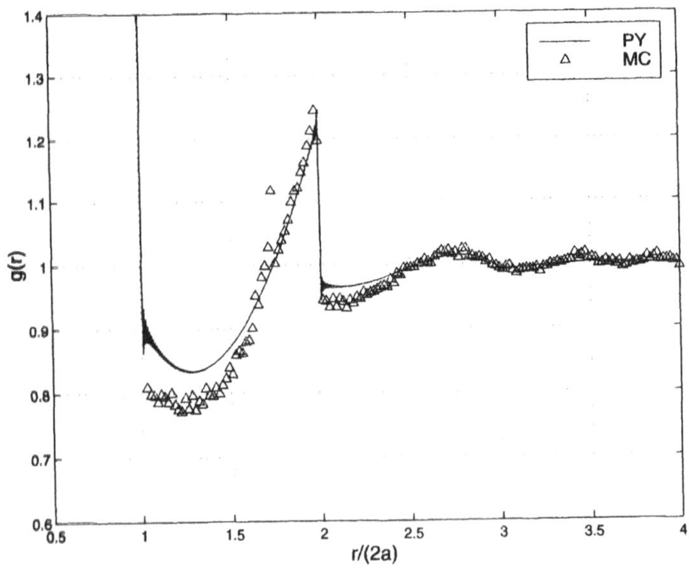

Figure 8.4.9 Pair distribution functions obtained from Monte Carlo simulations and Percus–Yevick approximation for a system of single size sticky spheres with $\tau = 0.2$ and $f_v = 0.3$. The symbols are Monte Carlo simulation results and the curve is for Percus–Yevick calculations.

where $d_{ik} = (d_i + d_k)/2$ is the cross-term hard sphere diameter, and τ_{ik} is a measure of the attraction between particles i and k. For the mixture case, a factor $(\frac{d_{ik}}{12\tau_{ik}})$ contributes to the bonding probability that depends on the distance and adhesion strength between particles of species i and k. Thus, $v_{\text{eff}}^{(1)i}$ is not a constant, it depends on the types of two particles that form a bond. The effective volume $v_{\text{eff}}^{(2)ij}$ of the test particle k having a double-bond with particles i and j is

$$\begin{aligned} v_{\text{eff}}^{(2)ij} &= \left(\frac{d_{ik}}{12\tau_{ik}}\right)\left(\frac{d_{jk}}{12\tau_{jk}}\right)\int_0^{2\pi} r\ d\phi \\ &= \left(\frac{d_{ik}}{12\tau_{ik}}\right)\left(\frac{d_{jk}}{12\tau_{jk}}\right)\int_0^{2\pi} \left[d_{ik}\ \sin\!\left(\cos^{-1}(\hat{e}_{ik}\cdot\hat{e}_{ij})\right)\right] d\phi \\ &= \left(\frac{d_{ik}}{12\tau_{ik}}\right)\left(\frac{d_{jk}}{12\tau_{jk}}\right) 2\pi \left[d_{ik}\ \sin\!\left(\cos^{-1}(\hat{e}_{ik}\cdot\hat{e}_{ij})\right)\right] \end{aligned} \tag{8.4.50}$$

where the unit vectors $\hat{e}_{ij}$ and $\hat{e}_{ik}$ are along the directions connecting the centers of particles i and j and particles i and k, respectively. The effective volume $v_{\text{eff}}^{(3)ijl}$ of the test particle k forming a triple-bond with particles i, j and l is

$$v_{\text{eff}}^{(3)ijl} = \left(\frac{d_{ik}}{12\tau_{ik}}\right)\left(\frac{d_{jk}}{12\tau_{jk}}\right)\left(\frac{d_{lk}}{12\tau_{lk}}\right)\frac{1}{|\hat{e}_{ik}\cdot(\hat{e}_{jk}\times\hat{e}_{lk})|} \tag{8.4.51}$$

here $\hat{e}_{jk}$ and $\hat{e}_{lk}$ are unit vectors from the center of the test particle k to the respective centers of particles j and l. The total effective volumes $V_{\text{eff}}^{(\alpha)}$, $\alpha = 0, 1, 2, 3$, are then given by

$$V_{\text{eff}}^{(0)} = V \tag{8.4.52}$$

$$V_{\text{eff}}^{(1)} = \sum_{i\neq k}^{N}\left(\frac{d_{ik}}{12\tau_{ik}}\right) 4\pi\ d_{ik}^2 \tag{8.4.53}$$

$$V_{\text{eff}}^{(2)} = \sum_{\substack{i<j\\ i,j\neq k}}^{N}\left(\frac{d_{ik}}{12\tau_{ik}}\right)\left(\frac{d_{jk}}{12\tau_{jk}}\right) 2\pi\left[d_{ik}\sin\!\left(\cos^{-1}(\hat{e}_{ik}\cdot\hat{e}_{ij})\right)\right] \tag{8.4.54}$$

$$V_{\text{eff}}^{(3)} = \sum_{\substack{i<j<l\\ i,j,l\neq k}}^{N}\left(\frac{d_{ik}}{12\tau_{ik}}\right)\left(\frac{d_{jk}}{12\tau_{jk}}\right)\left(\frac{d_{lk}}{12\tau_{lk}}\right)\frac{1}{|\hat{e}_{ik}\cdot(\hat{e}_{jk}\times\hat{e}_{lk})|} \tag{8.4.55}$$

The particle displacement algorithm for the mixture of adhesive particles is the same as that of the single species case. However, it is noted that the

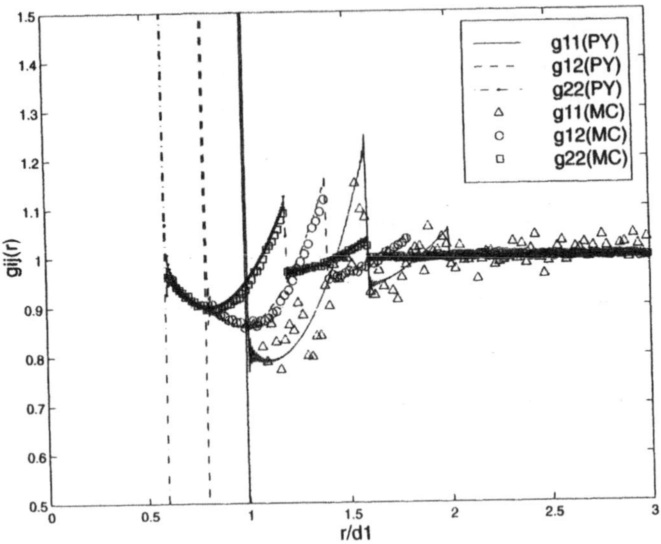

Figure 8.4.10 Pair distribution functions obtained from Monte Carlo simulations and Percus–Yevick approximation for a binary mixture of sticky spheres with $a_2 = 0.6a_1$, $f_1 = 0.1$, $f_2 = 0.15$, and $\tau_{11} = \tau_{12} = \tau_{22} = 0.2$. The symbols are Monte Carlo simulation results and the curves are for Percus–Yevick calculations.

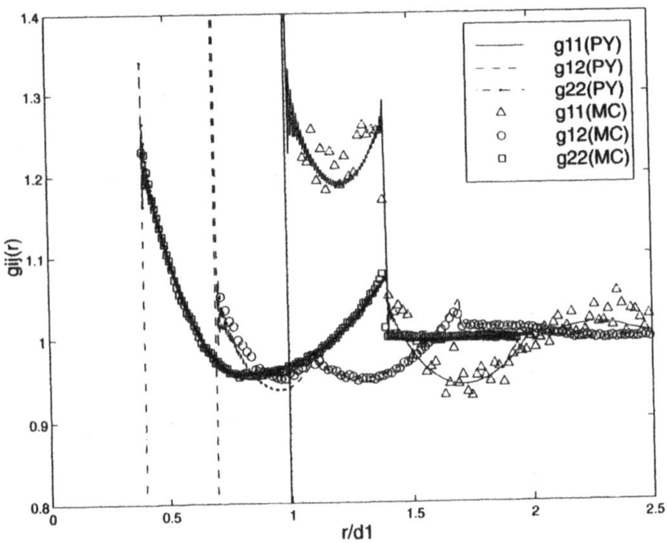

Figure 8.4.11 Pair distribution functions obtained from Monte Carlo simulations and Percus–Yevick approximation for a binary mixture of sticky spheres with $a_2 = 0.4a_1$, $f_1 = 0.2$, $f_2 = 0.04$, and $\tau_{11} = 1$, $\tau_{12} = 0.2$, and $\tau_{22} = \infty$. The symbols are Monte Carlo simulation results and the curves are for Percus–Yevick calculations.

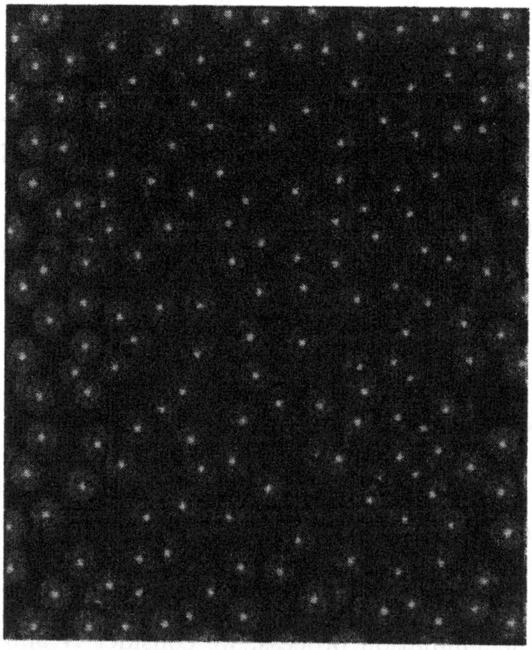

Figure 8.4.12 Three-dimensional sample of computed generated sticky spheres with $f = 0.35$ and $\tau = 0.2$.

single-bond effective volume $v_{\text{eff}}^{(1)i}$ is not a constant but depends on the types of two particles i and k that form such a single-bond. Thus, an integer random number with equal probability between 1 and $N - 1$ is required to select the candidate particle i (which cannot be the test particle itself) according to the probability

$$p^{(1)}(i) = \frac{v_{\text{eff}}^{(1)i}}{V_{\text{eff}}^{(1)}} \tag{8.4.56}$$

where $v_{\text{eff}}^{(1)i}$ is given by (8.4.49). Thus for the mixture case, we have to store three arrays of probabilities, (8.4.56), (8.4.47), and (8.4.48), to select a candidate particle i, pair (i, j), and triplet (i, j, l) to from single-, double-, and triple-bond, respectively, with the test particle.

In Figs. 8.4.10 and 8.4.11, we illustrate the Monte Carlo results of pair distribution functions for a medium with adhesive binary mixtures of particles. For Fig. 8.4.10 we placed $N = 1144$ spheres inside a unit cell in which $N_2 = 1000$ of them are smaller particles and $N_1 = 144$ are larger in size with $a_2 = 0.6a_1$. The volume fractions are $f_1 = 0.1$ and $f_2 = 0.15$. They have the stickiness with $\tau_{11} = \tau_{12} = \tau_{22} = 0.2$. In Fig. 8.4.11 we consider the case of particles having different adhesive forces. This system has a high concentration of larger particles and a much less volume fraction of smaller particles. The parameters are $N = 885$, $N_1 = 200$, $N_2 = 685$, $a_2 = 0.4a_1$, $f_1 = 0.2$, $f_2 = 0.04$, and the stickiness parameters are $\tau_{11} = 1$, $\tau_{12} = 0.2$, and $\tau_{22} = \infty$. The Monte Carlo results for both cases are obtained by averaging over 12,000 passes. In general, the Monte Carlo results are in good agreement with the Percus–Yevick approximation calculations. In Fig. 8.4.12, we show a typical configuration for a 3-D sample of sticky spheres with $\tau = 0.2$ and volume fraction $f = 0.35$ [Zurk, 1995]. The generated realizations of random media can be used for the numerical solutions of Maxwell's equations.

5 Particle Placement Algorithm for Spheroids

In this section we apply Metropolis shuffling process to generate the positions and orientations of a system of densely packed spheroids. For nonspherical particles such as spheroids, rotations need to be performed in addition to displacements.

To simplify the simulation process, we consider N spheroidal particles of the same shape that are randomly placed in a simple cubic cell. Periodic boundary conditions are employed to minimize the finite size effect. A central primary cell with side length L is set up to contain the N spheroids, and this central cell is surrounded by the periodic images of itself. The state of each spheroid is specified by the location $\overline{r} = (x, y, z)$ of its center and the orientation $\Omega = (\theta, \phi)$ of its symmetry axis. Given the particle concentration, size, and shape, the N spheroids are initially aligned along the $\hat{z}$-direction, and are placed in close contact with each other but without overlapping. In such an initial set-up, we are able to create configurations of higher fractional volumes. The shuffling process consists of a succession of trial moves in which we attempt to change the locations and orientations of particles by the combination of random displacements and rotations. For the spheroid j, the new location $\overline{r}'_j$ is determined from its present location $\overline{r}_j$ by

$$x'_j = x_j + \Delta\eta_x \tag{8.5.1}$$

$$y'_j = y_j + \Delta\eta_y \tag{8.5.2}$$

$$z'_j = z_j + \Delta\eta_z \tag{8.5.3}$$

where Δ is the maximum displacement allowed in each spatial movement, and η_x, η_y, and η_z are random numbers uniformly distributed between -1 and 1. Each spheroid has an axis of symmetry. We also perform random rotation. The new orientation Ω'_j of the particle is determined by

$$\theta'_j = \pi\eta_\theta \tag{8.5.4}$$

$$\phi'_j = 2\pi\eta_\phi \tag{8.5.5}$$

where η_θ and η_ϕ are random numbers with magnitudes between 0 and 1.

5.1 Contact Functions of Two Ellipsoids

During the course of simulation, in order to select the allowed new locations and orientations of particles, we have to examine whether the trial moves lead to overlap of spheroids. The checking of the overlap of spheroids is facilitated by the use of a contact function. The contact function $F_{AB}(\overline{r}_A, \Omega_A, \overline{r}_B, \Omega_B)$ is defined for a pair of particles A and B centered at $\overline{r}_A$ and $\overline{r}_B$, and with orientations Ω_A and Ω_B, respectively. For two ellipsoids A and B, Perram and Wertheim [1985] define a parametric function of the two particles as

$$F(\overline{r}, \lambda) = \lambda F_A(\overline{r} - \overline{r}_A, \Omega_A) + (1 - \lambda) F_B(\overline{r} - \overline{r}_B, \Omega_B) \tag{8.5.6}$$

where λ is a parameter and $F_i(\overline{r} - \overline{r}_i, \Omega_i)$, $i = A, B$, refers to an ellipsoid i centered at $\overline{r}_i$

$$F_i(\overline{r} - \overline{r}_i, \Omega_i) = (\overline{r} - \overline{r}_i)^T \cdot \overline{\overline{G}}_i^{-1}(\Omega_i) \cdot (\overline{r} - \overline{r}_i) \tag{8.5.7}$$

with T denoting the transpose, and $\overline{\overline{G}}_i$ being the matrix

$$\overline{\overline{G}}_i(\Omega_i) = a_i^2 \overline{a}_i \overline{a}_i^T + b_i^2 \overline{b}_i \overline{b}_i^T + c_i^2 \overline{c}_i \overline{c}_i^T \tag{8.5.8}$$

In (8.5.8), $\overline{a}_i$ is the column vector denoting the direction of the $\overline{a}_i$ vector with the given orientation Ω_i. Similar definitions apply to $\overline{b}_i$ and $\overline{c}_i$. For example $(\theta_A, \phi_A) = \Omega_A$, then $\overline{a}_A^T = [\sin\theta_A \cos\phi_A, \sin\theta_A \sin\phi_A, \cos\theta_A]$. Also a_i, b_i, and c_i are the length of the semi-axis $\overline{a}_i$, $\overline{b}_i$, and $\overline{c}_i$, respectively. The matrix $\overline{\overline{G}}_i$ depends on the shape and orientation of the ellipsoid. $\overline{\overline{G}}_i$ and $\overline{\overline{G}}_i^{-1}$ are symmetric and positive definite. The parameter λ is restricted within the interval $0 \le \lambda \le 1$, so that $F(\overline{r}, \lambda) \ge 0$. For fixed λ, the function $F(\overline{r}, \lambda)$ has a unique minimum as a function of $\overline{r}$. The contact function $F_{AB}(\overline{r}_A, \Omega_A, \overline{r}_B, \Omega_B)$ is thus constructed in terms of the object function $S(\lambda)$.

$$F_{AB}(\overline{r}_A, \Omega_A, \overline{r}_B, \Omega_B) = \max_{0 \le \lambda \le 1} \{S(\lambda)\} \tag{8.5.9}$$

where

$$S(\lambda) = \min_{\overline{r}} [F(\overline{r}, \lambda)] \tag{8.5.10}$$

The contact function takes the values

$$F_{AB}(\overline{r}_A, \Omega_A, \overline{r}_B, \Omega_B) \begin{cases} < 1 & \text{if } A \text{ and } B \text{ overlap} \\ = 1 & \text{if } A \text{ and } B \text{ are externally tangent} \\ & \text{just touching each other} \\ > 1 & \text{if } A \text{ and } B \text{ are not overlapping} \end{cases} \qquad (8.5.11)$$

The minimization in (8.5.10) can be easily solved, which will be illustrated in the next section. The maximization in (8.5.9) is carried out numerically using the Method of Brent [1973].

This process of randomly moving and rotation, and checking overlap is repeated for each particle inside the test volume. A single attempt to change the state of each particle is called a Monte Carlo step (MCS). The configurations used for the simulation of scattering have been selected at intervals of 10^3N MCSs apart. This is in order to establish a Markov chain of configurations, thereby eliminating correlations among configurations. The adjustments of random displacements and rotations are such that 50% of the attempted Monte Carlo moves are accepted.

5.2 Illustrations of Contact Functions

The contact function described in (8.5.11) plays an important role in the Monte Carlo simulation for checking the overlap of particles. The parametric function $F(\overline{r}, \lambda)$, defined in (8.5.6) for two ellipsoids A and B, has the minimum of $F(\overline{r}, \lambda) = 0$ at $\overline{r} = \overline{r}_B$ for $\lambda = 0$, and it is $F(\overline{r}, \lambda) = 0$ at $\overline{r} = \overline{r}_A$ for $\lambda = 1$. For $0 < \lambda < 1$, the location of the minimum of $F(\overline{r}, \lambda)$ is determined by

$$\nabla F(\overline{r}, \lambda) = 0 \qquad (8.5.12)$$

or

$$\lambda \overline{\overline{A}}^{-1} \cdot (\overline{r} - \overline{r}_A) + (1 - \lambda) \overline{\overline{B}}^{-1} \cdot (\overline{r} - \overline{r}_B) = 0 \qquad (8.5.13)$$

where $\overline{\overline{A}}^{-1} = (\overline{r} - \overline{r}_A)^T \cdot \overline{\overline{G}}_A^{-1}(\Omega_A) \cdot (\overline{r} - \overline{r}_A)$ and $\overline{\overline{B}}^{-1} = (\overline{r} - \overline{r}_B)^T \cdot \overline{\overline{G}}_B^{-1}(\Omega_B) \cdot (\overline{r} - \overline{r}_B)$ as defined in (8.5.7). From (8.5.13), the location of the minimum $\overline{r}(\lambda)$ can be obtained as

$$\overline{r}(\lambda) - \overline{r}_A = (1 - \lambda)\, \overline{\overline{A}} \cdot \overline{\overline{C}} \cdot \overline{r}_{BA} \qquad (8.5.14a)$$

$$\overline{r}(\lambda) - \overline{r}_B = -\lambda\, \overline{\overline{B}} \cdot \overline{\overline{C}} \cdot \overline{r}_{BA} \qquad (8.5.14b)$$

where $\overline{r}_{BA} = \overline{r}_B - \overline{r}_A$ and the matrix $\overline{\overline{C}}$ is given by

$$\overline{\overline{C}} = \left[(1 - \lambda) \overline{\overline{A}} + \lambda \overline{\overline{B}} \right]^{-1} \qquad (8.5.15)$$

The matrix $\overline{\overline{C}}$ is symmetric and its existence is guaranteed by the positive definite property of matrices $\overline{\overline{A}}$ and $\overline{\overline{B}}$. Insertion of (8.5.14a) and (8.5.14b) in (8.5.6) yields an equation for the object function

$$S(\lambda) \equiv \min_{\overline{r}} F(\overline{r}, \lambda) = \lambda(1-\lambda)\, \overline{r}_{BA}^T \cdot \overline{\overline{C}} \cdot \overline{r}_{BA} \tag{8.5.16}$$

which does not contain $\overline{r}$ explicitly.

In computer simulations, it is straightforward to perform the numerical inversion of the 3×3 matrix on the right-hand side of (8.5.15) and to compute the object function $S(\lambda)$ using (8.5.16). However, for some special cases explicit expressions of $S(\lambda)$ may be obtained. If the two particles are spheres of radii a and b, the respective matrices $\overline{\overline{A}}$ and $\overline{\overline{B}}$ are

$$\overline{\overline{A}} = a^2\, \overline{\overline{I}} \tag{8.5.17a}$$

$$\overline{\overline{B}} = b^2\, \overline{\overline{I}} \tag{8.5.17b}$$

where $\overline{\overline{I}}$ is the identity matrix. Then the matrix $\overline{\overline{C}}$ is

$$\overline{\overline{C}} = \frac{1}{(1-\lambda)a^2 + \lambda b^2}\, \overline{\overline{I}} \tag{8.5.18}$$

and the object function $S(\lambda)$ is

$$S(\lambda) = \frac{\lambda(1-\lambda)}{(1-\lambda)a^2 + \lambda b^2}\, |\overline{r}_{BA}|^2 \tag{8.5.19}$$

$\max S(\lambda)$ occurs at $\lambda = \lambda_M = a/(a+b)$ and $F_{AB} = S(\lambda_M) = |\overline{r}_{BA}|^2/(a+b)^2$ which clearly obeys (8.5.11).

Another useful case is for two spheroids of the same size, but have different orientations. Let $a_A = a_B = ea$, and $b_A = c_A = b_B = c_B = a$ with e being the elongation of spheroidal particle. Then the respective matrices $\overline{\overline{A}}$ and $\overline{\overline{B}}$ are given by

$$\overline{\overline{A}} = a^2\, \overline{\overline{I}} + a^2(e^2-1)\, \hat{a}_A\, \hat{a}_A \tag{8.5.20}$$

$$\overline{\overline{B}} = a^2\, \overline{\overline{I}} + a^2(e^2-1)\, \hat{a}_B\, \hat{a}_B \tag{8.5.21}$$

where $\hat{a}_A$ and $\hat{a}_B$ are unit vectors along their respective symmetry axes. In matrix notations, $\hat{a}_A$ and $\hat{a}_B$ are column vectors containing the components of $\hat{a}_A$ and $\hat{a}_B$ along the $\hat{x}$, $\hat{y}$, and $\hat{z}$ directions. Then the matrix $\overline{\overline{C}}$ is given as

$$\begin{aligned} \overline{\overline{C}} &= \frac{1}{a^2}\left[\overline{\overline{I}} - u\, \hat{a}_A\, \hat{a}_A - v\, \hat{a}_B\, \hat{a}_B\right]^{-1} \\ &= \frac{1}{a^2}\left[\overline{\overline{I}} + \alpha\, \hat{a}_A\, \hat{a}_A + \beta\, \hat{a}_B\, \hat{a}_B + \gamma\,(\hat{a}_A\, \hat{a}_B + \hat{a}_B\, \hat{a}_A)\right] \end{aligned} \tag{8.5.22}$$

with

$$u = (1 - \lambda)\ (1 - e^2) \tag{8.5.23}$$

$$v = \lambda\ (1 - e^2) \tag{8.5.24}$$

$$\alpha = \frac{u(1 - v)}{(1 - u)(1 - v) - uv(\hat{a}_A \cdot \hat{a}_B)^2} \tag{8.5.25}$$

$$\beta = \frac{v(1 - u)}{(1 - u)(1 - v) - uv(\hat{a}_A \cdot \hat{a}_B)^2} \tag{8.5.26}$$

$$\gamma = \frac{uv\ (\hat{a}_A \cdot \hat{a}_B)}{(1 - u)(1 - v) - uv(\hat{a}_A \cdot \hat{a}_B)^2} \tag{8.5.27}$$

Insertion of (8.5.22) into (8.5.16) yields the object function for the case of two identical spheroids

$$S(\lambda) = \frac{\lambda(1 - \lambda)}{a^2} \Big[|\overline{r}_{BA}|^2 + \alpha(\overline{r}_{BA} \cdot \hat{a}_A)^2 + \beta(\overline{r}_{BA} \cdot \hat{a}_B)^2 + 2\gamma(\overline{r}_{BA} \cdot \hat{a}_A)(\overline{r}_{BA} \cdot \hat{a}_B) \Big] \tag{8.5.28}$$

The algorithm of spheroidal particle placement will be applied in the next chapter of dense media models and simulations.

It is noted that $S(\lambda)$ is non-negative over the interval $0 \leq \lambda \leq 1$, and vanishes at both end points when $\lambda = 0$ and 1. As λ changes from 0 to 1, the location $\overline{r}(\lambda)$ will trace a path starting at $\overline{r}_B$ and ending at $\overline{r}_A$. The following three situations for two ellipsoids A and B can be distinguished:

(i) A and B non-overlapping: The path $\overline{r}(\lambda)$ pass through the region outside both particles, where both F_A and F_B are greater than 1, and $F(\overline{r}, \lambda) > 1$. Thus, $\max_{0 \leq \lambda \leq 1} S(\lambda)$ attains values greater than 1.

(ii) A and B overlapping: The minimum value of $S(\lambda)$ is less than 1 inside the overlapped region of two particles, where both F_A and F_B are less than 1. This precludes the path $\overline{r}(\lambda)$ from entering the region outside both particles, and hence $\max_{0 \leq \lambda \leq 1} S(\lambda)$ is less than 1.

(iii) A and B externally tangent: At the touching point $F(\overline{r}, \lambda)$ is equal to 1, which implies $S(\lambda) \leq 1$ for all λ. This means that the path $\overline{r}(\lambda)$ cannot go outside both particles, so that $\max_{0 \leq \lambda \leq 1} S(\lambda)$ is equal to 1.

Based on these observations, Perram and Wertheim [1985] proposed the contact function as given in (8.5.11). They also proved that the object function

$S(\lambda)$ has a unique maximum in the interval $0 \leq \lambda \leq 1$. Various optimization methods can be applied to obtain the maximum value of $S(\lambda)$. For example, Brent's method [1973], due to the concavity property of $S(\lambda)$, converges quickly to the maximum. The explicit expression of the derivative of the object function $S(\lambda)$ is given as

$$S'(\lambda) = \bar{r}_{BA}^T \cdot \overline{\overline{C}} \cdot \left[(1-\lambda)^2 \, \overline{\overline{A}} - \lambda^2 \, \overline{\overline{B}}\right] \cdot \overline{\overline{C}} \cdot \bar{r}_{BA} \tag{8.5.29}$$

which might be used to speed up the optimization scheme applied.

REFERENCES AND ADDITIONAL READINGS

Allen, M. P. and D. J. Tildesley (1989), *Computer Simulation of Liquids*, Oxford University Press, New York.

Baxter, R. J. (1968a), Percus–Yevick equation for hard spheres with surface adhesion, *J. Chem. Phys.*, **49(6)**, 2770–2773.

Baxter, R. J. (1968b), Ornstein-Zernike relation for a disordered fluid, *Aust. J. Phys.*, **21**, 563–569.

Baxter, R. J. (1970), Ornstein-Zernike relation and Percus–Yevick approximation for fluid mixtures, *J. Chem. Phys.*, **52**, 4559–4562.

Baxter, R. J. (1971), Distribution functions, in *Physical Chemistry: An Advanced Treatise. Vol. IIIA. The Liquid State*, edited by D. Henderson, Chapter 4, 267–334, Academic Press, New York.

Brent, R. P. (1973), *Algorithms for Minimization without Derivatives*, Prentice-Hall, Englewood Cliffs, NJ.

Ding, K. H. (1989), Electromagnetic wave propagation and scattering in dense media, Ph.D. thesis, University of Washington, Seattle.

Ding, K. H., C. Mandt, L. Tsang, and J. A. Kong (1992), Monte Carlo simulations of pair distribution functions with multiple sizes in dense discrete random media, *J. Electromag. Waves and Appl.*, **6(1)**, 1015–1029, 1992.

Ding, K. H. and L. Tsang (1988), Effective propagation constants of dense nontenuous media with multi-species of particles, *J. Electromag. Waves and Appl.*, **2(8)**, 757–777.

Ding, K. H. and L. Tsang (1989), Effective propagation constants in media with densely distributed dielectric particles of multiple sizes and permittivities, in *Progress in Electromagnetic Research*, **PIER 1**, (edited by J. A. Kong), Chapter 3, 241–295, Elsevier, New York.

Ding, K. H., L. M. Zurk, and L. Tsang (1994), Pair distribution functions and attenuation rates for sticky particles in dense media, **8(12)**, 1585–1604.

Frenkel, D. and B. Smit (1996), *Understanding Molecular Simulations: From Algorithms to Applications*, Academic Press, San Diego.

Hansen, J. P. and I. R. McDonald (1986), *Theory of Simple Liquids*, Academic Press, New York.

Kranendonk, W. G. T. and D. Frenkel (1988), Simulation of the adhesive-hard-sphere model, *Molecular Phys.*, **64(3)**, 403–424.

McQuarrie, D. A. (1976), *Statistical Mechanics*, Harper and Row, New York.

Metropolis, N., A. W. Rosenbluth, N. Rosenbluth, A. H. Teller, and E. Teller (1953), Equation of state calculation by fast computing machines, *J. Chem. Phys.*, **21(6)**, 1087–1092.

Penders, M. H. G. M. and A. Vrij (1990), A turbidity study on colloidal silica particles in concentrated suspensions sing the polydisperse adhesive hard sphere model, *J. Chem. Phys.*, **93(5)**, 3704–3711.

Percus, J. K. and G. J. Yevick (1958), Analysis of classical statistical mechanics by means of collective coordinates, *Phys. Rev.*, **110**, 1–13.

Perla, R., J. Dozier, and R. E. Davis (1986), Preparation of serial subsections in dry snow specimens, *J. of Microscopy*, **141**, 111-114.

Perram, J. W. and M. S. Wertheim (1985), Statistical mechanics of hard ellipsoids. I. overlap algorithm and the contact function, *J. of Comp. Phys.*, **58**, 409–416.

Perram, J. W., M. S. Wertheim, J. L. Lebowitz, and G. O. Williams (1984), Monte Carlo simulations of hard spheroids, *Chem. Phys. Lett.*, **105**, 277–280.

Seaton, N. A. and E. D. Glandt (1986), Monte Carlo simulation of adhesive disks, *J. Chem. Phys.*, **84(8)**, 4595–4601.

Seaton, N. A. and E. D. Glandt (1987), Monte Carlo simulation of adhesive spheres, *J. Chem. Phys.*, **87(3)**, 1785–1790.

Shi, J., R. E. Davis, and J. Dozier (1993), Stereological determination of dry-snow parameters for discrete-scatterer microwave modeling, *Annals of Glaciology*, **17**.

Waseda, Y. (1980), *The Structure of Non-Crystalline Materials, Liquids and Amorphous Solids*, McGraw-Hill, New York.

Wertheim, M. S. (1963), Exact solution of the Percus–Yevick integral equation for hard spheres, *Phys. Rev. Lett.*, **20**, 321–323.

Wertheim, M. S. (1964), Analytical solution of the Percus–Yevick equation, *Equilibrium Theory of Classical Fluids*, H. L. Frisch and J. L. Lebowitz, Eds., W. A. Benjamin, Inc., New York.

Ziman, J. M. (1979), *Models of Disorder*, Cambridge University Press, New York.

Zurk, L. M. (1995), Electromagnetic wave propagation and scattering in dense, discrete random media with application to remote sensing of snow, University of Washington, Seattle.

Zurk, L. M., L. Tsang, D. P. Winebrenner, J. Shi, and R. E. Davis (1997), Electromagnetic scattering calculated from pair distribution functions retrieved from planar snow sections, *IEEE Trans. Geosci. Remote Sens.*, **35(6)**, 1419–1428.

Chapter 9

SIMULATIONS OF TWO-DIMENSIONAL DENSE MEDIA

1 Introduction

In Chapter 7, we have discussed analytical theory and Monte Carlo simulations of three-dimensional point scatterers. In Chapter 8, we have examined the generation of dense media that consist of three-dimensional particles densely packed together. We study both analytical theory and Monte Carlo simulations of pair distribution functions and structure factors. In Chapter 10, we will study 3-D simulations of scattering by dense media. Volume scattering is a computationally intensive problem having one more dimension than surface integral equation used in rough surface scattering. Thus in this chapter, we study extensively the simpler problem of the simulations of 2-D scattering problems that consist of infinitely long cylinders. We make use of vector cylindrical wave functions described in Chapter 1 of Volume I.

In Section 2, we describe Monte Carlo generations of positions of hard (non-sticky) cylinders and also sticky cylinders. We also study the Percus–Yevick approximation of pair distribution functions for two-dimensional hard disks.* In Section 3, we perform Monte Carlo simulations of scattering by densely packed non-sticky cylinders and sticky cylinders. We study the frequency dependence of scattering by dense media. In Section 4, we apply the sparse-matrix canonical-grid (SMCG) method to fast computations of scattering by cylinders.

1.1 Extinction as a Function of Concentration

In a dense medium, the particles are densely packed together occupying high fractional volumes. In the three-dimensional case, for a medium consisting of spheres with radius a, the volume of the sphere v_o and the fractional volume f occupied by the particles are respectively

$$v_o = \frac{4\pi}{3}a^3 \tag{9.1.1}$$

$$f = n_o\frac{4\pi}{3}a^3 \tag{9.1.2}$$

where n_o is the number of spheres per unit volume. Let the particles be of permittivity ϵ_p, and they are embedded in a background of permittivity ϵ. As discussed in Chapter 8, in a dense medium, the pair distribution functions differ significantly from that of particles with independent positions. This will affect the scattering phase functions and the extinction coefficients as well as the effective permittivity of the medium.

---

* As far as the generation of cylinder positions is concerned, disks and infinite circular cylinders are equivalent. We shall use the terms *disk* and *cylinder* interchangeably.

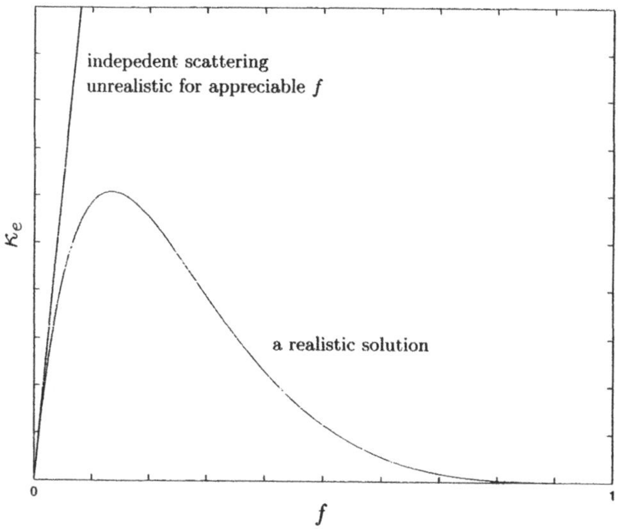

Figure 9.1.1 Scattering attenuation as a function of fractional volume of particles: Independent scattering and a more realistic solution.

In Volume I, we have employed the radiative transfer theory to evaluate the scattering of waves by discrete scatterers. The extinction rate and phase matrix are constructed by assuming that the particles scatter independently. Hence, the extinction rate will be linearly proportional to the number of particles per unit volume and the fractional volume of particles $f = n_o v_o$ (Fig. 9.1.1). However, physical intuition indicates that the linear relation cannot be correct for arbitrary f. For example, at $f = 1$, when the entire volume is occupied by scatterers, the medium becomes a homogeneous medium. Hence, in the absence of absorption, scattering should be equal to zero in the limit $f = 1$ (Fig. 9.1.1). Independent scattering is not valid for materials with an appreciable fractional volume of scatterers. This has been verified by controlled laboratory experiments [Ishimaru and Kuga, 1982; Mandt et al. 1992; West et al. 1994].

When a coherent wave propagates through a scattering medium, it is attenuated by both absorption and scattering. The coherent wave propagation constant is denoted by K, where

$$K = K_r + iK_i \tag{9.1.3}$$

The attenuation K_i is a summation of the absorption K_a and scattering K_s

$$K_i = K_a + K_s \tag{9.1.4}$$

The scattering part of the attenuation is dependent on the particle size.

Usually

$$K_r \gg K_i \tag{9.1.5}$$

The effective permittivity ϵ_{eff} is defined as

$$\epsilon_{\text{eff}} = \frac{K^2}{\omega^2 \mu} \tag{9.1.6}$$

Equation (9.1.5) gives the approximate relation

$$K^2 = (K_r + iK_a)^2 + 2iK_s K_r \tag{9.1.7}$$

We use $\kappa_a = 2K_a$, $\kappa_s = 2K_s$, $\kappa_e = \kappa_a + \kappa_s$, where κ_a, κ_s, and κ_e are the absorption coefficient, scattering coefficient, and extinction coefficient, respectively. The effective permittivity includes absorption and scattering. The classical mixing formulas for effective permittivity applies when scattering attenuation can be ignored [Bottcher, 1952; Maxwell-Garnett, 1904; Polder and van Santern, 1946]. Scattering is generally size dependent. For example, Rayleigh scattering gives a scattering cross section that is proportional to k^4a^6. Hence, at very low frequencies, scattering attenuation can be neglected. In such a limit, the effective permittivity will be the same as the effective permittivity from classical mixture formula. Thus classical mixture formula can be regarded as a very low frequency limiting case of the effective permittivity and wavenumber of dense media.

Random discrete scatterers can be classified as in Table 9.1.1. Particles are described as tenuous if their dielectric properties are only slightly different from the background medium. The characteristics of wave propagation in such media will be quite similar to those of continuous random media. The properties of K_r and K_i for the four classes of particles are described in Table 9.1.1, where k is used to denote the propagation constant of the background medium. Of the four classes of random discrete particles, class D is the subject of dense media.

1.2 Extinction as a Function of Frequency

In experiments of scattering by random distribution of particles, the scattering and extinction are usually measured as a function of frequency. For example, in satellite passive microwave remote sensing, the brightness temperatures are measured at 10 GHz, 19 GHz, 37 GHz, and 94 GHz. In satellite active microwave remote sensing backscattering measurements, the frequencies are 1.5 GHz, 5 GHz, and 10 GHz. The objectives of these sensors are to extract the snow parameters from these multi-frequency measurements. Thus it is important to study the frequency dependence of scattering and

Class	Dielectric property of particles	Particle concentration	Particle positions	K_r	Relation of K_i with f
A	tenuous	sparse	independent	$\simeq k$	linear
B	non-tenuous	sparse	independent	$\simeq k$	linear
C	tenuous	dense	correlated	$\simeq k$	nonlinear
D	non-tenuous	dense	correlated	significantly different from k	nonlinear

Table 9.1.1 Classification of random discrete scatterers.

extinction by random distribution of particles. For a small particle of radius a, it is well known that the dependence of scattering is k^4a^6 for 3-D particles, and k^3a^4 for 2-D scattering. The dependence of frequency is strong. When the frequency is doubled, scattering increases by 16 times and 8 times respectively for 3-D and 2-D scattering. For particle sizes comprable to the wavelength, then the freqeuncy dependence is weaker and becomes independent of frequency at the geometric optics limit. For sparse concentration of random distribution of particles, because independent scattering is valid, the frequency dependence follows that of single particles. If the particles have a size distribution with some particles smaller than the wavelength and others comparable to the wavelength, then the medium can exhibit varied frequency dependence of scattering. However, does the frequency dependence of scattering change with the concentration of particles? In this section, we will show that the frequency dependence of scattering of dense media is different from that of sparse concentration of particles. When the particles are densely packed together, they can adhere to form aggregates. In the sticky particle dense media model, we study densely packed aggregates of particles. We shall show that for the case of densely packed small particles, the sticky property makes them exhibit frequency dependence of scattering that is much weaker than $(\text{frequency})^4$ and $(\text{frequency})^3$ respectively of the 3-D case and the 2-D case. For passive microwave remote sensing of snow, this is particularly important. The ice grain in snow is of the order of 0.04 cm in radius (0.8 mm in diameter). The wavelengths are 1.58 cm and 0.81 cm respectively at 19 GHz and 37 GHz. This means that $ka = 0.16$ and $ka = 0.31$ respectively at 19 GHz and 37 GHz. Although the ice grains are small at these two frequencies, the extinction measurements in snow seldom exhibit the 16 times difference between the two frequencies as predicted for classical Rayleigh scattering of small particles [Hallikainen et al. 1987]. In Monte Carlo simulations presented in this chapter, we will use the sticky particle model to position the particles. Then we calculate rigorously the solutions of Maxwell's equations at several frequencies to study the frequency dependence of scattering.

2 Random Positions of Cylinders

2.1 Monte Carlo Simulations of Positions of Hard Cylinders

We describe the Monte Carlo procedure of shuffling the positions of randomly positioned two-dimensional particles in a square area. Let there be N particles. For simplicity, let

$$N = N_1^2 \tag{9.2.1}$$

where N_1 is an integer. The particles are placed in a unit square of $[0,1] \times [0,1]$. If n_o is the number of particles per unit area, then

$$n_o = N \tag{9.2.2}$$

Let b be the diameter of the particle, with $b = 2a$, where a is the radius of the particle. Let f be the fractional area occupied by the particles. In the computer code, we specify N and f. Then the diameter b is

$$b = \sqrt{\frac{4f}{\pi N}} \tag{9.2.3}$$

Inside the square, we first placed the particles periodically in both $\hat{x}$ and $\hat{y}$ directions with spacing s given by

$$s = \frac{1}{N_1} \tag{9.2.4}$$

For example, the first row of N_1 particles can have coordinates

$$\left(\frac{s}{2}, \frac{s}{2}\right), \left(\frac{3s}{2}, \frac{s}{2}\right), \left(\frac{5s}{2}, \frac{s}{2}\right), \ldots, \left(\left(N_1 - \frac{1}{2}\right)s, \frac{s}{2}\right) \tag{9.2.5}$$

the second row has coordinates

$$\left(\frac{s}{2}, \frac{3s}{2}\right), \left(\frac{3s}{2}, \frac{3s}{2}\right), \left(\frac{5s}{2}, \frac{3s}{2}\right), \ldots, \left(\left(N_1 - \frac{1}{2}\right)s, \frac{3s}{2}\right) \tag{9.2.6}$$

and so on.

To generate the 0th realization, we use shuffling. We perform N_{pass} passes. For each pass, there is an attempt to move each particle once even though the move may not be accepted. The move is as follows. Let

$$\Delta = c_{st} b \tag{9.2.7}$$

where c_{st} is an adjustable constant for the displacement. For a particle j with coordinates (x_j, y_j) we generate its possible new coordinates (x'_j, y'_j) from two random numbers r_1 and r_2 that are randomly and uniformly distributed between -1 and 1. In MATLAB, the function **rand** can be used to generate r_1 and r_2.

$$x'_j = x_j + r_1 \Delta \tag{9.2.8}$$

$$y'_j = y_j + r_2 \Delta \tag{9.2.9}$$

Next we need to check whether the new coordinates, (x'_j, y'_j), are acceptable by making sure that they do not overlap with other particles. Note that the periodic boundary conditions are used. Thus each particle has images in other squares. To check for overlap with particle l, $(l = 1, 2, \ldots, N$ and $l \neq j)$, we calculate

$$r_x = x_l - x'_j \tag{9.2.10}$$

However, if $r_x > 0.5$, then the image of x_l is closer to x'_j. Thus we do the following

$$\text{if } r_x > 0.5, \quad \text{replace } r_x \text{ by } r_x - 1;$$
$$\text{if } r_x < -0.5, \quad \text{replace } r_x \text{ by } r_x + 1$$

Do the same for $r_y = y_l - y'_j$, with the appropriate replacements.

Then we calculate the distance between particles j and l

$$r_d = \sqrt{r_x^2 + r_y^2} \tag{9.2.11}$$

If

$$r_d \geq b \text{ for all } l \neq j,$$

then the move is accepted for particle j

$$\text{replace } \; x_j \text{ by } x'_j;$$
$$\text{replace } \; y_j \text{ by } y'_j.$$

If

$$r_d < b \text{ for any } l \neq j,$$

then the move is rejected and the original coordinates (x_j, y_j) are kept. If $r_d > b$ for all $l \neq j$, we take (x'_j, y'_j) and call it temporarily the new coordinates (x_j, y_j). Before we finally accept the new (x_j, y_j), we need to make sure that particle j is within the unit square. If it is outside, it has become an image, and we need to translate it back into the unit square. Thus for the new x_j

$$\text{if } x_j < 0, \quad \text{replace } x_j \text{ by } x_j + 1;$$
$$\text{if } x_j > 1, \quad \text{replace } x_j \text{ by } x_j - 1$$

Do the same for y_j, with the appropriate replacements. Then we accept (x_j, y_j) as the new position for particle j.

The process is applied to all the particles $j = 1, 2, 3, \ldots, N$. When that is done, one pass is completed. A new realization is generated after N_{pass} passes to make sure that the particles are sufficiently randomized. To get the 0th realization from the initial periodic configuration, it is better to do several times of N_{pass} passes. From the 0th realization, we generate new realizations for a total N_r realizations, with each one created after N_{pass} passes. For each accepted new position for a particle, we count it as an accepted move. Let N_{acc} be the total number of accepted moves after N_r realizations. Then the acceptance rate is

$$\text{accept rate} = \frac{N_{acc}}{N N_{pass} N_r} \tag{9.2.12}$$

If the dsipacement Δ is too small, then all the moves will be accepted, and the realizations will be dependent on each other. On the other hand, if the displacement Δ is too large, then the moves will not be accepted. A good acceptance rate is between 30% and 70% by choosing an appropriate Δ through adjusting c_{st} in (9.2.7).

2.2 Simulations of Pair Distribution Functions

We calculate the pair distribution function from the realizations generated. Let the separation of the disks be counted in intervals of Δr. For the nth interval, the center separations of the particles are bewteen $r_{sn} = b + (n-1)\Delta r$ and $r_{un} = b + n\Delta r$. Since a pair of disks is considered to be uncorrelated when their separation is greater than $5b$ (i.e., $g(r) = 1$ when $r > 5b$), and they cannot interpenetrate, the interval Δr is determined by

$$\Delta r = \frac{4b}{M} \tag{9.2.13}$$

where M is the total number of intervals and determines the resolution of the Monte Carlo pair distribution function. For each relaization, we count the $N(N-1)$ pairs of separation. In the counting process, for the jth particle, $j = 1, 2, 3, \ldots, N$, we calculate its separation with the lth particle ($l = 1, 2, 3, \ldots, N$ and $l \neq j$).

$$r_x = x_l - x_j; \qquad r_y = y_l - y_j;$$

$$\text{if } r_x > 0.5, \text{ replace } r_x \text{ by } r_x - 1;$$
$$\text{if } r_x < -0.5, \text{ replace } r_x \text{ by } r_x + 1;$$

Do the same for r_y, with the appropriate replacements. Then we compute the separation r_d

$$r_d = \sqrt{r_x^2 + r_y^2}$$

and find out the interval where r_d falls. That will increase by one the counting of separations for that interval.

Let C_n be the number of counts of pair separations that fall in the nth interval. Then

$$\frac{C_n}{N(N-1)} \tag{9.2.14}$$

is the probability of finding a particle at the nth interval of separation. Thus, using the definition of the pair distribution function and the conditional probability

$$\frac{C_n}{N(N-1)} = \int_{nth} d\overline{r} p(\overline{r}|0) \tag{9.2.15}$$

where $p(\overline{r}|0)$ is the conditional probability of finding a particle at $\overline{r}$ given a particle at the origin. Since

$$p(\overline{r}|0) = \frac{g(r)}{A} \tag{9.2.16}$$

we have

$$\frac{C_n}{N(N-1)} = \int_{r_{sn}}^{r_{un}} dr 2\pi r \frac{g(r)}{A} = \frac{g(\widetilde{r}_n)}{A} \pi \left(r_{un}^2 - r_{sn}^2\right) \tag{9.2.17}$$

where $\widetilde{r}_n$ is a weighted position in the nth interval. For example, $\widetilde{r}_n = \sqrt{\left(r_{un}^2 + r_{sn}^2\right)/2}$. Thus

$$g(\widetilde{r}_n) = \frac{\langle C_n \rangle}{n_o(N-1)\pi \left(r_{un}^2 - r_{sn}^2\right)} \tag{9.2.18}$$

where angular bracket refers to average over realizations.

2.3 Percus–Yevick Approximation of Pair Distribution Functions

In Chapter 8, we have studied three-dimensional pair distribution functions based on the Percus–Yevick approximation. These results are based on the analytic closed form solution for the direct correlation function. However, for the case of two-dimensional problems, there are no closed form analytic solution for the direct correlation function. There are numerical slutions of Percus–Yevick integral equation for the case of hard disks [Lado, 1968]. There are also approximate analytic expressions for the direct correlation functions of hard disks. We summarize the results Ripoll and Tejero [1995] on the approximate analytic expressions for the two-dimensional case.

Let

$$x = \frac{r}{b} \tag{9.2.19}$$

where r is the radial distance between the two disks and b is the diameter of the disk. Also, the two-dimensional fractional volume is

$$f = n_o \pi a^2 \tag{9.2.20}$$

where $a = b/2$ is the radius of the disk and n_0 is the number of particles per unit area. Then, under the Percus–Yevick approximation,

$$c(x) = 0 \quad \text{for } x > 1 \tag{9.2.21}$$

For $x \leq 1$, direct correlation $c(x)$ assumes the form,

$$c(x) = c(0)\left[1 - 4f + 4f\omega_2\left(\frac{x}{2}\right) + s_2(f)x\right] \tag{9.2.22}$$

where

$$\omega_2(x) = \frac{2}{\pi}\left[\cos^{-1} x - x\sqrt{1 - x^2}\right] \tag{9.2.23}$$

$$s_2(f) = \frac{3f^2}{8}\left[\frac{8(1 - 2p) + (25 - 9p)pf - (7 - 3p)pf^2}{1 + f + 3pf^2 - pf^3}\right] \tag{9.2.24}$$

$$c(0) = -\frac{1 + f + 3pf^2 - pf^3}{(1 - f)^3} \tag{9.2.25}$$

$$p = \frac{7}{3} - \frac{4\sqrt{3}}{\pi} \tag{9.2.26}$$

Given the direct correlation function, we can use the Ornstein-Zernike equation to calculate the pair distribution function $g(r)$

$$h(r) = g(r) - 1 \tag{9.2.27}$$

where $h(r)$ is the total correlation which is related to $c(r)$ by means of the Ornstein-Zernike equation

$$h(\overline{r}_{12}) = c(\overline{r}_{12}) + n_o \int d\overline{r}_3 c(\overline{r}_{13})\, h(\overline{r}_{32}) \tag{9.2.28}$$

Define the two-dimensional Fourier transforms of $h(r)$ and $c(r)$

$$H(\overline{p}) = \frac{1}{4\pi^2}\int_{-\infty}^{\infty} d\overline{r} e^{-i\overline{p}\cdot\overline{r}} h(\overline{r}) \tag{9.2.29}$$

$$C(\overline{p}) = \frac{1}{4\pi^2}\int_{-\infty}^{\infty} d\overline{r} e^{-i\overline{p}\cdot\overline{r}} c(\overline{r}) \tag{9.2.30}$$

Then from (9.2.28)–(9.2.30), we have

$$H(\overline{p}) = C(\overline{p}) + n_o 4\pi^2 C(\overline{p}) H(\overline{p}) \tag{9.2.31}$$

$$H(\overline{p}) = \frac{C(\overline{p})}{1 - n_o 4\pi^2 C(\overline{p})} \tag{9.2.32}$$

For the case of isotropic total and direct correlation functions, (9.2.29) and (9.2.30) take the form of the Fourier-Bessel transform

$$H(p) = \frac{1}{2\pi}\int_0^\infty dr\ h(r) J_0(pr) \tag{9.2.33}$$

$$C(p) = \frac{1}{2\pi}\int_0^\infty dr\ c(r) J_0(pr) \tag{9.2.34}$$

where J_0 is the Bessel function of zeroth order. By changing the variable to $x = r/b$, (9.2.34) becomes

$$C(p) = \frac{b^2}{2\pi}\int_0^\infty dx\ xc(bx) J_0(pbx) \tag{9.2.35}$$

After $C(p)$ is obtained, $H(p)$ can be computed from (9.2.32). The pair distribution function $g(\overline{r})$ is given by

$$g(\overline{r}) = 1 + \int_{-\infty}^{\infty} d\overline{p}\ e^{i\overline{p}\cdot\overline{r}} H(\overline{p}) \tag{9.2.36}$$

or

$$g(bx) = 1 + 2\pi \int_0^\infty dp\ pH(p) J_0(pbx) \tag{9.2.37}$$

2.4 Results of Simulations

In Figs. 9.2.1, 9.2.2, and 9.2.3, we compare the results of the pair distribution functions from the Monte Carlo simulations and the analytic Percus–Yevick approximation for fractional volumes of 10%, 30%, and 50%, respectively. A total of 200 cylinders are used in the simulations. The number of passes N_{pass}, the displacement Δ, and the number of realizations used are given in the figure captions. Monte Carlo simulations agree well with the approximate PY pair distribution functions.

2.5 Monte Carlo Simulations of Sticky Disks

In order to capture the finite binding probabilities of adhesive disk pairs, Seaton and Glandt [1986] have developed new methods which allow the change of a sticky disk's number of bonds with other disks during its Monte

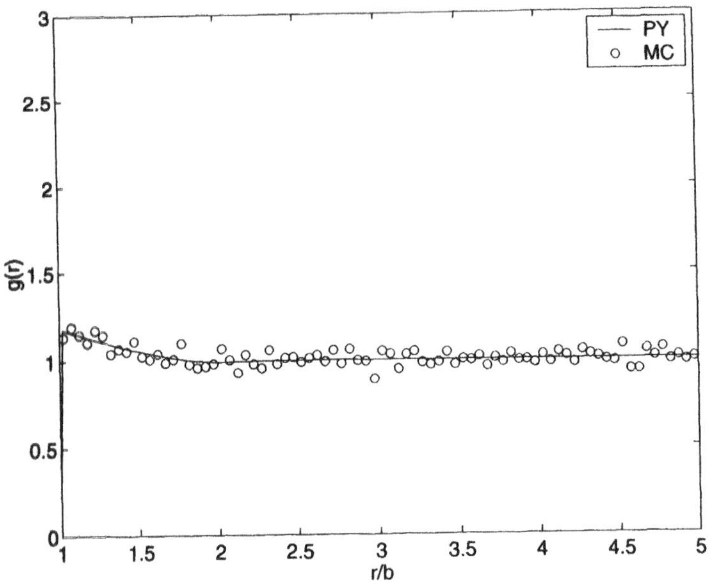

Figure 9.2.1 Pair distribution function for cylinders with $f = 0.1$. Results from the Monte Carlo (MC) simulations are obtained using 200 cylinders with $\Delta = 0.020$ ($b = 0.025$), $N_{pass} = 2000$, and $N_r = 50$. Results of Percus–Yevick (PY) approximation are also shown.

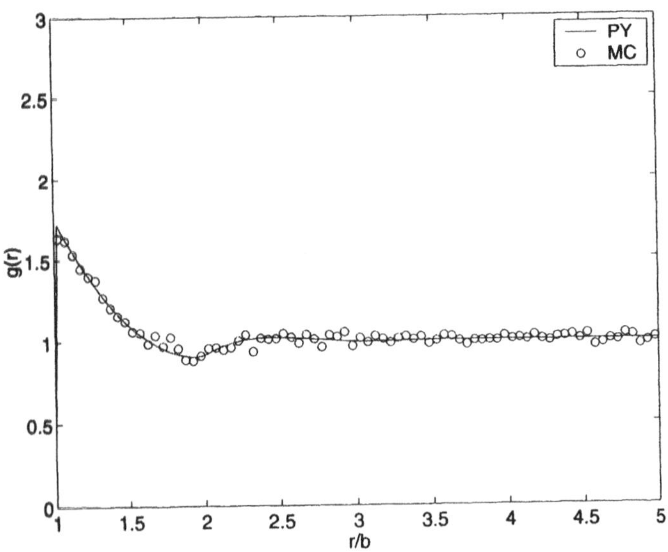

Figure 9.2.2 Pair distribution function for cylinders with $f = 0.3$. Results from the Monte Carlo (MC) simulations are obtained using 200 cylinders with $\Delta = 0.015$ ($b = 0.044$), $N_{pass} = 2000$, and $N_r = 50$. Results of Percus–Yevick (PY) approximation are also shown.

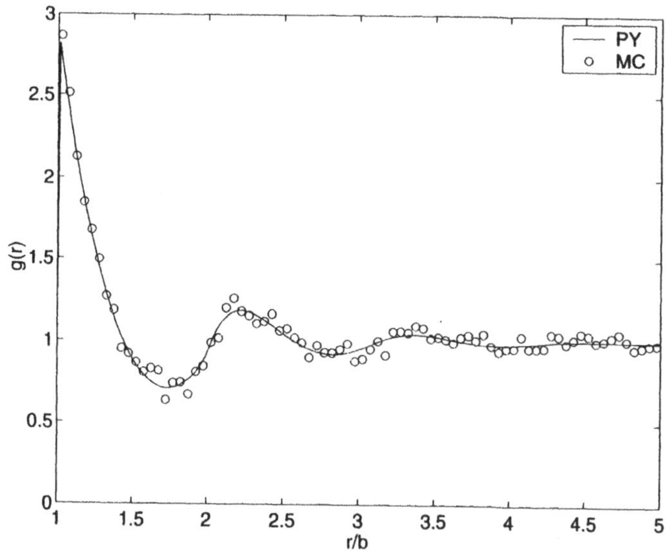

Figure 9.2.3 Pair distribution function for cylinders with $f = 0.5$. Results from the Monte Carlo (MC) simulations are obtained using 200 cylinders with $\Delta = .023$ ($b = 0.056$), $N_{pass} = 1000$, and $N_r = 10$. Results of Percus–Yevick (PY) approximation are also shown.

Carlo displacement. They have taken into account transitions between three binding states $\alpha = 0, 1, 2$. The four binding states of a particle are illustrated Fig. 9.2.4. They also call the number of bonds as "particle energy state" (PES) to denote the number of potential wells in which the particle sits. The maximum number of bonds with other particles is 6 for two-dimension. In Fig. 9.2.4, we show 4 of the 7 binding states. In short, the 7 binding states are the configurations where the number of particles that stick together is 0, 1, 2, ..., 6. In numerical simulations of this section, we use the first three binding states only, corresponding to (a), (b), and (c) of Fig. 9.2.4.

Seaton and Glandt also derive the "unnormalized" total transition probability p_α for changing the displaced particle k's binding state to the α-bond state. The calculation of the probability p_α involves integrating over all possible contact sites available to this particular α-bond state, allowing particle overlap to occur. Let $\overline{r}_k$ be the position vector of particle k and A be the total area of the system. The unnormalized total transition probability p_0 for the test disk k to form no bond with all other disks is

$$p_0 = \int_A d\overline{r}_k = A \tag{9.2.38}$$

Let d be the diameter of the particle and τ be the sticky paramter. The

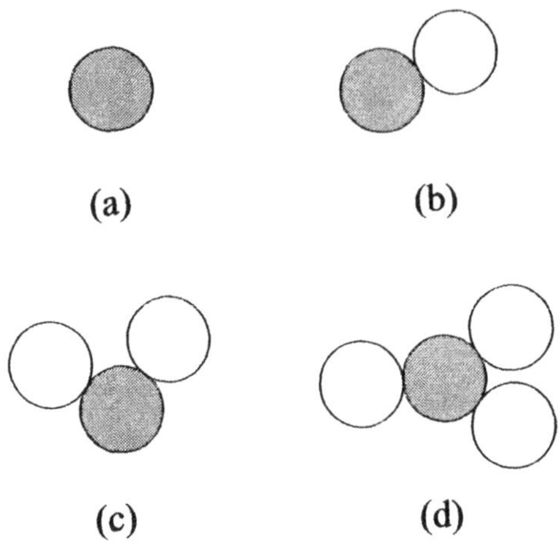

Figure 9.2.4 Four binding states of sticky disks: (a) unbounded, (b) single-bond, (c) double-bond, and (d) triple-bond.

unnormalized total transition probability p_1 for the test particle k to be in the single-bond state is given by

$$p_1 = \sum_{i \neq k}^{N} p_1^i \tag{9.2.39}$$

where the summation is over all other disks and

$$p_1^i = \int_A \left(\frac{d}{4\tau}\right) \delta(r_{ik} - d)\, d\overline{r}_k = \left(\frac{d}{4\tau}\right) 2\pi d \tag{9.2.40}$$

Thus

$$p_1 = \left(\frac{d}{4\tau}\right)(N-1)2\pi d \tag{9.2.41}$$

which is equal to the total circular perimeter available to the test particle k times a factor $\frac{d}{4\tau}$. Note that the difference between the three-dimensional and two-dimensional adhesive particles, the probability is proportional to a factor $(\frac{d}{4\tau})^\alpha$ for an α-bond sticky disk instead of $(\frac{d}{12\tau})^\alpha$ for an α-bond sticky sphere. This has been indicated by Stell [1991]. To calculate the double-bond unnormalized total transition probability p_2 for the test particle k, a double summation with $i < j \neq k$, where i, j are particle indices, is required. The probability p_2 is given as

$$p_2 = \sum_{\substack{i<j \\ i,j \neq k}}^{N} p_2^{ij} \tag{9.2.42}$$

where

$$\begin{aligned} p_2^{ij} &= \int_A \left(\frac{d}{4\tau}\right)^2 \delta(r_{ik}-d)\,\delta(r_{jk}-d)\,d\overline{r}_k \\ &= \left(\frac{d}{4\tau}\right)^2 \iint \delta(r_{ik}-d)\,\delta(r_{jk}-d)\,\frac{r_{jk}}{r_{ij}\sin\theta_{jk}}\,dr_{ik}\,dr_{jk} \\ &= \left(\frac{d}{4\tau}\right)^2 \frac{2d}{r_{ij}\sqrt{1-\left(\frac{r_{ij}}{2d}\right)^2}} \end{aligned} \tag{9.2.43}$$

where $d\overline{r}_k = r_{ik}dr_{ik}d\theta_{jk}$ and $r_{jk}^2 = r_{ik}^2 + r_{ij}^2 - 2r_{ik}r_{ij}\cos\theta_{jk}$ have been used to obtain the second equality in (9.2.43). r_{ij} is the distance between particles i and j, and θ_{jk} is the angle between the line segments $\overline{r}_{ij}$ and $\overline{r}_{ik}$ connecting particles i with particle j and with particle k, respectively. Note that $p_2=0$ when $r_{ij} > 2d$, since particle k cannot form a double-bond with praticles i and j simultaneously when their distance is greater than two diameters. p_2 has an integrable singularity at $r_{ij}=2d$ for any pair of disks i and j. Note that the total "transition probability" defined by Seaton and Glandt has the dimension of "volume" which is similar to the "effective volume" used by Kranendonk and Frenkel [1988].

The definitions of unnormalized total transition probabilities do not depend on the previous binding state of the displaced particle k. However, during an actual Monte Carlo step the binding state of a displaced disk may change from its present β-bond configuration to a new β'-bond configuration. Similarly, in order to calculate the unnormalized total transition probabilities, a list of all pairs of particles with which the test particle k could form a double-bond, has to be established. Then, the new binding state β' for the test disk k is selected according to the normalized probability

$$P(\beta') = \frac{p_{\beta'}}{\sum_{\alpha=0}^{2} p_\alpha} \tag{9.2.44}$$

It is noted that the magnitude of $P(\beta')$ depends on the binding structure of the system prior to the disk k being moved. The transitions involve all three binding states. In the following, we describe the Monte Carlo process.

Step 1. The Monte Carlo simulations for sticky disks also begin with N unbounded disks regularly placed within a unit square without overlapping. To create the 0th (initial) realization of random distribution, the periodic particle positions are randomized by using the shuffling method described in Section 2.1.

Step 2. To generate sticky disks, we do not shuffle the test disk within a small square of $[0, \Delta] \times [0, \Delta]$. Instead, the disk k makes a transition according to the breaking of its current binding state and the formation of new binding state with other disks. We evaluate the transition probability of particle k. Before changing the test disk k's binding state, we have to build a list of all possible pairs with which the test disk k could form a double-bond. We compute the normalized probabilities $P(\beta')$ according to the respective unnormalized total transition probability $p_{\beta'}$ for $\beta' = 0, 1, 2$. We also store the array of probabilities for the pair of disks (i, j)

$$p^{(2)}(i,j) = \frac{p_2^{ij}}{p_2} \tag{9.2.45}$$

Step 3. A random number generator, between 0 and 1, is used to generate a random number r and compared with the obtained $P(\beta')$ of the current system to decide the new binding state of the test disk. Three obvious possibilities are: $\beta' = 0, 1, 2$ corresponding to unbounded, single-bond, and double-bond state, respectively. If $0 \le r \le P(0)$, then $\beta' = 0$ is chosen. If $P(0) < r \le P(0) + P(1)$, then $\beta' = 1$ is chosen. If $P(0) + P(1) < r \le 1$, then $\beta' = 2$ is chosen.

Step 4. The possible new location of disk k, (x'_k, y'_k) is then determined as follows, depending on which β' is chosen is Step 3. If $\beta' = 0$, the coordinates (x'_k, y'_k) of the test disk is determined using two random number generators, each between 0 and 1, for its x and y coordinates. If $\beta' = 1$, an integer random number i with equal probabilities, $i = 1, 2, \ldots, N$, $i \neq k$, is generated to select the potential single-bond candidate disk. A random angle generator, between 0 and 2π, is employed to determine ϕ. Then

$$x'_k = x_i + d \cos\phi$$
$$y'_k = y_i + d \sin\phi$$

where (x_i, y_i) are the coordinates of the candidate disk i and d is the diameter of the disk. If $\beta' = 2$, a random number, between 0 and 1, is generated. This is then used to select a candidate pair of disks (i, j) from the established pair-list according to the stored array of probabilities $p^{(2)}(i, j)$ to form a double-bond with the test disk k. For this double-bond state, there are only two choices for the test disk k's locations, either above or below the line segment joining the centers of disks i and j. Then we determine (x'_k, y'_k) as follows. Let

$$x_d = x_j - x_i$$

$$y_d = y_j - y_i$$

Calculate the polar coordinates (r_d, γ) from (x_d, y_d) by

$$x_d = r_d \cos\gamma$$
$$y_d = r_d \sin\gamma$$

where $r_d = \sqrt{x_d^2 + y_d^2}$ and $\gamma = \tan^{-1}(y_d/x_d)$. Then calculate $\alpha = \cos^{-1} 2r_d/d$. Generate a random number u between 0 and 1. If $0 \le u < 0.5$, $\phi = \gamma - \alpha$. If $0.5 \le u \le 1$, $\phi = \gamma + \alpha$. The coordinates (x_k', y_k') are then

$$x_k' = x_i + d\cos\phi$$
$$y_k' = y_i + d\sin\phi$$

Step 5. It is possible that the new position of the displaced disk may overlap with other disks in the system. We check overlap similar to what is done in Section 2.1 for hard cylinders. If there is no overlap, we will accept the move and update the coordinates of the displaced disk. Otherwise, reject the displacement and return the test disk to its original position. As in the case of sticky sphere, we have to maintain an updated catalog of the system's binding configuration, so that we need to rebuild the list of possible double-bond candidates whenever a trial move of the test disk is successful.

In Figs. 9.2.5 and 9.2.6, we illustrate the Monte Carlo simulations of the pair distribution functions for the two-dimensional sticky disks. The result in Fig. 9.2.5 is for the case of 200 sticky disks and fractional volume of 20%. For Fig. 9.2.6, we use 300 sticky disks for the fractional volume of 30%. The Monte Carlo results are obtained by averaging over 12000 and 10000 passes, respectively. For both cases, the sticky paramter is $\tau = 0.1$. Compared with the 3-D results shown in Chapter 8, the 2-D sticky disks exhibit discontinuities in the pair distribution functions at pair separations of $r = \sqrt{3}d$, $\sqrt{7}d$, and $3d$ in addition to $r = d$ and $2d$. These features have also been shown in the study of Seaton and Glandt [1986] for adhesive disks.

3 Monte Carlo Simulations of Scattering by Cylinders

3.1 Scattering by a Single Cylinder

Consider a cylinder of permittivity ϵ_p and radius a centered at the origin (Fig. 9.3.1). Let the incident wave have $\exp(ik_{iz}z)$ dependence. The incident wave is given by

$$\overline{E}_i = \sum_{n=-\infty}^{\infty} \left[a_n^{(M)} Rg\overline{M}_n(k_{i\rho}, k_{iz}, \overline{r}) + a_n^{(N)} Rg\overline{N}_n(k_{i\rho}, k_{iz}, \overline{r}) \right] \qquad (9.3.1)$$

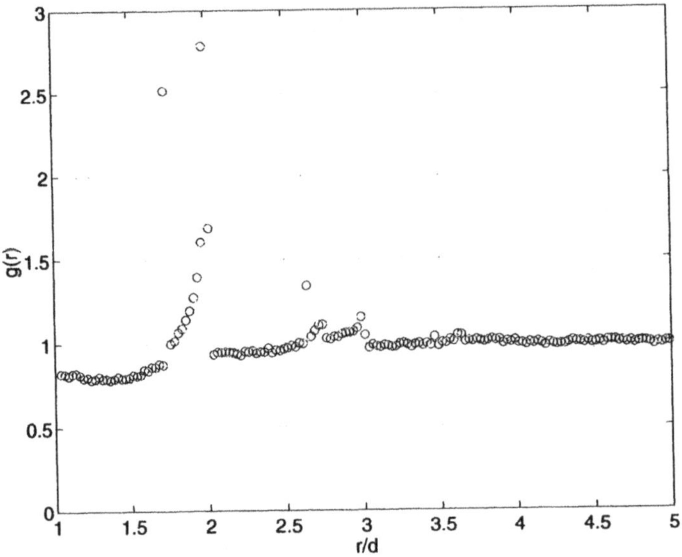

Figure 9.2.5 Pair distribution function obtained from Monte Carlo simulations for a system of single size sticky disks with $\tau = 0.1$ and $f = 0.2$. 200 disks are used in the simulations, and the results are obtained by averaging over 12000 passes.

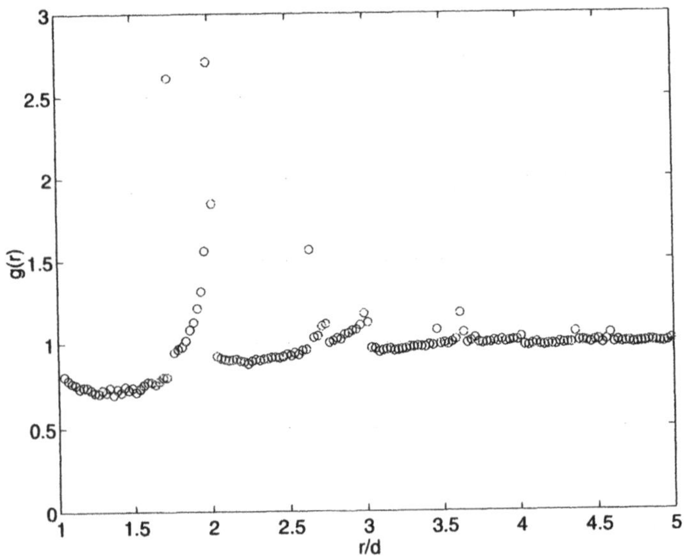

Figure 9.2.6 Pair distribution function obtained from Monte Carlo simulations for a system of single size sticky disks with $\tau = 0.1$ and $f = 0.3$. 300 disks are used in the simulations, and the results are obtained by averaging over 10000 passes.

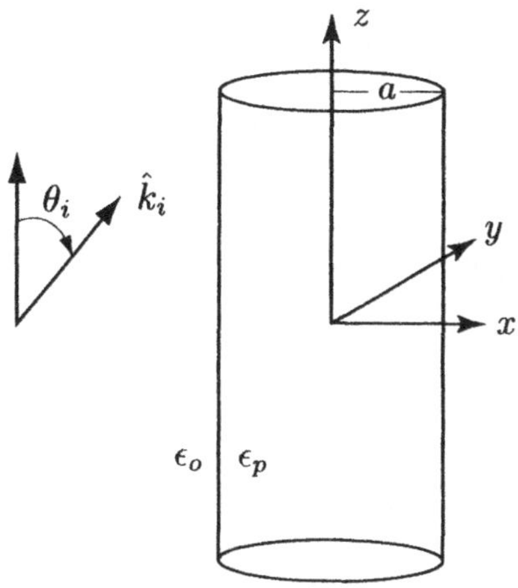

Figure 9.3.1 Scattering by a single cylinder with permittivity ϵ_p and radius a.

where $a_n^{(M)}$ and $a_n^{(N)}$ are the TE and TM wave coefficients and the regular vector cylinderical waves $Rg\overline{M}_n$ and $Rg\overline{N}_n$ are defined in Chapter 1, Section 4.2 of Volume I. In (9.3.1),

$$k_{i\rho} = \sqrt{k^2 - k_{iz}^2} \tag{9.3.2}$$

The scattered field is given by

$$\overline{E}_s = \sum_{n=-\infty}^{\infty} \left[a_n^{(M)s} \overline{M}_n(k_{i\rho}, k_{iz}, \overline{r}) + a_n^{(N)s} \overline{N}_n(k_{i\rho}, k_{iz}, \overline{r}) \right] \tag{9.3.3}$$

where $a_n^{(M)s}$ and $a_n^{(N)s}$ are the TE and TM scattered wave expansion coefficients and the vector cylinderical waves $\overline{M}_n$ and $\overline{N}_n$ are defined in Chapter 1, Section 4.2 of Volume I.

The internal field is given by

$$\overline{E}_{int} = \sum_{n=-\infty}^{\infty} \left[c_n^{(M)} Rg\overline{M}_n(k_{ip\rho}, k_{iz}, \overline{r}) + c_n^{(N)} Rg\overline{N}_n(k_{ip\rho}, k_{iz}, \overline{r}) \right] \tag{9.3.4}$$

where $c_n^{(M)}$ and $c_n^{(N)}$ are the TE and TM internal field coefficients and

$$k_{ip\rho} = \sqrt{k_p^2 - k_{iz}^2} \tag{9.3.5}$$

with $k_p = \omega\sqrt{\mu\epsilon_p}$. Because of phase matching, both the scattered field and the internal field have the same $\exp(ik_{iz}z)$ dependence.

From Chapter 1, Section 6.2 of Volume I, there are four equations relating the scattered field coefficients $a_n^{(M)s}, a_n^{(N)s}$ and the internal field coefficients $c_n^{(M)}, c_n^{(N)}$ to the incident field coefficients $a_n^{(M)}, a_n^{(N)}$. From equations (1.6.58)–(1.6.61) of Volume I,

$$\begin{aligned} & - k_{i\rho} a_n^{(M)} J_n'(k_{i\rho}a) - k_{i\rho} a_n^{(M)s} H_n^{(1)\prime}(k_{i\rho}a) - \frac{nk_{iz}}{ka} a_n^{(N)} J_n(k_{i\rho}a) \\ & - \frac{nk_{iz}}{ka} a_n^{(N)s} H_n^{(1)}(k_{i\rho}a) \\ = & - k_{ip\rho} c_n^{(M)} J_n'(k_{ip\rho}a) - \frac{nk_{iz}}{k_p a} c_n^{(N)} J_n(k_{ip\rho}a) \end{aligned} \tag{9.3.6}$$

$$\frac{k_{i\rho}^2}{k} J_n(k_{i\rho}a) a_n^{(N)} + \frac{k_{i\rho}^2}{k} H_n^{(1)}(k_{i\rho}a) a_n^{(N)s} = \frac{k_{ip\rho}^2}{k_p} J_n(k_{i\rho}a) c_n^{(N)} \tag{9.3.7}$$

$$\begin{aligned} & k\Big[-k_{i\rho} a_n^{(N)} J_n'(k_{i\rho}a) - k_{i\rho} a_n^{(N)s} H_n^{(1)\prime}(k_{i\rho}a) \\ & \quad - \frac{nk_{iz}}{ka} a_n^{(M)} J_n(k_{i\rho}a) - \frac{nk_{iz}}{ka} a_n^{(M)s} H_n^{(1)}(k_{i\rho}a)\Big] \\ & = k_p \left[-k_{ip\rho} c_n^{(N)} J_n'(k_{ip\rho}a) - \frac{nk_{iz}}{k_p a} c_n^{(M)} J_n(k_{ip\rho}a)\right] \end{aligned} \tag{9.3.8}$$

$$k_{i\rho}^2 J_n(k_{i\rho}a) a_n^{(M)} + k_{i\rho}^2 H_n^{(1)}(k_{i\rho}a) a_n^{(M)s} = k_{ip\rho}^2 J_n(k_{i\rho}a) c_n^{(M)} \tag{9.3.9}$$

Eliminating the scattered field coefficients gives the equations relating the internal field coefficients to the incident field coefficients. From Volume I, (1.6.62) to (1.2.67), we have

$$A_n^{MM}(k_{iz}) c_n^{(M)} + A_n^{MN}(k_{iz}) c_n^{(N)} = a_n^{(M)} \tag{9.3.10}$$

$$A_n^{NM}(k_{iz}) c_n^{(M)} + A_n^{NN}(k_{iz}) c_n^{(N)} = a_n^{(N)} \tag{9.3.11}$$

where

$$A_n^{MM}(k_{iz}) = -\frac{ia\pi}{2k_{i\rho}^2}\left[k_{p\rho}^2 k_{i\rho} J_n(k_{p\rho}a) H_n^{(1)\prime}(k_{i\rho}a) - k_{p\rho} k_{i\rho}^2 J_n'(k_{p\rho}a) H_n^{(1)}(k_{i\rho}a)\right] \tag{9.3.12}$$

$$A_n^{MN}(k_{iz}) = -\frac{ia\pi}{2k_{i\rho}^2}\left[\frac{n}{k_p a} k_{iz}(k_{p\rho}^2 - k_{i\rho}^2) J_n(k_{p\rho}a) H_n^{(1)}(k_{i\rho}a)\right] \tag{9.3.13}$$

$$A_n^{NM}(k_{iz}) = \frac{k_p}{k} A_n^{MN}(k_{iz}) \tag{9.3.14}$$

$$A_n^{NN}(k_{iz}) = -\frac{ia\pi}{2k_{i\rho}^2}\left[\frac{kk_{p\rho}^2 k_{i\rho}}{k_p} J_n(k_{p\rho}a) H_n^{(1)\prime}(k_{i\rho}a) - \frac{k_p k_{p\rho} k_{i\rho}^2}{k} J_n'(k_{p\rho}a) H_n^{(1)}(k_{i\rho}a)\right] \tag{9.3.15}$$

After the internal field coefficients and scattered field coefficients are obtained, we can write the relations between the internal field and the incident field as

$$c_n^{(M)} = B_n^{MM} a^{(M)} + B_n^{MN} a^{(N)} \tag{9.3.16}$$
$$c_n^{(N)} = B_n^{NM} a^{(M)} + B_n^{NN} a^{(N)} \tag{9.3.17}$$

and the relations between the scattered field and incident field as

$$a_n^{(M)s} = T_n^{MM} a^{(M)} + T_n^{MN} a^{(N)} \tag{9.3.18}$$
$$a_n^{(N)s} = T_n^{NM} a^{(M)} + T_n^{NN} a^{(N)} \tag{9.3.19}$$

Generally TE and TM waves are coupled for scattering by dielectric cylinders. The TE and TM waves are decoupled and the equations simplify for the following two cases:

(A) Scattering by dielectric cylinders with incidence in the x-y plane. For this case, the TE and TM waves are decoupled.

$$k_{iz} = 0; \quad k_{i\rho} = k, \ k_{p\rho} = k_p \tag{9.3.20}$$

$$A_n^{MM}(k_{iz}) = -\frac{ia\pi}{2k^2}\left[k_p^2 k J_n(k_p a) H_n^{(1)\prime}(ka) - k_p k^2 J_n'(k_p a) H_n^{(1)}(ka)\right] \tag{9.3.21}$$

$$A_n^{MN}(k_{iz}) = 0 \tag{9.3.22}$$

$$A_n^{NM}(k_{iz}) = 0 \tag{9.3.23}$$

$$A_n^{NN}(k_{iz}) = -\frac{ia\pi}{2k^2}\left[k^2 k_p J_n(k_p a) H_n^{(1)\prime}(ka) - k_p^2 k J_n'(k_p a) H_n^{(1)}(ka)\right] \tag{9.3.24}$$

Then from (9.3.16)–(9.3.19), we have

$$c_n^{(M)} = B_n^{(M)} a^{(M)} \tag{9.3.25}$$
$$c_n^{(N)} = B_n^{(N)} a^{(N)} \tag{9.3.26}$$
$$a_n^{(M)s} = T_n^{(M)} a^{(M)} \tag{9.3.27}$$
$$a_n^{(N)s} = T_n^{(N)} a^{(N)} \tag{9.3.28}$$

where

$$B_n^{(M)} = B_{-n}^{(M)} = \frac{\frac{2ik}{\pi a k_p}}{k_p H_n^{(1)\prime}(ka) J_n(k_p a) - k H_n^{(1)}(ka) J_n'(k_p a)} \tag{9.3.29}$$

$$B_n^{(N)} = B_{-n}^{(N)} = -\frac{\frac{2ik}{\pi a k_p}}{k_p H_n^{(1)}(ka) J_n'(k_p a) - k H_n^{(1)\prime}(ka) J_n(k_p a)} \tag{9.3.30}$$

$$T_n^{(M)} = T_{-n}^{(M)} = -\frac{x_n^{(M)}}{x_n^{(M)} + iy_n^{(M)}} \tag{9.3.31}$$

$$T_n^{(N)} = T_{-n}^{(N)} = -\frac{x_n^{(N)}}{x_n^{(N)} + iy_n^{(N)}} \tag{9.3.32}$$

$$x_n^{(M)} = k_p J_n'(ka) J_n(k_p a) - k J_n(ka) J_n'(k_p a) \tag{9.3.33}$$

$$y_n^{(M)} = k_p Y_n'(ka) J_n(k_p a) - k Y_n(ka) J_n'(k_p a) \tag{9.3.34}$$

$$x_n^{(N)} = k_p J_n(ka) J_n'(k_p a) - k J_n'(ka) J_n(k_p a) \tag{9.3.35}$$

$$y_n^{(N)} = k_p Y_n(ka) J_n'(k_p a) - k Y_n'(ka) J_n(k_p a) \tag{9.3.36}$$

For the case of small cylinders, we use small argument approximations of Hankel and Bessel functions. We also keep the leading term of the real part and the leading term of the imaginary part.

Analogous to the sphere case treated in Eqs. (2.8.54)–(2.8.59) from Chapter 2 of Volume I, since $x_n^{(N)} \ll y_n^{(N)}, x_n^{(M)} \ll y_n^{(M)}$, we have

$$T_n^{(M)} = T_{nr}^{(M)} + iT_{ni}^{(M)} \tag{9.3.37}$$

$$T_{ni}^{(M)} = \frac{x_n^{(M)}}{y_n^{(M)}} \tag{9.3.38}$$

$$T_{nr}^{(M)} = -\left(\frac{x_n^{(M)}}{y_n^{(M)}}\right)^2 \tag{9.3.39}$$

$$T_n^{(N)} = T_{nr}^{(N)} + iT_{ni}^{(N)} \tag{9.3.40}$$

$$T_{ni}^{(N)} = \frac{x_n^{(N)}}{y_n^{(N)}} \tag{9.3.41}$$

$$T_{nr}^{(N)} = -\left(\frac{x_n^{(N)}}{y_n^{(N)}}\right)^2 \tag{9.3.42}$$

Neglecting T-matrix coefficients that are less than $\mathcal{O}(k^2a^2)$, it can be shown that for TM waves, only $n = 0$ term contributes while for TE waves, only $n = \pm 1$ contribute.

Making small argument approximations, we obtain for TM waves,

$$x_0^{(N)} = -\frac{a}{2}\left(k_p^2 - k^2\right) \tag{9.3.43}$$

$$y_0^{(N)} = -\frac{2\pi}{a} \tag{9.3.44}$$

Thus

$$T_0^{(N)} = iT_{0i}^{(N)} - \left(T_{0i}^{(N)}\right)^2 \tag{9.3.45}$$

$$T_{0i}^{(N)} = \frac{\pi a^2}{4}\left(k_p^2 - k^2\right) \tag{9.3.46}$$

For TE waves

$$x_1^{(M)} = k_p \frac{k_p a}{4} - k\frac{ka}{4} \tag{9.3.47}$$

$$y_1^{(M)} = \frac{1}{a\pi}\left(\frac{k_p^2}{k^2} + 1\right) \tag{9.3.48}$$

$$T_1^{(M)} = T_{-1}^{(M)} = iT_{1i}^{(M)} - \left(T_{1i}^{(M)}\right)^2 \tag{9.3.49}$$

$$T_{1i}^{(M)} = \frac{\pi a^2 k^2\left(k_p^2 - k^2\right)}{4\left(k_p^2 + k^2\right)} \tag{9.3.50}$$

(B) Scattering by perfect electric conducting (PEC) cylinders. For this case, the TE and TM waves are decoupled, and we have

$$c_n^{(M)} = c_n^{(N)} = 0 \tag{9.3.51}$$

$$a_n^{(M)s} = T_n^{(M)} a^{(M)} \tag{9.3.52}$$

$$a_n^{(N)s} = T_n^{(N)} a^{(N)} \tag{9.3.53}$$

where

$$T_n^{(M)} = T_{-n}^{(M)} = -\frac{J_n'(k_{i\rho}a)}{H_n^{(1)\prime}(k_{i\rho}a)} \tag{9.3.54}$$

$$T_n^{(N)} = T_{-n}^{(N)} = -\frac{J_n(k_{i\rho}a)}{H_n^{(1)}(k_{i\rho}a)} \tag{9.3.55}$$

For plane wave incidence in the x-y plane, the incident field is

$$\overline{E}_i(\overline{r}) = \left(\hat{v}_i E_{v_i} + \hat{h}_i E_{h_i}\right) e^{i\overline{k}_i \cdot \overline{r}} \tag{9.3.56}$$

with the incident direction vector

$$\overline{k}_i = k\left(\hat{x}\cos\phi_i + \hat{y}\sin\phi_i\right) \tag{9.3.57}$$

Then the incident wave is, from Eq. (1.6.54) in Chapter 1 of Volume I and setting $k_{i\rho} = k$ and $k_{iz} = 0$,

$$\overline{E}_i(\overline{r}) = \sum_{n=-\infty}^{\infty} \frac{i^n e^{-in\phi_i}}{k}\left[iE_{h_i} Rg\overline{M}_n(k, 0, \overline{r}) - E_{v_i} Rg\overline{N}_n(k, 0, \overline{r})\right] \tag{9.3.58}$$

Thus for this case, the incident field coefficients are

$$a_n^{(M)} = \frac{i^{n+1} e^{-in\phi_i}}{k} E_{h_i} \tag{9.3.59}$$

$$a_n^{(N)} = -\frac{i^n e^{-in\phi_i}}{k} E_{v_i} \tag{9.3.60}$$

3.2 Foldy-Lax Multiple Scattering Equations for Cylinders

The cylinders are centered at $\overline{r}_1, \overline{r}_2, \ldots, \overline{r}_N$. Let the incident direction be in the x-y plane so that $\theta_i = 90°, k_{iz} = 0, k_{i\rho} = k$. We use the Foldy-Lax multiple scattering equations which states that

$$\overline{E}^{ex(q)} = \overline{E}_i + \sum_{p=1, p\neq q}^{N} \overline{E}^{s(p)} \tag{9.3.61}$$

The equation states that the final exciting field of particle q is equal to the incident wave plus the scattered waves to particle q from all other particles p except q itself. The incident wave can also be expressed in terms of cylindrical waves centered at the cylinder q.

$$\begin{aligned}\overline{E}_i(\overline{r}) &= \left(\hat{v}_i E_{v_i} + \hat{h}_i E_{h_i}\right) e^{i\overline{k}_i \cdot \overline{\rho}} = \left(\hat{v}_i E_{v_i} + \hat{h}_i E_{h_i}\right) e^{i\overline{k}_i \cdot \overline{\rho}_q} e^{i\overline{k}_i \cdot (\overline{\rho} - \overline{\rho}_q)} \\ &= e^{i\overline{k}_i \cdot \overline{\rho}_q} \sum_{n=-\infty}^{\infty} \frac{i^n e^{-in\phi_i}}{k} \left[iE_{h_i} Rg\overline{M}_n(k, \overline{\rho} - \overline{\rho}_q) \right. \\ &\qquad \left. - E_{v_i} Rg\overline{N}_n(k, \overline{\rho} - \overline{\rho}_q) \right]\end{aligned} \tag{9.3.62}$$

The exciting field of cylinder q is equal to

$$\overline{E}^{ex(q)} = \sum_{n=-\infty}^{\infty} \left[w_n^{(M)(q)} \overline{M}_n(k, \overline{\rho} - \overline{\rho}_q) + w_n^{(N)(q)} \overline{N}_n(k, \overline{\rho} - \overline{\rho}_q) \right] \tag{9.3.63}$$

where $w_n^{(M)(q)}$ and $w_n^{(N)(q)}$ are the final exciting field coefficients. The exciting field of cylinder p is

$$\overline{E}^{ex(p)} = \sum_{n'=-\infty}^{\infty} \left[w_{n'}^{(M)(p)} Rg\overline{M}_{n'}(k, \overline{\rho} - \overline{\rho}_p) + w_{n'}^{(N)(p)} Rg\overline{N}_{n'}(k, \overline{\rho} - \overline{\rho}_p) \right] \tag{9.3.64}$$

The scattered field from particle p is,

$$\overline{E}^{s(p)} = \sum_{n'=-\infty}^{\infty} \left[T_{n'}^{(M)} w_{n'}^{(M)(p)} \overline{M}_{n'}(k, \overline{\rho} - \overline{\rho}_p) + T_{n'}^{(N)} w_{n'}^{(N)(p)} \overline{N}_{n'}(k, \overline{\rho} - \overline{\rho}_p) \right] \tag{9.3.65}$$

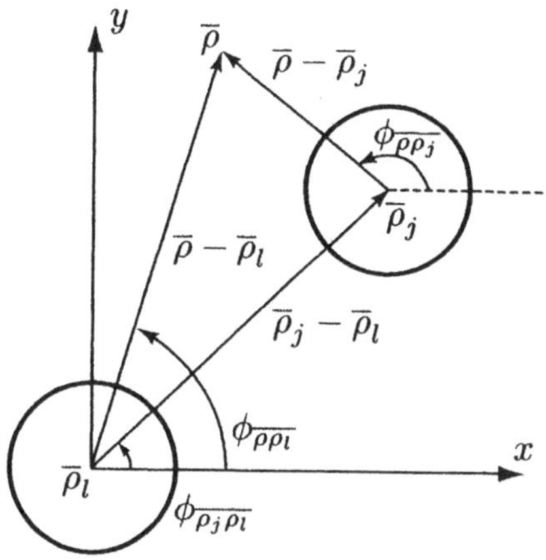

Figure 9.3.2 Translational addition theorem in the cylindrical coordinate system.

where $T_{n'}^{(M)}$ and $T_{n'}^{(N)}$ are the T-matrix coefficients.

To obtain equations for the exciting field coefficents from the Foldy-Lax multiple scattering equations, we need the translational addition theorem of

$$H_{n'}^{(1)}\left(k\left|\overline{\rho}-\overline{\rho}'\right|\right)e^{in'\phi_{\overline{\rho\rho'}}} \tag{9.3.66}$$

where $\phi_{\overline{\rho\rho'}}$ is the aziumthal angle that $\overline{\rho}-\overline{\rho}'$ makes with the x-axis (see Fig. 9.3.2). Let

$$\overline{\rho}-\overline{\rho}'=\overline{\rho}-\overline{\rho}_0-\left(\overline{\rho}'-\overline{\rho}_0\right) \tag{9.3.67}$$

$$\left|\overline{\rho}'-\overline{\rho}_0\right|\geq\left|\overline{\rho}-\overline{\rho}_0\right| \tag{9.3.68}$$

Then

$$H_{n'}^{(1)}\left(k\left|\overline{\rho}-\overline{\rho}'\right|\right)e^{in'\phi_{\overline{\rho\rho'}}}$$
$$=\sum_{n=-\infty}^{\infty}J_n\left(k\left|\overline{\rho}-\overline{\rho}_0\right|\right)e^{in\phi_{\overline{\rho\rho_0}}}H_{n-n'}^{(1)}\left(k\left|\overline{\rho}'-\overline{\rho}_0\right|\right)e^{-i(n-n')\phi_{\overline{\rho'\rho_0}}} \tag{9.3.69}$$

For the vector cylindrical wave functions with

$$\left|\overline{\rho}_p-\overline{\rho}_q\right|\geq\left|\overline{\rho}-\overline{\rho}_q\right| \tag{9.3.70}$$

we have

$$\overline{M}_{n'}\left(k,\overline{\rho}-\overline{\rho}_p\right)$$
$$=\sum_{n=-\infty}^{\infty}Rg\overline{M}_n\left(k,\overline{\rho}-\overline{\rho}_q\right)H_{n-n'}^{(1)}\left(k\left|\overline{\rho}_p-\overline{\rho}_q\right|\right)e^{-i(n-n')\phi_{\overline{\rho_p\rho_q}}} \tag{9.3.71}$$

$$\overline{N}_{n'}\left(k,\overline{\rho}-\overline{\rho}_p\right) = \sum_{n=-\infty}^{\infty} Rg\overline{N}_n\left(k,\overline{\rho}-\overline{\rho}_q\right) H^{(1)}_{n-n'}\left(k\left|\overline{\rho}_p-\overline{\rho}_q\right|\right) e^{-i(n-n')\phi_{\overline{\rho_p\rho_q}}} \tag{9.3.72}$$

Putting the cylindrical wave translational formula into Foldy-Lax multiple scattering equations, we have

$$\begin{aligned}
&\sum_{n=-\infty}^{\infty}\left[w_n^{(M)(q)} Rg\overline{M}_n\left(k,\overline{\rho}-\overline{\rho}_q\right) + w_n^{(N)(q)} Rg\overline{N}_n\left(k,\overline{\rho}-\overline{\rho}_q\right)\right] \\
&= e^{i\overline{k}_i\cdot\overline{\rho}_q} \sum_{n=-\infty}^{\infty} \frac{i^n e^{-in\phi_i}}{k}\left[iE_{h_i} Rg\overline{M}_n\left(k,\overline{\rho}-\overline{\rho}_q\right) - E_{v_i} Rg\overline{N}_n\left(k,\overline{\rho}-\overline{\rho}_q\right)\right] \\
&+ \sum_{\substack{p=1\\p\neq q}}^{N} \sum_{n'=-\infty}^{\infty} \sum_{n=-\infty}^{\infty}\left[T_{n'}^{(M)} w_{n'}^{(M)(p)} Rg\overline{M}_n\left(k,\overline{\rho}-\overline{\rho}_q\right)\right. \\
&\quad \left. + T_{n'}^{(N)} w_{n'}^{(N)(p)} Rg\overline{N}_n\left(k,\overline{\rho}-\overline{\rho}_q\right)\right] \times \\
&\quad H^{(1)}_{n-n'}\left(k\left|\overline{\rho}_p-\overline{\rho}_q\right|\right) e^{-i(n-n')\phi_{\overline{\rho_p\rho_q}}}
\end{aligned} \tag{9.3.73}$$

Balancing the coefficients for the TE waves gives

$$\begin{aligned}
w_n^{(M)(q)} &= e^{i\overline{k}_i\cdot\overline{\rho}_q} \frac{i^{n+1} e^{-in\phi_i}}{k} E_{h_i} \\
&+ \sum_{n'=-\infty}^{\infty} \sum_{\substack{p=1\\p\neq q}}^{N} H^{(1)}_{n-n'}\left(k\left|\overline{\rho}_p-\overline{\rho}_q\right|\right) e^{-i(n-n')\phi_{\overline{\rho_p\rho_q}}} T_{n'}^{(M)} w_{n'}^{(M)(p)}
\end{aligned} \tag{9.3.74}$$

and for TM waves

$$\begin{aligned}
w_n^{(N)(q)} &= -e^{i\overline{k}_i\cdot\overline{\rho}_q} \frac{i^{n} e^{-in\phi_i}}{k} E_{v_i} \\
&+ \sum_{n'=-\infty}^{\infty} \sum_{\substack{p=1\\p\neq q}}^{N} H^{(1)}_{n-n'}\left(k\left|\overline{\rho}_p-\overline{\rho}_q\right|\right) e^{-i(n-n')\phi_{\overline{\rho_p\rho_q}}} T_{n'}^{(N)} w_{n'}^{(N)(p)}
\end{aligned} \tag{9.3.75}$$

The above equations can be written in a matrix form. For example, for TE case with $N_{\max} = 1$ and $N = 2$, we have

$$
\begin{bmatrix}
1 & 0 & 0 & -H_0^{(1)}\left(k\left|\overline{\rho}_2-\overline{\rho}_1\right|\right)\cdot T_{-1}^{(M)} & -H_{-1}^{(1)}\left(k\left|\overline{\rho}_2-\overline{\rho}_1\right|\right)\cdot e^{i\phi_{\overline{\rho_2\rho_1}}}T_0^{(M)} & -H_{-2}^{(1)}\left(k\left|\overline{\rho}_2-\overline{\rho}_1\right|\right)\cdot e^{i2\phi_{\overline{\rho_2\rho_1}}}T_1^{(M)} \\
0 & 1 & 0 & -H_1^{(1)}\left(k\left|\overline{\rho}_2-\overline{\rho}_1\right|\right)\cdot e^{-i\phi_{\overline{\rho_2\rho_1}}}T_{-1}^{(M)} & -H_0^{(1)}\left(k\left|\overline{\rho}_2-\overline{\rho}_1\right|\right)\cdot T_0^{(M)} & -H_{-1}^{(1)}\left(k\left|\overline{\rho}_2-\overline{\rho}_1\right|\right)\cdot e^{i\phi_{\overline{\rho_2\rho_1}}}T_1^{(M)} \\
0 & 0 & 1 & -H_2^{(1)}\left(k\left|\overline{\rho}_2-\overline{\rho}_1\right|\right)\cdot e^{-i2\phi_{\overline{\rho_2\rho_1}}}T_{-1}^{(M)} & -H_1^{(1)}\left(k\left|\overline{\rho}_2-\overline{\rho}_1\right|\right)\cdot e^{-i\phi_{\overline{\rho_2\rho_1}}}T_0^{(M)} & -H_0^{(1)}\left(k\left|\overline{\rho}_2-\overline{\rho}_1\right|\right)\cdot T_1^{(M)} \\
-H_0^{(1)}\left(k\left|\overline{\rho}_1-\overline{\rho}_2\right|\right)\cdot T_{-1}^{(M)} & -H_{-1}^{(1)}\left(k\left|\overline{\rho}_1-\overline{\rho}_2\right|\right)\cdot e^{i\phi_{\overline{\rho_1\rho_2}}}T_0^{(M)} & -H_{-2}^{(1)}\left(k\left|\overline{\rho}_1-\overline{\rho}_2\right|\right)\cdot e^{i2\phi_{\overline{\rho_1\rho_2}}}T_1^{(M)} & 1 & 0 & 0 \\
-H_1^{(1)}\left(k\left|\overline{\rho}_1-\overline{\rho}_2\right|\right)\cdot e^{-i\phi_{\overline{\rho_1\rho_2}}}T_{-1}^{(M)} & -H_0^{(1)}\left(k\left|\overline{\rho}_1-\overline{\rho}_2\right|\right)\cdot T_0^{(M)} & -H_{-1}^{(1)}\left(k\left|\overline{\rho}_1-\overline{\rho}_2\right|\right)\cdot e^{i\phi_{\overline{\rho_1\rho_2}}}T_1^{(M)} & 0 & 1 & 0 \\
-H_2^{(1)}\left(k\left|\overline{\rho}_1-\overline{\rho}_2\right|\right)\cdot e^{-i2\phi_{\overline{\rho_1\rho_2}}}T_{-1}^{(M)} & -H_1^{(1)}\left(k\left|\overline{\rho}_1-\overline{\rho}_2\right|\right)\cdot e^{-i\phi_{\overline{\rho_1\rho_2}}}T_0^{(M)} & -H_0^{(1)}\left(k\left|\overline{\rho}_1-\overline{\rho}_2\right|\right)\cdot T_1^{(M)} & 0 & 0 & 1
\end{bmatrix}
$$

$$
\times
\begin{bmatrix}
w_{-1}^{(M)(1)} \\ w_0^{(M)(1)} \\ w_1^{(M)(1)} \\ w_{-1}^{(M)(2)} \\ w_0^{(M)(2)} \\ w_{-1}^{(M)(2)}
\end{bmatrix}
=
\begin{bmatrix}
e^{i\overline{k}_i\cdot\overline{\rho}_1}\dfrac{e^{i\phi_i}}{k}E_{h_i} \\
e^{i\overline{k}_i\cdot\overline{\rho}_1}\dfrac{i}{k}E_{h_i} \\
e^{i\overline{k}_i\cdot\overline{\rho}_1}\dfrac{i^2e^{-i\phi_i}}{k}E_{h_i} \\
e^{i\overline{k}_i\cdot\overline{\rho}_2}\dfrac{e^{i\phi_i}}{k}E_{h_i} \\
e^{i\overline{k}_i\cdot\overline{\rho}_2}\dfrac{i}{k}E_{h_i} \\
e^{i\overline{k}_i\cdot\overline{\rho}_2}\dfrac{i^2e^{-i\phi_i}}{k}E_{h_i}
\end{bmatrix}
\tag{9.3.76}
$$

After the exciting field coefficients are solved, the "final" scattered field from cylinder q and the final internal field of the cylinder q can be obtained.

$$
\overline{E}^{s(q)} = \sum_{n=-\infty}^{\infty}\left[a_n^{(M)s(q)}Rg\overline{M}_n\left(k,\overline{\rho}-\overline{\rho}_q\right)+a_n^{(N)s(q)}Rg\overline{N}_n\left(k,\overline{\rho}-\overline{\rho}_q\right)\right]
\tag{9.3.77}
$$

$$a_n^{(M)s(q)} = T_n^{(M)} w_n^{(M)(q)} \tag{9.3.78a}$$

$$a_n^{(N)s(q)} = T_n^{(N)} w_n^{(M)(q)} \tag{9.3.78b}$$

The total scattered field is

$$\overline{E}_s = \sum_{q=1}^{N} \overline{E}^{s(q)} \tag{9.3.79}$$

For the final internal field of cylinder q

$$\overline{E}_{int}^{(q)} = \sum_{n=-\infty}^{\infty} \left[c_n^{(M)(q)} Rg\overline{M}_n \left(k_p, \overline{\rho} - \overline{\rho}_q\right) + c_n^{(N)(q)} Rg\overline{N}_n \left(k_p, \overline{\rho} - \overline{\rho}_q\right) \right] \tag{9.3.80}$$

where

$$c_n^{(M)(q)} = B_n^{(M)} w_n^{(M)(q)}$$

$$c_n^{(N)(q)} = B_n^{(N)} w_n^{(M)(q)} \tag{9.3.81}$$

After the "final" exciting field coeffcients are obtained, we can calculate the "final" scattered field in the direction $\hat{k}_s = k(\cos\phi_s\hat{x} + \sin\phi_s\hat{y})$. Using the asymptotic form of Hankel functions and derivatives, we have, in the limit of $\rho \to \infty$,

$$\overline{m}_n \left(k_\rho, k_z, \overline{\rho}\right) = -\hat{\phi}\, i \sqrt{\frac{2k_\rho}{\pi\rho}}\, e^{i\left(k_\rho\rho - \frac{\pi}{4}\right) + in\frac{\pi}{2}} \tag{9.3.82}$$

$$\overline{n}_n \left(k_\rho, k_z, \overline{\rho}\right) = \sqrt{\frac{2k_\rho}{\pi\rho}}\, e^{i\left(k_\rho\rho - \frac{\pi}{4}\right) + in\frac{\pi}{2}} \left(-\frac{k_z}{k}\hat{\rho} + \frac{k_\rho}{k}\hat{z} \right) \tag{9.3.83}$$

$$\overline{E}_s = \sqrt{\frac{2k}{\pi\rho}} e^{i\left(k\rho - \frac{\pi}{4}\right)} \sum_{q=1}^{N} \sum_{n=-\infty}^{\infty} \left[-\hat{\phi}\, i T_n^{(M)} w_n^{(M)(q)} + \hat{z}\, T_n^{(N)} w_n^{(N)(q)} \right] \times e^{in\left(\phi_s - \frac{\pi}{2}\right)} e^{-i\hat{k}_s \cdot \overline{\rho}_q} \tag{9.3.84}$$

3.3 Coherent Field, Incoherent Field and Scattering Coefficient

Let A be the area in which the cylinders are placed. The fractional area occupied by the cylinders is

$$f = \frac{N\pi a^2}{A} \tag{9.3.85}$$

The Monte Carlo simulations of scattering by the cylinders are performed for N_r realizations. Let $\overline{E}_{sr}$ be the scattered field $\overline{E}_s$ for the rth realization. The coherent scattered field $\langle \overline{E}_s \rangle$ is obtained by averaging over N_r realizations

$$\langle \overline{E}_s \rangle = \frac{1}{N_r} \sum_{r=1}^{N_r} \overline{E}_{sr} \tag{9.3.86}$$

The averaged intensity is given by

$$\langle |\overline{E}_s|^2 \rangle = \frac{1}{N_r} \sum_{r=1}^{N_r} |\overline{E}_{sr}|^2 \tag{9.3.87}$$

and the incoherent intensity is equal to

$$\langle |\overline{E}_s|^2 \rangle - |\langle \overline{E}_s \rangle|^2 \tag{9.3.88}$$

The scattering rate is the scattering cross section per unit volume. To obtain unit volume, we can consider a unit length in the $\hat{z}$-direction. Then the scattering coefficient is

$$\kappa_s = \frac{\displaystyle\int_0^{2\pi} d\phi_s \rho \left(\left\langle |\overline{E}_s|^2 \right\rangle - |\langle \overline{E}_s \rangle|^2 \right)}{A \left(|E_{vi}|^2 + |E_{hi}|^2 \right)} \tag{9.3.89}$$

Only the incoherent intensity is used in calculating the scattering coefficient for reasons as discussed in Chapter 7. If the cylinders are not absorptive, then $\kappa_e = \kappa_s$.

3.4 Scattered Field and Internal Field Formulations

Instead of using equations for the exciting field coefficients, the scattered field coefficients can also be used in Foldy-Lax equations. For TE waves

$$\begin{aligned} a_n^{(M)s(q)} &= e^{i\overline{k}_i \cdot \overline{\rho}_q} \frac{i^{n+1} e^{-in\phi_i}}{k} T_n^{(M)} E_{h_i} \\ &+ T_n^{(M)} \sum_{n'=-\infty}^{\infty} \sum_{\substack{p=1 \\ p \neq q}}^{N} H_{n-n'}^{(1)} \left(k \left| \overline{\rho}_p - \overline{\rho}_q \right| \right) e^{-i(n-n')\phi_{\overline{\rho_p \rho_q}}} a_{n'}^{(M)s(p)} \end{aligned} \tag{9.3.90}$$

and for TM waves

$$\begin{aligned} a_n^{(N)s(q)} &= -T_n^{(N)} e^{i\overline{k}_i \cdot \overline{\rho}_q} \frac{i^{n} e^{-in\phi_i}}{k} E_{v_i} \\ &+ T_n^{(N)} \sum_{n'=-\infty}^{\infty} \sum_{\substack{p=1 \\ p \neq q}}^{N} H_{n-n'}^{(1)} \left(k \left| \overline{\rho}_p - \overline{\rho}_q \right| \right) e^{-i(n-n')\phi_{\overline{\rho_p \rho_q}}} a_{n'}^{(N)s(p)} \end{aligned} \tag{9.3.91}$$

The internal field coefficients can also be used in Foldy-Lax equations leading to the internal field formulation. For TE waves

$$c_n^{(M)(q)} = e^{i\bar{k}_i\cdot\bar{\rho}_q}\frac{i^{n+1}e^{-in\phi_i}}{k}B_n^{(M)}E_{h_i}$$
$$+ B_n^{(M)}\sum_{n'=-\infty}^{\infty}\sum_{\substack{p=1\\p\neq q}}^{N} H_{n-n'}^{(1)}\left(k\left|\bar{\rho}_p-\bar{\rho}_q\right|\right)e^{-i(n-n')\phi_{\overline{\rho_p\rho_q}}}S_{n'}^{(M)}c_{n'}^{(M)(q)} \tag{9.3.92}$$

where

$$S_n^{(M)} = \frac{T_n^{(M)}}{B_n^{(M)}} = \frac{i\pi a k_p}{2k}x_n^{(M)} \tag{9.3.93}$$

that relates the scattered field to the internal field. For TM waves

$$c_n^{(N)(q)} = -e^{i\bar{k}_i\cdot\bar{\rho}_q}\frac{i^{n}e^{-in\phi_i}}{k}E_{v_i}B_n^{(N)}$$
$$+ B_n^{(N)}\sum_{n'=-\infty}^{\infty}\sum_{\substack{p=1\\p\neq q}}^{N} H_{n-n'}^{(1)}\left(k\left|\bar{\rho}_p-\bar{\rho}_q\right|\right)e^{-i(n-n')\phi_{\overline{\rho_p\rho_q}}}S_{n'}^{(N)}c_{n'}^{(N)(p)} \tag{9.3.94}$$

where

$$S_n^{(N)} = \frac{T_n^{(N)}}{B_n^{(N)}} = -\frac{i\pi a k_p}{2k}x_n^{(N)} \tag{9.3.95}$$

3.5 Low Frequency Formulas

We next consider the low frequency limit of the formulas for the Foldy-Lax multiple scattering equations.

TM case

For the low frequency limit, only $m = 0$ (monopole) contributes to $T_m^{(N)}$. Then

$$T_0^{(N)} = \frac{i\left(k_p^2 - k^2\right)}{4}\sigma \tag{9.3.96}$$

$$B_0^{(N)} = \frac{k}{k_p} \tag{9.3.97}$$

where $\sigma = \pi a^2$ is the cross section area of the cylinder. The exciting field equations then become

$$w_0^{(N)(q)} = -e^{i\overline{k}_i\cdot\overline{\rho}_q}e^{i\overline{k}_i\cdot\overline{\rho}_q}\frac{E_{v_i}}{k} + \sum_{\substack{p=1\\p\neq q}}^{N} H_0^{(1)}\left(k\left|\overline{\rho}_p-\overline{\rho}_q\right|\right)\frac{i\left(k_p^2-k^2\right)}{4}\sigma w_0^{(N)(p)} \tag{9.3.98}$$

and the internal field equations become

$$c_0^{(N)(q)} = -e^{i\overline{k}_i\cdot\overline{\rho}_q}\frac{E_{v_i}}{k_p} + \sum_{\substack{p=1\\p\neq q}}^{N} H_0^{(1)}\left(k\left|\overline{\rho}_p-\overline{\rho}_q\right|\right)\frac{i\left(k_p^2-k^2\right)}{4}\sigma c_0^{(N)(p)} \tag{9.3.99}$$

Note that in the low frequency limit, only the z component of the electric field exists, we also have for the internal field of cylinder q as

$$E_z^{int(q)} = k_p c_0^{(N)(q)} \tag{9.3.100}$$

Thus the internal field equation can be put in the form

$$E_z^{int(q)} = -e^{i\overline{k}_i\cdot\overline{\rho}_q}E_{v_i} + \sum_{\substack{p=1\\p\neq q}}^{N}\frac{i}{4}H_0^{(1)}\left(k\left|\overline{\rho}_p-\overline{\rho}_q\right|\right)\left(k_p^2-k^2\right)\sigma E_z^{int(p)} \tag{9.3.101}$$

The above matrix equation is consistent with that of a volume integral equation using pulse basis function for each cylinder by recognizing that the 2-D Green's function is $\frac{i}{4}H_0^{(1)}\left(k\left|\overline{\rho}_p-\overline{\rho}_q\right|\right)$.

TE case

Only $n=\pm1$ contribute. Thus we have

$$w_1^{(M)(q)} = -e^{i\overline{k}_i\cdot\overline{\rho}_q}\frac{e^{-i\phi_i}}{k}E_{h_i} + \sum_{\substack{p=1\\p\neq q}}^{N} H_0^{(1)}\left(k\left|\overline{\rho}_p-\overline{\rho}_q\right|\right)T_1^{(M)}w_1^{(M)(p)} + \sum_{\substack{p=1\\p\neq q}}^{N} H_2^{(1)}\left(k\left|\overline{\rho}_p-\overline{\rho}_q\right|\right)e^{-i2\phi_{\overline{\rho_p\rho_q}}}T_1^{(M)}w_{-1}^{(M)(p)} \tag{9.3.102}$$

$$w_{-1}^{(M)(q)} = e^{i\overline{k}_i\cdot\overline{\rho}_q}\frac{e^{i\phi_i}}{k}E_{h_i} + \sum_{\substack{p=1\\p\neq q}}^{N} H_0^{(1)}\left(k\left|\overline{\rho}_p-\overline{\rho}_q\right|\right)T_1^{(M)}w_{-1}^{(M)(p)} + \sum_{\substack{p=1\\p\neq q}}^{N} H_2^{(1)}\left(k\left|\overline{\rho}_p-\overline{\rho}_q\right|\right)e^{i2\phi_{\overline{\rho_p\rho_q}}}T_1^{(M)}w_1^{(M)(p)} \tag{9.3.103}$$

3.6 Independent Scattering

The results of independent scattering can be readily derived. From (9.3.75), we approximate for TM waves,

$$w_n^{(N)(q)} = -e^{i\overline{k}_i \cdot \overline{\rho}_q} \frac{i^n e^{-in\phi_i}}{k} E_{v_i} \tag{9.3.104}$$

Then

$$\overline{E}_s = \sqrt{\frac{2k}{\pi\rho}} e^{ik\rho - \frac{\pi}{4}} \hat{z} \sum_{q=1}^{N} \sum_{n=-\infty}^{\infty} T_n^{(N)} w_n^{(N)(q)} e^{in(\phi_s - \frac{\pi}{2})} e^{-i\overline{k}_s \cdot \rho_q} \tag{9.3.105}$$

Assuming independent scattering of the scatterers, $|\overline{E}_s|^2$ becomes

$$|\overline{E}_s|^2 = \frac{2k}{\pi\rho} N \left| \sum_{n=-\infty}^{\infty} T_n^{(N)} w_n^{(N)} e^{in(\phi_s - \frac{\pi}{2})} \right|^2 \tag{9.3.106}$$

The the scattering coefficient is defined as

$$\kappa_s = \frac{\int_0^{2\pi} d\phi_s \, \rho \, |\overline{E}_s|^2}{A |E_{v_i}|^2} \tag{9.3.107}$$

Using (9.3.104)–(9.3.106) in (9.3.107), we have

$$\kappa_s = \frac{4n_o}{k} \sum_{n=-\infty}^{\infty} |T_n^{(N)}|^2 \tag{9.3.108}$$

Similarly, for TE waves,

$$\kappa_s = \frac{4n_o}{k} \sum_{n=-\infty}^{\infty} |T_n^{(M)}|^2 \tag{9.3.109}$$

At low frequency for TM waves, using (9.3.96), we have

$$\kappa_s = \frac{4n_o}{k} \left| \frac{k_p^2 - k^2}{4} \right|^2 \pi^2 a^4 \tag{9.3.110}$$

This gives

$$\begin{aligned} \frac{\kappa_s}{k} &= \frac{n_o}{4} k^2 a^2 \pi^2 a^2 \left| \frac{\epsilon_p}{\epsilon} - 1 \right|^2 \\ &= \frac{f\pi}{4} k^2 a^2 \left| \frac{\epsilon_p}{\epsilon} - 1 \right|^2 \end{aligned} \tag{9.3.111}$$

Thus, from (9.3.110), κ_s has a (frequency)^3 dependence.

3.7 Simulation Results for Sticky and Non-Sticky Cylinders

In the following, we present simulation results of TM waves. We find that for the case of small cylinders, the condition number of the matrix equation with internal field formulation is lower than that of the exciting field formulation. Thus the results shown in this section are computed based on the internal field formulation of Eq. (9.3.94). Matrix equation is solved by direct matrix inversion. We illustrate results for $ka = 0.2$ to $ka = 0.8$. Multipoles are truncated at $N_{\max} = 3$.

In Fig. 9.3.3, we plot the normalized extinction coefficient κ_s/k for non-sticky and sticky cylinders for $ka = 0.2,\ 0.4,\ 0.8$ and $\epsilon_p = 3.2\epsilon_o$, $\epsilon = \epsilon_o$. The results show convergence with respect to the number of realizations. We use up to 50 realizations. The fractional volume is $f = 0.2$, and we use 200 cylinders in the simulations. For sticky cylinders, we use a sticky parameter of $\tau = 0.1$. The independent scattering result is computed based on (9.3.108). In Table 9.3.1, we tabulate the results for κ_s/k. The frequency dependence of scattering can be derived from Table 9.3.1 as follows. We consider three frequencies at ω_1, $\omega_2 = 2\omega_1$, and $\omega_4 = 4\omega_1$. Let $k_1 = 1$ be the wavenumber of ω_1. Then $k_2 = 2$ and $k_4 = 4$ respectively of ω_2 and ω_4. Let a be such that $k_1 a = 0.2$. Then $k_2 a = 0.4$ and $k_4 a = 0.8$. For non-sticky particles, extinction coefficient κ_{s1} at ω_1 has $\kappa_{s1}/k_1 = 0.0116$ so that $\kappa_{s1} = 0.0116$. For ω_2, $k_2 a = 0.4$. From Table 9.3.1, the extinction coefficient κ_{s2} has $\kappa_{s2}/k_2 = 0.0335$ so that $\kappa_{s2} = 0.0670$. At ω_4, $k_4 a = 0.8$, then the extinction coefficient at ω_4 has $\kappa_{s4}/k_4 = 0.0281$ so that $\kappa_{s4} = 0.1124$. The case of sticky particles and the case of independent scattering are determined similarly. Table 9.3.2 gives the results of extinction. It is clear that sticky cylinders have a weaker frequency dependence than non-sticky cylinders. Sticky cylinders also have a weaker frequency dependence than independent scattering. For example, $\kappa_{s2}/\kappa_{s1} = 5.78$ for non-sticky cylinders, $\kappa_{s2}/\kappa_{s1} = 2.93$ for sticky cylinders, and $\kappa_{s2}/\kappa_{s1} = 9.2$ for independent scattering.

ka	κ_s/k (non-sticky)	κ_s/k (sticky)	κ_s/k (independent)
0.2	0.0116	0.0327	0.0355
0.4	0.0335	0.0480	0.1633
0.8	0.0281	0.0268	0.2706

Table 9.3.1 Extinction coefficients for non-sticky and sticky cylinders.

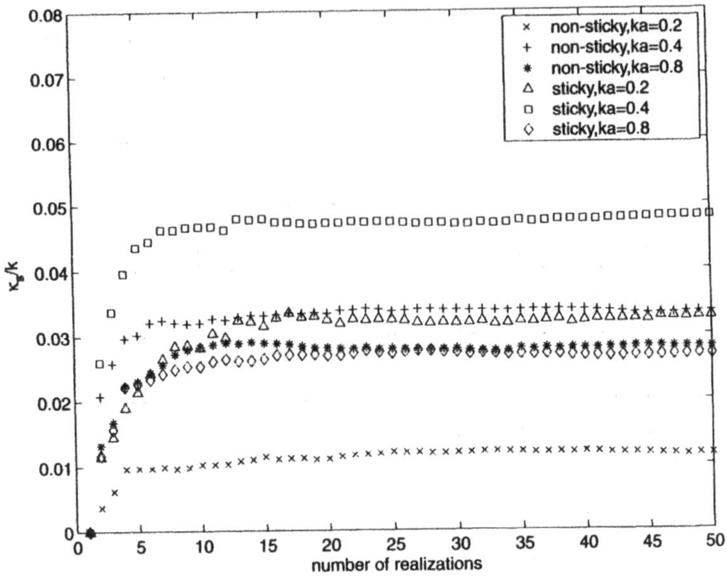

Figure 9.3.3 Extinction coefficients for non-sticky and sticky cylinders.

	κ_{s1}	κ_{s2}	κ_{s4}
non-sticky	0.0116	0.0670	0.1124
sticky	0.0327	0.0960	0.1072
independent	0.0355	0.3266	1.0824

Table 9.3.2 Extinction coefficients based on Table 9.3.1 with $a = 0.2$, $k_1a = 0.2$, $k_2a = 0.4$, $k_4a = 0.8$.

4 Sparse-Matrix Canonical-Grid Method for Scattering by Many Cylinders

4.1 Introduction

In Chapters 5 and 6, we applied the sparse-matrix canonical-grid (SMCG) method to solve rough surface scattering problem. For the case of rough surface scattering, equally spaced points on a flat surface was used on the canonical grid. In this section, we illustrate how the method is applied to volume scattering [Chan and Tsang, 1995; Chew et al. 1995]. In the SMCG method, we carry out the decomposition of strong and weak interactions. The weak interactions, which account for the majority of the required CPU time and memory, are calculated using a canonical grid with a translational

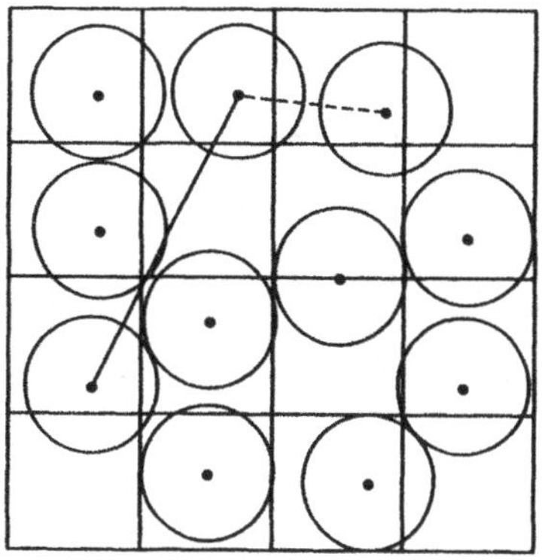

Figure 9.4.1 Decomposition of strong and weak interactions.

addition theorem. This facilitates the use of FFT and results in an $N \log N$--type efficiency for CPU and $\mathcal{O}(N)$ for memory. We use the concept of the sparse-matrix canonical grid method for scattering by many discrete scatterers. The near- and intermediate-range interactions are represented by the strong interaction matrix. For the weak part of the interaction among many discrete scatterers, we utilize a uniform grid as the canonical grid. By using a translational addition theorem, each scatterer is translated to a grid point in the uniform grid. We are making use of the equivalence principle to state that the field scattered by a scatterer is equivalent to the radiation from a grid point in the uniform grid. We further illustrate the method by considering scattering by N_p dielectric cylinders (two-dimensional scattering problem), the positions of cylinders are randomly placed in a square box (Fig. 9.4.1). The square box is next discretized by a two-dimensional uniformly spaced grid with N_g points. The grid size is chosen according to the cylinder distributions. Each cylinder is associated with a grid point that is closest to the center of the cylinder. Several cylinders may be associated with the same grid point.

The strong interactions between two cylinders in the vicinity of each other are calculated directly and represent the strong-interaction matrix. On the other hand, as depicted in Fig. 9.4.2, the weak interactions between two distant cylinders centered at $\overline{\rho}_\alpha$ and $\overline{\rho}_\beta$ are calculated indirectly. We first translate from $\overline{\rho}_\alpha$ to its associated grid point at $\overline{\rho}_{o\alpha}$. Then we calculate radiation from the grid point $\overline{\rho}_{o\alpha}$ to the grid point $\overline{\rho}_{o\beta}$. The points $\overline{\rho}_{o\alpha}$ and $\overline{\rho}_{o\beta}$ are on the uniform grid. Finally we translate from $\overline{\rho}_{o\beta}$ to the center of the cylinder at $\overline{\rho}_\beta$. All the operations can be performed by the use of translational addition theorem. The indirect calculation is cast into a form

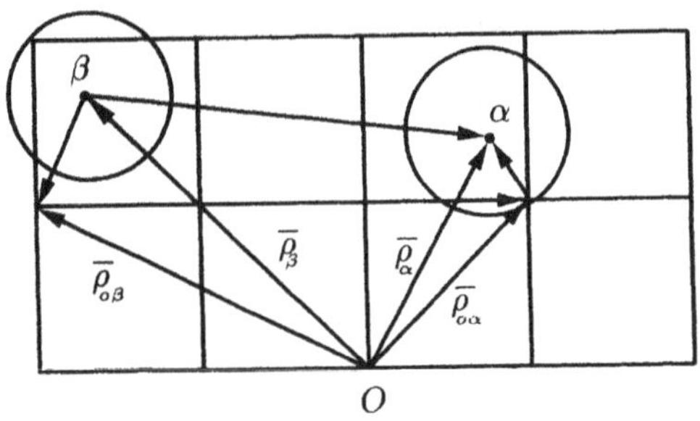

Figure 9.4.2 Direct and indirect interactions from cylinder β to cylinder α.

such that the weak interactions of all the cylinders, which account for the majority of the CPU time, can be evaluated simultaneously using FFT, thus achieving an $N_g \log N_g$-type efficiency. Note that N_g can be less than N_p when several scatterers associate with one grid point. This method only requires translation of a short or moderate distance to the nearest grid point so that the translational addition theorem only requires a low order. Using the matrix notation of the impedance matrix, the strategy can be represented as follows. By discretizing integral equations, a matrix equation

$$\overline{\overline{Z}}\,\overline{X} = \overline{b} \tag{9.4.1}$$

can be obtained. In (9.4.1), $\overline{\overline{Z}}$ is the impedance matrix, $\overline{X}$ is the unknown column vector, and $\overline{b}$ is the known vector corresponding to the excitation. Next the impedance matrix is decomposed into strong and weak interaction matrices

$$\overline{\overline{Z}} = \overline{\overline{Z}}^S + \overline{\overline{Z}}^W \tag{9.4.2}$$

In (9.4.2) $\overline{\overline{Z}}^S$ represents strong neighborhood interaction and $\overline{\overline{Z}}^W$ is the rest of $\overline{\overline{Z}}$. In $\overline{\overline{Z}}^W$, we use translational addition theorem to find equivalent sources at the uniform grid points, which are further translated to the actual observation points. In matrix notation

$$\overline{\overline{Z}}^W\overline{X} = \overline{\overline{T}}_r\overline{\overline{Z}}_p\overline{\overline{T}}_s\overline{X} \tag{9.4.3}$$

where $\overline{\overline{T}}_s$ is a diagonal translation matrix for the scatterers, $\overline{\overline{Z}}_p$ is the matrix representing the canonical uniform grid problem of radiating from grid points to grid points, and $\overline{\overline{T}}_r$ is the diagonal translation matrix to the observation points. Thus the product of $\overline{\overline{T}}_s$ and $\overline{X}$ requires $\mathcal{O}(N_p)$ steps. Next, multiplication by $\overline{\overline{Z}}_p$ can be performed by discrete convolution using 2-D FFTs, as

$\overline{\overline{Z}}_p$ is a translationally invariant matrix. This requires $\mathcal{O}(4N_g \log 4N_g)$ steps. Further multiplication by $\overline{\overline{T}}_r$ requires $\mathcal{O}(N_p)$ steps. In the memory, $\overline{\overline{T}}_r$ and $\overline{\overline{T}}_s$ are of the order $\mathcal{O}(N_p)$, because they are diagonal matrices. The $\overline{\overline{Z}}_p$ matrix is translationally invariant and consists of points that are equally spaced apart requires memory of the order of $\mathcal{O}(4N_g)$. Iterative solutions such as the conjugate gradient can be adopted for the iterative solution of the matrix equation.

4.2 The Two-Dimensional Scattering Problem of Many Dielectric Cylinders

Consider an incident wave with wave vector $\overline{k}_i$ scattered by a medium consisting of N_p circular dielectric cylinders centered at $\overline{\rho}_\alpha$ where $\alpha = 1, 2, \ldots, N_p$. Each cylinder has a radius of a, permittivity ϵ_p. We consider the algorithm for TE wave. The Foldy-Lax multiple scattering equations are, from (9.3.74),

$$w_n^{(\alpha)} = e^{i\overline{k}_i \cdot \overline{\rho}_\alpha} i^{n+1} e^{-in\phi_i} E_{hi} + \sum_{\substack{\beta=1 \\ \beta \neq \alpha}}^{N_p} w_n^{\alpha\beta} \tag{9.4.4}$$

where we have defined $w_n^{\alpha\beta}$ as the exciting field of α due to the scattered field from β. It is given by

$$w_n^{\alpha\beta} = \sum_{m=-\infty}^{\infty} T_m^{(\beta)} w_m^{(\beta)} H_{n-m}^{(1)}(k|\overline{\rho}_\beta - \overline{\rho}_\alpha|) \exp\left[-i(n-m)\phi_{\overline{\rho_\beta \rho_\alpha}}\right] \tag{9.4.5}$$

We have suppressed the (M) superscript in (9.4.5). Equation (9.4.5) is termed a direct calculation. Let $\overline{\rho}_{o\alpha}$ and $\overline{\rho}_{o\beta}$ be the grid points in the grid associated with the cylinders at $\overline{\rho}_\alpha$ and $\overline{\rho}_\beta$, respectively. After the successive use of the translational addition theorem to (9.4.5) to translate from $\overline{\rho}_\alpha$ and $\overline{\rho}_\beta$ to $\overline{\rho}_{o\alpha}$ and $\overline{\rho}_{o\beta}$, respectively, we have

$$\begin{aligned} w_n^{\alpha\beta} = & \sum_{m=-\infty}^{\infty} T_m^{(\beta)} w_m^{(\beta)} \cdot \\ & \sum_{m'=-\infty}^{\infty} \sum_{n'=-\infty}^{\infty} J_{m'-m}(k|\overline{\rho}_\beta - \overline{\rho}_{o\beta}|) \exp\left[-i(m'-m)\phi_{\overline{\rho_\beta \rho_{o\beta}}}\right] \\ & \qquad \times H_{n'-m'}^{(1)}(k|\overline{\rho}_{o\beta} - \overline{\rho}_{o\alpha}|) \exp\left[-i(n'-m')\phi_{\overline{\rho_{o\beta} \rho_{o\alpha}}}\right] \\ & \qquad \times J_{n-n'}(k|\overline{\rho}_{o\alpha} - \overline{\rho}_\alpha|) \exp\left[-i(n-n')\phi_{\overline{\rho_{o\alpha} \rho_\alpha}}\right] \end{aligned} \tag{9.4.6}$$

This procedure is termed indirect calculation. It is then clear that the exciting field coefficients among all cylinders can be computed simultaneously.

First, let us define a pre-multiplication process as

$$\tilde{w}^{\beta}_{m'} = \sum_{m=-\infty}^{\infty} J_{m'-m}(k|\overline{\rho}_{\beta} - \overline{\rho}_{o\beta}|) \exp\left[-i(m'-m)\phi_{\overline{\rho_{\beta}\rho_{o\beta}}}\right] T^{(\beta)}_{m} w^{(\beta)}_{m} \quad (9.4.7)$$

This step requires $\mathcal{O}(N_p)$ operations, where N_p is the number of scatterers. Next we sum over all scatterers to obtain

$$\tilde{\tilde{w}}^{\alpha}_{n'} = \sum_{m'=-\infty}^{\infty} H^{(1)}_{n'-m'}(k|\overline{\rho}_{o\beta} - \overline{\rho}_{o\alpha}|) \exp\left[-i(n'-m')\phi_{\overline{\rho_{o\beta}\rho_{o\alpha}}}\right] \tilde{w}^{\beta}_{m'} \quad (9.4.8)$$

Because the points $\overline{\rho}_{o\alpha}$ and $\overline{\rho}_{o\beta}$ are equally spaced on the uniform grid, this process can be performed simultaneously for all α and β combinations via the use of 2-D FFTs. This step requires $\mathcal{O}(4N_g \log N_g)$ operations. Finally, we perform post-multiplication to complete the indirect calculation so that

$$w^{\alpha\beta}_{n} = \sum_{n'=-\infty}^{\infty} J_{n-n'}(k|\overline{\rho}_{o\alpha} - \overline{\rho}_{\alpha}|) \exp\left[-i(n-n')\phi_{\overline{\rho_{o\alpha}\rho_{\alpha}}}\right] \tilde{\tilde{w}}^{\alpha}_{n'} \quad (9.4.9)$$

The final step requires $\mathcal{O}(N_p)$ operations. In this way, we obtain the N_p^2 combinations of $w^{\alpha\beta}_n$ simultaneously. Note that if the scatterers are of small to moderate size, only a few orders in the T-matrix coefficients are needed. If the center of the cylinder is not too far from the grid point, only a few terms in the translational formula are needed in the above infinite series. The fact that a majority of the interactions are weak and can be evaluated simultaneously using FFTs means that this algorithm is efficient when the number of scatterers is large, for example, on the order of several thousand. We define a neighborhood distance r_d. For a distance between scatterers less than the neighborhood distance, we exercise the direct calculation corresponding to the strong interaction matrix $\overline{\overline{Z}}^{S}$. For a distance between scatterers larger than r_d, we exercise the indirect calculation corresponding to the weak interaction $\overline{\overline{Z}}^{W}$ utilizing FFTs.

4.3 Numerical Results of Scattering and CPU Comparisons

Figure 9.4.3 shows the distribution of $N_p = 4000$ cylinders with a total area of 40% of a unit square box. This is generated by discretizing the region with a 64×64 grid ($N_g = 4096$) where a cylinder is assigned to each grid point. We then randomly removed 96 cylinders from the region. The remaining 4000 cylinders are allowed to move away from the grid point with

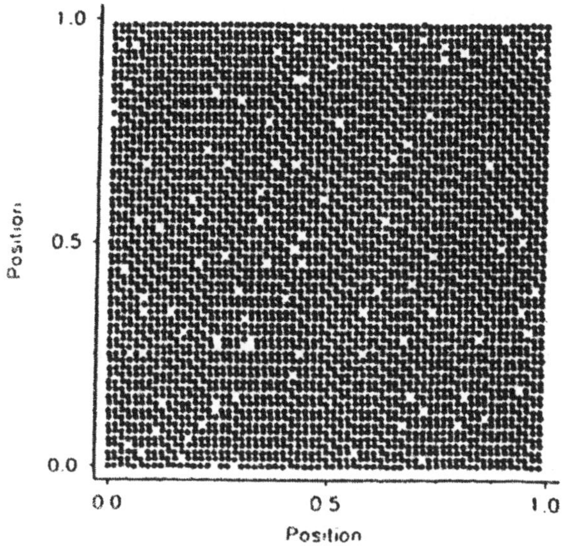

Figure 9.4.3 4000 cylinders randomly placed in a unit square box.

a random displacement between 0 and 50% of the radius of the cylinder. The average distance from the grid point for all the cylinders is 0.27% of the radius. The cylinder has a complex dielectric constant of 20+i3 with a radius of 1/20 of a free-space wavelength (λ_o). The total area occupied by the square box is scaled to $78.54\lambda_o^2$. The transverse electric (TE) case is considered with the angle of incidence $\phi = 30°$. Three cylindrical modes are needed to represent the scattered field of each cylinder; that is, $n = -1, 0,$ and 1 in (9.4.5). Therefore, we have a total of 12,000 unknowns. We vary the summation indices (m) in (9.4.5) and (9.4.7)–(9.4.9) until the solution converges. We also vary the neighborhood distance r_d to observe the changes in the solution. In (9.4.4), $\overline{w}^\alpha$ is calculated for $\alpha = 1, 2,$ and 3 whose centers are located at $(x, y) = (-0.0004, -0.0011), (0.2496, 0.2498)$, and $(0.5161, 0.5164)$ in Fig. 9.4.3, respectively. For the indirect summation of (9.4.7) using the SMCG method, we need to identify the neighboring cylinders associated with each cylinder, to compute the Bessel functions and Hankel functions, and to perform the convolutions using FFTs. We compare both the accuracy and the CPU time of the SMCG and those of direct summation for one iteration. For the direct summation, we extrapolate the CPU for the 4000 cylinders from that of the first 40 multiplied by 100. Table 9.4.1 shows the accuracy of the SMCG method versus direct summation, and Table 9.4.2 shows the

Cylinder			Sparse-Marix Canonical-Grid Method		
α		Direct Summation	$m = 3$, $n_d = 1$	$m = 2$, $n_d = 0$	$m = 1$, $n_d = 0$
1	$n = -1$	$-307.618 + i309.548$	$-307.618 + i309.550$	$-307.478 + i309.747$	$-294.812 + i310.465$
	0	$442.489 + i111.032$	$442.489 + i111.032$	$442.497 + i111.013$	$443.113 + i111.381$
	1	$-127.204 - i452.282$	$-127.204 - i452.848$	$-127.023 - i452.967$	$-141.035 - i459.339$
2	$n = -1$	$-181.995 + i295.316$	$-181.995 + i295.316$	$-181.997 + i295.355$	$-178.900 + i297.484$
	0	$350.850 + i28.2337$	$350.850 + i28.2336$	$350.853 + i28.2186$	$351.379 + i28.3405$
	1	$-146.117 - i322.364$	$-146.116 - i322.363$	$-146.148 - i322.377$	$-150.337 - i321.097$
3	$n = -1$	$162.856 + i155.520$	$162.856 + i155.519$	$162.932 + i155.537$	$162.169 + i160.916$
	0	$42.4817 - i219.952$	$42.4816 - i219.951$	$42.4738 - i219.951$	$42.5461 - i220.327$
	1	$-217.190 + i77.7958$	$-217.191 + i77.7962$	$-217.269 + i77.7984$	$-216.326 + i71.9728$

Table 9.4.1 Accuracy of the Sparse-Matrix Canonical-Grid Method Versus Direct Summation.

	Sparse-Marix Canonical-Grid Method			
	$m = 3$, $n_d = 2$	$m = 2$, $n_d = 1$	$m = 1$, $n_d = 1$	Direct Sum.
Neighborhood sorting time	23	0	0	0
Bessel function	3	2	2	0
Hankel function	31	21	11	0
Strong int.	46	0	0	28,700
Weak int. pre-multipl.	1	1	1	0
Weak int. convolution	58	30	11	0
Weak int. post-multipl.	1	1	1	0
Total CPU (second)	163	55	26	28,700

Table 9.4.2 CPU time in seconds (based on a Sun SPARC 10) of the sparse-matrix canonical-grid method vs. direct summation.

comparison of CPU time. It is observed that the SMCG can be faster by two orders of magnitude for the example presented in this section. This method can be extended to multiple scattering of many spheres.

REFERENCES AND ADDITIONAL READINGS

Allen, M. P. and D. J. Tildesley (1989), *Computer Simulation of Liquids*, Oxford University Press, New York.

Bottcher, C. J. F. (1952), *Theory of Electric Polarization*, Elsevier, Amsterdam.

Chan, C. H. and L. Tsang (1995), A sparse-matrix canonical-grid method for scattering by many scatterers, *Microwave Opt. Technol. Lett.*, **8(2)**, 114–118.

Chew, W. C. (1990), *Waves and Fields in Inhomogeneous Media*, Van Nostrand Reinhold, New York.

Chew, W. C., J. H. Lin, and X. G. Yang (1995), An FFT T-matrix method for 3D microwave scattering solution from random discrete scatterers, *Microwave Opt. Technol. Lett.*, **9**, 194–196.

Colbeck, S. C. (1982), An overview of snow metamorphism, *Reviews of Phys. and Space Phys.*, **20(1)**, 45–61.

Ding, K. H., L. M. Zurk, and L. Tsang (1994), Pair distribution functions and attenuation rates for sticky particles in dense media, **8(12)**, 1585–1604.

Frenkel, D. and B. Smit (1996), *Understanding Molecular Simulation: From Algorithms to Applications*, Academic Press, San Diego.

Hansen, J. P. and I. R. McDonald (1986), *Theory of Simple Liquids*, Academic Press, New York.

Hallikainen, M. T., F. T. Ulaby, and T. E. V. Deventer (1987), Extinction behavior of dry snow in the 18- to 90-GHz range, *IEEE Trans. Geosci. Remote Sens.*, **25(6)**, 737–750.

Ishimaru, A. and Y. Kuga (1982), Attenuation constant of a coherent field in a dense distribution of particles, *J. Opt. Soc. Am.*, **72**, 1317–1320.

Jin, Y.-Q. (1994), *Electromagnetic Scattering Modelling for Quantitative Remote Sensing*, World Scientific, London.

Koh, G. (1992), Experimental study of electromagnetic wave propagation in dense random media, *Waves in Random Media*, **2**, 39–48.

Kranendonk, W. G. T. and D. Frenkel (1988), Simulation of the adhesive-hard-sphere model, *Molecular Phys.*, **64(3)**, 403–424.

Kuga, Y., F. T. Ulaby, T. F. Haddock, and R. D. DeRoo (1991), Millimeter-wave radar scattering from snow: I. Radiative transfer model, *Radio Sci.*, **26(2)**, 239–341.

Lado, F (1968), Equation of state of the hard-disk fluid from approximate integral equations, *J. Chem. Phys.*, **49(7)**, 3092–3096.

Lu, C. C., W. C. Chew, and L. Tsang (1995), The application of recursive aggregate T-matrix algorithm in the Monte Carlo simulations of the extinction rate of random distribution of particles, *Radio Sci.*, **30(1)**, 25–28.

Mandt, C., Y. Kuga, L. Tsang, and A. Ishimaru (1992), Microwave propagation and scattering in a dense distribution of spherical particles: experiment and theory, *Waves in Random Media*, **2(3)**, 225–234.

Maxwell-Garnett, J. C. (1904), Colours in metal glasses and in metallic films, *Trans. Roy. Soc. London*, **203**, 385–420.

Metropolis, N., A. W. Rosenbluth, N. Rosenbluth, A. H. Teller, and E. Teller (1953), Equation of state calculation by fast computing machines, *J. Chem. Phys.*, **21(6)**, 1087–1092.

Polder, D. and J. H. van Santern (1946), The effective permeability of mixture of solids, *Physica*, **12**, 257–271.

Ripoll, M. S. and C.F. Tejero (1995), Approximate analytical expression for the direct correlation function of hard discs within the Percus–Yevick equation, *Molecular Phys.*, **85(2)**, 423–428.

Rosenfeld, S. and N. C. Grody (2000), Metamorphic signature of snow revealed in SSM/I measurements, *IEEE Trans. Geosci. Remote Sens.*, **38(1)**, 53–63.

Sarabandi, K. and P. Siqueira (1997), Numerical scattering analysis for two-dimensional dense random media: Characterization of effective permittivity, *IEEE Trans. Antennas Propagat.*, **45(5)**, 858–867.

Seaton, N. A. and E. D. Glandt (1986), Monte Carlo simulation of adhesive disks, *J. Chem. Phys.*, **84(8)**, 4595–4601.

Seaton, N. A. and E. D. Glandt (1987), Monte Carlo simulation of adhesive spheres, *J. Chem. Phys.*, **87(3)**, 1785–1790.

Siqueira, P. and K. Sarabandi (1996), Method of moments evaluation of the two-dimensional quasicrystalline approximation, *IEEE Trans. Antennas Propagat.*, **44(8)**, 1067–1077.

Stell, G. (1991), Sticky spheres and related systems, *J. Statistical Phys.*, **63**, 1203–1221.

Tsang, L., C.-T. Chen, A. T. C. Chang, J. Guo, and K.-H. Ding (2000), Dense media radiative transfer based on quasicrystalline approximation with applications to passive microwave remote sensing of snow, *Radio Sci.*, **35(3)**, 731–749.

Tsang, L. and J. A. Kong (1992), Scattering of electromagnetic waves from a dense medium consisting of correlated Mie scatterers with size distributions and applications to dry snow, *J. Electromag. Waves and Appl.*, **6(3)**, 265–286.

Tsang, L., C. Mandt, and K. H. Ding (1992), Monte Carlo simulations of the extinction rate of dense media with randomly distributed dielectric spheres based on solution of Maxwell's equations, *Optics Lett.*, **17(5)**, 314–316.

West, R., D. Gibbs, L. Tsang, and A. K. Fung (1994), Comparison of optical scattering experiments and the quasicrystalline approximation for dense media, *J. Opt. Soc. Am. A*, **11(6)**, 1854–1858.

Veysoglu, M. E. and J. A. Kong (1996), Multi-scale correlation function for random medium models, *Progress in Electromag. Res.*, **PIER 14**, 279–315.

Zurk, L. M. (1995), Electromagnetic wave propagation and scattering in dense, discrete random media with application to remote sensing of snow, University of Washington, Seattle.

Chapter 10

DENSE MEDIA MODELS AND THREE-DIMENSIONAL SIMULATIONS

1 Introduction

In scattering of waves by densely packed media, the particles do not scatter independently. The correlated scattering between particles has to be taken into account. This was verified by controlled experiments [Ishimaru and Kuga, 1982; Mandt et al. 1992; West et al. 1994]. In Volume III, we study extensively the analytic theory of dense media scattering. These include the Feynman diagrammatic methods leading to the Dyson equation and the Bethe–Salpeter equation. To make analytic theory numerically tractable, approximations have to be made on the mass and intensity operators. In Volume III, we study the quasi-crystalline approximation (QCA), the quasi-crystalline approximation with coherent potential (QCA–CP), the correlated ladder approximation and the dense media radiative transfer theory (DMRT). These approaches take into account the pair-distribution functions of particle positions for which we have used the Percus–Yevick pair-distribution functions as illustrated in Chapter 8. Near-field coherent wave interactions are also included. In this chapter, we study Monte Carlo simulations of three-dimensional dense media. In Section 2, we first formulate a simple analytic model for scattering by dense media. In Section 3, we use a volume integral equation approach for Monte Carlo simulations of spheres and spheroids. In Section 4, we use the T-matrix approach. Scattering simulations include up to 5000 particles. We simulate the extinction rates and the phase matrices. The scattering results compare well with dense media analytic theory for the case of spheres. However, for complicated cases of densely packed media such as densely packed nonspherical particles, numerical simulations of Maxwell's equations become more attractive.

2 Simple Analytical Models For Scattering From a Dense Medium

2.1 Effective Permittivity

Consider a distribution of spheres with permittivity ϵ_s embedded in a background medium of permittivity ϵ. Let the medium have polarization $\overline{P}$ (Fig. 10.2.1).

We first determine the field created by polarization $\overline{P}$ inside a sphere on itself. Consider a single sphere with polarization $\overline{P} = \hat{z}P$ and let the permittivity outside the sphere be ϵ (Fig. 10.2.2).We next calculate the electric field due to the polarization $\overline{P}$. Let Φ_{out} and Φ_{in} be the scalar potentials

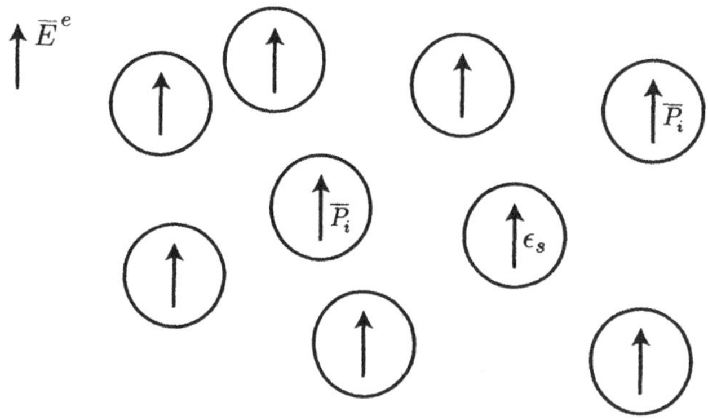

Figure 10.2.1 Collection of dielectric spheres.

outside and inside the sphere, respectively. Solving Laplace's equation

$$\Phi_{out} = \frac{A}{r^2}\cos\theta \tag{10.2.1}$$

$$\Phi_{in} = Br\cos\theta \tag{10.2.2}$$

The boundary conditions at $r = a$ are

$$\Phi_{out} = \Phi_{in}$$

$$-\epsilon\hat{n}\cdot\nabla\Phi_{out} = -\epsilon\hat{n}\cdot\nabla\Phi_{in} + \hat{n}\cdot\overline{P} \tag{10.2.3}$$

Thus with $\hat{n} = \hat{r}$ and $\hat{z} = \cos\theta\hat{r} - \sin\theta\hat{\theta}$

$$\frac{A}{a^2} = Ba \tag{10.2.4}$$

$$\epsilon\frac{2A}{a^3} = -\epsilon B + P \tag{10.2.5}$$

Solving (10.2.4) and (10.2.5) gives

$$B = \frac{P}{3\epsilon} \tag{10.2.6}$$

The electric field inside the sphere $\overline{E}_p$ due to this $\overline{P}$ is

$$\overline{E}_p = -\nabla\Phi_{in} = -B\hat{z} = -\frac{\overline{P}}{3\epsilon} \tag{10.2.7}$$

Note that $\overline{E}_p$ is *not* the total field of the problem. It is only the field due to polarization $\overline{P}$ (Fig. 10.2.2). Next we consider the collection of induced dipoles shown in Fig. 10.2.1. Let the electric field inside the medium be $\overline{E}$. Consider the electric field $\overline{E}^e$ exciting a particle. It is equal to the electric field $\overline{E}$ minus the electric field created by $\overline{P}$. Thus

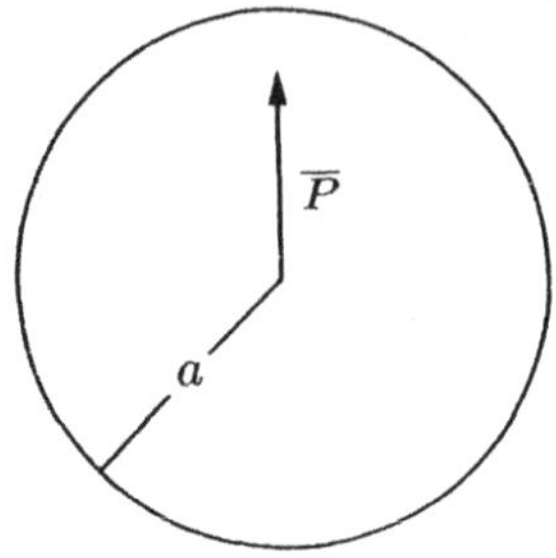

Figure 10.2.2 Polarization in a dielectric sphere.

$$\overline{E}^e = \overline{E} + \frac{\overline{P}}{3\epsilon} \tag{10.2.8}$$

Under the exciting electric field $\overline{E}^e$, the dipole moment induced inside the sphere is $\overline{p}$, with

$$\overline{p} = \alpha \overline{E}^e \tag{10.2.9}$$

where α is the polarizability of the sphere. The polarization $\overline{P}$ is the dipole moment per unit volume. Thus

$$\overline{P} = n_o \overline{p} \tag{10.2.10}$$

where n_o is the number of spheres per unit volume. Putting (10.2.8), (10.2.9), and (10.2.10) together gives

$$\overline{P} = n_o \alpha \overline{E}^e = n_o \alpha \left(\overline{E} + \frac{\overline{P}}{3\epsilon} \right) \tag{10.2.11}$$

Solving (10.2.11) gives

$$\overline{P} = \frac{n_o \alpha}{1 - \frac{n_o \alpha}{3\epsilon}} \overline{E} \tag{10.2.12}$$

The electric flux density $\overline{D}$ is

$$\overline{D} = \epsilon \overline{E} + \overline{P} \tag{10.2.13}$$

Putting (10.2.12) in (10.2.13) gives

$$\overline{D} = \epsilon \overline{E} + \frac{n_o \alpha \overline{E}}{1 - \dfrac{n_o \alpha}{3\epsilon}} = \epsilon \left[\frac{1 + \dfrac{2 n_o \alpha}{3\epsilon}}{1 - \dfrac{n_o \alpha}{3\epsilon}} \right] \overline{E} \tag{10.2.14}$$

The effective permittivity is ϵ_{eff} such that

$$\overline{D} = \epsilon_{\text{eff}} \overline{E} \tag{10.2.15}$$

Comparing (10.2.14) and (10.2.15) gives

$$\epsilon_{\text{eff}} = \epsilon \left[\frac{1 + \dfrac{2n_o\alpha}{3\epsilon}}{1 - \dfrac{n_o\alpha}{3\epsilon}} \right] \tag{10.2.16}$$

Next we determine the polarizability α of a sphere of radius a and permittivity ϵ_p, consider $\overline{E}^e = E^e\hat{z}$ to be the exciting field upon the sphere. Let Φ_{out} and Φ_{in} be the scalar potentials outside and inside the sphere, respectively. Solving Laplace's equation by letting

$$\Phi_{out} = -E^e r\cos\theta + \frac{A}{r^2}\cos\theta \tag{10.2.17}$$

$$\Phi_{in} = Br\cos\theta \tag{10.2.18}$$

The boundary conditions are the continuity of the potential and the continuity of the normal component of the displacement. At $r = a$

$$\Phi_{out} = \Phi_{in} \tag{10.2.19}$$

$$\epsilon\frac{\partial\Phi_{out}}{\partial r} = \epsilon_p\frac{\partial\Phi_{in}}{\partial r} \tag{10.2.20}$$

Using (10.2.17)–(10.2.18) in (10.2.19)–(10.2.20) gives

$$-E^e a + \frac{A}{a^2} = Ba \tag{10.2.21}$$

$$-\epsilon E^e - \frac{2A}{a^3}\epsilon = \epsilon_p B \tag{10.2.22}$$

Solving (10.2.21) and (10.2.22) gives

$$B = -\frac{3\epsilon E^e}{\epsilon_p + 2\epsilon} \tag{10.2.23}$$

Thus the induced electric field inside the particle is

$$\overline{E}_{in} = -\nabla\Phi_{in} = \frac{3\epsilon\overline{E}^e}{\epsilon_p + 2\epsilon} \tag{10.2.24}$$

The induced dipole moment inside the particle will be

$$\overline{p} = v_o\overline{P}_{in} \tag{10.2.25}$$

where $\overline{P}_{in}$ is the induced polarization inside the particle. Thus

$$\overline{P}_{in} = (\epsilon_p - \epsilon)\overline{E}_{in} \tag{10.2.26}$$

Using (10.2.24)–(10.2.26)

$$\overline{p} = \frac{v_o(\epsilon_p - \epsilon)3\epsilon}{\epsilon_p + 2\epsilon}\overline{E}^e \tag{10.2.27}$$

Comparing (10.2.27) and (10.2.9) gives the polarizability

$$\alpha = 3\epsilon v_o \frac{(\epsilon_p - \epsilon)}{\epsilon_p + 2\epsilon} \tag{10.2.28}$$

Putting α from (10.2.28) in (10.2.16) gives the effective permittivity of a medium consisting of particles of permittivity ϵ_p embedded in a background medium of permittivity ϵ.

$$\epsilon_{\text{eff}} = \epsilon \left(\frac{1 + 2fy}{1 - fy} \right) \tag{10.2.29}$$

where

$$f = n_o v_o \tag{10.2.30}$$

is the fractional volume occupied by the particles and

$$y = \frac{\epsilon_p - \epsilon}{\epsilon_p + 2\epsilon} \tag{10.2.31}$$

Equation (10.2.16) is known as the Clausius-Mossoti formula or the Lorentz-Lorenz formula. On the other hand, (10.2.29) is known as the Maxwell-Garnett mixing formula. The Maxwell-Garnett mixing formula can be put in the following symmetric form

$$\frac{\epsilon_{\text{eff}} - \epsilon}{\epsilon_{\text{eff}} + 2\epsilon} = f \frac{\epsilon_p - \epsilon}{\epsilon_p + 2\epsilon} \tag{10.2.32}$$

which is also known as the Rayleigh mixing formula. Given the effective permittivity, we can also calculate the effective propagation constant K with

$$K = \omega \sqrt{\mu \epsilon_{\text{eff}}} \tag{10.2.33}$$

so that

$$K^2 = k^2 \frac{1 + 2fy}{1 - fy} \tag{10.2.34}$$

However the imaginary part of the effective propagation constant K in (10.2.33) only accounts for absorption. To include the attenuation due to scattering, we have to take into account incoherent scattering which we will study in the next section.

2.2 Scattering Attenuation and Coherent Propagation Constant

In this section, the coherent propagation constant is calculated by including the attenuation due to scattering. The scattered intensity can be attributed to the radiation of the induced dipole that is induced by the exciting field.

From (10.2.8) and (10.2.12)

$$\overline{E}^e = \frac{\overline{E}}{1 - \frac{n_o\alpha}{3\epsilon}} = \frac{\overline{E}}{1 - fy} \tag{10.2.35}$$

From (10.2.9) and (10.2.35), we have

$$\overline{p} = \frac{\alpha\overline{E}}{1 - fy} = \frac{3\epsilon v_o y \overline{E}}{1 - fy} \tag{10.2.36}$$

Since the electric field $\overline{E}$ is in the medium that has effective wavenumber K. Let $K_r = \text{Re}(K)$. Thus we can write

$$\overline{E} = E_o \hat{e} e^{i\overline{K}_r \cdot \overline{r}} \tag{10.2.37}$$

where $\hat{e}$ is the polarization of the electric field and $\overline{K}_r$ is the wavevector with $|\overline{K}_r| = K_r$. At the position $\overline{r}_i$ of the ith particle, we have induced dipole moment

$$\overline{p}_i = \frac{3v_o\epsilon y}{1 - fy} E_o\, e^{i\overline{K}_r \cdot \overline{r}_i} \hat{e} \tag{10.2.38}$$

The radiation field of the ith particle is

$$\overline{E}_{si}(\overline{r}) = \frac{\omega^2 \mu e^{ikR_i}}{4\pi R_i} (\hat{R}_i \times \overline{p}_i) \times \hat{R}_i \tag{10.2.39}$$

and $\overline{R}_i = \overline{r} - \overline{r}_i$ is the vector pointing from the ith dipole to the observation point.

In the far field direction $\hat{k}_s$, we have $|\overline{R}_i| = r - \hat{k}_s \cdot \overline{r}_i$ and using (10.2.38) and (10.2.39)

$$\overline{E}_{si}(\overline{r}) = \overline{A} J_i \tag{10.2.40}$$

where

$$J_i = \exp(i\overline{\alpha} \cdot \overline{r}_i) \tag{10.2.41}$$

$$\overline{\alpha} = \overline{K}_r - \overline{k}_s \tag{10.2.42}$$

and

$$\overline{A} = \frac{k^2 a^3 e^{ikr}}{r} (\hat{k}_s \times \hat{e}) \times \hat{k}_s \frac{y}{1 - fy} E_o \tag{10.2.43}$$

is a vector independent of i. The total scattered field is the sum of the radiation fields of all the dipoles. Hence,

$$\overline{E}_s = \sum_{i=1}^{N} \overline{A} J_i \tag{10.2.44}$$

The radiation intensity I_s is

$$I_s = \frac{|\overline{E}_s|^2}{2\eta} = \frac{|\overline{A}|^2}{2\eta}\left\{\sum_{i=1}^{N}|J_i|^2 + \sum_{i=1}^{N}\sum_{\substack{j=1\\ j\neq i}}^{N} J_i J_j^*\right\}$$

$$= \frac{|\overline{A}|^2}{2\eta}\left\{N + \sum_{i=1}^{N}\sum_{\substack{j=1\\ j\neq i}}^{N} J_i J_j^*\right\} \tag{10.2.45}$$

We decompose I_s into coherent and incoherent intensity,

$$\langle \mathcal{I}_s\rangle = \langle I_s\rangle - \frac{|\langle \overline{E}_s\rangle|^2}{2\eta} = \frac{\langle|\overline{E}_s|^2\rangle}{2\eta} - \frac{|\langle \overline{E}_s\rangle|^2}{2\eta} \tag{10.2.46}$$

$$\langle I_s\rangle = \frac{|\overline{A}|^2}{2\eta}\left\{N + \sum_{i=1}^{N}\sum_{\substack{j=1\\ j\neq i}}^{N} \langle J_i J_j^*\rangle\right\} \tag{10.2.47}$$

Calculating the configuration average of the second term in (10.2.47) gives

$$\langle I_s\rangle = \frac{|\overline{A}|^2}{2\eta}\left\{N + N(N-1)\int d\overline{r}_i \int d\overline{r}_j p_2(\overline{r}_i, \overline{r}_j) e^{i\overline{\alpha}\cdot(\overline{r}_i - \overline{r}_j)}\right\} \tag{10.2.48}$$

where $p_2(\overline{r}_i, \overline{r}_j)$ is the joint probability density function of particles located at $\overline{r}_i$ and $\overline{r}_j$. On the other hand, the coherent field is, from (10.2.44)

$$\langle \overline{E}_s\rangle = N\overline{A}\int d\overline{r}_i p(\overline{r}_i) e^{i\overline{\alpha}\cdot\overline{r}_i} \tag{10.2.49}$$

Letting

$$p(\overline{r}_i) = \frac{1}{V} \tag{10.2.50}$$

$$p_2(\overline{r}_i, \overline{r}_j) = \frac{g(\overline{r}_i, \overline{r}_j)}{V^2} \tag{10.2.51}$$

where $g(\overline{r})$ is the pair distribution function, we have

$$\langle \overline{E}_s\rangle = n_o\overline{A}\int d\overline{r}_i e^{i\overline{\alpha}\cdot\overline{r}_i} \tag{10.2.52}$$

$$\langle I_s\rangle = \frac{|\overline{A}|^2}{2\eta}\left\{N + n_o^2\int d\overline{r}_i \int d\overline{r}_j g(\overline{r}_i - \overline{r}_j) e^{i\overline{\alpha}\cdot(\overline{r}_i - \overline{r}_j)}\right\} \tag{10.2.53}$$

Coherent scattered intensity is, from (10.2.52)

$$\frac{|\langle \overline{E}_s\rangle|^2}{2\eta} = \frac{n_o^2|\overline{A}|^2}{2\eta}\int d\overline{r}_i \int d\overline{r}_j e^{i\overline{\alpha}\cdot(\overline{r}_i - \overline{r}_j)} \tag{10.2.54}$$

By taking the difference of (10.2.53) and (10.2.54), we have the incoherent intensity as

$$\begin{aligned}\langle \mathcal{I}_s \rangle &= \frac{|\overline{A}|^2}{2\eta} \left\{ N + n_o^2 \int d\overline{r}_i \int d\overline{r}_j e^{i\overline{\alpha}\cdot(\overline{r}_i - \overline{r}_j)} \Big(g(\overline{r}_i - \overline{r}_j) - 1 \Big) \right\} \\ &= \frac{|\overline{A}|^2}{2\eta} \left\{ N + V n_0^2 \int d\overline{r} e^{i\overline{\alpha}\cdot\overline{r}} \Big(g(\overline{r}) - 1 \Big) \right\} \\ &= n_o \frac{|\overline{A}|^2}{2\eta} V \left\{ 1 + n_o \int d\overline{r} e^{i\overline{\alpha}\cdot\overline{r}} \Big(g(\overline{r}) - 1 \Big) \right\} \end{aligned} \tag{10.2.55}$$

We further assume spherical symmetry of pair distribution functions.

$$g(\overline{r}) = g(r) \tag{10.2.56}$$

Also

$$g(r) = 0 \qquad \text{for } r < b = 2a$$

As $r \to \infty$

$$g(r) = 1 \tag{10.2.57}$$

As shown in Chapter 8, $g(r)$ is practically equal to unity for r larger than a few diameters. We assume that $ka < 1$. However, if, say $ka = 0.2$, $3kb = 1.2$ is not much less than unity. Thus, we may need to keep the $e^{i\overline{\alpha}\cdot\overline{r}}$ term in (10.2.55) Let

$$H(\overline{p}) = \frac{1}{(2\pi)^3} \int_{-\infty}^{\infty} d\overline{r} (g(\overline{r}) - 1) e^{-i\overline{p}\cdot\overline{r}} \tag{10.2.58}$$

be the structure factor. Then

$$\langle \mathcal{I}_s \rangle = n_o \frac{|A|^2}{2\eta} V \left\{ 1 + (2\pi)^3 n_o H(\overline{k}_s - \overline{K}_r) \right\} \tag{10.2.59a}$$

Let $\hat{e}$ be in $\hat{y}$ direction and $\overline{K}_r$ be in $\hat{z}$ direction, and

$$\hat{k}_s = \sin\theta_s \cos\phi_s \hat{x} + \sin\theta_s \sin\phi_s \hat{y} + \cos\theta_s \hat{z}$$

Then

$$\left| (\hat{k}_s \times \hat{e}) \times \hat{k}_s \right|^2 = (\sin^2\theta_s \cos^2\phi_s + \cos^2\theta_s)$$

The total incoherent scattered power is obtained by integration of the incoherent intensity over 4π solid angles.

$$\begin{aligned} P_s &= \int_{4\pi} d\Omega_s r^2 \langle \mathcal{I}_s \rangle \\ &= \int_0^{\pi} d\theta_s \sin\theta_s \int_0^{2\pi} d\phi_s k^4 a^6 \left| \frac{y}{1 - fy} \right|^2 \frac{E_o^2}{2\eta} \\ &\cdot n_o V \left\{ 1 + (2\pi)^3 n_o H(\overline{k}_s - K_r \hat{z}) \right\} (\sin^2\theta_s \cos^2\phi_s + \cos^2\theta_s) \end{aligned} \tag{10.2.59b}$$

Since by spherical symmetry

$$H\left(\overline{k}_s - K_r\hat{z}\right) = H\left(|\overline{k}_s - K_r\hat{z}|\right) = H\left(\sqrt{k^2\sin^2\theta_s + (k\cos\theta_s - K_r)^2}\right) \tag{10.2.60}$$

is independent of ϕ_s. Then

$$P_s = 2\pi\int_0^{\pi} d\theta_s \sin\theta_s \left(\frac{\sin^2\theta_s}{2} + \cos^2\theta_s\right) k^4 a^6 \left|\frac{y}{1-fy}\right|^2 \frac{E_o^2}{2\eta}$$
$$n_o V\left\{1 + (2\pi)^3 n_o H\left(|\overline{k}_s - K_r\hat{z}|\right)\right\} \tag{10.2.61a}$$

Suppose we have a cylindrical volume of area A and length l containing n_o particles per unit volume. The input power due to the wave in (10.2.37) is

$$P_{in} = \frac{1}{2}E_0^2 A\left(\frac{\epsilon_{\text{eff}}}{\mu}\right)^{\frac{1}{2}} \tag{10.2.61b}$$

The number of particles in the cylindrical volume is $N = n_o Al$. Let κ_s denote the scattering rate and κ_a denote the absorption rate. The scattering attenuation rate denoted by κ_s is given by

$$\kappa_s = \frac{P_s}{lP_{in}} = \frac{P_s}{\frac{1}{2}E_o^2 V\left(\frac{\epsilon_{\text{eff}}}{\mu}\right)^{\frac{1}{2}}} \tag{10.2.61c}$$

Putting (10.2.61a) into (10.2.61c) gives and using $\sqrt{\frac{\mu}{\epsilon_{\text{eff}}}} = \frac{k}{K_r}$, $f = 4\pi n_o a^3/3$

$$\kappa_s = \frac{3k^5a^3}{2K_r} f\left|\frac{y}{1-fy}\right|^2 \int_0^{\pi} d\theta_s \sin\theta_s\left(\frac{\sin^2\theta_s}{2} + \cos^2\theta_s\right)$$
$$\left\{1 + (2\pi)^3 n_o H\left(|\overline{k}_s - K_r\hat{z}|\right)\right\} \tag{10.2.62}$$

For the special case that the correlation is short ranged so that

$$H\left(|\overline{k}_s - K_r\hat{z}|\right) \simeq H(0) = \frac{1}{(2\pi)^3}\int_{-\infty}^{+\infty} d\overline{r}\,(g(r) - 1)$$
$$= \frac{1}{(2\pi)^3}\left\{-\frac{4\pi}{3}b^3 + 4\pi b^3 H_0\right\} \tag{10.2.63a}$$

where

$$H_0 = \int_1^{\infty} d\sigma\sigma^2\,(g(\sigma b) - 1) \tag{10.2.63b}$$

For the short ranged case, putting (10.2.63a) in (10.2.62), we can integrate over θ_s to get

$$\kappa_s = \frac{2k^5a^3}{K_r} f\left|\frac{y}{1-fy}\right|^2 (1 - 8f + 24fH_0) \tag{10.2.64a}$$

The scattering attenuation can be added to the effective propagation constant of (10.2.34) as follows

$$K^2 \cong K_r^2 + iK_r(2K_i) = K_r^2 + iK_r(\kappa_a + \kappa_s) = K_r^2 + iK_r\kappa_a + iK_r\kappa_s$$
$$= (K^2)_{\text{absorption only}} + iK_r\kappa_s \quad (10.2.64b)$$

Using (10.2.62), (10.2.64*a*), and (10.2.34)

$$K^2 = k^2\frac{(1+2fy)}{1-fy} + i\frac{3}{2}k^5a^3f\left|\frac{y}{1-fy}\right|^2 \times \int_0^\pi d\theta_s \sin\theta_s$$
$$\cdot\left(\frac{\sin\theta_s}{2} + \cos^2\theta_s\right)\left\{1 + (2\pi)^3 n_o H\left(|\overline{k}_s - K_r\hat{z}|\right)\right\} \quad (10.2.65a)$$

Equation (10.2.65*a*) is the general result for small particle but longer ranged pair distribution function. If the pair distribution function is short ranged, we have

$$K^2 = k^2\frac{(1+2fy)}{1-fy} + i2fk^5a^3\left|\frac{y}{1-fy}\right|^2(1 - 8f + 24fH_0) \quad (10.2.65b)$$

2.3 Coherent Reflection and Incoherent Scattering From a Half-Space of Scatterers

Consider a plane wave incident on a half-space of dielectric scatterers (Fig. 10.2.3). We shall first consider vertically polarized incidence. The case of horizontally polarized incidence can be treated in a similar manner. The incident wave is in the direction $(\theta_i, \phi_i)$and the incident field is $\overline{E}_{inc} = \hat{\theta}_{id}E_o\exp(i\overline{k}_{id}\cdot\overline{r})$ with $\hat{k}_{id} = \sin\theta_i\cos\phi_i\hat{x} + \sin\theta_i\sin\phi_i\hat{y} - \cos\theta_i\hat{z}$, $\hat{\phi}_{id} = -\sin\phi_i\hat{x} + \cos\phi_i\hat{y}$, and $\hat{\theta}_{id} = \hat{\phi}_{id}\times\hat{k}_{id}$ forming the orthonormal triad. The subscript d is to denote the fact that the wave is propagating downward. In the lower region, the coherent effective propagation constant is K. The transmitted macroscopic field is $\overline{E}_t$ with propagation direction following Snell's law is:

$$\overline{E}_t = \hat{\theta}_{td}T_{01}^{TM}\frac{k}{K}e^{i\overline{K}_d\cdot\overline{r}} \quad (10.2.66)$$

with K given by (10.2.65), $\overline{K}_d = k\sin\theta_i\cos\phi_i\hat{x} + k\sin\theta_i\sin\phi_i\hat{y} - \sqrt{K^2 - k^2\sin^2\theta_i}\hat{z} = K_x\hat{x} + K_y\hat{y} - K_z\hat{z}$, $\hat{\theta}_{td} = -\cos\theta_t\cos\phi_i\hat{x} - \cos\theta_t\sin\phi_i\hat{y} - \sin\theta_t\hat{z}$ and θ_t is the transmitted angle obeying Snell's law

$$\sin\theta_t = \frac{k}{K_r}\sin\theta_i \quad (10.2.67)$$

The Fresnel transmission coefficient for TM waves is T_{01}^{TM} for the magnetic

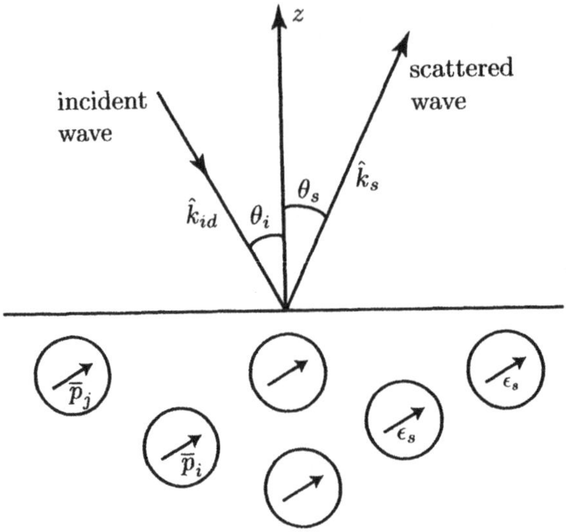

Figure 10.2.3 Plane wave incident on a half-space of dielectric scatterers. Induced dipoles.

field with

$$T_{01}^{TM} = \frac{2K^2 k_{iz}}{K^2 k_{iz} + k^2 K_z} \tag{10.2.68}$$

where $k_{iz} = k\cos\theta_i$ and $K_z = \sqrt{K^2 - k^2\sin^2\theta_i}$.

Dipoles are induced in the dielectric spheres and the dipole moment $\overline{p}_j$ of the jth particle, centered at $\overline{r}_j$ is, using (10.2.38),

$$\overline{p}_j = \hat{\theta}_{td}\frac{3v_o\epsilon y}{1 - fy}T_{01}^{TM}\frac{k}{K}e^{i\overline{K}_d\cdot\overline{r}} \tag{10.2.69}$$

The total radiation field of the dipoles is then

$$\overline{E}_s(\overline{r}) = \sum_{j=1}^{N}\overline{W}J_j \tag{10.2.70}$$

where $J_j = \exp\left[i(\overline{K}_d - \overline{k}_s)\cdot\overline{r}_j\right]$,

$$\begin{aligned}\overline{W} &= \frac{\omega^2\mu e^{ikr}}{4\pi r}(\hat{k}_s \times \hat{\theta}_{td}) \times \hat{k}_s\frac{3v_o\epsilon y}{1 - fy}T_{01}^{TM}\frac{k}{K} \\ &= \frac{k^2 a^3 e^{ikr}}{r}(\hat{k}_s \times \hat{\theta}_{td}) \times \hat{k}_s\frac{y}{1 - fy}T_{01}^{TM}\frac{k}{K}\end{aligned} \tag{10.2.71}$$

$\hat{k}_s = \sin\theta_s\cos\phi_s\hat{x} + \sin\theta_s\sin\phi_s\hat{y} + \cos\theta_s\hat{z} = k_{sx}\hat{x} + k_{sy}\hat{y} + k_{sz}\hat{z}$ is the propagation direction of the scattered wave with $\overline{k}_s = k\hat{k}_s$.

Coherent Reflected Field

To calculate the coherent scattered field, the configuration average of (10.2.71) is taken

$$\langle \overline{E}_s(\overline{r}) \rangle = \overline{W} \sum_{j=1}^{N} \langle J_j \rangle = \overline{W} N \langle J_j \rangle \tag{10.2.72}$$

Since $p(\overline{r}_j) = \frac{1}{V}$, we have, on evaluating $\langle J_j \rangle$ by integrating over the lower half-space,

$$\langle J_j \rangle = \frac{1}{V} \int_{-\infty}^{\infty} dx_j \int_{-\infty}^{\infty} dy_j \int_{-\infty}^{0} dz_j e^{i(\overline{K}_d - \overline{k}_s)\cdot \overline{r}_j}$$

giving rise to $\delta(K_x - k_{sx})\,\delta(K_y - k_{sy})$ indicating that the coherent scattered field is in the specular direction.

$$\begin{aligned} \langle \overline{E}_s \rangle = n_o \overline{W} \frac{4\pi^2 i}{(K_z + k_{iz})} \delta(k \sin\theta_s \cos\phi_s - k \sin\theta_i \cos\phi_i) \\ \cdot\, \delta(k \sin\theta_s \sin\phi_s - k \sin\theta_i \sin\phi_i) \end{aligned} \tag{10.2.73}$$

For $\hat{k}_s$ in specular reflected direction

$$(\hat{k}_s \times \hat{\theta}_{td}) \times \hat{k}_s = -\cos(\theta_i + \theta_t)\hat{\theta}_i \tag{10.2.74a}$$

We make use of the relation of (10.2.34) so that

$$n_o \overline{W} = \frac{e^{ikr}}{4\pi r} \left(-\cos(\theta_i + \theta_t)\hat{\theta}_t \right) (K^2 - k^2) T_{01}^{TM} \frac{k}{K} \tag{10.2.74b}$$

Also $K^2 - k^2 = (K_z + k_{iz})(K_z - k_{iz})$, so that

$$\frac{4\pi^2 i n_o \overline{W}}{K_z + k_{iz}} = \frac{\pi i e^{ikr}}{r} \left(-\cos(\theta_i + \theta_t)\hat{\theta}_t \right) (K_z - k_{iz}) T_{01}^{TM} \frac{k}{K} \tag{10.2.74c}$$

We make use of the relation

$$(K_z + k_{iz}) \cos(\theta_i - \theta_t) = \frac{k_{iz} K^2 + k^2 K_z}{kK} \tag{10.2.74d}$$

so that

$$T_{01}^{TM} \frac{k}{K} = \frac{2k_{iz}}{(K_z + k_{iz}) \cos(\theta_i - \theta_t)} \tag{10.2.75a}$$

Also we have

$$\left(\frac{K_z - k_{iz}}{K_z + k_{iz}} \right) \frac{\cos(\theta_i + \theta_t)}{\cos(\theta_i - \theta_t)} = \frac{K^2 k_{iz} - k^2 K_z}{K^2 k_{iz} + k^2 K_z} \equiv R_{01}^{TM} \tag{10.2.75b}$$

Putting (10.2.75) into (10.2.74c) and (10.2.73), we get

$$\langle \overline{E}_s \rangle = -\hat{\theta}_i i 2\pi k_{iz} R_{01}^{TM} \frac{e^{ikr}}{r} \delta(k \sin\theta_s \cos\phi_s - k\sin\theta_i \cos\phi_i) \cdot \delta(k\sin\theta_s \sin\phi_s - k\sin\theta_i \sin\phi_i) \tag{10.2.76}$$

Thus the coherent reflected wave is in the specular direction, containing no depolarization and obeying the Fresnel reflection formula for TM waves.

The case of horizontally polarized incident waves is treated in a similar manner. For an incident field of $\overline{E}_{inc} = \hat{\phi}_i E_o \exp(i\overline{k}_{id} \cdot \overline{r})$. The result for the coherent reflected field is

$$\langle \overline{E}_s(\overline{r}) \rangle = -\hat{\phi}_i i 2\pi k_{iz} R_{01}^{TE} \frac{e^{ikr}}{r} \delta(k \sin\theta_s \cos\phi_s - k\sin\theta_i \cos\phi_i) \cdot \delta(k\sin\theta_s \sin\phi_s - k\sin\theta_i \sin\phi_i) \tag{10.2.77}$$

where

$$R_{01}^{TE} = \frac{k_{iz} - K_z}{k_{iz} + K_z} \tag{10.2.78}$$

is the Fresnel reflection coefficient for TE waves.

Incoherent Scattered Field

The incoherent scattered field $\overline{\mathcal{E}}_s$ is $\overline{\mathcal{E}}_s(\overline{r}) = \overline{E}_s(\overline{r}) - \langle \overline{E}_s(\overline{r}) \rangle$. Hence, the incoherent intensity is

$$\langle \overline{\mathcal{E}}_s(\overline{r}) \cdot \overline{\mathcal{E}}_s^*(\overline{r}) \rangle = |\overline{W}|^2 \left\{ \sum_{i=1}^{N} \langle |J_i|^2 \rangle + \sum_{i=1}^{N} \sum_{\substack{j=1 \\ i\neq j}}^{N} \langle J_i J_j^* \rangle - N^2 |\langle J_1 \rangle|^2 \right\} \tag{10.2.79}$$

Following a similar procedure as in Section 2.2, we have

$$\langle \overline{\mathcal{E}}_s^* \cdot \overline{\mathcal{E}}_s \rangle = |\overline{W}|^2 \left\{ \frac{n_o A_0}{2\mathrm{Im}(K_z)} + n_o^2 \int d\overline{r}_i \int d\overline{r}_j \, [g(\overline{r}_i - \overline{r}_j) - 1] \cdot e^{i\overline{K}_d \cdot \overline{r}_i - i\overline{K}_d^* \cdot \overline{r}_j} e^{-i\overline{k}_s \cdot \overline{r}_i + i\overline{k}_s \cdot \overline{r}_j} \right\} \tag{10.2.80}$$

where A_0 is the area of the target area. The integral in (10.2.80) can be carried out readily. Hence,

$$\langle \overline{\mathcal{E}}_s^*(\overline{r}) \cdot \overline{\mathcal{E}}_s(\overline{r}) \rangle = |\overline{W}|^2 \frac{n_o A_0}{2\mathrm{Im}(K_z)} \left\{ 1 + n_o \int d\overline{r} \, [g(\overline{r}) - 1] \, e^{i(\mathrm{Re}\overline{K}_d - \overline{k}_s)\cdot \overline{r}} \right\} \tag{10.2.81}$$

where for vertically polarized incidence, $\overline{W}$ is given by (10.2.71) and for horizontally polarized incidence

$$\overline{W} = \frac{\omega^2 \mu e^{ikr}}{4\pi r}(\hat{k}_s \times \hat{\phi}_i) \times \hat{k}_s \frac{3v_o \epsilon y}{1 - fy}(1 + R_{01}^{TE})$$

$$= \frac{k^2 a^3 e^{ikr}}{r}(\hat{k}_s \times \hat{\phi}_i) \times \hat{k}_s \frac{y}{1 - fy} \frac{2k_{iz}}{K_z + k_{iz}} \qquad (10.2.82a)$$

so that

$$n_o \overline{W} = \frac{e^{ikr}}{4\pi r}\left((\hat{k}_s \times \hat{\phi}_i) \times \hat{k}_s\right)(K^2 - k^2)\frac{2k_{iz}}{K_z + k_{iz}} \qquad (10.2.82b)$$

In backscattering direction,

$$\hat{k}_s \times (\hat{k}_s \times \hat{\phi}_i) = \hat{\phi}_s.$$

Thus

$$\langle|\overline{\mathcal{E}}_s(\overline{r})|^2\rangle = \left|\frac{(K_z - k_{iz})k_{iz}}{n_o}\right|^2 \frac{n_o A_0}{2\mathrm{Im}(K_z) r^2} \frac{1}{4\pi^2}$$

$$\cdot \left\{1 + n_o \int_{-\infty}^{\infty} d\overline{r}(g(\overline{r}) - 1)e^{i(\mathrm{Re}\overline{K}_d - \overline{k}_s)\cdot\overline{r}}\right\} \qquad (10.2.83)$$

The backscattering coefficient is

$$\sigma_{hh}(\theta_i) = \frac{4\pi r^2 \langle|\overline{\mathcal{E}}_s(\overline{r})|^2\rangle}{A_0}$$

so that

$$\sigma_{hh} = \left|\frac{(K_z - k_{iz})k_{iz}}{n_o}\right|^2 \frac{n_o}{2\pi \mathrm{Im}(K_z)} \left\{1 + n_o \int_{-\infty}^{\infty} d\overline{r}(g(\overline{r}) - 1)e^{i(\mathrm{Re}\overline{K}_d - \overline{k}_s)\cdot\overline{r}}\right\} \qquad (10.2.84)$$

For the case of vertical polarized incidence, we can perform similar calculations. In the backscattering direction, $\theta_s = \theta_i$ and $\phi_s = \pi + \phi_i$, $\hat{k}_s \times (\hat{k}_s \times \hat{\theta}_{td}) = -\cos(\theta_i - \theta_t)\hat{\theta}_s$. Hence, the backscattering coefficients are the same for σ_{vv} and σ_{hh}. We have

$$\sigma_{vv}(\theta_i) = \sigma_{hh}(\theta_i) = \left|\frac{(K_z - k_{iz})k_{iz}}{n_o}\right|^2 \frac{n_o}{2\pi \mathrm{Im}(K_z)}$$

$$\times \left\{1 + n_o \int_{-\infty}^{\infty} d\overline{r}\,[g(\overline{r}) - 1]\, e^{i(\mathrm{Re}\overline{K}_d - \overline{k}_s)\cdot\overline{r}}\right\} \qquad (10.2.85)$$

The results of coherent reflection in this section agree with the QCA results in the low frequency approximation. The bistatic intensity agrees with QCA combined with distorted Born approximation. The QCA approximation will be treated in Volume III.

2.4 A Simple Dense Media Radiative Transfer Theory

A simple dense media radiative transfer theory can be developed based on the derivations of Sections 2.1 and 2.2. The model is based on the assumption that the particles are small so that $ka \ll 1$. However because the particle positions can have long-range correlation (e.g., sticky particles), the correlation distance can be comparable to or larger than a wavelength. Specifically, say $ka = 0.2$ and the correlation distance is up to $5kb$, where $b = 2a$ is the diameter. Then $5kb = 10ka = 2$ is larger than unity.

First we calculate the effective propagation constant which is determined by (10.2.65a)

$$K^2 = k^2\frac{(1+2fy)}{1-fy} + i\frac{3}{2}k^5a^3f\left|\frac{y}{1-fy}\right|^2 \int_0^\pi d\theta \sin\theta \left(\frac{\sin^2\theta}{2} + \cos^2\theta\right)$$
$$\left\{1 + (2\pi)^3 n_o H(\sqrt{k^2\sin^2\theta + (k\cos\theta - K_r)^2})\right\} \tag{10.2.86}$$

The effective propagation constant is

$$K = K_r + iK_i \tag{10.2.87}$$

The extinction rate is

$$\kappa_e = 2K_i \tag{10.2.88}$$

The structure factor is

$$H(\overline{p}) = \frac{1}{(2\pi)^3}\int_{-\infty}^{\infty} d\overline{r}(g(\overline{r}) - 1)e^{-i\overline{p}\cdot\overline{r}} \tag{10.2.89}$$

Because of spherical symmetry, this can be expressed as a sine transform.

$$H(p) = H(\overline{p}) = \frac{1}{2\pi^2}\int_{-\infty}^{\infty} dr r^2(g(r) - 1)\frac{\sin pr}{pr} \tag{10.2.90}$$

In practice, it is often easier to calculate the structure factor than calculating the pair distribution function. The scattering rate is

$$\kappa_s = \frac{3}{2}\frac{k^5a^3}{K_r}f\left|\frac{y}{1-fy}\right|^2 \int_0^\pi d\theta \sin\theta \left(\frac{\sin^2\theta}{2} + \cos^2\theta\right)$$
$$\left\{1 + (2\pi)^3 n_o H\left(\sqrt{k^2\sin^2\theta + (k\cos\theta - K_r)^2}\right)\right\} \tag{10.2.91}$$

To calculate the phase function, we first consider scattering using the 1-2 system. Consider propagation in the $\hat{i}$ direction and incident polarization in the $\hat{1}_i$ or $\hat{2}_i$ as done in Chapter 1, Section 1.2 of Volume I. The scattered direction is $\hat{s} = \hat{k}_s$ which makes an angle Θ with the incident direction. The polarizations of the scattered wave are in the $\hat{1}_s$ and $\hat{2}_s$ directions. Then, we can apply the analysis as in Section 2.2. The polarization dependence of $(\hat{k}_s \times \hat{e}) \times \hat{k}_s$ of the scattered wave is that of Rayleigh polarization dependence

$$\langle |f_{11}|^2 \rangle = |f_0|^2 V \left[1 + (2\pi)^3 n_o H \left(\sqrt{k^2 \sin^2 \Theta + (k \cos \Theta - K_r)^2} \right) \right] \quad (10.2.92)$$

where V is the volume containing the scatterers and the number of particles in V is $n_o V$. In (10.2.92)

$$|f_0|^2 = n_o \frac{k^5 a^6}{K_r} \left| \frac{y}{1 - fy} \right|^2 \quad (10.2.93)$$

Following similar analysis

$$\langle |f_{12}|^2 \rangle = 0 \quad (10.2.94)$$

$$\langle |f_{21}|^2 \rangle = 0 \quad (10.2.95)$$

$$\langle |f_{22}|^2 \rangle = |f_0|^2 V \left[1 + (2\pi)^3 n_o H \left(\sqrt{k^2 \sin^2 \Theta + (k \cos \Theta - K_r)^2} \right) \right] \quad (10.2.96)$$

For the phase matrix in the $\hat{v}$, $\hat{h}$ system, results are similar to that of Rayleigh phase matrix except with the inclusion of the $1 + (2\pi)^3 n_o H$ factor. Thus, in the vertical and horizontal polarization system, for incident direction (θ', ϕ') and scattered direction (θ, ϕ) as done in Chapter 7, Section 2.1 of Volume I, we have the phase matrix

$$\overline{\overline{P}}(\theta, \phi; \theta', \phi') = \begin{bmatrix} P_{11} & P_{12} & P_{13} & 0 \\ P_{21} & P_{22} & P_{23} & 0 \\ P_{31} & P_{32} & P_{33} & 0 \\ 0 & 0 & 0 & P_{44} \end{bmatrix} \quad (10.2.97)$$

where

$$\begin{aligned} P_{11} &= w \left[\sin^2 \theta \sin^2 \theta' + 2 \sin \theta \sin \theta' \cos \theta \cos \theta' \cos(\phi - \phi') \right. \\ &\quad \left. + \cos^2 \theta \cos^2 \theta' \cos^2(\phi - \phi') \right] \end{aligned} \quad (10.2.98)$$

$$P_{12} = w \cos^2 \theta \sin^2(\phi - \phi') \quad (10.2.99)$$

$$\begin{aligned} P_{13} &= w \left[\cos \theta \sin \theta \sin \theta' \sin(\phi - \phi') \cos^2 \theta \right. \\ &\quad \left. \cdot \cos \theta' \sin(\phi - \phi') \cos(\phi - \phi') \right] \end{aligned} \quad (10.2.100)$$

$$P_{21} = w \cos^2 \theta' \sin^2(\phi - \phi') \quad (10.2.101)$$

$$P_{22} = w \cos^2(\phi - \phi') \quad (10.2.102)$$

$$P_{23} = - w \cos \theta' \sin(\phi - \phi') \cos(\phi - \phi') \quad (10.2.103)$$

$$\begin{aligned} P_{31} &= w \left[-2 \sin \theta \sin \theta' \cos \theta' \sin(\phi - \phi') \right. \\ &\quad \left. - 2 \cos \theta \cos^2 \theta' \cos(\phi - \phi') \sin(\phi - \phi') \right] \end{aligned} \quad (10.2.104)$$

$$P_{32} = 2w \cos \theta \sin(\phi - \phi') \cos(\phi - \phi') \quad (10.2.105)$$

$$P_{33} = w\left[\sin\theta\sin\theta'\cos(\phi-\phi') + \cos\theta\cos\theta'(\cos^2(\phi-\phi') - \sin^2(\phi-\phi'))\right] \quad (10.2.106)$$

$$P_{44} = w[\sin\theta\sin\theta'\cos(\phi-\phi') + \cos\theta\cos\theta'] \quad (10.2.107)$$

where

$$w = |f_0|^2\left[1 + (2\pi)^3 n_o H\left(\sqrt{k^2\sin^2\Theta + (k\cos\Theta - K_r)^2}\right)\right] \quad (10.2.108)$$

$$\cos\Theta = \cos\theta\cos\theta' + \sin\theta\sin\theta'\cos(\phi-\phi') \quad (10.2.109)$$

3 Simulations Using Volume Integral Equations

Simulations of particle positions by packing several thousands of particles randomly in a box was done in Chapter 8. In this section, we solve Maxwell's equations for scattering by these realizations of particle positions. The simulations are performed over many samples (realizations) and the scattering results are averaged over these realizations. The results give information about the collective scattering effects of many particles packed together.

In this section, we use volume integral equation to simulate the scattering by densely packed spheres. We still restrict our attention to the case of small particles. In Section 4, we shall use multipole expansions to study scattering by spheres.

3.1 Volume Integral Equation

Consider waves in an inhomogeneous medium with permittivity $\epsilon_p(\overline{r})$. For the case of N discrete scatterers with permittivity ϵ_p, occupying regions $V_1, V_2, \ldots, V_N$, we have discontinuous permittivity $\epsilon_p(\overline{r})$ with

$$\epsilon_p(\overline{r}) = \begin{cases} \epsilon_p & \text{for } \overline{r} \in V_j,\ j = 1, 2, \ldots, N \\ \epsilon & \text{otherwise} \end{cases} \quad (10.3.1)$$

Let the inhomogeneous electric susceptibility be

$$\chi(\overline{r}) = \frac{\epsilon_p(\overline{r})}{\epsilon} - 1 = \epsilon_r(\overline{r}) - 1 \quad (10.3.2)$$

where $\epsilon_r(\overline{r})$ is the relative permittivity. The electric field is, from (2.1.37) of Chapter 2,

$$\overline{E}(\overline{r}) = \overline{E}_{inc}(\overline{r}) + k^2\int_{\text{all space}} d\overline{r}'\, g(\overline{r},\overline{r}')\chi(\overline{r}')\overline{E}(\overline{r}') + \nabla\int_{\text{all space}} d\overline{r}'\, g(\overline{r},\overline{r}')\nabla'\cdot\left(\chi(\overline{r}')\overline{E}(\overline{r}')\right) \quad (10.3.3)$$

Note that the integration space of (10.3.3) is over all space. The first integral is over regions where $\chi(\overline{r})$ is nonzero, that is, where $\epsilon_p(\overline{r}) \neq \epsilon$. The second integral is over space where $\nabla \cdot (\chi(\overline{r})\overline{E}(\overline{r}))$ is nonzero. For the case of discontinuous permittivity like discrete scatterers, the jump of permittivity across the scatterer boundary can have nonzero divergence at the boundary and give rise to a surface integral. Thus $\nabla \cdot (\chi(\overline{r})\overline{E}(\overline{r}))$ in (10.3.3) can be both zero outside and inside the scatterer and yet have a surface integral contribution.

We can also use polarization density $\overline{P}(\overline{r})$. Let

$$\overline{P}(\overline{r}) = \epsilon\chi(\overline{r})\overline{E}(\overline{r}) \tag{10.3.4}$$

Then (10.3.3) becomes

$$\begin{aligned}\overline{E}(\overline{r}) = {} & \overline{E}_{inc}(\overline{r}) + k^2 \int_{\text{all space}} d\overline{r}' g(\overline{r}, \overline{r}') \frac{\overline{P}(\overline{r}')}{\epsilon} \\ & + \nabla \int_{\text{all space}} d\overline{r}' g(\overline{r}, \overline{r}') \nabla' \cdot \left(\frac{\overline{P}(\overline{r}')}{\epsilon} \right)\end{aligned} \tag{10.3.5}$$

We can recast the last term of (10.3.5) by using the divergence theorem

$$\begin{aligned}&\int_{\text{all space}} d\overline{r}' g(\overline{r}, \overline{r}') \nabla' \cdot \left(\frac{\overline{P}(\overline{r}')}{\epsilon} \right) \\ &= \int_{\text{all space}} d\overline{r}' \nabla' \cdot \left(g(\overline{r}, \overline{r}') \frac{\overline{P}(\overline{r}')}{\epsilon} \right) - \int_{\text{all space}} d\overline{r}' \left(\nabla' g(\overline{r}, \overline{r}') \right) \cdot \frac{\overline{P}(\overline{r}')}{\epsilon} \\ &= \oint_{S_\infty} d\overline{S}' \left(g(\overline{r}, \overline{r}') \frac{\overline{P}(\overline{r}')}{\epsilon} \right) - \int_{\text{all space}} d\overline{r}' \left(\nabla' g(\overline{r}, \overline{r}') \right) \cdot \frac{\overline{P}(\overline{r}')}{\epsilon}\end{aligned} \tag{10.3.6}$$

The first term in (10.3.6) vanishes because of radiation condition at infinity. Putting (10.3.6) in (10.3.5) gives

$$\begin{aligned}\overline{E}(\overline{r}) = {} & \overline{E}_{inc}(\overline{r}) + k^2 \int_{\text{all space}} d\overline{r}' g(\overline{r}, \overline{r}') \frac{\overline{P}(\overline{r}')}{\epsilon} \\ & - \nabla \int_{\text{all space}} d\overline{r}' \left(\nabla' g(\overline{r}, \overline{r}') \right) \cdot \frac{\overline{P}(\overline{r}')}{\epsilon}\end{aligned} \tag{10.3.7}$$

Equation (10.3.7) has an advantage in that one only needs to integrate over regions where $\overline{P}(\overline{r})$ is nonzero (i.e., $\chi(\overline{r})$ nonzero) and does not have surface integral contributions as in (10.3.3).

3.2 Simulation of Densely Packed Dielectric Spheres

Let an incident electric field $\overline{E}_{inc}(\overline{r})$ be impinging upon N number of randomly positioned dielectric spheres. Sphere j is centered at $\overline{r}_j$ and has permittivity $\epsilon_j, j = 1, 2, 3, \ldots, N$. The discrete scatterers are embedded in a background with permittivity ϵ. Particle j occupies region V_j. Let $\epsilon_p(\overline{r})$ be the permittivity as a function of $\overline{r}$. Thus

$$\epsilon_p(\overline{r}) = \begin{cases} \epsilon_j & \text{for } \overline{r} \text{ in } V_j \\ \epsilon & \text{for } \overline{r} \text{ in background} \end{cases} \tag{10.3.8}$$

The induced polarization $\overline{P}(\overline{r})$is nonzero over the particles only. Let $\overline{P}_j(\overline{r})$ be the polarization density inside particle j. Then the volume integral equation gives

$$\overline{E}(\overline{r}) = \overline{E}_{inc}(\overline{r}) + k^2 \sum_{j=1}^{N} \int_{V_j} d\overline{r}' g(\overline{r}, \overline{r}') \frac{\overline{P}_j(\overline{r}')}{\epsilon} - \sum_{j=1}^{N} \nabla \int_{V_j} d\overline{r}' \nabla' g(\overline{r}, \overline{r}') \cdot \frac{\overline{P}_j(\overline{r}')}{\epsilon} \tag{10.3.9}$$

To solve (10.3.9) by the method of moments, we assume that the dielectric spheres of radii $a_j, j = 1, 2, \ldots, N$, are small so that only dipoles are induced inside the spheres. Let

$$\overline{E}(\overline{r}) = \overline{E}_j(\overline{r}) \qquad \text{for } \overline{r} \in V_j \tag{10.3.10}$$

We expand the internal field $\overline{E}_j(\overline{r})$ into electrostatic dipole functions

$$\overline{E}_j(\overline{r}) = \sum_{\alpha=1}^{3} c_{j\alpha} \overline{f}_{j\alpha}(\overline{r}) \tag{10.3.11}$$

where

$$\overline{f}_{j1}(\overline{r}) = \hat{x} \frac{1}{\sqrt{v_j}} \tag{10.3.12}$$

$$\overline{f}_{j2}(\overline{r}) = \hat{y} \frac{1}{\sqrt{v_j}} \tag{10.3.13}$$

$$\overline{f}_{j3}(\overline{r}) = \hat{z} \frac{1}{\sqrt{v_j}} \tag{10.3.14}$$

where $v_j = \frac{4\pi a_j^3}{3}$ is the volume of the jth sphere. The normalization factor is put in (10.3.12)–(10.3.14) so that

$$\int_{V_j} d\overline{r} \overline{f}_{j\alpha} \cdot \overline{f}_{j\beta} = \delta_{\alpha\beta} \tag{10.3.15}$$

Substituting (10.3.11) into (10.3.9) results in

$$\overline{E}(\overline{r}) = \overline{E}_{inc}(\overline{r}) + \sum_{j=1}^{N}\sum_{\alpha=1}^{3} c_{j\alpha}\overline{q}_{j\alpha}(\overline{r}) \tag{10.3.16}$$

where

$$\overline{q}_{j\alpha}(\overline{r}) = k^2 \int_{V_j} d\overline{r}' g(\overline{r},\overline{r}')\overline{f}_{j\alpha}(\epsilon_{rj}-1) - \nabla \int_{V_j} d\overline{r}' \nabla' g(\overline{r},\overline{r}') \cdot \overline{f}_{j\alpha}(\epsilon_{rj}-1) \tag{10.3.17}$$

and $\epsilon_{rj} = \frac{\epsilon_j}{\epsilon}$ is the relative permittivity of the jth particle.

Of particular importance is the expression for $\overline{q}_{j\alpha}(\overline{r})$ when $\overline{r}$ is in V_j (the self term). The first term in (10.3.17) is a wave field while the second term in (10.3.17) is dominated by the electrostatic field for $\overline{r}$ in V_j, assuming a small particle. Thus for $\overline{r}$ in V_j

$$\overline{q}_{j\alpha}(\overline{r}) \simeq -\nabla \int_{V_j} d\overline{r}' \nabla' g(\overline{r},\overline{r}') \cdot \overline{f}_{j\alpha}(\epsilon_{rj}-1) \tag{10.3.18}$$

However, the right hand side of (10.3.18) is the electrostatic field produced inside V_j by polarization $\overline{P}_{j\alpha} = \overline{f}_{j\alpha}\epsilon(\epsilon_{rj}-1)$ which by (10.2.7) is $-\frac{\overline{P}_{j\alpha}}{3\epsilon} = -\frac{\overline{f}_{j\alpha}}{3}(\epsilon_{rj}-1)$. Thus for $\overline{r}$ in V_j,

$$\overline{q}_{j\alpha}(\overline{r}) \simeq -\frac{\overline{f}_{j\alpha}}{3}(\epsilon_{rj}-1) \tag{10.3.19}$$

To solve (10.3.16), we apply testing functions of $\overline{f}_{l\beta}(\overline{r})$ and dot product it with (10.3.16) and integrate it over region V_l.

$$\int_{V_l} d\overline{r}\overline{f}_{l\beta}(\overline{r}) \cdot \overline{E}(\overline{r}) = \int_{V_l} d\overline{r}\overline{f}_{l\beta}(\overline{r}) \cdot \overline{E}_{inc}(\overline{r}) + \sum_{j=1}^{N}\sum_{\alpha=1}^{3} c_{j\alpha} \int_{V_l} d\overline{r}\overline{f}_{l\beta}(\overline{r}) \cdot \overline{q}_{j\alpha}(\overline{r}) \tag{10.3.20}$$

Using (10.3.10)–(10.3.11) indicates that the left hand side of (10.3.20) is equal to $c_{l\beta}$. The second term in (10.3.20) on the right hand side is decomposed into $j \neq l$ and $j = l$ terms. We thus have

$$c_{l\beta} = \int_{V_l} d\overline{r}\overline{f}_{l\beta} \cdot \overline{E}_{inc}(\overline{r}) + \sum_{\substack{j=1 \\ j\neq l}}^{N}\sum_{\alpha=1}^{3} c_{j\alpha} \int_{V_l} d\overline{r}\overline{f}_{l\beta} \cdot \overline{q}_{j\alpha}(\overline{r}) + \sum_{\alpha=1}^{3} c_{l\alpha} \int_{V_l} d\overline{r}\overline{f}_{l\beta} \cdot \overline{q}_{l\alpha}(\overline{r}) \tag{10.3.21}$$

Using (10.3.19) in the last term of (10.3.21) gives

$$c_{l\beta} = \int_{V_l} d\overline{r}\overline{f}_{l\beta} \cdot \overline{E}_{inc}(\overline{r}) + \sum_{\substack{j=1 \\ j\neq l}}^{N}\sum_{\alpha=1}^{3} c_{j\alpha} \int_{V_l} d\overline{r}\overline{f}_{l\beta} \cdot \overline{q}_{j\alpha}(\overline{r}) + c_{l\beta}\left(-\frac{1}{3}\right)(\epsilon_{rl}-1)$$

so that

$$c_{l\beta} = \frac{3}{2+\epsilon_{rl}}\left\{\int_{V_l} d\overline{r}\,\overline{f}_{l\beta}\cdot\overline{E}_{inc}(\overline{r}) + \sum_{\substack{j=1\\ j\neq l}}^{N}\sum_{\alpha=1}^{3} c_{j\alpha}\int_{V_l} d\overline{r}\,\overline{f}_{l\beta}\cdot\overline{q}_{j\alpha}(\overline{r})\right\} \quad (10.3.22)$$

with $l = 1, 2, \ldots, N$, and $\beta = 1, 2, 3$. Equation (10.3.22) provides the matrix equations of $3N$ equations for the $3N$ unknowns $c_{l\beta}$, $l = 1, 2, \ldots, N$ and $\beta = 1, 2, 3$.

By using the small particle assumption, the matrix elements in (10.3.22) can be simplified.

$$\int_{V_l} d\overline{r}\,\overline{f}_{l\beta}\cdot\overline{E}_{inc}(\overline{r}) = v_l\overline{f}_{l\beta}\cdot\overline{E}_{inc}(\overline{r}_l) \quad (10.3.23)$$

For $j \neq l$

$$\begin{aligned}\int_{V_l} d\overline{r}\,\overline{f}_{l\beta}(\overline{r})\cdot\overline{q}_{j\alpha}(\overline{r}) &= \int_{V_l} d\overline{r}\,\overline{f}_{l\beta}\cdot\left\{k^2\int_{V_j} d\overline{r}'g(\overline{r},\overline{r}')\overline{f}_{j\alpha}(\epsilon_{rj}-1)\right.\\ &\quad \left. -\nabla\int_{V_j} d\overline{r}'\nabla' g(\overline{r},\overline{r}')\cdot\overline{f}_{j\alpha}(\epsilon_{rj}-1)\right\}\\ &\approx k^2(\epsilon_{rj}-1)v_l v_j\overline{f}_{l\beta}\cdot\overline{\overline{G}}(\overline{r}_l,\overline{r}_j)\cdot\overline{f}_{j\alpha}\end{aligned} \quad (10.3.24)$$

where the dyadic Green's function is

$$\overline{\overline{G}}(\overline{r},\overline{r}) = \left(\overline{\overline{I}} + \frac{\nabla\nabla}{k^2}\right)g(\overline{r},\overline{r}') \quad (10.3.25)$$

Explicit expressions of $\overline{\overline{G}}(\overline{r},\overline{r}')$ are given in (2.3.5)–(2.3.7) of Chapter 2. Substituting (10.3.23) and (10.3.24) into (10.3.22), we have

$$c_{l\beta} = \frac{3}{2+\epsilon_{rl}}\left\{v_l\overline{f}_{l\beta}\cdot\overline{E}_{inc}(\overline{r}_l) + \sum_{\substack{j=1\\ j\neq l}}^{N}\sum_{\alpha=1}^{3} c_{j\alpha}k^2(\epsilon_{rj}-1)v_l v_j\overline{f}_{l\beta}\cdot\overline{\overline{G}}(\overline{r}_l,\overline{r}_j)\cdot\overline{f}_{j\alpha}\right\} \quad (10.3.26)$$

Equation (10.3.26) is the Foldy-Lax multiple scattering equations based on volume integral equation. After the matrix equations are solved, the final scattered field is

$$\overline{E}_s(\overline{r}) = k^2\sum_{j=1}^{N}\int_{V_j} d\overline{r}'\overline{\overline{G}}(\overline{r},\overline{r}')\cdot\frac{\overline{P}_j(\overline{r}')}{\epsilon} \quad (10.3.27)$$

We calculate the final scattered field $\overline{r}$ in the far field in the direction $\hat{k}_s = (\theta_s, \phi_s)$. Let $\hat{v}_s$ and $\hat{h}_s$ be the vertical and horizontal polarization vectors respectively. The polarization in particle j is

$$\overline{P}_j(\overline{r}) = \sum_{\alpha=1}^{3} \epsilon(\epsilon_{rj} - 1)c_{j\alpha}\overline{f}_{j\alpha}(\overline{r}) \tag{10.3.28}$$

We have, putting (10.3.28) in (10.3.27) and taking the result in the far-field,

$$\overline{E}_s(\overline{r}) = \frac{k^2 e^{ikr}}{4\pi r}(\hat{v}_s\hat{v}_s + \hat{h}_s\hat{h}_s)\sum_{j=1}^{N}\sum_{\alpha=1}^{3} v_j e^{-i\overline{k}_s\cdot\overline{r}_j}(\epsilon_{rj} - 1)c_{j\alpha}\overline{f}_{j\alpha} \tag{10.3.29}$$

This can be written in terms of the horizontal and vertical polarized components

$$\overline{E}_s = (E_{vs}\hat{v}_s + E_{hs}\hat{h}_s)\frac{e^{ikr}}{r} \tag{10.3.30}$$

with

$$E_{vs} = \frac{k^2}{4\pi}\sum_{j=1}^{N}\sum_{\alpha=1}^{3} v_j e^{-i\overline{k}_s\cdot\overline{r}_j}(\epsilon_{rj} - 1)c_{j\alpha}(\hat{v}_s \cdot \overline{f}_{j\alpha}) \tag{10.3.31a}$$

$$E_{hs} = \frac{k^2}{4\pi}\sum_{j=1}^{N}\sum_{\alpha=1}^{3} v_j e^{-i\overline{k}_s\cdot\overline{r}_j}(\epsilon_{rj} - 1)c_{j\alpha}(\hat{h}_s \cdot \overline{f}_{j\alpha}) \tag{10.3.31b}$$

The scattered power is

$$P_s = \int_0^{2\pi} d\phi_s \int_0^{\pi} d\theta_s \sin\theta_s \frac{(|E_{vs}|^2 + |E_{hs}|^2)}{2\eta} \tag{10.3.32}$$

To get the incoherent power, we have to subtract out the coherent intensity. Thus,

$$P_s^{\text{incoherent}} = \int_0^{2\pi} d\phi_s \int_0^{\pi} d\theta_s \sin\theta_s \frac{1}{2\eta}\left\{\langle|E_{vs} - \langle E_{vs}\rangle|^2\rangle + \langle|E_{hs} - \langle E_{hs}\rangle|^2\rangle\right\} \tag{10.3.33}$$

Equation (10.3.26) can also be put in the more familiar form of exciting fields. From (10.3.26), we can also have

$$\sum_{\beta=1}^{3} c_{l\beta}\overline{f}_{l\beta} = \frac{3}{2 + \epsilon_{rl}}\left\{\overline{E}_{inc}(\overline{r}_l) + \sum_{\substack{j=1\\ j\neq l}}^{N} k^2(\epsilon_{rj} - 1)v_j\overline{\overline{G}}(\overline{r}_l, \overline{r}_j)\cdot\sum_{\alpha=1}^{3} c_{j\alpha}\overline{f}_{j\alpha}\right\} \tag{10.3.34}$$

The dipole moment is

$$\overline{p}_j(\overline{r}) = v_j\epsilon(\epsilon_{rj} - 1)\overline{E}_j = v_j\epsilon(\epsilon_{rj} - 1)\sum_{\alpha=1}^{3} c_{j\alpha}\overline{f}_{j\alpha} \tag{10.3.35}$$

Dipole moment is also equal to

$$\overline{p}_j = \alpha_j \overline{E}_j^e = 3\epsilon v_j \frac{\epsilon_{rj} - 1}{\epsilon_{rj} + 2} \overline{E}_j^e \tag{10.3.36}$$

where $\overline{E}_j^e$ is exciting field of particle j. Thus we have from (10.3.36) and (10.3.35)

$$3\epsilon v_j \frac{\epsilon_{rj} - 1}{\epsilon_{rj} + 2} \overline{E}_j^e = v_j \epsilon (\epsilon_{rj} - 1) \sum_{\alpha=1}^{3} c_{j\alpha} \overline{f}_{j\alpha}$$

so that

$$\overline{E}_j^e = \frac{\epsilon_{rj} + 2}{3} \sum_{\alpha=1}^{3} c_{j\alpha} \overline{f}_{j\alpha} \tag{10.3.37}$$

Multiplying by $(\epsilon_{rl} + 2)/3$ in (10.3.34) gives

$$\overline{E}_l^e = \overline{E}_{inc}(\overline{r}_l) + \sum_{\substack{j=1 \\ j \neq l}}^{N} k^2 (\epsilon_{rj} - 1) v_j \frac{3}{\epsilon_{rj} + 2} \overline{\overline{G}}(\overline{r}_l, \overline{r}_j) \cdot \overline{E}_j^e \tag{10.3.38}$$

We recognize the scattering transition matrix element of

$$t_j = 3k^2 v_j \frac{\epsilon_{rj} - 1}{\epsilon_{rj} + 2} \tag{10.3.39}$$

This puts the matrix equation in the form of the multiple scattering equation of Foldy-Lax form based on exciting fields

$$\overline{E}_l^e = \overline{E}_{inc}(\overline{r}_l) + \sum_{\substack{j=1 \\ j \neq l}}^{N} \overline{\overline{G}}(\overline{r}_l, \overline{r}_j) \cdot t_j \overline{E}_j^e \tag{10.3.40}$$

3.3 Densely Packed Spheroids

In this section, we perform Monte Carlo simulations of non-spherical particles that are non-sticky. We describe the formulation of scattering by many spheroids in terms of volume integral equation. The method of moments is used with choices of basis functions that are appropriate for spheroids. The shuffling process to generate the positions of densely packed spheroids was described in Chapter 8. Both cases of aligned orientation and random orientation are considered. We illustrate the numerical results using several thousands of spheroids. Salient features of numerical results indicate that (1) the extinction rates of densely packed small spheroids are smaller than those of independent scattering; (2) for aligned spheroids, the extinction

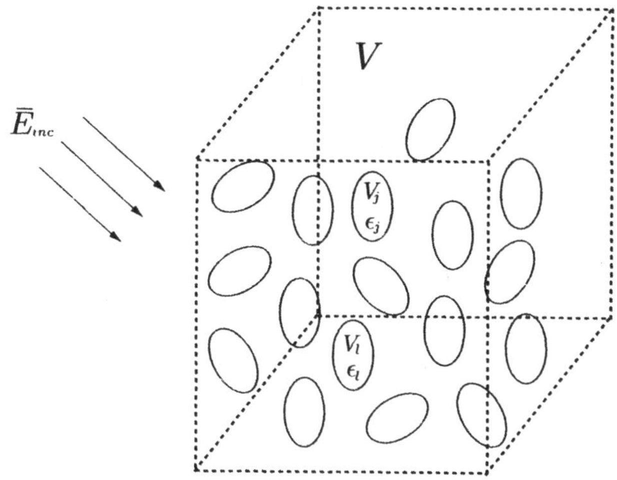

Figure 10.3.1 An electric field $\overline{E}_{inc}(\overline{r})$ incidents upon N non-overlap, small spheroids that are randomly positioned and oriented in a volume V.

rates are polarization dependent, while for completely random orientation, the extinction rates are polarization independent; (3) the co-polarized part of the phase matrix is smaller than that of independent scattering, while the depolarized part is larger than that of independent scattering. This means that the ratio of cross-polarization to co-polarization is significantly higher than that of independent scattering.

Let an incident electric field $\overline{E}_{inc}(\overline{r})$ be impinging upon N number of randomly positioned small spheroids (Fig. 10.3.1). Spheroid j is centered at $\overline{r}_j$ and has permittivity ϵ_j, $j = 1, 2, 3, \ldots, N$. The discrete scatterers are embedded in a background with permittivity ϵ. Particle j occupies region V_j. Let $\epsilon_p(\overline{r})$ be the permittivity as a function of $\overline{r}$,

$$\epsilon_p(\overline{r}) = \begin{cases} \epsilon_j & \text{for } \overline{r} \text{ in } V_j \\ \epsilon & \text{for } \overline{r} \text{ in the background} \end{cases} \tag{10.3.41}$$

Then the volume integral equation is, from (10.3.9)

$$\overline{E}(\overline{r}) = \overline{E}_{inc}(\overline{r}) + k^2 \sum_{j=1}^{N} \int_{V_j} d\overline{r}' \, \frac{g(\overline{r}, \overline{r}')}{\epsilon} \overline{P}_j(\overline{r}') - \sum_{j=1}^{N} \nabla \int_{V_j} d\overline{r}' \, \frac{\nabla' g(\overline{r}, \overline{r}')}{\epsilon} \cdot \overline{P}_j(\overline{r}') \tag{10.3.42}$$

where $\overline{P}_j(\overline{r}) = \epsilon \chi_j(\overline{r}) \overline{E}_j(\overline{r})$. To solve (10.3.42) by the method of moments, we expand the electric field $\overline{E}_j(\overline{r})$ inside the jth spheroid in a set of N_b basis functions

$$\overline{E}_j(\overline{r}) = \sum_{\alpha=1}^{N_b} a_{j\alpha}\overline{f}_{j\alpha}(\overline{r}) \tag{10.3.43}$$

Here the spheroid is assumed to be small, we choose the basis functions in (10.3.43) to be the electrostatic solution of that of a spheroid.

Let the jth spheroid be centered at $\overline{r}_j$ with

$$\overline{r}_j = x_j\hat{x} + y_j\hat{y} + z_j\hat{z} \tag{10.3.44}$$

and the symmetry axes of the spheroid be $\hat{x}_{bj}$, $\hat{y}_{bj}$, and $\hat{z}_{bj}$ with respective semi-axes lengths be a_j, a_j, and c_j. The orientation of the symmetry axis $\hat{z}_{bj}$ is

$$\hat{z}_{bj} = \sin\beta_j\cos\alpha_j\hat{x} + \sin\beta_j\sin\alpha_j\hat{y} + \cos\beta_j\hat{z} \tag{10.3.45}$$

The first eight normalized basis functions for the electric fields are

$$\overline{f}_{j1} = \hat{z}_{bj}\frac{1}{\sqrt{v_{0j}}} \tag{10.3.46}$$

$$\overline{f}_{j2} = \hat{x}_{bj}\frac{1}{\sqrt{v_{0j}}} \tag{10.3.47}$$

$$\overline{f}_{j3} = \hat{y}_{bj}\frac{1}{\sqrt{v_{0j}}} \tag{10.3.48}$$

$$\overline{f}_{j4} = \frac{1}{\sqrt{A_{4j}}}(x_{bj}\hat{x}_{bj} + y_{bj}\hat{y}_{bj} - 2z_{bj}\hat{z}_{bj}) \tag{10.3.49}$$

$$\overline{f}_{j5} = \frac{1}{\sqrt{A_{5j}}}(z_{bj}\hat{x}_{bj} + x_{bj}\hat{z}_{bj}) \tag{10.3.50}$$

$$\overline{f}_{j6} = \frac{1}{\sqrt{A_{5j}}}(z_{bj}\hat{y}_{bj} + y_{bj}\hat{z}_{bj}) \tag{10.3.51}$$

$$\overline{f}_{j7} = \frac{1}{\sqrt{A_{7j}}}(x_{bj}\hat{x}_{bj} - y_{bj}\hat{y}_{bj}) \tag{10.3.52}$$

$$\overline{f}_{j8} = \frac{1}{\sqrt{A_{7j}}}(y_{bj}\hat{x}_{bj} + x_{bj}\hat{y}_{bj}) \tag{10.3.53}$$

where A_α, $\alpha = 4, 5, \ldots, 8$, are normalization factors such that

$$\int_V d\overline{r}\,\overline{f}_\alpha \cdot \overline{f}_\beta = \delta_{\alpha\beta} \tag{10.3.54}$$

In (10.3.54), α, $\beta = 1, 2, \ldots, 8$, and V is the volume domain of the spheroid. There are three basis functions for electric dipole and five basis functions for electric quadrupoles.

Next we describe the derivation of the quadrupole basis functions. Consider a single small prolate spheroid of semi-axis a in $\hat{x}$ and $\hat{y}$ directions and semi-axis c in $\hat{z}$ direction. The volume of the small prolate spheroid is $v_o = 4\pi a^2 c/3$ and let the permittivity of the spheroid be ϵ_p.

For prolate spheroids with $c \geq a$, let

$$f_o = \sqrt{c^2 - a^2} \tag{10.3.55}$$

be the semifocal length. The prolate spheroidal coordinates, ξ, η, ϕ, are related to the Cartesian coordinates by

$$x = f_o\sqrt{(\xi^2 - 1)(1 - \eta^2)}\cos\phi \tag{10.3.56}$$

$$y = f_o\sqrt{(\xi^2 - 1)(1 - \eta^2)}\sin\phi \tag{10.3.57}$$

$$z = f_o\xi\eta \tag{10.3.58}$$

Then the five quadrupole solutions to Laplace's equation are

$$\begin{aligned} P_2^2(\xi)P_2^2(\eta)\exp(2i\phi) &= 9(\xi^2 - 1)(1 - \eta^2)\exp(2i\phi) \\ &= \frac{9}{f_o^2}(x^2 - y^2 + i2xy) \end{aligned} \tag{10.3.59}$$

$$\begin{aligned} P_2^1(\xi)P_2^1(\eta)\exp(i\phi) &= -9\xi\eta\sqrt{(\xi^2 - 1)(1 - \eta^2)}\exp(i\phi) \\ &= -\frac{9z}{f_o^2}(x + iy) \end{aligned} \tag{10.3.60}$$

$$P_2^{-1}(\xi)P_2^{-1}(\eta)\exp(-i\phi) = -\frac{z}{4f_o^2}(x - iy) \tag{10.3.61}$$

$$P_2^{-2}(\xi)P_2^{-1}(\eta)\exp(-2i\phi) = \frac{1}{64f_o^2}(x^2 - y^2 - 2ixy) \tag{10.3.62}$$

$$\begin{aligned} P_2^0(\xi)P_2^0(\eta) &= \frac{1}{4}(9\xi^2\eta^2 - 3\eta^2 - 3\xi^2 + 1) \\ &= -\frac{3}{4}\frac{x^2 + y^2}{f_o^2} + \frac{3}{2}\frac{z^2}{f_o^2} - \frac{1}{2} \end{aligned} \tag{10.3.63}$$

where P_n^m is the associated Legendre polynomial. We can obtain the five quadrupole basis functions for the potential functions by linear combinations of Φ_4, Φ_5, Φ_6, Φ_7, and Φ_8. Let

$$\Phi_4 = -\frac{1}{2}(x^2 + y^2) + z^2 - \frac{f_o^2}{3} \tag{10.3.64}$$

$$\Phi_5 = -zx \tag{10.3.65}$$

$$\Phi_6 = -zy \tag{10.3.66}$$

$$\Phi_7 = -\frac{1}{2}(x^2 - y^2) \tag{10.3.67}$$

$$\Phi_8 = -xy \tag{10.3.68}$$

The five electric field quadrupole basis functions are calculated by taking the negative gradient of the above and then normalizing the basis functions. We obtain (10.3.49)–(10.3.53) in this manner.

The polarization is

$$\overline{P}_\alpha = (\epsilon_p - \epsilon)\overline{f}_\alpha \tag{10.3.69}$$

for $\alpha = 1, 2, 3, \ldots, 8$. The polarization $\overline{P}_\alpha$ will induce an internal field $\overline{E}_{ind}^{(\alpha)}$, $\alpha = 1, 2, \ldots, N_b$, inside the spheroid. They are as follows

$$\overline{E}_{ind}^{(\alpha)} = C_\alpha \overline{f}_\alpha \tag{10.3.70}$$

$\alpha = 1, 2, \ldots, 8$, where

$$C_1 = -\frac{(\epsilon_p - \epsilon)}{\epsilon}\left[\frac{\xi_o}{2}\ln\left(\frac{\xi_o + 1}{\xi_o - 1}\right) - 1\right](\xi_o^2 - 1) \tag{10.3.71}$$

$$C_2 = C_3 = -\frac{1}{2}\left(\frac{\epsilon_p - \epsilon}{\epsilon}\right)\xi_o\left[\xi_o - \frac{\xi_o^2 - 1}{2}\ln\left(\frac{\xi_o + 1}{\xi_o - 1}\right)\right] \tag{10.3.72}$$

$$C_4 = -\frac{3\xi_o}{2}(\xi_o^2 - 1)\left(\frac{\epsilon_p - \epsilon}{\epsilon}\right)\left\{\frac{1}{2}(3\xi_o^2 - 1)\ln\left(\frac{\xi_o + 1}{\xi_o - 1}\right) - 3\xi_o\right\} \tag{10.3.73}$$

$$C_5 = C_6 = -\frac{(\xi_o^2 - 1)}{2}\left(\frac{\epsilon_p - \epsilon}{\epsilon}\right)(2\xi_o^2 - 1) \cdot \left\{\frac{3\xi_o^2 - 2}{\xi_o^2 - 1} - \frac{3\xi_o}{2}\ln\left(\frac{\xi_o + 1}{\xi_o - 1}\right)\right\} \tag{10.3.74}$$

$$C_7 = C_8 = -\left(\frac{\xi_o^2 - 1}{4}\right)\left(\frac{\epsilon_p - \epsilon}{\epsilon}\right)\xi_o \cdot \left\{\frac{3}{2}(\xi_o^2 - 1)\ln\left(\frac{\xi_o + 1}{\xi_o - 1}\right) - \frac{3\xi_o^3 - 5\xi_o}{\xi_o^2 - 1}\right\} \tag{10.3.75}$$

with

$$\xi_o = \frac{1}{\sqrt{1 - \frac{a^2}{c^2}}} \tag{10.3.76}$$

We shall illustrate the derivation of C_2 and C_5. The rest can be derived in a similar fashion.

Let the electric field inside the spheroid be

$$\overline{E} = \overline{f}_2 = \hat{x}\frac{1}{\sqrt{v_o}} \tag{10.3.77}$$

The polarization inside the spheroid is

$$\overline{P} = (\epsilon_p - \epsilon)\frac{\hat{x}}{\sqrt{v_o}} \tag{10.3.78}$$

To determine the electrostatic solution as a result of polarization of (10.3.78), we construct solutions based on Legendre functions of the first and second kind and spheroidal coordinates ξ, η, and ϕ. Let Φ_{in} and Φ_{out} be the scalar potentials inside and outside the spheroid. Both Φ_{in} and Φ_{out} satisfy Laplace equation. They obey the boundary condition that at the surface of the spheroid $\xi = \xi_o$,

$$\Phi_{out} = \Phi_{in} \tag{10.3.79}$$

$$-\hat{\xi} \cdot \nabla \Phi_{out} + \hat{\xi} \cdot \nabla \Phi_{in} = \hat{\xi} \cdot \frac{\overline{P}}{\epsilon} \tag{10.3.80}$$

To solve Φ_{in} and Φ_{out}, let

$$\Phi_{in} = A\sqrt{(\xi^2 - 1)(1 - \eta^2)} \cos\phi \tag{10.3.81}$$

$$\Phi_{out} = BQ_1^1(\xi)P_1^1(\eta) \cos\phi \tag{10.3.82}$$

where $P_n^m(\eta)$ is the associated Legendre polynomial and $Q_n^m(\xi)$ is the Legendre function of second kind. In (10.3.82)

$$P_1^1(\eta) = \sqrt{1 - \eta^2} \tag{10.3.83}$$

$$Q_1^1(\xi) = \sqrt{\xi^2 - 1} \left[\frac{\xi}{\xi^2 - 1} - \frac{1}{2} \ln \left(\frac{\xi + 1}{\xi - 1} \right) \right] \tag{10.3.84}$$

Applying boundary conditions (10.3.79) and (10.3.82) gives

$$A = \frac{1}{2} \left(\frac{\epsilon_p - \epsilon}{\epsilon} \right) \frac{f_o \xi_o}{\sqrt{v_o}} \left[\xi_o - \frac{\xi_o^2 - 1}{2} \ln \left(\frac{\xi_o + 1}{\xi_o - 1} \right) \right] \tag{10.3.85}$$

where $f_o = \sqrt{c^2 - a^2}$. The internal induced electric field is $\overline{E}_{ind} = -\nabla \Phi_{in}$. This gives

$$\overline{E}_{ind}^{(2)} = C_2 \overline{f}_2 \tag{10.3.86}$$

with C_2 and $\overline{f}_2$ as given in (10.3.72) and (10.3.47), respectively.

For the case of $\overline{f}_5$, let the electric field inside the spheroid be

$$\overline{E} = \overline{f}_5 \tag{10.3.87}$$

Then

$$\overline{P} = \frac{(\epsilon_p - \epsilon)}{\sqrt{A_5}} (z\hat{x} + x\hat{z}) \tag{10.3.88}$$

The potentials Φ_{in} and Φ_{out} are proportional to $P_2^1(\xi)P_2^1(\eta)\cos\phi$ and $Q_2^1(\xi)P_2^1(\eta)\cos\phi$ respectively, where $P_n^m(\xi)$ is Legendre function of the first kind. Let

$$\Phi_{in} = -Af_o^2 \xi\eta \sqrt{(\xi^2 - 1)(1 - \eta^2)} \cos\phi \tag{10.3.89}$$

$$\Phi_{out} = B\sqrt{\xi^2 - 1} \left[\frac{3\xi^2 - 2}{\xi^2 - 1} - \frac{3\xi}{2} \ln \left(\frac{\xi + 1}{\xi - 1} \right) \right] 3\eta \sqrt{1 - \eta^2} \cos\phi \tag{10.3.90}$$

Applying boundary conditions of (10.3.79) and (10.3.80) gives

$$A = \frac{1}{\sqrt{A_5}} \left\{ -\frac{(\xi_o^2 - 1)}{2} \left(\frac{\epsilon_p - \epsilon}{\epsilon} \right) (2\xi_o^2 - 1) \left[\frac{3\xi_o^2 - 2}{\xi_o^2 - 1} - \frac{3\xi_o}{2} \ln \left(\frac{\xi_o + 1}{\xi_o - 1} \right) \right] \right\} \tag{10.3.91}$$

Using (10.3.89), we get $\overline{E}_{ind}^{(5)} = -\nabla \Phi_{in} = C_5 \overline{f}_5$ where C_5 and $\overline{f}_5$ are as given in (10.3.74) and (10.3.50), respectively.

If the particles are closely packed, the near field interactions have large spatial variations over the size of a spheroid that may induce quadrupole fields inside the spheroid. However, the non-near field interactions have small spatial variations over the size of a spheroid and only induce dipole fields inside the spheroid.

Substituting (10.3.43) into (10.3.42), we get

$$\overline{E}(\overline{r}) = \overline{E}_{inc}(\overline{r}) + \sum_{j=1}^{N} \sum_{\alpha=1}^{N_b} a_{j\alpha} \overline{q}_{j\alpha}(\overline{r}) \tag{10.3.92}$$

where

$$\overline{q}_{j\alpha}(\overline{r}) = k^2 \int_{V_j} d\overline{r}' \; g(\overline{r}, \overline{r}') \overline{f}_{j\alpha}(\overline{r}') (\epsilon_{rj} - 1) - \nabla \int_{V_j} d\overline{r}' \; \nabla' g(\overline{r}, \overline{r}') \cdot \overline{f}_{j\alpha}(\overline{r}') (\epsilon_{rj} - 1) \tag{10.3.93}$$

is the electric field induced by the polarization $\overline{P}_j(\overline{r})$ of the spheroid j. Of particular importance is the internal field created by $\overline{P}_j(\overline{r})$ on itself. For this self term contribution, the second term in (10.3.93) dominates. Because of the smallness of the spheroid, an electrostatic solution can be sought for the second term in (10.3.93). We have from (10.3.70), for $\overline{r}$ in V_j, the self term becomes

$$\overline{q}_{j\alpha}(\overline{r}) \cong C_{j\alpha} \overline{f}_{j\alpha}(\overline{r}) \tag{10.3.94}$$

where j is the particle index, α is the basis function index, and $C_{j\alpha}$'s are given by (10.3.71)–(10.3.75) for each particle j. In the expressions of (10.3.70)–(10.3.75), we need to replace ϵ_p and ξ_o by the value for each particle. An approximation sign is used in (10.3.94) to indicate the low frequency approximation.

Next, we apply Galerkin's method by testing (10.3.92) with $\overline{f}_{l\beta}(\overline{r})$,

$$\begin{aligned} a_{l\beta} &= \int_{V_l} d\overline{r}\, \overline{f}_{l\beta}(\overline{r}) \cdot \overline{E}(\overline{r}) = \int_{V_l} d\overline{r}\, \overline{f}_{l\beta}(\overline{r}) \cdot \sum_{\alpha=1}^{N_b} a_{l\alpha} \overline{f}_{l\alpha}(\overline{r}) \\ &= \int_{V_l} d\overline{r}\, \overline{f}_{l\beta}(\overline{r}) \cdot \overline{E}_{inc}(\overline{r}) + \sum_{\substack{j=1 \\ j\neq l}}^{N} \sum_{\alpha=1}^{N_b} a_{j\alpha} \int_{V_l} d\overline{r}\, \overline{f}_{l\beta}(\overline{r}) \cdot \overline{q}_{j\alpha}(\overline{r}) \\ &\quad + \sum_{\alpha=1}^{N_b} a_{l\alpha} \int_{V_l} d\overline{r}\, \overline{f}_{l\beta}(\overline{r}) \cdot \overline{q}_{l\alpha}(\overline{r}) \\ &= \int_{V_l} d\overline{r}\, \overline{f}_{l\beta}(\overline{r}) \cdot \overline{E}_{inc}(\overline{r}) + \sum_{\substack{j=1 \\ j\neq l}}^{N} \sum_{\alpha=1}^{N_b} a_{j\alpha} \int_{V_l} d\overline{r}\, \overline{f}_{l\beta}(\overline{r}) \cdot \overline{q}_{j\alpha}(\overline{r}) \\ &\quad + a_{l\beta} C_{l\beta} \end{aligned} \tag{10.3.95}$$

This gives

$$a_{l\beta} = \frac{1}{(1 - C_{l\beta})} \left\{ \int_{V_l} d\overline{r}\, \overline{f}_{l\beta}(\overline{r}) \cdot \overline{E}_{inc}(\overline{r}) + \sum_{\substack{j=1 \\ j\neq l}}^{N} \sum_{\alpha=1}^{N_b} a_{j\alpha} \int_{V_l} d\overline{r}\, \overline{f}_{l\beta}(\overline{r}) \cdot \overline{q}_{j\alpha}(\overline{r}) \right\} \tag{10.3.96}$$

Because of the small spheroid assumption, only the dipole term contributes to the first term in (10.3.96) which is the polarization induced by the incident field. Thus

$$\int_{V_l} d\overline{r}\, \overline{f}_{l\beta}(\overline{r}) \cdot \overline{E}_{inc}(\overline{r}) = \begin{cases} v_{0l} \overline{f}_{l\beta} \cdot \overline{E}_{inc}(\overline{r}_l) & \text{for } \beta = 1,2,3 \\ 0 & \text{for } \beta > 3 \end{cases} \tag{10.3.97}$$

After the coefficients $a_{l\beta}$, $l = 1, 2, \ldots, N$, and $\beta = 1, 2, \ldots, N_b$ are solved, the far field scattered field in the direction (θ_s, ϕ_s) is expressed as

$$\overline{E}_s(\overline{r}) = k^2 \frac{e^{ikr}}{4\pi r} \left(\hat{v}_s \hat{v}_s + \hat{h}_s \hat{h}_s \right) \cdot \sum_{j=1}^{N} \sum_{\alpha=1}^{N_b} a_{j\alpha} (\epsilon_{rj} - 1) \int_{V_j} d\overline{r}'\, e^{-i\overline{k}_s \cdot \overline{r}'} \overline{f}_{j\alpha}(\overline{r}') \tag{10.3.98}$$

where $\epsilon_{rj} = \epsilon_j/\epsilon$. Under the small spheroid assumption, only the dipole fields will contribute to the far field radiation in (10.3.98). Thus, we have

$$\overline{E}_s(\overline{r}) \simeq k^2 \frac{e^{ikr}}{4\pi r} \left(\hat{v}_s \hat{v}_s + \hat{h}_s \hat{h}_s \right) \cdot \sum_{j=1}^{N} \sum_{\alpha=1}^{N_b} a_{j\alpha} (\epsilon_{rj} - 1)\, v_{0j} \overline{f}_{j\alpha} e^{-i\overline{k}_s \cdot \overline{r}_j} \tag{10.3.99}$$

We next illustrate the results of the numerical simulations by using $N = 2000$ spheroids and up to $f = 30\%$ by volume fraction. The relative permittivity used for the spheroids is 3.2 and the size parameter of the spheroids used is such that $ka = 0.2$. At this volume fraction, permittivity, and size, we did not include the quadrupole effects in the simulations. For dipole interactions, we replace the integral in the last term of (10.3.93) as follows.

$$\int_{V_l} d\overline{r}\ \overline{f}_{l\beta}(\overline{r}) \cdot \overline{q}_{j\alpha}(\overline{r}) = (\epsilon_{rj} - 1)\, v_{0j} v_{0l} k^2 \overline{f}_{l\beta} \cdot \overline{\overline{G}}(\overline{r}_l, \overline{r}_j) \cdot \overline{f}_{j\alpha} \qquad (10.3.100)$$

In the simulations, all the spheroids are prolate and are identical in size with $c = ea$, where e is the elongation ratio of the prolate spheroid. The size of the box in which the spheroids are placed is

$$V = \frac{Nv}{f} \qquad (10.3.101)$$

where f is the fractional volume, and $v = 4\pi a^2 c/3$ is the volume of one spheroid.

An incident electric field of

$$\overline{E}_{inc}(\overline{r}) = \hat{y} e^{ikz} \qquad (10.3.102)$$

is launched onto the box containing the N spheroids. The matrix equation of (10.3.96) is solved by iteration. After the matrix equation is solved, the scattered field is calculated by (10.3.99). The scattered field is decomposed into vertical and horizontal polarization

$$\overline{E}_s = E_{vs}\hat{v}_s + E_{hs}\hat{h}_s \qquad (10.3.103)$$

We performed $N_r = 50$ realizations in the numerical illustrations. Let σ be the realization index. Decomposition of the field into coherent and incoherent scattered fields is also made as in Section 3.2. The incoherent scattered field is decomposed into vertical and horizontal polarization

$$\overline{\mathcal{E}}^{\sigma}_s = \mathcal{E}^{\sigma}_{vs}\hat{v}_s + \mathcal{E}^{\sigma}_{hs}\hat{h}_s \qquad (10.3.104)$$

The averaged N-particle bistatic scattering cross sections are

$$\sigma_{v_s N}(\theta_s, \phi_s) = \frac{1}{N_r} \sum_{\sigma=1}^{N_r} |\mathcal{E}^{\sigma}_{vs}|^2 \qquad (10.3.105)$$

$$\sigma_{h_s N}(\theta_s, \phi_s) = \frac{1}{N_r} \sum_{\sigma=1}^{N_r} |\mathcal{E}^{\sigma}_{hs}|^2 \qquad (10.3.106)$$

For the simulations, the particles are not absorptive. Thus the extinction rate is the same as the scattering rate. The extinction rate is

$$\kappa_s = \kappa_e = \frac{1}{V} \int_0^{\pi} d\theta_s\ \sin\theta_s \int_0^{2\pi} d\phi_s\ (\sigma_{v_s N} + \sigma_{h_s N}) \qquad (10.3.107)$$

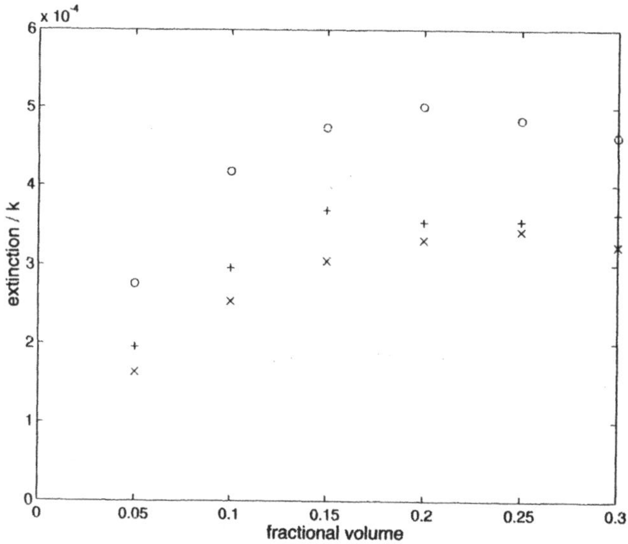

Figure 10.3.2 Extinction rate as a function of fractional volume of particles. Relative permittivity of particles $\epsilon_r = 3.2$. For spheroids $ka = 0.2$ and $e = 1.8$. for spheres $ka = 0.2$. The dotted curve is for the medium with spheres, the symbol "+" is for the medium with randomly oriented spheroids. The symbols "o" and "x" are for the medium with aligned spheroids but with incident wave being vertically polarized and horizontally polarized, respectively.

In Fig. 10.3.2, we illustrate the extinction coefficients normalized by the wavenumber k as a function of fractional volume. We consider the case consisting of aligned prolate spheroids with $ka = 0.2$ and $e = 1.8$. In such a medium, a vertically polarized incident wave with the incident polarization aligned with the symmetry axis of the prolate spheroids has a higher extinction rate than that of the horizontally polarized incident wave. The extinction rate is polarization dependent. In the same figure, we also show the extinction rate for the case when the spheroids are randomly oriented. For random orientation, the probability density function of orientation $p(\beta, \alpha)$ is

$$p(\beta, \alpha) = \frac{\sin \beta}{4\pi} \tag{10.3.108}$$

for $0 \leq \beta \leq \pi$, and $0 \leq \alpha \leq 2\pi$.

The result of the attenuation for the randomly oriented case is between those of vertically and horizontally polarized incidence of the aligned case. The extinction rates are also compared with those of a medium with spherical particles of $ka = 0.2$ and $e = 1$. The spherical case predicts a smaller attenuation than the spheroidal case, even though the medium has the same fractional volume. Figure 10.3.3 shows the extinction rates as a function

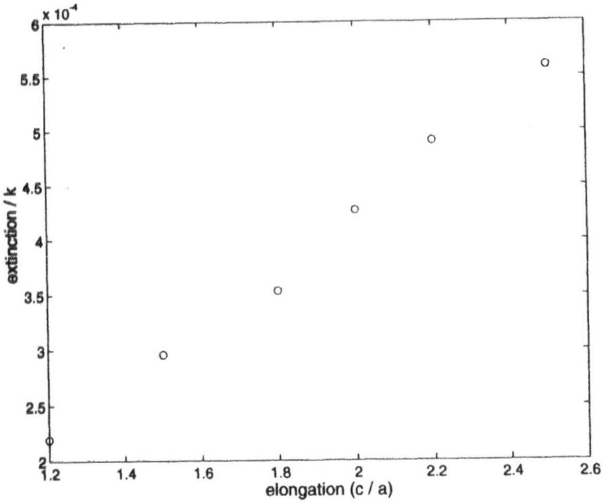

Figure 10.3.3 Extinction rate as a function of elongation ratio of spheroids. The spheroids are randomly oriented. For spheroids $ka = 0.2$ and fractional volume $f = 20\%$.

of particle elongation for a medium with randomly oriented spheroids. The size parameter is $ka = 0.2$ and fractional volume $f = 20\%$. It shows that the spheroid with longer elongation ratio predicts a larger attenuation due to its larger particle volume.

Next we illustrate the scattering phase matrices. The phase matrices are bistatic scattering cross sections per unit volume of a conglomeration of particles. We consider the incident wave and polarization as given by (10.3.102). The spheroids are randomly oriented in the following illustrations. We also compare with the results of independent scattering. The independent scattering results are obtained by including only the first term inside the curly bracket of (10.3.96). That is

$$a_{l\beta} = \frac{1}{(1 - C_{l\beta})} \int_{V_l} d\overline{r}\ \overline{f}_{l\beta}(\overline{r}) \cdot \overline{E}_{inc}(\overline{r}) \tag{10.3.109}$$

We represent the phase matrix by using the $(\hat{k}, \hat{1}, \hat{2})$ orthogonal system for polarization as treated in Section 1.2 of Chapter 1 of Volume I. We consider the incident wave as given by (10.3.102) with $\hat{k}_i = \hat{z}$ and incident polarization $\hat{e}_i = \hat{y}$.

Case (a): $\phi_s = 0°$ and $\phi_s = 180°$. For $\phi_s = 0°$,

$$\hat{k}_s = \sin\theta_s\hat{x} + \cos\theta_s\hat{z}; \quad \hat{v}_s = \cos\theta_s\hat{x} - \sin\theta_s\hat{z}; \quad \hat{h}_s = \hat{y}$$

Then

$$\hat{1}_s = \hat{1}_i = (\hat{k}_s \times \hat{k}_i)/|\hat{k}_s \times \hat{k}_i| = -\hat{y} = -\hat{h}_s; \quad \hat{2}_s = \hat{k}_s \times \hat{1}_s = \hat{v}_s$$

Thus the incident wave with $\hat{e}_i = \hat{y}$ is of polarization $\hat{1}_i$. Similar polarization vector formulas can be obtained for $\phi_s = 180°$. Hence we define the phase matrix elements as

$$P_{11}(\theta_s) = \frac{\sigma_{h_s N}}{V} \tag{10.3.110}$$

$$P_{21}(\theta_s) = \frac{\sigma_{v_s N}}{V} \tag{10.3.111}$$

The quantities $P_{11}(\theta_s)$ and $P_{21}(\theta_s)$ correspond to co-polarization and cross-polarization, respectively. In Figs. 10.3.4a and 10.3.4b, we plot P_{11} and P_{21}, respectively, as a function of scattering angle for scattering angle between 0° and 360°. We give the results of $\phi_s = 0°$ and $\phi_s = 180°$ in the same figure. The following definition is used. For $\phi_s = 0°$, we have scattering angle between 0° and 180° and the scattering angle is equal to θ_s with $0 \le \theta_s \le 180°$. For $\phi_s = 180°$, the scattering angle is equal to $360° - \theta_s$ with $0 \le \theta_s \le 180°$, covering the range of scattering angles between 180° and 360°.

Case (b): $\phi_s = 90°$ and $\phi_s = 270°$. For $\phi_s = 90°$,

$$\hat{k}_s = \sin\theta_s\hat{y} + \cos\theta_s\hat{z}; \quad \hat{v}_s = \cos\theta_s\hat{y} - \sin\theta_s\hat{z}; \quad \hat{h}_s = -\hat{x}$$

Then

$$\hat{1}_s = \hat{1}_i = (\hat{k}_s \times \hat{k}_i)/|\hat{k}_s \times \hat{k}_i| = \hat{x} = -\hat{h}_s$$

$$\hat{2}_i = \hat{k}_i \times \hat{1}_i = \hat{y}; \quad \hat{2}_s = \hat{k}_s \times \hat{1}_s = \hat{v}_s$$

Thus the incident wave with $\hat{e}_i = \hat{y}$ is of polarization $\hat{2}_i$. Similar polarization vector formulas can be obtained for $\phi_s = 270°$. Hence we define the phase matrix elements as

$$P_{12}(\theta_s) = \frac{\sigma_{h_s N}}{V} \tag{10.3.112}$$

$$P_{22}(\theta_s) = \frac{\sigma_{v_s N}}{V} \tag{10.3.113}$$

The quantities $P_{22}(\theta_s)$ and $P_{12}(\theta_s)$ correspond to co-polarization and cross-polarization, respectively. In Figs. 10.3.4c and 10.3.4d, we plot P_{12} and P_{22}, respectively, as a function of scattering angle. The scattering angle is between 0° and 360°. The following definition is used. For $\phi_s = 90°$, we have the scattering angle between 0° and 180°. The scattering angle is equal to θ_s. For $\phi_s = 270°$, the scattering angle is equal to $360° - \theta_s$ with $0 \le \theta_s \le 180°$, covering the range of scattering angles between 180° and 360°.

In Fig. 10.3.4, we show the results of P_{11}, P_{21}, P_{12}, and P_{22} for the fractional volume of 10%. The results of independent scattering are also shown for comparison. The unit is such that wavelength is equal to unity.

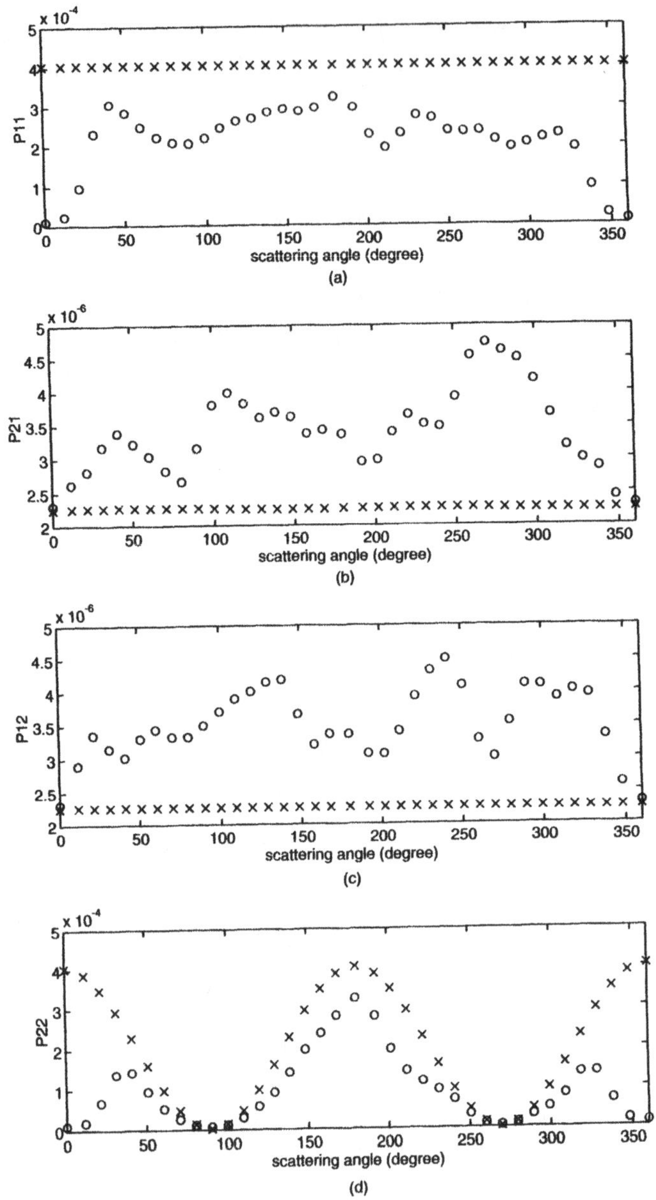

Figure 10.3.4 Phase matrix as a function of scattering angle for $ka = 0.2$, fractional volume $f = 10\%$, elongation ratio $e = 1.8$, relative permittivity of particles $\epsilon_r = 3.2$. The spheroids are randomly oriented. In the simulations, $N = 2000$ particles are used and the results are averaged over $N_r = 50$ realizations. (a) P_{11}, (b) P_{21}, (c) P_{12}, and (d) P_{22}. The symbol "o" represents the dense medium results, and the symbol "x" represents the independent scattering results.

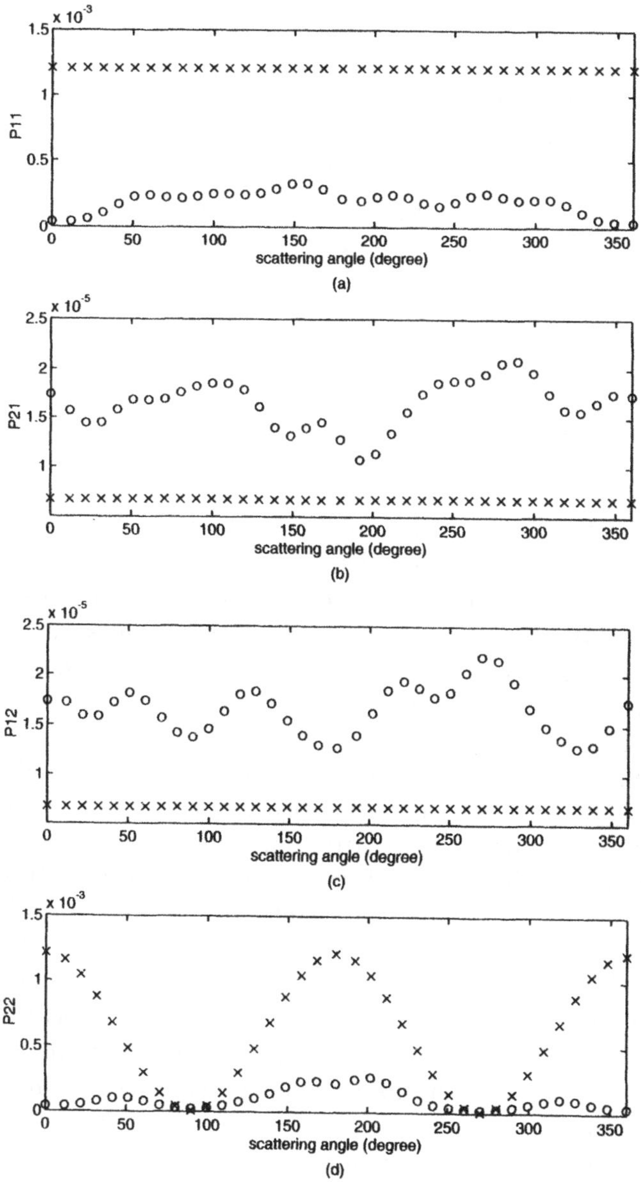

Figure 10.3.5 Phase matrix as a function of scattering angle for $ka = 0.2$, fractional volume $f = 30\%$, elongation ratio $e = 1.8$, relative permittivity of particles $\epsilon_r = 3.2$. The spheroids are randomly oriented. In the simulations, $N = 2000$ particles are used and the results are averaged over $N_r = 50$ realizations. (a) P_{11}, (b) P_{21}, (c) P_{12}, and (d) P_{22}. The symbol "o" represents the dense medium results, and the symbol "x" represents the independent scattering results.

f	P_{11}(dense)	P_{21}(dense)	P_{11}(indep)	P_{21}(indep)
10%	3.24×10^{-4}	4.73×10^{-6}	4.03×10^{-4}	2.27×10^{-6}
20%	3.90×10^{-4}	1.46×10^{-5}	8.06×10^{-4}	4.52×10^{-6}
30%	3.30×10^{-4}	2.09×10^{-5}	1.21×10^{-3}	6.80×10^{-6}

f	$\dfrac{P_{11}(\text{indep})}{P_{11}(\text{dense})}$	$\dfrac{P_{11}(\text{dense})}{P_{21}(\text{dense})}$	$\dfrac{P_{11}(\text{indep})}{P_{21}(\text{indep})}$
10%	1.25	68.50	177.64
20%	2.07	26.64	178.24
30%	3.67	15.80	178.08

Table 10.3.1 Values of phase matrix for various fractional volumes. The elements depend on scattering angles. The maximum values are tabulated.

The dimension of phase matrix is inverse of distance. We note that the co-polarization, P_{11} and P_{22}, are smaller than those of independent scattering, while the cross-polarization, P_{21} and P_{12}, are higher than those of independent scattering. We also note that the simulation results fluctuate because of random phase situation while that of independent scattering are smooth curves. We also note that P_{22} has angular dependence that is the characteristic of Rayleigh scattering.

In Fig. 10.3.5, we show the corresponding results for a higher fractional volume of 30%. In this case, the co-polarization of P_{11} and P_{22} become much lower than those of independent scattering, while cross-polarization P_{21} and P_{12} are much higher than those of independent scattering. To compare the result quantitatively, we tabulate the values of P_{11} and P_{21} in Table 10.3.1. The values are dependent on scattering angles. We tabulate the maximum values only. We also list the results for those of fractional volume of 20%. We note that the ratio of co-polarization of independent scattering to that of dense media scattering increases from 1.25 to 3.67 from fractional volume of 10% to 30%. This means that independent scattering significantly overestimates the total amount of scattering at high fractional volume for non-sticky particles. The cross-polarization is a small fraction of the total scattering. However, it has the opposite trend. The ratio of cross-polarization to co-polarization stays around 178 for independent scattering. However, the ratio decreases from 68.5 to 15.8 when fractional volume increases from 10% to 30%. Compared with 178, the ratio of 15.8 is ten times less, meaning that the ratio of cross-polarization to co-polarization for dense media can be ten times different from that of independent scattering.

4 Numerical Simulations Using T-Matrix Formalism

4.1 Multiple Scattering Equations

In this section, we use the multiple scattering equations based on the T-matrix formalism. Consider N spherical scatterers bounded by surfaces S_1, $S_2, \ldots, S_N$ (Fig. 10.4.1) occupying regions $V_1, V_2, \ldots, V_N$. The scatterers are centered at $\overline{r}_1, \overline{r}_2, \ldots, \overline{r}_N$ and have radii a_1, a_2, $\ldots$, a_N, respectively. Let the background region be denoted by V_o. The ith scatterer has permittivity equal to ϵ_i, wavenumber k_i, and permeability μ. Consider an incident field expressed in terms of vector spherical wave functions

$$\overline{E}^{inc}(\overline{r}) = \sum_{m,n} \left[a_{mn}^{(M)} Rg\overline{M}_{mn}(kr,\theta,\phi) + a_{mn}^{(N)} Rg\overline{N}_{mn}(kr,\theta,\phi) \right] \tag{10.4.1}$$

The Foldy-Lax equations are of the following form

$$\overline{E}_l^{ex} = \overline{E}^{inc} + \overline{\overline{G}}_o \sum_{\substack{j=1 \\ j\neq l}}^{N} \overline{\overline{T}}_j \overline{E}_j^{ex} \tag{10.4.2}$$

The interpretation is that the field exciting particle l is the sum of the incident field and scattered field from all particles j except particle l. The term $\overline{\overline{G}}_o \overline{\overline{T}}_j \overline{E}_j^{ex}$ is the field exciting particle j that is scattered by particle j and then propagates to particle l.

Let $\overline{r}$ be in the vicinity of particle l. The exciting field of particle l is

$$\overline{E}_l^{ex}(\overline{r}) = \sum_{mn} \left[w_{mn}^{(M)(l)} Rg\overline{M}_{mn}(k\overline{r}\overline{r}_l) + w_{mn}^{(N)(l)} Rg\overline{N}_{mn}(k\overline{r}\overline{r}_l) \right] \tag{10.4.3}$$

where $w_{mn}^{(M)(l)}$ and $w_{mn}^{(N)(l)}$ are unknown exciting field coefficients to be determined.

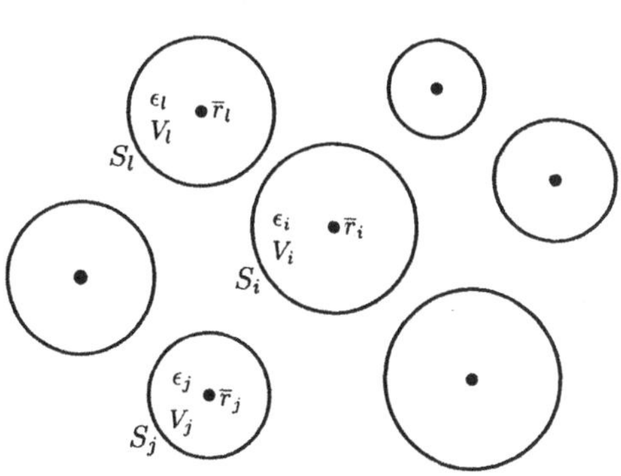

Figure 10.4.1 Particles $1, 2, \ldots, N$ occupying regions $V_1, V_2, \ldots, V_N$ and bounded by surfaces $S_1, S_2, \ldots, S_N$, respectively.

For $\overline{r}$ in the vicinity of particle j, the exciting field of particle j is

$$\overline{E}_j^{ex}(\overline{r}) = \sum_{mn} \left[w_{mn}^{(M)(j)} Rg\overline{M}_{mn}(k\overline{r}\overline{r}_j) + w_{mn}^{(N)(j)} Rg\overline{N}_{mn}(k\overline{r}\overline{r}_j) \right] \tag{10.4.4}$$

The particle j will scatter $\overline{E}_j^{ex}$ and give rise to outgoing vector spherical waves. The scattered wave by particle j is

$$\overline{E}_j^{s}(\overline{r}) = \sum_{mn} \left[w_{mn}^{(M)(j)} T_n^{(M)(j)} \overline{M}_{mn}(k\overline{r}\overline{r}_j) + w_{mn}^{(N)(j)} T_n^{(N)(j)} \overline{N}_{mn}(k\overline{r}\overline{r}_j) \right] \tag{10.4.5}$$

where $T_n^{(M)}$ and $T_n^{(N)}$ are T-matrix elements of spheres as given in (2.8.46)–(2.8.47) of Chapter 2, Volume I. Putting (10.4.5) and (10.4.1) in (10.4.2), we have for $|\overline{r} - \overline{r}_l| \gtrsim a_l$,

$$\begin{aligned} &\sum_{mn} \left[w_{mn}^{(M)(l)} Rg\overline{M}_{mn}(k\overline{r}\overline{r}_l) + w_{mn}^{(N)(l)} Rg\overline{N}_{mn}(k\overline{r}\overline{r}_l) \right] \\ &= \sum_{\mu\nu} \left[a_{\mu\nu}^{(M)} Rg\overline{M}_{\mu\nu}(k\overline{r}) + a_{\mu\nu}^{(N)} Rg\overline{N}_{\mu\nu}(k\overline{r}) \right] \\ &+ \sum_{\substack{j=1 \\ j\neq l}}^{N} \sum_{\mu\nu} \left[w_{\mu\nu}^{(M)(j)} T_\nu^{(M)(j)} \overline{M}_{\mu\nu}(k\overline{r}\overline{r}_j) + w_{\mu\nu}^{(N)(j)} T_\nu^{(N)(j)} \overline{N}_{\mu\nu}(k\overline{r}\overline{r}_j) \right] \end{aligned} \tag{10.4.6}$$

In order to balance coefficients in (10.4.6), we make the transformation of the vector spherical waves so that the spherical waves are all centered at $\overline{r}_l$.

The translational addition theorem for vector spherical waves can be expressed as follows. Consider the triangle with vectors $\overline{r}_o$, $\overline{r}$, and $\overline{r}'$ forming the three sides. The vector spherical wave centered at O can be expressed in terms of linear combinations of vector spherical waves at O' (Fig. 10.4.2). Note that in Fig. 10.4.2,

$$\overline{r} = \overline{r}_o + \overline{r}' \tag{10.4.7}$$

Then the translation addition theorem is [Cruzan, 1961, 1962]

$$\begin{aligned} Rg\overline{M}_{mn}(k\overline{r}) = \sum_{\nu\mu} \Big\{ & Rg A_{\mu\nu mn}(k\overline{r}_o) Rg\overline{M}_{\mu\nu}(k\overline{r}') \\ & + Rg B_{\mu\nu mn}(k\overline{r}_o) Rg\overline{N}_{\mu\nu}(k\overline{r}') \Big\} \end{aligned} \tag{10.4.8}$$

$$\begin{aligned} Rg\overline{N}_{mn}(k\overline{r}) = \sum_{\nu\mu} \Big\{ & Rg B_{\mu\nu mn}(k\overline{r}_o) Rg\overline{M}_{\mu\nu}(k\overline{r}') \\ & + Rg A_{\mu\nu mn}(k\overline{r}_o) Rg\overline{N}_{\mu\nu}(k\overline{r}') \Big\} \end{aligned} \tag{10.4.9}$$

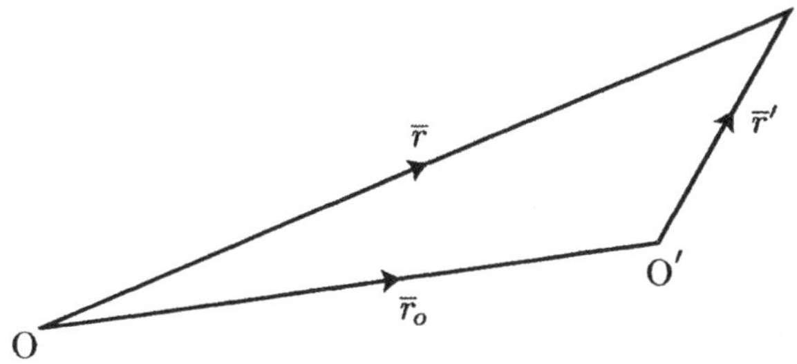

Figure 10.4.2 Translational addition theorem for vector spherical waves. Three vectors $\overline{r}$, $\overline{r}'$, and $\overline{r}_o$ form a triangle.

and for $r_o > r'$

$$\overline{M}_{mn}(k\overline{r}) = \sum_{\nu\mu} \left\{ A_{\mu\nu mn}(k\overline{r}_o) Rg\overline{M}_{\mu\nu}(k\overline{r}') + B_{\mu\nu mn}(k\overline{r}_o) Rg\overline{N}_{\mu\nu}(k\overline{r}') \right\} \tag{10.4.10}$$

$$\overline{N}_{mn}(k\overline{r}) = \sum_{\nu\mu} \left\{ B_{\mu\nu mn}(k\overline{r}_o) Rg\overline{M}\mu\nu(k\overline{r}') + A_{\mu\nu mn}(k\overline{r}_o) Rg\overline{N}_{\mu\nu}(k\overline{r}') \right\} \tag{10.4.11}$$

where

$$A_{\mu\nu mn}(k\overline{r}_o) = \frac{\gamma_{mn}}{\gamma_{\mu\nu}}(-1)^{\mu} \sum_p a(m,n|-\mu,\nu|p)a(n,\nu,p)h_p(kr_o)Y_p^{m-\mu}(\theta_o,\phi_o) \tag{10.4.12}$$

$$B_{\mu\nu mn}(k\overline{r}_o) = \frac{\gamma_{mn}}{\gamma_{\mu\nu}}(-1)^{\mu+1} \sum_p a(m,n|-\mu,\nu|p,p-1)b(n,\nu,p)h_p(kr_o)Y_p^{m-\mu}(\theta_o,\phi_o) \tag{10.4.13}$$

and γ_{mn} is defined in Eq. (1.4.59) of Chapter 1, Volume I. The expressions $RgA_{\mu\nu mn}(k\overline{r}_o)$ and $RgB_{\mu\nu mn}(k\overline{r}_o)$ are respectively those in (10.4.12) and (10.4.13) with $h_p(k\overline{r}_o)$ replaced by $j_p(k\overline{r}_o)$. In (10.4.12) and (10.4.13)

$$a(m,n|\mu,\nu|p) = (-1)^{m+\mu}(2p+1)\left[\frac{(n+m)!(\nu+\mu)!(p-m-\mu)!}{(n-m)!(\nu-\mu)!(p+m+\mu)!}\right]^{1/2} \cdot \begin{pmatrix} n & \nu & p \\ m & \mu & -(m+\mu) \end{pmatrix} \begin{pmatrix} n & \nu & p \\ 0 & 0 & 0 \end{pmatrix} \tag{10.4.14}$$

$$a(m,n|\mu,\nu|p,q) = (-1)^{m+\mu}(2p+1)\left[\frac{(n+m)!(\nu+\mu)!(p-m-\mu)!}{(n-m)!(\nu-\mu)!(p+m+\mu)!}\right]^{1/2}$$
$$\cdot \begin{pmatrix} n & \nu & p \\ m & \mu & -(m+\mu) \end{pmatrix} \begin{pmatrix} n & \nu & q \\ 0 & 0 & 0 \end{pmatrix} \tag{10.4.15}$$

$$a(n,\nu,p) = \frac{i^{\nu-n+p}}{2\nu(\nu+1)}\Big[2\nu(\nu+1)(2\nu+1) + (\nu+1)(n+\nu-p)$$
$$\cdot (n+p-\nu+1) - \nu(n+\nu+p+2)(\nu+p-n+1)\Big] \tag{10.4.16}$$

$$b(n,\nu,p) = -\frac{(2\nu+1)}{2\nu(\nu+1)} i^{\nu+p-n}\Big[(n+\nu+p+1)(\nu+p-n)$$
$$\cdot (n+p-\nu)(n+\nu-p+1)\Big]^{1/2} \tag{10.4.17}$$

and

$$\begin{pmatrix} j_1 & j_2 & j_3 \\ m_1 & m_2 & -(m_1+m_2) \end{pmatrix} \tag{10.4.18}$$

are the Wigner $3j$ symbols [Edmonds, 1957]. We note the following:

(1) The m index consists of $m = 0, \pm 1, \pm 2, \ldots, \pm n$ and μ index consists of $\mu = 0, \pm 1, \pm 2, \ldots, \pm \nu$.

(2) $A_{\mu\nu mn}$, $B_{\mu\nu mn}$ may not be of the same dimension as $\overline{M}_{mn}$ and $\overline{N}_{mn}$. For example, we may truncate $\overline{M}_{mn}$ and $\overline{N}_{mn}$ at a multipole order of $n = N_M$. However, for $A_{\mu\nu mn}$ and $B_{\mu\nu mn}$, the summations of (10.4.8)–(10.4.11), we may need to sum over $\nu = 1, 2, \ldots,$ to beyond N_M to have sufficient accuracy. This means $A_{\mu\nu mn}$ and $B_{\mu\nu mn}$ can be non-square matrices.

(3) For the Wigner $3j$ symbols in (10.4.18),

$$m_1 = 0, \pm 1, \ldots, \pm j_1$$
$$m_2 = 0, \pm 1, \ldots, \pm j_2$$

Also j_3 is bounded by

$$|j_1 - j_2| \le j_3 \le j_1 + j_2 \tag{10.4.19}$$

The symmetry relations for Wigner $3j$ symbols are, for even permutations,

$$\begin{pmatrix} j_1 & j_2 & j_3 \\ m_1 & m_2 & m_3 \end{pmatrix} = \begin{pmatrix} j_2 & j_3 & j_1 \\ m_2 & m_3 & m_1 \end{pmatrix} = \begin{pmatrix} j_3 & j_1 & j_2 \\ m_3 & m_1 & m_2 \end{pmatrix} \tag{10.4.20a}$$

and for odd permutations

$$\begin{pmatrix} j_2 & j_1 & j_3 \\ m_2 & m_1 & m_3 \end{pmatrix} = \begin{pmatrix} j_1 & j_3 & j_2 \\ m_1 & m_3 & m_2 \end{pmatrix} = \begin{pmatrix} j_3 & j_2 & j_1 \\ m_3 & m_2 & m_1 \end{pmatrix}$$
$$= (-1)^{j_1+j_2+j_3} \begin{pmatrix} j_1 & j_2 & j_3 \\ m_1 & m_2 & m_3 \end{pmatrix} \tag{10.4.20b}$$

(4) In view of (10.4.20a) and (10.4.20b),

$$\begin{pmatrix} n & \nu & p \\ m & \mu & -(m+\mu) \end{pmatrix} \begin{pmatrix} n & \nu & p \\ 0 & 0 & 0 \end{pmatrix}$$

is only nonzero for

$$p = |n-\nu|, |n-\nu|+2, |n-\nu|+4, \ldots, n+\nu \tag{10.4.21a}$$

and

$$\begin{pmatrix} n & \nu & p \\ m & \mu & -(m+\mu) \end{pmatrix} \begin{pmatrix} n & \nu & p-1 \\ 0 & 0 & 0 \end{pmatrix}$$

is only nonzero for

$$p = |n-\nu|+1, |n-\nu|+3, \ldots, n+\nu-1 \tag{10.4.21b}$$

Equations (10.4.21a) and (10.4.21b) can be applied to the summations in (10.4.19) and (10.4.20), respectively.

Next we apply the translation theorem to (10.4.6). Note that

$$\overline{rr_j} = \overline{r} - \overline{r}_j = \overline{r} - \overline{r}_l + \overline{r}_l - \overline{r}_j = \overline{rr_l} + \overline{r_l r_j} \tag{10.4.22}$$

and

$$|\overline{rr_l}| \le |\overline{r_l r_j}| \tag{10.4.23}$$

Also $\overline{r} = \overline{rr_l} + \overline{r}_l$. We have on using translation addition theorem in (10.4.6),

$$\sum_{mn} \left[w_{mn}^{(M)(l)} Rg\overline{M}_{mn}(k\overline{rr_l}) + w_{mn}^{(N)(l)} Rg\overline{N}_{mn}(k\overline{rr_l}) \right] =$$
$$\sum_{\mu\nu}\sum_{mn} \Big\{ a_{\mu\nu}^{(M)} \Big[RgA_{mn\mu\nu}(k\overline{r}_l) Rg\overline{M}_{mn}(k\overline{rr_l}) + RgB_{mn\mu\nu}(k\overline{r}_l) Rg\overline{N}_{mn}(k\overline{rr_l}) \Big]$$
$$+ a_{\mu\nu}^{(N)} \Big[RgB_{mn\mu\nu}(k\overline{r}_l) Rg\overline{M}_{mn}(k\overline{rr_l}) + RgA_{mn\mu\nu}(k\overline{r}_l) Rg\overline{N}_{mn}(k\overline{rr_l}) \Big] \Big\}$$

$$+\sum_{\substack{j=1\\ j\neq l}}^{N}\sum_{\mu\nu}\sum_{mn}\Big\{w_{\mu\nu}^{(M)(j)}T_{\nu}^{(M)(j)}\Big[A_{mn\mu\nu}(k\overline{r_l r_j})Rg\overline{M}_{mn}(k\overline{r r_l})$$

$$+B_{mn\mu\nu}(k\overline{r_l r_j})Rg\overline{N}_{mn}(k\overline{r r_l})\Big]+w_{\mu\nu}^{(N)(j)}T_{\nu}^{(N)(j)}$$

$$\cdot\Big[B_{mn\mu\nu}(k\overline{r_l r_j})Rg\overline{M}_{mn}(k\overline{r r_l})+A_{mn\mu\nu}(k\overline{r_l r_j})Rg\overline{N}_{mn}(k\overline{r r_l})\Big]\Big\} \quad (10.4.24)$$

Balancing the coefficients of $Rg\overline{M}_{mn}(k\overline{r r_l})$ and $Rg\overline{N}_{mn}(k\overline{r r_l})$, we obtain

$$w_{mn}^{(M)(l)}=\sum_{\mu\nu}\Big[RgA_{mn\mu\nu}(k\overline{r}_l)a_{\mu\nu}^{(M)}+RgB_{mn\mu\nu}(k\overline{r}_l)a_{\mu\nu}^{(N)}\Big]$$

$$+\sum_{\substack{j=1\\ j\neq l}}^{N}\sum_{\mu\nu}\Big[A_{mn\mu\nu}(k\overline{r_l r_j})T_{\nu}^{(M)(j)}w_{\mu\nu}^{(M)(j)}$$

$$+B_{mn\mu\nu}(k\overline{r_l r_j})T_{\nu}^{(N)(j)}w_{\mu\nu}^{(N)(j)}\Big] \quad (10.4.25a)$$

$$w_{mn}^{(N)(l)}=\sum_{\mu\nu}\Big[RgB_{mn\mu\nu}(k\overline{r}_l)a_{\mu\nu}^{(M)}+RgA_{mn\mu\nu}(k\overline{r}_l)a_{\mu\nu}^{(N)}\Big]$$

$$+\sum_{\substack{j=1\\ j\neq l}}^{N}\sum_{\mu\nu}\Big[B_{mn\mu\nu}(k\overline{r_l r_j})T_{\nu}^{(M)(j)}w_{\mu\nu}^{(M)(j)}$$

$$+A_{mn\mu\nu}(k\overline{r_l r_j})T_{\nu}^{(N)(j)}w_{\mu\nu}^{(N)(j)}\Big] \quad (10.4.25b)$$

Equations (10.4.25*a*)–(10.4.25*b*) can be conveniently put in matrix form.

Let $\overline{\overline{A}}$ and $\overline{\overline{B}}$ represent respectively $A_{mn\mu\nu}$ and $B_{mn\mu\nu}$, and the matrices are of dimension $L_{\max}\times L_{\max}$, using the notation from Table 2.7.1 in Chapter 2 of Volume I. Also let $\overline{\overline{T}}^{(M)(j)}$ represent $T_{\nu}^{(M)(j)}$. Note that $\overline{\overline{T}}^{(M)}$ is of dimension $L_{\max}\times L_{\max}$. Also let column vectors $\overline{w}^{(M)(l)}$ and $\overline{a}^{(M)}$ represent $w_{mn}^{(M)(l)}$ and $a_{mn}^{(M)}$, etc. These column vectors are of dimension $L_{\max}\times 1$. Then we have

$$\overline{w}^{(M)(l)}=Rg\overline{\overline{A}}(k\overline{r}_l)\overline{a}^{(M)}+Rg\overline{\overline{B}}(k\overline{r}_l)\overline{a}^{(N)}$$

$$+\sum_{\substack{j=1\\ j\neq l}}^{N}\Big[\overline{\overline{A}}(k\overline{r_l r_j})\overline{\overline{T}}^{(M)(j)}\overline{w}^{(M)(j)}+\overline{\overline{B}}(k\overline{r_l r_j})\overline{\overline{T}}^{(N)(j)}\overline{w}^{(N)(j)}\Big] \quad (10.4.26a)$$

$$\overline{w}^{(N)(l)} = Rg\overline{\overline{B}}(k\overline{r}_l)\overline{a}^{(M)} + Rg\overline{\overline{A}}(k\overline{r}_l)\overline{a}^{(N)} + \sum_{\substack{j=1 \\ j\neq l}}^{N}\left[\overline{\overline{B}}(k\overline{r_l r_j})\overline{\overline{T}}^{(M)(j)}\overline{w}^{(M)(j)} + \overline{\overline{A}}(k\overline{r_l r_j})\overline{\overline{T}}^{(N)(j)}\overline{w}^{(N)(j)}\right] \quad (10.4.26b)$$

Let

$$\overline{a}_{inc} = \begin{bmatrix} \overline{a}^{(M)} \\ \overline{a}^{(N)} \end{bmatrix} \quad (10.4.27)$$

be of dimension $2L_{\max} \times 1$,

$$\overline{\overline{\sigma}}(k\overline{r}) = \begin{bmatrix} \overline{\overline{A}}(k\overline{r}) & \overline{\overline{B}}(k\overline{r}) \\ \overline{\overline{B}}(k\overline{r}) & \overline{\overline{A}}(k\overline{r}) \end{bmatrix} \quad (10.4.28)$$

be of dimension $2L_{\max} \times 2L_{\max}$, and

$$\overline{\overline{T}} = \begin{bmatrix} \overline{\overline{T}}^{(M)} & 0 \\ 0 & \overline{\overline{T}}^{(N)} \end{bmatrix} \quad (10.4.29)$$

be of dimension $2L_{\max} \times 2L_{\max}$. Then, the multiple scattering equations of (10.4.26*a*) and (10.4.26*b*) can be expressed in compact matrix form.

$$\overline{w}^{(l)} = \sum_{\substack{j=1 \\ j\neq l}}^{N} \overline{\overline{\sigma}}(k\overline{r_l r_j})\overline{\overline{T}}^{(j)}\overline{w}^{(j)} + Rg\overline{\overline{\sigma}}(k\overline{r}_l)\overline{a}_{inc} \quad (10.4.30)$$

for $l = 1, 2, \ldots, N$.

After the coefficients $\overline{w}^{(j)}$'s are solved, then the scattered wave is given by

$$\overline{E}^S(\overline{r}) = \sum_{j=1}^{N}\left\{\sum_{mn} a_{mn}^{S(M)(j)}\overline{M}_{mn}(k\overline{r r_j}) + a_{mn}^{S(N)(j)}\overline{N}_{mn}(k\overline{r r_j})\right\} \quad (10.4.31)$$

Let $\overline{a}^{S(j)}$ be a $2L_{\max} \times 1$ column matrix defined by

$$\overline{a}^{S(j)} = \begin{bmatrix} \overline{a}^{S(M)(j)} \\ \overline{a}^{S(N)(j)} \end{bmatrix} \quad (10.4.32)$$

and $\overline{a}^{S(M)(j)}$ and $\overline{a}^{S(N)(j)}$ are $L_{\max} \times 1$ matrices representing $a_{mn}^{S(M)(j)}$ and $a_{mn}^{S(N)(j)}$. From (10.4.5), the scattered field coefficients $\overline{a}^{S(j)}$ satisfy the relation

$$\overline{a}^{S(j)} = \overline{\overline{T}}^{(j)}\overline{w}^{(j)} \quad (10.4.33)$$

Hence, once the exciting field coefficients $\overline{w}^{(j)}$ are solved from (10.4.30), the coefficients of the scattered field are calculated by (10.4.33). The final scattered field is then calculated via (10.4.31).

For the case of plane wave incidence, we can further simplify (10.4.30). Consider an incident wave with $\hat{k}_i = \sin\theta_i \cos\phi_i \hat{x} + \sin\theta_i \sin\phi_i \hat{y} + \cos\theta_i \hat{z}$. Let

$$\overline{E}^{inc}(\overline{r}) = \overline{A} e^{i\overline{k}_i \cdot \overline{r}} = \sum_{n,m} \left[a_{mn}^{(M)} Rg\overline{M}_{mn}(kr) + a_{mn}^{(N)} Rg\overline{N}_{mn}(kr) \right] \tag{10.4.34}$$

Given polarization vector $\overline{A}$, $a_{mn}^{(M)}$ and $a_{mn}^{(N)}$ can be calculated readily. We note that $\overline{E}^{inc}(\overline{r})$ can be written in the following alternate form

$$\overline{E}^{inc}(\overline{r}) = e^{i\overline{k}_i \cdot \overline{r}_l} \overline{A} e^{i\overline{k}_i \cdot (\overline{r} - \overline{r}_l)} = e^{i\overline{k}_i \cdot \overline{r}_l} \sum_{n,m} \left[a_{mn}^{(M)} Rg\overline{M}_{mn}(k\overline{r}\overline{r}_l) + a_{mn}^{(N)} Rg\overline{N}_{mn}(k\overline{r}\overline{r}_l) \right] \tag{10.4.35}$$

Hence $e^{i\overline{k}_i \cdot \overline{r}_l} a_{mn}^{(M)}$ and $e^{i\overline{k}_i \cdot \overline{r}_l} a_{mn}^{(N)}$ are the contributions of the incident wave to the exciting field coefficients $w_{mn}^{(M)(l)}$ and $w_{mn}^{(N)(l)}$ respectively of particle l. Thus

$$Rg\overline{\overline{\sigma}}(k\overline{r}_l)\overline{a}_{inc} = e^{i\overline{k}_i \cdot \overline{r}_l} \overline{a}_{inc} \tag{10.4.36}$$

and we have the following simplified form of the multiple scattering equations for plane wave incidence

$$\overline{w}^{(l)} = \sum_{\substack{j=1 \\ j \neq l}}^{N} \overline{\overline{\sigma}}(k\overline{r_l r_j}) \overline{\overline{T}}^{(j)} \overline{w}^{(j)} + e^{i\overline{k}_i \cdot \overline{r}_l} \overline{a}_{inc} \tag{10.4.37}$$

for $l = 1, 2, \ldots, N$. After the exciting field coefficients are calculated. The scattered field is calculated by (10.4.31). Let $\overline{a}^{S(j)}$ be a $2L_{\max} \times 1$ column matrix defined by

$$\overline{a}^{S(j)} = \begin{bmatrix} \overline{a}^{S(M)(j)} \\ \overline{a}^{S(N)(j)} \end{bmatrix} \tag{10.4.38}$$

and $\overline{a}^{S(M)(j)}$ and $\overline{a}^{S(N)(j)}$ are $L_{\max} \times 1$ matrices representing $a_{mn}^{S(M)(j)}$ and $a_{mn}^{S(N)(j)}$. The scattered field coefficients $\overline{a}^{S(j)}$ satisfy the relation

$$\overline{a}^{S(j)} = \overline{\overline{T}}^{(j)} \overline{w}^{(j)} \tag{10.4.39}$$

The final scattered field in the far field is

$$\overline{E}^S = \frac{e^{ikr}}{kr} \sum_{mn} \gamma_{mn} \left[a_{mn}^{S(M)} \overline{C}_{mn}(\theta_s, \phi_s) i^{-n-1} + a_{mn}^{S(N)} \overline{B}_{mn}(\theta_s, \phi_s) i^{-n} \right] \tag{10.4.40}$$

in the observation direction of $\hat{k}_s = \sin\theta_s \cos\phi_s \hat{x} + \sin\theta_s \sin\phi_s \hat{y} + \cos\theta_s \hat{z}$.

4.2 Computational Considerations

In implementing the numerical simulation of scattering, several items have to be computed. We consider their computations in the following.

Wigner 3j Symbols

The Wigner $3j$ symbols are related to the Clebsch-Gordan coefficients by

$$\begin{pmatrix} j_1 & j_2 & j \\ m_1 & m_2 & m \end{pmatrix} = (-1)^{j_1-j_2-m}(2j+1)^{-\frac{1}{2}}(j_1j_2m_1m_2 \mid j_1j_2j(-m)) \tag{10.4.41}$$

where $(j_1j_2m_1m_2 \mid j_1j_2j,(-m))$ are the Clebsch-Gordan coefficients [Edmonds, 1957; Abramowitz and Stegun, 1964]. It is given by the following formula

$$\begin{aligned}
&(j_1j_2m_1m_2 \mid j_1j_2j,(-m)) \\
&= \delta(m, m_1+m_2)\cdot\sqrt{\frac{(j_1+j_2-j)!(j+j_1-j_2)!(j+j_2-j_1)!(2j+1)}{(j+j_1+j_2+1)!}} \\
&\cdot\sum_k\left(\frac{(-1)^k\sqrt{(j_1+m_1)!(j_1-m_1)!(j_2+m_2)!(j_2-m_2)!(j+m)!(j-m)!}}{k!(j_1+j_2-j-k)!(j_1-m_1-k)!(j_2+m_2-k)!}\right. \\
&\left.\cdot\frac{1}{(j-j_2+m_1+k)!(j-j_1-m_2+k)!}\right)
\end{aligned} \tag{10.4.42}$$

with $\delta(i,k) = \delta_{ik}$ being the Kronecker delta. Thus

$$\begin{pmatrix} j_1 & j_2 & j \\ m_1 & m_2 & m \end{pmatrix} = 0 \tag{10.4.43}$$

if $m \neq -(m_1+m_2)$. The values of j are within the range

$$|j_1 - j_2| \leq j \leq j_1 + j_2 \tag{10.4.44}$$

Also

$$m_1 = 0, \pm 1, \pm 2, \ldots, \pm j_1 \tag{10.4.45a}$$

$$m_2 = 0, \pm 1, \pm 2, \ldots, \pm j_2 \tag{10.4.45b}$$

In the summation over k in (10.4.42), the summation is over integers of k such that none of the factorial is negative. Note $1! = 1$, $0! = 1$, $(-1)! =$ undefined. Thus summation over k in (10.4.42) is $k = 0, 1, \ldots$ such that none of the factorial is negative. This means for a given j_1, j_2, m_1, m_2, j, m, there are only a finite number of terms in the summation.

To prevent the factorial to acquire too large an integer value, it is computationally advantageous to work with the logarithm of factorial. Thus let

$$A = \sqrt{\frac{(j_1+j_2-j)!(j+j_1-j_2)!(j+j_2-j_1)!(2j+1)}{(j+j_1+j_2+1)!}} \times \sqrt{(j_1+m_1)!(j_1-m_1)!} \times \sqrt{(j_2+m_2)!(j_2-m_2)!(j+m)!(j-m)!} \tag{10.4.46}$$

$$\begin{aligned} A_1 &= \log A \\ &= \frac{1}{2}\Big[\log((j_1+j_2-j)!) + \log((j+j_1-j_2)!) + \log((j+j_2-j_1)!) \\ &\quad + \log(2j+1) + \log((j_1+m_1)!) + \log((j_1-m_1)!) + \log((j_2+m_2)!) \\ &\quad + \log((j_2-m_2)!) + \log((j+m)!) + \log((j-m)!) \\ &\quad - \log((j+j_1+j_2+1)!)\Big] \end{aligned} \tag{10.4.47}$$

$$B = \frac{1}{k!(j_1+j_2-j-k)!(j_1-m_1-k)!(j_2+m_2-k)!} \cdot \frac{1}{(j-j_2+m_1+k)!(j-j_1-m_2+k)!} \tag{10.4.48}$$

$$\begin{aligned} B_1 &= \log B \\ &= -\Big[\log(k!) + \log((j_1+j_2-j-k)!) \\ &\quad + \log((j_1-m_1-k)!) + \log((j_2+m_2-k)!) \\ &\quad + \log((j-j_2+m_1+k)!) + \log((j-j_1-m_2+k)!)\Big] \end{aligned} \tag{10.4.49}$$

Thus we calculate A_1 and B_1 instead of A and B.

The Clebsch-Gordan coefficient is

$$(j_1 j_2 m_1 m_2 \mid j_1 j_2 j m) = \delta_{m,m_1+m_2} e^{A_1} \sum_k (-1)^k e^{B_1} \tag{10.4.50}$$

Computation of T-Matrix Coefficients

We need to compute $j_n(x)$ and $y_n(x)$. The recurrence relations are, for increasing orders

$$j_{n+1}(x) = \frac{(2n+1)}{x} j_n(x) - j_{n-1}(x) \tag{10.4.51}$$

$$y_{n+1}(x) = \frac{(2n+1)}{x} y_n(x) - y_{n-1}(x) \tag{10.4.52}$$

The lowest two orders are

$$j_0(x) = \frac{\sin x}{x} \tag{10.4.53}$$

$$j_1(x) = \frac{\sin x}{x^2} - \frac{\cos x}{x} \tag{10.4.54}$$

$$y_0(x) = -\frac{\cos x}{x} \tag{10.4.55}$$

$$y_1(x) = -\frac{\cos x}{x^2} - \frac{\sin x}{x} \tag{10.4.56}$$

Using initial values of (10.4.53)–(10.4.56), the values of higher order spherical Bessel and Neumann functions can be calculated by (10.4.51)–(10.4.52). However, for small argument x, the recurrence relation of (10.4.51) for spherical bessel functions may not provide numerically accurate values as higher order spherical Bessel functions have smaller and smaller values. For this case we can use the backward recurrence formula for spherical Bessel functions.

$$j_{n-1}(x) = \frac{(2n+1)}{x} j_n(x) - j_{n+1}(x) \tag{10.4.57}$$

For smaller argument and a higher order n,

$$j_n(x) = \frac{x^n}{1 \cdot 3 \cdot 5 \cdots (2n+1)} \tag{10.4.58}$$

Thus one start with high order values of bessel function and use (10.4.57) to work backward to $j_1(x)$ and $j_0(x)$.

Computation of Vector Spherical Harmonics and Vector Spherical Wave Functions

We need to compute associated Legendre polynomials defined by $\sin\theta$ and derivatives of associated Legendre polynomials. We define t_n^m and s_n^m functions

$$t_n^m(\cos\theta) = \sqrt{2\pi n(n+1)}\gamma_{mn} m \frac{P_n^m(\cos\theta)}{\sin\theta} \tag{10.4.59}$$

$$s_n^m(\cos\theta) = \sqrt{2\pi n(n+1)}\gamma_{mn} \frac{dP_n^m(\cos\theta)}{d\theta} \tag{10.4.60}$$

To compute t_n^m and s_n^m for $n \geq 1$, we use the following recurrence relations

$$t_{m+1}^{m+1}(\cos\theta) = -\sin\theta \left(\frac{m+1}{m}\right)\left(\frac{2m+3}{2m+2}\right)^{1/2} t_m^m(\cos\theta) \tag{10.4.61}$$

$$t_{m+1}^m(\cos\theta) = \sqrt{2m+3}\cos\theta t_m^m(\cos\theta) \tag{10.4.62}$$

$$t_n^m(x) = \sqrt{\frac{2n+1}{n^2-m^2}}\left[\sqrt{2n-1}\,x t_{n-1}^m(x) - \left(\frac{(n-1)^2-m^2}{(2n-3)}\right)^{1/2} t_{n-2}^m(x)\right] \tag{10.4.63}$$

for $n \geq m+2$. The recurrence relation can be initialized by

$$t_1^1(\cos\theta) = -\sqrt{3}/2 \tag{10.4.64}$$

and compute t_2^2, t_3^3, ... by using (10.4.61). Then we compute t_2^1, t_3^2, ... by using (10.4.62). Also

$$t_n^0(\cos\theta) = 0 \tag{10.4.65}$$

We then use (10.4.63) to compute t_n^m for increasing n.

The computation of the $s_n^m(\cos\theta)$ function can be carried out by using the following relations.

(i) For $m > 0$:

$$s_m^m(\cos\theta) = \cos\theta\, t_m^m(\cos\theta) \tag{10.4.66}$$

$$m s_n^m(\cos\theta) = n\cos\theta\, t_n^m(\cos\theta) - t_{n-1}^m(\cos\theta)\left(\frac{(2n+1)(n^2-m^2)}{2n-1}\right)^{1/2}, \quad (n \geq m+1) \tag{10.4.67}$$

(ii) For $m = 0$:

$$s_n^0(\cos\theta) = \frac{\cos\theta}{n-1}(4n^2-1)^{1/2} s_{n-1}^0(\cos\theta) - \frac{n}{n-1}\left(\frac{2n+1}{2n-3}\right)^{1/2} s_{n-2}^0(\cos\theta) \tag{10.4.68}$$

The computation is initialized by the relations

$$s_0^0(\cos\theta) = 0 \tag{10.4.69}$$

$$s_1^0(\cos\theta) = -\sqrt{3/2}\sin\theta \tag{10.4.70}$$

To calculate t_n^m and s_n^m for negative m, we can use the relations

$$t_n^{-m}(\cos\theta) = (-1)^{m+1} t_n^m(\cos\theta) \tag{10.4.71}$$

$$s_n^{-m}(\cos\theta) = (-1)^m s_n^m(\cos\theta) \tag{10.4.72}$$

Thus, to set up a computer code, we first decide on a maximum value of n equal to $N_{\max}$ and initialize the values by using (10.4.64) and (10.4.65). Next the values of t_m^m and t_{m+1}^m for $m = 1, 2, \ldots, N_{\max}$ are calculated by (10.4.61) and (10.4.62). The values for negative degrees m are obtained by using (10.4.71). The values of the functions s_n^m for $m > 0$ are obtained from the t_n^m values by (10.4.66) and (10.4.67). The function s_n^0 is computed by using the recurrence relation of (10.4.68) which is initialized by (10.4.69) and (10.4.70). Negative degrees of s_n^m are calculated by using (10.4.72).

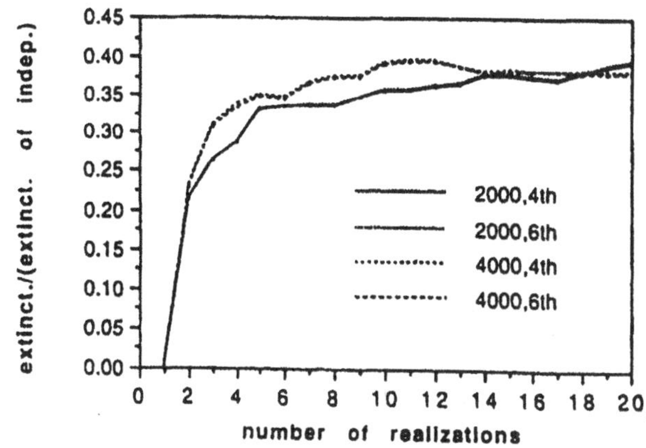

Figure 10.4.3 Convergence of extinction rate versus number of realizations and number of iterations for fractional volume $f = 15\%$ and $N = 2000$ and 4000. The extinction rate is normalized to the independent scattering case. Other parameters are $\epsilon_s = 3.2\epsilon_o$ and $ka = 0.2$.

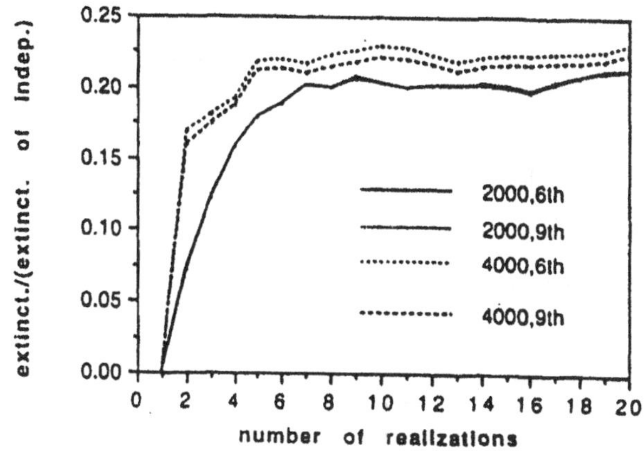

Figure 10.4.4 Convergence of extinction rate versus number of realizations and number of iterations for fractional volume $f = 25\%$ and $N = 2000$ and 4000. The extinction rate is normalized to the independent scattering case. Other parameters are $\epsilon_s = 3.2\epsilon_o$ and $ka = 0.2$.

4.3 Results and Comparisons with Analytic Theory

We solve (10.4.37) by iterations. In this section we present results for dielectric spheres of permittivity $\epsilon_s = 3.2\epsilon_o$. In Fig. 10.4.3 we show the results of $\kappa_e/(\kappa_e)_{\text{ind}}$ at a fractional volume of 15% as a function of realization by using 4 and 6 iterations. We note that for 2000 spheres there is practically no difference between 4 and 6 iterations. For 4000 spheres, the results of 4 and 6 iterations are practically identical. This demonstrates numerically the convergence with iterations for the simulated extinction rate. The results in

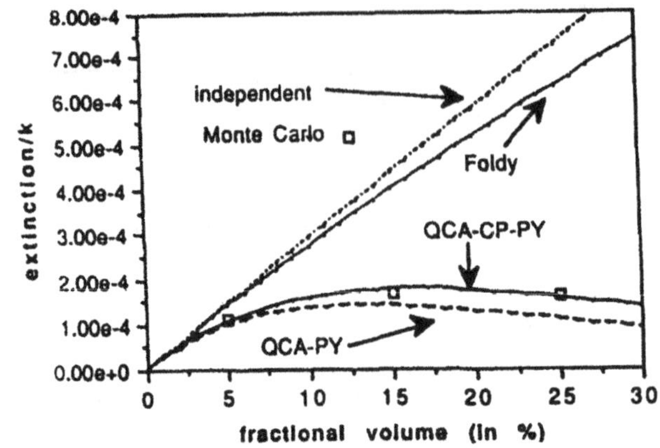

Figure 10.4.5 Extinction rate normalized to the free-space wave number as a function of the fractional volume of scatterers. The plots show calculations based on independent scattering, Foldy's formula, QCA–PY, QCA–CP–PY, and Monte Carlo simulations. Other parameters are $\epsilon_s = 3.2\epsilon_o$ and $ka = 0.2$.

	Fractional Volume		
Method	5%	15%	25%
Monte Carlo	0.748	0.381	0.224
4000 spheres	(4th)*	(6th)	(9th)
Monte Carlo	0.732	0.398	0.213
2000 spheres	(4th)	(6th)	(9th)
Independent scattering	1	1	1
Foldy	0.969	0.915	0.870
QCA–PY	0.674	0.318	0.150
QCA–CP–PY	0.732	0.402	0.215

Table 10.4.1 Numerical values of the ratio $\kappa_e/(\kappa_e)_{\text{ind}}$. (*The numbers in parentheses below the Monte Carlo values denote the numbers of iterations used.)

Fig. 10.4.3 also demonstrate convergence with the number of realizations. However, there is a small difference between the results with 2000 and 4000 spheres. The small difference can be attributed to the difference in the number of spheres that lie close to the edge of the cubic box in the cases of 2000 and 4000 spheres. The case of 25% is shown in Fig. 10.4.4 with 6 and 9 iterations. In Fig. 10.4.5 we compare the simulated results of 5% (30 realizations), 15% (20 realizations), and 25% (20 realizations) with analytic approximations of independent scattering, Foldy's approximation, QCA–PY, and QCA–CP–PY. We plot κ_e/k as a function of fractional volume f. We note that independent scattering predicts a linear increase with fractional volume, while Foldy's approximation predicts a monotonic increase with fractional volume at a slower rate than that of independent scattering. The QCA–PY and the QCA–CP–PY predict saturation and a decrease with further increase of fractional volume and are in good agreement with the simulations. Note that these simulations are for hard spheres without interparticle force. The results are tabulated in Table 10.4.1.

4.4 Simulation of Absorption Coefficient

For particles with absorption, the absorption coefficient can also be simulated. Upon solving the Foldy-Lax multiple scattering equations of (10.4.37), we obtain the coefficients $w_{mn}^{(M)(l)}$ and $w_{mn}^{(N)(l)}$ which gives the final exciting field of particle l. The power absorbed by particle l is then, from (2.8.84) in Chapter 2 of Volume I,

$$W_a^{(l)} = -\frac{1}{2\eta k^2}\sum_{mn}\left\{|w_{mn}^{(M)(l)}|^2(\mathrm{Re}T_n^{(M)(l)} + |T_n^{(M)(l)}|^2) + |w_{mn}^{(N)(l)}|^2(\mathrm{Re}T_n^{(N)(l)} + |T_n^{(N)(l)}|^2)\right\} \tag{10.4.73}$$

The power absorbed by N particles is

$$W_{aN} = \sum_{l=1}^{N} W_a^{(l)} \tag{10.4.74}$$

The absorption coefficient is equal to the absorption cross section per unit volume of space. Let S_i be the incident wave Poynting vector. Then

$$\kappa_a = \frac{1}{S_i V} W_{aN} \tag{10.4.75}$$

is the absorption cross section per unit volume of space. In (10.4.75), V is the total volume in which the sample of spheres is placed. For plane wave with unit magnitude of electric field, $S_i = 1/(2\eta)$. Then we have

$$\kappa_a = -\frac{1}{Vk^2}\sum_{l=1}^{N}\sum_{mn}\left\{|w_{mn}^{(M)(l)}|^2(\mathrm{Re}T_n^{(M)(l)} + |T_n^{(M)(l)}|^2) + |w_{mn}^{(N)(l)}|^2(\mathrm{Re}T_n^{(N)(l)} + |T_n^{(N)(l)}|^2)\right\} \tag{10.4.76}$$

Simulations by particles continue to be an important problem [Tsang et al. 1992; Zurk et al. 1995; Siqueira and Sarabandi, 2000]. Other computation methods used included the recursive T-matrix method [Lu et al. 1995; Chew et al. 1990], the discrete dipole approximation (DDA) [Purcell and Pennypacker, 1973; Draine and Flatau, 1994] discussed in Chapter 2, and the finite-difference time-domain (FDTD) method [Karkkainen et al. 2000].

REFERENCES AND ADDITIONAL READINGS

Abramowitz, M. and J. A. Stegun (1964), *Handbook of Mathematical Functions*, Dover Publications, New York.

Chew, W. C. (1990), *Waves and Fields in Inhomogeneous Media*, Van Nostrand Reinhold, New York.

Chew, W. C., J. A. Friedrich, and R. Geiger (1990), A multiple scattering solution for the effective permittivity of a sphere mixture, *IEEE Trans. Geosci. Remote Sens.*, **28(2)**, 207–214.

Cruzan, O. R. (1961), Translational addition theorems for spherical vector wave functions, **TR-906**, Diamond Ordinance Fuse Laboratories, Department of the Army, Washington DC.

Cruzan, O. R. (1962), Translational addition theorems for spherical vector wave functions, *Quart. J. Appl. Math.*, **20**, 33–40.

Draine, B. T. and P. J. Flatau (1994), Discrete-dipole approximation for scattering calculations, *J. Opt. Soc. Am. A*, **11**, 1491–1499.

Edmonds, A. R. (1957), *Angular Momentum in Quantum Mechanics*, Princeton University, Princeton, NJ.

Ishimaru, A. and Y. Kuga (1982), Attenuation constant of a coherent field in a dense distribution of particles, *J. Opt. Soc. Am.*, **72**, 1317–1320.

Karkkainen, K. K., A. H. Sihvola, and K. I. Nikoskinen (2000), Effective permittivity of mixtures: numerical validation by the FDTD method, *IEEE Trans. Geosci. Remote Sens.*, **38(3)**, 1303–1308.

Lu, C. C., W. C. Chew, and L. Tsang (1995), The application of recursive aggregate T-matrix algorithm in the Monte Carlo simulations of the extinction rate of random distribution of particles, *Radio Sci.*, **30(1)**, 25–28.

Mandt, C. (1992), Multiple scattering in random meda: Backscattering enhancement in a sparse distribution of large scatterers and Monte Carlo simulations of the extinction rate in dense media, University of Washington, Seattle.

Mandt, C., Y. Kuga, L. Tsang, and A. Ishimaru (1992), Microwave propagation and scattering in a dense distribution of spherical particles: experiment and theory, *Waves in Random Media*, **2(3)**, 225–234.

Metropolis, N., A. W. Rosenbluth, N. Rosenbluth, A. H. Teller, and E. Teller (1953), Equation of state calculation by fast computing machines, *J. Chem. Phys.*, **21(6)**, 1087–1092.

Peterson, B. and S. Strom (1973), T matrix for electromagnetic scattering from an arbitrary number of scatterers and representation of E(3), *Phys. Rev. D*, **8**, 3661–3678.

Purcell, E. M. and C. R. Pennypacker (1973), Scattering and absorption of light by non-spherical dielectric grains, *Astrophys. J.*, **186**, 705–714.

Siqueira, P. and K. Sarabandi (1996), Method of moments evaluation of the two-dimensional quasicrystalline approximation, *IEEE Trans. Antennas Propagat.*, **44(8)**, 1067–1077.

Siqueira, P. R. and K. Sarabandi (2000), T-matrix determination of effective permittivity for three-dimensional dense random media, *IEEE Trans. Antennas Propagat.*, **48(2)**, 317–327.

Tsang, L., K. H. Ding, S. E. Shih, and J. A. Kong (1998), Scattering of electromagnetic waves from dense distributions of spherical particles based on Monte Carlo simulations, *J. Opt. Soc. Am. A*, **15(12)**, 2660–2670.

Tsang, L., J. A. Kong, and R. T. Shin (1985), *Theory of Microwave Remote Sensing*, Wiley-Interscience, New York.

Tsang, L., C. Mandt, and K. H. Ding (1992), Monte Carlo simulations of the extinction rate of dense media with randomly distributed dielectric spheres based on solution of Maxwell's equations, *Optics Lett.*, **17(5)**, 314–316.

West, R., D. Gibbs, L. Tsang, and A. K. Fung (1994), Comparison of optical scattering experiments and the quasicrystalline approximation for dense media, *J. Opt. Soc. Am. A*, **11(6)**, 1854–1858.

Zurk, L. M., L. Tsang, K. H. Ding, and D. P. Winebrenner (1995), Monte Carlo simulations of the extinction rate of densely packed spheres with clustered and non-clustered geometries, *J. Opt. Soc. Am. A*, **12(8)**, 1772–1781.

Chapter 11

ANGULAR CORRELATION FUNCTION AND DETECTION OF BURIED OBJECT

1 Introduction

Besides using radar cross section, it is important to study statistical moments of the field. It is also useful to study object buried in random media. In remote sensing applications, investigators have used angular correlation functions [Chan et al. 1999] and frequency correlation functions [Sarabandi and Nashashibi, 1999]. Angular correlation function has also been used in detection of object that is buried in random media [Tsang et al. 1996; Zhang and Tsang, 1997, 1998; Zhang et al. 1998a,b]. Experiments have also been performed [Chan et al. 1999].

Studies show that the angular correlation function of scattering by random rough surface is generally small away from the memory line. The angular correlation function is the correlation function of two scattered fields in directions θ_{s2} and θ_{s1} corresponding to two incident waves in the θ_{i2} and θ_{i1} directions, respectively (Fig. 11.1.1). The correlation function is obtained by taking an average over realizations (samples) of rough surfaces. The memory line obeys the angular relation of $\sin\theta_{s2} - \sin\theta_{s1} = \sin\theta_{i2} - \sin\theta_{i1}$. The memory line is a result of statistical translational invariance of the random rough surface. This is analogous to phase matching. For a planar boundary separating two homogeneous media, the problem is horizontal translational invariant. The horizontal translational invariance gives rise to Snell's law as a result of phase matching. For periodic structures, the translational invariance of the structure holds when the translation is equal to a period. In this case, phase matching gives Floquet's theorem. For random medium, assuming statistical translational invariance of the medium, the memory effect is a consequence of the phase matching of the statistical moments of the fields.

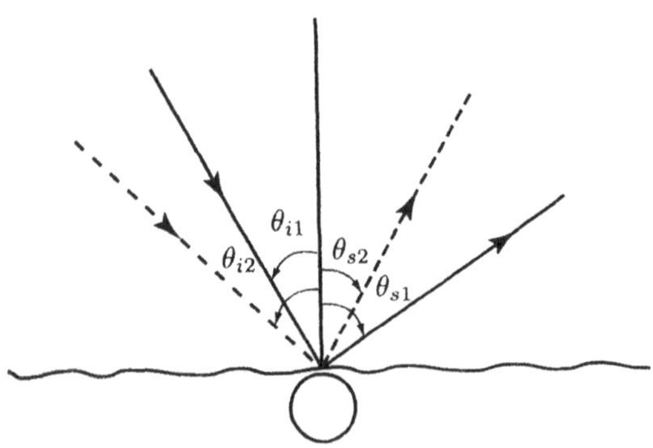

Figure 11.1.1 Configuration of angular correlation of wave scattering by a buried object under a rough surface. Scales are in wavelengths.

In this chapter, we study the angular correlation function, the memory effect [Feng et al. 1988] and the detection of object buried in random media. In Section 2, we study two-dimensional simulations and compare the angular correlation function and radar cross section. In Section 3, we perform three-dimensional simulations. Both angular and polarization correlations are utilized. In Section 4, we study SAR imaging. Angular correlation function and angular-frequency correlation function are used. It is shown that using angular correlation function in circular SAR improves the imaging of the buried object in random media.

2 Two-Dimensional Simulations of Angular Memory Effect and Detection of Buried Object

2.1 Introduction

In Section 2.2, we first give a simple derivation of the memory effect. In Section 2.3, we study the detection of buried object. In the real-life problem of buried object detection, the buried object is under a single random rough surface. If the wave is incident on another part of the random rough surface (another sample of rough surface), the buried object may no longer be within the incident beam. Thus, averaging over realizations of random rough surface is not applicable. In this section we study the electromagnetic wave scattering of a tapered incident wave by a buried object under a single random rough surface. The angular correlation function is calculated by taking the average over frequencies instead of taking the average over realizations of random rough surfaces. It is shown that for a sufficiently long random rough surface, the memory line still exists on averaging over frequencies. We also show that away from the memory line, the contribution of random rough surface scattering to the angular correlation function is small. This means that scattering by the buried object becomes more conspicuous.

2.2 Simple and General Derivation of Memory Effect

The memory effect is a condition of generalized phase matching due to statistical translational invariance of clutter (volume scattering and rough surface scattering). This is the only condition that is needed to derive the memory effect condition for the angular correlation function. The memory effect was first described by Feng et al. [1988].

The angular correlation function is the correlation function of two scattered fields in directions θ_{s2} and θ_{s1} corresponding to two incident waves in the θ_{i2} and θ_{i1} direction, respectively (Fig. 11.1.1). The correlation function is obtained by taking an average over realizations (samples) of rough surfaces.

If the clutter is statistically translationally invariant in a horizontal direction, the statistical phase-matching condition for the two-dimensional scattering problem is

$$k_{sx1} - k_{sx2} = k_{ix1} - k_{ix2} \tag{11.2.1}$$

or

$$\sin\theta_{s1} - \sin\theta_{s2} = \sin\theta_{i1} - \sin\theta_{i2} \tag{11.2.2}$$

The relations of (11.2.1)–(11.2.2) are proved as follows: Let J_{i1} and J_{i2} be the two antenna sources that launch the incident waves in direction $\overline{k}_{i1}$ and $\overline{k}_{i2}$, respectively. The wavevector components in x direction are respectively k_{ix1} and k_{ix2}. Note that the antenna currents J_{i1} and J_{i2} that launch the incident waves are not random. Let $G(\overline{\rho}, \overline{\rho}')$ be the exact Green's function for the boundary value problem that includes volume scattering and rough surface scattering effects. Assume statistical translational invariance in the horizontal x-direction and let $\overline{\Delta x} = \hat{x}\Delta x$. Then the following statistical properties hold for the exact Green's function

$$\langle G(\overline{\rho} + \overline{\Delta x}, \overline{\rho}' + \overline{\Delta x})\rangle = \langle G(\overline{\rho}, \overline{\rho}')\rangle \tag{11.2.3}$$

$$\langle G(\overline{\rho} + \overline{\Delta x}, \overline{\rho}'' + \overline{\Delta x}) G^*(\overline{\rho}' + \overline{\Delta x}, \overline{\rho}''' + \overline{\Delta x})\rangle = \langle G(\overline{\rho}, \overline{\rho}'') G^*(\overline{\rho}', \overline{\rho}''')\rangle \tag{11.2.4}$$

where the angular brackets denote the ensemble average. Note that the horizontal translational invariance holds when averages are taken. It does not hold for each realization.

Let the scattered field be ψ_{s1} and ψ_{s2} for the two antenna currents J_{i1} and J_{i2}, respectively. Then

$$\psi_{\left[\substack{s1\\s2}\right]}(\overline{\rho}) = \int d\overline{\rho}' G(\overline{\rho}, \overline{\rho}') J_{\left[\substack{i1\\i2}\right]}(\overline{\rho}') \tag{11.2.5}$$

Because the incident wave is a plane wave with $e^{ik_{ix}x}$ dependence, the antenna that launches the incident plane wave obeys the condition

$$J_{\left[\substack{i1\\i2}\right]}(\overline{\rho}' + \overline{\Delta x}) = \exp\left\{ik_{\left[\substack{ix1\\ix2}\right]}\Delta x\right\} J_{\left[\substack{i1\\i2}\right]}(\overline{\rho}') \tag{11.2.6}$$

The second moment, as given by the average of the product of two scattered fields, is

$$\langle \psi_{s1}(\overline{\rho})\psi_{s2}{}^*(\overline{\rho}')\rangle = \int d\overline{\rho}'' \int d\overline{\rho}''' \langle G(\overline{\rho}, \overline{\rho}'') G^*(\overline{\rho}', \overline{\rho}''')\rangle J_{i1}(\overline{\rho}'') J_{i2}^*(\overline{\rho}''') \tag{11.2.7a}$$

The two incident field directions are k_{i1} and k_{i2}. Displacement of $\overline{\rho}$ and $\overline{\rho}'$ by $\overline{\Delta x}$ gives

$$\begin{aligned}
&\langle \psi_{s1}(\overline{\rho}+\overline{\Delta x})\psi_{s2}^*(\overline{\rho}'+\overline{\Delta x})\rangle \\
&= \int d\overline{\rho}'' \int d\overline{\rho}''' \langle G(\overline{\rho}+\overline{\Delta x},\overline{\rho}'')G^*(\overline{\rho}'+\overline{\Delta x},\overline{\rho}''')\rangle J_{i1}(\overline{\rho}'')J_{i2}^*(\overline{\rho}''') \\
&= \int d\overline{\rho}'' \int d\overline{\rho}''' \langle G(\overline{\rho}+\overline{\Delta x},\overline{\rho}''+\overline{\Delta x})G^*(\overline{\rho}'+\overline{\Delta x},\overline{\rho}'''+\overline{\Delta x})\rangle \\
&\quad \cdot J_{i1}(\overline{\rho}''+\overline{\Delta x})J_{i2}^*(\overline{\rho}'''+\overline{\Delta x}) \\
&= \int d\overline{\rho}'' \int d\overline{\rho}''' \langle G(\overline{\rho},\overline{\rho}'')G^*(\overline{\rho}',\overline{\rho}''')\rangle J_{i1}(\overline{\rho}''+\overline{\Delta x})J_{i2}^*(\overline{\rho}'''+\overline{\Delta x})
\end{aligned} \tag{11.2.7b}$$

The second equality is a result of transformation of dummy variables. The third equality is a result of using the statistical translational invariance property of (11.2.4). Next we make use of the translational property (11.2.6) of the antenna current in (11.2.7). Then

$$\begin{aligned}
&\langle \psi_{s1}(\overline{\rho}+\overline{\Delta x})\psi_{s2}^*(\overline{\rho}'+\overline{\Delta x})\rangle \\
&\quad = e^{ik_{ix1}\Delta x}e^{-ik_{ix2}\Delta x}\int d\overline{\rho}''\int d\overline{\rho}'''\langle G(\overline{\rho},\overline{\rho}'')G^*(\overline{\rho}',\overline{\rho}''')\rangle J_{i1}(\overline{\rho}'')J_{i2}^*(\overline{\rho}''') \\
&\quad = e^{ik_{ix1}\Delta x}e^{-ik_{ix2}\Delta x}\langle \psi_{s1}(\overline{\rho})\psi_{s2}^*(\overline{\rho}')\rangle
\end{aligned} \tag{11.2.8}$$

In the scattered directions k_{sx1} and k_{sx2} respectively of ψ_{s1} and ψ_{s2}, we have

$$\langle \psi_{s1}(\overline{\rho}+\overline{\Delta x})\psi_{s2}^*(\overline{\rho}'+\overline{\Delta x})\rangle = e^{i(k_{sx1}-k_{sx2})\Delta x}\langle \psi_{s1}(\overline{\rho})\psi_{s2}^*(\overline{\rho}')\rangle \tag{11.2.9}$$

Substituting (11.2.9) into (11.2.8) gives

$$e^{i(k_{sx1}-k_{sx2})\Delta x}\langle \psi_{s1}(\overline{\rho})\psi_{s2}^*(\overline{\rho}')\rangle = e^{i(k_{ix1}-k_{ix2})\Delta x}\langle \psi_{s1}(\overline{\rho})\psi_{s2}^*(\overline{\rho}')\rangle \tag{11.2.10}$$

Thus balancing the two sides of (11.2.10) gives either $\langle \psi_{s1}(\overline{\rho})\psi_{s2}^*(\overline{\rho}')\rangle = 0$ or the memory effect condition

$$k_{ix1} - k_{ix2} = k_{sx1} - k_{sx2} \tag{11.2.11}$$

In terms of angles,

$$\sin\theta_{i1} - \sin\theta_{i2} = \sin\theta_{s1} - \sin\theta_{s2} \tag{11.2.12}$$

2.3 ACF of Random Rough Surfaces with Different Averaging Methods

In the simulation of wave scattering from rough surfaces, it is customary as shown in Chapters 4 to 6 to use the realization average to calculate the scattering cross section. However, when the object is buried under a single rough surface, realization averaging is not applicable, and other means of taking coherent averages must be investigated. We examine the ACF based on three methods of averaging: realization, frequency, and angular. The normalized scattered field is defined by

$$\psi^s(\bar{r}) = \frac{e^{ikr}}{\sqrt{r}}\psi_s^N(\hat{k}_s, \hat{k}_N) \tag{11.2.13}$$

Realization averaging: The ensemble average is obtained by taking averaging over different samples (rough surfaces) with the same statistics.

$$\Gamma_r(\theta_{s1}, \theta_{i1}; \theta_{s2}, \theta_{i2}) = \frac{1}{N_r}\sum_{n=1}^{N_r} \psi_s^N(\theta_{s1}, \theta_{i1}, n)\psi_s^{N*}(\theta_{s2}, \theta_{i2}, n)/\sqrt{P_1 P_2} \tag{11.2.14}$$

where n denotes the realization index, and N_r is the number of realizations. P_1 and P_2 are total power flux of the two incident waves, respectively. In general, many independent realizations must be generated to get smooth averages.

Frequency averaging: Frequency averaging takes an ensemble average over a frequency bandwidth $2\Delta f$ centered f_o. This can be done if scattering data are available over a wide frequency band. The excitation frequency varies between $-\Delta f + f_o$ to $\Delta f + f_o$

$$\Gamma_f(\theta_{s1}, \theta_{i1}; \theta_{s2}, \theta_{i2}) = \frac{1}{N_f}\sum_{n=1}^{N_f} \psi_s^N(\theta_{s1}, \theta_{i1}; f_n)\psi_s^{N*}(\theta_{s2}, \theta_{i2}; f_n)/\sqrt{P_1 P_2} \tag{11.2.15}$$

where N_r is the number of frequencies over the frequency range $f_o - \Delta f < f_n < f_o + \Delta f$, and n is the frequency index for f_n.

Angular averaging: The angular averaging is defined by the small changes of incident and scattering angles around the fixed angles.

$$\begin{aligned}&\Gamma_a(\theta_{s1}, \theta_{i1}; \theta_{s2}, \theta_{i2})\\ &= \frac{1}{N_a}\sum_{n=1}^{N_a} \psi_s^N(\theta_{s1} + \delta_n; \theta_{i1} + \delta_n)\psi_s^{N*}(\theta_{s2} + \delta_n, \theta_{i2} + \delta_n)/\sqrt{P_1 P_2}\end{aligned} \tag{11.2.16}$$

where N_a is the number of the angles, and δ_n is the small angular difference for index n.

Figure 11.2.1 shows the ACF magnitude with the three different ensemble averaging methods. The random rough surface is generated by using the spectrum method with a Gaussian height distribution and a Gaussian correlation function. The tapering parameter g chosen to be $L/4$. We plot the ACF as a function of $\sin\theta_{i2}$ and $\sin\theta_{s2}$ with fixed angles $\theta_{i1} = 20°$ and $\theta_{s1} = -20°$. Figure 11.2.1a is the result for one rough surface without any averaging. Figure 11.2.1b is that of realization averaging taken over 100 realizations. It is clear that the existence of the memory line becomes apparent if a sufficient number of independent samples are included. Figure 11.2.1c shows the ACF magnitude of a single rough surface based on the frequency averaging method. The single rough surface profile is as shown in Fig. 11.1.1. There are 50 equally spaced frequency samples over the frequency bandwidth of $0.5f_o$ to $1.5f_o$. Although the ACF magnitude by frequency averaging is noisier than that of realization averaging, a distinct memory line is clearly visible. To suppress fluctuation in Fig. 11.2.1c, more independent samples must be included in the averaging process. It was estimated that only about 10 independent samples can be obtained with the bandwidth of $0.5f_o$ to $1.5f_o$. For the frequency averaging to be effective, a wide bandwidth may be required. Figure 11.2.1d shows the ACF magnitude by the angular averaging method given in (11.2.16). The results are smoother than those without averaging (Fig. 11.2.1a), but the memory line is not as clearly visible as that of frequency averaging.

2.4 Scattering by a Buried Object Under a Rough Surface

In this section we study the scattered field from an object placed below a rough surface. The incident wave is a horizontally polarized (TE) wave, and we assume that the buried object is a perfect conductor as shown in Fig. 11.1.1.

Let ψ_o and ψ_1 be the fields in regions 0 and 1, respectively, and let the boundary conditions be $\psi_o = \psi_1$, $\frac{\partial \psi_o}{\partial_n} = \frac{\partial \psi_1}{\partial n}$ at the rough surface and $\psi_1 = 0\cdot$ on the surface of the object. We make use of the surface integral equations from Chapter 4. In region 0 we have the integral equation given by, for $\overline{r}$ on S_r,

$$\frac{1}{2}\psi_o(\overline{r}) = \psi^i(\overline{r}) + \int\!\!\!\!\!-_{S_r} \left[\psi_o(\overline{r}')\frac{\partial G_o(\overline{r},\overline{r}')}{\partial n'} - G_o(\overline{r},\overline{r}')\frac{\partial \psi_o(\overline{r}')}{\partial n'}\right] ds' \qquad (11.2.17)$$

where $\psi^i(\overline{r})$ is the incident wave field, $G_o(\overline{r},\overline{r}')$ is the Green's function for region 0, and $\int\!\!\!\!\!-_{S_r}$ denotes the principal value integral over the rough surface.

In region 1 we have the integral equation given by

$$\frac{1}{2}\psi_1(\overline{r}) = \psi_b^s(\overline{r}) - \int\!\!\!\!\!-_{S_r} \left[\psi_1(\overline{r}')\frac{\partial G_1(\overline{r},\overline{r}')}{\partial n'} - G_1(\overline{r},\overline{r}')\frac{\partial \psi_1(\overline{r}')}{\partial n'}\right] ds' \qquad (11.2.18)$$

where $\psi_b^s(\overline{r})$ is the scattered field from the buried object and is given by

$$\psi_b^s(\overline{r}) = -\int_{S_b} G_1(\overline{r},\overline{r}')\frac{\partial \psi_1(\overline{r}')}{\partial n'}\, ds' \qquad (11.2.19)$$

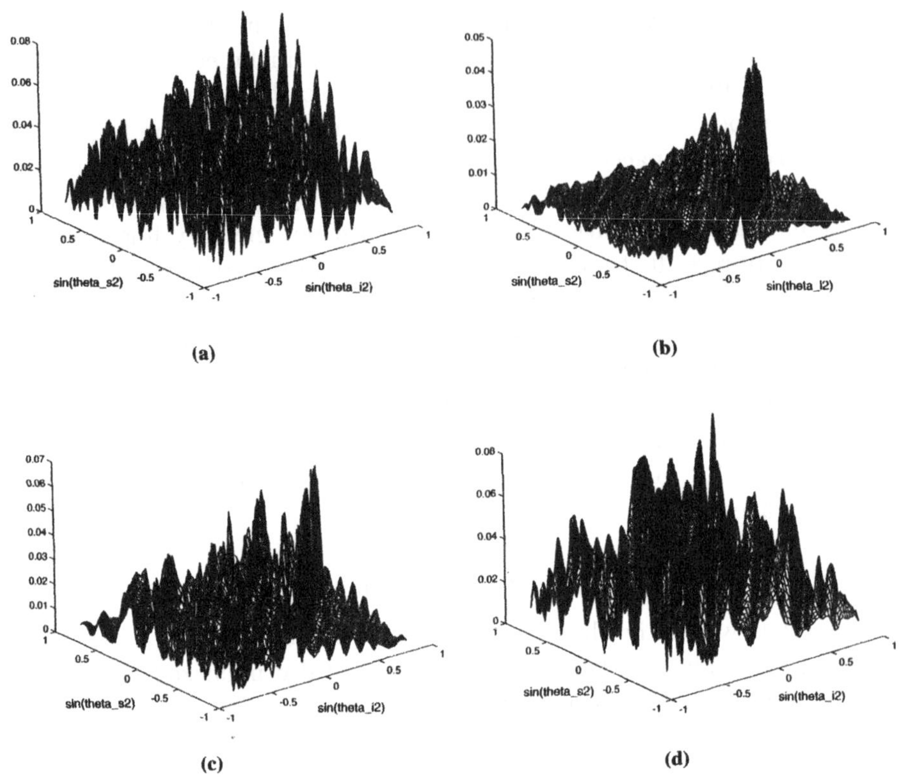

Figure 11.2.1 Three-dimensional plots of ACF magnitude by different averaging methods. Reference angles are $(\theta_{i1} = 20°,\ \theta_{s1} = -20°)$. Dielectric constant of region 1 is $\epsilon_r = 3.7 + i0.13$. $h = 0.35\lambda_o$, $l - 1.0\lambda_o$, $L = 40\lambda_o$, $g = L/4$. (a) One realization. (b) Realization averaging over 100 rough surfaces. (c) Frequency averaging over a frequency band of $0.5f_o$ to $1.5f_o$. (d) Angular averaging over an angular range of $\theta - 10°$ to $\theta + 10°$.

and s_b is the surface of the buried object.

For $\overline{r}$ on the surface of the buried object, we have

$$\psi_r^s(\overline{r}) - \int_{s_b} G_1(\overline{r}, \overline{r}') \frac{\partial \psi_1(\overline{r}')}{\partial n'}\, ds' = 0 \tag{11.2.20}$$

where $\psi_r^s(\overline{r})$ is the scattered field from the rough surface, given by

$$\psi_r^s(\overline{r}) = \int_{s_r} \left[\psi_1(\overline{r}') \frac{\partial G_1(\overline{r}, \overline{r}')}{\partial n'} - G_1(\overline{r}, \overline{r}') \frac{\partial \psi_1(\overline{r}')}{\partial n'} \right] ds' \tag{11.2.21}$$

Thus Eqs. (11.2.18) and (11.2.20) become, respectively,

$$\frac{1}{2}\psi_1(\overline{r}) + \int_{s_b} G_1(\overline{r},\overline{r}')\frac{\partial \psi_1(\overline{r}')}{\partial n'}\, ds' + \fint_{s_r}\left[\psi_1(\overline{r}')\frac{\partial G_1(\overline{r},\overline{r}')}{\partial n'} - G_1(\overline{r},\overline{r}')\frac{\partial \psi_1(\overline{r}')}{\partial n'}\right] ds' = 0 \quad (11.2.22)$$

for $\overline{r}$ on s_r.

$$\int_{s_b} G_1(\overline{r},\overline{r}')\frac{\partial \psi_1(\overline{r}')}{\partial n'}\, ds' - \fint_{s_r}\left[\psi_1(\overline{r}')\frac{\partial G_1(\overline{r},\overline{r}')}{\partial n'} - G_1(\overline{r},\overline{r}')\frac{\partial \psi_1(\overline{r}')}{\partial n'}\right] ds' = 0 \quad (11.2.23)$$

for $\overline{r}$ on s_b.

Equations (11.2.17), (11.2.22), and (11.2.23) are three coupled surface integral equations and can be put into a matrix form with discretization. The surface unknowns are $(\psi)_{s_r}$, $\left(\frac{\partial\psi}{\partial n}\right)_{s_r}$, and $\left(\frac{\partial\psi}{\partial n}\right)_{s_b}$, which denote the values of ψ, its normal derivative on the rough surface s_r, and the values of the normal derivative of ψ_1 on an object surface s_b, respectively.

$$\begin{bmatrix} A & B & 0 \\ C & D & E \\ F & G & H \end{bmatrix} \begin{bmatrix} (\psi)_{s_r} \\ \left(\dfrac{\partial\psi}{\partial n}\right)_{s_r} \\ \left(\dfrac{\partial\psi_1}{\partial n}\right)_{s_b} \end{bmatrix} = \begin{bmatrix} \psi_i \\ 0 \\ 0 \end{bmatrix} \quad (11.2.24)$$

The quantities A through H are impedance matrices.

After the matrix (11.2.24) is solved, the scattered field in region 0 is calculated by

$$\psi^s(\overline{r}) = \int_{s_r}\left[\psi(\overline{r}')\frac{\partial G(\overline{r},\overline{r}')}{\partial n'} - G(\overline{r},\overline{r}')\frac{\partial \psi(\overline{r}')}{\partial n'}\right] ds' \quad (11.2.25)$$

The normalized scattered field is defined in (11.2.13), and the ACF based on frequency averaging can be obtained using (11.2.15). When $\theta_{i1} = \theta_{i2}$ and $\theta_{s1} = \theta_{s2}$, the ACF becomes the scattered intensity. The average scattered intensity can be defined based on the frequency averaging method similar to that of ACF.

$$I_f(\theta_s, \theta_i) = \frac{1}{N_f}\sum_{n=1}^{N_f} |\psi_s^N(\theta_s, \theta_i; f_n)|^2 / P \quad (11.2.26)$$

We conducted simulations using a circular cylinder as the buried object. Frequency averaging is used. The cylinder is placed at a depth $d = 2\lambda_o$, and a surface length of $L = 40\lambda_o$ is used (Fig. 11.1.1). Unless specified otherwise, all distance units are based on wavelength λ_o of the center frequency.

In Fig. 11.2.2a and 11.2.2b, both ACF magnitude and scattered intensity with and without the buried object are shown as a function of the incident angle θ_{i2}. The reference angles are $\theta_{i1} = 30^\circ$ $\theta_{s1} = -50^\circ$, and the variable angles are set to $\theta_{s2} = -\theta_{i2}$. This configuration provides the backscattering cross section (RCS) for intensity, but the ACF magnitude intercepts the memory line only at $\theta_{i2} = 39^\circ$ at which a peak can be observed in Fig. 11.2.2a, since $2\sin 39^\circ = \sin 30^\circ + \sin 50^\circ$. In general, the RCS varies slowly as a function of angle, whereas the ACF becomes small away from the memory line due to the destructive phase interference in the coherent averaging process. We note that ACF in Fig. 11.2.2 is small because the memory line is avoided.

The ACF in Fig. 11.2.2b is much bigger than Fig. 11.2.2a because ACF of buried object is much bigger than ACF of rough surface away from the memory line. We calculated the ratios of ACF with and without the buried object. A similar definition is used for the intensity. The definition of the respective ratios are

$$\text{ratio of } |\text{ACF}| = \frac{|\text{ACF}| \text{ with buried object}}{|\text{ACF}| \text{ without buried object}} = \frac{|\Gamma_{sc}(\theta_{s1}, \theta_{i1}; \theta_{s2}, \theta_{i2})|}{|\Gamma_s(\theta_{s1}, \theta_{i1}; \theta_{s2}, \theta_{i2})|} \tag{11.2.27}$$

$$\text{ratio of } I = \frac{I \text{ with buried object}}{I \text{ without buried object}} = \frac{I_{sc}(\theta_s, \theta_i)}{I_s(\theta_s, \theta_i)} \tag{11.2.28}$$

The comparisons of these two ratios are shown in Fig. 11.2.2c. Although the ratio of ACF contains a large amount of fluctuation due to a limited number of independent samples, the presence of the buried object is clearly identifiable in the ACF data because the ACF ratio is considerably higher than that of the ratio of intensity. For example, for $-40^\circ < \theta_{i2} < 40^\circ$, the ratio of intensity is about 1 dB while the ratio of ACF is up to 15 dB. In Fig. 11.2.2a–c, the intensity and its ratio are that for $(\theta_s, \theta_i) = (\theta_{s2}, \theta_{i2})$ at back directions $(\theta_s = -\theta_i)$.

The large difference between the ratio of ACF and that of RCS in Fig. 11.2.2c is partially due to the small scattering from the rough surface at the reference angles ($\theta_{i1} = 30^\circ$, $\theta_{s1} = -50^\circ$). For a fair comparison, two ratios of intensities are shown in Fig. 11.2.2d. One is for the reference angles with $(\theta_s, \theta_i) = (\theta_{s1}, \theta_{i1}) = (-50^\circ, 30^\circ)$ used in (11.2.28), and one is for the observation angles $(\theta_s, \theta_i) = (\theta_{s2}, \theta_{i2}) = (-\theta_{i2}, \theta_{i2})$ in (11.2.28). It is to be noted that the results shown in Fig. 11.2.2d are obtained by an addi-

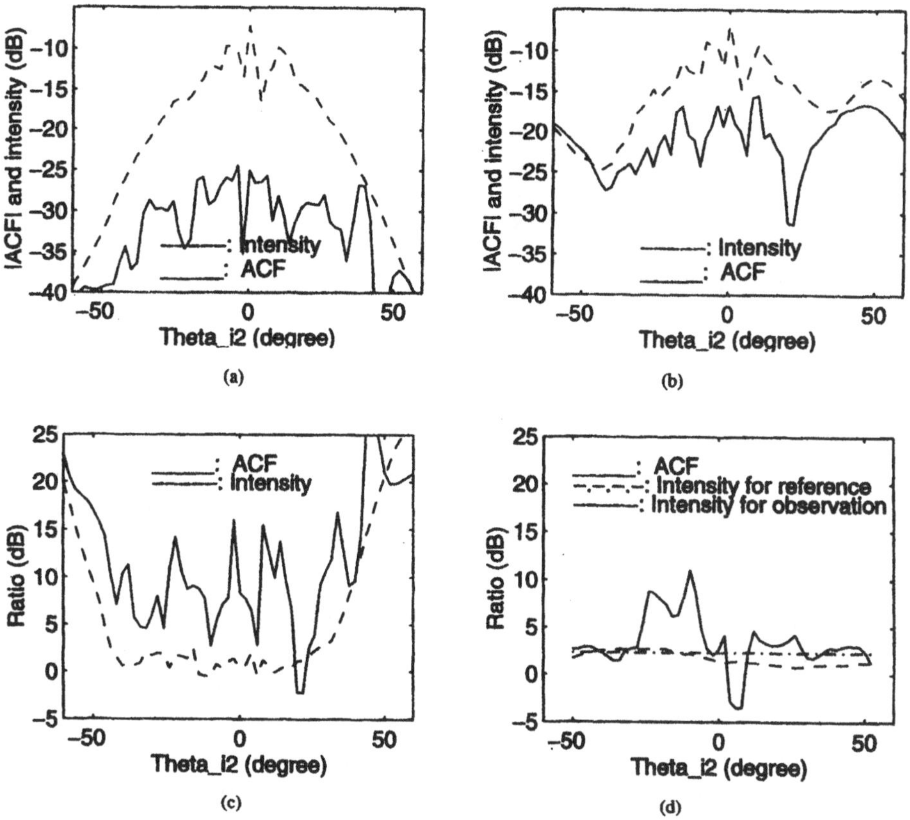

Figure 11.2.2 Comparisons of ACF magnitude (solid line) and intensity (dashed line). Averaging band is $(0.5f_o$ to$1.5f_o)$. Reference angles are $\theta_{i1} = 30^\circ$ and $\theta_{s1} = -50^\circ$. $\epsilon_r = 3.7+i0.13$. $L = 40\lambda_o$, $g = L/4$. (a) Without the object. (b) With the object ($a = 1.5\lambda_o$, $x_p = 0$, $d = 2\lambda_o$). (c) Ratios of results in (a) and those in (b). In (a), (b), and (c), the intensity is the backscattered intensity for the observation angles ($\theta_{s2} = -\theta_{i2}$) and a rough surface with $h = 0.18\lambda_o$, and $l = 1.0\lambda_o$ is used. (d) Ratios for a rougher surface case of $h = 0.5\lambda_o$ and $l = 0.5\lambda_o$. Two ratios of intensities are shown. One is for the reference angles with $(\theta_s, \theta_i) = (\theta_{s1}, \theta_{i1}) = (-50^\circ, 30^\circ)$, and one is for the observation angles with $(\theta_s, \theta_i) = (\theta_{s2}, \theta_{i2}) = (-\theta_{i2}, \theta_{i2})$.

tional averaging. We further take the angular averaging over the frequency averaged results as given by

$$\Gamma_{fa}(\theta_{s1}, \theta_{i1}; -\theta_{i2}, \theta_{i2}) = \frac{1}{N_a} \sum_{n=1}^{N_a} \Gamma_f(\theta_{s1}, \theta_{i1}; -\theta_{i2} - \delta_n, \theta_{i2} + \delta_n) \quad (11.2.29)$$

The angular averaging is taken over a 20° angular range. We can see that both the ratio of the intensity at reference angles and that at observation

θ_{i2} ($\theta_{s2} = -\theta_{i2}$)	$\lvert\Gamma_{sc}\rvert/\lvert\Gamma_s\rvert$ (20°, −40°)	(30°, −30°)	(30°, −40°)	(30°, −50°)	I_{sc}/I_s
−30°	2.776	3.924	−2.674	6.000	1.699
−10°	3.409	5.542	3.320	2.741	−0.573
10°	1.912	3.723	13.575	12.735	0.593
30°	2.463	2.819	9.489	10.715	2.818

Table 11.2.1 Ratios of ACF magnitudes and intensities in decibel scale.

angles are small. However, the ratio of ACF can be 10 dB. Parameters for Fig. 11.2.2d are the same as those used in Fig. 11.2.2c except that changes are made for the parameters of the rough surface with $h = 0.5\lambda_o$ and $l = 0.5\lambda_o$ instead of $h = 0.18\lambda_o$ and $l = 1.0\lambda_o$. A rougher surface produces large clutter scattering at large angles. In this case, the ratio of intensity at both reference angles and observation angles are small, but the ratio of ACF can be larger by 10 dB.

In Table 11.2.1, the ratios obtained with (11.2.27) and (11.2.28) are listed for the reference angles $(\theta_{i1}, \theta_{s1})$ of $(20°, -40°)$, and $(30°, -30°)$, $(30°, -40°)$, and $(30°, -50°)$. In most cases, the ratio of ACF is higher than that of the intensity.

Figure 11.2.3 shows the ACF magnitude for various rms heights and soil moisture conditions. Results are shown for the reference angles in the backscattering direction as a function of incident angles θ_{i2}, $\Gamma_f(\theta_{s1} = -20°, \theta_{i1} = 20°; -\theta_{i2}, \theta_{i2})$. In Fig. 11.2.3a, the ACF magnitudes for two surface rms heights ($h = 0.35\lambda_o$ and $h = 0.18\lambda_o$) are obtained with and without an object. The memory line is at $\theta_{i2} = 20°$ where the values are at maximum. For rough surface scattering only, the ACF magnitude of $h = 0.35\lambda_o$ is substantially different from that of $h = 0.18\lambda_o$ because the rough surface scattering depends on the surface characteristics. However, results with the buried object for the two different rms heights are similar because, at these angles, the ACF is dominated by the scattering from the buried object and the contribution from the rough surface is small. Figure 11.2.3b shows results of two soil moisture conditions. Moisture contents of 5% and 30% in soil corresponds to the dielectric constants of $\epsilon_r = (3.7 + i0.13)$ and $\epsilon_r = (16.16 + i1.15)$, respectively. We note that for the large moisture case (30%), the effect of the buried object becomes less because of the heavy attenuation in the soil.

Figure 11.2.4a shows the ACF magnitude for three different depths of the object. Although the ACF magnitude decreases with increasing depth,

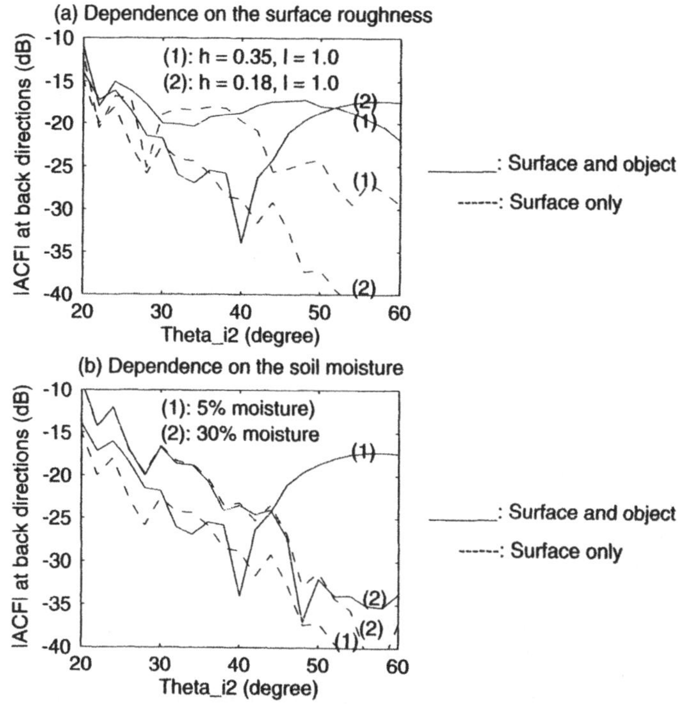

Figure 11.2.3 ACF magnitude for various surface properties. Averaging band is ($0.5f_o$ to $1.5f_o$). Reference angles are ($\theta_{i1} = 20°$, $\theta_{s1} = -20°$). $a = 1.5\lambda_o$, $L = 40\lambda_o$, $g = L/4$. (a) Dependence on the roughness, $\epsilon_r = 3.7 + 0.13i$ for two cases (1) $h = 0.35\lambda_o$, $l = 1.0\lambda_o$ and (2) $h = 0.18\lambda_o$, $l = 1.0\lambda_o$. (b) Dependence on the moisture, $h = 0.18\lambda_o$, $l = 1.0\lambda_o$, $\epsilon_r = 3.7 + 0.13i$ for 5% moisture, $\epsilon_r = 16.67 + 1.15i$ for 30% moisture.

it is still much higher than that of rough surface scattering if the angles are chosen carefully. Figure 11.2.4b shows the dependence of ACF magnitude on the horizontal position of the object. When x_p is less than $2\lambda_o$ (curves 1 and 2 in Fig. 11.2.4b), the object is still within the incident beam and the scattered wave contains both rough surface and object contributions. This accounts for the much larger value of the ACF magnitude for curves 1 and 2. For $x_p = 4\lambda_o$ and $6\lambda_o$ (curves 3 and 4), the object is on the edge of the incident beam, so that the ACF magnitude with and without the buried object are comparable. Figure 11.2.4c shows the ACF magnitude for three different object sizes: $a = 1.5\lambda_o$, $1\lambda_o$, and $0.5\lambda_o$. As expected, the ACF magnitude decreases as the size of the object decreases. However, if the incident angle θ_{i2} is far away from the memory line of the rough surface, a significant difference in the ACF magnitude is apparent even with the smallest size ($a = 0.5\lambda_o$).

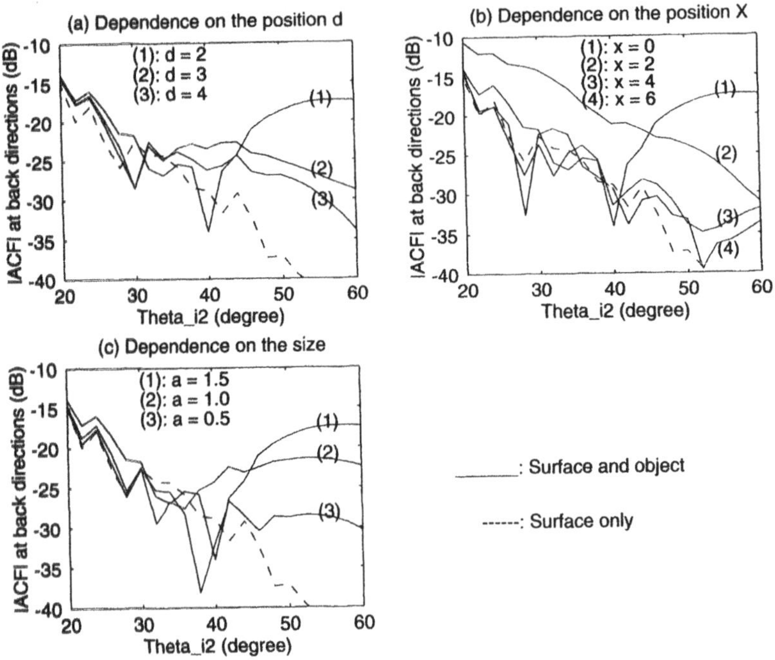

Figure 11.2.4 ACF magnitude of different object positions and sizes. Averaging band is $0.5f_o$ to $1.5f_o$. Reference angles are $\theta_{i1} = 20°$ and $\theta_{s1} = -20°$. $\epsilon_r = 3.7 + 0.13i$. $h = 0.18\lambda_o$, $l = 1.0\lambda_o$, $L = 40\lambda_o$, $g = L/4$. (a) Various depth dependence d ($x_p = 0$, $a = 1.5\lambda_o$). (b) Various horizontal position dependence x_p ($d = 2\lambda_o$, $a = 1.5\lambda_o$). (c) Various size dependence a ($x_p = 0$, $d = 2\lambda_o$).

3 Angular Correlation Function of Scattering by a Buried Object Under a 2-D Random Rough Surface (3-D Scattering)

3.1 Introduction

Since the practical problems involve a 3-D object buried under a 2-D rough surface, a solution of electromagnetic wave scattering from a 2-D random rough surface with a 3-D buried object is needed. To speed up the solution of surface integral equations for rough surface with MoM, fast computational methods of Chapters 5 and 6 are used.

Angular correlation functions are obtained by taking averages over a product of two signals. Thus, a key step of calculating the ACF is taking averages. For random media scattering, the average is usually taken over realizations of random media or rough surfaces, which is not applicable for the detection of the object buried under a rough surface. For 2-D scattering problems of a target embedded in clutter in Section 2, we have used frequency averaging and zenith angular averaging to obtain the ACF in Section 2. Numerical results have shown that detection of targets by the ACF with frequency and angular averaging has advantages over the radar cross section (RCS). Since the scattering characteristics of the object is frequency dependent, the frequency averaging may smear the result. In this section, we study 3-D scattering. In 3-D scattering, there is an additional degree of freedom, that of varying the azimuthal angle. Therefore, we will use azimuthal angular averaging instead.

We study electromagnetic wave scattering by a 3-D buried object under a 2-D random rough surface. We formulate the problem based on the Stratton-Chu surface integral equations for the rough surface and the surface of the object. Next the scattered wave fields from the object onto the rough surface are treated as additional incident fields on the rough surface. Then, the SMCG method is used for the solution of the matrix equation. Numerical results are calculated for a perfectly conducting sphere under the rough surface. Both the ACF and scattering coefficient are calculated. To take averages of the statistical results, we use azimuthal averaging. It is found that ACF is more effective in suppressing the effects of the rough surface scattering. Also, for 3-D scattering, we have cross-polarization. We utilize angular correlation and polarization correlation together giving the PACF (Polarization-angular correlation function). It is shown that the cross-polarization components of ACF can be more useful than co-polarization components for the detection of the buried object.

3.2 Formulation of Integral Equations

Consider an electromagnetic wave with fields $\overline{E}^i(x,y,z)$ and $\overline{H}^i(x,y,z)$ impinging on a 2-D rough surface with a random height profile $z = f(x,y)$. Above the rough surface is a free space (region 0) while the subsurface is characterized by permittivity ϵ_1 and permeability μ_1 (region 1). The incident direction is $\hat{k}_i = \sin\theta_i \cos\phi_i \hat{x} + \sin\theta_i \sin\phi_i \hat{y} - \cos\theta_i \hat{z}$. The incident electric and magnetic fields are given as

$$\overline{E}^i(x,y,z) = \int_{-\infty}^{+\infty} dk_x \int_{-\infty}^{+\infty} dk_y \exp(ik_x x + ik_y y - ik_z z) E_i(k_x, k_y) \cdot \left[a^{TE}\hat{e}(-k_z) + a^{TM}\hat{h}(-k_z)\right] \tag{11.3.1}$$

$$\overline{H}^i(x,y,z) = \frac{1}{\eta}\int_{-\infty}^{+\infty} dk_x \int_{-\infty}^{+\infty} dk_y \exp(ik_x x + ik_y y - ik_z z)E_i(k_x, k_y)$$
$$\cdot \left[-a^{TE}\hat{h}(-k_z) + a^{TM}\hat{e}(-k_z)\right] \tag{11.3.2}$$

where $(a^{TE} = 1, a^{TM} = 0)$ indicates incident TE (horizontal) polarization while $(a^{TE} = 0, a^{TM} = 1)$ indicates TM (vertical polarization). The polarization unit vectors $\hat{e}(-k_z)$ and $\hat{(}-k_z)$ are as defined in (2.1.18) of Volume I.

In (11.3.1) and (11.3.2), $E_i(k_x, k_y)$ is the spectrum of the incident wave. We use the following spectrum

$$E_i(k_x, k_y) = \frac{1}{4\pi^2}\int_{-\infty}^{+\infty} dx \int_{-\infty}^{+\infty} dy \exp(-ik_x x - ik_y y)$$
$$\times \exp(i(k_{ix}x + k_{iy}y)(1+w))\exp(-t) \tag{11.3.3}$$

where $t = t_x + t_y = (x^2 + y^2)/g^2$ and

$$t_x = \frac{(\cos\theta_i \cos\phi_i x + \cos\theta_i \sin\phi_i y)^2}{g^2\cos^2\theta_i} \tag{11.3.4}$$

$$t_y = \frac{(-\sin\phi_i x + \cos\phi_i y)^2}{g^2} \tag{11.3.5}$$

$$w = \frac{1}{k^2}\left[\frac{(2t_x - 1)}{g^2\cos^2\theta_i} + \frac{(2t_y - 1)}{g^2}\right] \tag{11.3.6}$$

The parameter g controls the tapering of the incident wave. The tapering is done in the spectral domain.

We apply Stratton-Chu surface integral equations on the rough surface and the surface of the buried object. The boundary conditions for the dielectric rough surface are

$$\begin{aligned} &\hat{n}\times\overline{E}_0 = \hat{n}\times\overline{E}_1, \quad \hat{n}\times\overline{H}_0 = \hat{n}\times\overline{H}_1 \\ &\epsilon_0\hat{n}\cdot\overline{E}_0 = \epsilon_1\hat{n}\cdot\overline{E}_1, \quad \mu_0\hat{n}\cdot\overline{H}_0 = \mu_1\hat{n}\cdot\overline{H}_1 \end{aligned} \tag{11.3.7}$$

By using $\mu_0 = \mu_1$, we use Stratton-Chu equations approaching from both sides of the rough surface and the boundary conditions. Let $\overline{E}_b^s$ and $\overline{H}_b^s$ denote the scattered fields from the buried target. The integral equations on the rough surface are, with $\overline{r}$ on the rough surface

$$\hat{n}\cdot\overline{E}^i(\overline{r}) = \frac{\hat{n}\cdot\overline{E}(\overline{r})}{2} - \hat{n}\cdot\left[\int\!\!\!\int \hat{n}'\times\overline{H}(\overline{r'})i\omega\mu_0 g_0 dS'\right.$$
$$\left. + \fint_{S_r}\{(\hat{n}'\times\overline{E}(\overline{r'}))\times\nabla' g_0 + \nabla' g_0\hat{n}'\cdot\overline{E}(\overline{r'})\}dS'\right] \tag{11.3.8}$$

$$\hat{n} \times \overline{H}^i(\overline{r}) = \frac{\hat{n} \times \overline{H}(\overline{r})}{2} - \hat{n} \times \left[\int -i\omega \hat{n}' \times \overline{E}(\overline{r'}) \epsilon_0 g_0 dS' \right.$$
$$\left. + \fint_{S_r} \{ \hat{n}' \times \overline{H}(\overline{r'})) \times \nabla' g_0 + \hat{n}' \cdot \overline{H}(\overline{r'}) \nabla' g_0 \} dS' \right] \tag{11.3.9}$$

$$- \hat{n} \times \overline{E}_b^s(\overline{r}) = -\frac{\hat{n} \times \overline{E}(\overline{r})}{2} - \hat{n} \times \left[\int i\omega \hat{n} \times \overline{H}(\overline{r'}) \mu_1 g_1 dS' \right.$$
$$\left. + \fint_{S_r} \{ (\hat{n}' \times \overline{E}(\overline{r'})) \times \nabla' g_1 + (\hat{n}' \cdot \overline{E}(\overline{r'}) \frac{\epsilon_1}{\epsilon_o} \nabla' g_1) \} dS' \right] \tag{11.3.10}$$

$$- \hat{n} \cdot \overline{H}_b^s(\overline{r}) = -\frac{\hat{n} \cdot \overline{H}(\overline{r})}{2} - \hat{n} \cdot \left[\int -\hat{n}' \times \overline{E}(\overline{r'}) i\omega \epsilon_1 g_1 dS' \right.$$
$$\left. + \fint_{S_r} \{ (\hat{n}' \times \overline{H}(\overline{r'})) \times \nabla' g_1 + \nabla' g_1 \hat{n}' \cdot \overline{H}(\overline{r'}) \} dS' \right] \tag{11.3.11}$$

where the integral $\fint$ represents the principal-value integral, S_r is the rough surface, and g_0 and g_1 are the scalar Green's function in region 0 (air) and region 1, respectively

$$g_0 = \frac{\exp(ik_0 R)}{4\pi R} \quad \text{and} \quad g_1 = \frac{\exp(ik_1 R)}{4\pi R} \tag{11.3.12}$$

The distance between a field point $\overline{r}$ and a source point $\overline{r}'$ is $R = \sqrt{(x - x')^2 + (y - y')^2 + (f(x, y) - f(x', y'))^2}$.

In Eqs. (11.3.10) and (11.3.11), $\overline{E}_b^s$ and $\overline{H}_b^s$ are scattered fields from the buried object onto the rough surface. They are calculated as follows.

We assume that the buried object is a perfectly conducting sphere. It is convenient to use the magnetic field integral equation (MFIE) to solve the surface current on the object. For an exciting field of $\overline{H}_b^e$ incoming on the buried object, the MFIE for the surface current $\overline{J}_b = \hat{n}_b \times \overline{H}_b$ on the buried object is

$$\hat{n}_b \times \overline{H}_b^e(\overline{r}) = \frac{\overline{J}_b(\overline{r})}{2} - \hat{n}_b \times \int_{S_b} \overline{J}_b(\overline{r'}) \times \nabla' g_1 dS' \tag{11.3.13}$$

where the unit normal vector of the surface of the buried object is $\hat{n}_b$, and S_b is the surface of the buried object. In (11.3.13), the exciting field of $\overline{H}_b^e$ for the buried object is the scattered field from the rough surface, i.e.

$$\hat{n}_b \times \overline{H}_b^e(\overline{r}) = -\hat{n}_b \times \int_{S_r} \left[-\hat{n}' \times \overline{E}(\overline{r'}) i\omega \epsilon_1 g_1 + (\hat{n}' \times \overline{H}(\overline{r'})) \times \nabla' g_1 \right.$$
$$\left. + \nabla' g_1 \hat{n}' \cdot \overline{H}(\overline{r'}) \right] dS' \tag{11.3.14}$$

The right hand sides of (11.3.13) and (11.3.14) are to be equated with $\overline{r}$ on S_b. Then, the scattered fields from the buried object are expressed in terms of the surface current as

$$\overline{H}_b^s(\overline{r}) = \int_{S_b} \overline{J_b}(\overline{r'}) \times \nabla' g_1 dS' \tag{11.3.15}$$

$$\overline{E}_b^s(\overline{r}) = \frac{-i}{\omega\epsilon_1} \nabla \times \int_{S_b} \overline{J_b}(\overline{r'}) \times \nabla' g_1 dS' \tag{11.3.16}$$

The expressions of (11.3.15) and (11.3.16) are substituted in (11.3.10) and (11.3.11) for $\overline{r}$ on S_r. Equations (11.3.8)–(11.3.11) combined with (11.3.13)–(11.3.16) constitute the coupled integral equations between the rough surface and the buried object. There are six equations in (11.3.8)–(11.3.11), two equations in (11.3.13)–(11.3.14). Equations (11.3.15) and (11.3.16) are to be substituted into the left hand sides of (11.3.10) and (11.3.11). The unknowns are the six field components on the rough surface and the components of the surface current on the object.

To solve the coupled integral equations, we discretize the rough surface and the surface of the object into small patches. For a patch at the rough surface, we use 6 knowns for the surface fields as follows:

$$J_1(\overline{r}) = \overline{n} \times \overline{E}_r(\overline{r}) \cdot \hat{x} \tag{11.3.17}$$

$$J_2(\overline{r}) = \overline{n} \times \overline{E}_r(\overline{r}) \cdot \hat{y} \tag{11.3.18}$$

$$J_3(\overline{r}) = \overline{n} \cdot \overline{E}_r(\overline{r}) \tag{11.3.19}$$

$$J_4(\overline{r}) = \overline{n} \times \overline{H}_r(\overline{r}) \cdot \hat{x} \tag{11.3.20}$$

$$J_5(\overline{r}) = \overline{n} \times \overline{H}_r(\overline{r}) \cdot \hat{y} \tag{11.3.21}$$

$$J_6(\overline{r}) = \overline{n} \cdot \overline{H}_r(\overline{r}) \tag{11.3.22}$$

where $\overline{r}$ is on S_r.

Using the above definitions for unknowns and the MoM, Eqs. (11.3.8)–(11.3.11) become a matrix equation as

$$\overline{\overline{Z}}\,\overline{J} = \overline{b}_i + \overline{b}_b \tag{11.3.23}$$

where the impedance matrix of rough surface is $\overline{\overline{Z}}$, $\overline{b}_i$ represents incident fields, and $\overline{b}_b$ represents scattered fields from the buried object.

Let $\overline{\overline{Z}}_b$ be the impedance matrix for the buried object, $\overline{\overline{Z}}_{rb}$ be impedance matrix that characterize scattering from the buried object to the rough surface, and $\overline{\overline{Z}}_{br}$ be the impedance matrix that characterize scattering from the rough surface to the buried object. Let the surface currents on the buried object be

$$J_{b1}(\overline{r}) = \hat{n}_b \times \overline{H}_b(\overline{r}) \cdot \hat{x} \tag{11.3.24}$$

$$J_{b2}(\overline{r}) = \hat{n}_b \times \overline{H}_b(\overline{r}) \cdot \hat{y} \tag{11.3.25}$$

$$J_{b3}(\overline{r}) = \hat{n}_b \times \overline{H}_b(\overline{r}) \cdot \hat{z} \tag{11.3.26}$$

where $\overline{r}$ is on S_b. From Eqs. (11.3.15)–(11.3.16) , we have

$$\overline{b}_b = \overline{\overline{Z}}_{rb}\overline{J}_b \tag{11.3.27}$$

From (11.3.13), we have

$$\overline{J}_b^e = \overline{\overline{Z}}_b\overline{J}_b \tag{11.3.28}$$

where $\overline{J}_b^e$ is the column vector that represents the values of $\hat{n}_b \times \overline{H}_b^e(\overline{r})$ on S_b, and from (11.3.14)

$$\overline{J}_b^e = \overline{\overline{Z}}_{br}\overline{J} \tag{11.3.29}$$

Thus

$$\overline{b}_b = \overline{\overline{Z}}_{rb}\,\overline{J}_b = \overline{\overline{Z}}_{rb}\,\overline{\overline{Z}}_b^{-1}\overline{J}_b^e = \overline{\overline{Z}}_{rb}\,\overline{\overline{Z}}_b^{-1}\overline{\overline{Z}}_{br}\,\overline{J} \tag{11.3.30}$$

We assume that the number of buried object surface unknowns is far less than that of the rough surface so that calculating the inverse of $\overline{\overline{Z}}_b$ does not present large CPU requirement. Substituting Eq. (11.3.30) into (11.3.23), we get

$$(\overline{\overline{Z}} - \overline{\overline{Z}}_{rb}\,\overline{\overline{Z}}_b^{-1}\overline{\overline{Z}}_{br})\overline{J} = \overline{b}_i \tag{11.3.31}$$

To speed up the solution of Eq. (11.3.31), we decompose the impedance matrix for the rough surface $\overline{\overline{Z}}$ into three parts: a block Toeplitz flat-surface part $\overline{\overline{Z}}^{FS}$, strong interaction part $\overline{\overline{Z}}^{S}$, and the weak remainder $\overline{\overline{Z}}^{W}$. Thus

$$\overline{\overline{Z}} = \overline{\overline{Z}}^{S} + \overline{\overline{Z}}^{FS} + \overline{\overline{Z}}^{W} \tag{11.3.32}$$

With the weak remainder part moved to the right-hand side, we have

$$(\overline{\overline{Z}}^{FS} + \overline{\overline{Z}}^{S} - \overline{\overline{Z}}_{rb}\,\overline{\overline{Z}}_b^{-1}\overline{\overline{Z}}_{br})\overline{J} = \overline{b}_i - \overline{\overline{Z}}^{W}\overline{J} \tag{11.3.33}$$

For a small object, the solution of Eq. (11.3.33) can be further speeded up by moving the buried object term to the right-hand side,

$$(\overline{\overline{Z}}^{FS} + \overline{\overline{Z}}^{S})\overline{J} = \overline{b}_i + \overline{\overline{Z}}_{rb}\,\overline{\overline{Z}}_b^{-1}\overline{\overline{Z}}_{br}\overline{J} - \overline{\overline{Z}}^{W}\overline{J} \tag{11.3.34}$$

The matrix equation is then iteratively

$$(\overline{\overline{Z}}^{FS} + \overline{\overline{Z}}^{S})\overline{J}^{(0)} = \overline{b}_i \tag{11.3.35}$$

$$\overline{b}^{(n+1)} = \overline{b}_i + \overline{\overline{Z}}_{rb}\overline{\overline{Z}}_b^{-1}\overline{\overline{Z}}_{br}\overline{J}^{(n)} - \overline{\overline{Z}}^{W}\overline{J}^{(n)} \tag{11.3.36}$$

$$(\overline{\overline{Z}}^{FS} + \overline{\overline{Z}}^{S})\overline{J}^{(n+1)} = \overline{b}^{(n+1)} \tag{11.3.37}$$

for $n = 0, 1, 2, \ldots$. Equations (11.3.35) and (11.3.37) are solved by conjugate gradient method (CGM). We also use the sparse matrix canonical grid method to calculate $\overline{\overline{Z}}^{W}\overline{J}^{(n)}$. The product of $\overline{\overline{Z}}^{(FS)}$ with $\overline{J}$ can be computed using a 2-D fast FFT algorithm as described in Chapter 6. Updating the right-hand side is also quickly calculated. In the numerical calculations shown in Section 3.4, the iteration is terminated when the error norm criterion is less than 0.2%.

3.3 Statistics of Scattered Fields

After the surface currents are solved, the scattered fields in medium 0 can be calculated. The scattering amplitudes for both the co-polarized and cross-polarized polarizations $F_{\beta\alpha}$ are respectively

$$
\begin{aligned}
F_{h\alpha} = \frac{ik}{4\pi\sqrt{2\eta P_\alpha^i}} \int_{S_r} dx'dy' \exp(-ik\gamma') \Bigg[\Bigg\{ J_1(x',y') \cos\theta_s \cos\phi_s \\
+ J_2(x',y') \cos\theta_s \sin\theta_s - J_1(x',y') \frac{\partial f(x',y')}{\partial x'} \sin\theta_s \\
- J_2(x',y') \frac{\partial f(x',y')}{\partial y'} \sin\theta s \Bigg\} - \eta \{ J_4(x',y') \sin\phi_s - J_5(x',y') \cos\phi_s \} \Bigg]
\end{aligned}
\tag{11.3.38}
$$

and

$$
\begin{aligned}
F_{v\alpha} = \frac{ik}{4\pi\sqrt{2\eta P_\alpha^i}} \int_{S_r} dx'dy' \exp(-ik\gamma') \Big[\{ J_1(x',y') \sin\phi_s - J_2(x',y') \cos\phi_s \} \\
+ \eta \{ J_4(x',y') \cos\theta_s \cos\phi_s + J_5 \cos\theta_s \sin\phi_s - J_4(x',y') \frac{\partial f(x',y')}{\partial x'} \sin\theta_s \\
- J_5(x',y') \frac{\partial f(x',y')}{\partial y'} \sin\theta_s \} \Big]
\end{aligned}
\tag{11.3.39}
$$

where $\gamma' = x' \sin\theta_s \cos\phi_s + y' \sin\theta_s \sin\phi_s + f(x',y') \cos\theta_s$. The scattering amplitudes in (11.3.39) are normalized by the square root of the incident power $2\eta P_{i\alpha}$ where

$$
P_\alpha^i = \frac{2\pi^2}{\eta} \int_{k_\rho < k} dk_x dk_y |E_{i\alpha}(k_x, k_y)|^2 \frac{k_z}{k}
\tag{11.3.40}
$$

As discussed in Section 2, since the buried object is under a single random rough surface, the realization averaging that is usually done in random rough surface scattering simulations is not applicable. In the numerical results, both the bistatic scattering coefficient (normalized RCS) and ACF are calculated based on azimuthal averaging. Let N_ϕ be the number of azimuthal angles. The bistatic scattering coefficient is

$$\sigma_{\beta\alpha}(\theta_s, \theta_i) = \frac{1}{N_\phi} \sum_{n=1}^{N_\phi} |F_{\beta\alpha}(\theta_s, \phi_{sn}; \theta_i, \phi_{in})|^2 \tag{11.3.41}$$

and

$$\Gamma_{\beta\alpha}(\theta_{s2}, \theta_{i2}; \theta_{s1}, \theta_{i1}) = \frac{1}{N_\phi} \sum_{n=1}^{N_\phi} F_{\beta\alpha}(\theta_{s2}, \phi_{s2n}; \theta_{i2}, \phi_{i2n}) F^*_{\beta\alpha}(\theta_{s1}, \phi_{s1n}; \theta_{i1}, \phi_{i1n}) \tag{11.3.42}$$

where ϕ_{in} and ϕ_{sn} are incident and scattering azimuthal angles. In the scattering plane, they are related to each other by the relations of (i) $\phi_{sn} = \phi_{in}$ for θ_s having the same sign as θ_i, (ii) $\phi_{sn} = \phi_{in} + 180°$ for θ_s having an opposite sign of θ_i.

For θ_s having the same sign as θ_i, the scattering is in the forward direction, and we have set $\phi_{sn} = \phi_{in}$. We also let the two incident ϕ's to be the same, $\phi_{i2n} = \phi_{i1n}$. The azimuthal averaging over ϕ's means that we keep the source and receiving directions on opposite sides of the scattering plane and rotating both by the same amount in the azimuthal direction. For θ_s having the opposite sign as θ_i, the scattering is in the backward direction. We set $\phi_{sn} = \phi_{in} + 180°$, meaning that the source and the observation are in the same side of the scattering plane. Azimuthal averaging means rotating source and observation by the same amount in the azimuthal direction.

Equation (11.3.42) can be extended to calculate the polarization-angular correlation function (PACF) as follows

$$\Gamma_{\beta_1\alpha_1\beta_2\alpha_2}(\theta_{s1}, \theta_{i1}; \theta_{s2}, \theta_{i2})$$
$$= \frac{1}{N_\phi} \sum_{n=1}^{N_\phi} F_{\beta_1\alpha_1}(\theta_{s1}, \phi_{s1n}; \theta_{i1}, \phi_{i1n}) F^*_{\beta_2\alpha_2}(\theta_{s2}, \phi_{s2n}; \theta_{i2}, \phi_{i2n}) \tag{11.3.43}$$

Equation (11.3.43) includes effects of both angular correlations and polarization correlations.

3.4 Numerical Illustrations of ACF and PACF

The numerical simulation is conducted for a perfectly electrical conductor (PEC) sphere buried under a 2-D Gaussian random rough surface. The rough surface is generated by using the spectrum method with an assumption of a Gaussian spectrum. The sizes of the rough surface in the x and y directions are $L_x = L_y = 8.0\lambda$. The surface rms heights are $h_x = h_y = 0.02\lambda$, and the correlation lengths are $l_x = l_y = 0.5\lambda$. The relative dielectric constant of

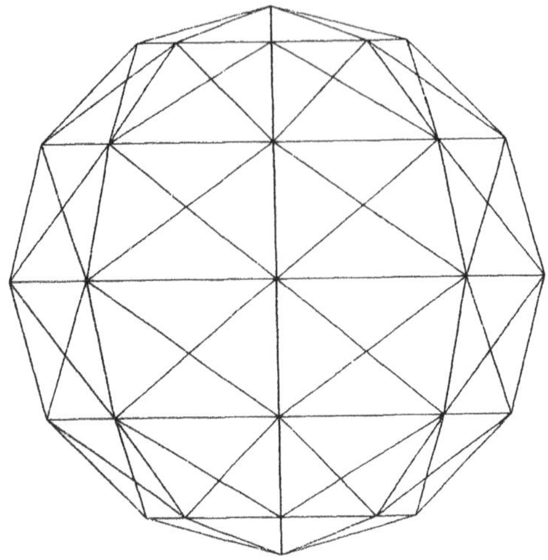

Figure 11.3.1 Discretization of a sphere into 80 triangle patches.

the lower medium is $\epsilon_r = (2.0 + i0.2)$. The surface is sampled at 64 points per λ^2 giving 4096 points on the rough surface and 24576 surface unknowns. The neighborhood distance in the implementation of SMCG is $r_d = 3.5\lambda$. The sphere of radius of $a = 0.3\lambda$ is buried under the rough surface at a depth of $d = 0.6\lambda$. The sphere surface is discretized into 80 triangle patches as shown in Fig. 11.3.1 for which the surface currents are represented by 240 unknowns. The impedance matrix of the sphere $\overline{\overline{Z}}_b$ is calculated and tested by calculating the scattering cross section of a sphere in free space. The numerical results agree with that obtained by Mie scattering as shown in Fig. 11.3.2.

We solve the matrix equations with the buried object contribution on the left-hand side (11.3.33) and that on the right side (11.3.34). Both give the same result. The scattering coefficients are shown in Fig. 11.3.3. The CPU with the target term on the right-hand side is five times faster than the CPU with the target term on the left-hand side.

We calculate the scattering amplitudes for 10 azimuthal angles at 0°, 36°,..., and 324°, respectively. There is only one realization of the random rough surface. The RCS and ACF are calculated by using azimuthal angular averaging as given in Eq. (11.3.41) and (11.3.42). We plot the results as functions of the scattering angle θ_{s2}. Parameters for other angles are $\theta_{i1} = 20°$, $\theta_{s1} = -40°$, and $\theta_{i2} = 20°$. The memory line is at $\theta_{s2} = -40°$ which shows a moderate peak for the ACF without the target. Figure 11.3.4 shows the results for hh polarization component. Both the results with and without

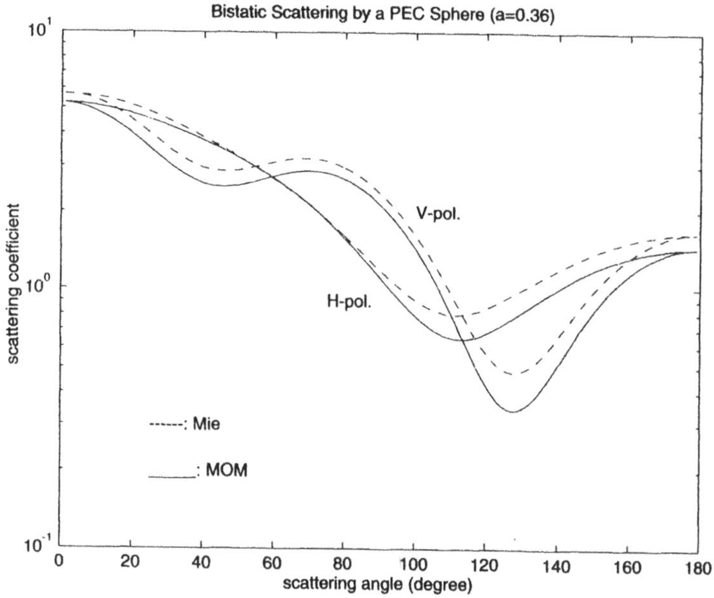

Figure 11.3.2 Comparison of MoM and Mie scattering for the radar cross section of a PEC sphere.

the target sphere are shown for comparison. Figure 11.3.4b is for RCS. As expected, there is a peak in the specular direction, which is due to the slightly rough surface. The difference of RCS with and that without the target is large only for large scattering angles, since the rough surface scattering is small in these cases. As shown in Fig. 11.3.4a, however, the difference of ACF can be 7dB, even for angles closed to the nadir direction. This is because the memory effect is avoided and rough surface scattering is minimized in the ACF.

The fully polarimetric results of RCS and ACF are calculated and shown in Figs. 11.3.5 and 11.3.6. Figure 11.3.5 shows the results of RCS. We see that the differences of RCS between with and without the target for co-polarizations are larger than those for cross-polarizations. This is because the cross-polarization components are mainly due to the rough surface scattering. Because the target is a sphere, it has only a small cross-polarization contribution in RCS. It is also found that there are larger differences for the vv component than for the hh component, since the vertical polarization wave has better penetration through the rough surface. Figure 11.3.5 shows the results of ACF. We can see the large difference of ACF between with and without a target in both the co-polarization and cross-polarization result.

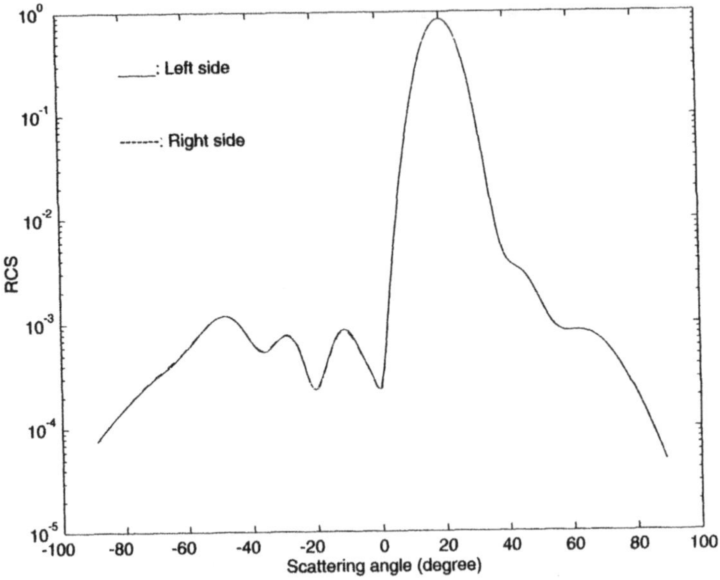

Figure 11.3.3 Comparison between the solution of the matrix equation for the target object term on left-hand side and that on right-hand side.

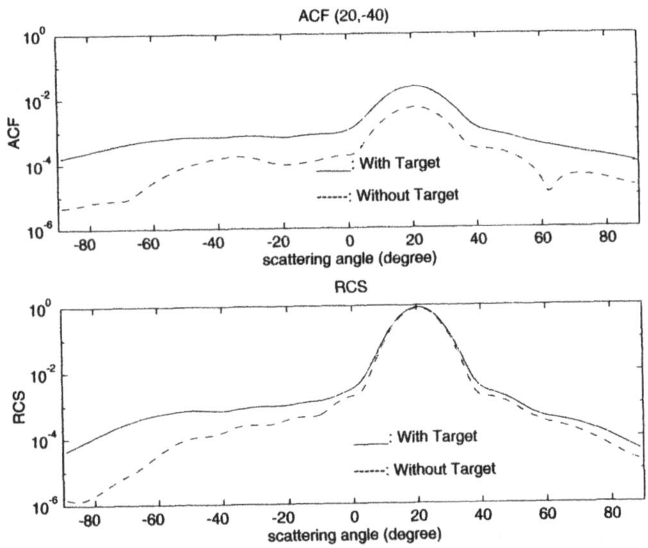

Figure 11.3.4 ACF and RCS of EM wave scattering by a PEC sphere buried under a 2-D rough surface for hh polarization component. Parameters are: $a = 0.3\lambda$, $d = 0.6\lambda$; $L_x = L_y = 8.0\lambda$, $h = 0.02\lambda$, $l_x = l_y = 0.5\lambda$, $r_d = 3.5\lambda$, $\theta_{i1} = 20°$, $\theta_{s1} = -40°$, and $\theta_{i2} = 20°$. Solid line with target; dashed line without target.

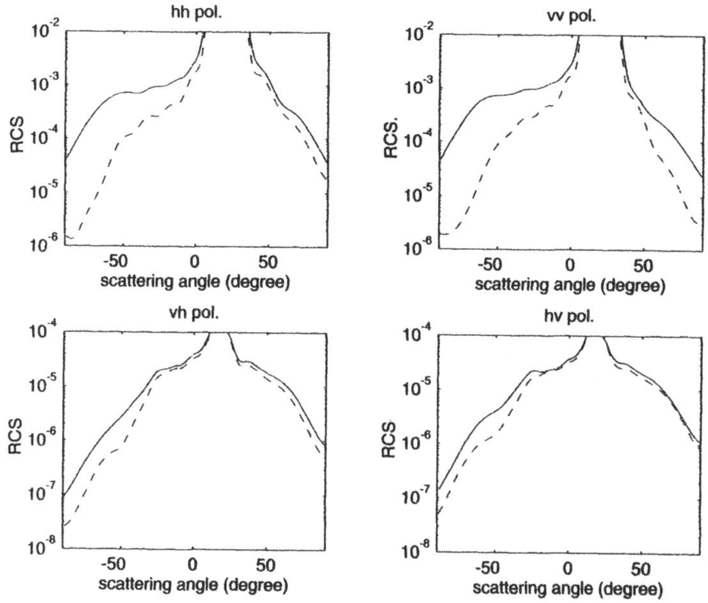

Figure 11.3.5 RCS of EM wave scattering by a PEC sphere buried under a 2-D rough surface for co-pol. and cross-pol. components. Parameters are: $a = 0.3\lambda$, $d = 0.6\lambda$; $L_x = L_y = 8.0\lambda$, $h = 0.02\lambda$, $l_x = l_y = 0.5\lambda$, $r_d = 3.5\lambda$, $\theta_{i1} = 20°$, $\theta_{s1} = -40°$, and $\theta_{i2} = 20°$. Solid line with target; dashed line without target.

This is because of the random phase of rough surface scattering that causes cancellation in the ACF calculation. The results of polarization-angular correlation function (PACF) are also calculated and shown in Fig. 11.3.7. Figure 11.3.7a is the PACF between hh components and vh components. Figure 11.3.7b is the PACF between vv components and hv components. We can see an even larger difference up to 10 to 20dB. This is due to the fact that the PACF of rough surface scattering has little polarization correlation.

4 Angular Correlation Function Applied to Correlation Imaging in Target Detection

4.1 Introduction

SAR imaging is an important topic in remote sensing [Soumekh, 1996; Axelsson, 1995]. Resolution and signal-to-noise ratio are two important criteria in image processing. A radar system of fine resolution usually requires high operating frequency. At high frequencies, however, waves are also scattered by clutter such as rough surface and random media, causing low signal-noise

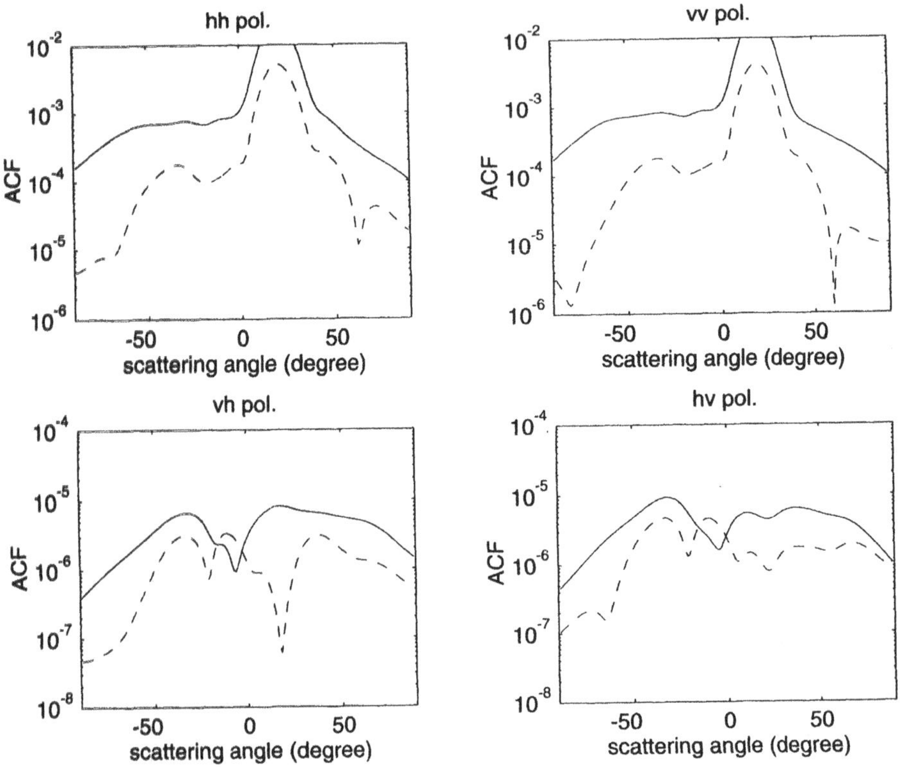

Figure 11.3.6 ACF of EM wave scattering by a PEC sphere buried under a 2-D rough surface for co-pol. and cross-pol. components. Parameters are: $a = 0.3\lambda$, $d = 0.6\lambda$; $L_x = L_y = 8.0\lambda$, $h = 0.02\lambda$, $l_x = l_y = 0.5\lambda$, $r_d = 3.5\lambda$, $\theta_{i1} = 20°$, $\theta_{s1} = -40°$, and $\theta_{i2} = 20°$.

ratio. Therefore, developing data processing methods that would give fine resolution and effective clutter suppression is an important goal.

Imaging is to obtain the detailed information from a wide-band and a large range of angular measurements. The conventional SAR imaging method can be called field imaging, in which the target function is obtained from the measured fields by inverse Fourier transform [Soumekh, 1996] or by focusing [Axelsson, 1995; Moore, 1996]. The focusing method is also called correlation imaging since the field is correlated with a reference signal. The focusing method was described in Chapter 6, Section 2 of Volume I for linear and circular SAR. The field focusing method can be improved by introducing filter function such as matched filtering if the scattering property of the target is known [Moore, 1996]. However, if the target scattering function is not known, the matched filtering method cannot be applied. In this section,

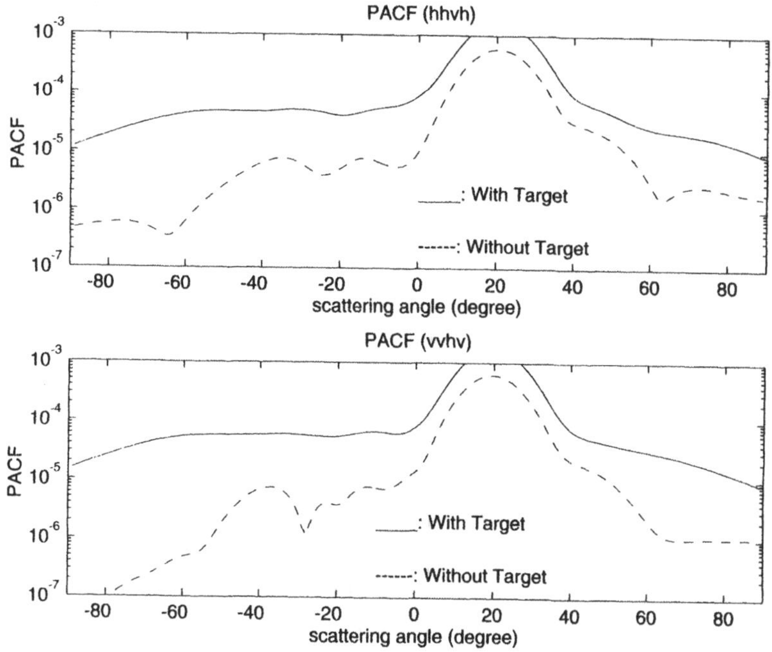

Figure 11.3.7 PACF of EM wave scattering by a PEC sphere buried under a 2-D rough surface between co-pol. and cross-pol. components. Parameters are: $a = 0.3\lambda$, $d = 0.6\lambda$; $L_x = L_y = 8.0\lambda$, $h = 0.02\lambda$, $l_x = l_y = 0.5\lambda$, $r_d = 3.5\lambda$, $\theta_{i1} = 20°$, $\theta_{s1} = -40°$, and $\theta_{i2} = 20°$. (a): hhvh, (b) vvhv.

we study the ACF imaging method that is based on the calculation of mutual correlation function of two received signals at two different angles.

Correlation imaging is to calculate the correlation function with focusing on every position in a region and then a image is obtained for the entire region. The memory effect of the mutual correlation function for random scattering is avoided. The correlation imaging uses the phase difference (or sum) of two wave propagation paths, while the field imaging is to use the phase of one path. Therefore, correlation imaging uses a larger spectrum domain information which gives a finer resolution and can suppress clutter. The correlation imaging can be realized by calculating angular correlation function (ACF), frequency correlation function (FCF), and the combination of both frequency angular correlation function (FACF).

In this section, we perform numerical simulations of SAR imaging. In Section 4.2, we discuss the fundamentals of correlation imaging. The similarities and differences between field imaging and correlation imaging are compared. Then, in Section 4.3, the angular correlation imaging method is illustrated by using Monte Carlo simulations for circular SAR and linear SAR. Circular SAR gives finer resolution and better signal-clutter ratio of 100%. For linear SAR, it is found that the frequency angular correlation imaging gives better performance.

4.2 Formulation of Imaging

Imaging is to obtain the detailed information of a region by measuring the scattered field from many view angles and frequencies. Under the far-field approximation, the received signal is a set of data $E(\overline{k})$ which can be written as

$$E(\overline{k}) = \int f(\overline{r}, \overline{k}) e^{i\overline{k}\cdot\overline{r}} d\overline{r} \tag{11.4.1}$$

where $\overline{k} = \overline{k}_i - \overline{k}_s$ is the difference between the incident wave vector $\overline{k}_i$ and the scattered wave vector $\overline{k}_s$. Thus, $\overline{k}$ is a function of frequency and view angle. The target function $f(\overline{r}, \overline{k})$ is generally frequency and angular dependent.

In the ideal case, $f(\overline{r}, \overline{k})$ is not dependent on $\overline{k}$. Thus

$$f(\overline{r}, \overline{k}) \approx f_{id}(\overline{r}) \tag{11.4.1a}$$

Then in this ideal case, the target function is the inverse Fourier transform of received signals.

$$f_{id}(\overline{r}) = \frac{1}{(2\pi)^3} \int E(\overline{k}) e^{-i\overline{k}\cdot\overline{r}} d\overline{k} \tag{11.4.2}$$

A second way to obtain the target function is by focusing. The received signal $E(\overline{k})$ is correlated with a reference response E_o focused on $\overline{r}_o$. From Eq. (6.2.25) of Chapter 6, Volume I,

$$C_F(\overline{r}_o) = \frac{1}{(2\pi)^3} \int d\overline{k} E(\overline{k}) E_o^* e^{-i\overline{k}\cdot\overline{r}_o} \tag{11.4.3}$$

Putting (11.4.1) in (11.4.3) and using $E_o = 1$, we have

$$C_F(\overline{r}_o) = \frac{1}{(2\pi)^3} \int d\overline{k} \int d\overline{r} f_{id}(\overline{r}) e^{i\overline{k}\cdot\overline{r}} E_o^* e^{-i\overline{k}\cdot\overline{r}_o} = f_{id}(\overline{r}_o) \tag{11.4.4}$$

We will use the notation that

$$\langle w(\overline{k}) \rangle_{\overline{k}} = \frac{1}{(2\pi)^3} \int d\overline{k}\; w(\overline{k}) \tag{11.4.5a}$$

where $\langle \cdots \rangle_{\overline{k}}$ denotes an averaging over $\overline{k}$ space. For functions that are dependent on two $\overline{k}$ vectors

$$\langle w(\overline{k}_1, \overline{k}_2) \rangle_{\overline{k}_1 \overline{k}_2} = \frac{1}{(2\pi)^6} \int d\overline{k}_1 \int d\overline{k}_2\; w(\overline{k}_1, \overline{k}_2) \tag{11.4.5b}$$

This operation can be interpreted as spectrum averaging, with E_o named the filter function.

Both methods (inverse Fourier transform and focusing) of finding target function in (11.4.2) and (11.4.3) are based on the received field, which we shall call field imaging.

In correlation imaging, we calculate the correlation function of two received signals. We further perform focusing by spectrum averaging as

$$\begin{aligned} C_\Gamma(\overline{r}_o) &= \langle E(\overline{k}_1)e^{-i\overline{k}_1\cdot\overline{r}_o}E^*(\overline{k}_2)e^{i\overline{k}_2\cdot\overline{r}_o}\rangle_{\overline{k}_1\overline{k}_2} \\ &= \frac{1}{(2\pi)^3}\int d\overline{k}_1\frac{1}{(2\pi)^3}\int d\overline{k}_2E(\overline{k}_1)e^{-i\overline{k}_1\cdot\overline{r}_o}E^*(\overline{k}_2)e^{i\overline{k}_2\cdot\overline{r}_o} \end{aligned} \tag{11.4.6}$$

Putting (11.4.1) in (11.4.6), we have, for the ideal case

$$\begin{aligned} &C_\Gamma(\overline{r}_o) \\ &= \frac{1}{(2\pi)^3}\int d\overline{k}_1\frac{1}{(2\pi)^3}\int d\overline{k}_2e^{-i\overline{k}_1\cdot\overline{r}_o}e^{i\overline{k}_2\cdot\overline{r}_o}\int d\overline{r}_1f_{id}(\overline{r}_1)e^{i\overline{k}_1\cdot\overline{r}_1}\int d\overline{r}_2f_{id}^*(\overline{r}_2)e^{-i\overline{k}_2\cdot\overline{r}_2} \\ &= \int d\overline{r}_1\int d\overline{r}_2f_{id}(\overline{r}_1)f_{id}^*(\overline{r}_2)\delta(\overline{r}_1-\overline{r}_o)\delta(\overline{r}_2-\overline{r}_o) = |f_{id}(\overline{r}_o)|^2 \end{aligned} \tag{11.4.7}$$

Thus ideally, $C_\Gamma(\overline{r}_o)$ gives $|f_{id}(\overline{r}_o)|^2$. The correlation imaging can be related to frequency angular correlation function of scattering.

We shall show that correlation imaging also has better performance for the nonideal case when $f(\overline{r},\overline{k})$ is dependent on $\overline{k}$.

First, we let $\overline{k}=\overline{k}_1$ and $\overline{k}_d=\overline{k}_2-\overline{k}_1$. Then

$$C_\Gamma(\overline{r}_o) = \frac{1}{(2\pi)^6}\int d\overline{k}_d\int d\overline{k}E(\overline{k})E^*(\overline{k}+\overline{k}_d)e^{i\overline{k}_d\cdot\overline{r}_o} \tag{11.4.8}$$

The frequency angular correlation function is

$$\Gamma_a(\overline{k}_d) = \langle E(\overline{k})E^*(\overline{k}+\overline{k}_d)\rangle_{\overline{k}} = \frac{1}{(2\pi)^3}\int d\overline{k}E(\overline{k})E^*(\overline{k}+\overline{k}_d) \tag{11.4.9}$$

Using (11.4.9) in (11.4.8), we have

$$C_\Gamma(\overline{r}_o) = \langle\Gamma_a(\overline{k}_d)e^{i\overline{k}_d\cdot\overline{r}_o}\rangle_{\overline{k}_d} = \frac{1}{(2\pi)^3}\int d\overline{k}_d\Gamma_a(\overline{k}_d)e^{i\overline{k}_d\cdot\overline{r}_o} \tag{11.4.10}$$

$\Gamma_a(\overline{k}_d)$ is known as the frequency angular correlation function (FACF) of scattering. Note that in (11.4.9), we have taken the averaging over $\overline{k}$ to obtain $\Gamma_a(\overline{k}_d)$.

For random media scattering, very often, averaging is taken over realizations

$$\Gamma_{a,r}(\overline{k}_d) = \langle E(\overline{k})E^*(\overline{k}+\overline{k}_d)\rangle_{\text{realization}} \tag{11.4.11}$$

For the case of realization averaging, the angular correlation function (ACF) of scattering by random media (clutter) has been studied. It has been shown in Section 2 that the scattering by random media does not contribute to ACF except along the memory directions of $\overline{k}_d = 0$. That is, the ACF of random scattering is small except for $\overline{k}_d = 0$.

For imaging purpose, random scattering can be minimized by avoiding the memory effect by choosing $\overline{k}_d \neq 0$. We next define a more general expression for correlation imaging function by generalizing (11.4.6)

$$C_\Gamma(\overline{r}_o) = \frac{1}{(2\pi)^6} \int d\overline{k}_1 \int_{\overline{k}_d \neq 0} d\overline{k}_2 E(\overline{k}_1) e^{-2ik_1|\overline{R}(\hat{k}_1) - \overline{r}_o|} \cdot E^*(\overline{k}_2) e^{2ik_2|\overline{R}(\hat{k}_2) - \overline{r}_o|} W(\overline{k}_1; \overline{k}_2) \tag{11.4.12}$$

where $\overline{R}(\overline{k}_1)$ and $\overline{R}(\overline{k}_2)$ are distances between the focusing position and the receiver positions. In Eq. (11.4.12), $W(\overline{k}_1; \overline{k}_2)$ is a weighting function, which can be chosen to further suppress clutter effect if the clutter scattering characteristics are known. We shall assume that it is equal to unity for simplicity. We will use (11.4.8) to process (simulated) SAR data.

The correlation imaging may be done by calculating angular correlation function (ACF), frequency correlation function (FCF) and frequency angular correlation function (FACF). The central idea of correlation imaging is to use a large phase difference for focusing and to avoid the memory effect so that clutter is suppressed. The large phase difference, however, may not be obtained using ACF imaging in some practical problems, like linear SAR. In this case, frequency angular correlation imaging (FACF imaging) gives better performance. In practice, the integration in (11.4.12) is discretized and approximated by finite sums.

4.3 Simulations of SAR Data and ACF Processing

In this section, we use the Monte Carlo simulations of wave scattering to simulate SAR data. Then we use ACF to process the simulated SAR data.

Consider an incident wave impinging upon the targets that are embedded in random medium (Fig. 11.4.1). Let the incident wave be given by

$$E_i(\overline{k}_i, \overline{r}) = e^{ik|\overline{R}_t - \overline{r}|} \tag{11.4.13}$$

where $\overline{R}_t$ is antenna position, $\overline{r}$ is field position.

In the numerical simulation, the random medium is modeled by a collection of randomly distributed small scatterers. Let the scattering amplitude of the targets be denoted by $F_{tn}(\hat{k}_s, \hat{k}_i)$, $n = 1, 2, \ldots, N_t$ where N_t is the number of targets. Let $f_n(\hat{k}_s, \hat{k}_i)$ be the scattering amplitude of the small

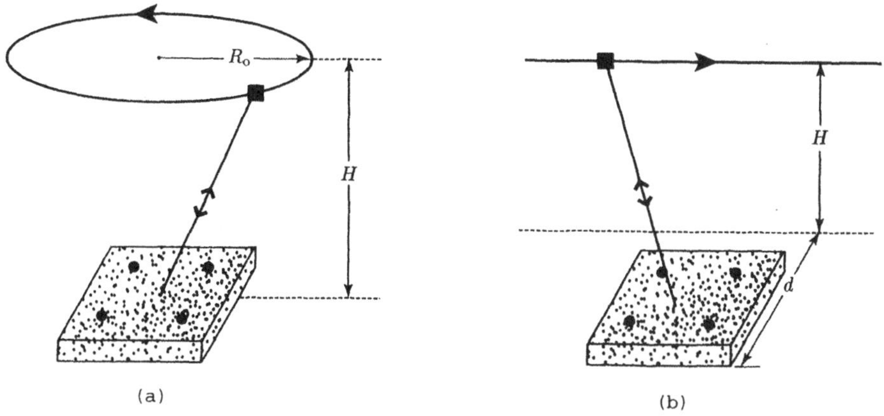

Figure 11.4.1 Configurations of SAR imaging for targets embedded in small scatterers. (a) Circular SAR, (b) linear SAR.

random scatterers with $n = 1, 2, \ldots, N_p$ where N_p is the number of small scatterers.

Assuming single scattering, the scattered field at the receiver is given by

$$E(\overline{k}, \overline{R}) = \sum_{n=1}^{N_t} e^{ik|\overline{R}(\hat{k}_i)-\overline{r}_n|-|\overline{R}(\hat{k}_s)-\overline{r}_n|} F_{tn}(\hat{k}_s, \hat{k}_i)$$
$$+ \sum_{n=1}^{N_p} e^{ik|\overline{R}(\hat{k}_i)-\overline{r}_n|-|\overline{R}(\hat{k}_s)-\overline{r}_n|} f_n(\hat{k}_s, \hat{k}_i) \qquad (11.4.14)$$

where the received field is expressed in terms of the scattering amplitude of the targets $F_{tn}(k)$ and the scattering amplitudes of randomly distributed particles $f_n(k)$

In this manner, we generate the simulated data of the scattered field for many combinations of incident and scattered field directions and frequencies. Next we describe how correlation imaging is used to process the simulated circular SAR and the linear SAR data.

Circular SAR Imaging

Let a mono-static radar be moving along a circular path of radius of R_o and at height H above the target region (Figure 11.4.1a). The radar position is a function of the azimuthal angle ϕ as $\overline{R}(\phi) = R_o \cos\phi\hat{x} + R_o \sin\phi\hat{y} + H\hat{z}$.

We write the total scattered field at the radar

$$E(k, \phi) = \sum_{n=1}^{N_t} e^{2ik|\overline{R}(\phi) - \overline{r}_n|} F_{tn}(k) + \sum_{n=1}^{N_p} e^{2ik|\overline{R}(\phi) - \overline{r}_n|} f_n(k) \tag{11.4.15}$$

where the received signal is expressed in terms of the scattering amplitude of targets $F_{tn}(k)$ and the scattering amplitudes of randomly distributed particles $f_n(k)$ $(n = 1, 2, \cdots, N_p)$. The factor of 2 in the phase accounts for the round trip phase shift for a monostatic radar.

The conventional SAR imaging, which we call field imaging, can be realized by correlating the received signal with a reference signal E_o and then summing over frequency and angle. From (11.4.3),

$$C_F(\overline{r}_o) = \sum_{m=1}^{N_k} \sum_{n=1}^{N_\phi} E(k_m, \phi_n) E_o^* e^{-2ik_m|\overline{R}(\phi_n) - \overline{r}_o|} \tag{11.4.16}$$

The reference signal E_o is chosen to be a constant. As shown in (11.4.3)–(11.4.4), (11.4.16) has well defined peaks when $\overline{r}_o$ approaches the location of the scatterers. To have a fair comparison with ACF imaging which is proportional to the square of the signal, we also take the absolute valued square of (11.4.16). That is

$$C_{2F}(\overline{r}_o) = |C_F(\overline{r}_o)|^2 \tag{11.4.17}$$

The result of (11.4.16)–(11.4.17) is known as field imaging. Next we calculate ACF imaging. The ACF imaging is to correlate two signals received at different angles and to sum over frequency and angles. From (11.4.12),

$$C_\Gamma(\overline{r}_o) = \sum_{m=1}^{N_k} \sum_{n_1=1}^{N_\phi} \sum_{n_2=1, n_2 \neq n_1}^{N_\phi} E(k_m, \phi_{n_1}) e^{-2ik_m|\overline{R}(\phi_{n_1}) - \overline{r}_o|} \cdot E^*(k_m, \phi_{n_2}) e^{2ik_m|\overline{R}(\phi_{n_2}) - \overline{r}_o|} \tag{11.4.18}$$

Note that the integrand in (11.4.10) is the angular correlation function. The memory effect of random scattering is avoided by $n_2 \neq n_1$. In practical simulations, we let $|n_1 - n_2| > N_o$ instead of just $n_2 \neq n_1$ in (11.4.18) so that the angular difference $\phi_d = |\phi_{n_1} - \phi_{n_2}|$ is larger than the correlation angle of the angular correlation function. The choice of N_o depends on the correlation angle of random scattering. The correlation angle of random scattering is usually small, which is in the order of λ/L where L is the size of the medium. By having $|n_2 - n_1| > N_o$, only a small portion of information is removed for the target. Therefore, target information is essentially preserved in (11.4.18) while the clutter due to random scattering is significantly suppressed.

In the numerical simulations, 80,000 small particles with radius of $a = 0.04\lambda_o$ are used to model clutter. They are randomly distributed in a layer region of $40\lambda_o \times 40\lambda_o \times 0.5\lambda_o$. Four target spheres with radius of $a = 0.3\lambda_o$ are placed at positions of $(10,14,0)\lambda_o$, $(10,28,0)\lambda_o$, $(30,10,0)\lambda_o$, and $(30,32,0)\lambda_o$. The dielectric constant of both the targets and the small particles are (3.23+i0.36). All distances are in terms of the wavelength λ_o of the center frequency. The circular path of the receiver has a radius of $R_o = 1732\lambda_o$ at height of $H = 1000\lambda_o$. The backscattering amplitudes for targets and particles are calculated based on Mie scattering. The received signal is calculated by Eq. (11.4.14). The frequency band is from $0.5f_o$ to $1.5f_o$ with an increment of $0.01f_o$. Thus $N_k = 100$, $k_1 = 0.5f_o$, $k_2 = 0.51f_o$, ..., etc. The azimuth angle is from 0° to 360° at an interval of 3.6 degrees. Thus $N_\phi = 100$, $\phi_1 = 0°$, $\phi_2 = 3.6°$, $\phi_3 = 7.2°$, etc.

The simulated data is processed by field imaging and ACF imaging. The results are normalized by the maximum of the target. The normalized results are shown in Fig. 11.4.2 and 11.4.3. Figure 11.4.2 shows the results of the square of (11.4.16), i.e., the results of (11.4.17) of field imaging. The 4 targets are spread out to a two wavelength size and are obscured by the background clutter. Figure 11.4.3 shows the results of ACF imaging. We see that the spot size of targets is smaller, which is a result of finer resolution. The background clutter is significantly suppressed. We use the condition of $|n_1 - n_2| > 5$, i.e., $N_o = 5$, to ensure that the angular pairs ϕ_{n_1} and ϕ_{n_2} are away from the memory dot. The better performance of ACF imaging is due to (1) the clutter effect is minimized by avoiding the memory effect of random scattering and (2) the spreading due to the frequency dependence of scattering is compensated by the cross-range resolution in ACF imaging. The angular correlation function with focusing (ACF focusing) is also shown in Fig. 11.4.4. The ACF focusing $C_\Gamma(\overline{r}_o, \phi_d)$ is obtained by using Eq. (11.4.18) with the summation over $\sum_{n_2=1}^{N_\phi}$ replaced by $n_2 = n_1 + n_d$, as given by

$$C_\Gamma(\overline{r}_o, \phi_d) = \sum_{m=1}^{N_k} \sum_{n=1}^{N_\phi} E(k_m, \phi_n) e^{-2ik_m|\overline{R}(\phi_n) - \overline{r}_o|} \cdot E^*(k_m, \phi_{n+n_d}) e^{2ik_m|\overline{R}(\phi_{n+n_d}) - \overline{r}_o|} \tag{11.4.19}$$

Therefore, ACF imaging is the summation over many images of ACF focusing. Because ACF focusing is a random function of correlation angle ϕ_d when $\overline{r}_o$ is not at a target. Comparing (11.4.18) and (11.4.19), we note that the clutter is further suppressed by having one more averaging in (11.4.18) than in (11.4.19).

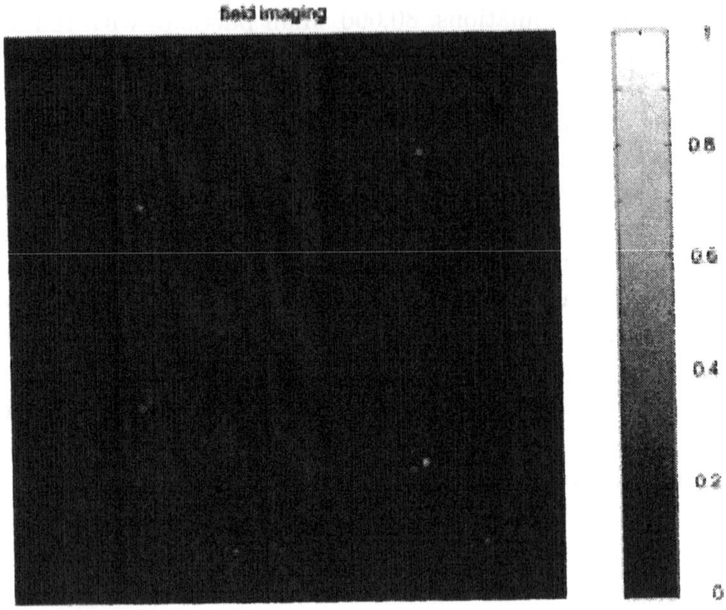

Figure 11.4.2 Simulated image of targets embedded in clutter by field imaging for circular SAR.

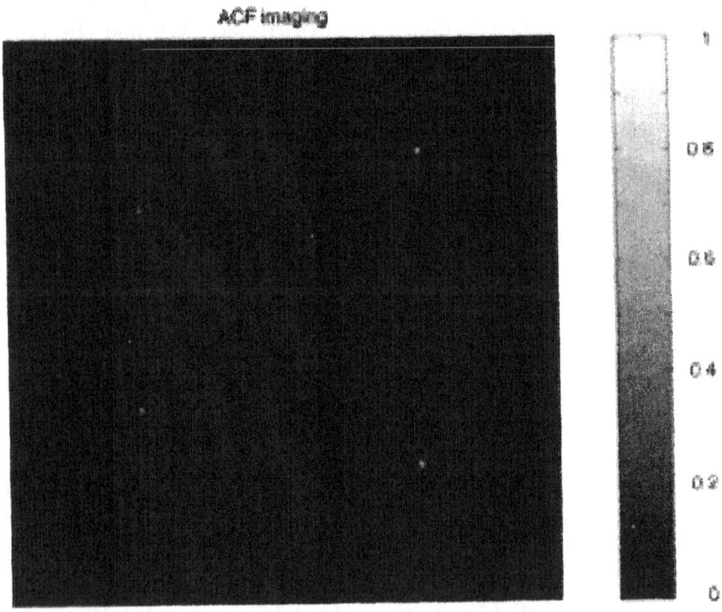

Figure 11.4.3 Simulated image of targets embedded in clutter by ACF imaging for circular SAR.

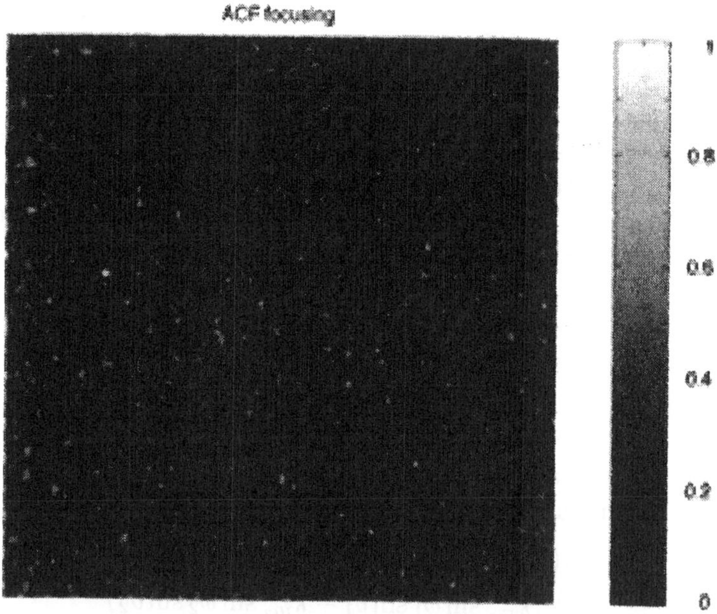

Figure 11.4.4 Simulated image of targets embedded in clutter by ACF focusing for circular SAR.

Linear SAR Imaging

For linear SAR as shown in Fig. 11.4.1, a monostatic radar moves along a linear path at height H above the target region. The radar position is function of the antenna position x as $\overline{R}(x) = x\hat{x} - d\hat{y} + H\hat{z}$. Therefore, the received signals can be written as

$$E(k,x) = \sum_{n=1}^{N_t} e^{2ik|\overline{R}(x)-\overline{r}_n|} F_{tn}(k) + \sum_{n=1}^{N_p} e^{2ik|\overline{R}(x)-\overline{r}_n|} f_n(k) \qquad (11.4.20)$$

where the received signal is a function of wave number k (frequency) and antenna position x.

The field imaging is to correlate the received signal with a reference signal E_o and then sum over frequency k_m and azimuthal position x_m.

$$C_F(\overline{r}_o) = \sum_{m=1}^{N_k} \sum_{n=1}^{N_x} E(k_m, x_n) E_o(k_m)^* e^{-2ik_m|\overline{R}(x_n)-\overline{r}_o|} \qquad (11.4.21a)$$

and

$$C_{2F}(\overline{r}_o) = |C_F(\overline{r}_o)|^2 \qquad (11.4.21b)$$

In the numerical simulations, the reference signal E_o is chosen to be unity for simplicity. The correlation defined above has well defined peaks when $\overline{r}_o$ approaches the locations of the scatterers.

It was shown that in circular SAR, the angular correlation imaging can give clear images. On the other hand, in linear SAR, spatial correlation imaging gives fine cross-range resolution and frequency correlation imaging gives fine range resolution. In general, frequency spatial correlation will give better performance in correlation imaging for linear SAR. The frequency angular correlation function (FACF) imaging is defined by

$$C_\Gamma(\overline{r}_o) = \sum_{m_1=1}^{N_k} \sum_{n_1=1}^{N_x} \sum_{m_2=1}^{N_k} \sum_{n_2=1,k_d>K}^{N_x} E(k_{m_1}, x_{n_1}) e^{-2ik_{m_1}|\overline{R}(x_{n_1})-\overline{r}_o|}$$
$$E^*(k_{m_2}, x_{n_2}) e^{2ik_{m_2}|\overline{R}(x_{n_2})-\overline{r}_o|} \tag{11.4.22}$$

The memory effect is avoided in the summation by choosing $k_d > K$. The magnitude of the difference of the two wave vectors k_d is given by

$$\begin{aligned} k_d = \ & \big[(k_{m_1}\cos\phi_1 \sin\theta_1 - k_{m_2}\cos\phi_2 \sin\theta_2)^2 \\ & + (k_{m_1}\sin\phi_1 \sin\theta_1 - k_{m_2}\sin\phi_2 \sin\theta_2)^2 \\ & + (k_{m_1}\cos\theta_1 - k_{m_2}\cos\theta_2)^2\big]^{1/2} \end{aligned} \tag{11.4.23}$$

where the angular pairs (θ_1, ϕ_1) and (θ_2, ϕ_2) are the orientation angles for vectors $\overline{R}(x_{n_1})-\overline{r}_o$ and $\overline{R}(x_{n_2})-\overline{r}_o$, respectively. In the numerical simulations of correlation focusing, k_d is chosen to be larger than $k_o/16$, i.e., $k_d > K = \frac{k_o}{16}$, where k_o is the center wave number. Equation (11.4.22) reduces to FCF imaging with $n_2 = n_1$ and to ACF imaging by letting $m_2 = m_1$. The FCF imaging is then given by

$$C_\Gamma^{(FCF)}(\overline{r}_o) = \sum_{n=1}^{N_x} \sum_{m_1=1}^{N_k} \sum_{m_2=1,k_d>K}^{N_k} E(k_{m_1}, x_n) e^{-2ik_{m_1}|\overline{R}(x_n)-\overline{r}_o|}$$
$$\cdot E^*(k_{m_2}, x_n) e^{2ik_{m_2}|\overline{R}(x_n)-\overline{r}_o|} \tag{11.4.24}$$

and ACF imaging is written as

$$C_\Gamma(\overline{r}_o) = \sum_{m=1}^{N_k} \sum_{n_1=1}^{N_x} \sum_{n_2=1,k_d>K}^{N_x} E(k_m, x_{n_1}) e^{-2ik_m|\overline{R}(x_{n_1})-\overline{r}_o|}$$
$$\cdot E^*(k_m, x_{n_2}) e^{2ik_m|\overline{R}(x_{n_2})-\overline{r}_o|} \tag{11.4.25}$$

For the numerical simulations of linear SAR imaging, 80,000 small particles are used to model clutter. They are randomly distributed in a layer region of $40\lambda_o \times 40\lambda_o \times 0.5\lambda_o$. Four targets are placed at positions of $(10,14,0)\lambda_o$,

$(10,28,0)\lambda_o$, $(30,10,0)\lambda_o$, and $(30,32,0)\lambda_o$. The dielectric constant of the targets and the scatterers are equal to $(3.23 + i0.36)$. The back-scattering amplitudes for the targets and the particles are calculated based on Mie scattering. The receiver moves from $x_s = -d/2$ to $x_s = d/2$ with an increment of $d/100$. The horizontal position is $y_s = d$ with $d = 1732\lambda_o$ and the height is $z_s = H = 1000\lambda_o$. The received signal is calculated over a frequency band $0.5f_o$ to $1.5f_o$ with an increment of $0.01f_o$ and 100 azimuthal positions with equal space.

The simulated data is processed by the methods of field imaging and FACF imaging. The normalized results are shown in Fig. 11.4.5, and 11.4.6. Figure 11.4.5 shows field imaging of $C_{2F}(\overline{r}_o) = |C_F(\overline{r}_o)|^2$. The 4 targets are obscured by the background clutter. Figure 11.4.6 shows the results of FACF imaging with the memory effect avoided, which has lower clutter level than that in Fig. 11.4.5. The results of FCF imaging and ACF imaging are also shown in Fig. 11.4.7 and 11.4.8. We see that ACF imaging loses range resolution and FCF does not have good cross-range resolution. Therefore, FACF imaging gives better performance than ACF imaging and FCF imaging for linear SAR.

To compare results quantitatively, we define the visibility of targets in clutter as a ratio of target signal and the average signal strength for the entire image covering the region. The visibility is calculated by

$$v = \frac{\text{Max}(|I(\overline{r}_o)|)}{\int d\overline{r}_o |I(\overline{r}_o)| \Big/ \int d\overline{r}_o} \tag{11.4.26}$$

In Eq. (11.4.26), the numerator corresponds to the I of the targets, and the denominator is the mean value of background clutter.

The visibility of targets in clutter for images shown in Figs. 11.4.2, 11.4.3, 11.4.5 and 11.4.6 are calculated. We find that correlation imaging gives better visibility than the conventional field imaging. The image in Fig. 11.4.3 obtained for the ACF imaging has a visibility of 2 times that of field imaging shown in Fig. 11.4.2. The visibility of the image in Fig. 11.4.6 of FACF is 1.3 times that in Fig. 11.4.5. The results are summarized in Table 11.4.1

Imaging Method	Circular		Linear	
	Field	ACF	Field	FACF
Visibility	16.5	33.8	15.2	19.2

Table 11.4.1 Comparison of visibility of target in clutter.

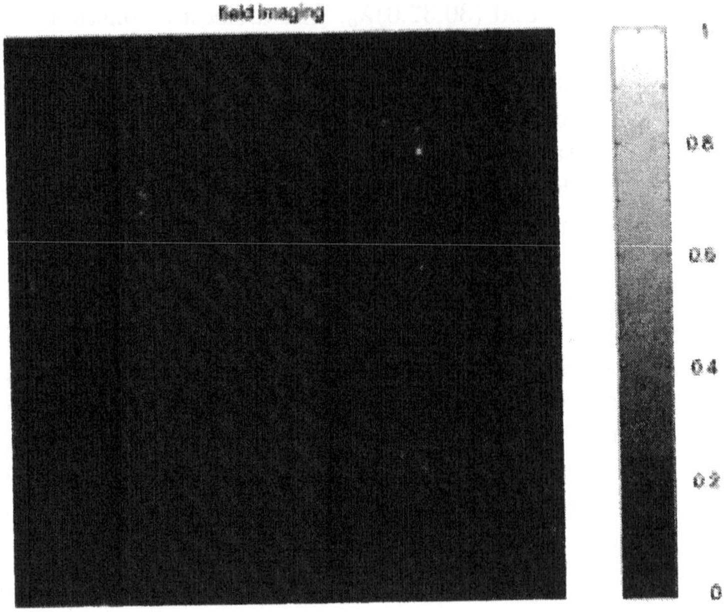

Figure 11.4.5 Simulated image of targets embedded in clutter by field imaging for linear SAR.

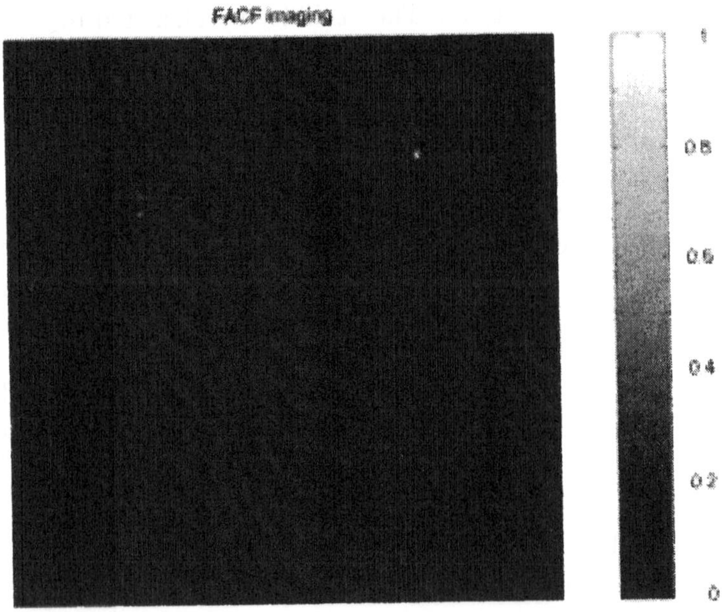

Figure 11.4.6 Simulated image of targets embedded in clutter by FACF imaging for linear SAR.

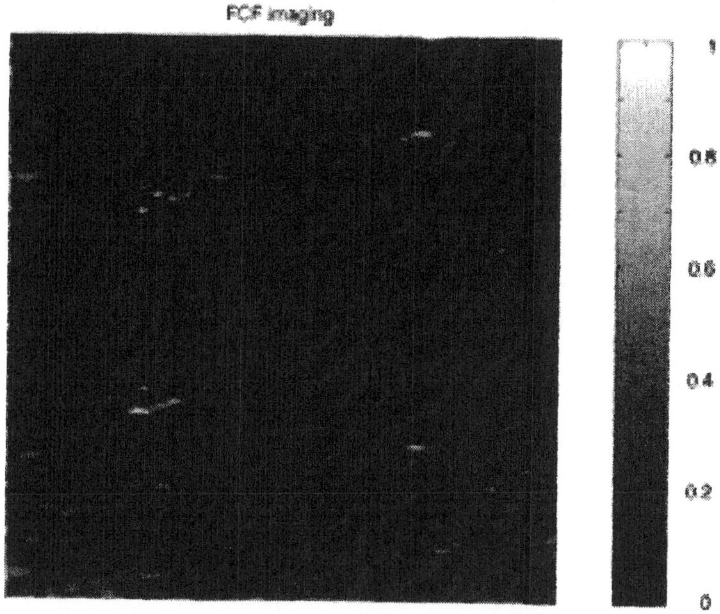

Figure 11.4.7 Simulated image of targets embedded in clutter by FCF imaging for linear SAR.

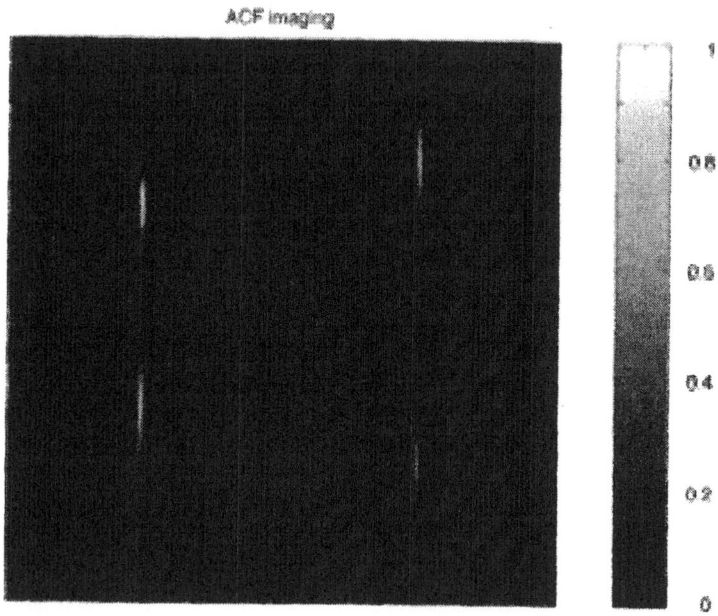

Figure 11.4.8 Simulated image of targets embedded in clutter by ACF imaging for linear SAR.

From the formulation and numerical results, we found that correlation imaging gives better results for the detection of targets embedded in clutter than field imaging when the size of random particles is smaller than a wavelength. For the same received signals, the image obtained by correlation imaging has a finer resolution and a larger signal-clutter ratio than that by field imaging. This is because the clutter effect is minimized in correlation imaging by avoiding the memory effect of random scattering. It is shown that the clutter effect can be substantially reduced for both circular and linear configurations. In a similar manner, a 3-D image can be obtained with focusing on different layers. The frequency dependence can be compensated by using a weighting function. The image processing algorithm can be speeded up by using FFT with the far-field approximation for the phase.

REFERENCES AND ADDITIONAL READINGS

Axelsson, S. (1995), Frequency and azimuth variation and their influence upon low-frequency SAR imaging, *IEEE Trans. Geosci. Remote Sens.*, **33**, 1258–1265.

Berkovits, R., M. Kaveh, and S. Feng (1989), Memory effect of waves in disordered systems: a real-space approach, *Phys. Rev. B*, **40**, 737–740.

Berkovits, R. and S. Feng (1994), Correlations in coherent multiple scattering, *Phy. Rep.*, **238**, 135–172.

Chan, T.-K., Y. Kuga, and A. Ishimaru (1999), Experimental studies on circular SAR imaging in clutter using angular correlation function technique, *IEEE Trans. Geosci. Remote Sens.*, **37(5)**, 2192–2197.

Chan, T.-K., Y. Kuga, and A. Ishimaru (1997), Subsurface detection of a buried object using angular correlation function measurement, *Waves in Random Media*, **7(3)**, 457–465.

Chan, T.-K., Y. Kuga, and A. Ishimaru (1996), Angular memory effect of millimeter-wave scattering from two-dimensional conducting random rough surfaces, *Radio Sci.*, **31**, 1067–1076.

Feng, S., C. Kane, P. A. Lee, and A. D. Stone (1988), Correlations and fluctuations of coherent wave transmission through disordered media, *Phys. Rev. Lett.*, **61**, 834–837.

Freund, I. (1990), Correlation imaging through multiply scattering media, *Phys. Lett. A*, **147(8/9)**, 502–506.

Freund, I., M. Rosenbluh, and S. Feng (1988), Memory effects in propagation of optical waves through disordered media, *Phys. Rev. Lett.*, **61**, 2328–2331.

Kawanishi, T., Z. L. Wang, M. Izutsu, H. Ogura (1999), Conjugate memory effect of random scattered waves, *J. Opt. Soc. Am. A*, **16**, 1342–1349.

Knotts, M. E., T. R. Michel, and K. A. O'Donnell (1992), Angular correlation functions of polarized intensities scattered from a one-dimensionally rough surface, *J. Opt. Soc. Am. A*, **9**, 1882–1831.

Kuga, Y., C. T. C. Le, A. Ishimaru, and L. Ailes-Sengers (1996), Analytical experimental, and numerical studies of angular memory signatures of waves scattered from one-dimensional rough surfaces, *IEEE Trans. Geosci. Remote Sens.*, **34**, 1300–1307.

Le, C. T. C., Y. Kuga, and A. Ishimaru (1996), Angular correlation function based on the second-order Kichhoff approximation and comparison with experiments, *J. Opt. Soc. Am. A*, **13**, 1057–1067.

Lu, J. Q. and Z. H. Gu (1997), Angular correlation function of speckle patterns scattered from a one-dimensional rough dielectric film on a glass substrate, *Appl. Optics*, **36**, 4562–4570.

Michel, T. R. and K. A. O'Donnell (1992), Angular correlation functions of amplitudes scattered from a one-dimensional, perfectly conducting rough surface, *J. Opt. Soc. Am. A*, **9(8)**, 1374–1384.

Moore (1996), A new algorithm for the formation of ISAR images, *IEEE Trans. Aerosp. Electron. Syst.*, **32(2)**, 714–721.

Nieto-Vesperinas, M. and J. M. Soto-Crespo (1987), Monte-Carlo simulations for scattering of electromagnetic waves from perfectly conducting random rough surfaces, *Optics Lett.*, **12**, 979–981.

O'Neill, K. (2000), Broadband bistatic coherent and incoherent detection of buried objects beneath randomly rough surfaces, *IEEE Trans. Geosci. Remote Sens.*, **38(2)**, 891–898.

Sarabandi, K. and A. Nashashibi (1999), Analysis and applications of backscattered frequency correlation function, *IEEE Trans. Geosci. Remote Sens.*, **37**, 1895–1906.

Tsang, L., G. Zhang, and K. Pak (1996), Detection of a buried object under a single random rough surface with angular correlation function in EM wave scattering, *Microwave Opt. Technol. Lett.*, **11(6)**, 300–304. Corrections to "Detection of a buried object under a single random rough surface with angular correlation function in EM wave scattering", *Microwave Opt. Technol. Lett.*, **12**, 375.

Soumekh, M. (1996), Reconnaissance with slant plant circular SAR imaging, *IEEE Trans. Image Processing*, **35**, 45–51.

Soumekh, M. (1999), *Synthetic Aperture Radar Signal Processing with MATLAB Algorithms*, Wiley-Interscience, New York.

Zhang, G. and L. Tsang (1997), Angular correlation function of wave scattering by a random rough surface and discrete scatterers and its application in the detection of a buried object, *Waves in Random Media*, **7(3)**, 467–478.

Zhang, G. and L. Tsang (1998), Application of angular correlation function of clutter scattering and correlation imaging in target detection, *IEEE Trans. Geosci. Remote Sens.*, **36**, 1485–1493.

Zhang, G., L. Tsang, and Y. Kuga (1997a), Studies of the angular correlation function of scattering by random rough surfaces with and without a buried object, *IEEE Trans. Geosci. Remote Sens.*, **35**, 444–453.

Zhang, G., L. Tsang, and Y. Kuga (1997b), The angular correlation function of wave scattering by a buried object embedded in random discretic scatterers under a random rough surface, *Microwave Opt. Technol. Lett.*, **14**, 144–151.

Zhang, G., L. Tsang, and Y. Kuga (1998a), Numerical studies of the detection of targets embedded in clutter by using angular correlation functions and angular correlation imaging, *Microwave Opt. Technol. Lett.*, **17(2)**, 82–86.

Zhang, G., L. Tsang, and K. Pak (1998b), Angular correlation function and scattering coefficients of electromagnetic waves scattered by a buried object under a two-dimensional rough surface, *J. Opt. Soc. Am. A*, **15(12)**, 2995–3002.

Chapter 12

MULTIPLE SCATTERING BY CYLINDERS IN THE PRESENCE OF BOUNDARIES

1 Introduction

In Chapters 7 to 11, we have studied the multiple scattering problems with particles in an infinite medium. In this chapter, we study the case when the cylinders are placed on top of reflective boundaries. In Section 2, we study scattering by dielectric cylinders lying above a dielectric half-space. Applications to scattering by vegetation and forests are illustrated. We discuss the uniform randomly distributed cylinders and also cylinders in clusters. The two cases have very different scattering properties. The study of backscattering enhancement for the case of cylinders in the presence of reflective boundary is also illustrated. This is a case of backscattering enhancement that is exhibited in volume scattering in the presence of reflective boundary. The subject of backscattering enhancement will be treated in more details in Chpater 8 of Volume III. In Section 3, we study the problem of metallic cylinders placed between two perfect conductors. For this case, the parallel waveguide geometry facilitates the use of waveguide modes for propagation and scattering among cylinders.

2 Scattering by Dielectric Cylinders Above a Dielectric Half-Space

Scattering of waves from a layer of vertical dielectric cylinders overlying a homogeneous half-space is an important remote sensing problem in microwave propagation, scattering, and emission of vegetation. For example, in forests and corn canopy, the scattering can be dominated by the trunks and corn stalks. In this section we illustrate analytically the first-order scattering solution and the concepts of g_s (pair distribution functions among scatterers in a cluster) and g_p (pair distribution function between clusters) in the two-dimensional random positioning of the vertical cylinders. Analytical solution is derived for single scattering. We shall present the forms of the analytical expressions for the first-order solutions in Section 2.1. Both first- and second-order solutions will then be rigorously derived in Section 2.2.

2.1 Scattering from a Layer of Vertical Cylinders: First-Order Solution

Consider a plane electromagnetic wave incident upon a layer of vertical dielectric cylinders overlying a homogeneous dielectric half-space (Fig. 12.2.1). The boundary of the dielectric half-space is at $z = -d$. There are N cylin-

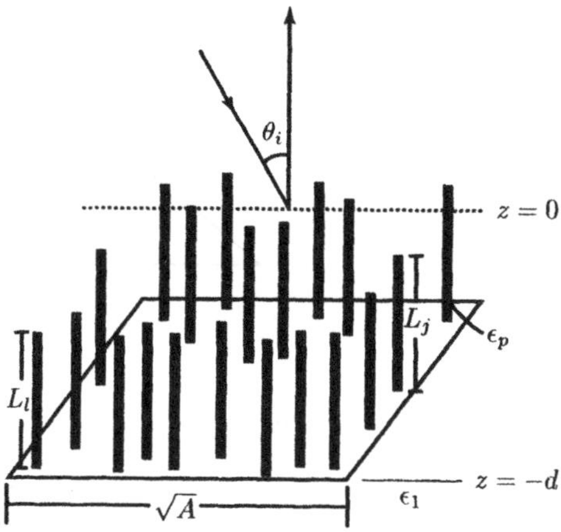

Figure 12.2.1 Scattering by vertical cylinders of length L and permittivity ϵ_p overlying a dielectric half-space of permittivity ϵ_1.

ders in the canopy planted in an area A. The centers of the cylinders are located at $\overline{r}_1, \overline{r}_2, \ldots, \overline{r}_N$. The αth cylinder is of length L_α and radius a_α with $\alpha = 1, 2, \ldots, N$, where N is a large number. The permittivities of the cylinders and the homogeneous half-space are ϵ_p and ϵ_1, respectively. The incident wave is in direction $(\pi - \theta_i, \phi_i)$ with incident electric field as

$$\overline{E}^{inc} = (E_{vi}\hat{v}_i + E_{hi}\hat{h}_i)\, e^{ik_{ix}x + ik_{iy}y - ik_{iz}z} \tag{12.2.1}$$

where $k_{ix} = k \sin\theta_i \cos\phi_i$, $k_{iy} = k \sin\theta_i \sin\phi_i$, $k_{iz} = k\cos\theta_i$, $\hat{v}_i = \hat{\theta}(\pi - \theta_i, \phi_i)$, and $\hat{h}_i = \hat{\phi}(\pi - \theta_i, \phi_i)$. We consider observation direction $\hat{k}_s$ with

$$\hat{k}_s = k_{sx}\hat{x} + k_{sy}\hat{y} + k_{sz}\hat{z} \tag{12.2.2}$$

where $k_{sx} = k \sin\theta_s \cos\phi_s$, $k_{sy} = k \sin\theta_s \sin\phi_s$, and $k_{sz} = k\cos\theta_s$. The single scattering solution will have four contributions. By single scattering we mean single volume scattering that can also include bouncing from the boundary of the homogeneous half-space at $z = -d$. Let $\overline{r}_j = \overline{\rho}_j + z_j\hat{z}$ be the position of the jth cylinder. The four contributions are as follows (Fig. 12.2.2):

(a) Direct volume scattering by cylinder j that has a dependence of

$$e^{i\overline{k}_{d\rho}\cdot\overline{\rho}_j} e^{-ik_{iz}z_j - ik_{sz}z_j} \frac{L_j}{2\pi} \operatorname{sinc}\left[(k_{iz} + k_{sz})\frac{L_j}{2}\right]$$

where

$$\begin{aligned} \overline{\rho}_j &= x_j\hat{x} + y_j\hat{y} \\ \overline{k}_d &= k(\hat{k}_i - \hat{k}_s) \end{aligned} \tag{12.2.3}$$

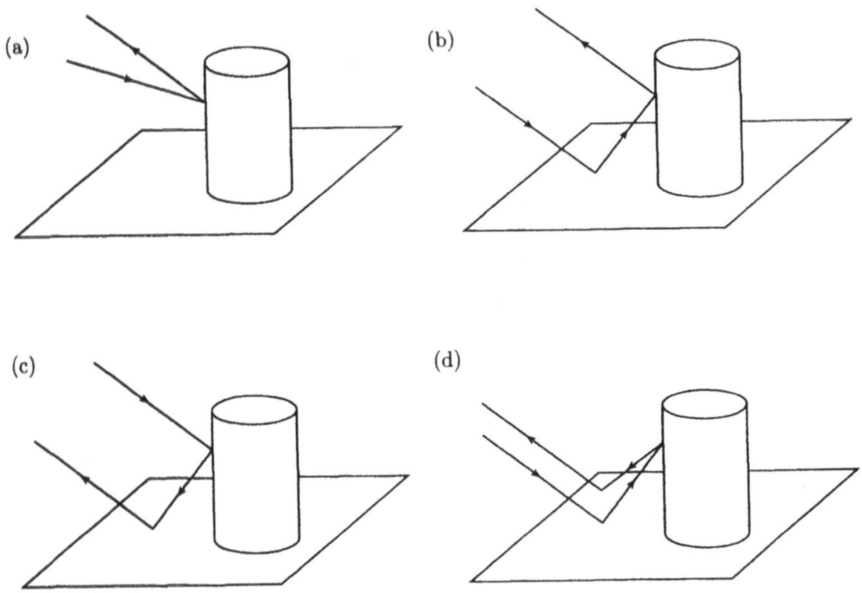

Figure 12.2.2 Four scattering contributions for single scattering of the cylinder in the presence of a reflective boundary.

$$\overline{k}_{d\rho} = \overline{k}_{i\rho} - \overline{k}_{s\rho} = (k_{ix} - k_{sx})\hat{x} + (k_{iy} - k_{sy})\hat{y} \tag{12.2.4}$$

is a vector in the x-y-plane and

$$\overline{k}_{d\rho} \cdot \overline{\rho}_j = k_{dx} x_j + k_{dy} y_j \tag{12.2.5}$$

The phase dependence can be directly traced from the figure. The phase dependence $e^{-ik_{iz}z_j - ik_{sz}z_j}$ is due to the vertical z_j position of the cylinder. The phase dependence of $e^{i\overline{k}_{d\rho}\cdot\overline{\rho}_j}$ is due to the horizontal position $\overline{\rho}_j$ of the cylinder j, where $\overline{k}_{d\rho}$ is the wave vector difference between the incident and scattered wave vectors in the horizontal direction. The amplitude dependence is as illustrated by the conical pattern of the cylinder that was discussed in Chapter 1, Section 6.2 of Volume I.

(b) Reflection by boundary followed by volume scattering by cylinder j, which has a dependence of

$$e^{i\overline{k}_{d\rho}\cdot\overline{\rho}_j} e^{ik_{iz}(z_j+2d) - ik_{sz}z_j} \frac{L_j}{2\pi} \operatorname{sinc}\left[\frac{(k_{iz} - k_{sz})L_j}{2}\right]$$

The phase factor $\exp(ik_{iz}2d)$ is a result of the incident wave traveling through the canopy.

(c) Volume scattering followed by reflection that has a dependence of

$$e^{i\overline{k}_{d\rho}\cdot\overline{\rho}_j} e^{-ik_{iz}z_j} e^{+ik_{sz}(z_j+2d)} \frac{L_j}{2\pi} \text{sinc}\left[\frac{(k_{iz}-k_{sz})L_j}{2}\right]$$

(d) Reflection–volume–reflection scattering. Reflection followed by volume scattering that is further followed by reflection. The dependence is

$$e^{i\overline{k}_{d\rho}\cdot\overline{\rho}_j} e^{ik_{iz}(z_j+2d)} e^{ik_{sz}(z_j+2d)} \frac{L_j}{2\pi} \text{sinc}\left[\frac{(k_{iz}+k_{sz})L_j}{2}\right]$$

In this case both incident waves and scattered waves traveled through the canopy. Thus

$$\begin{aligned}
&\overline{E}^{(s)(1)}(\overline{r}) \\
&= \frac{e^{ikr}}{r}\sum_{j=1}^{N}\frac{L_j}{2\pi}e^{i\overline{k}_{d\rho}\cdot\overline{\rho}_j}\left\{\text{sinc}\left[(k_{iz}+k_{sz})\frac{L_j}{2}\right]\overline{f}_v(\hat{k}_s,\hat{k}_i)e^{-ik_{sz}z_j}e^{-ik_{iz}z_j}\right. \\
&\quad + \text{sinc}\left[(k_{iz}-k_{sz})\frac{L_j}{2}\right]\overline{f}_{vr}(\hat{k}_s,\hat{k}_i)e^{ik_{iz}z_j-ik_{sz}z_j+2ik_{iz}d} \\
&\quad + \text{sinc}\left[(k_{iz}-k_{sz})\frac{L_j}{2}\right]\overline{f}_{rv}(\hat{k}_s,\hat{k}_i)e^{-ik_{sz}z_j-ik_{iz}z_j}e^{2ik_{sz}(z_j+d)} \\
&\quad \left. + \text{sinc}\left[(k_{iz}+k_{sz})\frac{L_j}{2}\right]\overline{f}_{rvr}(\hat{k}_s,\hat{k}_i)e^{ik_{sz}z_j}e^{ik_{iz}z_j}e^{2i(k_{iz}+k_{sz})d}\right\}
\end{aligned} \quad (12.2.6)$$

where $\overline{f}_v, \overline{f}_{vr}, \overline{f}_{rv}$, and $\overline{f}_{rvr}$ are field vectors that depend on cylinder radius a_j and is polarization dependent on the scattering characteristics of the cylinders and the reflection by the half-space boundary. They will be derived rigorously in Section 2.2. The subscripts v and r denote volume and reflection, respectively.

In the following, we shall assume that (1) the position of z_j is $z_j = L_j/2 - d$ (that is, the cylinder is attached to the boundary of the dielectric half-space) and (2) the length of the cylinder is Gaussian distribution with mean L_o and standard deviation σ_L. The probability density function (pdf) is

$$p(L) = \frac{1}{\sqrt{2\pi}\sigma_L}\exp\left[-\frac{(L-L_o)^2}{2\sigma_L^2}\right] \quad (12.2.7)$$

The probability distribution of length will smooth out side lobes that may exist in the conical scattering pattern. The radius of the cylinder is equal to constant a for all the cylinders. The pdf and joint pdf of horizontal positions are independent of lengths of cylinders. Thus, letting $z_j = L_j/2 - d$ in (12.2.6) gives

$$\overline{E}^{s(1)}(\overline{r}) = \frac{e^{ikr}}{r}\overline{F}^{(1)} \tag{12.2.8a}$$

where

$$\overline{F}^{(1)} = \sum_{j=1}^{N} e^{i\overline{k}_{dp}\cdot\overline{\rho}_j}\overline{f}(L_j) \tag{12.2.8b}$$

is the first-order scattering amplitude of the canopy, and

$$\begin{aligned}\overline{f}(L_j) = \frac{L_j}{2\pi}e^{i(k_{sz}+k_{iz})d}\Bigg\{ &\operatorname{sinc}\left[(k_{iz}+k_{sz})\frac{L_j}{2}\right]\overline{f}_v e^{-i(k_{sz}+k_{iz})L_j/2} \\ &+ \operatorname{sinc}\left[(k_{iz}-k_{sz})\frac{L_j}{2}\right]\overline{f}_{vr} e^{i(k_{iz}-k_{sz})L_j/2} \\ &+ \operatorname{sinc}\left[(k_{iz}-k_{sz})\frac{L_j}{2}\right]\overline{f}_{rv} e^{i(k_{sz}-k_{iz})L_j/2} \\ &+ \operatorname{sinc}\left[(k_{iz}+k_{sz})\frac{L_j}{2}\right]\overline{f}_{rvr} e^{i(k_{sz}+k_{iz})L_j/2}\Bigg\}\end{aligned} \tag{12.2.9}$$

The pdf and joint pdf of horizontal positions of cylinders are as follows. Let the cylinders be in clusters with N_s cylinders per cluster. Hence the number of clusters is N_c with

$$N_c = \frac{N}{N_s} \tag{12.2.10}$$

Note that N and N_c are large numbers while N_s may not be a large number. Let the cluster center be at (x_α, y_α) with $\alpha = 1, \ldots, N_c$. Each cluster lies within a radius R_c. Thus the pdf and joint pdf of clusters are

$$p(x_\alpha, y_\alpha) = \frac{1}{A} \tag{12.2.11}$$

where $A = L_x L_y$ is the area under observation and $L_x \gg \lambda$, $L_y \gg \lambda$. The joint pdf of clusters is

$$p_2(\overline{\rho}_\alpha, \overline{\rho}_\beta) = \frac{g_p(\overline{\rho}_\alpha, \overline{\rho}_\beta)}{A^2} \tag{12.2.12}$$

where g_p is the pair distribution function. We also disallow interpenetration of clusters so that

$$g_p(\overline{\rho}_\alpha, \overline{\rho}_\beta) = 0 \quad \text{for} \quad |\overline{\rho}_\alpha - \overline{\rho}_\beta| < d_c \tag{12.2.13}$$

where d_c is the minimum separation of the centers of two clusters. Within each cluster α, the positions of the secondary scatterers are at $\overline{\rho}_\alpha + \overline{\rho}_{\alpha j}$,

$j = 1, \ldots, N_s$.

$$p(x_{\alpha j}, y_{\alpha j}) = \frac{1}{A_c} \tag{12.2.14}$$

where $A_c = \pi R_c^2$ is the area occupied by a cluster.

$$p_2(\overline{\rho}_{\alpha j}, \overline{\rho}_{\alpha l}) = \frac{g_s(\overline{\rho}_{\alpha j}, \overline{\rho}_{\alpha l})}{A_c^2} \tag{12.2.15}$$

Note that for the random media problem, N, N_c, and $A \to \infty$ while N_s and A_c may not be large number.

The pair functions are normalized quantities so that the magnitudes are of the order of unity. The summation of (12.2.8b) is then replaced by $\sum_{\alpha=1}^{N_c}\sum_{j=1}^{N_s}$.

$$\overline{F}^{(1)}(\hat{k}_s, \hat{k}_i) = \sum_{\alpha=1}^{N_c}\sum_{j=1}^{N_s} e^{i\overline{k}_{d\rho}\cdot(\overline{\rho}_\alpha + \overline{\rho}_{\alpha j})}\overline{f}(L_{\alpha j}) \tag{12.2.16}$$

We first evaluate the coherent field by taking the average of (12.2.16)

$$\langle \overline{F}^{(1)}(\hat{k}_s, \hat{k}_i)\rangle = N_c N_s \langle e^{i\overline{k}_{d\rho}\cdot(\overline{\rho}_\alpha + \overline{\rho}_{\alpha j})}\rangle\langle \overline{f}(L_{\alpha j})\rangle \tag{12.2.17}$$

where

$$\langle e^{i\overline{k}_{d\rho}(\overline{\rho}_\alpha + \overline{\rho}_{\alpha j})}\rangle = \frac{1}{A}\int d\overline{\rho}_\alpha\, e^{i\overline{k}_{d\rho}\cdot\overline{\rho}_\alpha}\frac{1}{A_c}\int_{A_c} d\overline{\rho}_{\alpha j}\, e^{i\overline{k}_{d\rho}\cdot\overline{\rho}_{\alpha j}} \tag{12.2.18}$$

Assuming a rectangular area $A = L_x L_y$, we have

$$\langle e^{i\overline{k}_{d\rho}\cdot(\overline{\rho}_\alpha + \overline{\rho}_{\alpha j})}\rangle = \operatorname{sinc}\left[\frac{k_{dx}L_x}{2}\right]\operatorname{sinc}\left[\frac{k_{dy}L_y}{2}\right]X(\overline{k}_{d\rho}) \tag{12.2.19}$$

where

$$X(\overline{k}_{d\rho}) = \frac{1}{A_c}\int_{A_c} d\overline{\rho}_{\alpha j}\, e^{i\overline{k}_{d\rho}\cdot\overline{\rho}_{\alpha j}} \tag{12.2.20}$$

Depending on clustering size A_c; the integral of the exponential term in (12.2.20) may not undergo large phase fluctuation over the cluster area. From (12.2.20), $X(0) = 1$. For L_x and $L_y \to \infty$, the sinc terms in (12.2.19) are sharply peaked at $k_{dx} = k_{dy} = 0$ so that we can replace $k_{d\rho}$ by 0 in $X(k_{d\rho})$ in (12.2.19). Thus

$$\langle e^{i\overline{k}_{d\rho}\cdot(\overline{\rho}_\alpha + \overline{\rho}_{\alpha j})}\rangle = \operatorname{sinc}\left[\frac{k_{dx}L_x}{2}\right]\operatorname{sinc}\left[\frac{k_{dy}L_y}{2}\right] \tag{12.2.21}$$

Hence

$$\langle \overline{F}^{(1)}(\hat{k}_s, \hat{k}_i)\rangle = N \text{sinc}\left[\frac{k_{dx}L_x}{2}\right] \text{sinc}\left[\frac{k_{dy}L_y}{2}\right] \langle \overline{f}(L)\rangle \tag{12.2.22}$$

Using (12.2.7) and (12.2.9), we have

$$\langle \overline{f}(L)\rangle = \frac{1}{2\pi}\Big\{W(-(k_{sz}+k_{iz}))\overline{f}_v + W(k_{iz}-k_{sz})\overline{f}_{vr} + W(k_{sz}-k_{iz})\overline{f}_{rv} + W(k_{sz}+k_{iz})\overline{f}_{rvr}\Big\} e^{i(k_{iz}+k_{sz})d} \tag{12.2.23}$$

$$W(\alpha) = \frac{1}{i\alpha}\left\{e^{i\alpha L_o} e^{-\alpha^2\sigma_L^2/2} - 1\right\} = \langle L \,\text{sinc}\frac{\alpha L}{2}\, e^{i\alpha L/2}\rangle \tag{12.2.24}$$

The bistatic scattering coefficient is

$$\gamma(\hat{k}_s, \hat{k}_i) = \frac{4\pi\langle|\overline{F}|^2\rangle}{A\cos\theta_i}$$

The contribution of the bistatic coefficient is the sum of the coherent field γ^c contribution and the incoherent field contribution, γ^i :

$$\gamma = \gamma^c + \gamma^i \tag{12.2.25}$$

For coherent field contribution,

$$\gamma^c = \frac{4\pi|\langle\overline{F}\rangle|^2}{A\cos\theta_i} = \frac{4\pi n_A}{\cos\theta_i} N\text{sinc}^2\left(\frac{k_{dx}L_x}{2}\right)\text{sinc}^2\left(\frac{k_{dy}L_y}{2}\right)|\langle\overline{f}(L)\rangle|^2 \tag{12.2.26}$$

where $n_A = N/A$ is the number of cylinders per unit area. Thus the coherent intensity is sharply peaked in the specular direction and depends on the total number of scatterers N, the size of the area L_x and L_y, and the shape of the area under observation. The incoherent field is

$$\overline{\mathcal{F}} = \overline{F} - \langle\overline{F}\rangle \tag{12.2.27}$$

The incoherent field can contain partial coherent effects. The bistatic scattering coefficient of the incoherent field is γ^i with

$$\gamma^i(\hat{k}_s, \hat{k}_i) = \frac{4\pi\langle|\overline{\mathcal{F}}(\hat{k}_s, \hat{k}_i)|^2\rangle}{A\cos\theta_i} \tag{12.2.28}$$

The incoherent bistatic scattering coefficient, on the other hand, as shown by following calculation, depends only on the intrinsic properties of the random media (e.g., n_A) and does not depend on parameters such as N, L_x, and L_y. Thus the γ^i obtained with Monte Carlo simulations is a physically meaningful quantity that can be used for practical applications.

Next we form $|\overline{F}^{(1)}(\hat{k}_s, \hat{k}_i)|^2$, which gives a fourfold summation of $\sum_\alpha \sum_j \sum_\beta \sum_l$. This can be separated into three terms: (i) $\alpha = \beta$, $j = l$, scattered field and field conjugate from the same cylinder, (ii) $\alpha = \beta$, $j \neq l$, scattered field and field conjugate from the same cluster but from different cylinders in the cluster, (iii) $\alpha \neq \beta$, scattered field and field conjugate from cylinders of two different clusters. Thus,

$$
\begin{aligned}
&\langle |\overline{F}^{(1)}(\hat{k}_s, \hat{k}_i)|^2 \rangle \\
&= \sum_{\alpha=1}^{N_c} \sum_{j=1}^{N_s} \langle |\overline{f}(L_{\alpha j})|^2 \rangle + \sum_{\alpha=1}^{N_c} \sum_{\substack{j=1 \\ j \neq l}}^{N_s} \sum_{l=1}^{N_s} \langle \exp\left(i\overline{k}_{d\rho} \cdot (\overline{\rho}_{\alpha j} - \overline{\rho}_{\alpha l})\right) \rangle |\langle \overline{f}(L) \rangle|^2 \\
&+ \sum_{\substack{\alpha=1 \\ \alpha \neq \beta}}^{N_c} \sum_{\beta=1}^{N_c} \sum_{j=1}^{N_s} \sum_{l=1}^{N_s} \langle \exp\left(i\overline{k}_{d\rho} \cdot ((\overline{\rho}_\alpha + \overline{\rho}_{\alpha j}) - (\overline{\rho}_\beta + \overline{\rho}_{\beta l}))\right) \rangle \cdot |\langle \overline{f}(L) \rangle|^2
\end{aligned}
\tag{12.2.29}
$$

Using the definitions of pair distribution functions of cluster-cluster g_p and cylinder-cylinder in the same cluster g_s, we have

$$
\begin{aligned}
&\langle |\overline{F}^{(1)}(\hat{k}_s, \hat{k}_i)|^2 \rangle = \\
&\quad N \langle |\overline{f}(L)|^2 \rangle + N_c N_s (N_s - 1) \int_{A_c} d\overline{\rho}_{\alpha j} \int_{A_c} d\overline{\rho}_{\alpha l} \, \frac{g_s(\overline{\rho}_{\alpha j}, \overline{\rho}_{\alpha l})}{A_c^2} \\
&\qquad \cdot \exp\left(i\overline{k}_{d\rho} \cdot (\overline{\rho}_{\alpha j} - \overline{\rho}_{\alpha l})\right) |\langle \overline{f}(L) \rangle|^2 \\
&\quad + N_c (N_c - 1) N_s^2 \int_A d\overline{\rho}_\alpha \int_A d\overline{\rho}_\beta \, \frac{g_p(\overline{\rho}_\alpha, \overline{\rho}_\beta)}{A^2} \\
&\qquad \cdot \exp\left(i\overline{k}_{d\rho} \cdot (\overline{\rho}_\alpha - \overline{\rho}_\beta)\right) |X(\overline{k}_{d\rho})|^2 \, |\langle \overline{f}(L) \rangle|^2
\end{aligned}
\tag{12.2.30}
$$

The pair functions over clusters $g_p(\overline{\rho}_\alpha, \overline{\rho}_\beta)$ are to be integrated over a large area A as indicated in (12.2.30). Besides the nonpenetrating condition of (12.2.13), the function must asymptotically approach unity because the cluster positions must be independent if they are far apart. Thus,

$$
\lim_{|\overline{\rho}_\alpha - \overline{\rho}_\beta| \to \infty} g_p(\overline{\rho}_\alpha, \overline{\rho}_\beta) = 1 \tag{12.2.31}
$$

Furthermore, we assume that it is translational invariant. $g_p(\overline{\rho}_\alpha, \overline{\rho}_\beta) = g_p(\overline{\rho}_\alpha - \overline{\rho}_\beta)$. On the other hand, g_s, the pair function within a cluster, extends over a smaller area A_c and does not share the same properties as g_p. We write $g_p = (g_p - 1) + 1$ in (12.2.30) and let N_c and $N \to \infty$. We also make use of the translational invariant property of g_p :

$$\langle|\overline{F}^{(1)}(\hat{k}_s,\hat{k}_i)|^2\rangle = N\langle|\overline{f}(L)|^2\rangle$$
$$+N(N_s-1)\int_{A_c} d\overline{\rho}_{\alpha j}\int_{A_c} d\overline{\rho}_{\alpha l}\frac{g_s(\overline{\rho}_{\alpha j},\overline{\rho}_{\alpha l})}{A_c^2}\, e^{i\overline{k}_{d\rho}\cdot(\overline{\rho}_{\alpha j}-\overline{\rho}_{\alpha l})}\,|\langle\overline{f}(L)\rangle|^2$$
$$+\frac{N^2}{A}\int_A d\overline{\rho}(g_p(\overline{\rho})-1)\exp(i\overline{k}_{d\rho}\cdot\overline{\rho})\,|X(\overline{k}_{d\rho})|^2\,|\langle\overline{f}(L)\rangle|^2$$
$$+N^2\int d\overline{\rho}_\alpha\int d\overline{\rho}_\beta\frac{\exp\left(i\overline{k}_{d\rho}\cdot(\overline{\rho}_\alpha-\overline{\rho}_\beta)\right)}{A^2}\,|X(\overline{k}_{d\rho})|^2\,|\langle\overline{f}(L)\rangle|^2 \qquad (12.2.32)$$

The last term in (12.2.32) is the coherent intensity. Subtracting it from (12.2.32) gives the incoherent intensity, which, however, can contain partial coherent effects. Thus the bistatic intensity of the incoherent field, under the first-order approximation, is

$$\gamma^{i(1)}(\hat{k}_s,\hat{k}_i) = \frac{4\pi}{\cos\theta_i}\Bigg\{n_A\langle|\overline{f}(L)|^2\rangle$$
$$+n_A(N_s-1)\int d\overline{\rho}_{\alpha j}\int d\overline{\rho}_{\alpha l}\frac{g_s(\overline{\rho}_{\alpha j},\overline{\rho}_{\alpha l})}{A_c^2}\exp\left(i\overline{k}_{d\rho}\cdot(\overline{\rho}_{\alpha j}-\overline{\rho}_{\alpha l})\right)|\langle\overline{f}(L)\rangle^2$$
$$+n_A^2\int_A d\overline{\rho}(g_p(\overline{\rho})-1)e^{i\overline{k}_{d\rho}\cdot\overline{\rho}}|X(\overline{k}_{d\rho})|^2|\langle\overline{f}(L)\rangle|^2\Bigg\} \qquad (12.2.33)$$

As is clear from (12.2.33), the bistatic intensity of the incoherent field only depends on intrinsic properties of the random media such as quantities like n_A, L, g_s, and g_p. The value of γ^i obtained by Monte Carlo simulations can be used for practical applications. The integral of g_s is only over the cluster size and can be performed numerically. The integral of g_p is only over the region of A that g_p is not equal to 1. That is an area much smaller than A, and the integral of $g_p - 1$ can be performed numerically.

We note that the first term in (12.2.33) is the independent scattering result. The second term is the correlation due to clustering effects for a single cluster. The last term is due to correlation between clusters. We note that for $N_s \geq 2$ the second term due to g_s gives a positive contribution, giving rise to enhancement of scattering due to clustering. The larger the N_s, the stronger the effect. Physically, this can easily be interpreted by the fact that when scatterers form a cluster, they form a larger particle and can give rise to a larger cross section. The last term in (12.2.33) is due to correlation effects between clusters. It is usually negative (e.g., $g_p - 1 = -1$ for $|\overline{\rho}| < d_{min}$), giving rise to less incoherent scattering. We illustrate the results with the following simple example. Let $g_s = 1$, and A_c is a circular area with radius

R_c.

$$g_p(\overline{\rho}) = \begin{cases} 0 & \text{if } |\overline{\rho}| < 2R_c \\ 1 & \text{if } |\overline{\rho}| \geq 2R_c \end{cases} \tag{12.2.34}$$

Equation (12.2.34) is known as the hole correction and is valid when the density of clusters is not too large. Using the integral identity

$$\int dx\, x\, J_0(x) = x J_1(x) \tag{12.2.35}$$

we then have

$$X(\overline{k}_{d\rho}) = \frac{2}{k_{d\rho} R_c} J_1(k_{d\rho} R_c) \tag{12.2.36}$$

Hence

$$\gamma^{i(1)}(\hat{k}_s, \hat{k}_i) = \frac{4\pi}{\cos\theta_i} \left\{ n_A \langle |\overline{f}(L)|^2 \rangle + \left| \frac{2}{k_{d\rho} R_c} J_1(k_{d\rho} R_c) \langle \overline{f}(L) \rangle \right|^2 \right.$$
$$\left. \cdot \left[n_A (N_s - 1) - n_A^2 \frac{4\pi R_c J_1(k_{d\rho} 2R_c)}{k_{d\rho}} \right] \right\} \tag{12.2.37}$$

where the $2R_c$ argument in the last term of (12.2.37) arises from the fact that the hole correction has radius of $2R_c$ for the hole. It will be shown in the next section that the analytical result of (12.2.37) agrees well with Monte Carlo simulations.

2.2 First- and Second-Order Solutions

In this section we formulate the Foldy-Lax multiple scattering equations of the problem. A half-space Green's function is used that is also expressed in vector cylindrical waves. We also derive the first-order solution and give expressions of $\overline{f}_v$, $\overline{f}_{vr}$, $\overline{f}_{rv}$, and $\overline{f}_{rvr}$ that were needed in Section 2.1. The incident field is given by (12.2.1). The sum of the incident field and reflected field is

$$\overline{E}_{inc} + \overline{E}_{ref}$$
$$= \left(E_{vi}\hat{v}_i + E_{hi}\hat{h}_i \right) e^{i\overline{K}_i \cdot \overline{r}}$$
$$+ \left(S_{01}(k_{iz}) E_{vi} \hat{\theta}(\theta_i, \phi_i) + R_{01}(k_{iz}) E_{hi} \hat{\phi}(\theta_i, \phi_i) \right) e^{2ik_{iz}d} e^{i\overline{k}_i \cdot \overline{r}} \tag{12.2.38}$$

where

$$\overline{K}_i = k_{ix}\hat{x} + k_{iy}\hat{y} - k_{iz}\hat{z} \tag{12.2.39}$$
$$\overline{k}_i = k_{ix}\hat{x} + k_{iy}\hat{y} + k_{iz}\hat{z} \tag{12.2.40}$$

Both the incident fields and the reflected fields are expressed plane waves in (12.2.38).

Consider cylinder l centered at $\overline{r}_l$. The field exciting cylinder l is the sum of incident field, reflected field, and scattered wave from other cylinders j with $j \neq l$. First we express the incident field and scattered field in terms of vector cylindrical waves centered at $\overline{r}_l$.

$$\begin{aligned}
&\overline{E}_{inc} + \overline{E}_{ref} \\
&\quad = -e^{i\overline{K}_i \cdot \overline{r}_l} \left[E_{vi} \frac{1}{k_{i\rho}} \sum_n (-1)^n e^{-in\phi_i - in\pi/2} Rg\,\overline{N}_n(k_{i\rho}, -k_{iz}, \overline{r} - \overline{r}_l) \right. \\
&\quad \left. + E_{hi} \frac{1}{ik_{i\rho}} \sum_n (-1)^n e^{-in\phi_i - in\pi/2} Rg\,\overline{M}_n(k_{i\rho}, -k_{iz}, \overline{r} - \overline{r}_l) \right] \\
&\quad - e^{i\overline{k}_i \cdot \overline{r}_l} e^{2ik_{iz}d} \left[S_{01i} \frac{E_{vi}}{k_{i\rho}} \sum_n (-1)^n e^{-in\phi_i - in\pi/2} Rg\,\overline{N}_n(k_{i\rho}, k_{iz}, \overline{r} - \overline{r}_l) \right. \\
&\quad \left. + R_{01i} \frac{E_{hi}}{ik_{i\rho}} \sum_n (-1)^n e^{-in\phi_i - in\pi/2} Rg\,\overline{M}_n(k_{i\rho}, k_{iz}, \overline{r} - \overline{r}_l) \right] \qquad (12.2.41)
\end{aligned}$$

To determine the exciting field of cylinder l and to find the scattered field from cylinder j to cylinder l, we use a procedure that includes the following steps:

(*i*) Write down the expression of internal field for cylinder j in terms of vector cylindrical waves with unknown coefficients.

(*ii*) Use Green's function to find the scattered field from cylinder j. The Green's function is the half-space Green's function. The half-space Green's function includes the direct scattered field and the scattered field that is reflected by the half-space boundary at $z = -d$. The scattered field is expressed in terms of vector cylindrical waves centered at $\overline{r}_j$.

(*iii*) Use translation addition theorem to express the vector cylindrical waves from cylinder j in terms of vector cylindrical waves centered at cylinder l.

(*iv*) Equate exciting field coefficients of cylinder l to an incoming wave on cylinder l that includes incident, reflected, and scattered field from all cylinders j, $j = 1, 2, \ldots, N$, except $j = l$. Then obtain self-consistent multiple scattering equations.

Step i: The internal field inside cylinder j is assumed to have the following form:

$$\overline{E}_p^{(j)} = \sum_{n'=-\infty}^{\infty} \int_{-\infty}^{\infty} dk'_z \Big[c_{n'}^{(M)(j)}(k'_z) Rg\,\overline{M}_{n'}(k'_{p\rho}, k'_z, \overline{r} - \overline{r}_j) + c_{n'}^{(N)(j)}(k'_z) Rg\,\overline{N}_{n'}(k'_{p\rho}, k'_z, \overline{r} - \overline{r}_j) \Big] \tag{12.2.42}$$

$$i\omega\mu\overline{H}_p^{(j)} = \sum_{n'=-\infty}^{\infty} k_p \int_{-\infty}^{\infty} dk'_z \Big[c_{n'}^{(M)(j)}(k'_z) Rg\,\overline{N}_{n'}(k'_{p\rho}, k'_z, \overline{r} - \overline{r}_j) + c_{n'}^{(N)(j)}(k'_z) Rg\,\overline{M}_{n'}(k'_{p\rho}, k'_z, \overline{r} - \overline{r}_j) \Big] \tag{12.2.43}$$

where $c_n^{(M)(j)}$ and $c_n^{(N)(j)}$ are unknown internal field coefficients to be determined self-consistently. In (12.2.42) and (12.2.43), $k'_{p\rho} = \sqrt{k_p^2 - k_z'^2}$.

Step ii: The scattered field from cylinder j is

$$\overline{E}^{s(j)}(\overline{r}') = \int_{-L_j/2+z_j}^{-L_j/2+z_j} dz \int_0^{2\pi} d\phi_{\rho\rho_j} a_j \Big\{ i\omega\mu\hat{\rho}_{\rho\rho_j} \times \overline{H}_p^{(j)}(\overline{r}) \cdot \Big(\overline{\overline{G}}_o(\overline{r}, \overline{r}') + \overline{\overline{G}}_{ref}(\overline{r}, \overline{r}') \Big) + \hat{\rho}_{\rho\rho_j} \times \overline{E}_p^{(j)}(\overline{r}) \cdot \nabla \times \Big(\overline{\overline{G}}_o(\overline{r}, \overline{r}') + \overline{\overline{G}}_{ref}(\overline{r}, \overline{r}') \Big) \Big\} \tag{12.2.44}$$

where $\rho_{\rho\rho_j}$ and $\phi_{\rho\rho_j}$ are polar coordinates with center at $\overline{\rho}_j$ (Fig. 12.2.3). In (12.2.44), $\overline{\overline{G}}_o + \overline{\overline{G}}_{ref}$ is the half-space Green's function which contains the reflected part that denotes a scattered field that is further reflected by boundary at $z = -d$. From Chapter 5, Section 2.2 of Volume I, we have the reflected part of the Dyadic Green's function in plane wave representation

$$\overline{\overline{G}}_{ref}(\overline{r}, \overline{r}') = \frac{i}{8\pi^2} \int d\overline{k}_\perp \frac{1}{k_z} \Big\{ R^{TE} \hat{e}(k_z) e^{i\overline{k}\cdot\overline{r}} \hat{e}(-k_z) e^{-i\overline{K}\cdot\overline{r}'} + R^{TM} \hat{h}(k_z) e^{i\overline{k}\cdot\overline{r}} \hat{h}(-k_z) e^{-i\overline{K}\cdot\overline{r}'} \Big\} \tag{12.2.45}$$

where

$$R^{TE}(k_z) = R_{01}(k_z) e^{2ik_z d} \tag{12.2.46a}$$

$$R^{TM}(k_z) = S_{01}(k_z) e^{2ik_z d} \tag{12.2.46b}$$

for the half-space case with boundary at $z = -d$. Using the transformation between vector plane waves and vector cylindrical waves as given by (2.1.26) and (2.1.25), we have

$$\overline{\overline{G}}_{ref}(\overline{r},\overline{r}')$$
$$= \frac{i}{4\pi}\int_0^\infty dk_\rho k_\rho \frac{1}{k_{oz}k_\rho{}^2} e^{2ik_{oz}z_j} e^{2ik_{oz}d}\sum_n (-1)^n$$
$$\cdot \Big[R_{01}(k_{oz})Rg\,\overline{M}_{-n}(k_\rho,k_{oz},\overline{r}-\overline{r}_j)Rg\,\overline{M}_n(k_\rho,k_{oz},\overline{r}'-\overline{r}_j)$$
$$+ S_{01}(k_{oz})Rg\,\overline{N}_{-n}(k_\rho,k_{oz},\overline{r}-\overline{r}_j)Rg\,\overline{N}_n(k_\rho,k_{oz},\overline{r}'-\overline{r}_j)\Big] \quad (12.2.47a)$$

For the free-space Green's function $\overline{\overline{G}}_o$, we have

$$\overline{\overline{G}}_o(\overline{r},\overline{r}')$$
$$= \frac{i}{8\pi}\sum_{n=-\infty}^{\infty}(-1)^n\int_{-\infty}^{\infty} dk_z \frac{1}{k_{o\rho}^2}\Big[R_g\overline{M}_n(k_{o\rho},k_z,\overline{r}-\overline{r}_j)\overline{M}_{-n}(k_{o\rho},-k_z,\overline{r}'-\overline{r}_j)$$
$$+ R_g\overline{N}_o(k_{o\rho},k_z,\overline{r}-\overline{r}_j)\overline{N}_{-n}(k_{o\rho},-k_z,\overline{r}'-\overline{r}_j)\Big] \quad (12.2.47b)$$

Integration in $\overline{\overline{G}}_o$ is over $\int_{-\infty}^{\infty} dk_z$ because the direct scattered wave from cylinder j to cylinder l includes both upward and downward going waves. However, integration in (12.2.47a) for $\overline{\overline{G}}_{ref}$ is over $\int_0^\infty dk_\rho k_\rho$ because the reflected wave only has upward going waves. Substitute (12.2.42), (12.2.43), (12.2.47a), and (12.2.47b) in (12.2.44). Integration over $d\phi_{\rho\rho_j}$ gives $n' = n$. Also since $\overline{r}$ is on the surface of the cylinder, we have $|\overline{\rho} - \overline{\rho}_j| = a$. We then have

$$\overline{E}^{s(j)}(\overline{r}') =$$
$$\sum_{n'=-\infty}^{\infty}\left(-\frac{L_j}{2\pi}\right)\int_{-\infty}^{\infty} dk_z'\int_{-\infty}^{\infty} dk_z \text{sinc}\left((k_z'-k_z)\frac{L_j}{2}\right)$$
$$\cdot\Big\{\overline{M}_{n'}(k_{o\rho},k_z,\overline{r}'-\overline{r}_j)\Big[c_{n'}^{(M)(j)}(k_z')Rg\,A_{n'}^{MM}(k_{o\rho},k_z,k_{p\rho}',k_z',a_j)$$
$$+ c_{n'}^{(N)(j)}(k_z')Rg\,A_{n'}^{MN}(k_{o\rho},k_z,k_{p\rho}',k_z',a_j)\Big]$$
$$+ \overline{N}_n'(k_{o\rho},k_z,\overline{r}'-\overline{r}_j)\Big[c_{n'}^{(M)(j)}(k_z')Rg\,A_{n'}^{NM}(k_{o\rho},k_z,k_{p\rho}',k_z',a_j)$$
$$+ c_{n'}^{(N)(j)}(k_z')Rg\,A_{n'}^{NN}(k_{o\rho},k_z,k_{p\rho}',k_z',a_j)\Big]\Big\}$$
$$+ \sum_{n'=-\infty}^{\infty}\left(-\frac{L_j}{\pi}\right)\int_{-\infty}^{\infty} dk_z'\int_0^\infty dk_\rho \frac{k_\rho}{k_{oz}} e^{2ik_{oz}(z_j+d)}\text{sinc}\left(\left(\frac{k_{oz}+k_z'}{2}\right)L_j\right)$$
$$\cdot\Big\{Rg\,\overline{M}_{n'}(k_\rho,k_{oz},\overline{r}'-\overline{r}_j)R_{01}(k_{oz})\Big[c_{n'}^{(M)(j)}(k_z')Rg\,A_{n'}^{MM}$$

$$\cdot (k_\rho, -k_{oz}, k'_{p\rho}, k'_z, a_j) + c_{n'}^{(N)(j)}(k'_z) Rg\, A_{n'}^{MN}(k_\rho, -k_{oz}, k'_{p\rho}, k'_z, a_j)\Big]$$
$$+ Rg\, \overline{N}_{n'}(k_\rho, k_{oz}, \overline{r}' - \overline{r}_j) S_{01}(k_{oz}) \Big[c_{n'}^{(M)(j)}(k'_z) Rg\, A_{n'}^{NM}(k_\rho, -k_{oz}, k'_{p\rho}, k'_z, a_j)$$
$$+ c_{n'}^{(N)(j)}(k'_z) Rg\, A_{n'}^{NN}(k_{o\rho}, -k_{oz}, k'_{p\rho}, k'_z, a_j)\Big]\Big\} \tag{12.2.48}$$

where the coupling coefficients $Rg\, A_{n'}$ are as given in (1.6.64)–(1.6.67) from Chapter 1 of Volume I, with Hankel functions replaced by Bessel functions. The integral in the first term of (12.2.48) is over dk_z with $k_{o\rho} = (k^2 - k_z^2)^{1/2}$ while the second term is over dk_ρ with $k_{oz} = (k^2 - k_\rho^2)^{1/2}$. This gives

Step iii: We next use the translation addition theorem to express vector cylindrical waves centered at $\overline{r}_j$ in terms of vector cylindrical waves centered at $\overline{r}_l$. We can then derive the expression of an exciting field at cylinder l due to a scattered field from cylinder j.

$$\overline{E}^{s(j)}(\overline{r}') =$$
$$\sum_{n=-\infty}^{\infty} \sum_{n'=-\infty}^{\infty} \left(-\frac{L_j}{2\pi}\right) \int_{-\infty}^{\infty} dk'_z \int_{-\infty}^{\infty} dk_z \,\mathrm{sinc}\left((k'_z - k_z)\frac{L_j}{2}\right)$$
$$\cdot H_{n-n'}^{(1)}(k_{o\rho}|\overline{\rho}_j - \overline{\rho}_l|) e^{-i(n-n')\phi_{\rho_j \rho_l} - ik_z(z_j - z_l)}$$
$$\cdot \Big\{ Rg\, \overline{M}_n(k_{o\rho}, k_z, \overline{r}' - \overline{r}_l) \Big[c_{n'}^{(M)(j)}(k'_z) Rg\, A_{n'}^{MM}(k_{o\rho}, k_z, k'_{p\rho}, k'_z, a_j)$$
$$+ c_{n'}^{(N)(j)}(k'_z) Rg\, A_{n'}^{MN}(k_{o\rho}, k_z, k'_{p\rho}, k'_z, a_j)\Big]$$
$$+ Rg\, \overline{N}_n(k_{o\rho}, k_z, \overline{r}' - \overline{r}_l) \Big[c_{n'}^{(M)(j)}(k'_z) Rg\, A_{n'}^{NM}(k_{o\rho}, k_z, k'_{p\rho}, k'_z, a_j)$$
$$+ c_{n'}^{(N)(j)}(k'_z) Rg\, A_{n'}^{NN}(k_{o\rho}, k_z, k'_{p\rho}, k'_z, a_j)\Big]\Big\}$$
$$+ \sum_{n=-\infty}^{\infty} \sum_{n'=-\infty}^{\infty} \left(-\frac{L_j}{\pi}\right) \int_{-\infty}^{\infty} dk'_z \int_{0}^{\infty} dk_\rho \frac{k_\rho}{k_{oz}} e^{2ik_{oz}(z_j + d)}$$
$$\cdot \mathrm{sinc}\left((k'_z + k_{oz})\frac{L_j}{2}\right) J_{n-n'}(k_\rho|\overline{\rho}_j - \overline{\rho}_l|) e^{-i(n-n')\phi_{\rho_j \rho_l}} e^{-ik_{oz}(z_j - z_l)}$$
$$\cdot \Big\{ Rg\, \overline{M}_n(k_\rho, k_{oz}, \overline{r}' - \overline{r}_l) R_{01}(k_{oz}) \Big[c_{n'}^{(M)(\alpha)}(k'_z) Rg\, A_{n'}^{MM}(k_\rho, -k_{oz}, k'_{p\rho}, k'_z, a_j)$$
$$+ c_{n'}^{(N)(\alpha)}(k'_z) Rg\, A_{n'}^{MN}(k_\rho, -k_z, k'_{p\rho}, k'_z, a_j)\Big]$$
$$+ Rg\, \overline{N}_n(k_\rho, k_{oz}, \overline{r}' - \overline{r}_l) S_{01}(k_{oz}) \Big[c_{n'}^{(M)(j)}(k'_z) Rg\, A_{n'}^{NM}(k_\rho, -k_{oz}, k'_{p\rho}, k'_z, a_j)$$
$$+ c_{n'}^{(N)(j)}(k'_z) Rg\, A_{n'}^{NN}(k_\rho, -k_{oz}, k'_{p\rho}, k'_z, a_j)\Big]\Big\} \tag{12.2.49}$$

The first term in (12.2.49) is due to $\overline{\overline{G}}_o$, and the second term in (12.2.49) is due to the reflected Green's function $\overline{\overline{G}}_{ref}$.

Step iv: To obtain self-consistent multiple scattering equations, we express the exciting field of cylinder l as follows:

$$\overline{E}^{ex(l)} = \sum_{n=-\infty}^{\infty} \int_{-\infty}^{\infty} dk_z \Big[a_n^{(M)(l)}(k_z) Rg\,\overline{M}_n(k_{o\rho}, k_z, \overline{r} - \overline{r}_l) + a_n^{(N)}(k_z) Rg\,\overline{N}_n(k_{o\rho}, k_z, \overline{r} - \overline{r}_l) \Big] \quad (12.2.50)$$

In matching the coefficients to scattered field from cylinder j, a problem exists. That is, the reflected term in (12.2.49) is integration of dk_ρ from 0 to ∞ while that of (12.2.50) is dk_z from $-\infty$ to ∞. This is because in the reflected term, if dk_ρ integration is performed in the complex k_ρ-plane, it will have to include a contribution of a branch cut of $k_{1z} = (k_1^2 - k_\rho^2)^{1/2}$ of the bottom half-space ϵ_1 as well as possibly a Zenneck surface wave pole [Kong, 2000]. However, when we examine the exponential terms, dependence involving k_{oz} in the reflected term of (12.2.48), we find that, with $z_j = -d + L_j/2$,

$$\exp(2ik_{oz}(z_j + d)) \exp(-ik_{oz}L_j/2) \exp\big(ik_{oz}(z' - z_j)\big) = \exp\big(ik_{oz}(z' + d)\big) \quad (12.2.51)$$

The $\operatorname{sinc}\big((k_{oz} + k_z')L_j/2\big)$ shows the conical pattern with k_{oz} close to $-k_z'$ which is real, while z' in (12.2.51) is from $-d$ to $-d + L_j$. Thus the contribution of evanescent waves in dk_ρ integration will be small. Hence we approximate the second term in (12.2.49) as follows.

$$\int_0^\infty dk_\rho = \int_0^k dk_\rho + \int_k^\infty dk_\rho \simeq \int_0^k dk_\rho = \int_0^k \frac{dk_z\, k_z}{k_{o\rho}} \quad (12.2.52)$$

Integration is only over dk_z from 0 to k for reflected term. Once the approximation is made, the exciting field coefficients $a_n^{(M)(l)}$ and $a_n^{(N)(l)}$ can be equated to the sum of incident, reflected, and scattered fields. We obtain the following self-consistent Foldy-Lax multiple scattering equations in equating the exciting field of cylinder l to its internal field as represented by (1.6.75) and (1.6.76). Thus, for $-\infty < k_z < \infty$, $l = 1, 2, \ldots, N$, we obtain

$$a_n^{(M)(l)}(k_z) = A_n^{MM}(k_{o\rho}, k_z, k_{p\rho}, k_z, a_l) c_n^{(M)(l)}(k_z) + A_n^{MN}(k_{o\rho}, k_z, k_{p\rho}, k_z, a_l) c_n^{(N)(l)}(k_z)$$

$$
\begin{aligned}
&= - e^{i\overline{K}_i\cdot\overline{r}_l}\frac{E_{hi}}{ik_{i\rho}}(-1)^n e^{-in\phi_i - in\pi/2}\delta(k_z + k_{iz}) \\
&\quad - e^{i\overline{k}_i\cdot\overline{r}_l}e^{2ik_{iz}d}\frac{R_{01}(k_{iz})E_{hi}}{ik_{i\rho}}(-1)^n e^{-in\phi_i - in\pi/2}\delta(k_z - k_{iz}) \\
&\quad + \sum_{\substack{j=1 \\ j\neq l}}^{N}\left(-\frac{L_j}{2\pi}\right)\sum_{n'=-\infty}^{\infty}\int_{-\infty}^{\infty} dk_z' \text{sinc}\left((k_z' - k_z)\frac{L_j}{2}\right) \\
&\quad \times H_{n-n'}^{(1)}(k_{o\rho}|\overline{\rho}_j - \overline{\rho}_l|)e^{-i(n-n')\phi_{\rho_j\rho_l}}e^{-ik_z(z_j - z_l)} \\
&\quad \times \Big[c_{n'}^{(M)(j)}(k_z')Rg\, A_{n'}^{MM}(k_{o\rho}, k_z, k_{p\rho}', k_z', a_j) \\
&\quad + c_{n'}^{(N)(j)}(k_z')Rg\, A_{n'}^{MN}(k_{o\rho}, k_z, k_{p\rho}', k_z', a_j)\Big] \\
&\quad + \Big(U(k_z) - U(k_z - k)\Big) \\
&\quad \times \sum_{\substack{j=1 \\ j\neq l}}^{N}\left(-\frac{L_j}{\pi}\right)\sum_{n'=-\infty}^{\infty}\int_{-\infty}^{\infty} dk_z' \text{sinc}\left((k_z' + k_z)\frac{L_j}{2}\right)e^{2ik_z(z_j + d)} \\
&\quad \times J_{n-n'}(k_{o\rho}|\overline{\rho}_j - \overline{\rho}_l|)e^{-i(n-n')\phi_{\rho_j\rho_l} - ik_z(z_j - z_l)}R_{01}(k_z) \\
&\quad \times \Big[c_{n'}^{(M)(j)}(k_z')Rg\, A_{n'}^{MM}(k_{o\rho}, -k_z, k_{p\rho}', k_z', a_j) \\
&\quad + c_{n'}^{(N)(j)}(k_z')Rg\, A_{n'}^{MN}(k_{o\rho}, -k_z, k_{p\rho}', k_z', a_j)\Big]
\end{aligned} \tag{12.2.53a}
$$

$$
\begin{aligned}
a_n^{(N)(l)}(k_z) &= A_n^{NM}(k_{o\rho}, k_z, k_{p\rho}, k_z, a_l)c_n^{(M)(l)}(k_z) \\
&\quad + A_n^{NN}(k_{o\rho}, k_z, k_{p\rho}, k_z, a_l)c_n^{(N)(l)}(k_z) \\
&= - e^{i\overline{K}_i\cdot\overline{r}_l}\frac{E_{vi}}{k_{i\rho}}(-1)^n e^{-in\phi_i - in\pi/2}\delta(k_z + k_{iz}) \\
&\quad - e^{i\overline{k}_i\cdot\overline{r}_l}S_{01}(k_{iz})e^{2ik_{iz}d}\frac{E_{v_i}}{k_{i\rho}}(-1)^n e^{-in\phi_i - in\pi/2}\delta(k_z - k_{iz}) \\
&\quad + \sum_{\substack{j=1 \\ j\neq l}}^{N}\left(-\frac{L_j}{2\pi}\right)\sum_{n'=-\infty}^{\infty}\int_{-\infty}^{\infty} dk_z' \text{sinc}\left((k_z' - k_z)\frac{L_j}{2}\right) \\
&\quad \times H_{n-n'}^{(1)}(k_{o\rho}|\overline{\rho}_j - \overline{\rho}_l|)e^{-i(n-n')\phi_{\rho_j\rho_l} - ik_z(z_j - z_l)} \\
&\quad \times \Big[Rg\, A_{n'}^{NM}(k_{o\rho}, k_z, k_{p\rho}', k_z', a_j)c_{n'}^{(M)(j)}(k_z') \\
&\quad + Rg\, A_{n'}^{NN}(k_{o\rho}, k_z, k_{p\rho}', k_z', a_j)c_{n'}^{(N)(j)}(k_z')\Big]
\end{aligned}
$$

$$+ \Big(U(k_z) - U(k_z - k)\Big)$$

$$+ \sum_{\substack{j=1 \\ j\neq 1}}^{N} \left(-\frac{L_j}{\pi}\right) \sum_{n'=-\infty}^{\infty} \int_{-\infty}^{\infty} dk_z' \text{sinc}\left((k_z' + k_z)\frac{L_j}{2}\right) e^{2ik_z(z_j+d)}$$

$$\times J_{n-n'}(k_{o\rho}|\overline{\rho}_j - \overline{\rho}_l|) e^{-i(n-n')\phi_{\rho_j \rho_l} - ik_z(z_j - z_l)} S_{01}(k_z)$$

$$\times \Big[Rg\, A_{n'}^{NM}(k_{o\rho}, -k_z, k'_{p\rho}, k'_z, a_j) c_{n'}^{(M)(j)}(k'_z)$$

$$+ Rg\, A_{n'}^{NN}(k_{o\rho}, -k_z, k'_{p\rho}, k'_z, a_j) c_{n'}^{(N)(j)}(k'_z) \Big] \qquad (12.2.53b)$$

where U is Heaviside step function. Equations (12.2.53a) and (12.2.53b) contain $2N$ integral equations for the $2N$ unknowns $c_n^{(M)(l)}(k_z)$ and $c_n^{(N)(l)}(k_z)$, $l = 1, 2, \ldots, N$. Once they are solved numerically, these will give the "final" internal field coefficients $c_n^{(M)(l)}(k_z)$ and $c_n^{(N)(l)}(k_z)$. These can then be substituted back into (12.2.48) to find the "final" scattered field from each cylinder j. Summation over all $\overline{E}_s^{(j)}$ then gives the final scattered field $\overline{E}_s$ (which can be taken in the far field) in the scattered direction of (θ_s, ϕ_s),

$$\overline{E}^s(\overline{r}) = \frac{e^{ikr}}{r} \sum_{j=1}^{N} \sum_{n=-\infty}^{\infty} (-i)^n e^{in\phi_s} \frac{iL_j}{\pi} k \sin\theta_s e^{-i\overline{k}_s \cdot \overline{r}_j}$$

$$\cdot \int_{-\infty}^{\infty} dk_z' \text{sinc}\left((k_z' - k_{sz})\frac{L_j}{2}\right)$$

$$\Big\{ -i\hat{h}_s \Big[Rg\, A_n^{MM}(k_{s\rho}, k_{sz}, k'_{p\rho}, k'_z, a_j) c_n^{(M)(j)}(k'_z)$$

$$+ Rg\, A_n^{MN}(k_{s\rho}, k_{sz}, k'_{p\rho}, k'_z, a_j) c_n^{(N)(j)}(k'_z) \Big]$$

$$- \hat{v}_s \Big[Rg\, A_n^{NM}(k_{s\rho}, k_{sz}, k'_{p\rho}, k'_z, a_j) c_n^{(M)(j)}(k'_z)$$

$$+ Rg\, A_n^{NN}(k_{s\rho}, k_{sz}, k'_{p\rho}, k'_z, a_j) c_n^{(N)(j)}(k'_z) \Big] \Big\}$$

$$+ \frac{e^{ikr}}{r} \sum_{j=1}^{N} \sum_{n=-\infty}^{\infty} (-i)^n e^{in\phi_s} \frac{iL_j}{\pi} k \sin\theta_s e^{-i\overline{k}_s \cdot \overline{r}_j}$$

$$\cdot \int_{-\infty}^{\infty} dk_z{}' e^{2ik_{sz}(z_j+d)} \text{sinc}\left((k_z' + k_{sz})\frac{L_j}{2}\right)$$

$$\cdot \Big\{ -i\hat{h}_s R_{01}(k_{sz}) \Big[c_n^{(M)(j)}(k'_z) Rg\, A_n^{MM}(k_{s\rho}, -k_{sz}, k'_{p\rho}, k'_z, a_j)$$

$$+ c_n^{(N)(j)}(k_z')Rg\, A_n^{MN}(k_{s\rho}, -k_{sz}, k_{p\rho}', k_z', a_j)\Big]$$
$$- \hat{v}_s S_{01}(k_{sz})\Big[c_n^{(M)(j)}(k_z')Rg\, A_n^{NM}(k_{s\rho}, -k_{sz}, k_{p\rho}', k_z', a_j)$$
$$+ c_n^{(N)(j)}(k_z')Rg\, A_n^{NN}(k_{s\rho}, -k_{sz}, k_{p\rho}', k_z', a_j)\Big]\Big\} \qquad (12.2.54)$$

where $k_{s\rho} = k\sin\theta_s$ and $k_{sz} = k\cos\theta_s$.

A convenient way of solving the Foldy-Lax multiple scattering equations of (12.2.53a) and (12.2.53b) is by iteration. The first-order solution can also be derived analytically. Let

$$c_n^{(M)(j)}(k_z) = c_n^{(M)(j)d}(-k_{iz})\delta(k_z + k_{iz})e^{i\overline{K}_i\cdot\overline{r}_j}$$
$$+ c_n^{(M)(j)u}(k_{iz})\delta(k_z - k_{iz})e^{i\overline{k}_i\cdot\overline{r}_j}e^{2ik_{iz}d} + c_n^{(N)(j)H}(k_z) \qquad (12.2.55)$$

$$c_n^{(N)(j)}(k_z) = c_n^{(N)(j)d}(-k_{iz})\delta(k_z + k_{iz})e^{i\overline{K}_i\cdot\overline{r}_j}$$
$$+ c_n^{(N)(j)u}(k_{iz})\delta(k_z - k_{iz})e^{i\overline{k}_i\cdot\overline{r}_j}e^{2ik_{iz}d} + c_n^{(N)(j)H}(k_z) \qquad (12.2.56)$$

where superscripts u and d denote upward- and downward-going first-order solutions, respectively, and superscript H denotes higher-order solution. On substituting (12.2.55)–(12.2.56) into (12.2.53) and balancing the coefficients of the delta functions, the first-order solution of the internal field obeys the equation

$$A_n^{MM}(k_{i\rho}, -k_{iz}, k_{ip\rho}, -k_{iz}, a_j)c_n^{(M)(j)d}(-k_{iz})$$
$$+ A_n^{MN}(k_{i\rho}, -k_{iz}, k_{ip\rho}, -k_{iz}, a_j)c_n^{(N)(j)d}(-k_{iz})$$
$$= -\frac{E_{hi}}{ik_{i\rho}}(-1)^n e^{-in\phi_i - in\pi/2} \qquad (12.2.57a)$$

$$A_n^{NM}(k_{i\rho}, -k_{iz}, k_{ip\rho}, -k_{iz}, a_j)c_n^{(M)(j)d}(-k_{iz})$$
$$+ A_n^{NN}(k_{i\rho}, -k_{iz}, k_{ip\rho}, -k_{iz}, a_j)c_n^{(N)(j)d}(-k_{iz})$$
$$= -\frac{E_{vi}}{k_{i\rho}}(-1)^n e^{-in\phi_i - in\pi/2} \qquad (12.2.57b)$$

$$A_n^{MM}(k_{i\rho}, k_{iz}, k_{ip\rho}, k_{iz}, a_j)c_n^{(M)(j)u}(k_{iz})$$
$$+ A_n^{MN}(k_{i\rho}, k_{iz}, k_{ip\rho}, k_{iz}, a_j)c_n^{(N)(j)u}(k_{iz})$$
$$= -\frac{E_{hi}}{ik_{i\rho}}(-1)^n e^{-in\phi_i - in\pi/2}R_{01}(k_{iz}) \qquad (12.2.58a)$$

$$A_n^{NM}(k_{i\rho}, k_{iz}, k_{ip\rho}, k_{iz}, a_j)c_n^{(M)(j)u}(k_{iz})$$
$$+ A_n^{NN}(k_{i\rho}, k_{iz}, k_{ip\rho}, k_{iz}, a_j)c_n^{(N)(j)u}(k_{iz})$$
$$= -\frac{E_{vi}}{k_{i\rho}}(-1)^n e^{-in\phi_i - in\pi/2}S_{01}(k_{iz}) \qquad (12.2.58b)$$

Equations (12.2.57a) and (12.2.57b) are the coupled equations for the downward-going waves of the first-order solution, and (12.2.58a) and (12.2.58b) are the coupled equations for the upward-going waves of the first-order solution. They can be solved readily.

Substituting (12.2.55)–(12.2.56) into (12.2.54) gives the far-field solution of the first-order scattered field in the form as given in (12.2.6), where

$$\begin{aligned}\overline{f}_v(\hat{k}_s,\hat{k}_i) = 2ik\sin\theta_s \sum_{n=-\infty}^{\infty} (-i)^n e^{in\phi_s}\Big\{-i\hat{h}_s \\ \cdot\Big[Rg\,A_n^{MM}(k_{s\rho},k_{sz},k_{ip\rho},-k_{iz},a_j)c_n^{(M)(j)d}(-k_{iz}) \\ + Rg\,A_n^{MN}(k_{s\rho},k_{sz},k_{ip\rho},-k_{iz},a_j)c_n^{(N)(j)d}(-k_{iz})\Big] \\ - \hat{v}_s\Big[Rg\,A_n^{NM}(k_{s\rho},k_{sz},k_{ip\rho},-k_{iz},a_j)c_n^{(M)(j)d}(-k_{iz}) \\ + Rg\,A_n^{NN}(k_{s\rho},k_{sz},k_{ip\rho},-k_{iz},a_j)c_n^{(N)(j)d}(-k_{iz})\Big]\Big\}\end{aligned} \tag{12.2.59a}$$

$$\begin{aligned}\overline{f}_{vr}(\hat{k}_s,\hat{k}_i) = 2ik\sin\theta_s \sum_{n=-\infty}^{\infty} (-i)^n e^{in\phi_s}\Big\{-i\hat{h}_s \\ \cdot\Big[Rg\,A_n^{MM}(k_{s\rho},k_{sz},k_{ip\rho},k_{iz},a_j)c_n^{(M)(j)u}(k_{iz}) \\ + Rg\,A_n^{MN}(k_{s\rho},k_{sz},k_{ip\rho},k_{iz},a_j)c_n^{(N)(j)u}(k_{iz})\Big] \\ - \hat{v}_s\Big[Rg\,A_n^{NM}(k_{s\rho},k_{sz},k_{ip\rho},k_{iz},a_j)c_n^{(M)(j)u}(k_{iz}) \\ + Rg\,A_n^{NN}(k_{s\rho},k_{sz},k_{ip\rho},k_{iz},a_j)c_n^{(N)(j)u}(k_{iz})\Big]\Big\}\end{aligned} \tag{12.2.59b}$$

$$\begin{aligned}\overline{f}_{rv}(\hat{k}_s,\hat{k}_i) = 2ik\sin\theta_s \sum_{n=-\infty}^{\infty} (-i)^n e^{in\phi_s}\Big\{-i\hat{h}_s R_{01}(k_{sz}) \\ \cdot\Big[c_n^{(M)(j)d}(-k_{iz})Rg\,A_n^{MM}(k_{s\rho},-k_{sz},k_{ip\rho},-k_{iz},a_j) \\ + c_n^{(N)(j)d}(-k_{iz})Rg\,A_n^{MN}(k_{s\rho},-k_{sz},k_{ip\rho},-k_{iz},a_j)\Big] \\ - \hat{v}_s S_{01}(k_{sz})[c_n^{(M)(j)d}(-k_{iz})Rg\,A_n^{NM}(k_{s\rho},-k_{sz},k_{ip\rho},-k_{iz},a_j) \\ + c_n^{(N)(j)d}(-k_{iz})Rg A_n^{NN}(k_{s\rho},-k_{sz},k_{ip\rho},-k_{iz},a_j)\Big]\Big\}\end{aligned} \tag{12.2.59c}$$

$$
\begin{aligned}
\overline{f}_{rvr}(\hat{k}_s, \hat{k}_i) = 2ik \sin\theta_s \sum_{n=-\infty}^{\infty} (-i)^n e^{in\phi_s} \Big\{ -i\hat{h}_s R_{01}(k_{sz}) \\
\Big[c_n^{(M)(j)u}(k_z') Rg\, A_n^{MM}(k_{s\rho}, -k_{sz}, k_{ip\rho}, k_{iz}, a_j) \\
+ c_n^{(N)(j)u}(k_z') Rg\, A_n^{MN}(k_{s\rho}, -k_{sz}, k_{ip\rho}, k_{iz}, a_j) \Big] \\
- \hat{v}_s S_{01}(k_{sz}) \Big[c_n^{(M)(j)u}(k_z') Rg\, A_n^{NM}(k_{s\rho}, -k_{sz}, k_{ip\rho}, k_{iz}, a_j) \\
+ c_n^{(N)(j)u}(k_z') Rg\, A_n^{NN}(k_{s\rho}, -k_{sz}, k_{ip\rho}, k_{iz}, a_j) \Big] \Big\}
\end{aligned}
\tag{12.2.59d}
$$

2.3 Results of Monte Carlo Simulations

In this section we illustrate the Monte Carlo simulation results of the bistatic scattering coefficient as a function of the observation angle. In the simulations, $N = 100$ vertical dielectric cylinders of radius $a = 1$ cm and average height L are arranged in two types of configurations as shown in Figs. 12.2.4 and 12.2.5. The values of $L = 60$ cm and 90 cm are chosen. In Fig. 12.2.4, the cylinders are uniformly and randomly distributed over a dielectric surface. The dielectric surface is infinite; however, the cylinders are only placed over a square area A with $\sqrt{A} = 560.5$ cm. In Fig. 12.2.5, the same number of cylinders are distributed in clusters. There are N_c clusters. In each cluster the N_s ($N = N_c N_s$) cylinders are randomly distributed inside a circle of radius a_c. The cylinders occupy 0.1% of fractional volume in all the cases considered. That is, $\frac{N\pi a^2}{A} = 0.001$. The fractional volumes are the same for the uniform case of Fig. 12.2.4 and the cluster case of Fig. 12.2.5. However, the cluster case has locally dense properties. The permittivity for the cylinder is $\epsilon_p = (20+i3)\epsilon_0$. The permittivity for the dielectric surface is $\epsilon_1 = (15+i5)\epsilon_0$. A vector plane wave of wavelength $\lambda = 20$ cm is incident at $\theta_i = 45.55°$ and $\phi_i = 0°$. The positions of cylinders are generated randomly for the uniformly random distribution case. For the clustered distribution case, the positions of cylinders are generated in two stages. First, N_c positions are generated randomly within a square area A. Centered at these N_c positions, we define cluster circles of radius a_c which do not overlap with each other. Then in the second stage, we randomly generated N_s cylinders of radius a inside each cluster circle of radius a_c. The lengths of cylinders are assumed to have a Gaussian distribution with average length L and standard deviation σ_L. The results are calculated and averaged for 50 realizations.

We illustrate the results for the incoherent bistatic scattering coefficients.

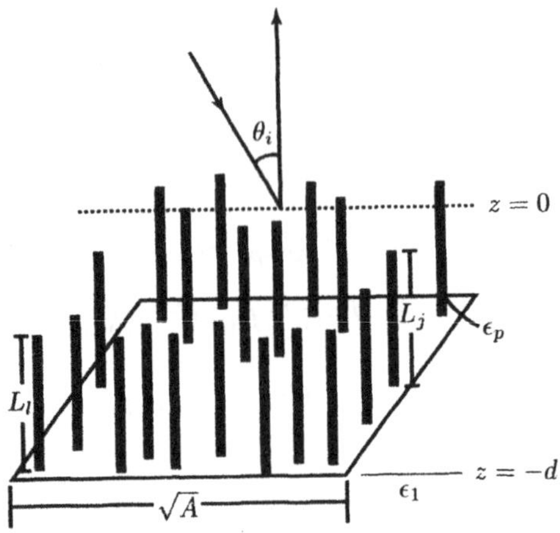

Figure 12.2.4 Monte Carlo simulation geometry. Vertical dielectric cylinders are randomly distributed over a dielectric surface at $z = -d$. The lower half-space is of permittivity ϵ_1. The lth and jth cylinders have length L_l and L_j, respectively. The illuminated area is a square with length $\sqrt{A}$.

The bistatic scattering coefficient is defined by

$$\gamma_{\beta\alpha}(\theta_s, \phi_s; \theta_i, \phi_i) = \lim_{r\to\infty} \frac{4\pi r^2 |E_\beta^s|^2}{|E_\alpha^i|^2 A \cos\theta_i} \tag{12.2.60}$$

We distinguish between the coherent part and the incoherent part. The incoherent bistatic scattering coefficient is defined as

$$\gamma_{\beta\alpha}^i = \lim_{r\to\infty} \frac{4\pi r^2 \langle |E_\beta^s - \langle E_\beta^s \rangle|^2 \rangle}{|E_\alpha^i|^2 A \cos\theta_i} \tag{12.2.61}$$

where angular brackets denote average over realizations. In the numerical results we plot the bistatic scattering coefficients in the plane of incidence. We also define the absorption coefficient as

$$\gamma_{abs}(\theta_i, \phi_i) = \frac{4\pi\sigma_{abs}^N}{A\cos\theta_i} \tag{12.2.62}$$

Backscattering Enhancement for Uniform Random Case

In Fig. 12.2.6 we plot the bistatic scattering coefficient in decibels for a horizontally polarized incident plane wave as a function of observation angle for the uniform random case. The average length and the standard deviation of cylinders are $L = 60$ cm and $\sigma_L = 1$ cm. The Monte Carlo first-order result is compared with that of independent scattering. It follows from Section 2.1

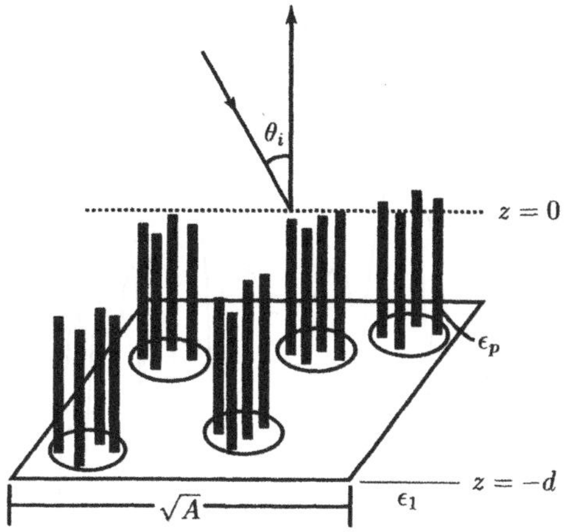

Figure 12.2.5 Monte Carlo simulation geometry for the cluster case. The clusters are randomly distributed over a dielectric square surface of side length $\sqrt{A}$. Within each cluster of radius a_c, the vertical cylinders are randomly and uniformly distributed.

that the first-order solution of the scattered electric field has four terms

$$E_s(\overline{r}) = \frac{e^{ikr}}{r}(f_v + f_{rv} + f_{vr} + f_{rvr}) \tag{12.2.63}$$

The first term is the direct volume scattering term. The second term is volume scattering followed by reflection, while the third term is reflection followed by volume scattering. The fourth term is reflection, followed by volume scattering and then a second reflection. We compare this with the independent scattering result which is defined by

$$\gamma_{ind} = \frac{4\pi(|f_v|^2 + |f_{rv}|^2 + |f_{vr}|^2 + |f_{rvr}|^2)}{A\cos\theta_i} \tag{12.2.64}$$

There is a conspicuous peak in the backscattering direction as observed in the Monte Carlo result. This is a backscattering enhancement result and is not in the independent scattering result. Backscattering enhancement is treated in Chapter 8 of Volume III. In the backscattering direction, $f_{rv} = f_{vr}$ in (12.2.63), giving rise to constructive interference. The peak of the independent scattering is due to the geometrical shape of the vertical cylinder so that, when combined with the dielectric surface, it has the effect of a corner reflector. The Monte Carlo result is about 3 dB higher than the independent scattering result. In the first-order scattering case, the 3 dB backscattering enhancement is primarily due to the constructive interference between the f_{rv} term and the f_{vr} term, which is also a result of reciprocity. On the other hand, the dip in the specular (forward) direction in the Monte Carlo result is due to subtraction of the coherent intensity from the total intensity. In

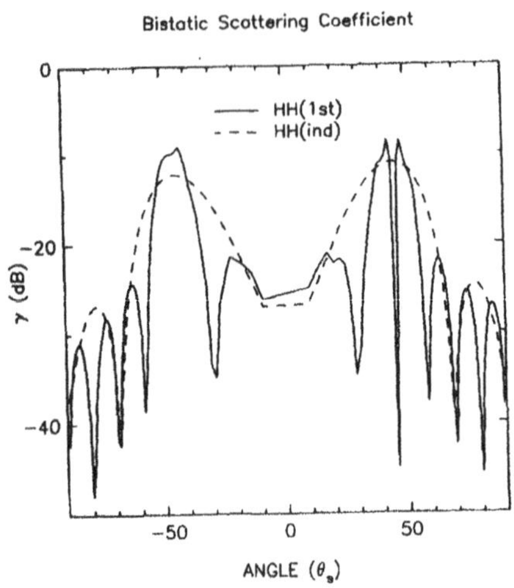

Figure 12.2.6 Comparison of HH bistatic scattering coefficients (dB) between the Monte Carlo simulation and the independent scattering approach (50 realizations). The parameters are $\epsilon_p = (20+i3)\epsilon_0$, $\epsilon_1 = (15+i5)\epsilon_0$. $N = 100$, $\sqrt{A} = 560.5$ cm, $a = 1$ cm, $L = 60$ cm, $\sigma_L = 1$ cm, $\lambda = 20$ cm, $\theta_i = 45.55°$, and $\phi_i = 0°$.

the Monte Carlo simulation, the scattered waves are in phase in the specular direction and gives rise to a large maximum in the specular direction. Subtraction of the coherent intensity gives a dip in the first- and second-order solution of Monte Carlo simulations. This is absent in the classical independent scattering approximation, which only includes incoherent scattering. If an infinite area is used, the coherent peak is a delta function; however, in Monte Carlo simulation, the cylinders are distributed over a finite area A so that the coherent peak in the specular direction has a finite angular width. Thus the dip in the specular direction is also of finite width.

Variation of Enhancement Angular Width with Layer Thickness

To illustrate the angular width for the backscattering peak, Fig. 12.2.7 compares the bistatic scattering coefficient of the first-order scattering result in linear scale with two different average lengths $L_1 = 60$ cm and $L_2 = 90$ cm and standard deviations $\sigma_{L_1} = 6$ cm and $\sigma_{L_2} = 9$ cm, respectively. For the case of a longer cylinder, the peak for backscattering enhancement appears higher while the angular width becomes narrower. The angular widths for the 60-cm case and the 90-cm case are, respectively, 28° and 18°. Thus the enhancement peak is of the order of a wavelength divided by the layer

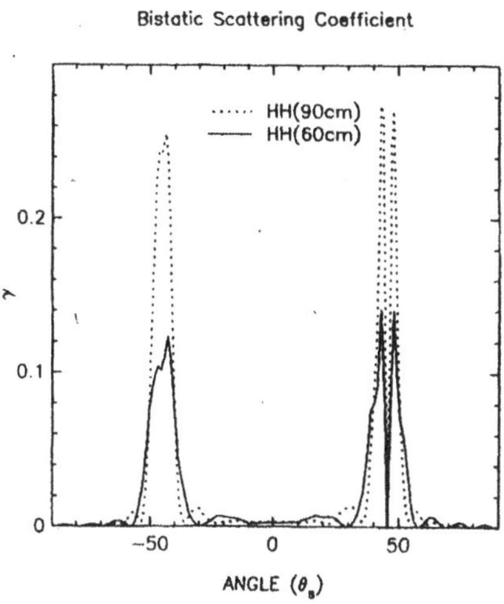

Figure 12.2.7 Comparison of HH bistatic scattering coefficients (linear scale) between the case of cylinders with $L = 90$ cm and $\sigma_L = 6$ cm and the case of cylinders with $L = 90$ cm and $\sigma_L = 9$ cm (50 realizations). The parameters are $\epsilon_p = (20+i3)\epsilon_0$, $\epsilon_1 = (15+i5)\epsilon_0$, $N = 100$, $\sqrt{A} = 560.5$ cm, $a = 1$ cm, $\lambda = 20$ cm, $\theta_i = 45.55°$, and $\phi_i = 0°$.

thickness. This angular width is substantially larger than the purely volume scattering case, which is of the order of a wavelength divided by the mean transport path (see Chapter 8, Volume III).

Second-Order Theory for Uniformly Random Case

In Fig. 12.2.8 we compare the results of second-order theory, first-order theory, and first-order independent scattering. However, there is only very little difference between the results of first-order and second-order solutions for the uniformly random distributions. This is because of the small optical thickness of the layer.

Second-Order Theory for Clustered Distribution Case

We next illustrate the clustering distribution case in Figs. 12.2.9–12.2.11. The local fractional volume is equal to $\frac{N_s \pi a^2}{\pi a_c^2}$ and is equal to 5% in Figs. 12.2.9 and 12.2.10, and 25% in Fig. 12.2.11. The second-order solution becomes important for the clustered distribution cases. We keep the same total fractional volume of 0.1%; however, the cylinders are placed in clusters with the cluster radius $a_c = 10$ cm in Fig. 12.2.9 (VV) and Fig. 12.2.10 (HH). We note that the clustering case gives much larger second-order scattering than

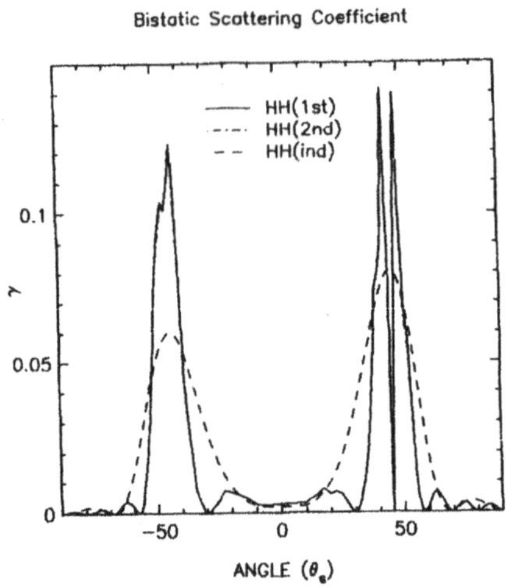

Figure 12.2.8 Comparison of HH bistatic scattering coefficients. Results of first-order, second-order, and independent scattering are shown (50 realizations). The parameters are $\epsilon_p = (20 + i3)\epsilon_0$, $\epsilon_1 = (15 + i5)\epsilon_0$, $N = 100$, $\sqrt{A} = 560.5$ cm, $a = 1$ cm, $L = 60$ cm, $\sigma_L = 1$ cm, $\lambda = 20$ cm, $\theta_i = 45.55°$, and $\phi_i = 0°$.

the case of uniformly random distribution. This can be attributed to the fact that the scatterers are in the vicinity of each other in a cluster. The second-order scattering of vertical polarization is more important than that of horizontal polarization. That is because vertical cylinders scatter vertically polarized waves more than horizontally polarized waves and create more coherent wave interaction among the cylinders in a cluster. For second-order radiative transfer theory, the conditional probability of a second order given a first order depends on the optical thickness, which is small in this case. Thus radiative transfer theory cannot account for the enhancement of scattering due to clustering effects. In Fig. 12.2.11 we have more cylinders in each cluster with higher fractional volume of 25%. There are $N_s = 10$ cylinders that are randomly distributed in each of $N_c = 10$ clusters with the cluster radius $a_c = 6.32$ cm. The other parameters are the same as in Fig. 12.2.6. In comparison with Fig. 12.2.6, significant enhancements are observed in both backscattering and forward-scattering directions. Since the local density is higher than Figs. 12.2.9 and 12.2.10, the second-order effects are stronger. Also, the clustered scattering predicts much larger second-order scattering effects than in the case of uniformly random distribution. This is due to the coherent mutual wave interactions between scatterers that are in close proximity with each other.

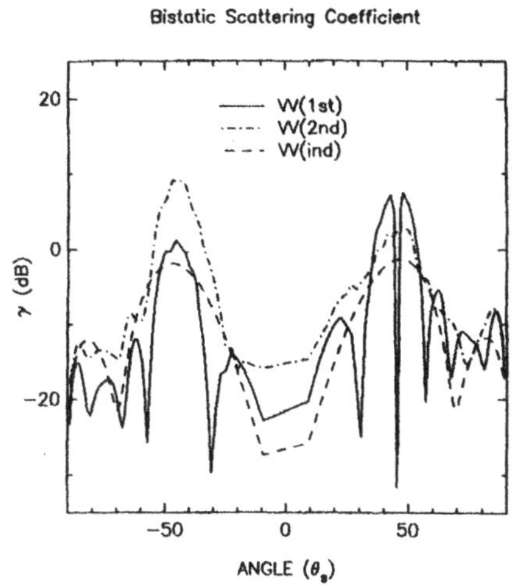

Figure 12.2.9 Comparison of VV bistatic scattering coefficients (dB). Results of first-order, second-order, and independent scattering are shown (50 realizations). The parameters are $N_s = 5$, $N_c = 20$, $a_c = 10$ cm, $\epsilon_p = (20 + i3)\epsilon_0$, $\epsilon_1 = (15 + i5)\epsilon_0$, $N = 100$, $\sqrt{A} = 560.5$ cm, $a = 1$ cm, $L = 60$ cm, $\sigma_L = 1$ cm, $\lambda = 20$ cm, $\theta_i = 45.55°$, and $\phi_i = 0°$.

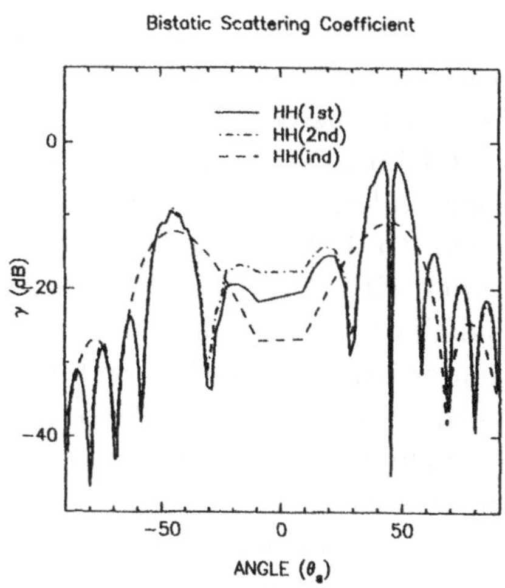

Figure 12.2.10 Comparison of HH bistatic scattering coefficients (dB). Results of first-order, second-order, and independent scattering are shown (50 realizations). The parameters are $N_s = 5$, $N_c = 20$, $a_c = 10$ cm, $\epsilon_p = (20 + i3)\epsilon_0$, $\epsilon_1 = (15 + i5)\epsilon_0$, $N = 100$, $\sqrt{A} = 560.5$ cm, $a = 1$ cm, $L = 60$ cm, $\sigma_L = 1$ cm, $\lambda = 20$ cm, $\theta_i = 45.55°$, and $\phi_i = 0°$.

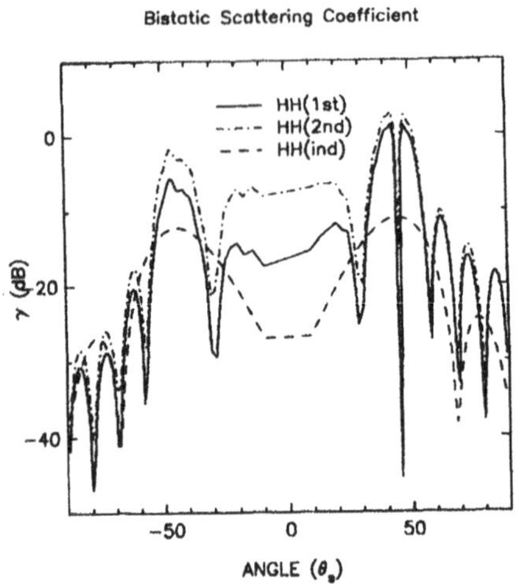

Figure 12.2.11 Comparison of HH bistatic scattering coefficients (dB). Results of first-order, second-order, and independent scattering as shown (50 realizations). The parameters are $N_s = 10$, $N_c = 10$, $a_c = 6.32$ cm, $\epsilon_p = (20 + i3)\epsilon_0$, $\epsilon_1 = (15 + i5)\epsilon_0$, $N = 100$, $\sqrt{A} = 560.5$ cm, $a = 1$ cm, $L = 60$ cm, $\sigma_L = 1$ cm, $\lambda = 20$ cm, $\theta_i = 45.55°$, and $\phi_i = 0°$.

Absorption Coefficients for Clustered Distribution Case

In Table 12.2.1 and Table 12.2.2 we give the absorption coefficients, as defined by (12.2.62), for the clustering case with $N_s = 5$ cylinders randomly distributed in each of $N_c = 20$ clusters with the cluster radius $a_c = 10$ cm, for vertically polarized and horizontally polarized incident waves, respectively. The local fractional volume is 5%. The azimuthal angle ϕ_i of the incident wave is kept equal to 0° for all cases. We compare the results of first-order and second-order with the independent absorption case. The independent absorption is defined as the sum of the absorption of the cylinders assuming that each cylinder exists by itself. For the wave theory, we have tabulated γ_{abs}^{1st}, the first-order result and γ_{abs}^{2nd}, the second-order theory result. The second-order theory includes the first-order and second-order scattering results. We note that for the case of clustering, there is considerable mutual interaction such that the absorption is significantly enhanced. Since the absorption cross section is proportional to the square of the magnitude of internal field, it is expected that the internal field inside each cylinder is greatly altered by the strong coherent mutual interaction between scatterers which are close to each other. The small variation of the absorption coefficient for horizontal polarization with incident angles is due to the small tangential component of the exciting field which induces a small internal

θ_i	γ_{abs}^{ind}	γ_{abs}^{1st}	γ_{abs}^{2nd}
9.33°	−19.74 dB	−18.43 dB	−16.54 dB
17.27°	−16.73 dB	−15.52 dB	−13.60 dB
22.62°	−14.55 dB	−13.33 dB	−11.57 dB
32.49°	−12.53 dB	−11.36 dB	−9.65 dB
41.58°	−12.48 dB	−11.50 dB	−9.31 dB
51.63°	−12.61 dB	−11.99 dB	−9.80 dB
61.48°	−12.23 dB	−11.89 dB	−9.87 dB
65.56°	−11.85 dB	−11.61 dB	−9.46 dB
70.47°	−11.16 dB	−11.03 dB	−8.90 dB

Table 12.2.1 Absorption coefficients for vertically polarized incident wave.

θ_i	γ_{abs}^{ind}	γ_{abs}^{1st}	γ_{abs}^{2nd}
9.33°	−21.40 dB	−20.06 dB	−18.12 dB
17.27°	−22.26 dB	−20.90 dB	−18.76 dB
22.62°	−22.61 dB	−21.19 dB	−19.00 dB
32.49°	−23.01 dB	−21.46 dB	−19.28 dB
41.58°	−23.15 dB	−21.46 dB	−19.56 dB
51.63°	−23.04 dB	−21.17 dB	−19.86 dB
61.48°	−22.54 dB	−20.38 dB	−19.53 dB
65.56°	−22.15 dB	−19.97 dB	−19.16 dB
70.47°	−21.48 dB	−19.15 dB	−18.51 dB

Table 12.2.2 Absorption coefficients for horizontally polarized incident wave.

field of dielectric cylinder. The variation for vertical polarization is large for smaller incident angles; however, this is because the tangential component of field varies with incident angle for vertically polarized case. It is also found that the first-order results of absorption coefficients for both vertical and horizontal polarizations do not vary with the cluster radius a_c for this case, while the second-order results show large variation with the cluster radius.

3 Scattering by Cylinders in the Presence of Two Reflective Boundaries

In this section, we study the multiple scattering by cylinders placed inside a parallel plate waveguide. The problem is conveniently formulated in terms of the dyadic Green's function of vector cylindrical waves and expressed in terms of waveguide modes. The case considered here resembles the coupling problem among multiple vias in high frequency circuits. The Foldy-Lax multiple scattering equations are derived using vector cylindrical waves and waveguide mode indices as basis functions. It is important to recognize that the Foldy-Lax equations are decoupled among the waveguide modes. In terms of the matrix equation, this decoupling reduces significantly the dimension of the matrix equation as one only needs to solve Foldy-Lax equation for each waveguide mode separately. In this section we also consider ring-like magnetic sources as excitation.

3.1 Vector Cylindrical Wave Expansion of Dyadic Green's Function Between Two Perfect Conductors

Consider two perfect electric conductors (PEC) at $z = d/2$ and $z = -d/2$. Let the magnetic source z' be between the two PECs (Fig. 12.3.1). Then the primary Green's function is given by (2.1.28) of Chapter 2, Volume I. We consider the case when the magnetic current source sheet $\overline{M}_s$ flows in the horizontal plane. Let $\overline{\overline{I}}_t = \hat{x}\hat{x} + \hat{y}\hat{y}$ be the transverse dyad. Then

$$\overline{H} = i\omega\epsilon \iint dx'dy' \overline{\overline{G}}(\overline{r}, \overline{r}') \cdot \overline{M}_s(\overline{r}') \tag{12.3.1}$$

The Green's function is next decomposed into a primary part and a response part $\overline{\overline{G}} = \overline{\overline{G}}_P + \overline{\overline{G}}_R$. The primary Green's function is

$$\overline{\overline{G}}_P(\overline{r}, \overline{r}') \cdot \overline{\overline{I}}_t = \begin{cases} \dfrac{i}{4\pi} \displaystyle\int_0^\infty dk_\rho \frac{1}{k_\rho k_z} \sum_n (-1)^n \, e^{ik_z(z-z') + in(\phi-\phi')} \\ \quad \Big\{ Rg\overline{m}_n(k_\rho, k_z, \overline{\rho}) Rg\overline{m}_{-n}(k_\rho, -k_z, \overline{\rho}') \\ \quad + Rg\overline{n}_n(k_\rho, k_z, \overline{\rho}) Rg\overline{n}_{-n}(k_\rho, -k_z, \overline{\rho}') \Big\} \cdot \overline{\overline{I}}_t & \text{for } z > z' \\ \dfrac{i}{4\pi} \displaystyle\int_0^\infty dk_\rho \frac{1}{k_\rho k_z} \sum_n (-1)^n \, e^{-ik_z(z-z') + in(\phi-\phi')} \\ \quad \Big\{ Rg\overline{m}_n(k_\rho, -k_z, \overline{\rho}) Rg\overline{m}_{-n}(k_\rho, k_z, \overline{\rho}') \\ \quad + Rg\overline{n}_n(k_\rho, -k_z, \overline{\rho}) Rg\overline{n}_{-n}(k_\rho, k_z, \overline{\rho}') \Big\} \cdot \overline{\overline{I}}_t & \text{for } z < z' \end{cases} \tag{12.3.2}$$

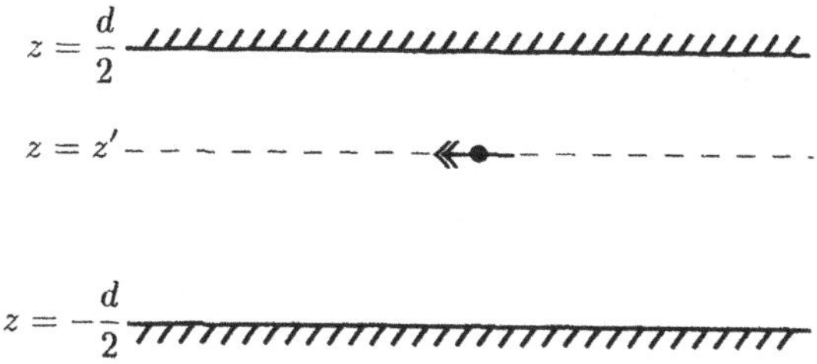

Figure 12.3.1 A horizontal magnetic source located between two perfect electric conductors.

where $\overline{m}_n$ and $\overline{n}_n$ are the vector cylindrical wave functions that are defined in (1.4.23)–(1.4.24) in Chapter 1 of Volume I. The response part of the Green's function is

$$\begin{aligned}\overline{\overline{G}}_R(\overline{r},\overline{r}')\cdot\overline{\overline{I}}_t = &\frac{i}{4\pi}\int_0^{\infty} dk_\rho \frac{1}{k_\rho k_z}\sum_n (-1)^n\, e^{in(\phi-\phi')}\\ &\left\{A_1 e^{ik_z(z-z')}Rg\overline{m}_n(k_\rho,k_z,\overline{\rho}) + B_1 e^{-ik_z(z-z')}Rg\overline{m}_n(k_\rho,-k_z,\overline{\rho})\right\}\\ &Rg\overline{m}_{-n}(k_\rho,k_z,\overline{\rho}')\cdot\overline{\overline{I}}_t\\ &+\frac{i}{4\pi}\int_0^{\infty} dk_\rho \frac{1}{k_\rho k_z}\sum_n (-1)^n\, e^{in(\phi-\phi')}\\ &\left\{C_1 e^{ik_z(z-z')}Rg\overline{n}_n(k_\rho,k_z,\overline{\rho}) + D_1 e^{-ik_z(z-z')}Rg\overline{n}_n(k_\rho,-k_z,\overline{\rho})\right\}\\ &Rg\overline{n}_{-n}(k_\rho,k_z,\overline{\rho}')\cdot\overline{\overline{I}}_t \end{aligned}\tag{12.3.3}$$

The total Green's function is

$$\overline{\overline{G}}(\overline{r},\overline{r}')\cdot\overline{\overline{I}}_t = \left[\overline{\overline{G}}_P(\overline{r},\overline{r}') + \overline{\overline{G}}_R(\overline{r},\overline{r}')\right]\cdot\overline{\overline{I}}_t \tag{12.3.4}$$

The boundary conditions is that the tangential electric field is equal to zero.

$$\hat{z}\times\nabla\times\overline{\overline{G}}(\overline{r},\overline{r}')\cdot\overline{\overline{I}}_t = \hat{z}\times\nabla\times\left[\overline{\overline{G}}_P(\overline{r},\overline{r}') + \overline{\overline{G}}_R(\overline{r},\overline{r}')\right]\cdot\overline{\overline{I}}_t = 0 \tag{12.3.5}$$

at $z = \pm d/2$. We note that for horizontal sources

$$Rg\overline{m}_n(k_\rho,k_z,\overline{\rho})\cdot\overline{\overline{I}}_t = Rg\overline{m}_n(k_\rho,-k_z,\overline{\rho})\cdot\overline{\overline{I}}_t \tag{12.3.6}$$

$$Rg\overline{n}_n(k_\rho,k_z,\overline{\rho})\cdot\overline{\overline{I}}_t = -Rg\overline{n}_n(k_\rho,-k_z,\overline{\rho})\cdot\overline{\overline{I}}_t \tag{12.3.7}$$

Also

$$\hat{z}\times Rg\overline{m}_n(k_\rho,k_z,\overline{\rho}) = \hat{z}\times Rg\overline{m}_n(k_\rho,-k_z,\overline{\rho}) \tag{12.3.8}$$

$$\hat{z}\times Rg\overline{n}_n(k_\rho,k_z,\overline{\rho}) = \hat{z}\times Rg\overline{n}_n(k_\rho,-k_z,\overline{\rho}) \tag{12.3.9}$$

The boundary condition at $z = d/2$ gives

$$e^{ik_z\frac{d}{2}}e^{-ik_zz'} + A_1e^{ik_z\frac{d}{2}}e^{-ik_zz'} - B_1e^{-ik_z\frac{d}{2}}e^{ik_zz'} = 0 \tag{12.3.10}$$

$$-e^{ik_z\frac{d}{2}}e^{-ik_zz'} + C_1e^{ik_z\frac{d}{2}}e^{-ik_zz'} + D_1e^{-ik_z\frac{d}{2}}e^{ik_zz'} = 0 \tag{12.3.11}$$

The boundary condition at $z = -d/2$ gives

$$-e^{ik_z(z'+\frac{d}{2})} + A_1e^{-ik_z(z'+\frac{d}{2})} - B_1e^{ik_z(z'+\frac{d}{2})} = 0 \tag{12.3.12}$$

$$e^{ik_z(z'+\frac{d}{2})} + C_1e^{-ik_z(z'+\frac{d}{2})} + D_1e^{ik_z(z'+\frac{d}{2})} = 0 \tag{12.3.13}$$

Solving (12.3.10)–(12.3.13) gives

$$A_1 = -C_1 = \frac{e^{2ik_z(z'+\frac{d}{2})}\left(1 + e^{-2ik_z(z'-\frac{d}{2})}\right)}{1 - e^{2ik_zd}} \tag{12.3.14}$$

$$B_1 = D_1 = \frac{e^{-2ik_z(z'-\frac{d}{2})}\left(1 + e^{2ik_z(z'+\frac{d}{2})}\right)}{1 - e^{2ik_zd}} \tag{12.3.15}$$

$$1 + A_1 = \frac{1 + e^{2ik_z(z'+\frac{d}{2})}}{1 - e^{2ik_zd}} \tag{12.3.16}$$

$$1 + B_1 = \frac{1 + e^{-2ik_z(z'-\frac{d}{2})}}{1 - e^{2ik_zd}} \tag{12.3.17}$$

Then adding the primary and the response Green's function as in (12.3.4), we get, for $z > z'$

$$\begin{aligned}
&\overline{\overline{G}}(\overline{r}, \overline{r}') \cdot \overline{\overline{I}}_t \\
&= \frac{i}{4\pi}\int_0^\infty dk_\rho \frac{1}{k_\rho k_z}\sum_n \frac{(-1)^n\, e^{in(\phi-\phi')}}{1 - e^{2ik_zd}}\left(1 + e^{2ik_z(z'+\frac{d}{2})}\right)e^{-ik_z(z'-\frac{d}{2})} \\
&\quad \left\{e^{ik_z(z-\frac{d}{2})}Rg\overline{m}_n(k_\rho, k_z, \overline{\rho}) + e^{-ik_z(z-\frac{d}{2})}Rg\overline{m}_n(k_\rho, -k_z, \overline{\rho})\right\} \\
&\quad Rg\overline{m}_{-n}(k_\rho, k_z, \overline{\rho}') \cdot \overline{\overline{I}}_t \\
&+ \frac{i}{4\pi}\int_0^\infty dk_\rho \frac{1}{k_\rho k_z}\sum_n \frac{(-1)^n\, e^{in(\phi-\phi')}}{1 - e^{2ik_zd}}\left(1 + e^{2ik_z(z'+\frac{d}{2})}\right)e^{-ik_z(z'-\frac{d}{2})} \\
&\quad \left\{-e^{ik_z(z-\frac{d}{2})}Rg\overline{n}_n(k_\rho, k_z, \overline{\rho}) + e^{-ik_z(z-\frac{d}{2})}Rg\overline{n}_n(k_\rho, -k_z, \overline{\rho})\right\} \\
&\quad Rg\overline{n}_{-n}(k_\rho, k_z, \overline{\rho}') \cdot \overline{\overline{I}}_t
\end{aligned} \tag{12.3.18}$$

and for $z < z'$

$$\begin{aligned}
&\overline{\overline{G}}(\overline{r}, \overline{r}') \cdot \overline{\overline{I}}_t \\
&= \frac{i}{4\pi}\int_0^\infty dk_\rho \frac{1}{k_\rho k_z}\sum_n \frac{(-1)^n\, e^{in(\phi-\phi')}}{1 - e^{2ik_zd}}\left(1 + e^{-2ik_z(z'-\frac{d}{2})}\right)e^{ik_z(z'+\frac{d}{2})}
\end{aligned}$$

$$\left\{e^{ik_z(z+\frac{d}{2})}Rg\overline{m}_n(k_\rho,k_z,\overline{\rho})+e^{-ik_z(z+\frac{d}{2})}Rg\overline{m}_n(k_\rho,-k_z,\overline{\rho})\right\}$$
$$Rg\overline{m}_{-n}(k_\rho,k_z,\overline{\rho}')\cdot\overline{\overline{I}}_t$$
$$+\frac{i}{4\pi}\int_0^\infty dk_\rho\frac{1}{k_\rho k_z}\sum_n\frac{(-1)^n e^{in(\phi-\phi')}}{1-e^{2ik_zd}}\left(1+e^{-2ik_z(z'-\frac{d}{2})}\right)e^{ik_z(z'+\frac{d}{2})}$$
$$\left\{-e^{ik_z(z+\frac{d}{2})}Rg\overline{n}_n(k_\rho,k_z,\overline{\rho})+e^{-ik_z(z+\frac{d}{2})}Rg\overline{n}_n(k_\rho,-k_z,\overline{\rho})\right\}$$
$$Rg\overline{n}_{-n}(k_\rho,k_z,\overline{\rho}')\cdot\overline{\overline{I}}_t \tag{12.3.19}$$

We next define magnetic field modal solutions and the electric field modal solutions as follows

$$Rg\overline{H}_n^{TE}(k_\rho,k_z,\overline{\rho},z)=$$
$$\frac{1}{2\eta}\left\{e^{ik_zz}Rg\overline{n}_n(k_\rho,k_z,\overline{\rho})-e^{-ik_zz}Rg\overline{n}_n(k_\rho,-k_z,\overline{\rho})\right\}e^{in\phi} \tag{12.3.20}$$
$$Rg\overline{H}_n^{TM}(k_\rho,k_z,\overline{\rho},z)=$$
$$\frac{-i}{2\eta}\left\{e^{ik_zz}Rg\overline{m}_n(k_\rho,k_z,\overline{\rho})+e^{-ik_zz}Rg\overline{m}_n(k_\rho,-k_z,\overline{\rho})\right\}e^{in\phi} \tag{12.3.21}$$

$$Rg\overline{E}_n^{TE}(k_\rho,k_z,\overline{\rho},z)=\frac{i}{\omega\epsilon}\nabla\times Rg\overline{H}_n^{TE}(k_\rho,k_z,\overline{\rho},z)$$
$$=\frac{i}{2}\left\{e^{ik_zz}Rg\overline{m}_n(k_\rho,k_z,\overline{\rho})-e^{-ik_zz}Rg\overline{m}_n(k_\rho,-k_z,\overline{\rho})\right\}e^{in\phi} \tag{12.3.22}$$
$$Rg\overline{E}_n^{TM}(k_\rho,k_z,\overline{\rho},z)=\frac{i}{\omega\epsilon}\nabla\times Rg\overline{H}_n^{TM}(k_\rho,k_z,\overline{\rho},z)$$
$$=\frac{1}{2}\left\{e^{ik_zz}Rg\overline{n}_n(k_\rho,k_z,\overline{\rho})+e^{-ik_zz}Rg\overline{n}_n(k_\rho,-k_z,\overline{\rho})\right\}e^{in\phi} \tag{12.3.23}$$

Then, using the modal solutions, we have, for $z>z'$

$$\overline{\overline{G}}(\overline{r},\overline{r}')\cdot\overline{\overline{I}}_t$$
$$=\frac{-\eta}{2\pi}\int_0^\infty dk_\rho\frac{i}{k_\rho 2k_z\sin(k_zd)}\sum_n(-1)^n\left(1+e^{2ik_z(z'+\frac{d}{2})}\right)e^{-ik_z(z'+\frac{d}{2})}$$
$$Rg\overline{H}_n^{TM}(k_\rho,k_z,\overline{\rho},z-\frac{d}{2})Rg\overline{m}_{-n}(k_\rho,k_z,\overline{\rho}')e^{-in\phi'}\cdot\overline{\overline{I}}_t$$
$$-\frac{i\eta}{2\pi}\int_0^\infty dk_\rho\frac{i}{k_\rho 2k_z\sin(k_zd)}\sum_n(-1)^n\left(1+e^{2ik_z(z'+\frac{d}{2})}\right)e^{-ik_z(z'+\frac{d}{2})}$$
$$Rg\overline{H}_n^{TE}(k_\rho,k_z,\overline{\rho},z-\frac{d}{2})Rg\overline{n}_{-n}(k_\rho,k_z,\overline{\rho}')e^{-in\phi'}\cdot\overline{\overline{I}}_t \tag{12.3.24}$$

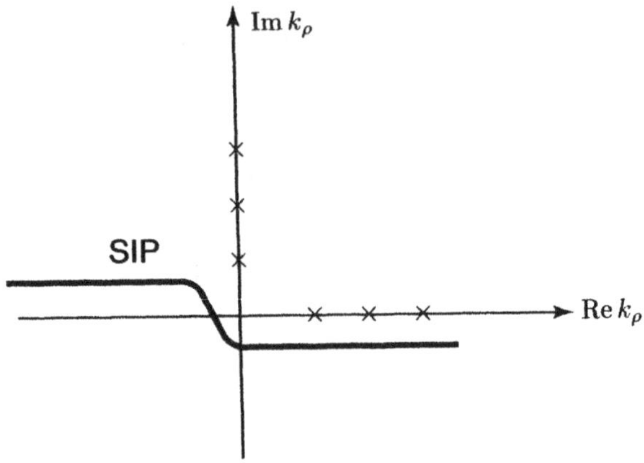

Figure 12.3.2 Sommerfeld integration path (SIP) and waveguide poles.

and for $z < z'$

$$\begin{aligned}
&\overline{\overline{G}}(\overline{r},\overline{r}') \cdot \overline{\overline{I}}_t \\
&= \frac{-\eta}{2\pi}\int_0^\infty dk_\rho \frac{i}{k_\rho 2k_z \sin(k_z d)} \sum_n (-1)^n \left(1 + e^{-2ik_z(z'-\frac{d}{2})}\right) e^{ik_z(z'-\frac{d}{2})} \\
&\quad Rg\overline{H}_n^{TM}(k_\rho, k_z, \overline{\rho}, z + \frac{d}{2}) Rg\overline{m}_{-n}(k_\rho, k_z, \overline{\rho}') e^{-in\phi'} \cdot \overline{\overline{I}}_t \\
&- \frac{i\eta}{2\pi}\int_0^\infty dk_\rho \frac{i}{k_\rho 2k_z \sin(k_z d)} \sum_n (-1)^n \left(1 + e^{-2ik_z(z'-\frac{d}{2})}\right) e^{ik_z(z'-\frac{d}{2})} \\
&\quad Rg\overline{H}_n^{TE}(k_\rho, k_z, \overline{\rho}, z + \frac{d}{2}) Rg\overline{n}_{-n}(k_\rho, k_z, \overline{\rho}') e^{-in\phi'} \cdot \overline{\overline{I}}_t
\end{aligned} \quad (12.3.25)$$

The integral in (12.3.24) and (12.3.25) is over the continuum of dk_ρ. We next convert to discrete waveguide modes using residue calculus. The poles are as indicated in Fig. 12.3.2. We have the poles for TM modes and TE modes

$$k_{zl} = \frac{l\pi}{d} \quad (12.3.26)$$

with $l = 0, 1, 2, \ldots$ for TM modes and $l = 1, 2, \ldots$ for TE modes. Note that $Rg\overline{E}_n^{TE}(k_\rho, k_z = 0, \overline{\rho}, z) = 0$, so that TE_l modes start with $l = 1$ rather than $l = 0$. On the other hand, TM_l modes start with $l = 0, 1, 2, \ldots$. The TM_0 mode is also called TEM mode.

First we need to extend dk_ρ integration for $-\infty$ to ∞. We use the usual transformation of Sommerfeld integral from Bessel functions to Hankel functions [Sommerfeld, 1962; Chew, 1990]. Let ( ) denotes terms that are even functions of k_ρ For $\rho < \rho'$, the transformation is as follows.

$$\int_0^\infty dk_\rho \frac{1}{k_\rho} J_n(k_\rho\rho) J_{-n}(k_\rho\rho')(\;) =$$

$$\frac{1}{2}\int_0^\infty dk_\rho \frac{1}{k_\rho} J_n(k_\rho\rho)\left(H_n^{(1)}(k_\rho\rho') + H_n^{(2)}(k_\rho\rho')\right)(\;)$$

$$= \frac{1}{2}\int_0^\infty dk_\rho \frac{1}{k_\rho} J_n(k_\rho\rho) H_n^{(1)}(k_\rho\rho')(\;)$$

$$+ \frac{1}{2}\int_{-\infty}^0 dk_\rho \frac{1}{(-k_\rho)} J_n(-k_\rho\rho) H_n^{(2)}(-k_\rho\rho')(\;)$$

$$= \frac{1}{2}\int_{-\infty,SIP}^\infty dk_\rho \frac{1}{k_\rho} J_n(k_\rho\rho) H_n^{(1)}(k_\rho\rho')(\;) \qquad (12.3.27)$$

where we have made use of the following relation

$$H_n^{(2)}\left(we^{-\pi i}\right) = -e^{n\pi i} H_n^{(1)}(w) \qquad (12.3.28)$$

In (12.3.27), SIP denotes Sommerfeld integration path (Fig. 12.3.2).

We illustrate for $z < z'$. We have

$$\overline{\overline{G}}(\overline{r}, \overline{r}') \cdot \overline{\overline{I}}_t$$

$$= \frac{-\eta}{4\pi}\int_{-\infty,SIP}^\infty dk_\rho \frac{i}{k_\rho 2k_z \sin(k_z d)} \sum_n (-1)^n \left(1 + e^{-2ik_z(z'-\frac{d}{2})}\right) e^{ik_z(z'-\frac{d}{2})}$$

$$Rg\overline{H}_n^{TM}(k_\rho, k_z, \overline{\rho}, z + \frac{d}{2})\overline{m}_{-n}(k_\rho, k_z, \overline{\rho}')e^{-in\phi'} \cdot \overline{\overline{I}}_t$$

$$- \frac{i\eta}{4\pi}\int_{-\infty,SIP}^\infty dk_\rho \frac{i}{k_\rho 2k_z \sin(k_z d)} \sum_n (-1)^n \left(1 + e^{-2ik_z(z'-\frac{d}{2})}\right) e^{ik_z(z'-\frac{d}{2})}$$

$$Rg\overline{H}_n^{TE}(k_\rho, k_z, \overline{\rho}, z + \frac{d}{2}) Rg\overline{n}_{-n}(k_\rho, k_z, \overline{\rho}')e^{-in\phi'} \cdot \overline{\overline{I}}_t \qquad (12.3.29)$$

For $\rho < \rho'$, we can deform upward and obtain the residues of the poles

$$\text{Residue}\frac{1}{k_z \sin(k_z d)} = -\frac{(-1)^l}{k_{\rho l} d} f_l \qquad (12.3.30)$$

$$f_l = \begin{cases} \frac{1}{2} & \text{for } l = 0 \\ 1 & \text{for } l = 1, 2, \ldots \end{cases} \qquad (12.3.31)$$

Thus for $z < z'$, $\rho < \rho'$

$$\overline{\overline{G}}(\overline{r},\overline{r}')\cdot\overline{\overline{I}}_t = -\frac{\eta}{4d}\sum_{n,l}\frac{(-1)^{n+l}}{k_{\rho l}^2}f_l\left(1+e^{-2ik_z(z'-\frac{d}{2})}\right)e^{ik_z(z'-\frac{d}{2})}$$
$$Rg\overline{H}_n^{TM}(k_{\rho l},k_{zl},\overline{\rho},z+\frac{d}{2})\overline{m}_{-n}(k_{\rho l},k_{zl},\overline{\rho}')e^{-in\phi'}\cdot\overline{\overline{I}}_t$$
$$-\frac{i\eta}{4d}\sum_{n,l}\frac{(-1)^{n+l}}{k_{\rho l}^2}f_l\left(1+e^{-2ik_z(z'-\frac{d}{2})}\right)e^{ik_z(z'-\frac{d}{2})}$$
$$Rg\overline{H}_n^{TE}(k_{\rho l},k_{zl},\overline{\rho},z+\frac{d}{2})\overline{n}_{-n}(k_{\rho l},k_{zl},\overline{\rho}')e^{-in\phi'}\cdot\overline{\overline{I}}_t \quad (12.3.32)$$

Similarly for $z < z', \rho > \rho'$

$$\overline{\overline{G}}(\overline{r},\overline{r}')\cdot\overline{\overline{I}}_t = -\frac{\eta}{4d}\sum_{n,l}\frac{(-1)^{n+l}}{k_{\rho l}^2}f_l\left(1+e^{-2ik_z(z'-\frac{d}{2})}\right)e^{ik_z(z'-\frac{d}{2})}$$
$$\overline{H}_n^{TM}(k_{\rho l},k_{zl},\overline{\rho},z+\frac{d}{2})Rg\overline{m}_{-n}(k_{\rho l},k_{zl},\overline{\rho}')e^{-in\phi'}\cdot\overline{\overline{I}}_t$$
$$-\frac{i\eta}{4d}\sum_{n,l}\frac{(-1)^{n+l}}{k_{\rho l}^2}f_l\left(1+e^{-2ik_z(z'-\frac{d}{2})}\right)e^{ik_z(z'-\frac{d}{2})}$$
$$\overline{H}_n^{TE}(k_{\rho l},k_{zl},\overline{\rho},z+\frac{d}{2})Rg\overline{n}_{-n}(k_{\rho l},k_{zl},\overline{\rho}')e^{-in\phi'}\cdot\overline{\overline{I}}_t \quad (12.3.33)$$

For the case $z > z', \rho < \rho'$, we have, similarly,

$$\overline{\overline{G}}(\overline{r},\overline{r}')\cdot\overline{\overline{I}}_t = -\frac{\eta}{4d}\sum_{n,l}\frac{(-1)^{n+l}}{k_{\rho l}^2}f_l\left(1+e^{2ik_z(z'+\frac{d}{2})}\right)e^{-ik_z(z'+\frac{d}{2})}$$
$$Rg\overline{H}_n^{TM}(k_{\rho l},k_{zl},\overline{\rho},z-\frac{d}{2})\overline{m}_{-n}(k_{\rho l},k_{zl},\overline{\rho}')e^{-in\phi'}\cdot\overline{\overline{I}}_t$$
$$-\frac{i\eta}{4d}\sum_{n,l}\frac{(-1)^{n+l}}{k_{\rho l}^2}f_l\left(1+e^{2ik_z(z'+\frac{d}{2})}\right)e^{-ik_z(z'+\frac{d}{2})}$$
$$Rg\overline{H}_n^{TE}(k_{\rho l},k_{zl},\overline{\rho},z-\frac{d}{2})\overline{n}_{-n}(k_{\rho l},k_{zl},\overline{\rho}')e^{-in\phi'}\cdot\overline{\overline{I}}_t \quad (12.3.34)$$

and for $z > z', \rho > \rho'$

$$\overline{\overline{G}}(\overline{r},\overline{r}')\cdot\overline{\overline{I}}_t = -\frac{\eta}{4d}\sum_{n,l}\frac{(-1)^{n+l}}{k_{\rho l}^2}f_l\left(1+e^{2ik_z(z'+\frac{d}{2})}\right)e^{-ik_z(z'+\frac{d}{2})}$$
$$\overline{H}_n^{TM}(k_{\rho l},k_{zl},\overline{\rho},z-\frac{d}{2})Rg\overline{m}_{-n}(k_{\rho l},k_{zl},\overline{\rho}')e^{-in\phi'}\cdot\overline{\overline{I}}_t$$

$$- \frac{i\eta}{4d} \sum_{n,l} \frac{(-1)^{n+l}}{k_{\rho l}^2} f_l \left(1 + e^{2ik_z(z'+\frac{d}{2})}\right) e^{-ik_z(z'+\frac{d}{2})}$$

$$\overline{H}_n^{TE}(k_{\rho l}, k_{zl}, \overline{\rho}, z - \frac{d}{2}) Rg\overline{n}_{-n}(k_{\rho l}, k_{zl}, \overline{\rho}') e^{-in\phi'} \cdot \overline{\overline{I}}_t \quad (12.3.35)$$

3.2 Dyadic Green's Function of a Cylindrical Scatterer Between Two PEC

Consider a perfectly conducting cylinder placed between two parallel plates placed at $z = d/2$ and $z = -d/2$. The cylinder is of radius a and is centered at $\overline{\rho} = 0$. We first consider the case when the source is at $z' = d/2$. Then we only need the solution for $z < z'$. The total Green's function of the previous section will be the primary Green's function for this case as it becomes the excitation field of the cylinder. It is the field of the source between two parallel plates and is the incident field on the cylinder. The response Green's function for this case is then due to the scattering by the cylinder. Let the source of the Green's function be at $\overline{\rho}'$ with $\rho' > a$. Thus, letting $z' = d/2$, we have, from (12.3.32) and (12.3.33), the following expressions:
for $z < d/2, \rho < \rho'$

$$\overline{\overline{G}}_P(\overline{r}, \overline{r}') \cdot \overline{\overline{I}}_t = -\frac{\eta}{2d} \sum_{n,l} \frac{(-1)^{n+l}}{k_{\rho l}^2} f_l Rg\overline{H}_n^{TM}(k_{\rho l}, k_{zl}, \overline{\rho}, z + \frac{d}{2})$$

$$\overline{m}_{-n}(k_{\rho l}, k_{zl}, \overline{\rho}') e^{-in\phi'} \cdot \overline{\overline{I}}_t$$

$$- \frac{i\eta}{2d} \sum_{n,l} \frac{(-1)^{n+l}}{k_{\rho l}^2} f_l Rg\overline{H}_n^{TE}(k_{\rho l}, k_{zl}, \overline{\rho}, z + \frac{d}{2})$$

$$\overline{n}_{-n}(k_{\rho l}, k_{zl}, \overline{\rho}') e^{-in\phi'} \cdot \overline{\overline{I}}_t \quad (12.3.36)$$

and for $\rho > \rho'$

$$\overline{\overline{G}}_P(\overline{r}, \overline{r}') \cdot \overline{\overline{I}}_t = -\frac{\eta}{2d} \sum_{n,l} \frac{(-1)^{n+l}}{k_{\rho l}^2} f_l \overline{H}_n^{TM}(k_{\rho l}, k_{zl}, \overline{\rho}, z + \frac{d}{2})$$

$$Rg\overline{m}_{-n}(k_{\rho l}, k_{zl}, \overline{\rho}') e^{-in\phi'} \cdot \overline{\overline{I}}_t$$

$$- \frac{i\eta}{2d} \sum_{n,l} \frac{(-1)^{n+l}}{k_{\rho l}^2} f_l \overline{H}_n^{TE}(k_{\rho l}, k_{zl}, \overline{\rho}, z + \frac{d}{2})$$

$$Rg\overline{n}_{-n}(k_{\rho l}, k_{zl}, \overline{\rho}') e^{-in\phi'} \cdot \overline{\overline{I}}_t \quad (12.3.37)$$

The electric field Green's function is for $\rho < \rho'$

$$\frac{i}{\omega\epsilon}\nabla \times \overline{\overline{G}}_P(\overline{r},\overline{r}') \cdot \overline{\overline{I}}_t = -\frac{\eta}{2d}\sum_{n,l} \frac{(-1)^{n+l}}{k_{\rho l}^2} f_l Rg\overline{E}_n^{TM}(k_{\rho l}, k_{zl}, \overline{\rho}, z + \frac{d}{2})$$
$$\overline{m}_{-n}(k_{\rho l}, k_{zl}, \overline{\rho}')e^{-in\phi'} \cdot \overline{\overline{I}}_t$$
$$-\frac{i\eta}{2d}\sum_{n,l} \frac{(-1)^{n+l}}{k_{\rho l}^2} f_l Rg\overline{E}_n^{TE}(k_{\rho l}, k_{zl}, \overline{\rho}, z + \frac{d}{2})$$
$$\overline{n}_{-n}(k_{\rho l}, k_{zl}, \overline{\rho}')e^{-in\phi'} \cdot \overline{\overline{I}}_t \qquad (12.3.38)$$

We can use the solution that we had in Chapter 9 because the $Rg\overline{E}_n^{TM}$ is a linear combination of $Rg\overline{N}_n$ and has a scattering coefficient of $T_n^{(N)}$ for scattering by a cylinder. The quantity $Rg\overline{E}_n^{TE}$ is a linear combination of $Rg\overline{M}_n$ and has a scattering coefficient of $T_n^{(M)}$. The expressions of $T_n^{(M)}$ and $T_n^{(N)}$ are as in (9.3.54) and (9.3.55) of Chapter 9. Thus the response Green's function of the magnetic field of the cylinder is

$$\overline{\overline{G}}_R(\overline{r},\overline{r}') \cdot \overline{\overline{I}}_t = -\frac{\eta}{2d}\sum_{n,l} \frac{(-1)^{n+l}}{k_{\rho l}^2} f_l T_n^{(N)}\overline{H}_n^{TM}(k_{\rho l}, k_{zl}, \overline{\rho}, z + \frac{d}{2})$$
$$\overline{m}_{-n}(k_{\rho l}, k_{zl}, \overline{\rho}')e^{-in\phi'} \cdot \overline{\overline{I}}_t$$
$$-\frac{i\eta}{2d}\sum_{n,l} \frac{(-1)^{n+l}}{k_{\rho l}^2} f_l T_n^{(M)}\overline{H}_n^{TE}(k_{\rho l}, k_{zl}, \overline{\rho}, z + \frac{d}{2})$$
$$\overline{n}_{-n}(k_{\rho l}, k_{zl}, \overline{\rho}')e^{-in\phi'} \cdot \overline{\overline{I}}_t \qquad (12.3.39)$$

We also define the surface current modal solutions as the current on the surface of the cylinder. These can be obtained from the modal solutions of the magnetic field and using the Wronskian on p. 44 of Volume I. The surface current modal solutions are

$$\overline{J}_{sn}^{TE}(k_\rho, k_z, z) = \hat{\rho} \times \left(Rg\overline{H}_n^{TE}(k_\rho, k_z, \overline{\rho}, z) + T_n^{(M)}\overline{H}_n^{TE}(k_\rho, k_z, \overline{\rho}, z)\right)_{\rho=a}$$
$$= \frac{\frac{2i}{\eta\pi k_\rho a}e^{in\phi}}{H_n^{(1)\prime}(k_\rho a)}\left[-\hat{z}\frac{nk_z}{ka}\cos(k_z z) - \hat{\phi}\frac{k_\rho^2}{k}i\sin(k_z z)\right] \qquad (12.3.40)$$

$$\overline{J}_{sn}^{TM}(k_\rho, k_z, z) = \hat{\rho} \times \left(Rg\overline{H}_n^{TM}(k_\rho, k_z, \overline{\rho}, z) + T_n^{(N)}\overline{H}_n^{TM}(k_\rho, k_z, \overline{\rho}, z)\right)_{\rho=a}$$
$$= \hat{z}\frac{\frac{2}{\eta\pi a}}{H_n^{(1)}(k_\rho a)}\{\cos k_z z\}\, e^{in\phi} \qquad (12.3.41)$$

3.3 Dyadic Green's Function with Multiple Cylinders

Consider N cylinders placed between the two parallel plates and centered at $\overline{\rho}_1, \overline{\rho}_2, \ldots, \overline{\rho}_N$. Let there be magnetic surface current density $\overline{M}_{su}$ at $(\overline{\rho}', z' = d/2)$ and $\overline{M}_s = \overline{M}_{sb}$ at $(\overline{\rho}', z' = -d/2)$ (Fig. 12.3.3). The multiple scattering can be formulated in terms of Foldy-Lax multiple scattering equations similar to that of Chapter 9, Section 3. In the following, we use the upper sign for equations and results with source $\overline{M}_{su}$ at $(\overline{\rho}', z' = d/2)$ and lower sign for equations and results with source $\overline{M}_{sb}$ at $(\overline{\rho}', z' = -d/2)$.

Within the Foldy-Lax formulation, the final exciting field of cylinder p is

$$\overline{H}_{ex}^{(p)} = \sum_{n,l} w_{lm}^{TM(p)} Rg\overline{H}_m^{TM}(k_{\rho l}, k_{zl}, \overline{\rho} - \overline{\rho}_p, z \pm d/2) + \sum_{n,l} w_{lm}^{TE(p)} Rg\overline{H}_m^{TE}(k_{\rho l}, k_{zl}, \overline{\rho} - \overline{\rho}_p, z \pm d/2) \qquad (12.3.42)$$

The final scattered field from cylinder p is

$$\overline{H}_s^{(p)} = \sum_{m,l} T_m^{(N)} w_{lm}^{TM(p)} \overline{H}_m^{TM}(k_{\rho l}, k_{zl}, \overline{\rho} - \overline{\rho}_p, z \pm d/2) + \sum_{m,l} T_m^{(M)} w_{lm}^{TE(p)} \overline{H}_m^{TE}(k_{\rho l}, k_{zl}, \overline{\rho} - \overline{\rho}_p, z \pm d/2) \qquad (12.3.43)$$

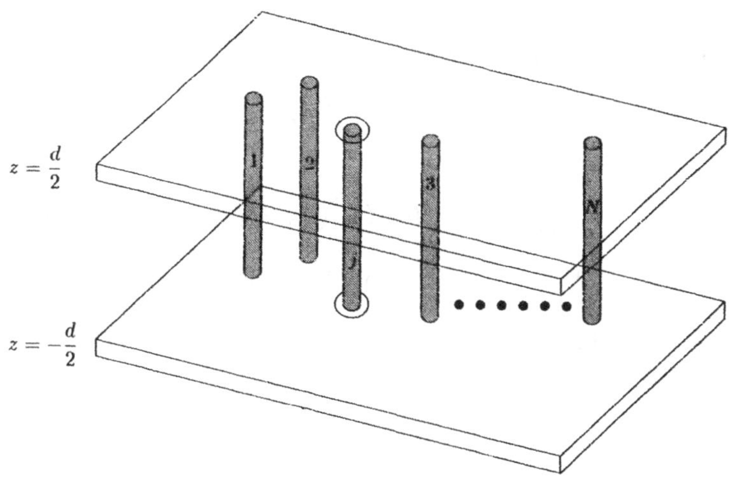

Figure 12.3.3 N cylinders are placed between two parallel plates with magnetic surface current density $\overline{M}_{su}$ at $(\overline{\rho}', z' = d/2)$ and $\overline{M}_s = \overline{M}_{sb}$ at $(\overline{\rho}', z' = -d/2)$. Magnetic ring source is at cylinder j.

On the cylinder p, the surface current is

$$\overline{J}_s^{(p)} = \sum_{m,l} w_{lm}^{TM(p)} \overline{J}_s^{TM}(k_{\rho l}, k_{zl}, z \pm d/2) + \sum_{m,l} w_{lm}^{TE(p)} \overline{J}_s^{TE}(k_{\rho l}, k_{zl}, z \pm d/2) \tag{12.3.44}$$

The scattered field will be incident on cylinder q. Since the mode functions are linear combinations of cylindrical wave functions of $\overline{M}_n$ and $\overline{N}_n$. The translation formula for the magnetic modal fields are the same as the vector cylindrical wave functions of (9.3.71) and (9.3.72) from Chapter 9.

$$\begin{aligned}\overline{H}_m^{TM}&(k_{\rho l}, k_{zl}, \overline{\rho} - \overline{\rho}_p, z \pm \frac{d}{2}) \\ &= \sum_{n=-\infty}^{\infty} Rg\overline{H}_n^{TM}(k_{\rho l}, k_{zl}, \overline{\rho} - \overline{\rho}_q, z \pm \frac{d}{2}) H_{n-m}^{(1)}(k_{\rho l} |\overline{\rho}_p - \overline{\rho}_q|) e^{-i(n-m)\phi_{\overline{\rho_p \rho_q}}}\end{aligned} \tag{12.3.45}$$

$$\begin{aligned}\overline{H}_m^{TE}&(k_{\rho l}, k_{zl}, \overline{\rho} - \overline{\rho}_p, z \pm \frac{d}{2}) \\ &= \sum_{n=-\infty}^{\infty} Rg\overline{H}_n^{TE}(k_{\rho l}, k_{zl}, \overline{\rho} - \overline{\rho}_q, z \pm \frac{d}{2}) H_{n-m}^{(1)}(k_{\rho l} |\overline{\rho}_p - \overline{\rho}_q|) e^{-i(n-m)\phi_{\overline{\rho_p \rho_q}}}\end{aligned} \tag{12.3.46}$$

The scattered field from p onto cylinder q is

$$\begin{aligned}\overline{H}_s^{(p)} = \sum_{m,l,n} \Big[& T_m^{(N)} w_{lm}^{TM(p)} Rg\overline{H}_n^{TM}(k_{\rho l}, k_{zl}, \overline{\rho} - \overline{\rho}_q, z \pm \frac{d}{2}) \\ & + T_m^{(M)} w_{lm}^{TE(p)} Rg\overline{H}_n^{TE}(k_{\rho l}, k_{zl}, \overline{\rho} - \overline{\rho}_q, z \pm \frac{d}{2}) \Big] \\ & H_{n-m}^{(1)}(k_{\rho l} |\overline{\rho}_p - \overline{\rho}_q|) e^{-i(n-m)\phi_{\overline{\rho_p \rho_q}}}\end{aligned} \tag{12.3.47}$$

Thus we get the Foldy-Lax multiple scattering equations, similar to Chapter 9,

$$\begin{aligned}w_{ln}^{TM(q)} &= a_{ln}^{TM(q)} \\ &+ \sum_{\substack{p=1 \\ p \neq q}}^{N} \sum_{m=-\infty}^{\infty} H_{n-m}^{(1)}(k_{\rho l} |\overline{\rho}_p - \overline{\rho}_q|) e^{-i(n-m)\phi_{\overline{\rho_p \rho_q}}} T_m^{(N)} w_{lm}^{TM(p)}\end{aligned} \tag{12.3.48}$$

$$w_{ln}^{TE(q)} = a_{ln}^{TE(q)} + \sum_{\substack{p=1 \\ p\neq q}}^{N} \sum_{m=-\infty}^{\infty} H_{n-m}^{(1)}(k_{\rho l}\left|\overline{\rho}_p - \overline{\rho}_q\right|) e^{-i(n-m)\phi_{\overline{\rho_p\rho_q}}} T_m^{(M)} w_{lm}^{TE(p)} \quad (12.3.49)$$

where $a_{ln}^{TM(q)}, a_{ln}^{TE(q)}$ are incident field of the current source onto cylinder q. The equations are basically identical to that of (9.3.74) and (9.3.75) in Chapter 9. This is because there is no coupling between different l because each l corresponds to a specific k_z. Neither is there coupling between TE and TM waves because the cylinders are perfect electric conductors as shown in (9.3.51)–(9.3.60) of Chapter 9. This is particularly important when one considers the dimension of the matrix equation associated with Foldy-Lax multiple scattering equations. The decoupling reduces significantly the dimension of the matrix equation, as one needs only to solve Foldy-Lax equation for each waveguide mode separately. This decoupling is useful when multiple waveguide modes are excited for waveguides of moderate thickness. For such cases, if discretization in the $\hat{z}$-direction is done using subsectional basis functions, a large number of basis functions are required and they are all coupled in multiple scattering. To determine $a_{ln}^{TM(q)}, a_{ln}^{TE(q)}$, we use the results of the primary Green's function of the previous section. Note that the primary Green's function is translational invariant in the horizontal plane.

$$\overline{\overline{G}}_P\left(\overline{\rho}, z, \overline{\rho}', z'\right) \cdot \overline{\overline{I}}_t = \overline{\overline{G}}_P\left(\overline{\rho} - \overline{\rho}_q, z, \overline{\rho}' - \overline{\rho}_q, z'\right) \cdot \overline{\overline{I}}_t \quad (12.3.50)$$

Thus the incident field from the magnetic surface current source $\overline{M}_s$ onto cylinder q for $|\overline{\rho} - \overline{\rho}_q| < |\overline{\rho}' - \overline{\rho}_q|$ is, from (12.3.32) and (12.3.34),

$$\begin{aligned}
i\omega\epsilon \int d\overline{\rho}' \overline{\overline{G}}_P(\overline{r}, \overline{r}') \cdot \overline{M}_s\left(\overline{\rho}'\right) &= i\omega\epsilon \int d\overline{\rho}' \overline{\overline{G}}_P\left(\overline{\rho} - \overline{\rho}_q, z, \overline{\rho}' - \overline{\rho}_q, z'\right) \cdot \overline{M}_s\left(\overline{\rho}'\right) \\
&= -\frac{\eta i\omega\epsilon}{2d} \sum_{n,l} \frac{(-1)^{n+l}}{k_{\rho l}^2} f_l Rg\overline{H}_n^{TM}(k_{\rho l}, k_{zl}, \overline{\rho} - \overline{\rho}_q, z \pm \frac{d}{2}) \\
&\quad \int d\overline{\rho}' \overline{m}_{-n}(k_{\rho l}, k_{zl}, \overline{\rho}' - \overline{\rho}_q) e^{-in\phi_{\overline{\rho'\rho_q}}} \cdot \overline{M}_s\left(\overline{\rho}'\right) \\
&+ \frac{\eta\omega\epsilon}{2d} \sum_{n,l} \frac{(-1)^{n+l}}{k_{\rho l}^2} f_l Rg\overline{H}_n^{TE}(k_{\rho l}, k_{zl}, \overline{\rho} - \overline{\rho}_q, z \pm \frac{d}{2}) \\
&\quad \int d\overline{\rho}' \overline{n}_{-n}(k_{\rho l}, k_{zl}, \overline{\rho}' - \overline{\rho}_q) e^{-in\phi'_{\overline{\rho'\rho_q}}} \cdot \overline{M}_s\left(\overline{\rho}'\right)
\end{aligned} \quad (12.3.51)$$

Thus the incident field coefficients onto cylinder q are

$$a_{ln}^{TM(q)} = -\frac{\eta i\omega\epsilon}{2d}\frac{(-1)^{n+l}}{k_{\rho l}^2} f_l \int d\overline{\rho}' \overline{m}_{-n}(k_{\rho l}, k_{zl}, \overline{\rho}' - \overline{\rho}_q) e^{-in\phi_{\overline{\rho' \rho_q}}} \cdot \overline{M}_s(\overline{\rho}') \tag{12.3.52}$$

$$a_{ln}^{TE(q)} = \frac{\eta\omega\epsilon}{2d}\frac{(-1)^{n+l}}{k_{\rho l}^2} f_l \int d\overline{\rho}' \overline{n}_{-n}(k_{\rho l}, k_{zl}, \overline{\rho}' - \overline{\rho}_q) e^{-in\phi'_{\overline{\rho' \rho_q}}} \cdot \overline{M}_s(\overline{\rho}') \tag{12.3.53}$$

After the exciting field coefficients are determined , we can find the far field scattered field. Asymptotically

$$\overline{m}_n(k_\rho, k_z, \overline{\rho}) = -\hat{\phi} i \sqrt{\frac{2k_\rho}{\pi\rho}} e^{i(k_\rho \rho - \frac{\pi}{4}) - in\frac{\pi}{2}} \tag{12.3.54}$$

$$\overline{n}_n(k_\rho, k_z, \overline{\rho}) = \left(-\frac{k_z}{k}\hat{\rho} + \frac{k_\rho}{k}\hat{z}\right) \sqrt{\frac{2k_\rho}{\pi\rho}} e^{i(k_\rho \rho - \frac{\pi}{4}) - in\frac{\pi}{2}} \tag{12.3.55}$$

$$\overline{H}_n^{TE}(k_\rho, k_z, \overline{\rho}, z) = \frac{1}{\eta}\left\{-\cos(k_z z)\frac{k_z}{k}\hat{\rho} + i\sin(k_z z)\frac{k_\rho}{k}\hat{z}\right\} e^{in\phi} \sqrt{\frac{2k_\rho}{\pi\rho}} e^{i(k_\rho \rho - \frac{\pi}{4}) - in\frac{\pi}{2}} \tag{12.3.56}$$

$$\overline{H}_n^{TM}(k_\rho, k_z, \overline{\rho}, z) = -\frac{1}{\eta}\cos(k_z z)\hat{\phi} e^{in\phi} \sqrt{\frac{2k_\rho}{\pi\rho}} e^{i(k_\rho \rho - \frac{\pi}{4}) - in\frac{\pi}{2}} \tag{12.3.57}$$

The total scattered field is

$$\begin{aligned}\overline{H}_s &= \sum_{p=1}^{N} \overline{H}_s^{(p)} \\ &= \sum_{p=1}^{N}\sum_{m,l} T_m^{(N)} w_{lm}^{TM(p)} \overline{H}_m^{TM}(k_{\rho l}, k_{zl}, \overline{\rho} - \overline{\rho}_p, z \pm \frac{d}{2}) \\ &+ \sum_{p=1}^{N}\sum_{m,l} T_m^{(M)} w_{lm}^{TE(p)} \overline{H}_m^{TE}(k_{\rho l}, k_{zl}, \overline{\rho} - \overline{\rho}_p, z \pm \frac{d}{2})\end{aligned} \tag{12.3.58}$$

The far field in direction $\hat{k}_s = \cos\phi_s \hat{x} + \sin\phi_s \hat{y}$ is

$$\overline{H}_s = -\frac{1}{\eta}\sum_{p=1}^{N}\sum_{n,l} T_n^{(N)} w_{ln}^{TM(p)} \cos(k_{zl}(z \pm \frac{d}{2}))\hat{\phi}_s$$

$$e^{in\phi_s} e^{-ik\hat{k}_s\cdot\overline{\rho_p}} \sqrt{\frac{2k_{\rho l}}{\pi\rho}} e^{i(k_{\rho l}\rho - \frac{\pi}{4}) - in\frac{\pi}{2}}$$

$$+\frac{1}{\eta}\sum_{p=1}^{N}\sum_{n,l} T_n^{(M)} w_{ln}^{TE(p)} \left\{-\cos(k_{zl}(z \pm \frac{d}{2}))\frac{k_{zl}}{k}\hat{\rho} + i\sin(k_{zl}(z \pm \frac{d}{2}))\frac{k_{\rho l}}{k}\hat{z}\right\}$$

$$e^{in\phi_s} e^{-ik\hat{k}_s\cdot\overline{\rho_p}} \sqrt{\frac{2k_{\rho l}}{\pi\rho}} e^{i(k_{\rho l}\rho - \frac{\pi}{4}) - in\frac{\pi}{2}} \tag{12.3.59}$$

3.4 Excitation of Magnetic Ring Currents

We consider the case when there is a ring source of magnetic surface current centered at cylinder j (Fig. 12.3.3). Specifically at $z' = d/2$,

$$\overline{M}_{su}\left(\overline{\rho}'\right) = -\frac{V_{ju}}{\left|\overline{\rho}' - \overline{\rho}_j\right| \ln\frac{b}{a}}\hat{\phi}'_{\overline{\rho\rho_j}} \quad \text{for } a \le \left|\overline{\rho}' - \overline{\rho}_j\right| \le b \tag{12.3.60}$$

and at $z' = -d/2$,

$$\overline{M}_{sb}\left(\overline{\rho}'\right) = -\frac{V_{jb}}{\left|\overline{\rho}' - \overline{\rho}_j\right| \ln\frac{b}{a}}\hat{\phi}'_{\overline{\rho\rho_j}} \quad \text{for } a \le \left|\overline{\rho}' - \overline{\rho}_j\right| \le b \tag{12.3.61}$$

We first consider the excitation terms of $a_{ln}^{TM(j)}$ and $a_{ln}^{TE(j)}$ for cylinder j.

$$\int d\overline{\rho}'\, \overline{m}_{-n}(k_{\rho l}, k_{zl}, \overline{\rho}' - \overline{\rho}_j) e^{-in\phi_{\overline{\rho'\rho_j}}} \cdot \overline{M}_s\left(\overline{\rho}'\right)$$

$$= \frac{2\pi V_j}{\ln\frac{b}{a}}\delta_{n0}\left[H_0^{(1)}\left(k_{\rho l}b\right) - H_0^{(1)}\left(k_{\rho l}a\right)\right] \tag{12.3.62}$$

where $V_j = V_{ju}$ for $\overline{M}_{su}$, and $V_j = V_{jb}$ for $\overline{M}_{sb}$. In (12.3.62), the Kronecker delta δ_{n0} arises because there is no azimuthal dependence of the source. Because $\overline{n}_0(k_{\rho l}, k_{zl}, \overline{\rho}' - \overline{\rho}_j) = 0$. Thus

$$\int d\overline{\rho}'\, \overline{n}_{-n}(k_{\rho l}, k_{zl}, \overline{\rho}' - \overline{\rho}_j) e^{-in\phi_{\overline{\rho'\rho_j}}} \cdot \overline{M}_s\left(\overline{\rho}'\right) = 0 \tag{12.3.63}$$

We then have

$$a_{ln}^{TM(j)} = -\frac{\eta i\omega\epsilon}{2d}\frac{(-1)^l}{k_{\rho l}^2} f_l \frac{2\pi V_j}{\ln\frac{b}{a}}\delta_{n0}\left[H_0^{(1)}\left(k_{\rho l}b\right) - H_0^{(1)}\left(k_{\rho l}a\right)\right] \tag{12.3.64}$$

$$a_{ln}^{TE(j)} = 0 \tag{12.3.65}$$

Next we calculate the excitation terms of $a_{ln}^{TM(q)}$ and $a_{ln}^{TE(q)}$ for cylinder $q \neq j$. We use translation addition theorem, since $|\overline{\rho}' - \overline{\rho}_q| < |\overline{\rho}_j - \overline{\rho}_q|$

$$
\begin{aligned}
a_{ln}^{TM(q)} &= -\frac{\eta i\omega\epsilon}{2d}\frac{(-1)^{n+l}}{k_{\rho l}^2} f_l \\
&\int d\overline{\rho}' \overline{m}_{-n}(k_{\rho l}, k_{zl}, \overline{\rho}' - \overline{\rho}_q) e^{-in\phi_{\overline{\rho' \rho_q}}} \cdot \overline{M}_s(\overline{\rho}') \\
&= -\frac{\eta i\omega\epsilon}{2d}\frac{(-1)^{n+l}}{k_{\rho l}^2} f_l \sum_{m=-\infty}^{\infty} H_{m+n}^{(1)}\left(k_{\rho l}\left|\overline{\rho}_q - \overline{\rho}_j\right|\right) e^{-i(m+n)\phi_{\overline{\rho_q \rho_j}}} \\
&\int d\overline{\rho}' Rg\overline{m}_m\left(k, \overline{\rho}' - \overline{\rho}_j\right) e^{im\phi_{\overline{\rho' \rho_q}}} \cdot \overline{M}_s(\overline{\rho}') \\
&= -\frac{\eta i\omega\epsilon}{2d}\frac{(-1)^{n+l}}{k_{\rho l}^2} f_l \sum_{m=-\infty}^{\infty} H_{m+n}^{(1)}\left(k_{\rho l}\left|\overline{\rho}_q - \overline{\rho}_j\right|\right) e^{-i(m+n)\phi_{\overline{\rho_q \rho_j}}} \\
&\left[\frac{2\pi V_j}{\ln\frac{b}{a}}\delta_{m0}\left[J_0(k_{\rho l}b) - J_0(k_{\rho l}a)\right]\right] \qquad (12.3.66)
\end{aligned}
$$

Thus for $q \neq j$

$$
\begin{aligned}
a_{ln}^{TM(q)} &= -\frac{ik}{2d}\frac{(-1)^{n+l}}{k_{\rho l}^2} f_l H_n^{(1)}\left(k_{\rho l}\left|\overline{\rho}_q - \overline{\rho}_j\right|\right) e^{-in\phi_{\overline{\rho_q \rho_j}}} \\
&\left[\frac{2\pi V_j}{\ln\frac{b}{a}}\left[J_0(k_{\rho l}b) - J_0(k_{\rho l}a)\right]\right] \qquad (12.3.67)
\end{aligned}
$$

$$
a_{ln}^{TE(q)} = 0 \qquad (12.3.68)
$$

We use these incident field coefficients of $a_{ln}^{TM(q)}$ in (12.3.48). After the exciting field coefficients $w_{ln}^{TM(q)}$ are determined, the surface current on cylinder p is

$$
\begin{aligned}
\overline{J}_s^{(p)} &= \sum_{m,l} w_{lm}^{TM(p)} \overline{J}_s^{TM}(k_{\rho l}, k_{zl}, z \pm d/2) \\
&= \sum_{m,l} w_{lm}^{TM(p)} (-\hat{z}) k_{\rho l} \frac{\frac{2i}{\pi k_{\rho l} a}}{H_m^{(1)}(k_{\rho l}a)} \frac{i}{\eta} \cos\left[k_{zl}(z \pm d/2)\right] e^{im\phi} \qquad (12.3.69)
\end{aligned}
$$

The far field scattered field is

$$\overline{H}_s = -\frac{1}{\eta}\sum_{p=1}^{N}\sum_{n,l} T_n^{(N)} w_{ln}^{TM(p)} \cos\left[k_{zl}(z \pm d/2)\right]\hat{\phi}_s$$
$$\times\, e^{in\phi_s} e^{-ik_{\rho l}\hat{k}_s\cdot\overline{\rho_p}} \sqrt{\frac{2k_{\rho l}}{\pi\rho}} e^{i\left(k_{\rho l}\rho - \frac{\pi}{4}\right) - in\frac{\pi}{2}} \quad (12.3.70)$$

The primary field from source j is

$$\overline{H}_P = -\frac{\eta i\omega\varepsilon}{2d}\sum_l \frac{(-1)^l}{k_{\rho l}^2} f_l \overline{H}_0^{TM}(k_{\rho l}, k_{zl}, \overline{\rho} - \overline{\rho}_j, z \pm d/2)$$
$$\times\, \frac{2\pi V_j}{\ln\frac{b}{a}}\left[J_0(k_{\rho l}b) - J_0(k_{\rho l}a)\right] \quad (12.3.71)$$

In the far field the primary field from source j is

$$\overline{H}_P = \frac{ik}{2\eta d}\sum_l \frac{(-1)^l}{k_{\rho l}^2} f_l \frac{2\pi V_j}{\ln\frac{b}{a}}\left[J_0(k_{\rho l}b) - J_0(k_{\rho l}a)\right]$$
$$\times \cos\left[k_{zl}(z \pm d/2)\right]\hat{\phi}_s e^{-ik_{\rho l}\hat{k}_s\cdot\overline{\rho_j}} \sqrt{\frac{2k_{\rho l}}{\pi\rho}} e^{i\left(k_{\rho l}\rho - \frac{\pi}{4}\right)} \quad (12.3.72)$$

The total field is

$$\overline{H} = \overline{H}_P + \overline{H}_S \quad (12.3.73)$$

In the far field, the power per unit length in the $\hat{z}$ direction per unit angle of ϕ_s is

$$\sigma_s(\phi_s) = \frac{\eta}{2}\left|\overline{H}\right|^2 \rho \quad (12.3.74)$$

and the total power per unit length in the $\hat{z}$ direction is

$$P = \int_0^{2\pi} d\phi_s \sigma_s(\phi_s) \quad (12.3.75)$$

3.4.1 First Order Solution

We calculate the first order solution. Since the cylinder j is close to the source at j, we combine the field from source j and the scattered field from cylinder j as the incident field on other cylinders. Let superscript (1) denote first order solution. Thus

$$w_{ln}^{TM(j)(1)} = a_{ln}^{TM(j)}$$
$$= -\frac{ki}{2d}\frac{(-1)^l}{k_{\rho l}^2} f_l \frac{2\pi V_j}{\ln\frac{b}{a}}\delta_{n0}\left[H_0^{(1)}(k_{\rho l}b) - H_0^{(1)}(k_{\rho l}a)\right] \quad (12.3.76)$$

For $q \neq j$,

$$w_{ln}^{TM(q)(1)} = a_{ln}^{TM(q)} + \sum_{m=-\infty}^{\infty} H_{n-m}^{(1)}(k_{\rho l}\left|\overline{\rho}_j - \overline{\rho}_q\right|)e^{-i(n-m)\phi_{\overline{\rho_j}\rho_q}} T_m^{(N)} a_{lm}^{TM(j)} \qquad (12.3.77)$$

Thus, putting (12.3.64) in (12.3.77), and noting that $\phi_{\overline{\rho_j}\rho_q} = \phi_{\overline{\rho_q \varrho_j}} + \pi$, we have

$$w_{ln}^{TM(q)(1)} = -\frac{ik}{2d}\frac{(-1)^{n+l}}{k_{\rho l}^2} f_l H_n^{(1)}\left(k\left|\overline{\rho}_q - \overline{\rho}_j\right|\right) e^{-in\phi_{\overline{\rho_q \varrho_j}}} \frac{2\pi V_j}{\ln\frac{b}{a}} \times \left\{\left[J_0(k_{\rho l}b) - J_0(k_{\rho l}a)\right] + T_0^{(N)}\left[H_0^{(1)}(k_{\rho l}b) - H_0^{(1)}(k_{\rho l}a)\right]\right\} \qquad (12.3.78)$$

Since $T_0^{(N)} = -J_0(k_{\rho l}a)/H_0^{(1)}(k_{\rho l}a)$, putting it in (12.3.78) gives

$$w_{ln}^{TM(q)(1)} = -\frac{ik}{2d}\frac{(-1)^{n+l}}{k_{\rho l}^2} f_l H_n^{(1)}\left(k\left|\overline{\rho}_q - \overline{\rho}_j\right|\right) e^{-in\phi_{\overline{\rho_q \varrho_j}}} \frac{2\pi V_j}{\ln\frac{b}{a}} \times \frac{J_0(k_{\rho l}b) H_0^{(1)}(k_{\rho l}a) - J_0(k_{\rho l}a) H_0^{(1)}(k_{\rho l}b)}{H_0^{(1)}(k_{\rho l}a)} \qquad (12.3.79)$$

3.4.2 Numerical Results

We illustrate numerical results of the radiation as given by $\sigma(\phi_s)$ of (12.3.74). The fractional volume of cylinders is at $f = 0.0005$, and the size of the cylinders is such that $ka = 0.05$, or $a = 0.00796\lambda$, where λ is the wavelength. The size of the aperture is $b = 2a = 0.0159\lambda$ (see Fig. 12.3.3). The separation between the parallel plates is such that $kd = 0.3$. In Figs. 12.3.4a–c, we show results of $\sigma(\phi_s)$ for $N = 16$, 64 and 256 cylinders, respectively. The positions of the cylinders are chosen stochastically using the shuffling algorithm in Section 2.1 of Chapter 9. For each case, the size of the square area in which the cylinders are placed are $(2.52\lambda)^2$, $(5.04\lambda)^2$, and $(10.08\lambda)^2$ respectively for $N = 16$, 64, and 256 cylinders. The source cylinder j is chosen that it is the cylinder that is closest to the center of the square. The voltage is $V_j = 1$ volt. The magnetic ring source is placed at $z = d/2$. We plot results for $\sigma(\phi_s)$ with the observation point at $z = -d/2$. The total power per unit length in $\hat{z}$ direction, $P = \int_0^{2\pi} d\phi_s \sigma_s(\phi_s)$ are 0.10, 0.11, and 0.11 respectively for 16, 64, and 256 cylinders. We note that when there are more cylinders, the radiation pattern has more angular fluctuations.

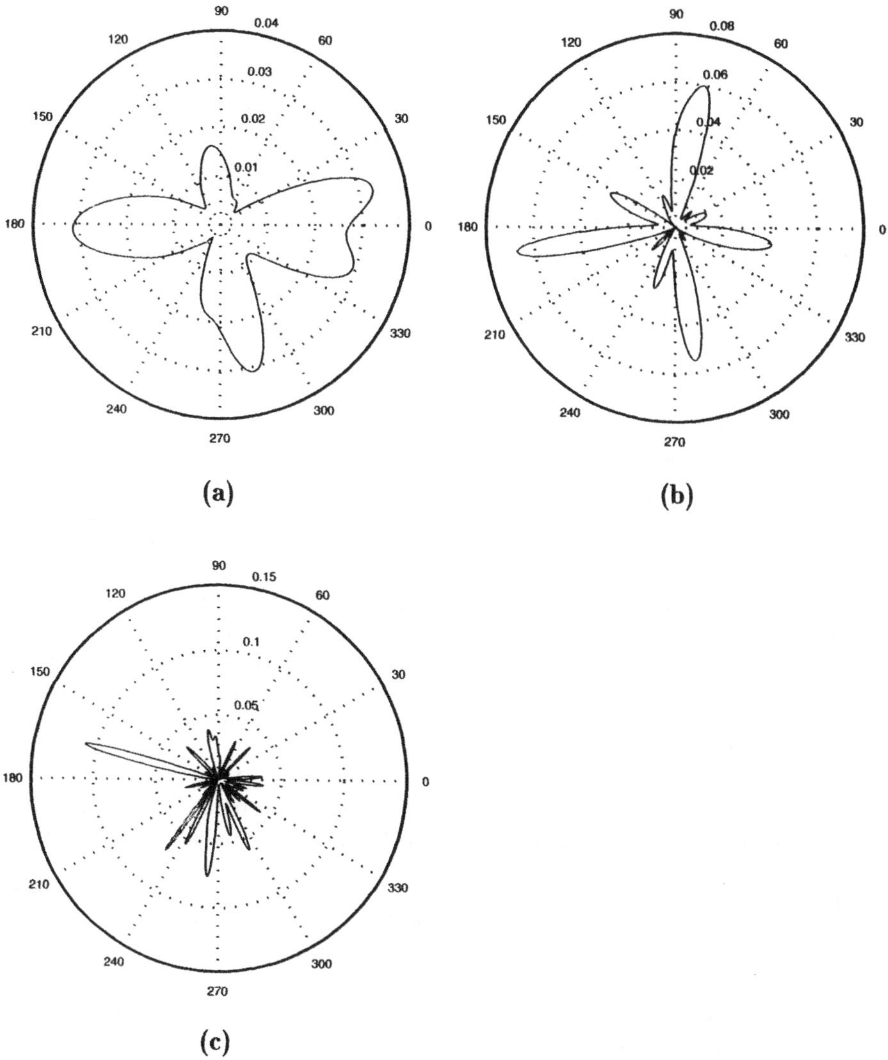

Figure 12.3.4 Radiation pattern $\sigma(\phi_s)$ from N cylinders with fractional volume $f = 0.005$ and size $ka = 0.05$. The separation of the parallel plates is such that $kd = 0.3$. (a) $N = 16$, (b) $N = 64$, and (c) $N = 256$.

REFERENCES AND ADDITIONAL READINGS

Au, W. C., J. A. Kong, and L. Tsang (1994), Absorption enhancement of scattering of electromagnetic waves by dielectric cylinder clusters, *Microwave Opt. Technol. Lett.*, **7(10)**, 454–457.

Au, W. C., L. Tsang, R. T. Shin, and J. A. Kong (1996), Collective scattering and absorption in microwave interaction with vegetation canopies, *Progress in Electromag. Res.*, **14**, 182–231, EMW Publishers, Cambridge, Massachusetts.

Chew, W. C. (1990), *Waves and Fields in Inhomogeneous Media*, Van Nostrand Reinhold, New York.

Chiu, T. and K. Sarabandi (2000), Electromagnetic scattering from short branching vegetation, *IEEE Trans. Geosci. Remote Sens.*, **38(2)**, 911–925.

Gu, Q., M. A. Tassoudji, S. Y. Poh, R. T. Shin, and J. A. Kong (1994), Coupled noise analysis for adjacent vias in multilayered digital circuits, *IEEE Trans. Circuits and Systems I: Fundamental Theory and Applications*, **41(12)**, 796–803.

Karam, M. A., A. K. Fung, and Y. M. M. Antar (1988), Electromagnetic wave scattering from some vegetation samples, *IEEE Trans. Geosci. Remote Sens.*, **26(6)**, 799–808.

Kong, J. A. (2000), *Electromagnetic Wave Theory*, EMW Publishing, Cambridge, MA.

Kuga, Y. and A. Ishimaru (1984), Retroreflectance from a dense distribution of spherical particles, *J. Opt. Soc. Am.*, **1**, 831–835.

Lin, Y.-C. and K. Sarabandi (1995), Electromagnetic scattering model for a tree trunk above a tilted ground plane, *IEEE Trans. Geosci. Remote Sens.*, **33(4)**, 1063–1070.

Sarabandi, K., P. F. Polatin, and F. T. Ulaby (1993), Monte Carlo simulation of scattering from a layer of vertical cylinders, *IEEE Trans. Antennas Propagat.*, **41(4)**, 465–475.

Seker, S. S. and A. Schneider (1988), Electromagnetic scattering from a dielectric cylinder of finite length, *IEEE Trans. Antennas Propagat.*, **36(2)**, 303–307.

Sommerfeld, A. (1962), *Partial Differential Equations*, Academic Press, New York.

Tsang, L., C. H. Chan, J. A. Kong, and J. Joseph (1992), Polarimetric signatures of a canopy of dielectric cylinders based on first and second order vector radiative transfer theory, *J. Electromag. Waves and Appl.*, **6(1)**, 19–51.

Tsang, L., K. H. Ding, G. Zhang, C. C. Hsu, and J. A. Kong (1995), Backscattering Enhancement and clustering effects of randomly distributed dielectric cylinders overlying a dielectric half space based on Monte-Carlo simulations, *IEEE Trans. Antennas Propagat.*, **43(5)**, 488–499.

Tsang, L. and A. Ishimaru (1984), Backscattering enhancement of random discrete scatterers, *J. Opt. Soc. Am. A*, **1**, 836–839.

Tsang, L., J. A. Kong, and R. T. Shin (1985), *Theory of Microwave Remote Sensing*, Wiley-Interscience, New York.

Chapter 13

ELECTROMAGNETIC WAVES SCATTERING BY VEGETATION

1 Introduction

Various theoretical models have been developed to characterize the electromagnetic wave scattering properties of vegetation canopies. In the past, wave scattering from vegetation has been studied extensively with vector radiative transfer theory as described in Volume I. Classical radiative transfer theory assumes that the particles scatter independently so that the scattering phase functions add. This assumption is based on the random phase of scattering by different particles and is valid if the particle positions are independent and the randomness of relative positions is comparable to or larger than a wavelength. However, such an assumption can be invalid for microwave scattering of certain cases of vegetation canopy where the randomness of their relative positions is less than a wavelength. For example, branches and leaves in a tree occur in clusters, and there are correlations between their relative positions. Scatterers with this kind of cluster structure can demonstrate collective scattering effects. In Chapter 7, we have shown by simulations of point scatterers the effects of clustering. Collective scattering effects include correlated scattering and take into account the relative phase of scattered waves from the scatterers and their neighbors. The mutual coherent wave interactions between scatterers are also to be included.

In numerical solutions of Maxwell's equations for random media, the positions and characteristics of the scatterers are randomly generated according to prescribed statistics. We have described procedures for generating 2-D and 3-D dense media in Chapters 8 and 9. For the case of vegetation, the scattering by correlated scatterers was first studied by using the coherent addition approximation (CAA). The probability density functions of positions are introduced. However, it is difficult to calculate the probability density functions and the pair distribution functions for natural vegetation. Coherent scattering effects including the relative structures of branches and leaves draw important attention [Yueh et al. 1992; Au et al. 1994, 1996]. To build realistic tree structures, Lindenmayer systems can be used [Prusinkiewicz and Lindenmayer, 1990]. Subsequently coherent scattering models are developed for these Lindenmayer systems. [Chen et al. 1995, 1996; Zhang et al. 1996; Le Toan et al. 1997; Lin and Sarabandi, 1999a,b].

In this chapter we study wave scattering by trees. The trees are grown by using stochastic L-systems. The correlation of scattering by different branches are included by using their relative positions as given by the growth procedure. The advantages of this method are that (1) the structure of trees is controlled by growth procedure and the calculations of the pair distribution functions and probability density function are not needed, (2) the trees grown by the stochastic Lindenmayer systems are quite realistic in appearance to natural trees, and (3) the use of stochastic process in the Lindenmayer systems ensures that every tree will be different though they are of the same generic structure.

In Section 2, we describe the L-systems, which allow the use of computer graphics for realistic visualization of plant structures and growth process. In Section 3, we first describe coherent scattering by a particle in the presence of a boundary. Then we calculate the scattering amplitude from a layer of trees overlaying a flat-ground surface by using coherent addition approximation and compare it to the independent scattering approximation and the tree-independent scattering approximation. The coherent addition approximation takes into account the relative phase shifts between scatterers in a realization of trees. The tree-independent scattering approximation considers every tree as an independent scatterer. Under the tree-independent approximation, the scattering cross section of a forest is the sum of scattering cross sections from each tree. On the other hand, under the independent scattering approximation, the scattering cross section of a forest is the sum of the scattering cross sections from each branch. The numerical results of backscattering coefficients for a layer of trees overlaying a flat-ground surface are illustrated as a function of frequency. The difference of the results by different approximations increases as frequency decreases. At the P-band, there can be a 20-dB difference between the result of coherent addition approximation and that of the independent scattering approximation. In Section 4, Foldy's approximation is invoked to account for the effects of scattering and absorption on coherent wave propagation within the vegetation canopy. In Section 5, we calculate the scattering coefficients from a layer of tree structures overlaying a flat-ground surface by using the discrete dipole approximation method and compare it to the independent scattering approximation and to the coherent addition approximation. The discrete dipole approximation is the volume integral approach described in Chapter 2, Section 3. The advantage of this approach is that the mutual interactions between the branches are included. It is found that there can be a 2-dB absorption enhancement and up to 17-dB scattering enhancement for cross-polarization. The difference between the discrete dipole approximation and the coherent addition approximation is due to the coherent mutual interactions among the branches within a tree. Lastly, the coherent addition approximation is applied to the numerical simulation of C-band spaceborne radar response for characterizing the rice growth and comparing with the ERS-1 backscatter data [Le Toan et al. 1997].

2 Plant Modeling by Using L-Systems

A vegetation canopy generally consists of many types of vegetative elements, such as leaves, branches, stems, and stalks. The distribution of leaves and branches is not uniformly random, they actually occur in random clusters locally in a plant or tree. A major problem in assessing their influence on the propagation of electromagnetic waves is to model their structurally complex and time-varying properties. However, recent mathematical advances have provided useful means for plant modeling based on fractals [Mandelbrot, 1983] or Lindenmayer systems (L-systems) [Prusinkiewicz and Lindenmayer, 1990]. In the modeling of vegetation, L-systems have proven to be successful from a practical perspective. In the remote sensing research, plant models generated by L-systems have been applied to study the transport of light in vegetation canopies [Goel et al. 1991]. For microwave and millimeter waves, the use of L-systems is important because one can then include coherent addition and coherent wave interaction effects in a vegetation canopy. In this section we describe the L-systems method and use this technique to generate trees with different structures.

The L-systems, introduced by Lindenmayer [1968], are a mathematical formalism for describing the development and growth of simple living systems. The original emphasis is on plant topology. Further development [Prusinkiewicz and Lindenmayer, 1990] takes into account the rules of describing plant development in time, and the distinctive feature of self-similarity, as characterized by Mandelbrot [1983], that is kept in the rewriting process of L-systems. Rewriting is the central idea in L-systems for defining complex objects by successively replacing parts of a simple initial object using a set of *rewriting rules* or *productions*. In L-systems the rewriting rules are applied in parallel and can be carried out recursively in the growth process.

L-systems have been utilized for a variety of applications. The description of L-systems presented in this section follows closely the work of Prusinkiewicz and Lindenmayer [1990]. For more details, we refer the readers to the existing literature [Herman et al. 1975; Mandelbrot, 1983; Prusinkiewicz and Hanan, 1989; Prusinkiewicz and Lindenmayer, 1990].

2.1 Lindenmayer Systems

In L-systems, a tree structure is specified by an ordered triplet $G = \langle V, \omega, P, \rangle$, where V, a collection of letters or symbols, is called the *alphabet* of the system, ω, a nonempty word (a sequence of letters) from the set V, is called the *axiom* (initial string of letters or symbols), and P is a finite *set of productions.* A pair $(a, \chi) \in P$ is called a production rule and is written as $a \rightarrow \chi$. The letter a and the word χ are called the *predecessor* and the *successor* of the production, respectively. There are many types of L-systems. The DOL-systems, deterministic and context-free, are the simplest class of L-systems where for each $a \in V$ there is exactly one χ such that $a \rightarrow \chi$.

Consider an example of the DOL-systems where the alphabet consists of two letters a and b, $V = \{a, b\}$. The axiom is the word b, and simple rewriting rules are specified for each letter a and b. This L-system is defined as follows:

Axiom

$$\omega : b$$

Production Rules

$$p_1 : b \longrightarrow a$$

$$p_2 : a \longrightarrow ab$$

where the rule $p_1 : b \longrightarrow a$ means that the letter b is to be replaced by the letter a, and the rule $p_2 : a \longrightarrow ab$ means that the letter a is to be replaced by the string ab. The application of production rules p_1 and p_2, starting from the axiom b, will produce the following sequence of symbols:

$$b \longrightarrow a \longrightarrow ab \longrightarrow aba \longrightarrow abaab \longrightarrow abaababa \longrightarrow \ldots$$

of length 1, 1, 2, 3, 5, 8, ... We note that in the above sequence, starting from the third step, both a and b in the strings are simultaneously replaced according to the rules p_1 and p_2.

Although it is possible to use DOL-systems to generate a variety of interesting objects, their modeling power is still quite limited. For example, a simple isosceles right triangle cannot be traced exactly, since the ratio of its hypotenuse length to the side length is an irrational number $\sqrt{2}$. In order to solve similar problems, Lindenmayer [1974] proposed that numerical parameters be associated with L-system symbols. Parametric L-systems operate on *parametric words,* which are strings of letters with associated parameters. The parameters are real numbers. A module with letter A and parameters $t_1, t_2, \ldots, t_n$ is denoted by $A(t_1, t_2, \ldots, t_n)$. The real-valued parameters appearing in the words correspond to *formal parameters* used in the specification of L-system productions. If Σ is a set of formal parameters, then $C(\Sigma)$ denotes a *logical expression* (condition) with parameters from Σ. A *parametric OL-system* is defined as an ordered quadruplet $G = \langle V, \Sigma, \omega, P, \rangle$, where V is the *alphabet* of the system, Σ is the *set of formal parameters,*

ω is a nonempty parametric word called the *axiom*, and P is a finite *set of productions*. A production is noted as

$$a \;:\; C \;\longrightarrow\; \chi$$

where the symbols : and $\longrightarrow$ are used to separate the three components of a production: the predecessor a, the condition C, and the successor χ. For example, the following production rule

$$A(t) \;:\; t > 5 \;\longrightarrow\; B(t+1)CD(t^{0.5}, t-2)$$

which has the predecessor $A(t)$, the condition $t > 5$, the successor $B(t + 1)CD(t^{0.5}, t-2)$, and t as the parameter. With the application of the above production rule, a module $A(9)$ results in a parametric word $B(10)CD(3,7)$.

All plants generated by the same deterministic L-systems are identical. In order to achieve flexibility and variety in the generation process, the stochastic application of productions may affect both the topology and the geometry of the plant. A *stochastic OL-systems* is an ordered quadruplet $G_\pi = \langle V, \omega, P, \pi \rangle$. The alphabet V, the axiom ω and the set of production P are defined as in an OL-system. Function π: P $\longrightarrow (0,1]$, called the *probability distribution*, maps the set of productions into the set of *production probabilities*. An example of a stochastic L-system is given below.

$$\begin{aligned}
\omega &: a \\
p_1 &: a \xrightarrow{0.2} \chi \\
p_2 &: a \xrightarrow{0.3} \xi \\
p_3 &: a \xrightarrow{0.5} \eta
\end{aligned}$$

where the letter $a \in V$ is the predecessor. The production probabilities, 0.2, 0.3, and 0.5, are listed above, and the production rules are applied with specified probabilities. The sum of all probabilities is equal to 1.

2.2 Turtle Interpretation of L-Systems

In order to generate various forms of leaves, branches, whole tree, or population of plants, symbols in the strings must be given appropriate interpretations graphically, for example, using 3D turtle graphics [Abelson and diSessa, 1982]. In this manner, one can map strings of symbols into graphic objects on a computer display. Moreover, by adding variations to the rewriting rules, realistic plant structures can be generated by this technique. Also, by controlling the rewriting rules in the generation of plant structure, it is possible to simulate the growth stages of a plant. A great variety of realistic plant structures generated using L-systems can be found in Prusinkiewicz and Lindenmayer [1990].

Prusinkiewicz uses a LOGO-style turtle [Abelson and diSessa, 1982] to interpret L-systems graphically. The state of the turtle is specified by its position and orientation. A vector $\overline{r}$ represents the turtle's position in the Cartesian coordinates (x, y, z). The orientation of the turtle is represented by three orthogonal unit vectors $\widehat{H}$, $\widehat{L}$, and $\widehat{U}$, indicating the turtle's *heading*, the direction to the *left*, and the direction *up*, respectively. These vectors satisfy the relation $\widehat{H} \times \widehat{L} = \widehat{U}$. The turtle can move from its current state, represented by a quadruplet: $\overline{r}$, $\widehat{H}$, $\widehat{L}$, $\widehat{U}$, to a new state with position vector $\overline{r}'$ and orientation vectors $\widehat{H}'$, $\widehat{L}'$, and and $\widehat{U}'$. The new and old orientations of turtle are related by a 3×3 rotation matrix $\overline{\overline{R}}$

$$\left[\widehat{H}' \; \widehat{L}' \; \widehat{U}'\right] = \left[\widehat{H} \; \widehat{L} \; \widehat{U}\right] \cdot \overline{\overline{R}} \tag{13.2.1}$$

Specifically, rotations by an angle α about vectors $\widehat{U}$, $\widehat{L}$ and $\widehat{H}$ are given as

$$\overline{\overline{R}}_U(\alpha) = \begin{bmatrix} \cos\alpha & \sin\alpha & 0 \\ -\sin\alpha & \cos\alpha & 0 \\ 0 & 0 & 1 \end{bmatrix} \tag{13.2.2}$$

$$\overline{\overline{R}}_L(\alpha) = \begin{bmatrix} \cos\alpha & 0 & -\sin\alpha \\ 0 & 1 & 0 \\ \sin\alpha & 0 & \cos\alpha \end{bmatrix} \tag{13.2.3}$$

$$\overline{\overline{R}}_H(\alpha) = \begin{bmatrix} 1 & 0 & 0 \\ 0 & \cos\alpha & -\sin\alpha \\ 0 & \sin\alpha & \cos\alpha \end{bmatrix} \tag{13.2.4}$$

Given the step size d and an angle of rotation δ, the turtle can respond to commands represented by the following symbols:

- F — Move forward a step d in the direction of heading. The state of the turtle changes to $\overline{r}'$ with $\overline{r}' = \overline{r} + d\widehat{H}$. A line segment between points $\overline{r}$ and $\overline{r}'$ is drawn.
- f — Move forward a step of length d as F but without drawing a line.
- $+$ — Turn left by angle δ. The orientation of the turtle changes according to (13.2.1) with $\overline{\overline{R}} = \overline{\overline{R}}_U(\delta)$.
- $-$ — Turn right by angle δ. The orientation of the turtle changes according to (13.2.1) with $\overline{\overline{R}} = \overline{\overline{R}}_U(-\delta)$.
- & — Pitch down by angle δ, using rotation matrix $\overline{\overline{R}} = \overline{\overline{R}}_L(\delta)$.
- $\wedge$ — Pitch up by angle δ, using rotation matrix $\overline{\overline{R}} = \overline{\overline{R}}_L(-\delta)$.
- \ — Roll left by angle δ, using rotation matrix $\overline{\overline{R}} = \overline{\overline{R}}_H(\delta)$.

/	Roll right by angle δ, using rotation matrix $\overline{\overline{R}} = \overline{\overline{R}}_H(-\delta)$.
|	Turn around, using rotation matrix $\overline{\overline{R}} = \overline{\overline{R}}_U(180^\circ)$.

With parametric L-systems, numerical parameters can be associated with L-systems symbols. If one or more parameters are associated with a symbol interpreted by the turtle, the value of the first parameter controls the turtle's state. If the symbol is not followed by any parameter, default values specified outside the L-systems are used as in the non-parametric case. The basic set of symbols affected by the introduction of parameters is listed below.

$F(a)$	Move forward a step of length $a > 0$. The state of the turtle changes to $\overline{r}'$, with $\overline{r}' = \overline{r} + a\widehat{H}$. A line segment between points $\overline{r}$ and $\overline{r}'$ is drawn.
$f(a)$	Move forward a step of length a as $F(a)$ but without drawing a line.
$+(a)$	Rotate around $\widehat{U}$ by an angle of a degrees. The turtle is turned to the left if $a > 0$, and it is turned to the right if $a < 0$.
$\&(a)$	Rotate around $\widehat{L}$ by an angle of a degrees. The turtle is pitched down if $a > 0$, and it is pitched up if $a < 0$.
$/(a)$	Rotate around $\widehat{H}$ by an angle of a degrees. The turtle is rolled to the right if $a > 0$, and it is rolled to the left if $a < 0$.

A few more symbols are introduced in the following to delimit a branch.

[	Push the current state of the turtle onto a pushdown stack. The information saved on the stack contains the turtle's position and orientation, and possibly other attributes such as the color and width of lines being drawn.
]	Pop a state from the stack and make it the current state of the turtle. No line is drawn, although in general the position of the turtle changes.
$!(\omega)$	Set the line width to ω.
$	Rolls the turtle around its own axis so that vector $\widehat{L}$ pointing to the left of the turtle is brought to a horizontal position.

It should be noticed that symbols $+$, $\&$, $\wedge$, and $/$ are used both as letters of the alphabet V and as operators in logical and arithmetic expressions. Their meaning depends on the context.

2.3 Computer Simulations of Stochastic L-Systems and Input Files

With the rules described in previous sections, modeling of plants becomes possible. A variety of commercial software and shareware are currently available for performing simulations using L-systems. For example, *lsys* is a plant generation program developed from *The Algorithmic Beauty of Plants* [Prusinkiewicz and Lindenmayer, 1990]. The code *lsys* was written by Jonathan Leech in c++ language and runs on UNIX systems. It is a freeware and is available by anonymous ftp from `ftp.cs.unc.edu`. The source code can be found in `pub/users/jon/lsys.tar.gz`. It is a software package designed for manipulating production L-systems. In addition to specifying the systems and applying production rules, it interprets the results graphically, producing Postscript output. A database containing the positions and widths of all branches generated by L-systems can be obtained by this tool. This software is capable of simulating various examples presented in *The Algorithmic Beauty of Plants.* In the following examples, we will demonstrate how to model plants by using this tool.

lsys accepts a grammar describing OL-systems augmented by brackets, stochastically applied productions, context-sensitivity, and parameters. All these features may be freely combined. Legal input is broken down into several classes: comments, assertions, definitions, the initial string, and productions. Syntactic structures common to all classes of input are described first, then each class. Unless escaped with or as otherwise specified, newline terminates all classes of input. Blank lines and comments are allowed.

Example 1: Generating tree-like structure. The input file is as follows.

`#define maxgen` 10

`#define` r_1 0.9 `/* contraction ratio for the trunk */`
`#define` r_2 0.6 `/* contraction ratio for the branches */`
`#define` a_0 45 `/* branching angle from the trunk */`
`#define` a_2 45 `/* branching angle for lateral axes */`
`#define` d 137.5 `/* divergence angle */`
`#define` w_r 0.707 `/* width decrease rate */`

`START` : $A(1, 10)$

$p_1 : A(l, w) : * \longrightarrow !(w)F(l)[\&(a_0)B(l * r_2, w * w_r)]/(d)A(l * r_1, w * w_r)$
$p_2 : B(l, w) : * \longrightarrow !(w)F(l)[-(a_2)\$C(l * r_2, w * w_r)]C(l * r_1, w * w_r)$
$p_3 : C(l, w) : * \longrightarrow !(w)F(l)[+(a_2)\$B(l * r_2, w * w_r)]B(l * r_1, w * w_r)$

Figure 13.2.1 Example of tree-like structure generated using L-systems.

In the definition section, **maxgen** specifies the maximum number of growth stage, r_1 for the contraction ratio of straight daughter branch, r_2 for the contraction ratio of lateral daughter branch, w_r for the width decrease rate, a_0 for the branching angle for branches arising from the trunk, a_2 for the branching angle for branches not directly attached to the trunk, and the divergence angle d specifies the angle between two successive branch planes for branches coming off the trunk. $A(1, 10)$ is the *axiom* (initial string) which specifies a starting trunk with one unit of length and ten units of width. According to production p_1, the apex of the main A produces an extension (internode F) of the trunk giving off a straight branch, and a lateral branch (apex B) in each derivation step. The application of rule p_1 decreases the length and width of the daughter segments by the factors r_2 and w_r, respectively. The subsequent development of lateral branches are described by rules p_2 and p_3. In each step, the straight apex, either B or C, issues a lateral apex of the next order at angle a_2 or $-a_2$ with respect to the mother axis. Two productions are employed to create lateral branches alternately to the left and right of the mother branch. Figure 13.2.1 shows the plant after 10 growth stages.

In Example 1, the parameters such as branching angles, divergence angles, and so on, are all fixed. All the plants generated using this input will be identical. In order to achieve randomization, we can apply random number generators in the growth process as shown below.

Example 2: Stochastic process. The input file is as follows.

```
#define maxgen 10
#define r1 0.9        /* contraction ratio for the trunk */
#define r2 0.6        /* contraction ratio for the branches */
#define a0 45         /* branching angle from the trunk */
#define a2 45         /* branching angle for lateral axes */
#define d 137.5       /* divergence angle */
#define wr 0.707      /* width decrease rate */
```

$\texttt{START} \; : \; A(0.9 + 0.1 * rand(2), 9 + rand(2))$

$$p_1 : A(l, w) : * \longrightarrow !(w * (0.9 + 0.1 * rand(2)))F(l * (0.9 + 0.1 * rand(2)))$$
$$[\&(a_0)B(l * r_2, w * w_r)]/(d)A(l * r_1, w * w_r)$$
$$p_2 : B(l, w) : * \longrightarrow !(w * (0.9 + 0.1 * rand(2)))F(l * (0.9 + 0.1 * rand(2)))$$
$$[-(a_2)\$C(l * r_2, w * w_r)]C(l * r_1, w * w_r)$$
$$p_3 : C(l, w) : * \longrightarrow !(w * (0.9 + 0.1 * rand(2)))F(l * (0.9 + 0.1 * rand(2)))$$
$$[+(a_2)\$B(l * r_2, w * w_r)]B(l * r_1, w * w_r)$$

where $rand(n)$ returns a uniformly distributed random number in the interval $[0, n)$. In this example, we substitute $!(w)$ in the previous example by $!(w * (0.9 + 0.1 * rand(2)))$, and $F(l)$ by $F(l * (0.9 + 0.1 * rand(2)))$. So the length and width for every segment will vary in each production. In general, parameters such as branching angles, divergence angles, and so on, can be randomized in this way.

Two other input files for generating tree structures, which will be used for the calculation of radar scattering by trees in later sections, are provided below.

A. Tree-like Structure with Ternary Branching

Input file A is shown below. Here $rand(2)$ returns a uniformly distributed number on [0,2). So the expression $(0.9 + 0.1 * rand(2))$ gives a uniformly distributed number on [0.9, 1.1) with a mean value of 1. The overall structure of the tree is defined by production p_1. In each derivation step, apex A produces three new branches terminated by their own apices. The new branches have a mean length of 12 with the length uniformly distributed in $[12 * (1 - 10\%), 12 * (1 + 10\%))$. The new branches have a mean width of 1.2

Input file A

```
/* Tree-like structure with ternary branching */
#define maxgen 5
#define d1 120.00      /* average divergence angle 1 */
#define a 22.5         /* average branching angle */
#define lr 1.2         /* average length increase rate */
#define vr 1.2         /* average width increase rate */
#define le 12          /* average initial length */
#define width 100

START : !(vr * (0.9 + 0.1 * rand(2)))
            F(20 * (0.9 + 0.1 * rand(2)))/(180 + rand(180))A

p1 : A ⟶ !(vr * (0.9 + 0.1 * rand(2)))
            [&(15+rand(15))F(le*(0.9+0.1*rand(2)))A]/(100+rand(40))
            [&(15+rand(15))F(le*(0.9+0.1*rand(2)))A]/(100+rand(40))
            [&(15 +rand(15))F(le * (0.9 + 0.1 * rand(2)))A]
p2 : F(l) ⟶ F(l * lr * (0.9 + 0.1 * rand(2)))
p3 : !(w) ⟶ !(w * vr * (0.9 + 0.1 * rand(2)))
```

with a uniform width distribution between $1.2*(1-10\%)$ and $1.2*(1+10\%)$. The branching angles have a mean value of 22.5° with a uniform distribution between 15° and 30°. The divergence angles have a mean value of 120° with a uniform distribution between 100° and 140°. Production p_2 shows that the growth rate of the length of each branch from one generation to the next has a mean value of 1.2 with a uniform distribution between $1.2*(1-10\%)$ and $1.2*(1+10\%)$. Production p_3 shows that the growth rate of the width of each branch from one generation to the next has a mean value of 1.2 with a uniform distribution between $1.2*(1-10\%)$ and $1.2*(1+10\%)$. The number of growth stages is **maxgen** = 5. After 5 generations, the total number of branches, including the main branch, is 324. Figure 13.2.2 shows one of the generated trees.

B. *Tree-like Structure with a Big Stem and Binary Branching*

Input file B is shown below. The overall structure of the plant is defined by production p_1. In each derivation step, apex A produces two new branches terminated by their own apices. The new branches have a mean length of 0.4 with the length uniformly distributed in $[0.4*(1-10\%), 0.4*(1+10\%))$. The new branches have a mean width of 0.04 with a uniform width dis-

Figure 13.2.2 Tree-like scattering object generated using L-systems.

Input file B

/* *Tree-like structure with a big stem and binary branching* */
`#define maxgen 5`
`#define` d_1 160.00 /* `divergence angle` */
`#define` a_0 40 /* `branching angle` */
`#define` l_m 0.04 /* `1/100 of average stem length` */
`#define` w_m $0.01 * (9 + rand(2))$ /* `stem width` */
`#define` l_b 0.04 /* `1/10 of average branch length` */
`#define` w_b 0.004 /* `1/10 of average branch width` */
`#define` *width* 100
`START` : $!(w_m)F(l_m * (9 + rand(2)))/(180 + rand(180))A$
$p_1 : A \longrightarrow$ $[\&(a_0 + rand(10))!(w_b * (9 + rand(2)))$ $F(l_b * (9 + rand(2)))]/(d_1 + rand(40))$ $[\&(a_0 + rand(10))!(w_b * (9 + rand(2)))$ $F(l_b * (9 + rand(2)))]/(d_1 + rand(40))!(w_m)$ $F(l_m * (9 + rand(2)))A$

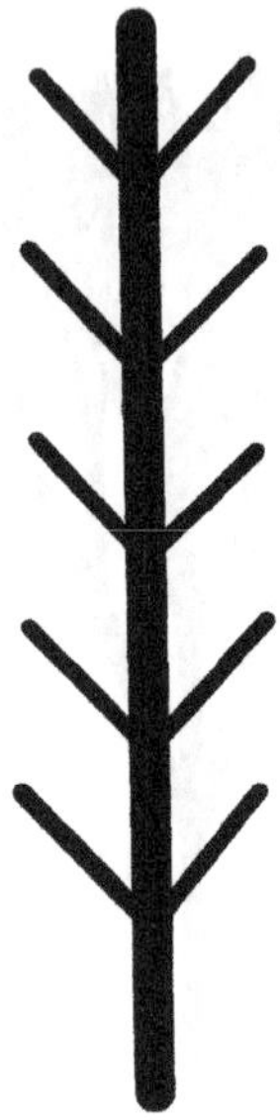

Figure 13.2.3 Configuration of the tree-like scattering object generated by L-systems.

tribution between $0.04 * (1 - 10\%)$ and $0.04 * (1 + 10\%)$. The branching angles have a mean value of 45° with a uniform distribution between 40° and 50°. Apex A also grows the main stem. The growth part of the main stem has a mean length of 0.4 with the length uniformly distributed in $[0.4 * (1 - 10\%), 0.4 * (1 + 10\%))$. The divergence angles have a mean value of 180° with a uniform distribution between 160° and 200°. The number of growth stages is `maxgen` = 5. After 5 generations, the total number of branches, including the main branch, is 11. Figure 13.2.3 shows one of the generated simple trees with binary branching structure.

3 Scattering from Trees Generated by L-Systems Based on Coherent Addition Approximation

In this section we study wave scattering from trees without leaves by using coherent addition approximation. A cylindrical model is employed to represent the tree trunks, stems, and branches. The scattering amplitude of each branch is calculated by using the infinite cylinder approximation discussed in Chapter 1, Section 6.2 of Volume I. The correlations of scattering by different branches are included by using their relative positions, which can be obtained after decoding the growth procedure.

We first consider in Section 3.1 coherent single scattering by a particle in the presence of a reflective boundary. Scattering by trees generated with L-systems is discussed in Section 3.2.

3.1 Single Scattering by a Particle in the Presence of Reflective Boundary

3.1.1 Electric Field and Dyadic Green's Function

Consider a particle located above a reflective boundary. The boundary is at $z = 0$ which separates region 0 and region 1. From Eqs. (5.2.15a–c) of Volume I, the dyadic Green's function is, for $z > z'$,

$$\begin{aligned}
&\overline{\overline{G}}(\overline{r}, \overline{r}') \\
&= \frac{i}{8\pi^2} \int d\overline{k}_\perp \frac{1}{k_z} e^{i\overline{k}_\perp \cdot (\overline{\rho} - \overline{\rho}')} e^{ik_z z} \hat{e}(k_z) \left[\hat{e}(k_z) e^{-ik_z z'} + R^{TE}(k_z) \hat{e}(-k_z) e^{ik_z z'} \right] \\
&+ \frac{i}{8\pi^2} \int d\overline{k}_\perp \frac{1}{k_z} e^{i\overline{k}_\perp \cdot (\overline{\rho} - \overline{\rho}')} e^{ik_z z} \hat{h}(k_z) \left[\hat{h}(k_z) e^{-ik_z z'} + R^{TM}(k_z) \hat{h}(-k_z) e^{ik_z z'} \right]
\end{aligned} \tag{13.3.1}$$

where, based on the notations from Chapter 5 of Volume I,

$$\overline{k} = k_x \hat{x} + k_y \hat{y} + k_z \hat{z} \tag{13.3.2a}$$

$$\overline{K} = k_x \hat{x} + k_y \hat{y} - k_z \hat{z} \tag{13.3.2b}$$

$$\overline{k}_\perp = k_x \hat{x} + k_y \hat{y} \tag{13.3.2c}$$

$$k_z = \sqrt{k^2 - k_x^2 - k_y^2} \tag{13.3.2d}$$

$$\overline{\rho} = x\hat{x} + y\hat{y} \tag{13.3.2e}$$

$$\overline{\rho}' = x'\hat{x} + y'\hat{y} \tag{13.3.2f}$$

For a far-field observation point $\overline{r}$ in the $\hat{k}_s$ direction, where

$$\hat{k}_s = \sin\theta_s \cos\phi_s \hat{x} + \sin\theta_s \sin\phi_s \hat{y} + \cos\theta_s \hat{z} \tag{13.3.3}$$

the far-field Green's function is

$$\begin{aligned}
&\overline{\overline{G}}(\overline{r}, \overline{r}') \\
&= \frac{e^{ikr}}{4\pi r} \hat{e}(k_{sz}) \left[\hat{e}(k_{sz}) e^{-ik_{sz} z'} + R^{TE}(k_{sz}) \hat{e}(-k_{sz}) e^{ik_{sz} z'} \right] e^{-i\overline{k}_{s\perp} \cdot \overline{\rho}'} \\
&+ \frac{e^{ikr}}{4\pi r} \hat{h}(k_{sz}) \left[\hat{h}(k_{sz}) e^{-ik_{sz} z'} + R^{TM}(k_{sz}) \hat{h}(-k_{sz}) e^{ik_{sz} z'} \right] e^{-i\overline{k}_{s\perp} \cdot \overline{\rho}'}
\end{aligned} \tag{13.3.4}$$

where

$$k_{sx} = k \sin\theta_s \cos\phi_s \tag{13.3.5a}$$

$$k_{sy} = k \sin\theta_s \sin\phi_s \tag{13.3.5b}$$

$$k_{sz} = k \cos\theta_s \tag{13.3.5c}$$

$$\overline{k}_{s\perp} = k_{sx}\hat{x} + k_{sy}\hat{y} \tag{13.3.5d}$$

Let the incident wave be a plane wave with the following wavevector components

$$k_{ix} = k \sin\theta_i \cos\phi_i \tag{13.3.6a}$$

$$k_{iy} = k \sin\theta_i \sin\phi_i \tag{13.3.6b}$$

$$k_{iz} = k \cos\theta_i \tag{13.3.6c}$$

$$\overline{k}_{i\perp} = k_{ix}\hat{x} + k_{iy}\hat{y} \tag{13.3.6d}$$

The incident wave is downward going in the direction $(\pi - \theta_i, \phi_i)$ while the reflected wave is upward going in the direction (θ_i, ϕ_i). Thus the total field in region 0 is

$$\begin{aligned}\overline{E} = &\left[E_{TE}\hat{e}(-k_{iz}) + E_{TM}\hat{h}(-k_{iz})\right] e^{i\overline{K}_i\cdot\overline{r}} \\ &+ \left[E_{TE}\hat{e}(k_{iz})R^{TE}(k_{iz}) + E_{TM}R^{TM}(k_{iz})\hat{h}(k_{iz})\right] e^{i\overline{k}_i\cdot\overline{r}}\end{aligned} \tag{13.3.7}$$

where E_{TE} and E_{TM} are the amplitudes of the TE and TM components of the incident wave. The TE and TM Fresnel reflection coefficients are denoted by R^{TE} and R^{TM}, respectively.

3.1.2 Scattering by a Single Particle

Consider a particle of permittivity ϵ_p and relative permittivity ϵ_r occupying the volume of V_p and is centered at $\overline{r}_p$. The scattered field is

$$\overline{E}_s = k^2 \int_{V_p} d\overline{r}'\,(\epsilon_r - 1)\,\overline{\overline{G}}\,(\overline{r}, \overline{r}') \cdot \overline{E}(\overline{r}') \tag{13.3.8}$$

Then in the far field and in the $\hat{k}_s$-direction, the scattered field becomes

$$\begin{aligned}\overline{E}_s = &\frac{e^{ikr}}{4\pi r}\hat{e}(k_{sz})k^2 \int_{V_p} d\overline{r}'\,(\epsilon_r - 1) \\ &\left[\hat{e}(k_{sz})e^{-ik_{sz}z'} + R^{TE}(k_{sz})\hat{e}(-k_{sz})e^{ik_{sz}z'}\right] e^{-i\overline{k}_{s\perp}\cdot\overline{\rho}'} \cdot \overline{E}(\overline{r}') \\ &+ \frac{e^{ikr}}{4\pi r}\hat{h}(k_{sz})k^2 \int_{V_p} d\overline{r}'\,(\epsilon_r - 1) \\ &\left[\hat{h}(k_{sz})e^{-ik_{sz}z'} + R^{TM}(k_{sz})\hat{h}(-k_{sz})e^{ik_{sz}z'}\right] e^{-i\overline{k}_{s\perp}\cdot\overline{\rho}'} \cdot \overline{E}(\overline{r}')\end{aligned} \tag{13.3.9}$$

In the Born approximation, we approximate $\overline{E}(\overline{r}')$ by the electric field of (13.3.7) given in the previous section. Thus

$$\overline{E}_s = \frac{e^{ikr}}{4\pi r}\hat{e}(k_{sz})k^2 \int_{V_p} d\overline{r}' \, (\epsilon_r - 1)$$
$$\left[\hat{e}(k_{sz})e^{-ik_{sz}z'} + R^{TE}(k_{sz})\hat{e}(-k_{sz})e^{ik_{sz}z'}\right] e^{-i\overline{k}_{s\perp}\cdot\overline{\rho}'} \cdot$$
$$\left\{\left[E_{TE}\hat{e}(-k_{iz}) + E_{TM}\hat{h}(-k_{iz})\right] e^{i\overline{K}_i\cdot\overline{r}'}\right.$$
$$\left. + \left[E_{TE}\hat{e}(k_{iz})R^{TE}(k_{iz}) + E_{TM}R^{TM}(k_{iz})\hat{h}(k_{iz})\right] e^{i\overline{k}_i\cdot\overline{r}'}\right\}$$
$$+ \frac{e^{ikr}}{4\pi r}\hat{h}(k_{sz})k^2 \int_{V_p} d\overline{r}' \, (\epsilon_r - 1)$$
$$\left[\hat{h}(k_{sz})e^{-ik_{sz}z'} + R^{TM}(k_{sz})\hat{h}(-k_{sz})e^{ik_{sz}z'}\right] e^{-i\overline{k}_{s\perp}\cdot\overline{\rho}'} \cdot$$
$$\left\{\left[E_{TE}\hat{e}(-k_{iz}) + E_{TM}\hat{h}(-k_{iz})\right] e^{i\overline{K}_i\cdot\overline{r}'}\right.$$
$$\left. + \left[E_{TE}\hat{e}(k_{iz})R^{TE}(k_{iz}) + E_{TM}R^{TM}(k_{iz})\hat{h}(k_{iz})\right] e^{i\overline{k}_i\cdot\overline{r}'}\right\} \quad (13.3.10)$$

In the above equation, there are four kinds of scattering processes that contribute to the scattered field (Fig. 13.3.1). Thus we express $\overline{E}_s$ as

$$\overline{E}_s = \overline{E}_{s1} + \overline{E}_{s2} + \overline{E}_{s3} + \overline{E}_{s4} \quad (13.3.11)$$

where $\overline{E}_{s1}$, $\overline{E}_{s2}$, $\overline{E}_{s3}$, and $\overline{E}_{s4}$ correspond to the four kinds of scattering processes. They are given by

$$\overline{E}_{s1} = \frac{e^{ikr}}{4\pi r}k^2 \int_{V_p} d\overline{r}' \, (\epsilon_r - 1) \left[\hat{e}(k_{sz})\hat{e}(k_{sz}) + \hat{h}(k_{sz})\hat{h}(k_{sz})\right]$$
$$\cdot \left[E_{TE}\hat{e}(-k_{iz}) + E_{TM}\hat{h}(-k_{iz})\right] e^{i(\overline{K}_i - \overline{k}_s)\cdot\overline{r}'} \quad (13.3.12)$$

$$\overline{E}_{s2} = \frac{e^{ikr}}{4\pi r}k^2 \int_{V_p} d\overline{r}' \, (\epsilon_r - 1) \left[\hat{e}(k_{sz})\hat{e}(k_{sz}) + \hat{h}(k_{sz})\hat{h}(k_{sz})\right]$$
$$\cdot \left[E_{TE}\hat{e}(k_{iz})R^{TE}(k_{iz}) + E_{TM}\hat{h}(k_{iz})R^{TM}(k_{iz})\right] e^{i(\overline{k}_i - \overline{k}_s)\cdot\overline{r}'} \quad (13.3.13)$$

$$\overline{E}_{s3} = \frac{e^{ikr}}{4\pi r}k^2 \int_{V_p} d\overline{r}' \, (\epsilon_r - 1)$$
$$\left[\hat{e}(k_{sz})R^{TE}(k_{sz})\hat{e}(-k_{sz}) + \hat{h}(k_{sz})R^{TM}(k_{sz})\hat{h}(-k_{sz})\right]$$
$$\cdot \left[E_{TE}\hat{e}(-k_{iz}) + E_{TM}\hat{h}(-k_{iz})\right] e^{i(\overline{K}_i - \overline{K}_s)\cdot\overline{r}'} \quad (13.3.14)$$

$$\overline{E}_{s4} = \frac{e^{ikr}}{4\pi r} k^2 \int_{V_p} d\overline{r}' \, (\epsilon_r - 1)$$
$$\left[\hat{e}(k_{sz}) R^{TE}(k_{sz}) \hat{e}(-k_{sz}) + \hat{h}(k_{sz}) R^{TM}(k_{sz}) \hat{h}(-k_{sz})\right]$$
$$\cdot \left[E_{TE}\hat{e}(k_{iz}) R^{TE}(k_{iz}) + E_{TM}\hat{h}(k_{iz}) R^{TM}(k_{iz})\right] e^{i(\overline{k}_i - \overline{K}_s)\cdot\overline{r}'} \qquad (13.3.15)$$

with the four respective phase factors governed by the following four corresponding difference wavevectors

$$\overline{k}_{d1} = \overline{K}_i - \overline{k}_s$$
$$= k(\sin\theta_i \cos\phi_i - \sin\theta_s \cos\phi_s)\hat{x}$$
$$+ k(\sin\theta_i \sin\phi_i - \sin\theta_s \sin\phi_s)\hat{y} - k\,(\cos\theta_i + \cos\theta_s)\,\hat{z} \qquad (13.3.16)$$

$$\overline{k}_{d2} = \overline{k}_i - \overline{k}_s$$
$$= k(\sin\theta_i \cos\phi_i - \sin\theta_s \cos\phi_s)\hat{x}$$
$$+ k(\sin\theta_i \sin\phi_i - \sin\theta_s \sin\phi_s)\hat{y} + k\,(\cos\theta_i - \cos\theta_s)\,\hat{z} \qquad (13.3.17)$$

$$\overline{k}_{d3} = \overline{K}_i - \overline{K}_s$$
$$= k(\sin\theta_i \cos\phi_i - \sin\theta_s \cos\phi_s)\hat{x}$$
$$+ k(\sin\theta_i \sin\phi_i - \sin\theta_s \sin\phi_s)\hat{y} + k\,(-\cos\theta_i + \cos\theta_s)\,\hat{z} \qquad (13.3.18)$$

$$\overline{k}_{d4} = \overline{k}_i - \overline{K}_s$$
$$= k(\sin\theta_i \cos\phi_i - \sin\theta_s \cos\phi_s)\hat{x}$$
$$+ k(\sin\theta_i \sin\phi_i - \sin\theta_s \sin\phi_s)\hat{y} + k\,(\cos\theta_i + \cos\theta_s)\,\hat{z} \qquad (13.3.19)$$

In the backscattering direction, $\theta_s = \theta_i, \phi_s = \pi + \phi_i$

$$\overline{k}_{d1} = 2k\sin\theta_i\cos\phi_i\hat{x} + 2k\sin\theta_i\sin\phi_i\hat{y} - 2k\cos\theta_i\hat{z} \qquad (13.3.20)$$
$$\overline{k}_{d2} = \overline{k}_{d3} = 2k\sin\theta_i\cos\phi_i\hat{x} + 2k\sin\theta_i\sin\phi_i\hat{y} \qquad (13.3.21)$$
$$\overline{k}_{d4} = 2k\sin\theta_i\cos\phi_i\hat{x} + 2k\sin\theta_i\sin\phi_i\hat{y} + 2k\cos\theta_i\hat{z} \qquad (13.3.22)$$

The expression of Born aproximation can be readily generalized to the case of a discrete scatterer. Let the discrete scatterer be centered at the position $\overline{r}_p$, and let $f_{\beta\alpha}(\theta_s, \phi_s; \theta_i, \phi_i)$ be the bistatic scattering amplitude from incident polarization α scattered into polarization β as defined in Chapter 1, Section 1 of Volume I. Then the bistaic scattering amplitude of the downward going incident wave scattered by the particle in the presence of the reflective boundary will include the four scattering processes as depicted in Fig. 13.3.1. Fig. 13.3.1a represents the scattering from the incident direction by a scatterer into the scattered direction. Fig. 13.3.1b represents the scattering of the reflected wave by a scatterer into the scattered direction. Fig. 13.3.1c represents the scattering from the incident direction by

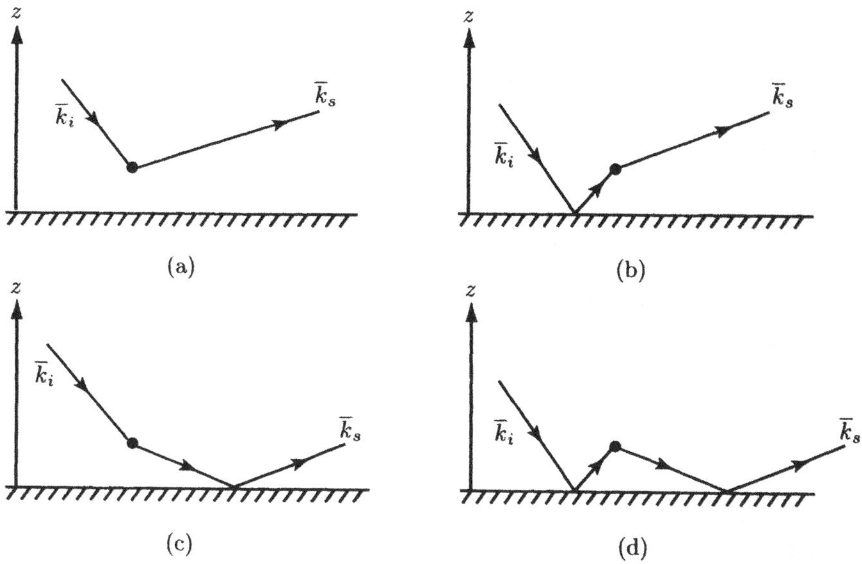

Figure 13.3.1 Four scattering mechanisms for first-order scattering in the presence of a reflective boundary.

a scatterer, and the wave is then reflected by the boundary before going into the scattered direction. Fig. 13.3.1d represents two reflections with one volume scattering in-between. We denote the total scattering amplitude by $\mathcal{F}_{\beta\alpha}(\theta_s, \phi_s; \pi - \theta_i, \phi_i)$ with

$$\begin{aligned}\mathcal{F}_{\beta\alpha}(\theta_s, \phi_s; \pi - \theta_i, \phi_i) &= f_{\beta\alpha}(\theta_s, \phi_s; \pi - \theta_i, \phi_i)e^{i\overline{k}_{d1}\cdot\overline{r}_p} \\ &+ f_{\beta\alpha}(\theta_s, \phi_s; \theta_i, \phi_i)R_\alpha(\theta_i)\ e^{i\overline{k}_{d2}\cdot\overline{r}_p} \\ &+ R_\beta(\theta_s)f_{\beta\alpha}(\pi - \theta_s, \phi_s; \pi - \theta_i, \phi_i)\ e^{i\overline{k}_{d3}\cdot\overline{r}_p} \\ &+ R_\beta(\theta_s)f_{\beta\alpha}(\pi - \theta_s, \phi_s; \theta_i, \phi_i)R_\alpha(\theta_i)\ e^{i\overline{k}_{d4}\cdot\overline{r}_p} \qquad (13.3.23)\end{aligned}$$

3.2 Scattering by Trees

Consider N_t trees generated by the L-systems in each realization. Each tree has N_b branches. The trunk is considered as one of the branches. Let $p = 1, 2, \ldots, N_t$ be the tree index and let $i_p = 1, 2, \ldots, N_b$ be the branch index of the pth tree. If the i_pth branch is centered at the origin, then the scattering amplitude of the i_pth branch is $f_{\beta\alpha}^{(i_p)}(\hat{k}_s, \hat{k}_i)$, where $\hat{k}_i$ and $\hat{k}_s$ are respectively the incident and scattered directions, and α and β are the incident and scattered polarizations, respectively. If we let the i_pth branch be centered at

$\overline{r}_{i_p}$, then the scattering amplitude is $f_{\beta\alpha}^{(i_p)}(\hat{k}_s, \hat{k}_i)\exp(i\overline{k}_d \cdot \overline{r}_{i_p})$, where $\overline{k}_d = k\hat{k}_i - k\hat{k}_s$. The coherent addition approximation assumes the total scattering amplitude of the N_t trees, denoted by $F_{\beta\alpha}(\hat{k}_s, \hat{k}_i)$, as the sum of the scattering amplitude of all individual branches.

$$F_{\beta\alpha}(\hat{k}_s, \hat{k}_i) = \sum_{p=1}^{N_t}\sum_{i_p=1}^{N_b} f_{\beta\alpha}^{(i_p)}(\hat{k}_s, \hat{k}_i)\exp(i\overline{k}_d \cdot \overline{r}_{i_p}) \tag{13.3.24}$$

Thus the coherent addition approximation, unlike the intensity approach, takes into account the phase shift due to the position of the scatterer. The average intensity of the scattered wave is obtained by taking the configurational average of the squared absolute value of (13.3.24):

$$\begin{aligned}&\langle|F_{\beta\alpha}(\hat{k}_s, \hat{k}_i)|^2\rangle \\ &= \sum_{q=1}^{N_t}\sum_{p=1}^{N_t}\sum_{j_q=1}^{N_b}\sum_{i_p=1}^{N_b}\langle f_{\beta\alpha}^{(j_q)*}(\hat{k}_s, \hat{k}_i) f_{\beta\alpha}^{(i_p)}(\hat{k}_s, \hat{k}_i)\exp(i\overline{k}_d \cdot (\overline{r}_{i_p} - \overline{r}_{j_q}))\rangle\end{aligned} \tag{13.3.25}$$

where angular brackets denote ensemble averaging. The analytic calculation of the average on the right-hand side of (13.3.25) depends on the pair distribution of branches that is difficult to obtain. Instead, we calculate wave scattering from the trees generated in Section 2 based on using L-systems and using Monte Carlo simulations. The results are averaged over many realizations.

From (13.3.25) the normalized backscattering coefficient is

$$\sigma_{\beta\alpha}^{(W)}(\hat{k}_s, \hat{k}_i) = \frac{4\pi}{A}\left\langle\left|\sum_{p=1}^{N_t}\sum_{i_p=1}^{N_b} f_{\beta\alpha}^{(i_p)}(\hat{k}_s, \hat{k}_i)\exp(i\overline{k}_d \cdot \overline{r}_{i_p})\right|^2\right\rangle \tag{13.3.26}$$

with A being the area of the pixel where the N_t trees are located.

Tree-independent scattering approximation assumes that the total scattering intensity can be obtained by summing the scattering intensity from each tree. This assumes that each tree scatters independently. For each tree the scattering field is obtained by adding the scattering fields from the branches in the same tree coherently.

$$\sigma_{\beta\alpha}^{(T,Ind)}(\hat{k}_s, \hat{k}_i) = \frac{4\pi}{A}\sum_{p=1}^{N_t}\left\langle\left|\sum_{i_p=1}^{N_b} f_{\beta\alpha}^{(i_p)}(\hat{k}_s, \hat{k}_i)\exp(i\overline{k}_d \cdot \overline{r}_{i_p})\right|^2\right\rangle \tag{13.3.27}$$

Tree-independent scattering should be a good approximation when the distances between trees are large compared with wavelength. In this case the phase factor $e^{i\overline{k}_d \cdot \overline{r}_{i_p}}$ from different trees in (13.3.27) fluctuates rapidly, and that makes the contribution of the cross terms from different trees very small on averaging.

Independent scattering approximation is a classical approximation method. It is used in conventional radiative transfer theory and states that the sum of scattered intensities from all branches add without considering any phase relation between scattering by different branches.

$$\sigma_{\beta\alpha}^{(Ind)}(\hat{k}_s, \hat{k}_i) = \frac{4\pi}{A} \sum_{p=1}^{N_t} \sum_{i_p=1}^{N_b} \langle |f_{\beta\alpha}^{(i_p)}(\hat{k}_s, \hat{k}_i)|^2 \rangle \tag{13.3.28}$$

The scattering cross section of a forest is equal to the sum of the scattering cross sections of the individual branches. Independent scattering can be valid only in the case where the separations between branches are comparable to or larger than a wavelength so that the branches can be treated as independent scatterers. As the frequency increases, the condition can be satisfied for some vegetation. We can also identify independent scattering as high-frequency approximation. The input file to the L-systems is the input file A given in Section 2.3. We also assume that the unit used in the tree-generating process is centimeter.

Three hundred trees are generated. The maximum height of these trees is 164.185 cm. The shadow of each tree can be put in a circle of minimum diameter 159.654 cm. We define a shadow cylinder for each tree as one which has a height equal to the maximum height of the tree, along with minimum diameter to cover the shadow of the tree. We define the local fractional volume as the total volume of all the branches of one tree divided by the volume of its shadow cylinder. The averaged local fractional volume f_{vl} for these 300 trees is 0.34%. In other words, the fractional volume occupied by the branches within the shadow cylinder is 0.34%.

In each realization of our calculations, ten of these trees are put into a pixel size of 756.3 cm × 756.3 cm. The positions of the trees are random but the shadow cylinders of the trees won't overlap with each other. The fractional area f_a, defined as the sum of the shadow areas of all these ten shadow cylinders divided by the area of the pixel, is 0.354. Thus the fractional volume occupied by the branches in the forest of trees is $f = f_a f_{vl} = 0.12\%$.

Each pixel is assumed to have a reflective boundary of permittivity $(16 + i4)\epsilon_o$. As shown in (13.3.23), the total scattering amplitude for each pixel include four terms. However, we only include three terms in the calculations here. These are the direct return, single reflection of the downgoing incident wave, and single reflection of the downgoing scattered wave from the scatterer. The backscattering coefficient of (13.3.26) then becomes

$$\sigma_{\beta\alpha}^{(W)}(\pi-\theta_i,\pi+\phi_i;\theta_i,\phi_i)$$
$$=\frac{4\pi}{A}\left\langle\left|\sum_{p=1}^{N_t}\sum_{i_p=1}^{N_b}\left[f_{\beta\alpha}^{(i_p)}(\theta_i,\pi+\phi_i;\pi-\theta_i,\phi_i)\exp(i\overline{k}_{d1}\cdot\overline{r}_{i_p})\right.\right.\right.$$
$$+f_{\beta\alpha}^{(i_p)}(\theta_i,\pi+\phi_i;\theta_i,\phi_i)\,R_\alpha(\theta_i)\,\exp(i\overline{k}_{d2}\cdot\overline{r}_{i_p})$$
$$\left.\left.\left.+R_\beta(\theta_i)\,f_{\beta\alpha}^{(i_p)}(\pi-\theta_i,\pi+\phi_i;\pi-\theta_i,\phi_i)\,\exp(i\overline{k}_{d3}\cdot\overline{r}_{i_p})\right]\right|^2\right\rangle \tag{13.3.29}$$

Here R_α is the reflection coefficient for the incident polarization, and R_β is the reflection coefficient for the scattered polarization. The expressions for k_{d1}, k_{d2}, k_{d3} for backscattering are given in (13.3.20) and (13.3.21).

Similarly, Eq. (13.3.27) becomes

$$\sigma_{\beta\alpha}^{(T,Ind)}(\pi-\theta_i,\pi+\phi_i;\theta_i,\phi_i)$$
$$=\frac{4\pi}{A}\sum_{p=1}^{N_t}\left\langle\left|\sum_{i_p=1}^{N_b}\left[f_{\beta\alpha}^{(i_p)}(\theta_i,\pi+\phi_i;\pi-\theta_i,\phi_i)\,\exp(i\overline{k}_{d1}\cdot\overline{r}_{i_p})\right.\right.\right.$$
$$+f_{\beta\alpha}^{(i_p)}(\theta_i,\pi+\phi_i;\theta_i,\phi_i)R_\alpha(\theta_i)\,\exp(i\overline{k}_{d2}\cdot\overline{r}_{i_p})$$
$$\left.\left.\left.+R_\beta(\theta_i)f_{\beta\alpha}^{(i_p)}(\pi-\theta_i,\pi+\phi_i;\pi-\theta_i,\phi_i)\,\exp(i\overline{k}_{d3}\cdot\overline{r}_{i_p})\right]\right|^2\right\rangle \tag{13.3.30}$$

Equation (13.3.28) becomes

$$\sigma_{\beta\alpha}^{(Ind)}(\pi-\theta_i,\pi+\phi_i;\theta_i,\phi_i)$$
$$=\frac{4\pi}{A}\sum_{p=1}^{N_t}\sum_{i_p=1}^{N_b}\left\langle|f_{\beta\alpha}^{(i_p)}(\theta_i,\pi+\phi_i;\pi-\theta_i,\phi_i)|^2\right.$$
$$+|f_{\beta\alpha}^{(i_p)}(\theta_i,\pi+\phi_i;\theta_i,\phi_i)\,R_\alpha(\theta_i)|^2$$
$$\left.+|R_\beta(\theta_i)\,f_{\beta\alpha}^{(i_p)}(\pi-\theta_i,\pi+\phi_i;\pi-\theta_i,\phi_i)|^2\right\rangle \tag{13.3.31}$$

We use a relative permittivity of $(11+i4)\epsilon_o$ for the branches in the following numerical simulations.

First we use a frequency of 5.3 GHz (C-band). Figures 13.3.2, 13.3.3, and 13.3.4 show the backscattering coefficients σ_{vv}, σ_{vh}, and σ_{hh} calculated by using the coherent addition approximation and compared with those of tree-independent approximation and independent scattering approximation. Both co-polarization and cross-polarization results of tree-independent

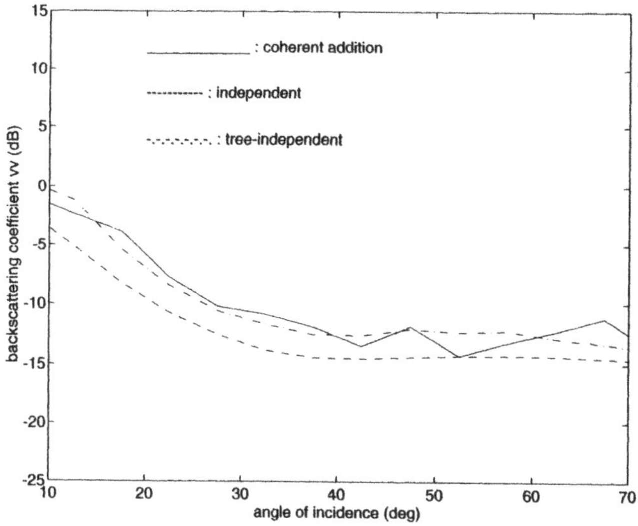

Figure 13.3.2 Comparison of backscattering coefficient σ_{vv} by coherent addition approximation, tree-independent scattering approximation, and independent scattering approximation. The frequency is 5.3 GHz. The fractional volume is $f = f_{vl} f_a = 0.12\%$. $\epsilon_s = (11+i4)\epsilon_o$. The scattering layer has a thickness of 29.32λ and the underlying half-space is flat and has a permittivity $\epsilon_{soil} = (16+i4)\epsilon_o$. The number of branches of each tree is 324.

approximation are close to those of coherent addition. This is because at this frequency the minimum distance between trees is much larger than the wavelength. We can see an approximately 3-dB difference between the co-polarization results of independent scattering approximation and coherent addition approximation in this case. To explain this, we have to look at the reflection of the three scattering mechanisms at the boundary. For coherent addition approximation, the total scattering amplitude for one scattering object is $f_{1\beta\alpha} + f_{2\beta\alpha} \cdot R_\alpha + f_{3\beta\alpha} \cdot R_\beta$ with 1, 2, and 3 indicating the first, the second, and the third scattering mechanism respectively. The incident and scattered polarizations are α and β, R_α is the reflection coefficient for incident polarization, and R_β is the reflection coefficient for scattered polarization. Note that in the second scattering mechanism, the wave reflects at the boundary first before being scattered from the object. So the reflection coefficient is R_α. In the third scattering mechanism, the wave scatters at the object first, then reflects at the boundary. Thus, the reflection coefficient is R_β. For independent scattering approximation, the scattering intensity is $|f_{1\beta\alpha}|^2 + |f_{2\beta\alpha}|^2 \cdot |R_\alpha|^2 + |f_{3\beta\alpha}|^2 \cdot |R_\beta|^2$. Because of reciprocity, the relation between $f_{2\beta\alpha}$ and $f_{3\beta\alpha}$ is as follows: $f_{2vv} = f_{3vv}$, $f_{2vh} = -f_{3hv}$, $f_{2hv} = -f_{3vh}$, $f_{2hh} = f_{3hh}$. If the first term is much smaller than the second and the

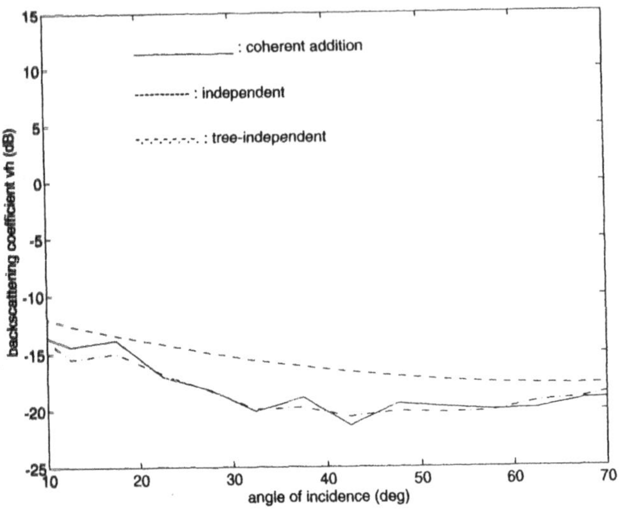

Figure 13.3.3 Comparison of backscattering coefficient σ_{vh} by coherent addition approximation, tree-independent scattering approximation, and independent scattering approximation. The frequency is 5.3 GHz. The fractional volume is $f = f_{vl} f_a = 0.12\%$. $\epsilon_s = (11+i4)\epsilon_o$. The scattering layer has a thickness of 29.32λ, and the underlying half-space is flat and has a permittivity $\epsilon_{soil} = (16+i4)\epsilon_o$. The number of branches of each tree is 324.

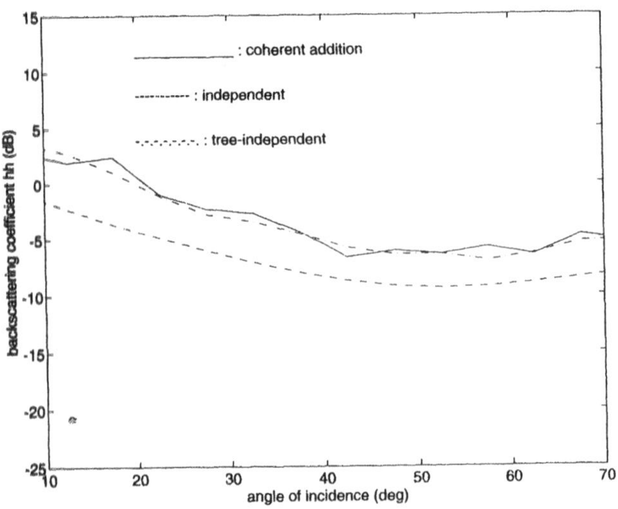

Figure 13.3.4 Comparison of backscattering coefficient σ_{hh} by coherent addition approximation, tree-independent scattering approximation, and independent scattering approximation. The frequency is 5.3 GHz. The fractional volume is $f = f_{vl} f_a = 0.12\%$. $\epsilon_s = (11+i4)\epsilon_o$. The scattering layer has a thickness of 29.32λ, and the underlying half-space is flat and has a permittivity $\epsilon_{soil} = (16+i4)\epsilon_o$. The number of branches of each tree is 324.

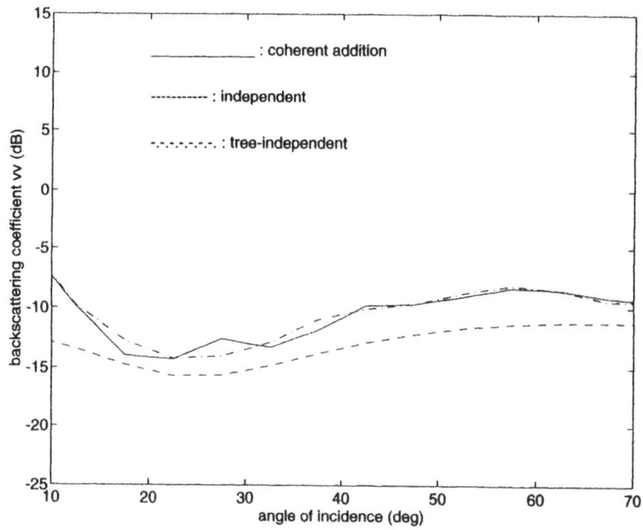

Figure 13.3.5 Comparison of backscattering coefficient σ_{vv} by coherent addition approximation, tree-independent scattering approximation, and independent scattering approximation. The frequency is 1.5 GHz. The fractional volume is $f = f_{vl} f_a = 0.12\%$. $\epsilon_s = (11+i4)\epsilon_o$. The scattering layer has a thickness of 8.21λ, and the underlying half-space is flat and has a permittivity $\epsilon_{soil} = (16+i4)\epsilon_o$. The number of branches of each tree is 324.

third, which is true when the branch is close to vertical, the co-polarization scattering amplitude is $2f_{2\alpha\alpha} \cdot R_\alpha$ by coherent addition approximation, giving an intensity of $4|f_{2\alpha\alpha}|^2|R_\alpha|^2$. The co-polarization scattering intensity is $2|f_{2\alpha\alpha}|^2 \cdot |R_\alpha|^2$ for independent scattering approximation. We see a difference of a factor of 2 between these two approaches for scattering from one single object. Therefore, even if there are no clustering effects at high frequencies, a 3-dB difference between coherent addition approximation and independent scattering approximation can still exist for co-polarization returns. But such claims may not be true for cross-polarization returns because $f_{2vh} \cdot R_h$ and $f_{3vh} \cdot R_v$, or $f_{2hv} \cdot R_v$ and $f_{3hv} \cdot R_h$, are usually not equal.

Next we apply an L-band frequency at 1.5GHz. Figures 13.3.5, 13.3.6, and 13.3.7 show the backscattering coefficients of σ_{vv}, σ_{vh}, and σ_{hh} respectively. The results for coherent addition approximation and tree-independent scattering approximation are still very close. The differences of the co-polarizations between coherent addition approximation and independent scattering approximation are larger than those at C-band, especially for h-polarization. We start to see some clustering effects since the distance between neighboring branches can be smaller than one wavelength and the positions are correlated.

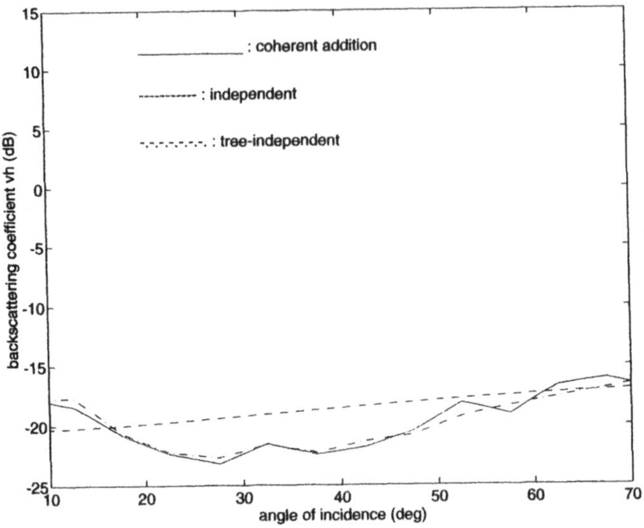

Figure 13.3.6 Comparison of backscattering coefficient σ_{vh} by coherent addition approximation, tree-independent scattering approximation, and independent scattering approximation. The frequency is 1.5 GHz. The fractional volume is $f = f_{vl} f_a = 0.12\%$. $\epsilon_s = (11+i4)\epsilon_o$. The scattering layer has a thickness of 8.21λ, and the underlying half-space is flat and has a permittivity $\epsilon_{soil} = (16 + i4)\epsilon_o$. The number of branches of each tree is 324.

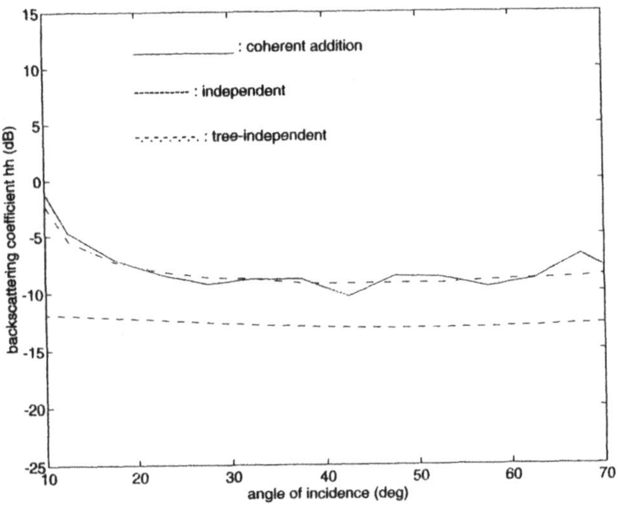

Figure 13.3.7 Comparison of backscattering coefficient σ_{vv} by coherent addition approximation, tree-independent scattering approximation, and independent scattering approximation. The frequency is 1.5 GHz. The fractional volume is $f = f_{vl} f_a = 0.12\%$. $\epsilon_s = (11+i4)\epsilon_o$. The scattering layer has a thickness of 8.21λ, and the underlying half-space is flat and has a permittivity $\epsilon_{soil} = (16 + i4)\epsilon_o$. The number of branches of each tree is 324.

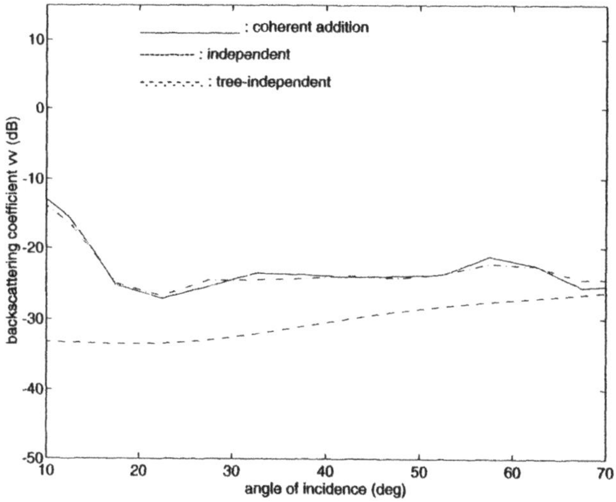

Figure 13.3.8 Comparison of backscattering coefficient σ_{vv} by coherent addition approximation, tree-independent scattering approximation, and independent scattering approximation. The frequency is 0.45 GHz. The fractional volume is $f = f_{vl} f_a = 0.12\%$. $\epsilon_s = (11+i4)\epsilon_o$. The scattering layer has a thickness of 2.46λ, and the underlying half-space is flat and has a permittivity $\epsilon_{soil} = (16 + i4)\epsilon_o$. The number of branches of each tree is 324.

Finally, we apply a frequency of 0.45GHz (P-band). Figures 13.3.8, 13.3.9, and 13.3.10 show the backscattering coefficients σ_{vv}, σ_{vh}, and σ_{hh} respectively. The results of coherent addition approximation still agrees well with those of tree-independent scattering approximation, but differ significantly from those of independent scattering approximation. There can be a 15-dB difference between the results of coherent addition approximation and that of independent scattering approximation. This is because the distances between trees are still large compared with wavelength, and the trees can still be treated as independent scatterers. But branches in the same tree can be very close to each other in terms of wavelength and will exhibit collective scattering effects.

4 Coherent Addition Approximation with Attenuation

For sparse media like vegetation, Foldy's approximation can be employed to account for the effects of absorption and scattering on the coherent wave caused by the inhomogeneities of the random medium. Rigorous derivation of the Foldy's approximation is treated in Volume III. As discussed in Chapter 7, Section 2.2 of Volume I, the propagation of coherent wave is governed by

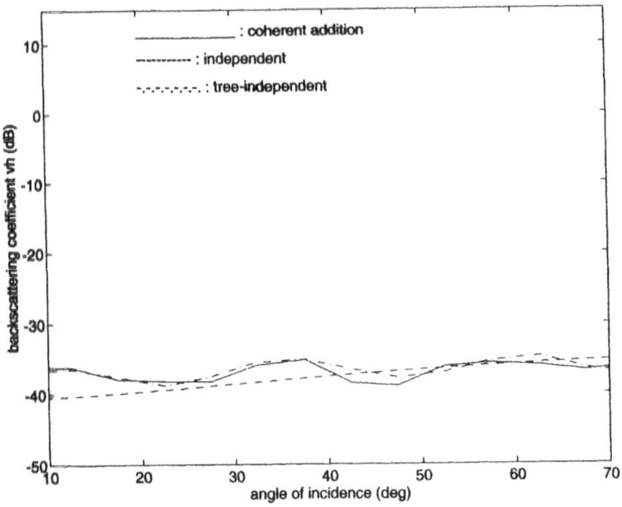

Figure 13.3.9 Comparison of backscattering coefficient σ_{vh} by coherent addition approximation, tree-independent scattering approximation, and independent scattering approximation. The frequency is 0.45 GHz. The fractional volume is $f = f_{vl} f_a = 0.12\%$. $\epsilon_s = (11+i4)\epsilon_o$. The scattering layer has a thickness of 2.46λ, and the underlying half-space is flat and has a permittivity $\epsilon_{soil} = (16+i4)\epsilon_o$. The number of branches of each tree is 324.

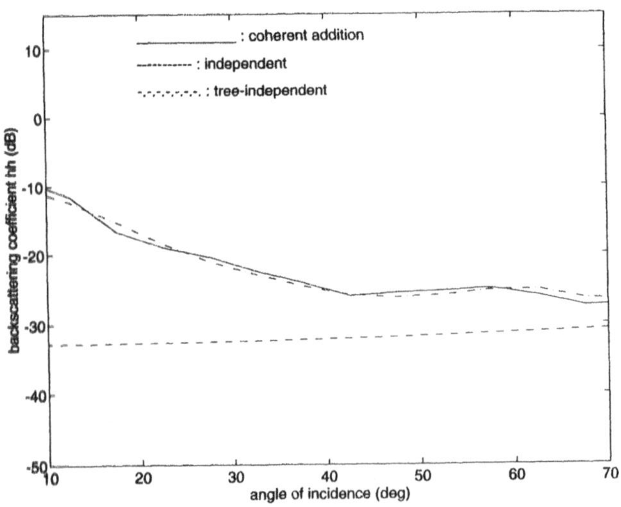

Figure 13.3.10 Comparison of backscattering coefficient σ_{vv} by coherent addition approximation, tree-independent scattering approximation, and independent scattering approximation. The frequency is 0.45 GHz. The fractional volume is $f = f_{vl} f_a = 0.12\%$. $\epsilon_s = (11+i4)\epsilon_o$. The scattering layer has a thickness of 2.46λ, and the underlying half-space is flat and has a permittivity $\epsilon_{soil} = (16+i4)\epsilon_o$. The number of branches of each tree is 324.

the following equation under Foldy's approximation

$$\frac{dE_h}{ds} = (ik_o + M_{hh})E_h + M_{hv}E_v \tag{13.4.1}$$

$$\frac{dE_v}{ds} = M_{vh}E_h + (ik_o + M_{vv})E_v \tag{13.4.2}$$

where E_h and E_v are the horizontal and vertical components of the electric field, s is the distance along the propagation direction (θ, ϕ), and

$$M_{jl} = \frac{i2\pi n_o}{k_o}\langle f_{jl}(\theta, \phi; \theta, \phi)\rangle \tag{13.4.3}$$

with $j, l = v, h$. Here n_o is the number density of scatterers, k_o is the wavenumber of the background medium, f_{jl} is the scattering amplitude matrix element, and the angular brackets denote the configuration average.

If the canopy structure exhibits statistically azimuthal symmetry, there will be no coupling between horizontal and vertical components of the coherent field. Thus

$$M_{hv} = M_{vh} = 0 \tag{13.4.4}$$

The effective propagation constants of horizontally and vertically polarized coherent waves are given respectively by

$$k_h = k_o - iM_{hh} \tag{13.4.5}$$

$$k_v = k_o - iM_{vv} \tag{13.4.6}$$

where k_o is the wavenumber of free space. The attenuation of coherent wave is accounted for by the real parts of M_{hh} and M_{vv}. These two equations of coherent wave propagation can be incorporated to take into account the attenuation and phase shift before and after the wave is scattered by a particular vegetative element.

5 Scattering from Plants Generated by L-Systems Based on Discrete Dipole Approximation

In this section we use the discrete dipole approximation (DDA) method to calculate the scattering from trees generated by stochastic L-systems. The discrete dipole approximation is a volume integral approach as described in Chapter 2. The advantage of this approach is that the mutual interactions between the branches are included. The scattering from a layer of trees overlaying ground is calculated by assuming each tree scatters independently. As shown in Section 3, this assumption has compared well with the coherent addition approximation through the C-band, the L-band, and the P-band.

5.1 Formulation of Discrete Dipole Approximation (DDA) Method

Using the DDA equations from (2.3.48) of Chapter 2, we have

$$\frac{\overline{p}_i}{\epsilon \Delta V} = \frac{\overline{\overline{\alpha}}_i}{\epsilon \Delta V} \cdot \overline{E}_i^{inc} - \frac{\overline{\overline{\alpha}}_i}{\epsilon \Delta V} \cdot \sum_{j=1, j\neq i}^{N} (\epsilon \Delta V) \overline{\overline{A}}_{ij} \cdot \frac{\overline{p}_j}{\epsilon \Delta V} \tag{13.5.1}$$

where ϵ is the permittivity of the background medium, and

$$\overline{\overline{\alpha}}_i = \alpha_{ix} \hat{x}\hat{x} + \alpha_{iy} \hat{y}\hat{y} + \alpha_{iz} \hat{z}\hat{z} = \left(\frac{\epsilon_{pi}}{\epsilon} - 1 \right) \overline{\overline{\beta}}_i \tag{13.5.2}$$

$$\frac{\beta_{ix}}{\epsilon \Delta V} = \frac{1}{1 + \left(\frac{\epsilon_{pi}}{\epsilon} - 1 \right) (L_x - D_x k^2)} \tag{13.5.3}$$

$$\frac{\beta_{iy}}{\epsilon \Delta V} = \frac{1}{1 + \left(\frac{\epsilon_{pi}}{\epsilon} - 1 \right) (L_y - D_y k^2)} \tag{13.5.4}$$

$$\frac{\beta_{iz}}{\epsilon \Delta V} = \frac{1}{1 + \left(\frac{\epsilon_{pi}}{\epsilon} - 1 \right) (L_z - D_z k^2)} \tag{13.5.5}$$

and $\Delta V = d_x d_y d_z$.

In the case of cells of circular cylindrical shape of radius a and length l, the corresponding results of L_x, L_y, L_z and D_x, D_y and D_z are

$$L_x = L_y = \frac{l}{2(4a^2 + l^2)^{1/2}} \tag{13.5.6a}$$

$$L_z = 1 - \frac{l}{(4a^2 + l^2)^{1/2}} \tag{13.5.6b}$$

$$D_x = D_y = \frac{ika^2 l}{6} + \frac{l}{8} \left\{ \sqrt{l^2 + 4a^2} - l \right\} + \frac{a^2}{4} \ln \left(\frac{l + \sqrt{l^2 + 4a^2}}{2a} \right) \tag{13.5.6c}$$

$$D_z = \frac{ika^2 l}{6} + \frac{a^2}{2} \ln \left(\frac{l + \sqrt{l^2 + 4a^2}}{2a} \right) \tag{13.5.6d}$$

and $\Delta V = \pi a^2 l$.

The equation for $\overline{\overline{A}}(\overline{r}, \overline{r}')$ is given in (2.3.8) of Chapter 2. The value of $\overline{\overline{A}}(\overline{r}, \overline{r}')$, for $\overline{r} = \overline{r}_i$ and $\overline{r}' = \overline{r}_j$, $(\overline{r}_i \neq \overline{r}_j)$ is expressed as

$$\overline{\overline{A}}_{ij} = \overline{\overline{A}}(\overline{r}_i, \overline{r}_j) \tag{13.5.7}$$

However, these may not be accurate enough when $\overline{r}_i$ and $\overline{r}_j$ are in the neighborhood of each other. Accuracy can be improved by numerical integration over the cell V_j centered at $\overline{r}_j$. Thus we can define a neighborhood distance

r_d so that

$$\overline{\overline{A}}_{ij} = \begin{cases} \dfrac{1}{\Delta V} \displaystyle\int_{V_j} d\overline{r}' \overline{\overline{A}}(\overline{r}_i, \overline{r}') & \text{for } |\overline{r}_i - \overline{r}_j| < r_d \\ \overline{\overline{A}}(\overline{r}_i, \overline{r}_j) & \text{for } |\overline{r}_i - \overline{r}_j| \geq r_d \end{cases} \tag{13.5.8}$$

The expression of $\overline{\overline{A}}_{ij}$ in (13.5.8) will still preserve the translational invariant property so that the FFT can still be taken when the matrix equation is solved by iterative method.

After the solution is obtained, we have the solution of the reduced dipole moment $\overline{p}_i$ for every cell. The electric field at every cell $\overline{E}_i$ is given by

$$\overline{E}_i = \frac{1}{\left(\frac{\epsilon_{pi}}{\epsilon} - 1\right)} \frac{\overline{p}_i}{\epsilon \Delta V_i} \tag{13.5.9}$$

Note that we only have to include those cells which have $\epsilon_{pi} \neq \epsilon$. The matrix equation (13.5.1) is of dimension $3N \times 3N$, where N is the number of small cylindrical subcells. The cylindrical subcells may have different lengths and radii in tree scattering. The factor 3 arises from the x, y, and z components of the polarization vector.

After the matrix equation (13.5.1) is solved, the far-field, scattered field in the direction $\hat{k}_s$ is

$$\overline{E}^s(\overline{r}) = \frac{e^{ikr}}{4\pi r} k^2 \left(\hat{v}_s \hat{v}_s + \hat{h}_s \hat{h}_s\right) \cdot \sum_{i=1}^{N} e^{-i\overline{k}_s \cdot \overline{r}_i} \frac{\overline{p}_i}{\epsilon \Delta V_i} \cdot \Delta V_i \tag{13.5.10}$$

Based on (13.5.10), the far-field scattering amplitude matrix can be computed readily.

The time-averaged power absorbed is equal to

$$\langle P_a \rangle = \frac{1}{2} \omega \sum_{i=1}^{N} \epsilon''_{pi} |\overline{E}_i|^2 \Delta V_i \tag{13.5.11}$$

where ϵ''_{pi} is the imaginary part of ϵ_{pi} for the ith cell. In terms of dipole moment of each cell, we have

$$\langle P_a \rangle = \frac{1}{2} \omega \sum_{i=1}^{N} \epsilon''_{pi} \Delta V_i \frac{1}{\left|\dfrac{\epsilon_{pi}}{\epsilon} - 1\right|^2} \left|\frac{\overline{p}_i}{\epsilon \Delta V_i}\right|^2 \tag{13.5.12}$$

We have verified the validity of the discrete dipole approximation method by calculating the backscattering coefficients from a vertical cylinder of length $l = 1\lambda$, radius $a = 0.05\lambda$, and permittivity $\epsilon_p = (3 + i0.5)\epsilon$ [Chen, 1994].

The method is further validated by comparing its results with those of another method of moment code for a body of revolution based on the surface integral formulation. The results agree well with each other. Energy conservation is checked by comparing the extinction coefficient obtained by the optical theorem and that by the sum of the total scattering coefficient and the absorption coefficient. The details of the validity of our numerical code can be found in [Chen, 1994].

5.2 Scattering by Simple Trees

In the following study, the trees to be used as our scattering objects are grown by using the stochastic L-systems discussed in Section 2. The input file to the L-systems is the input file B given in Section 2.3. Figure 13.2.3 shows one of the generated trees. Here, we assume that the unit used in the above process is one wavelength.

Three hundred trees are generated. The maximum height of these trees is 2.47λ. The shadow of each tree can be put in a circle of minimum diameter 0.632λ. We define a shadow cylinder for each tree as one which has a height of the maximum height of the tree and minimum diameter to cover the shadow of the tree. We define the local fractional volume as the total volume of all the branches of one tree divided by the volume of its shadow cylinder. The averaged local fractional volume for these 300 trees is $f_{vl} = 2.89\%$.

In each realization of our calculation, ten of these trees are put into a pixel of the size of $3\lambda \times 3\lambda$. The positions of the trees are random, but the shadow cylinders of the trees won't overlap each other. The fractional area f_a, defined as the sum of the shadow areas of all these ten shadow cylinders divided by the area of the pixel, is 0.346. Thus the fractional volume occupied by the branches in the forest is $f = f_{vl} f_a = 1.0\%$.

Each pixel is assumed to have a reflective boundary of permittivity $(16 + i4)\epsilon_o$. The following three scattering mechanisms are considered in the presence of the reflective boundary (Fig. 13.3.1a–c). The first term represents the scattering from the incident direction by a scatterer into the scattered direction. The second term represents the scattering of the reflected wave by a scatterer into the scattered direction. The third term represents the scattering from the incident direction by a scatterer, and the wave is then reflected by the boundary before going into the scattered direction.

The backscattering coefficients calculated using the discrete dipole approximation $\sigma_{\alpha\beta}^{(D)}$ can be expressed as follows:

$$\sigma_{\beta\alpha}^{(D)}(\pi-\theta_i,\pi+\phi_i;\theta_i,\phi_i)=\frac{4\pi}{A}\sum_{i_{tree}=1}^{N_t}\langle|f_{\beta\alpha}^{(i_{tree})}(\theta_i,\pi+\phi_i;\pi-\theta_i,\phi_i)$$
$$+f_{\beta\alpha}^{(i_{tree})}(\theta_i,\pi+\phi_i;\theta_i,\phi_i)\,R_\alpha(\theta_i)$$
$$+R_\beta(\theta_i)\,f_{\beta\alpha}^{(i_{tree})}(\pi-\theta_i,\pi+\phi_i;\pi-\theta_i,\phi_i)|^2\rangle \tag{13.5.13}$$

Here A is the area of the pixel where the N_t trees are located. The scattering amplitude is $f_{\beta\alpha}^{(i_{tree})}$ for the ith tree, R_α is the reflection coefficient for the incident polarization, and R_β is the reflection coefficient for the scattered polarization. Note that in the second scattering mechanism, the wave reflects at the boundary first, then scatters at the object. Thus, the reflection coefficient is R_α. In the third scattering mechanism, the wave scatters at the object first, then reflects at the boundary. Thus, the reflection coefficient is R_β.

In the coherent addition approximation (CAA), as given in (13.3.30), each branch is treated as a scatterer and the scattering amplitude is calculated. The total scattered field is obtained by adding the scattered fields from the branches coherently. In the independent scattering approximation, each branch is treated as an individual scatterer and the scattering amplitude is calculated. The scattered intensity for the independent scattering is assumed to be the sum of the scattered wave intensities from each scatterer. The expression of scattering coefficient is the same as that given in (13.3.31).

We use a relative permittivity of $11+i4$ for the branches in the following numerical simulations. Figures 13.5.1, 13.5.2, and 13.5.3 show the backscattering coefficients σ_{vv}, σ_{vh}, and σ_{hh}, respectively, which are calculated by discrete dipole approximation and are compared with those of coherent addition approximation and independent scattering approximation. It is observed that the coherent addition approximation gives good estimates of the copolarized backscattering coefficients (both vv and hh). The differences become larger and can be 17 dB for the case of cross-polarized backscattering coefficients. This is because the interactions between branches and the main stem give rise to cross-polarized backscattering coefficients which cannot be captured in the CAA. In the CAA, the internal fields are assumed to be the same as that of a single scatterer. The polarization current induced will be dictated by the incident field direction. When the polarization current radiates in the scattering direction, depolarization is usually small. However, in the volume integral equation with discrete dipole approximation, mutual coherent interaction, especially in the near field, can significantly change the direction of the internal field and the polarization current. This creates a much larger depolarized scattering cross section.

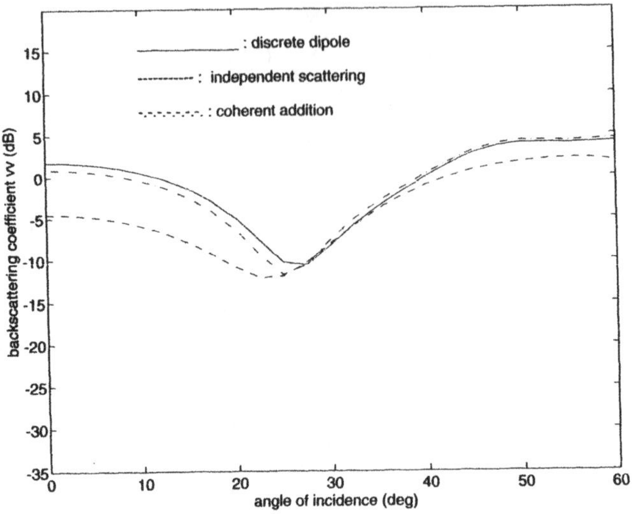

Figure 13.5.1 Comparison of backscattering coefficients σ_{vv} for a two-layer medium by discrete dipole approximation, coherent addition approximation, and independent scattering approximation. Fractional volume $f = f_{vl} f_a = 1.0\%$, $\epsilon_s = (11 + i4)\epsilon_o$. The scattering layer has a thickness of 2.47λ and the underlying half-space is flat and has a permittivity $\epsilon_{soil} = (16 + i4)\epsilon_o$. The number of branches for each simple tree is 11.

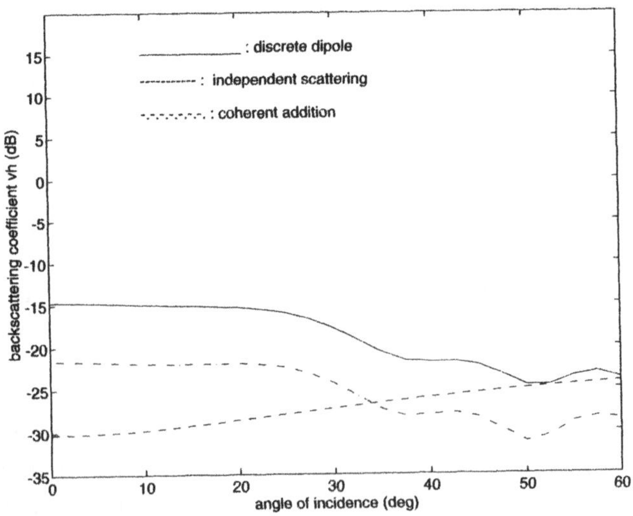

Figure 13.5.2 Comparison of backscattering coefficients σ_{vh} for a two-layer medium by discrete dipole approximation, coherent addition approximation, and independent scattering approximation. Fractional volume $f = f_{vl} f_a = 1.0\%$, $\epsilon_s = (11 + i4)\epsilon_o$. The scattering layer has a thickness of 2.47λ and the underlying half-space is flat and has a permittivity $\epsilon_{soil} = (16 + i4)\epsilon_o$. The number of branches for each simple tree is 11.

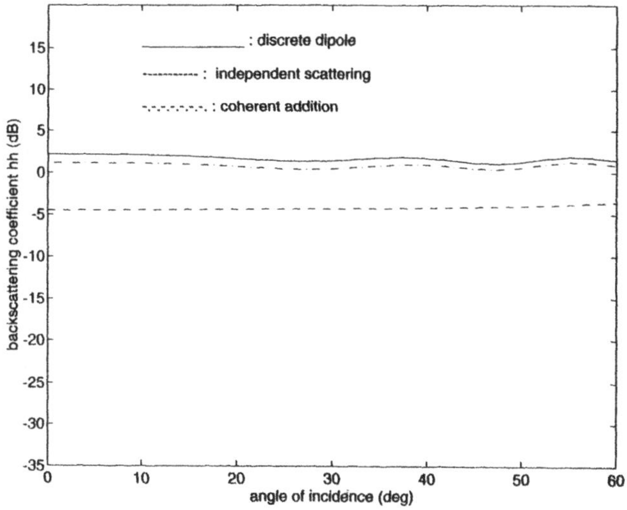

Figure 13.5.3 Comparison of backscattering coefficients σ_{hh} for a two-layer medium by discrete dipole approximation, coherent addition approximation, and independent scattering approximation. Fractional volume $f = f_{vl} f_a = 1.0\%$, $\epsilon_s = (11 + i4)\epsilon_o$. The scattering layer has a thickness of 2.47λ and the underlying half-space is flat and has a permittivity $\epsilon_{soil} = (16 + i4)\epsilon_o$. The number of branches for each simple tree is 11.

Figures 13.5.4, 13.5.5, and 13.5.6 show the bistatic scattering coefficients σ_{vv}, σ_{vh}, and σ_{hh} for an incident angle of 45°. Again, the results are in good agreement for co-polarizations (vv and hh) between discrete dipole approximation and coherent addition approximation. There are large differences for the cross-polarization (vh). It is also noted that there is a maximum at the backscattering direction for the bistatic scattering coefficient of vv but not for that of hh and vh. That is because the contribution of the central cylinder to the vv-polarized scattering wave is dominant and has a maximum at the conical direction which is reflected to the backscattering direction. However, for the hh-polarized wave, the contribution comes from the all branches which are randomly distributed along the $\hat{z}$-direction. The scattering pattern depends on the structure and does not necessarily have a maximum at the conical direction. The scattered wave of cross-polarization is due to the inclination of branches and mutual interaction and has a more complex scattering pattern.

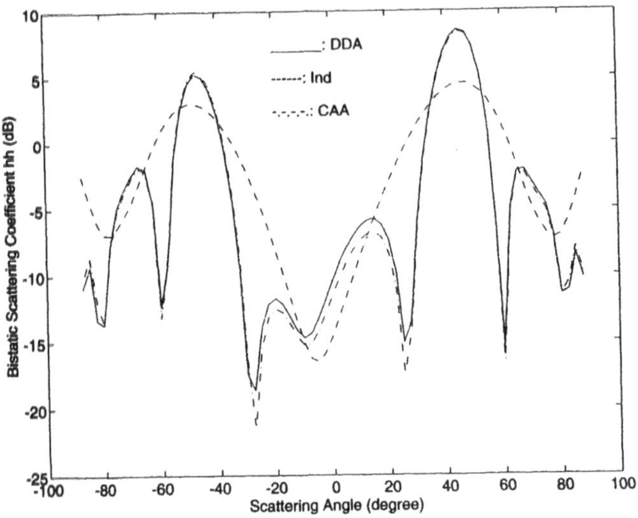

Figure 13.5.4 Comparison of bistatic scattering coefficients σ_{vv} for a two-layer medium by discrete dipole approximation, coherent addition approximation, and independent scattering approximation. Incident angle is 45°. Fractional volume $f = f_{vl} f_a = 1.0\%$, $\epsilon_s = (11 + i4)\epsilon_o$. The scattering layer has a thickness of 2.47λ and the underlying half-space is flat and has a permittivity $\epsilon_{soil} = (16 + i4)\epsilon_o$. The number of branches for each simple tree is 11.

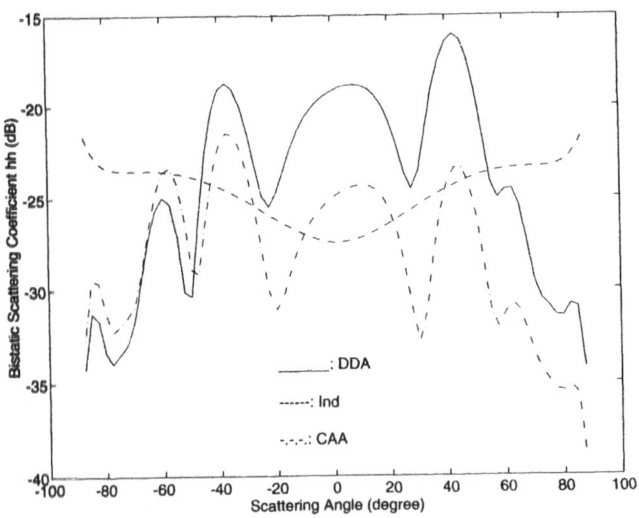

Figure 13.5.5 Comparison of bistatic scattering coefficients σ_{vh} for a two-layer medium by discrete dipole approximation, coherent addition approximation, and independent scattering approximation. Incident angle is 45°. Fractional volume $f = f_{vl} f_a = 1.0\%$, $\epsilon_s = (11 + i4)\epsilon_o$. The scattering layer has a thickness of 2.47λ and the underlying half-space is flat and has a permittivity $\epsilon_{soil} = (16 + i4)\epsilon_o$. The number of branches for each simple tree is 11.

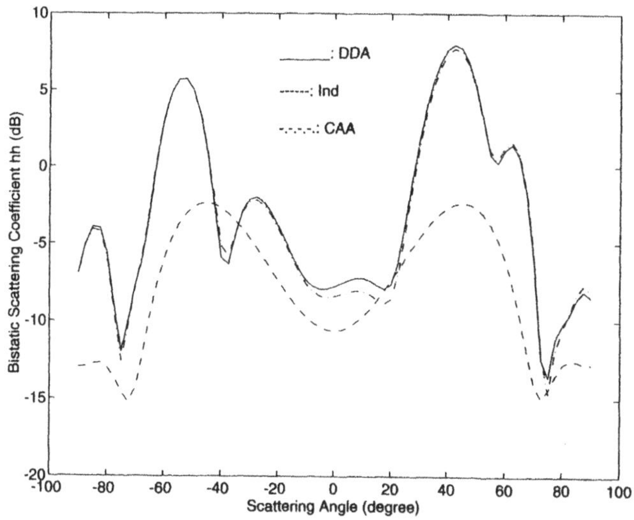

Figure 13.5.6 Comparison of bistatic scattering coefficients σ_{hh} for a two-layer medium by discrete dipole approximation, coherent addition approximation, and independent scattering approximation. Incident angle is 45°. Fractional volume $f = f_{vl} f_a = 1.0\%$, $\epsilon_s = (11 + i4)\epsilon_o$. The scattering layer has a thickness of 2.47λ and the underlying half-space is flat and has a permittivity $\epsilon_{soil} = (16 + i4)\epsilon_o$. The number of branches for each simple tree is 11.

5.3 Scattering by Honda Trees

Next we use DDA to calculate scattering by Honda trees. The Honda trees are generated based on L-systems. Figure 13.5.7 shows the growth process of a Honda tree up to the 5th generation. First, we have a cylinder. Then two shrunk cylinders are generated from the top. One is in the same direction as the mother branch. Another one has a branch angle of 45°. This process is repeated to each end branch, and the next generation of the tree is obtained.

Stochastic process can be introduced in the L-systems. For a Honda tree, the structure is controlled by the parameters such as contraction rate, branch angle, divergence angle, and width decrease rate. We use random number generators to randomize the parameters and achieve the randomization of the tree. Four Honda trees of the 6th generation are shown in Fig. 13.5.8. Each tree has 63 branches. The trees are different from each other while maintaining the same characteristics.

The scattered field in the direction $\hat{k}_s$ can be written as

$$\overline{E}^s(\overline{r}) = \frac{e^{ikr}}{4\pi r} \overline{\overline{F}}(\hat{k}_s, \hat{k}_i) \cdot \overline{E}_o^{inc} \tag{13.5.14}$$

where $\overline{E}_o^{inc}$ is the incident field in the local coordinate systems and $\overline{\overline{F}}$ is the

Figure 13.5.7 Growth process of a Honda tree.

Figure 13.5.8 Different trees but the same characteristics, each tree with 63 branches.

scattering amplitude matrix for the tree given by

$$\overline{\overline{F}}(\hat{k}_s, \hat{k}_i) = \frac{k_o{}^2}{\epsilon_o} \sum_{i=1}^{N} e^{-i\overline{k}_s \cdot \overline{r}_i} \begin{bmatrix} \hat{v}_s \cdot \overline{p}_{iv} & \hat{v}_s \cdot \overline{p}_{ih} \\ \hat{h}_s \cdot \overline{p}_{iv} & \hat{h}_s \cdot \overline{p}_{ih} \end{bmatrix} \tag{13.5.15}$$

where $\overline{p}_{iv}$ and $\overline{p}_{ih}$ are the dipole moments induced by a unit vertically polarized incident wave and by a unit horizontally polarized incident wave, respectively. N is the total number of discretized cylindrical cells of the tree.

The effects of ground surface are taken into account by introducing boundary reflections. The four scattering mechanisms are as follows: (a) the scattering from the incident direction by a scatterer into the scattered direction, (b) the reflected wave that is subsequently scattered by a scatterer into the scattered direction, (c) the scattered wave of a scatterer, which is subsequently reflected by the boundary into the scattered direction, and (d) the scattering of the reflected wave by a scatterer that is followed by a second boundary reflection (Fig. 13.3.1).

The bistatic scattering cross sections $\sigma_{\alpha\beta}$ of a single tree are calculated by three approaches:

Coherent Interaction Model (CIM): Each tree that consists of many branches is treated as a single scatterer. Maxwell's equations are solved using DDA for the entire tree.

$$\begin{aligned}\sigma_{\beta\alpha}^{(CIM)}(\theta_s,\phi_s;\theta_i,\phi_i) = \langle |F_{\beta\alpha}(\theta_s,\phi_s;\pi-\theta_i,\phi_i) + F_{\beta\alpha}(\theta_s,\phi_s;\theta_i,\phi_i)\, R_\alpha(\theta_i)\\ + R_\beta(\theta_s)\, F_{\beta\alpha}(\pi-\theta_s,\phi_s;\pi-\theta_i,\phi_i)\\ + R_\beta(\theta_s)\, F_{\beta\alpha}(\pi-\theta_s,\phi_s;\theta_i,\phi_i)\, R_\alpha(\theta_i)|^2\rangle \quad (13.5.16)\end{aligned}$$

where the phase shifts due to reflections associated with the four scattering processes are included in the definition of $F_{\beta\alpha}$.

Coherent Addition Approximation (CAA): Each branch is treated as an individual scatterer. The total scattered field is obtained by the summation of the scattered fields from each scatterer. Relative phase shifts are included. Mutual interaction is excluded.

$$\begin{aligned}\sigma_{\beta\alpha}^{(CAA)}(\theta_s,\phi_s;\theta_i,\phi_i) = \sum_{i_b=1}^{N_b}\langle |f_{\beta\alpha}^{(i_b)}(\theta_s,\phi_s;\pi-\theta_i,\phi_i)\exp(i\overline{k}_{d1}\cdot\overline{r}_{i_b})\\ + f_{\beta\alpha}^{(i_b)}(\theta_s,\phi_s;\theta_i,\phi_i)\, R_\alpha(\theta_i)\exp(i\overline{k}_{d2}\cdot\overline{r}_{i_b})\\ + R_\beta(\theta_s)\, f_{\beta\alpha}^{(i_b)}(\pi-\theta_s,\phi_s;\pi-\theta_i,\phi_i)\exp(i\overline{k}_{d3}\cdot\overline{r}_{i_b})\\ + R_\beta(\theta_s)\, f_{\beta\alpha}^{(i_b)}(\pi-\theta_s,\phi_s;\theta_i,\phi_i)\, R_\alpha(\theta_i)\exp(i\overline{k}_{d4}\cdot\overline{r}_{i_b})|^2\rangle \quad (13.5.17)\end{aligned}$$

where expressions for $\overline{k}_{d1}$, $\overline{k}_{d2}$, $\overline{k}_{d3}$, and $\overline{k}_{d4}$ are given in (13.3.16)–(13.3.19).

Independent Scattering Approximation (IND): Each branch is treated as an independent scatterer. The total scattered intensity is the sum of the scattered intensities from each scatterer.

$$\begin{aligned}\sigma_{\beta\alpha}^{(Ind)}(\theta_s,\phi_s;\theta_i,\phi_i) = \sum_{i_b=1}^{N_b}\langle |f_{\beta\alpha}^{(i_b)}(\theta_s,\phi_s;\pi-\theta_i,\phi_i)|^2\\ + |f_{\beta\alpha}^{(i_b)}(\theta_s,\phi_s;\theta_i,\phi_i)\, R_\alpha(\theta_i)|^2\\ + |R_\beta(\theta_s)\, f_{\beta\alpha}^{(i_b)}(\pi-\theta_s,\phi_s;\pi-\theta_i,\phi_i)|^2\\ + |R_\beta(\theta_s)\, f_{\beta\alpha}^{(i_b)}(\pi-\theta_s,\phi_s;\theta_i,\phi_i)\, R_\alpha(\theta_i)|^2\rangle \quad (13.5.18)\end{aligned}$$

In (13.5.16), $F_{\beta\alpha}$ is the scattering amplitude of the whole tree, while in (13.5.17)–(13.5.18), the scattering amplitude of the i_bth branch of the tree is $f_{\beta\alpha}^{(i_b)}$, R is the reflection coefficient, and N_b is the total number of branches. Equation (13.5.16) is the exact solution of wave scattering by a tree. Both

(13.5.17) and (13.5.18) are approximate methods. CAA gives a better approximation than independent scattering since it includes the relative phase shifts among branches.

Our scattering objects are the Honda trees (as shown in Fig. 13.5.8) with 63 branches generated by using stochastic L-systems. The central branch consists of six branches of decreasing sizes. Thirty trees are generated in this manner. The input file for the L-systems is given as follows:

```
#define maxgen 6
#define r1 0.85 + 0.1 * rand(1)    /* contraction ratio for trunk */
#define r2 0.55 + 0.1 * rand(1)    /* contraction ratio for branches */
#define a0 44 + rand(2)            /* branching angle from the trunk */
#define a2 44 + rand(2)            /* branching angle for lateral axes */
#define d 127.5 + rand(180)        /* divergence angle */
#define wr 0.85 + 0.1 * rand(1)    /* width decrease rate */

START : A(1, 10)
p1 : A(l, w) : * -> !(w)F(l)[&(a0)B(l * r2, w * wr)]/(d)A(l * r1, w * wr)
p2 : B(l, w) : * -> !(w)F(l)[-(a2)$C(l * r2, w * wr)]C(l * r1, w * wr)
p3 : C(l, w) : * -> !(w)F(l)[+(a2)$B(l * r2, w * wr)]B(l * r1, w * wr)
```

The maximum height of these trees is 5.18λ. The bistatic scattering cross sections for both co-polarization and cross-polarization are calculated by using (13.5.16)–(13.5.18) and are averaged over 30 realizations of trees. We plot the normalized cross section σ/λ^2 in the dB scale.

Figures 13.5.9 and 13.5.10 show the bistatic scattering cross sections for co-polarizations σ_{vv} and σ_{hh} for an incident angle of 45°. The results for CIM and CAA have a distinct scattering pattern. The result for IND, however, is a smooth curve. The results are in good agreement for both vv and hh between the coherent interaction model and coherent addition approximation. As noted for the case of simple trees in Section 5.2, there is a maximum at the backscattering direction for the bistatic scattering coefficient of vv, but not for that of hh. There is no maximum at the backscattering direction of vv for IND, because we treat all 63 branches as independent scatterers. For example, the central branch consists of six independent scattering branches of shrinking radius that is a result of L-systems structure.

Next, we compare the results by redefining the branches that continue in the same direction as one branch. To account for the decreasing radii, the length of the modified branch is the sum of the lengths of the branches. The

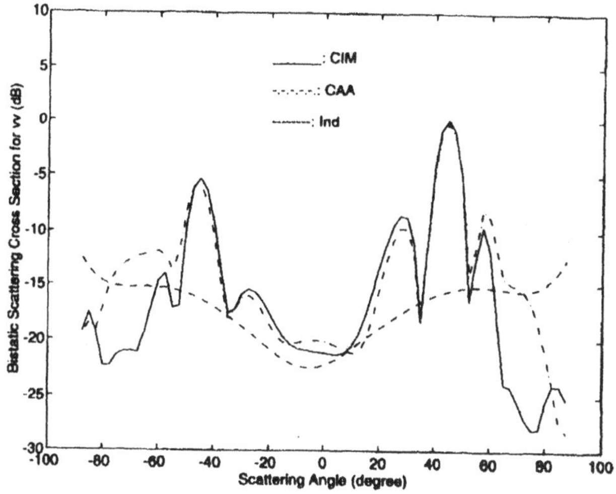

Figure 13.5.9 Comparison of bistatic scattering cross sections σ_{vv}/λ^2 for a Honda tree by coherent interaction model, coherent addition approximation, and independent scattering approximation. Incident angle is 45°. The tree is 5.18λ in height at maximum with 63 branches. $\epsilon_p = (11 + i4)\epsilon_o$. The underlying half-space is flat and has a permittivity $\epsilon_{soil} = (16 + i4)\epsilon_o$. They are averaged over 30 realizations.

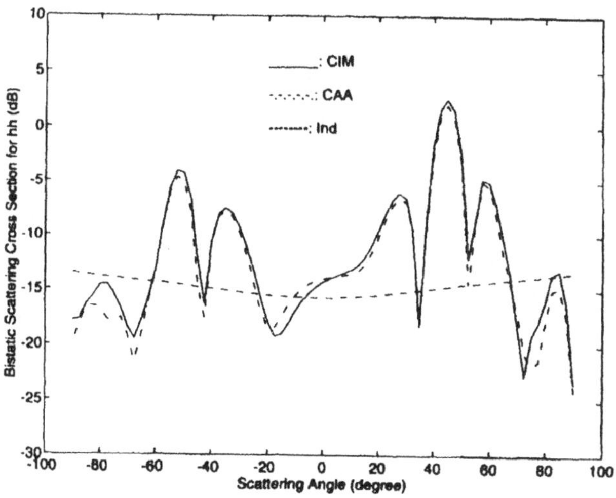

Figure 13.5.10 Comparison of bistatic scattering cross sections σ_{hh}/λ^2 for a Honda tree by coherent interaction model, coherent addition approximation, and independent scattering approximation. Incident angle is 45°. The tree is 5.18λ in height at maximum with 63 branches. $\epsilon_p = (11 + i4)\epsilon_o$. The underlying half-space is flat and has a permittivity $\epsilon_{soil} = (16 + i4)\epsilon_o$. They are averaged over 30 realizations.

radius of the modified branch is calculated by

$$a_j = \sqrt{\left(\sum_{i=1}^{N_j} a_i^2 l_i\right) \Bigg/ \left(\sum_{i=1}^{N_j} l_i\right)} \tag{13.5.19}$$

where N_j is the number of branches that continue in the same direction. The radius and length of the original ith branch are a_i and l_i, respectively. Thus, the modified central branch replaces the original six branches of different radius. Equation (13.5.19) keeps the volume of the modified branch to be the sum of individual branches. We rearrange the output data from the L-systems. The 63-branch tree then becomes a 32-branch tree-like structure. The results of wave scattering from the 32-branch tree-like structures are given below.

Figures 13.5.11 and 13.5.12 show the bistatic scattering cross sections for co-polarization σ_{vv} and σ_{hh} for an incident angle of 45°. The results of CIM and CAA in Figs. 13.5.11 and 13.5.12 are comparable to those in Figs. 13.5.9 and 13.5.10. This shows that since the wave approaches take into account the phase differences among branches, the results are not critically dependent on how the branches are labeled as independent scatterers. However, we can see large differences of the results of independent scattering when comparing Figs. 13.5.9–13.5.12. The results of independent scattering depend on how branches are labeled as independent scatterers. We also see large radar cross section in the backscattering direction for the independent scattering approximation, which does not appear in Figs. 13.5.9 and 13.5.10. This is due to the contribution of the modified central cylinder which has become much larger and contributes dominant scattering at the conical direction.

Figures 13.5.13 and 13.5.14 show the bistatic scattering cross sections for cross-polarization σ_{vh} and σ_{hv}. The results in Figs. 13.5.13 and 13.5.14 have the same parameters as Figs. 13.5.11 and 13.5.12. We can see differences between the results of the coherent interaction model and the coherent addition approximation. It is shown that the results of cross-polarization of CAA and CIM can be smaller than those of independent scattering. The cross-polarized component of the scattered wave is a result of the inclination angles of branches and the coherent wave interaction among the induced polarization within the branches. This field interaction depends strongly on how the branches are placed and oriented with respect to each other. It also creates a more complicated scattering pattern. The results of Fig. 13.5.14, show that the interactions between branches and the main stem give rise to cross-polarized backscattering coefficients which are different from those calculated by the CAA. In the CAA, the internal fields are assumed to be

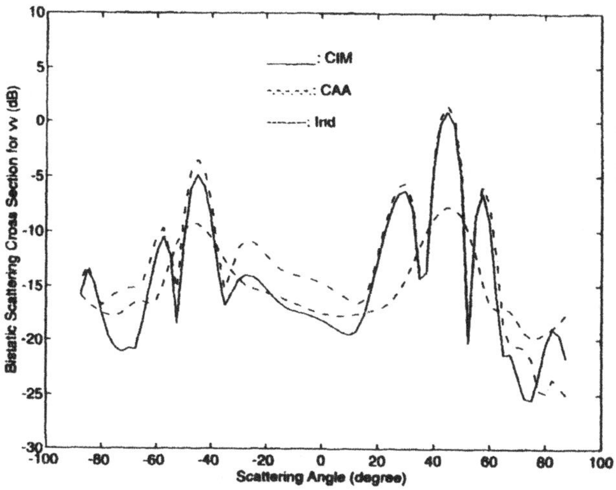

Figure 13.5.11 Comparison of bistatic scattering cross sections σ_{vv}/λ^2 for a modified Honda tree by coherent interaction model, coherent addition approximation, and independent scattering approximation. Incident angle is 45°. The tree is 5.18λ in height at maximum with 32 branches. $\epsilon_p = (11 + i4)\epsilon_o$. The underlying half-space is flat and has a permittivity $\epsilon_{soil} = (16 + i4)\epsilon_o$. They are averaged over 30 realizations.

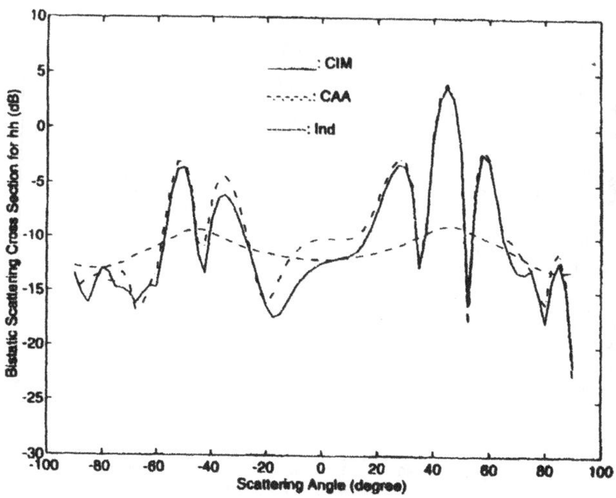

Figure 13.5.12 Comparison of bistatic scattering cross sections σ_{hh}/λ^2 for a modified Honda tree by coherent interaction model, coherent addition approximation, and independent scattering approximation. Incident angle is 45°. The tree is 5.18λ in height at maximum with 32 branches. $\epsilon_p = (11 + i4)\epsilon_o$. The underlying half-space is flat and has a permittivity $\epsilon_{soil} = (16 + i4)\epsilon_o$. They are averaged over 30 realizations.

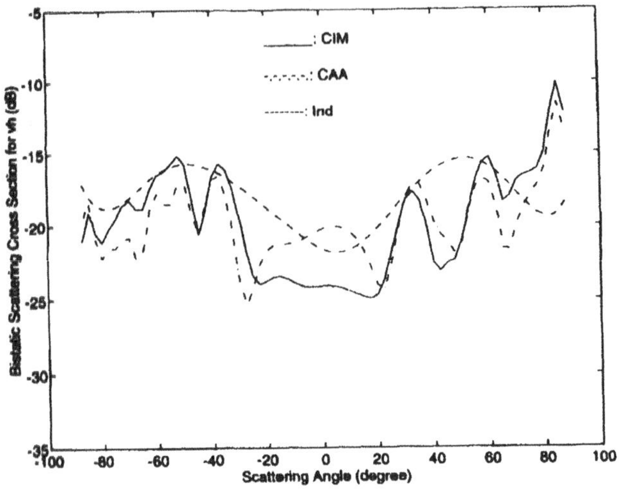

Figure 13.5.13 Comparison of bistatic scattering cross sections σ_{vh}/λ^2 for a modified Honda tree by coherent interaction model, coherent addition approximation, and independent scattering approximation. Incident angle is 45°. The tree is 5.18λ in height at maximum with 32 branches. $\epsilon_p = (11 + i4)\epsilon_o$. The underlying half-space is flat and has a permittivity $\epsilon_{soil} = (16 + i4)\epsilon_o$. They are averaged over 30 realizations.

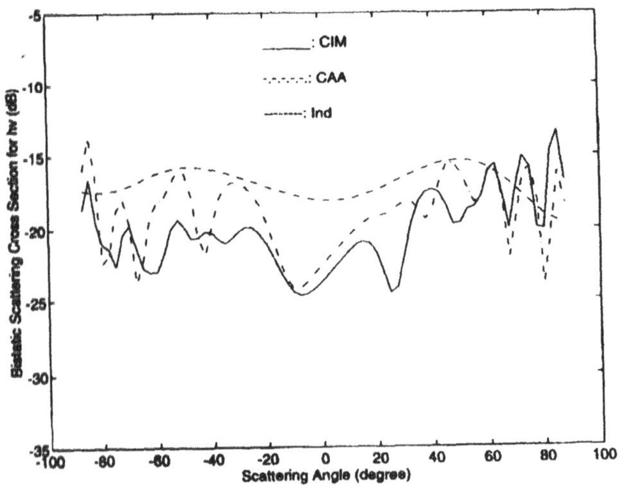

Figure 13.5.14 Comparison of bistatic scattering cross sections σ_{hv}/λ^2 for a modified Honda tree by coherent interaction model, coherent addition approximation, and independent scattering approximation. Incident angle is 45°. The tree is 5.18λ in height at maximum with 32 branches. $\epsilon_p = (11 + i4)\epsilon_o$. The underlying half-space is flat and has a permittivity $\epsilon_{soil} = (16 + i4)\epsilon_o$. They are averaged over 30 realizations.

the same as if each branch exists by itself. That internal field is dictated entirely by the polarization and the direction of the incident wave. However, the mutual coherent interaction, especially in the near field, can significantly change the direction of the internal field and the polarization current. This creates a larger difference of wave scattering for cross-polarization.

6 Rice Canopy Scattering Model

The application of space-borne sensors on monitoring rice crop growth is important to study the usage of earth lands. The possibility of using SAR measurements for the remote sensing of rice crop is based on a particular characteristic of the rice field that has a flooded ground surface during a large portion of its growing period, and a nearly vertical rice plant structure which is somewhat different from other types of vegetation. Satellite SAR data of rice field have shown strong temporal responses. At C-band, the ERS-1 backscatter data of rice fields increase significantly with the rice plant height and biomass until its fully-grown stage. The increase is also enhanced by the highly reflective underlying flooded surface through the volume-surface interaction. In this section, we describe a scattering model for microwave remote sensing of rice field. The model is developed based on the coherent addition approximation and Monte Carlo approach. Simulated backscattering coefficients employing ground truth characterization agree well with the temporal variations of SAR backscatter data from test rice field sites.

6.1 Model Description

The Monte Carlo configurations of rice fields used in the simulations are created on the basis of ground truth characteristics, such as volume fractions, sizes and shapes of rice plants. The locations, orientations, and distributions of rice plant components are generated using random number generators. Figure 13.6.1 shows a schematic diagram of the simulated rice field where rice branches are planted with nearly constant spacing over a square area A, but a small random variation in the spacing between rice branches are allowed. The lower half space is water with complex dielectric constant ϵ_1. Each rice plant contains a bunch of N_s dielectric cylindrical stems with height H, radius a, and complex dielectric constant ϵ_s. Each rice stem has attached N_ℓ leaves of elliptical disc shape with length ℓ, width w, thickness d, and complex dielectric constant ϵ_ℓ. Within a rice bunch, the stems are uniformly randomly placed inside a circle of radius a_c. There is a total number of N_c

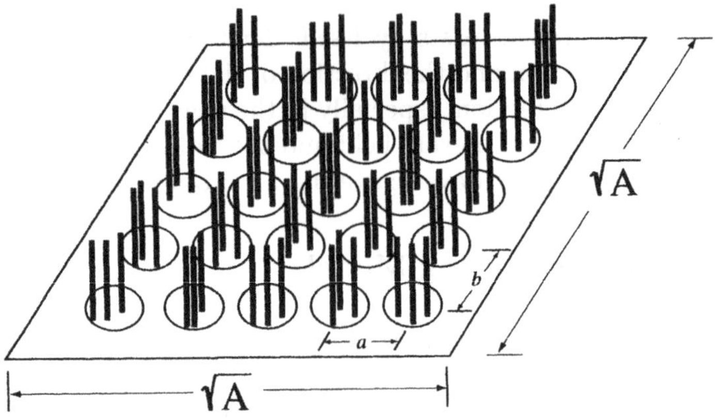

Figure 13.6.1 Rice model configuration. Rice bunches are planted over a square area A with spacing a in the x direction and spacing b in the y direction.

bunches inside the simulated rice field such that the total number of stems is $N_c \times N_s$ and $N_c \times N_s \times N_\ell$ for leaves. The rice stem height may assume a Gaussian distribution with the mean height given by the ground truth. The orientation of leaves can be described by Euler angles. The rice stems and leaves may tilt from the vertical direction to the maximum tilt angle θ_{ms} for stems and $\theta_{m\ell}$ for leaves.

Consider an incident wave $\overline{E}^i$, in the direction (θ_i, ϕ_i), impinging on a rice canopy of layer thickness h. As shown in Fig. 13.6.2, the first-order far-field solution of the backscattered field from a rice canopy can be expressed as the summation of four major scattering mechanisms,

$$E_q^s(\overline{r}) = \frac{e^{ikr}}{r}[\ S_1 + S_2 + S_3 + S_4\ ]E_p^i \tag{13.6.1}$$

where p and q denote the polarization of the incident and scattered waves, respectively. The first term S_1 describes the direct scattering from a scatterer, Fig. 13.6.2a,

$$S_1 = \sum_{\substack{t=\text{stem} \\ \text{or leaf}}} \sum_{j=1}^{N_t} f_{qp}^t(\pi - \theta_i, \pi + \phi_i; \theta_i, \phi_i)\ e^{i[\overline{k}_p^i(\theta_i,\phi_i) - \overline{k}_q^s(\pi-\theta_i,\pi+\phi_i)]\cdot\overline{r}_j^t} \tag{13.6.2}$$

where t denotes the scatterer type: stem or leaf, and $N_t = N_c \times N_s$ for stems and $N_t = N_c \times N_s \times N_\ell$ for leaves. f_{qp}^t is the scattering amplitude matrix element for an incident wave with polarization p and a scattered wave with polarization q. $\overline{k}_p^i$ is the incident wave propagation vector and $\overline{k}_q^s$ is for the scattered field. $\overline{r}_j^t = \hat{x}x_j^t + \hat{y}y_j^t - \hat{z}z_j^t$ is the location of the rice canopy element j of type t. The second term S_2 describes the single scattering from a scatterer

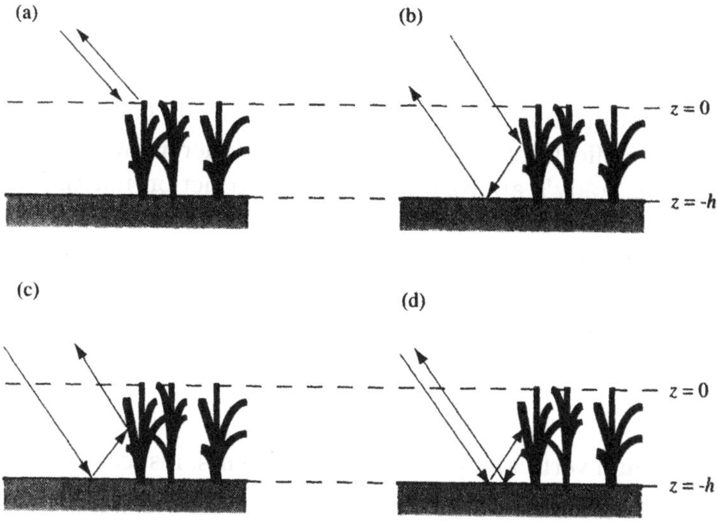

Figure 13.6.2 Four major backscattering mechanisms in a rice canopy: (a) direct scattering from the scatterer; (b) single scattering from the scatter followed by reflection off the ground surface; (c) ground surface reflection followed by single scattering from the scatterer; (d) reflection by ground surface followed by single scattering from the scatterer and further reflection off the ground surface.

followed by a reflection off the ground surface, Fig. 13.6.2b,

$$S_2 = \sum_{\substack{t=\text{stem} \\ \text{or leaf}}} \sum_{j=1}^{N_t} R_q(\theta_i)\ f^t_{qp}(\theta_i, \pi + \phi_i; \theta_i, \phi_i)\ e^{i[\overline{k}^i_p(\theta_i,\phi_i) - \overline{k}^s_q(\theta_i,\pi+\phi_i)]\cdot\overline{r}^t_j} \quad (13.6.3)$$

The third term S_3 is the reverse of the second scattering mechanism, Fig. 13.6.2c,

$$S_3 = \sum_{\substack{t=\text{stem} \\ \text{or leaf}}} \sum_{j=1}^{N_t} f^t_{qp}(\pi - \theta_i, \pi + \phi_i; \pi - \theta_i, \phi_i) R_p(\theta_i)\ e^{i[\overline{k}^i_p(\pi-\theta_i,\phi_i) - \overline{k}^s_q(\pi-\theta_i,\pi+\phi_i)]\cdot\overline{r}^t_j} \quad (13.6.4)$$

where $R_p(\theta_i)$ and $R_q(\theta_i)$ in (13.6.3) and (13.6.4) are the Fresnel reflection coefficients of the p and q-polarizations, respectively. The fourth term S_4 describes a reflection by the ground surface followed by the backscattering from the scatterer and further reflection off the ground surface, Fig. 13.6.2d,

$$S_4 = \sum_{\substack{t=\text{stem} \\ \text{or leaf}}} \sum_{j=1}^{N_t} R_q(\theta_i) f^t_{qp}(\theta_i, \pi + \phi_i; \pi - \theta_i, \phi_i) R_p(\theta_i) e^{i[\overline{k}^i_p(\pi-\theta_i,\phi_i) - \overline{k}^s_q(\theta_i,\pi+\phi_i)]\cdot\overline{r}^t_j} \quad (13.6.5)$$

To calculate the scattering amplitude matrix elements of rice plants, the infinite cylinder approximation can be applied for rice stems, and the infinite disk approximation can be employed for rice leaves. These two analytical approximations have been described in details in Chapter 1, Section 6 of Volume I. The effects of attenuation on the coherent wave is taken into account by using Foldy's approximation. The extinction matrix is obtained by averaging the forward scattering amplitude matrix over rice canopy scattering components. The rice plants are then assumed to be excited by this coherent wave. The two characteristic waves propagate along the direction (θ, ϕ) inside the canopy with the propagation constants

$$k_h = k_o - iM_{hh} \tag{13.6.6}$$

$$k_v = k_o - iM_{vv} \tag{13.6.7}$$

for horizontally and vertically polarized components, respectively. In (13.6.6) and (13.6.7),

$$M_{qp} = \frac{i2\pi}{k_o} \frac{N_c N_s}{Ah} \left[\langle f_{qp}^{stem}(\theta, \phi; \theta, \phi) \rangle + N_\ell \langle f_{qp}^{leaf}(\theta, \phi; \theta, \phi) \rangle \right] \tag{13.6.8}$$

where the angular brackets denote configurational average, h is the layer depth of rice canopy, and k_o is the wavenumber of free space. The real parts of M_{pp} and M_{qq} give the extinction. Since the calculated k_h and k_v are quite close to k_o, the effects of reflection and refraction at the top boundary of rice canopy layer are neglected. Based on equations (13.6.6) and (13.6.7), equation (13.6.1) can be expressed as

$$\begin{aligned}
E_q^s(\overline{r}) = \frac{e^{ikr}}{r} \sum_{\substack{t=stem \\ or\ leaf}} \sum_{j=1}^{N_t} \Bigg[& f_{qp}^t(\pi - \theta_i, \pi + \phi_i; \theta_i, \phi_i) e^{-M_{qq}\frac{z_j^t}{\cos\theta_i}} e^{-M_{pp}\frac{z_j^t}{\cos\theta_i}} e^{2i(k_x^i x_j^t + k_y^i y_j^t - k_z^i z_j^t)} \\
& + R_q(\theta_i) f_{qp}^t(\theta_i, \pi + \phi_i; \theta_i, \phi_i) e^{M_{qq}\frac{2h+z_j^t}{\cos\theta_i}} e^{-M_{pp}\frac{z_j^t}{\cos\theta_i}} e^{2i(k_x^i x_j^t + k_y^i y_j^t + k_z^i h)} \\
& + f_{qp}^t(\pi - \theta_i, \pi + \phi_i; \pi - \theta_i, \phi_i) R_p(\theta_i) \\
& \quad \times e^{-M_{qq}\frac{z_j^t}{\cos\theta_i}} e^{M_{pp}\frac{2h+z_j^t}{\cos\theta_i}} e^{2i(k_x^i x_j^t + k_y^i y_j^t + k_z^i h)} \\
& + R_q(\theta_i) f_{qp}^t(\theta_i, \pi + \phi_i; \pi - \theta_i, \phi_i) R_p(\theta_i) \\
& \quad \times e^{M_{qq}\frac{2h+z_j^t}{\cos\theta_i}} e^{M_{pp}\frac{2h+z_j^t}{\cos\theta_i}} e^{2i[k_x^i x_j^t + k_y^i y_j^t + k_z^i (2h+z_j^t)]} \Bigg] E_p^i
\end{aligned} \tag{13.6.9}$$

where $k_x^i = k_o \sin\theta_i \cos\phi_i$, $k_y^i = k_o \sin\theta_i \sin\phi_i$, and $k_z^i = k_o \cos\theta_i$.

In each realization of the simulated rice field, the center positions of N_c rice clusters are first created and then the positions of the N_s stems within every bunch are generated, both steps use random number generators with uniform distribution. The positions of rice stems are checked so that there is no overlapping between stems. The positions and orientations of the attached leaves on each stem are also generated using random number generators. The tilt angles of stems or leaves are similarly generated. The scattered electric field E_s from each realization is calculated according to (13.6.9). The backscattering coefficient is computed from

$$\sigma_{qp} = \frac{4\pi r^2}{A} \frac{\langle |E_q^s|^2 \rangle}{|E_p^i|^2} \tag{13.6.10}$$

where A is the simulated rice field area. The results are obtained by averaging over an ensemble of realizations. The angular brackets denote the configurational average.

6.2 Model Simulation

The rice canopy scattering model described in the previous section is applied to simulate the backscattering signatures at different rice growth stages and compared with ERS-1 data. The frequency is 5.3 GHz, the incidence angle is 23°, and the polarization is VV. The parameters used in the simulation are summarized in Table 13.6.1. The dielectric constants of rice plant at various growth stages are calculated from the gravimetric water content using an empirical formula given by Ulaby and El-Rays [1987]. Since the bottom part of the rice plants is in the water, the dielectric constant of the ground surface is that of water at the frequency of 5.3 GHz at 20°C, which is $\epsilon_1 = (74 + i21)\epsilon_o$. In the simulation, the average spacing between two rice bunches is about 22 cm. We also assume that the height of stem has a normal distribution with the standard deviation of 1 cm. The backscattering coefficients are obtained by averaging over 50 realizations. The parameters and data used in this comparison are obtained from the reference [Le Toan et al. 1997].

In Fig. 13.6.3, the model backscattering results of VV polarization at different growth stages are compared with the ERS-1 data. The comparison shows good agreement between model and measurements. The increasing trend of the temporal radar response is well captured by the Monte Carlo modeling. Since the bottom of rice plants are immersed in water, the backscattering returns are dominated by the volume-surface interactions.

age (days)	20	40	60	80	100
stem height H (cm)	20.0	35.0	50.0	68.4	76.7
stem radius a (cm)	0.1	0.12	0.14	0.16	0.18
# of stems per bunch N_s	10	10	10	10	10
# of bunches per m^2 (N_c/A)	20	20	20	20	20
gravimetric water content	0.7	0.72	0.74	0.71	0.71
dielectric constant ϵ_s, ϵ_ℓ	25.1+i7.9	26.3+i8.2	27.6+i8.6	25.7+i8.0	25.7+i8.0
leaf width w (cm)	0.41	0.74	0.98	1.15	1.24
leaf length ℓ (cm)	14.08	24.72	31.92	35.68	36
leaf thickness d (cm)	0.02	0.02	0.02	0.02	0.02
# leaves per stem N_ℓ	5	7	7	7	7
volume fraction (%)	0.06	0.09	0.123	0.161	0.204
bunch radius a_c (cm)	1.6	1.8	2.1	2.3	2.3
max stem tilt angle θ_{ms} (deg)	0	0	0	5	10
max leaf tilt angle $\theta_{m\ell}$ (deg)	5	10	20	30	40

Table 13.6.1 Input parameters for model simulation.

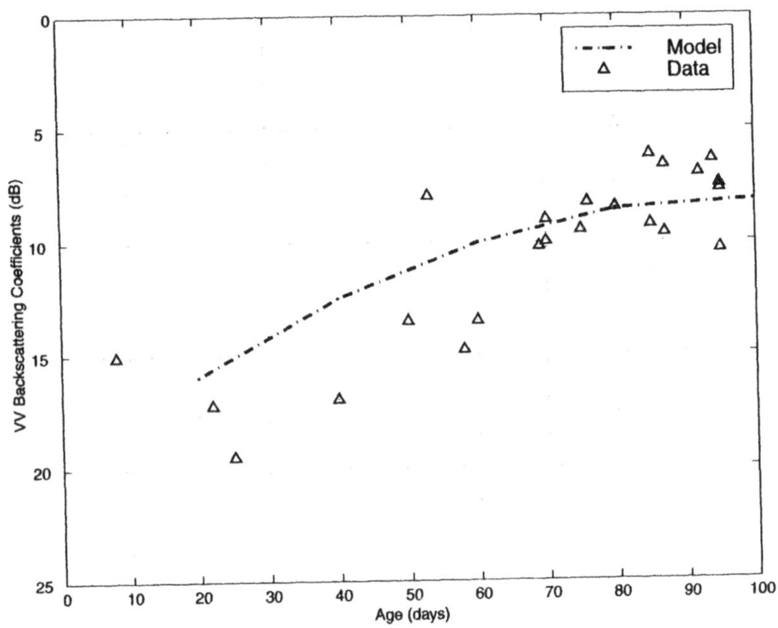

Figure 13.6.3 Comparison of measured and model backscattering coefficients at C-band versus rice age.

REFERENCES AND ADDITIONAL READINGS

Abelson, H. and A. A. diSessa (1982), *Turtle Geometry*, MIT Press, Cambridge, MA.

Au, W. C., J. A. Kong, and L. Tsang (1994), Absorption enhancement of scattering of electromagnetic waves by dielectric cylinder clusters, *Microwave Opt. Technol. Lett.*, **7(10)**, 454–457.

Au, W. C., L. Tsang, R. T. Shin, and J. A. Kong (1996), Collective scattering and absorption in microwave interaction with vegetation canopies, *Progress in Electromag. Res.*, **14**, 182–231, EMW Publishers, Cambridge, Massachusetts.

Chen, Z. (1994), Microwave remote sensing of vegetation: Stochastic Lindenmayer systems, collective scattering effects, and neural network inversions, Ph.D. thesis, University of Washington, Seattle.

Chen, Z., L. Tsang, and G. Zhang (1995), Scattering of electromagnetic waves by vegetation based on the wave approach and the stochastic Lindenmayer system, *Microwave Opt. Technol. Lett.*, **8(1)**, 30–33.

Chen, Z., L. Tsang, and G. Zhang (1996), Application of stochastic Lindenmayer systems to study collective and cluster scattering in microwave remote sensing of vegetation, *Progress in Electromag. Res.*, **14**, 233–277, EMW Publishers, Cambridge, Massachusetts.

Fung, A. K. (1994), *Microwave Scattering and Emission Models and Their Applications*, Artech House, Norwood, Massachusetts.

Goel, N. S., I. Rozehual, and R. L. Thompson (1991), A computer graphics based model for scattering from objects of arbitrary shapes in the optical region, *Remote Sens. Environ.*, **36**, 73–104.

Herman, G., A. Lindenmayer, and G. Rozenberg (1975), Description of developmental languages using recurrence systems, *Mathematical Systems Theory*, **8**, 316–341.

Le Toan, T., F. Ribbes, L. F. Wang, N. Floury, K. H. Ding J. A. Kong, M. Fujita, and T. Kurosa (1997), Rice crop mapping and monitoring using ERS-1 data based on experiment and modeling results, *IEEE Trans. Geosci. Remote Sens.*, **35**, 41–56.

Lin, Y. C. and K. Sarabandi (1999a), A Monte Carlo coherent scattering model for forest canopies using fractal-generated trees, *IEEE Trans. Geosci. Remote Sens.*, **37**, 440–451.

Lin, Y. C. and K. Sarabandi (1999b), Retrieval of forest parameters using a fractal-based coherent scattering model and a genetic algorithm, *IEEE Trans. Geosci. Remote Sens.*, **37**, 1415–1424.

Lindenmayer, A. (1968), Mathematical models for cellular interaction in development, Parts I and II, *Journal of Theoretical Biology*, **18**, 280–315.

Lindenmayer, A. (1974), Adding continuous components to L-systems, *L Systems,*, Lecture Notes in computer Science, **15**, 53–68, Springer-Verlag, Berlin.

Mandelbrot, B. B. (1983), *The Fractal Geometry of Nature*, W. H. Freeman & Co., New York.

Prusinkiewicz, P. and J. Hanan (1989), *Lindenmayer Systems, Fractals, and Plants*, Vol. 79 of Lecture Notes in Biomathematics, Springer-Verlag, Berlin.

Prusinkiewicz, P. and A. Lindenmayer (1990), *The Algorithmic Beauty of Plants*, Springer-Verlag, New York.

Stiles, J. M. and K. Sarabandi (2000), Electromagnetic scattering from grassland. I. A fully phase-coherent scattering model, *IEEE Trans. Geosci. Remote Sens.*, **38(1)**, 339–348.

Stiles, J. M., K. Sarabandi, and F. T. Ulaby (2000), Electromagnetic scattering from grassland. II. Measurement and model results, *IEEE Trans. Geosci. Remote Sens.*, **38(1)**, 349–356.

Tsang, L., K. H. Ding, G. Zhang, C. C. Hsu, and J. A. Kong (1995), Backscattering enhancement and clustering effects of randomly distributed dielectric cylinders overlying a dielectric half space based on Monte-Carlo simulations, *IEEE Trans. Antennas Propagat.*, **43(5)**, 488–499.

Ulaby, F. T. and C. Elachi, Eds. (1990), *Radar Polarimetry for Geoscience Applications*, Artech House, Norwood.

Ulaby, F. T. and M. A. El-Rays, (1987), Microwave dielectric spectrum of vegetation — Part II: dual-dispersion model, *IEEE Trans. Geosci. Remote Sens.*, **25**, 550–557.

Yueh, S. H., J. A. Kong, J. K. Jao, R. T. Shin, and T. Le Toan (1992), Branching model for vegetation, *IEEE Trans. Geosci. Remote Sens.*, **30**, 390–402.

Zhang, G., L. Tsang, and Z. Chen (1996), Collective scattering effects of trees generated by Stochastic Lindenmayer Systems, *Microwave Opt. Technol. Lett.*, **11(2)**, 107–111.

INDEX

Printed and bound by CPI Group (UK) Ltd, Croydon, CR0 4YY
18/09/2026
14965015-0003